VIKING FUND PUBLICATIONS IN ANTHROPOLOGY
edited by Sol Tax

Number Thirty-Six

AFRICAN ECOLOGY AND HUMAN EVOLUTION

AFRICAN ECOLOGY AND HUMAN EVOLUTION

edited by

F. CLARK HOWELL and FRANÇOIS BOURLIÈRE

ALDINE PUBLISHING COMPANY / *Chicago*

This volume comprises one of a series of publications on research in general anthropology published by the Wenner-Gren Foundation for Anthropological Research, Incorporated, a foundation created and endowed at the instance of Axel L. Wenner-Gren for scientific, educational, and charitable purposes. The reports, numbered consecutively as independent contributions, appear at irregular intervals.

Second printing 1966

Library of Congress Catalog Card Number 63-12473

First published 1963 by

ALDINE PUBLISHING COMPANY
320 West Adams Street
Chicago, Illinois 60606

Printed in the United States of America

PREFACE

This volume comprises the background papers and discussion transcription of a symposium, sponsored by the Wenner-Gren Foundation for Anthropological Research, Inc., which met at Burg Wartenstein, Austria, the European headquarters of the Foundation, July 10 to 22, 1961. The papers were largely prepared for circulation to other participants prior to the symposium; none of the papers was presented as such at the symposium. The paper by L. S. B. Leakey was prepared by him during the meeting from notes which formed the basis for several of his presentations to the participants of the status of recent investigations at Olduvai Gorge, northern Tanganyika. The papers by C. Arambourg, K. W. Butzer and L. Liben (not present at the symposium) were prepared after the symposium at the request of the chairmen. All of the other papers were submitted for publication, with some minor revisions, following the symposium. The papers by Arambourg, Biberson, Bourlière, de Heinzelin, Hiernaux, and Liben were translated by Marie Ann Honeywell, and that by Monod by Katherine Appleby and McGuire Gibson; each translation was edited by the particular author.

This symposium had its immediate genesis when we were present at an earlier symposium, on "The Social Life of Early Man," held at Burg Wartenstein, in 1959. Both of us had recent field experience in Africa, especially eastern and equatorial Africa, and were aware of the need to integrate the results of increasingly numerous field studies bearing on the biological-behavioral evolution of the higher primates (especially hominids) with other field studies on the paleoecology and the recent mammalian ecology of sub-Saharan Africa. At that time we outlined a plan for such a projected symposium and submitted it to Dr. Paul Fejos, President and Director of Research of the Foundation, who gave it his tentative approval. The plan was discussed further by us in Kenya that summer when we met en route to the Omo valley of southern Ethiopia (F. C. H.) and the Albert National Park of the Congo Republic (F. B.) in conjunction with our own research interests. The final plans for the symposium were made at a joint meeting in Paris in June, 1960.

We feel it is most fortunate that essentially all invited participants were able to attend the conference. We had hoped to include an experienced worker in tropical epidemiology, but our first choice was recently deceased and the number of participants was already at a maximum so that we chose to neglect this very important subject. Similarly we felt the need for an ecological botanist of wide experience in sub-Saharan Africa, but to find a single individual appeared so difficult that we chose to rely on our other participants, who together represented years of experience in varied parts of the African continent.

Our daily discussions were tape recorded and detailed notes were kept by Mrs. Betty Clark, assisted by the other participants. The summary transcript of these discussions, prepared by Mrs. Clark, and published here presents more adequately than any summary we might attempt the nature and scope of our conversations together at the great round table in the conference room at Wartenstein. Fortunately or unfortunately no one cared (or dared) to preserve for posterity the lively conversations over cocktails, at meals, or during our "free day" trips to Vienna, to a winery in eastern Austria, or our evening together at a Viennese performance of "The Merry Widow." And only the participants share the memory of the Bishop-Cooke duet of the "Paranthropus-Zinjanthropus" theme our last night together.

We will all remember the gracious hospitality at Wartenstein and the kindness shown us by all of its staff. We are deeply grateful to the Wenner-Gren Foundation, and to its President Dr. Paul Fejos, for the opportunity to hold this symposium at Burg Wartenstein and to publish the papers and discussion summary as a volume in this series.

F. Clark Howell
François Bourlière

TABLE OF CONTENTS

AFRICAN ECOLOGY AND HUMAN EVOLUTION

CLIMATIC-GEOMORPHOLOGIC INTERPRETATION OF PLEISTOCENE SEDIMENTS IN THE EURAFRICAN SUBTROPICS

KARL W. BUTZER

INTRODUCTION

EFFECTIVE UNDERSTANDING of Pleistocene stratigraphy in terms of palaeoclimatic units is dependent upon systematic interpretation of geomorphic features. Most existing literature however has emphasized cycles and denudation chronologies while the corresponding sediments are frequently ignored. The core of the problem of "pluvial" paleoclimates, viz. whether an increase of precipitation or a decrease in evaporation is involved, can only be reached by sedimentological investigation. Sedimentology cannot provide absolute answers and indeed small variations in rainfall intensity, amount, or variability within a single climatic province may render absolute or numerical generalizations inadequate. The size, slope, and relative elevation of a catchment area may also determine the details of sedimentation by fluviatile processes. Nevertheless, wherever tectonic factors can be discounted it is the sediment rather than the erosional form that is more elucidating—providing of course that the former is preserved.

Although some observed deposits or phenomena are more or less fossil, many can still be interpreted indirectly when compared to modern analogies. Transport capacity of the fluviatile agents which are responsible for the different specific classes of sediments can be compared by quantitative or semiquantitative evaluations of modern and Pleistocene constituents. Similarly, Pleistocene deposits of differing stratigraphical age can be comparatively analyzed and can on occasion be subjected to micro-stratigraphic analysis. The intensity of soil development at different periods may often be assessed by field observation alone. A knowledge of the vegetation type and vegetative mat associated with different degrees or types of denudation and deposition is also useful in process interpretation.

Comparative interpretation can be successfully applied to most Pleistocene sediments and the following discussion of sediments commonly found in the subtropical zone is based upon personal study in two distinct climatic provinces:

(1) the modern Mediterranean littoral climate (*Csa*) studied in a subhumid and semiarid variety (Catalonia and the Balearic Islands, respectively), and (2) the modern winter-rain desert climate (*BWhs*) studied in the arid zone of Egypt. Comparative observations in other parts of Spain and in Italy and Palestine have also been included although the writer does not claim general validity for the conclusions reached outside of the specific areas studied. Reference is specifically made to lowland areas which have a limited altitudinal variation in their catchment basins, and thus exclude sedimentation which may have been caused by cold climate agents of the Pleistocene. Tectonic activity was also absent or insignificant in the areas of detailed investigation during post-Villafranchian time.

The various gradational agents and related sediments will be outlined and briefly discussed.

I. AEOLIAN PROCESSES

Regressional Dunes (Aeolianites)

The shores of many subtropical and tropical lands, characterized by arid or semiarid climate today, have widespread calcareous littoral dunes. In areas of limestone bedrock these dunes are over 90% soluble in HCl and contain large quantities of organic materials such as molluscan debris and Foraminifera. In more humid climatic zones littoral dunes are composed of much greater proportions of quartz sands since calcareous materials have been extensively removed by solution. Such littoral dunes alone are not indicative of local aridity, but the calcareous facies is almost exclusively developed in semiarid landscapes where solution and leaching have been moderate if not ineffective. In these areas the greater part of such littoral dunes are consolidated and immobile, and consequently fossil. Such massive deposits are due to accelerated deposition following deflation of epicontinental lime sands and marine rubble during glacio-eustatic regressions. In a pioneer study on Bermuda, R. W. Sayles (1931) defined these fossil littoral or *regressional dunes* as *aeolianites*. They are frequent on rocky or cliff coasts where beach sands are no longer available for deflation today. [Synonyms: "*grès*" *dunaire* (French North Africa), *ramleh* (Lebanon), *kurkar* (Palestine-Israel).]

Morphology. Aeolianites developed in typical aeolian facies may be found either as (1) steeply inclined, foreset-bedded dunes embanked against coastal cliffs with typical seaward dip values of 40–60%, and landward values of 60–80%; (2) free longitudinal dunes of subdued morphology on coastal plains, where they often form littoral cordons. The local relief of one dunal generation may be in the order of 5–25m, while slopes are gentle, and seldom exceed 25%. Barkhan and transverse forms are also known in such deposits, particularly in desert areas with little or no vegetation; (3) undulating sand sheets with longitudinal affinities of subdued dunal topography, concentrated in face and lee of minor surface irregularities. These sheets are found beyond the rims of coastal cliffs and well inland on coastal plains or level uplands.

Sedimentology. 1.) It is difficult to give quantitative data on *grain size* distributions as many different gradations between fluviatile-colluvial and aeolian bedding may occur. Water-borne admixture can be invariably recognized by discoloration from the normal white or very pale brown (10 YR 7–8/2–4 according to the *Munsell Soil Color Charts*) due to addition of weathered materials. Aeolianites without such admixture were analyzed from Mallorcan samples and it was found necessary to subdivide them into a coastal (coarse-grained) and an interior (finer-grained) facies (Table 1). The typical coastal facies has an average composition 70% in the coarse sand (200–2000 microns) fraction, and 90% in the sand (20–2000 microns) fraction. The respective proportions of interior aeolianites may be reduced to 40% and 70%. Predominantly aeolian but highly mixed deposits will still retain a 75% sand fraction but the coarse sand component is reduced to 25–30%.

TABLE 1

Grain Sizes of Aeolianite Samples from Mallorca (in percent)

(A: typical Pleistocene coastal aeolianite; B: modern coastal dune; C: typical Pleistocene interior aeolianite; D: mixed Pleistocene coastal aeolianite; E: semicolluvial Pleistocene coastal aeolianite)

Sample	*200–2000μ*	*60–200μ*	*20–60μ*	*20–2000μ*	*6–20μ*	*2–6μ*	*<2μ*
A	66.6	22.6	3.6	92.8	2.2	1.3	3.7
B	80.7	5.5	0.2	86.4	3.1	1.9	8.6
C	42.7	22.9	6.8	72.4	8.9	6.2	12.5
D	59.1	28.5	3.4	91.0	1.1	1.4	6.5
E	28.3	41.6	6.1	76.0	9.9	4.3	9.8

2.) Aeolianites are composed of disintegrated molluscan rubble and clastic inorganic materials derived directly from coastal bedrock. They are then essentially a terrestrial form of calcarenite. Along the Miocene limestone coasts of Mallorca, unweathered littoral dunes, whether modern or fossil aeolianites, contain calcareous components as high as 95%. The second major component is quartz which averages 0.5–4.5%, whereas Fe_2O_3, Al_2O_3, MnO, etc., amount to as much as 1%. Mixed and particularly semicolluvial beds contain higher values of noncalcareous materials and quartz proportions rise to 30% or more in some beds.

3.) Analyses of *quartz-grain micromorphology* have disclosed that the percentage of rounded glossy water-worn, or frosted wind-worn grains are matched by a large percentage of angular unworn grains (Table 2). Furthermore, waterworn grains frequently out-number the wind-worn specimens but without an obvious explanation. This can be attributed to the fact that aeolian transport is over such small distances in many or most instances that quartz grains are only partly modified by aeolian agencies—unlike the continental dunes of desert regions where transport, which is often repeated after renewed deflation, involves great distances. Analysis of 300–1500 micron quartz components were of little more significance than those of 60–200 micron quartz for interpretative purposes, despite the reduction in the number of unworn grains. Wind-worn, smooth ellipsoidal calcite grains are in fact a more significant feature under the microscope.

TABLE 2
MICROMORPHOLOGIC ANALYSIS OF QUARTZ GRAINS
(of medium-sand fraction, in percent. Referring to Table 1)

Sample	*Waterworn*	*Windworn*	*Unworn*
A	7%	66%	27%
B	15	45	40
C	38	24	38
E	30	31	29

4.) Bedding features of many aeolianites are marked stratification surfaces occurring at certain intervals. These often show increased consolidation indicating temporary erosion and prolonged exposure to subaerial agents. Foreset bedding is common, crossbedding rare. Embanked aeolianites at the base of cliffs always exhibit marked stratification, whereas many coastal plain or interior deposits lack bedding. Poor or no stratification is usually associated with frequent root drip or with the root casts and calcified roots of halophile shrubs or conifers such as *Juniperus*. The writer attributes this class of aeolianites to deposition under vegetation, such as intermingling with the needle *førna* under coniferous woodland. Unbedded littoral dunes form in this fashion today. On the other hand, well-bedded coastal plain or interior aeolianites without root drip are certainly indicative of sparse vegetation and prevailing aridity.

Further palaeoclimatic information can be derived from bedding directions (directions of dip), which can be used to compare mean storm wind deviations of various stratigraphic units with modern conditions (cf. Butzer 1961, Butzer & Cuerda 1962).

Stratigraphic Significance. Well-developed aeolianites which occur along sandy or shallow-water coasts, or any form of aeolianite along rocky or cliff coasts without sandy beaches, may be regarded as evidence of marine regressions. Only under regressive conditions can sufficient unconsolidated shelf sediments be subjected to deflation by storm winds. In particular, as H. E. Wright (1962) has argued, such aeolianites are indicative of a marine regression actually in

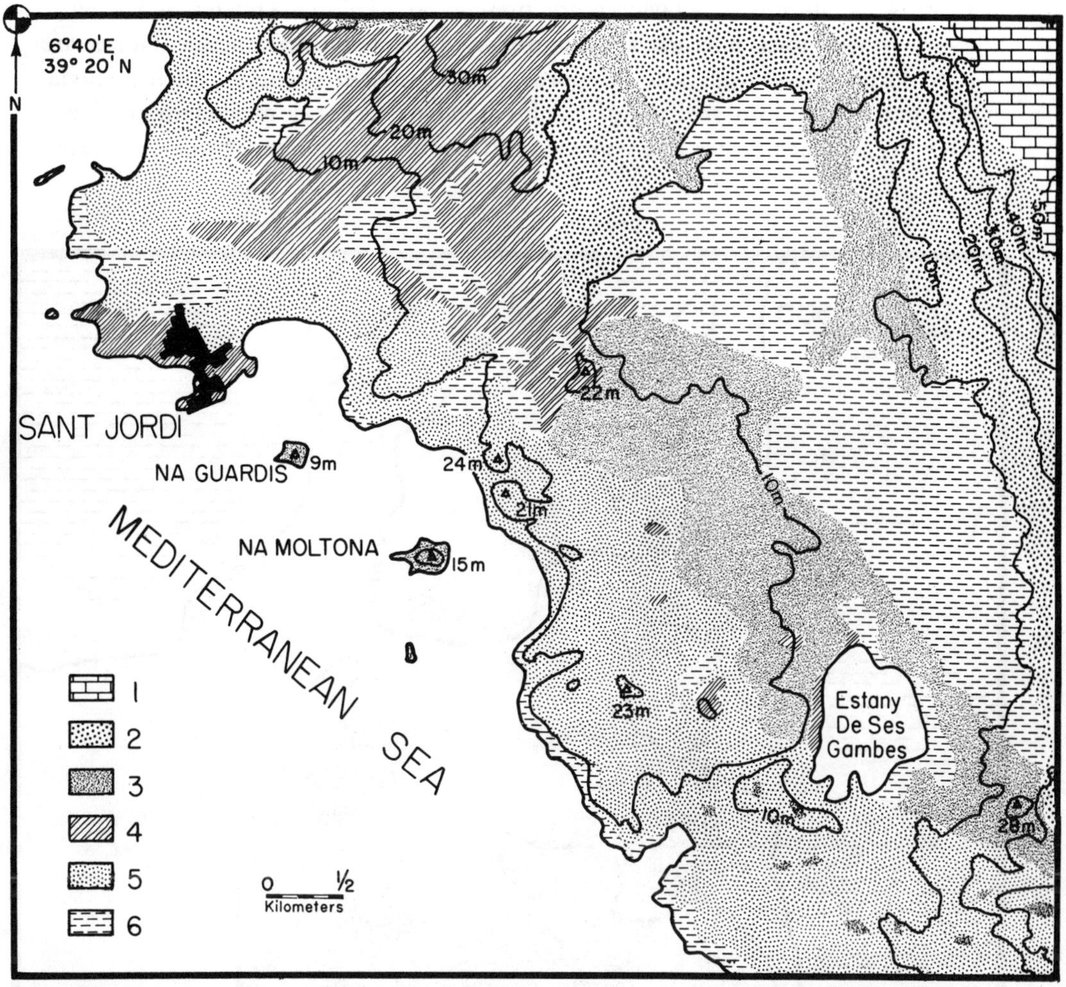

FIGURE 1

Littoral cordon of three generations of Pleistocene Aeolianites on the southern coast of Mallorca. 10 m. contour interval. 1: horizontal Miocene limestones; 2: cemented and weathered aeolianite of antepenultimate major regression; 3: consolidated and weathered aeolianite of penultimate major regression; 4: semiconsolidated aeolianite of last major regression; 5: unconsolidated littoral dunes (subrecent) or loose aeolian mantle; 6: unconsolidated and cemented colluvial silts of various ages. From 1:50,000 geological map by the writer.

progress. For once regression ceases or a renewed rise in sea level occurs, no new sands are exposed to deflation and consequently sedimentation stops. Conformable stratigraphic sequences are present on Mallorca (Butzer and Cuerda 1962) where final interglacial transgressive beds with thermophile faunas are succeeded by colluvial beds to approximately modern sea level. Such colluvial silts grade over into aeolianites which extend to well below modern sea level. Many typical aeolianites also contain small wind-worn mollusca which are often

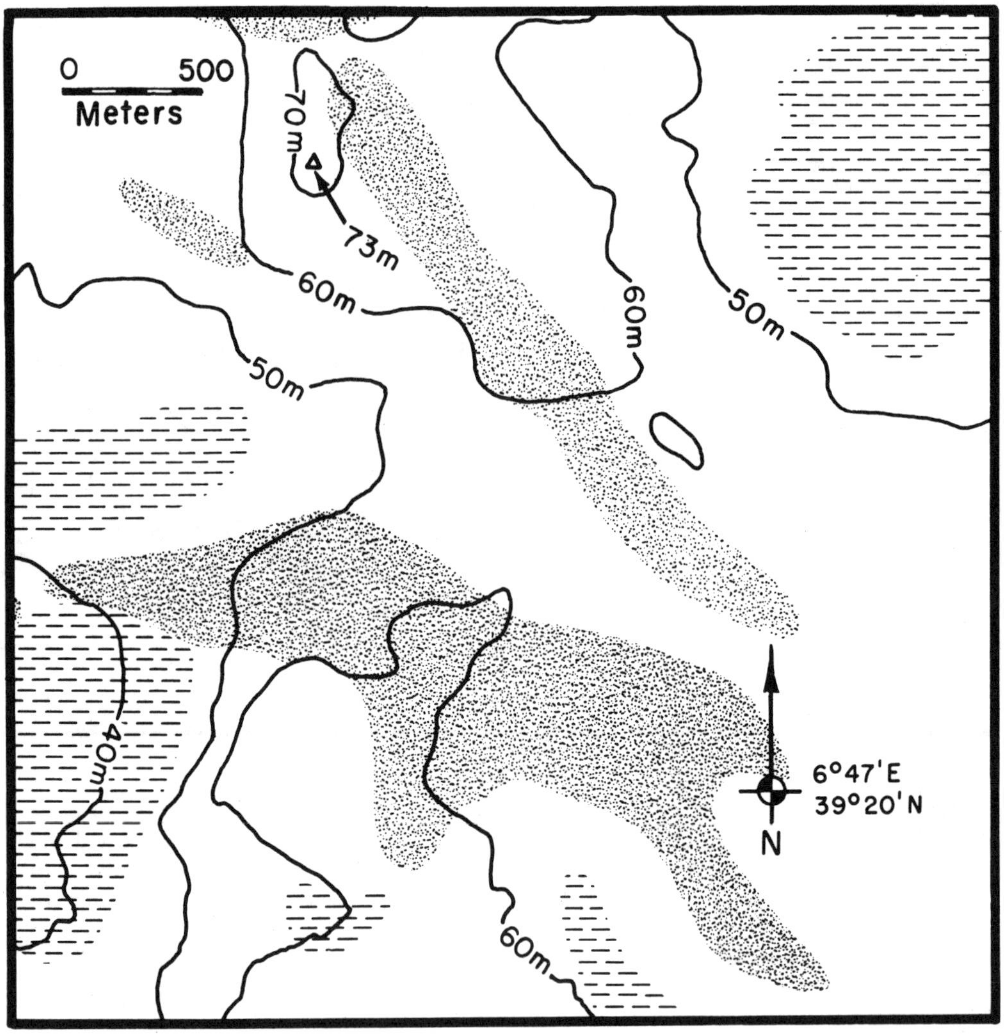

FIGURE 2

Interior aeolianites of subdued topography (from southern Mallorca). Dot pattern indicates shallow aeolianites of the penultimate regression, shaded pattern unconsolidated colluvial silts largely of last regressional date.

quite intact. Such coarse components are not present in later beds nor in those with interior situations. There can be absolutely no question of the regressive character of the aeolianites studied by the writer in Spain, or of those described by various French authors (e.g., J. Hilly 1957) for North Africa and by R. W. Hey (McBurney & Hey 1955) for Cyrenaica. Recent unfounded suggestions by K. H. Kaiser (1960) to the effect that Lebanese aeolianites are transgressive sediments must be discounted.

As interpreted here aeolianites are invaluable as an indirect means of correlation with higher latitude glaciation through the glacio-eustatic chronology. Interruptions of aeolianite deposition may indicate world-wide halts or oscillations of the continental glacier advances. Similarly aeolianite deposition must more or less cease when the maximum of a glacial regression has been attained (Wright, 1962). True aeolianite therefore is stratigraphically indicative of the earlier phases of continental glacial advances until the period of maximum ice extent.

The widespread occurrence of aeolianites provides a most valuable paleoclimatic and stratigraphic guide along subhumid to arid subtropical and tropical coasts.

Continental Dunes

Although continental dunes do adjoin coastal aeolianite fields in some areas, the former are generally a quite distinct feature in arid zones. Certainly dunes form under a wide range of thermal and precipitation conditions if a ready source of loose, unstabilized material is available. The extensive fields of parabolic or blow-out dunes of the Würm-Wisconsin tundras of Central Europe and the United States were deposited under a considerable herbaceous vegetation. Small active dune fields are not unknown along the sandy beds of seasonally overloaded streams in semiarid climatic zones. Continental dunes of the barchan, transverse and longitudinal type are however essentially a characteristic of the world's deserts. Within these deserts their distribution is determined by the presence of suitable materials for deflation. The majority of *ergs* moreover are not a product of present conditions.

Morphology. The dunal forms of the continental arid zone include free dunes of the well-known longitudinal, barchan, and transverse type, as well as tied dunes of the lee type. Sand or loess mantles of subdued or incoherent morphology are also not unknown within the same physiographic zone.

Sedimentology. The typical *grain size* distribution of continental dunes (Shepard and Young 1961, Bagnold 1954; J. Meckelein 1959, p. 176) differs appreciably from that of aeolianites. Medium-grained sands (60–200 microns) provide 20–90%, i.e., twice or more the medium sand components of aeolianites. Very coarse sands are less frequent since transport is over much greater distances than that involved in the formation of regressional littoral dunes. The most striking characteristic of these dunes is the almost complete absence of any component under 30 microns. Granulometric spectra are thus necessary to help to determine the mode of origin of fossil sands and particularly that of sands found in coastal proximity.

Fossil desert loess is reported from northern Tripolitania (C. Rathjens 1929) but no grain size analyses are available, and the proximity of the Mediterranean Sea is disturbing.

The material composition of continental dunes is characterized by a dominance of quartz although very appreciable calcite components may be present if the local bedrock is calcareous.

Quartz grain morphology of such aeolian sands has been adequately described by A. Cailleux (1952). Wind-worn frosted grains are quite dominant and well developed.

Paleoclimatic Significance. Fossil dune fields dominate the Saharan subcontinent today. Successive authors have described crusts, paleosols, and stabilizing oxidation horizons of incipient (B)-type from the great Algerian ergs and Fezzan *edeyens*. In fact several dune generations are obviously represented in these. The *goz* of the Sahel belt, which is a broad zone of apparently rather complex genesis, provides an example of fossil dunes that are now located well within semiarid or even subhumid climates. On the other hand the dune fields of the Libyan Desert are apparently mobile and this also applies to innumerable smaller fields in various parts of the Sahara.

In view of the occurrence of both fossil and mobile dunes in full desert country which is quite devoid of vegetation today, the climatic geomorphology of widespread aeolian deposition seems a little perplexing. In fact, J. Meckelein (1959, pp. 63 ff.) suggests that dunes are representative of marginal rather than full desert conditions, and that the Fezzan *edeyens* are really relics of a moister climate. This interpretation is open to question, but it seems certain that Meckelein's conclusion that former moist phases presumably provided large expanses of wadi sands and lacustrine sand plains, which were then available for subsequent deflation under quite arid conditions, is more convincing. Other authors have suggested that the Libyan Desert sands were originally derived from regressional dunes during various phases of the Pleistocene (G. Knetsch, personal communication).

The only solution to reliable paleoclimatic interpretation lies in comparative analysis (eg., Butzer 1959a, c). Local aeolian processes must be studied from the contemporary viewpoint and past and present source areas of deflation must be located and compared with fossil dune orientations. The possibilities of mechanical dune field migration and hence "accidental" sedimentation must be explored. But whatever the situation, provided that sedimentological characteristics are sufficiently distinctive or that the particular dunal morphology corresponds to barchan, transverse or longitudinal forms, the presence of little or no vegetation may be assumed.

II. FLUVIATILE PROCESSES

Alluvial Beds

Pleistocene sediments transported and laid down by more or less channelled watercourses are frequent throughout the subtropical arid and semiarid zones. They represent the effect of regular or irregular stream removal, transport and deposition of clays, silts, gravels or cobbles during waterflow of variable intensity or duration.

Morphology. Water-laid deposits of linear, channelled type are generally preserved in two major forms: undissected or dissected alluvial valley fill (fluviatile *terraces*), and radially deposited *fans* of alluvium at the mouths of steep-sided channels abutting on to open plains or broader valleys.

So-called non-cyclic or non-paired alluvial terraces are not known to the writer in the Mediterranean area or northern Africa. Tectonically controlled terraces are locally present. The overwhelming majority of alluvial terraces are however the result of local climatic change within the catchment area or are an indirect response to climatic change elsewhere expressed by changes in base level of the ocean or of major river valleys. An example of the latter is the relationship between the Nile River and its tributary wadis. Broader stream valleys of the river, arroyo, torrent, wadi, khor, etc., type very often display several terrace levels indicating several phases of climatic change, often susceptible to stratigraphic differentiation.

Alluvial fans are found in numerous situations where steep-sided streams emerge onto broad flat surfaces leading to radial spread of waters, loss of stream volume with distribution and braiding, and hence loss of velocity and carrying capacity. Reductions of stream gradient intensify the localized deposition of alluvium on the margins of the flat terrain. Such alluvial fans of varying size and development are to be found at the piedmont of various mountain ranges (e.g., Morocco, Mallorca) or at the mouths of various wadi channels emerging from dissected country onto open plains or into such large valleys as that of the Nile. Semiarid countrysides today, particularly in the vicinity of the Mediterranean Basin, often have complex alluviated valley systems with irregular valley gradients due to fan deposition at tributary confluences.

Sedimentology. On account of the tremendous range of intensity and duration of transporting capacity by running waters, granulometry, quartz grain morphometry, stratification, and sorting vary within an almost undefined range. According to the geology of the drainage basin in question the mineralogy of sediments will vary from place to place. Except for a basic characteristic of fluviatile wear (some rounding of coarser components) and stratification (crude horizontal bedding of either horizons or individual size components), which is necessary to determine the possible fluviatile nature of deposits, no diagnostic sedimentological characteristics can be cited. This does not render sedimentological study either ineffective or unnecessary, but makes such analysis essential.

Morphometric quartz grain analysis is capable of determining the degree of rounding as well as percentage of water-worn glossy grains (Cailleux 1952). Gravel size and morphometric gravel analyses are moreover essential to the characterization and eventual genetic interpretation of alluvial beds.

Gravel size studies are particularly useful in attempting an over-all interpretation of stream deposits. The standard units employed in size analysis are fine gravel (2–6 mm. diameter), medium gravel (6–20 mm. major axis), coarse gravel (20–60 mm. major axis), and cobbles (over 60 mm.). Completely angular materials such as grit (2–20 mm.) are also distinguished from detritus (over 20 mm. diameter).

Morphometric gravel analysis can only be briefly referred to in this paper but general references to the various indices, applications, and geomorphic significance can be obtained from M. Blenk (1960). After particular application of various indices to largely angular gravel samples of the Mediterranean area it seems that several indices can be most effectively employed. They are (1) the Szadeczky-Kordoss index of rounding (G. Lüttig 1956), which refers to the percentage of smoothed, convex circumference of the individual pebble. The following classes of mean values per sample are suggested by Butzer & Cuerda (1962):

0–10%	angular
11–20%	subangular
21–40%	subrounded
41–60%	rounded
over 60%	well rounded

Such divisions are quite useful for effective characterization of deposits. The Cailleux (1952) index of 2r/L, where *r* is the smallest radius of curvature and *L* the length of the major axis, was not considered satisfactory for analysis of the morphometry of arid zone samples. Further information on genetic homogeneity can be obtained by counting the pebbles of a sample (100 pebbles) with $\rho \leq 8\%$ as well as by the use of the coefficient of variability within a sample. (2) the index of flattening, whereby the ratios E/L and e/l (*E* breadth, *l* minor axis, *L* major axis) are simplest and most effective as comparative indices of dominant sliding (low ratios) or rolling (high ratios) motions (e.g., Blenk 1960). Pebble flattening is an index of mechanical action rather than climatic regime. The ratios indicated above can be more simply and accurately determined by such techniques as Lüttig's (1956), than by the more lengthy procedure involved in the evaluation of Cailleux's (l+L)/2E index.

The significance of morphometric gravel analyses is, however, not absolute.

Nevertheless, such analysis does enable (a) accurate, quantitative description of sediments; (b) comparative analysis of differences in precipitation effectiveness (this can be obtained by the study of fossil gravels of various ages as opposed to the modern bed materials of one catchment basin locality); and (c) differentiation of fluviatile, colluvial, or slope components within heterogeneous beds, or the identification of the dominant transport agent responsible for questionable sediments. Morphometric gravel analysis should always be used in comparison with modern samples from the same local area.

Paleoclimatic Significance. Climatic-geomorphologic interpretation of alluvial beds is only possible when the climatic or base-level stimuli to aggradation can be isolated. If these stimuli can be isolated, what kind of genetic interpretation can be made of "climatic" alluviation?

Local watercourses of the dry belts may be divided into three hydrological types: (1) perennial streams with considerable seasonal fluctuations in volume ("rivers"); (2) seasonal streams with protracted waterflow for a part or all of the rainy season, drying out for several weeks or months during the dry season (arroyos); and (3) episodic streams which experience irregular floods after major rains, often at intervals of many years, but which are otherwise permanently dry (torrents, wadis, or khors). On the basis of sediment study it is certainly possible to determine whether transporting capacity was greater or smaller than that of today, and very often possible to suggest whether a contemporary stream belongs to the same hydrological class as it did in the past. Greater rounding of gravel samples indicates greater transport distance and consequently not only either greater or longer waterflow or both, but also more runoff and a greater availability of moisture (see Table 5). Greater pebble size indicates greater erosive or transport capacity. Better stratification and sorting of the beds may indicate perennial or seasonal rather than episodic flow.

This comparative and partially quantitative approach renders the application of theoretical arguments to the problems of paleoclimates unnecessary. The familiar polemic as to whether alluviation or vertical incision indicates greater aridity or greater humidity, or whether alluviation upstream and downcutting downstream are representative of semiaridity as opposed to aridity, or vice versa, is both futile and unnecessary. The deposits themselves reflect the conditions of deposition and tell their own story. On the basis of sedimentary analysis it seems that climatically induced alluviation in the Eurafrican subtropics—in so far that it was solely determined by local moisture changes—can be largely associated with progressive increases of precipitation. Changes between the different forms of geomorphic equilibrium produce disruptions of the delicate balance of erosion and deposition in a stream system. Soils, residual mantles, and detritus provided by one climatic balance may be available for large scale areal denudation and subsequent alluviation of the entire stream channel during a climatic change to more intense or greater rainfall. Various forms of Pleistocene geomorphic equilibrium in the Mediterranean area have been outlined previously by the author (1961).

It is symptomatic of stream deposits of northern Africa and of the Mediterranean area that terraces usually accompany most or the whole length of rivers, so allowing no differentiation of "erosion upstream, deposition downstream" or vice versa. Instead, the differentiation of areal from linear erosion and deposition is more significant. The fact that many coastal streams aggraded their beds even during falling sea levels suggests that absolute loads are more significant than the longitudinal distribution of complementary agencies in the interpretation of the significance of alluvial beds.

Stratigraphic Significance. It is a well-known fact that alluvial terraces are quite useful as a means to establish relative stratigraphies in given localities especially when found in association with prehistoric assemblages, faunas, or floras. Of greater significance are alluvial deposits whose stratigraphic association to high or low sea levels can be determined, so that local continental chronologies can be related to glacio-eustatic fluctuations of sea level. Thus the complex problem of Nile and wadi terraces in Egypt (Butzer 1959b) is directly or indirectly associated with the Mediterranean chronology of sea level fluctuations. Hence the Pleistocene terraces of many smaller rivers draining to the sea may be of great stratigraphic importance.

Colluvial Beds (*Limons Rouges*)

A curious sediment, which has been described only from the Mediterranean area, western and southern Australia, is what French authors have called *limons rouges* or "red silts." There are predominantly fine beds with a varying admixture of gravel and detritus, and not infrequently, interbedded tufaceous crusts (cf. below). Their location as thin areal spreads on hillsides, in hollows, or on smooth surfaces with gentle gradients indicates that surface washing and colluvial deposition are the agencies responsible. Eroded soils of *terra rossa* type form the major constituent, along with a variable quantity of angular to subrounded local detritus. Aeolian components may also be present. Such sediments were first interpreted by G. Choubert (1948a, b) in Morocco, who showed that they were due to pluvial washing and sheet-flooding. C. Arambourg (1952) has also outlined analogous cave silts and surficial silts in North Africa and the Levant, where they were associated with a Levalloiso-Mousterian-type industry and an Upper Pleistocene fauna of warm affinities. A pedological study in Algeria by J. H. Durand (1959) represents the most recent analysis of more general interest. A synthetic analysis of the colluvial silts is still lacking while not one of the recent geomorphological texts (including French efforts) mentions such deposits.

Morphology. These deposits usually occur as relatively thin areal sheets, which seldom attain a meter in thickness. Thicknesses of 5 m. may be attained at the bottom of moderate or steep slopes or in original topographic depressions. Surface slopes seldom exceed 15% and characteristic examples can be cited from Mallorca where the general bedrock topography has slopes of 1-3%. In drainage channels the silts can often be observed to grade laterally into fluviatile conglomerates with an identical silt matrix. Clearly then the sediment morphology is nondescript, and consists of essentially shallow, irregular detrital mantles. It is not surprising that closely analogous alluvial and cave features should have been recognized first since the greater part of the typical spreads were simply considered as "soils."

Sedimentology. Corresponding to the colluvial character of these silts the sedimentology is *par force* highly variable from one instance to another. In general stratification of individual beds rather than of individual coarse components is characteristic while sorting is less common.

In the silt mantles on gentle or moderate slopes the coarse component of over 2 mm. diameter is largely confined to coarse angular to subrounded detritus or gravels embedded in finer sediments. The granulometry of the fines shows a spectrum in which the clay, silt, and sand (fine, medium, and coarse) fractions are approximately equal (Table 3). Frequently a moderate maximum may be

FIGURE 3

Topographic Situation of typical Colluvial Silts (from southern Mallorca). Heavy black line indicates silt deposits 50–200 cm. thick. Vertical exaggeration 5:1.

TABLE 3
GRAIN SIZES OF COLLUVIAL SILT SAMPLES FROM MALLORCA AND CABRERA
(in percent of fraction under 2 mm.)

(A–D: typical range of typical *limons rouges* samples derived mainly from ancient soils with a little aeolian admixture, primary or derived; E, F: semi-aeolian *limons rouges;* G: *limons rouges* derived from slightly weathered aeolianite; H: semi-aeolian silt resembling loess in appearance.)

Sample	*200–2000μ*	*60–200μ*	*20–60μ*	*6–20μ*	*2–6μ*	*2–60μ*	*<2μ*
A	7.8	20.1	4.1	21.6	5.6	31.3	40.8
B	6.9	4.4	21.7	18.1	14.7	54.5	34.2
C	18.7	17.9	12.8	14.4	12.7	39.9	23.5
D	25.0	16.7	7.9	22.7	12.3	42.9	15.4
E	42.2	25.4	9.5	7.8	6.9	24.2	8.2
F	13.0	37.2	13.6	11.2	11.0	35.8	14.0
G	50.9	25.1	12.4	1.3	3.5	17.2	6.8
H	13.7	20.7	19.5	32.6	8.9	61.0	4.6

found in the coarse silt and finer sand fraction, but the writer knows of no granulometric spectra in which over 70% of the sample occurs in the 2–60 micron range and consequently typical loess deposits are not known.[1] Semiaeolian beds are however not uncommon in association with regressional dunes, or specifically aeolian materials bedded by water action. More frequently however the grain-size spectrum is related to that of the source materials for surface washing, namely, soils and weathered older aeolian beds.[2] Particularly conclusive evidence of the dominance of water bedding in all but a very few coastal deposits is provided by the axial orientation of gravels, conformable tufaceous calcareous crust beds, and submicroscopic bedding of the fines. Such criteria are important when grain sizes are almost identical with those of aeolianites (e.g., Table 3 E, and Table 1 C).

Micromorphologic study of quartz grains (Table 4) does not provide any very significant information.

With the exception of unstratified breccia-like sediments, which are caused

1. C. Virgili & I. Zamarreño (1957), and L. Solé Sabarís (1960, personal communication) consider that true loess deposits occur at Sant Andriá near Barcelona. The writer visited the site together with Dr. Solé but does not consider them distinct from analogous beds on Mallorca whose granulometry is not truly "loessic" (Table 3, H). Grain size analyses of the Sant Andriá "loess" have been made, but the writer has not been able to obtain any details. In our opinion these beds are semi-aeolian silts and in fact innumerable water-laid sand, grit, or pebble bands interrupt the profile.

2. Another variety of red silt has been suggested by R. W. Hey (1962) from interior Tripolitania, which on the basis of unpublished laboratory analysis are thought to be "aeolian silts." Just how unequivocally "aeolian" the grain size spectrum may be remains to be seen, however.

TABLE 4
MICROMORPHOLOGIC ANALYSIS OF QUARTZ GRAINS
(of medium sand fraction, in percent. Refers to Table 3)

Sample	*Waterworn*	*Wind-worn*	*Unworn*
A	26	18	56
C	36	51	13
D	36	6	58
F	50	32	18
G	18	49	23
H	24	28	48

by the combined gravity and fluviatile activity of very steep slopes, *limons rouges* contain an interesting assortment of slightly rolled gravel wherever detritus (excluding aeolianite rubble) is locally available. A knowledge of the modifications of this gravel is essential to a determination of the conditions of transport and deposition.[3] Table 5 shows that the degree of rounding of gravel contained in colluvial beds is similar to that of gravel found in modern torrent beds, although it is usually somewhat better. Occasional bands of sub-rounded stream gravels (such as Sample I in Table 5) occur within the silt beds and emphasize the importance and duration of water flow in comparison to that of the present. The E/L ratio indicates that rolling in contrast to sliding motions were dominant during transport. The large number of gravel samples of colluvial silts analyzed from Mallorca and Soria indicate that both overall transport capacity and duration of effective water flow was equal or greater than that of the episodic and seasonal streams of today. The spectrum of homogeneity as expressed by the coefficient of variation is identical to that of modern torrents. Consequently violent, torrential sheetflooding as opposed to the gentler forms of surface washing (*ruissellement*) may be considered as the geomorphic agency responsible, aided by gravity action in areas of accentuated relief.

Paleoclimatic Significance. As indicated, colluviation of *limons rouges* is essentially a fossil process in subtropical latitudes, in spite of the atypical counterpart in southern Morocco and such analogous features as soil stripping induced by anthropogenic action in the Mediterranean world today. Modern analogies of greater pertinence may well exist in parts of the savanna belt today, but they are incompletely understood or have not yet been described. It therefore seems necessary to present an independent interpretation of the fossil features. Prolonged

3. Prof. R. Négre informed the writer that silt colluviation by surface washing is known to occur at very slow rates today, namely in southern Morocco in a zone with only 300 mm. precipitation. These are fine sediments, and cannot be considered genetically identical with Pleistocene *limons rouges*, for the gravel contained in Pleistocene beds of this type indicates greater water transport than today (in areas with 400–700 mm. precipitation at present).

TABLE 5

MORPHOMETRIC ANALYSES OF ALLUVIAL AND COLLUVIAL GRAVELS

(Samples A, D, F–I, and K from Mallorca; C, E from Gerona; B, J from Soria. C and E as well as G–I are from single stratigraphic sequences. K consists of colluvial silts of semi-alluvial type.)

	Sample Description	*Sample Size*	*Mean ρ in %*	*Percent ρ ≤ 8%*	*Coefficient of Variability of ρ (in %)*	*E/L (in %)*	*Lithology*
A	Episodic stream (recent)	100	12.1	57	111.3	52.1	Limestone
B	Episodic stream (Pleistocene)	100	10.9	55	95.3	58.3	Limestone
C	Seasonal stream (recent)	50	37.7	16	63.7	......	Granite
D	Seasonal stream (Pleistocene)	50	36.3	4	47.0	51.7	Limestone
E	Perennial stream (Pleistocene)	50	45.8	0	34.0	......	Granite
F	Perennial stream (Pleistocene)	100	57.6	0	36.5	55.2	Limestone
G	Episodic stream (recent)	100	8.4	68	106.2	55.4	Limestone
H	Pleistocene colluvium	100	14.0	52	104.7	53.0	Limestone
I	Pleistocene colluvium	100	23.0	38	78.3	58.9	Limestone
J	Pleistocene colluvium	100	12.9	43	83.7	58.3	Limestone
K	Mixed colluvium (Pleistocene)	100	26.6	10	55.0	......	Limestone

torrential rains, of some duration and not inconsiderable frequency, imply that annual rainfall amounts must have been at least as great as they are now. The possibilities of violent sheetflooding to strip soil and transport detritus would be unlimited under the vegetation type which exists in Mallorca, Catalonia, or Soria today. The garrigue has no sod grasses and little or no bunch grass or other soil-protecting herbs, while much bare soil is exposed. Even in the light oak or mixed oak-pine woodlands, dispersed coarse tuft grasses and the little or non-existant sod vegetation also provide only incomplete soil protection. Similar sod conditions also prevail under woodlands with 500 mm. or even with 1000 mm. precipitation today. Similarly, an increase in aridity would not significantly change the area of bare soil. Consequently no change in vegetation physiognomy or association is necessary to permit colluviation of the *limons rouges* type. The present seasonality of rainfall must therefore have been characteristic.

An explanation of more "fluvial" rather than absolutely moister conditions is untenable as present rainfall is rarely sufficient for coherent runoff today. (Butzer 1961). Morphologically effective sheetflooding under Mediterranean climates can only be understood as a consequence of pluvial erosion and transport. Only prolonged, intensive rainfalls will saturate the dominantly coarser textured *terra rossa* type soils to the point that they become impermeable. Accelerated runoff and soil stripping will then follow. These are the characteristic

FIGURE 4
Upper Pleistocene aeolianite embanked against Miocene limestone cliffs. Crude breccia in colluvial silts overlying aeolianite at right. Southern Mallorca, S'Estret d'es Temps.

deposits of the geomorphologically significant pluvials of the Mediterranean zone although they do not represent the wettest Pleistocene climates of this area, which seem to have occurred during phases of red soil development (Butzer 1961).

Stratigraphic Significance. Stratigraphic aspects of the colluvial silts are analogous to those of alluvial beds. The widespread occurrence of *limons rouges* particularly along coastlines enhances their stratigraphic value. Abundant snail faunas when studied statistically may be used not only to indicate ecological conditions but can be stratigraphically employed. Similarly, Munsell colours are often useful in the field since a relationship frequently exists between clay content and intensity of red colouration. Thus certain *limons rouges* are sufficiently distinctively coloured to be of some limited stratigraphic value in local correlations. Consequently, in spite of their nondescript morphology colluvial beds are of prime significance for total interpretation of past environments. It is interesting that major silt deposition of this type can be associated with the very beginning of the major glacial regressions.

Tufaceous and Travertine Crusts (*Croûtes zonaires*).

A widespread phenomenon of many subtropical and tropical lands are calcareous crusts or "caliche," which are mainly of Pleistocene, but also partly of Recent date. Interpretations of such crusts vary to an unbelievable degree and indicate that genetically distinct phenomena are too often grouped under the one misleading term "calcareous crust." In fact almost every author using this term implies something quite different from that described by another. The only comprehensive study for part of the Mediterranean area has been made by J. H. Durand (1959, pp. 75–136), who outlined several distinct types of "crusts" as follows: [4]

1. Powdery sediments (*formations pulvérulentes*). Powdery lime, gypsum, or salt inorganically precipitated at the bottom of lakes, lagoons, or playas.
2. Zonal crusts (*croûtes zonaires*). Calcareous surface deposits left by intermittent sheetflooding of highly calcareous waters. Each film of sediment is consolidated between successive water flows. Durand incorrectly assigns a Lower Pleistocene age to these crusts.
3. Chalky nodules (*nodules farineux*). Nodules formed at the same time as the deposition of a particular sediment by the almost instantaneous segregation of the carbonates within contraction fissures or hollows.
4. Concretions (*nodules concretionnés*). Nodules formed by—allegedly only modern—pedogenesis which involves calcareous precipitation at greater depth.

4. A lengthy lithological but unfortunately not pedologic or stratigraphic study by E. Rutte (1958) treats calcareous crusts in Spain. Both Durand's and the writer's interpretations disagree with Rutte's. Rutte considers the various genetic types of calcareous crust as genetically analogous.

FIGURE 5
Upper Pleistocene colluvial silt-breccia. Southern Mallorca, S'Estret d'es Temps.

This is a somewhat poor characterization of the typical Ca (calcium) or Sa (salt) horizons, due largely to carbonate (or salt) leaching of upper soil horizons. Such pedogenetic Ca horizons (as they should be called) form both nodules, irregular horizontal bands which often grade into massive honeycomb structures, as well as dense chalky horizons and chalky precipitation along contraction fissures. These deposits may occur in the (B) or C horizons. Nodules may also be chalky, and the genetic class of chalky nodules identified by Durand is therefore a little dubious and certainly difficult to identify.

5. Calcrete cementation (*encroûtements calcaires*). Precipitates or evaporites of calcium carbonate or salts found in shallow zones of aeration above a ground water table which is situated near the surface. These calcrete crusts are formed by the upward movement of capillary water with dissolved materials during periods of desiccation. Similar processes may also be involved in the formation of pedogenetic Ca horizons.

The most widespread and indicative of the various "calcareous crusts" are the so-called *croûtes zonaires* or tufaceous (and sometimes travertine-like) varieties. These will be considered in some detail here on the basis of Spanish samples studied.

Morphology. These deposits are widely found in association with colluvial silts, where they are either interbedded or cap horizons of a few millimeters to several tens of centimeters thick. Consequently they generally have no morphology whatever, except where independent areal spreads in hollows or shallow depressions occasionally attain 50–100 cm. These durable crusts have geomorphological significance when they form land surfaces.

Sedimentology. Sediments are composed of 75% or more of crypto-crystalline calcite with some detritus which varies from microscopic to pebble size. This detritus includes calcite crystals, often weathered; quartz grains, largely waterworn; manganese particles; feldspar crystals, if present in local bedrock; clay aggregates from microscopic to pebble size, representing stripped soil particles or lumps; and finally silicified root particles. Iron oxides are distributed within the free calcite and their presence leads to discoloration of the sediment in fine, wavy bands. A high colloidal SiO_2 content is present.

The macrosedimentology can be summarized under three headings.

1. Laminated, former surface crusts. These shallow, well-cemented crusts are most frequently if not almost invariably found in association with colluvial beds. The lamina can be easily recognized on the basis of the fine oxide bands. Although typically a pure white, these crusts have many rather discolored bands of pink, reddish yellow, or yellow color, which reflect the considerable silt content in the form of oxidized calcite crystals or of minute soil particles (clay aggregates). The discolored crusts have a transitional position between silts and tufaceous crusts, and the white laminated tufas indicate clear sedimentation with a reduction of mechanical erosion and a cessation of alluviation. As C. Virgili (1957) has already shown in the Barcelona area, concretionary zones are most frequently found below a capping of calcareous crusts. These are often but not necessarily conformable with the tufaceous crust, and are mainly of the calcrete cementation type caused by upward movement of capillary waters. They are however also formed by downward percolation of lime-charge waters but in many instances both movements are involved. It is difficult to decide whether an upward or a downward movement is dominant in this type of process.[5]

2. Massive, inconspicuously bedded calcite beds. These comparatively rare alluvial-type crusts which often possibly attain a meter or more in depth also contain coarse detritus in the form of crude, subangular chunks of former dehydrated soils of *terra rossa* or *terra fusca* type. Subsequent calcification imparts to such inclusions a resemblance to rock pebbles within a consolidated although less resistant calcite mass. Such massive beds may also occur in association with spring tufas of the soft, rather porous type which have frequent plant impressions.

3. Travertine beds. Massive tufaceous calcite sediments may grade locally into true travertines of a vertical, columnar structure.

Paleoclimatic Significance. Considerations of the genesis of the tufaceous or travertine-like calcareous crusts (*croûtes zonaires*) suggest that they are un-

5. The writer gratefully acknowledges stimulating discussion with L. Solé Sabarís on genetic differentiation of crusts. Solé outlined a rather significant and appealing sequence of calcareous segregation, concentration, and deposition during and after sedimentation of the sediment concerned. The writer looks forward to publication of the results and conclusions.

FIGURE 6
Colluvial silt grading into laminated, calcareous crusts. Middle Pleistocene age. Southern Mallorca, KM 5 of local route to Cap Salines.

doubtedly formed by the surface washing of lime-charged waters (Durand 1959, Butzer 1961). Durand believes that such slow calcareous precipitation is only feasible with warm waters. Certainly the lack of numerous coarse components such as sands and gravel presumably indicates that there is little loose detritus other than cracked, desiccated *terra rossa* soils which may have been available for stripping. The presence of the latter soils may suggest that these crusts were largely deposited after the first, severe late summer rains after the dry season. Also if parched *terra rossa* chunks were available in September, for example, then grain by grain soil stripping should have been possible later on during the rainy season. Many of the interbedded tufaceous crusts in colluvial silts may well be explained in this way.

The capping tufaceous crusts on silt deposits as well as the massive calcite or travertine beds must however be interpreted as representing a change in vegetation-soil-rainfall equilibrium. A rather complete mat of herbaceous vegetation must have sprung up after the first rains so that pluvial soil erosion was limited to removal and transport of odd fragments of dehydrated soil in late summer. The subsequent development of a sodgrass type vegetation, as is found today near seasonal or perennial sources of underground or spring waters, then effectively inhibited any further erosion. Such sodgrass along with the absence of much loose detritus suggests that there was a rather dense vegetative mat and very possibly incipient pedogenesis of the *terra rossa* or *terra fusca* type. The stripped soil fragments indicate that the dry season persisted, but over-all precipitation must have been greater than during the colluvial silt phases. For the Balearic Islands area—with 400–600 mm rainfall and with limestone bedrock—these fossil crusts represent wet phases with a precipitation several hundred millimeters greater than today. Rainfall reliability was probably greater, rainfall distribution more equitable, and over-all intensity of rainfall less or at least no greater than that of today. In other regions of the Mediterranean or the savanna belt analogous features may however be indicative of drier conditions. So for example, the limited carbonates present in areas with silicate bedrock are rapidly lost through solution under moist conditions. It is thus absolutely essential to specify the particular type of calcareous deposit concerned and to study its local significance. Generalizations over wide areas are dangerous and such frequent stereotype statements that "caliche deposits" are indicative of "aridity" without close investigation or specification are irresponsible.

Other than in local association with springs, tufaceous crusts no longer develop in the Mediterranean area (Durand 1959, cf. Rutte 1958). This raises a last outstanding problem of the original carbonate solution in water. This could of course be performed by the surface solution of humic acids which are derived from the biochemical activity of vegetation and soil fauna. Very little calcite however, would be released at the surface after a late summer rain. Strongly increased spring activity—seasonal or perennial—over larger areas must also have been involved. General surface washing (rather than violent sheetflooding) by activated, lime-charged underground waters after protracted rains seems the most probable source of these tufaceous and travertine-like crusts. The *terra rossa* fragments embedded in calcite may not even be related to these lime charged waters, but may well have been removed by earlier rain storms and then subsequently buried by lime precipitates. Innumerable fissures which show evidence of former spring activity can be observed on the cliff coasts of the Balearic Islands today and certainly support this interpretation. A climatically significant increase in spring activity must be unquestionably linked with travertines and thus stratigraphically associated with the tufaceous crust formation on the Balearic Islands.

In conclusion it is suggested (1) that tufaceous and travertine-like crusts or beds of the Mediterranean area are largely formed by precipitation from slowly moving, lime-charged waters of limited or moderate underground trajectory in soil or bedrock and also in part from sheetflooding rain waters; (2) that they are deposited under a dense vegetation mat with considerable spring activity and sufficient rainfall, with temperatures certainly no lower than those of today; (3) that their coarse clastic components were probably embedded *in situ* by such precipitates after their original transport by the first late summer downpours, before the grass mat had developed again after the intense dry season.

FIGURE 7
Massive, inconspicuously bedded calcite beds of Upper Pleistocene age, with embanked "pebbles" of *terra fusca* soil. Overlying semicemented aeolianite. Southern Mallorca, Cala Pi.

Stratigraphic Significance. Tufaceous crusts have only a limited stratigraphic significance, with exception of the massive calcite beds of late interglacial date sometimes found disconformably under regressional sequences (Butzer 1961; Butzer & Cuerda 1962). In their relationship to colluvial silts the *croûtes zonaires* proper occur too frequently and variably to be of anything but local paleoclimatic value.

III. LACUSTRINE AGENCIES

Subaqueous deposits of the arid and semiarid subtropics comprise a great number of genetic types:

A. Coastal-Lagoonal
   1. Salt water (marine)
   2. Brackish water (estuarine and semilacustrine)
   3. Fresh water (lacustrine)
B. Interior-Lacustrine
   1. Fresh and brackish water (swamps and lakes)
   2. Salt water (lakes and pans)

A wide variety of facies is also present:

1. evaporites, usually gypsum or halite;
2. calcareous beds, including chalky;
3. marls;
4. silts and clays;
5. sands;
6. organic deposits.

The following remarks are intended to describe no more than some of the typical aspects of these facies, since it is impossible to generalize about a multitude of local possibilities.

Evaporites, mainly gypsum (calcium sulfate) and other salts such as sodium, magnesium, and potassium chloride, which are frequently indicative of desiccation. Such desiccation may refer to periodic seasonal shrinkage during the dry season, or long term reduction of a larger lake to a lagoon or pan. So for example the Middle or Upper Pleistocene Jordan Sea, which was the ancestor of the Dead Sea, deposited some 50,000 varvelike alternations of silts (rainy season influx) and carbonates, sulfates, or chlorides (dry season evaporites) (cf. Butzer 1958, p. 78). Evaporites with the exception of open coastal lagoons are indicative of some degree of aridity.

Calcareous beds, usually *lacustrine chalks*, are not infrequent in French North Africa (Durand 1959, pp. 75–90) and in other parts of the Sahara, such as in

FIGURE 8
Travertines with intercalated massive calcite bed, consolidated aeolianites at base and top. Southern Mallorca, Cala Pi.

the depressions of the Kordofan *goz*. These chalks may or may not be fossiliferous. Such chalks usually indicate perennial lakes which were not subject to very great seasonal fluctuations of oxygen content.

Marls, or highly calcareous silts, may be deposited in both lakes and swamps. They are more frequent in the semiarid than in the arid zone where lateral streams are able to carry soil products into comparatively small water bodies.

Silts and *clays* are generally carried into standing waters in suspension by local streams. They may however also be at least partly aeolian in origin, as is the "lacustrine loess" of the Namakzar Kavir of the Persian Lut Desert (cf. Hückriede, 1962). Lacustrine silts and clays are again less typical of the Pleistocene deposits of the arid zone, and are generally limited to local situations such as spring-fed lakes in some of the Saharan oases.

Sands of lacustrine deposition are a comparatively widespread Pleistocene facies in the Saharan region. They are largely derived from sandy wadi deposits, and to a lesser extent also from wave action on local bedrock. The prehistoric Fayum and Chad lakes are striking examples.

Organic deposits, although comparatively rare in the subtropics are however not unknown, particularly in the temperate mountain areas of the Mediterranean zone. These may be gyttyas, peats or humic sands, silts, and clays. Few such deposits of Pleistocene date are known, however; their major significance is the possibility they provide for palynology (e.g., Menéndez-Amor & Florschütz 1961).

Some of these deposits are susceptible to different methods of interpretative study, but since local conditions will necessarily determine such methods, it does not seem warranted to discuss lacustrine deposits any further here.

IV. CONCLUSIONS

It is hoped that this outline of characteristic Pleistocene sediments and their paleoclimatic and stratigraphic significance may possibly serve towards a tentative sketch of Pleistocene sedimentary processes in the Old World sub-tropics. General textbooks and many specialized studies attempt to interpret lower latitude geomorphology summarily by analogy to higher latitude processes. Almost half of the existing literature on the lower middle latitude Pleistocene is devoted to glacial or "periglacial" features. No comprehensive study of Pleistocene processes equatorward of the former cold climate morphogenetic system exists.

Because of this lack of appreciation of the individuality of lower latitude processes, many a higher latitude specialist has erroneously ascribed glacial or "periglacial" attributes to features such as colluvial breccias. It is insufficient to have theoretical acquaintance with arid zone morphology and yet be unaware of the range of sedimentary processes present in semiarid and arid regions. Pleistocene geology in lower latitudes is as specialized as glacial geology or "periglacial" geomorphology.

A thorough investigation of contemporary and Pleistocene processes of the lower latitudes must be the object of many qualified specialists in different areas. Such process investigation should emphasize weathering, pedogenesis, and sedimentation. Above all, study of sites or regions should aim at comprehensive interpretation. Only in this way can the paleoecology of archaelogical sites, which is so crucial to our understanding of prehistoric time, be effectively reconstructed.

BIBLIOGRAPHY

ARAMBOURG, C.
1952. "The Red Beds of the Mediterranean Basin," *Proc. 1st Pan-Afr. Conf. Prehist.* (Nairobi 1947), Oxford: Blackwell, pp. 39–45.
BAGNOLD, R. A.
1954. *The Physics of Wind-blown Sand* (rev. ed.). London: Methuen.
BLENCK, M.
1960. "Ein Beitrag zur morphometrischen Schotteranalyse," *Zeit. Geomorph.*, 4:202–63.
BUTZER, K. W.
1958. "Quaternary Stratigraphy and Climate in the Near East," *Bonner Geogr. Abhandl.* 24.
1959a. "Some Recent Geological Deposits on the Egyptian Nile Valley," *Geog. J.*, 125:75–9.
1959b. "Contributions to the Pleistocene Geology of the Nile Valley," *Erdkunde*, 13:46–67.
1959c. "Die Naturlandschaft Agyptens während der Vorgeschichte und der Dynastischen Zeit," *Abhandl. Akad. Wiss. Liter. Mainz, Math.-naturwiss. Kl.*, No. 2.
1961. "Palaeoclimatic Implications of Pleistocene Stratigraphy in the Mediterranean Area," *Ann. New York Acad. Sci.*, 95:449–56.
BUTZER, K. W. and J. CUERDA
1962. "Coastal Stratigraphy of Southern Mallorca and its Implications for the Pleistocene Chronology of the Mediterranean Sea," *Jour. Geol.*, 70:398–416.
CAILLEUX, A.
1952. "Morphoskopische Analyse der Geschiebe und Sandkörner," *Geol. Rund.*, 40:11–19.
CHOUBERT, G.
1948a. "Sur l'age des limons rouges superficiels du Maroc," *C. R. Acad. Sci.* (Paris), 227:558–60.
1948b. "Sur la nature des limons rouges superficiels du Maroc," *ibid.*, pp. 639–41.
DURAND, J. H.
1959. "Les sols rouges et les croûtes en Algerie," *Serv. des Etudes Scien.*, Birmandreis.

Hey, R. W.
1962. "The Quaternary Geology and Palaeolithic Archaeology of Libya," *Quaternaria*, 6:435–49.
Hilly, J.
1951. "Les formations quaternaires du Massif de l'Edough et du Cap de Fer." Mimeographed extract of 1955 *Thèse.*
Hückriede, R.
1962. "Jung-Quartär und End-Mesolithikum in der Provinz Kerman," *Eiszeitalter und Gegenw.*, 12:25-42.
Kaiser, K. H.
1960. "Libanon." Colloquium at the Geological Institute, University of Cologne, November 30, 1960.
Lüttig, G.
1956. "Eine neue einfache geröllmorphometrische Methode," *Eiszeitalter and Gegenw.*, 7:13–20.
McBurney, C. B. M., and R. W. Hey
1955. *Prehistory and Pleistocene Geology in Cyrenaican Libya.* Cambridge: Cambridge University Press.
Meckelein, W.
1959. "Forschungen in der Zentralen Sahara, *Klimageomorphologie.*" Braunschweig: G. Westermann.
Menéndez Amor, J., and F. Florschütz
1961. "Contribucion al conocimiento de la historia de la vegetacion en España durante el Cuaternario," *Estudios geol.*, 17:83–99.
Mensching, H.
1955. "Karst und Terra rossa auf Mallorca," *Erdkunde*, 9:188–96.
Rathjens, C.
1928. "Löss in Tripolitanien," *Zeit. Ges. Erdkunde Berlin,* pp. 211–28.
Rutte, E.
1958. "Kalkkrusten in Spanien," *Neues Jb. Geol. u. Paläontol.*, Abh. B, 106:52–138.
Sayles, R. W.
1931. "Bermuda During the Ice Age," *Proc. Amer. Acad. Arts Sci.*, 66:382–467.
Shepard, F. P., and Young
1961. "Distinguishing between Beach and Dune Sands," *J. Sedimentary Petrol.*, 31:196-214.
Virgili, C., and I. Zamarreno
1957. In "Livret Guide de l'Excursion Environs de Barcelona et Montserrat," *5th Int. Cong. INQUA* (Madrid-Barcelona), pp. 7–16.
Wright, H. E.
1962. "Late Pleistocene Geology of Coastal Lebanon," *Quaternaria*, 6:525–39.

THE DISTRIBUTION OF TROPICAL AFRICAN BIRDS AS AN INDICATOR OF PAST CLIMATIC CHANGES

R. E. MOREAU

THE DISTRIBUTION of birds in Africa south of the Sahara is now comparatively well documented, both specifically and subspecifically.[1] Also, the taxonomy is more advanced than in other animal classes or in plants. There is no doubt that with few exceptions most of the bird species are associated very closely with individual vegetation types, but detailed study of this relationship awaits more complete data. Meanwhile, there are undoubtedly certain features in the distribution of birds which are so difficult to explain in terms of the present climatic picture that it is necessary to postulate past changes on a considerable scale. This is the subject of the present study. Attention will be concentrated on groups rather than individual species, which are more liable to be misleading. Some ancillary evidence will be adduced from other groups of animals and from plants.

First of all, there is one unexpected advantage and one specific disadvantage so far as this discussion is concerned. The advantage is that provided the biotope does not fluctuate enough to force the birds to move, they seem to remain remarkably sedentary. In montane forest this is evinced not merely by observation but also by the fact that well-marked subspeciation, implying reproductive isolation, can take place on mountains only a few miles apart and within easy sight of each other—for example, the white-eyes *(Zosterops)* and thrushes *(Turdus)* on Kilimanjaro, Teita, and North Pare, and *Francolinus* on Mt. Kenya and the Aberdares.

The disadvantage for the present discussion is the great lack of fossil material in Africa. Avian fossils are in any case difficult to identify accurately, particularly where passerine ("song") birds are concerned (see critical remarks by Moreau 1954a). The result is that, even with the more complete temperate-zone data, the best possible generalization about the ages of bird taxa is: "Many modern genera probably existed in the Pliocene, but many of the species differed from the recent ones" (Storer 1960). This may perhaps give an exaggerated im-

1. I am indebted to Mrs. B. P. Hall, R. W. Hayman, Dr. L. S. B. Leakey, Dr. B. Verdcourt, and L. T. Wigg for kindly providing data and for discussing problems with me; also to Dr. D. Lack for criticism of the final draft.

pression of the age of modern bird species. If the age of mammal species, comparatively well documented from fossil records, can be taken as a guide, it seems probable that climatic postulates based on the distribution of existing bird species relate to the latter half of the Pleistocene. Subspecific differentiation in birds, which has been well studied, can contribute little to the problems under discussion, since wide variation can exist in the rate at which such evolution takes place; apparently "good" subspecies can appear in less than 5000 years (cf. Mayr 1942).

The most significant features of avian geography in Africa concern the distribution of birds of the evergreen forest. Under present conditions this biotope is especially liable to appear as ecological islands. This applies particularly to the montane forest, which will be dealt with first. While colonization of montane forest islands as a result of individual movement cannot be ruled out, they appear rather to have been occupied by an already integrated community, one associated with a community of plants. This seems to require a breakdown of isolation, with whatever climatic changes that implies.

THE MONTANE AVIFAUNA

"Montane evergreen forest above about 1300 m. usually differs from its lowland counterpart in floristic composition, in the abundance of epiphytic bryophytes, and in the smaller height of the trees. *Olea, Ocotea, Juniperus, Podocarpus, Schefflera,* and *Pittosporum* are among the many genera characteristic of this type" (Keay 1959). In tropical Africa typical montane forest exists, at least vestigially, on most mountains that rise much above 5000 feet (at much lower altitudes near the sea). It may meet lowland evergreen forest (with a transition zone) but under existing conditions it very often does not. In any case, it is so distinct floristically that, in the opinion of botanists (quoted by Moreau 1952), no important element of it has evolved from neighboring lowland forest. The origin of this characteristic montane community, including its fauna, remains as much of an enigma as it was ten years ago.

In recent times, the areas of montane forest have been greatest on the Kenya highlands and on the mountains along the eastern rim of the Congo basin. In addition, these are in eastern Africa between northern Kenya and the Zambesi, numerous smaller areas effectively isolated one from the other by savanna, thorn steppe, and even subdesert. Some of these have been reduced by human activity to a few hundred acres, as on the Teita Hills of southern Kenya and on mountains in southern Nyasaland, while remaining typical (see especially Benson 1948 for Nyasaland). By contrast, the extensive Abyssinian plateau, while having some species in common with the Kenya highlands, is floristically so distinct as to constitute a separate "phytogeographical region" (Prof. R. Pichi-Sermolli, cited in Moreau 1952). In the vast area of western Africa, islands of montane vegetation are scarce. There are some small relict patches in the highest parts of southwestern Angola and a more important group around the head of the Gulf of Guinea, namely, Cameroon Mt., Kupe, Manenguba, the Bamenda-Banso highlands, and the Obudu plateau, all within a radius of about eighty miles. But isolated as these mountains are from the others in tropical Africa, the "vast majority of the [plant] species occurring in them" are "also in East Africa" (R. W. Hepper *in litt.*). West of this there appears to be nothing that is, ecologically speaking, typically montane, though several small massifs on the northern edge of Liberia and Sierra Leone reach the necessary height, and the Futa Jallon in the Republic of Guinea approaches it.

The small size and the grouping of many of the montane islands makes it

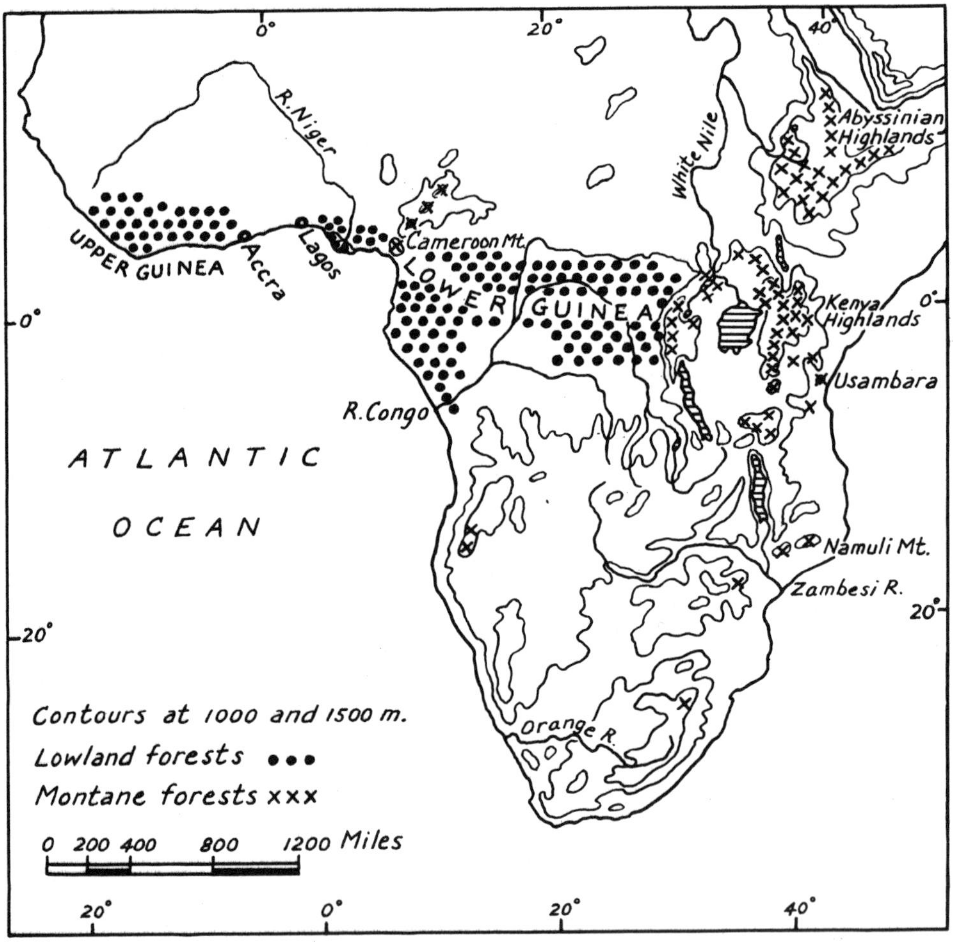

FIGURE 1
Sketch-map of the Ethiopian Region to show the distribution of the principal lowland and montane evergreen forests. (A large number of isolated montane forests in East Africa are too restricted to be shown on a small-scale map.)

difficult to map them adequately, even on the 1:10,000,000 scale adopted by the "Vegetation Map of Africa" (1959). The same applies to the relict lowland forest on the East African coast. All that the sketch map can do is to convey a general impression (Fig. 1).

While the altitude of 1300 m. (approximately 4200 ft.) has been quoted as indicating the transition between lowland and montane forest, this figure is subject to important modifications so far as the avifauna is concerned. In much of the interior of the continent the transition seems to come at about 5000 ft. (cf. Chapin 1932), but it is much lower close to the sea, e.g., around the head of the Gulf of Guinea and in the Usambara and Uluguru Mountains of Tanganyika Territory. In the forests of Usambara there are 35 species of passerine birds at 4500 ft., of which 26 occur down to 3000 ft., but only six at 2000 ft. (Moreau 1954b), so that in that area the main change in the avifauna takes place around 2500 ft. In general, the very marked change in the avifauna with altitude and the occurrence of this change throughout tropical Africa led Chapin (1923) in his study of African bird geography to designate a "Humid Montane District," the validity of which subsequent work has fully confirmed. (It is by no means only in Africa that such a division has been found applicable. In Colombia, for example, Chapman (1917) drew the line between the Tropical Zone and the Subtropical Zone at 1600 m., or 5200 ft. "The forest being continuous, an altitudinal difference of 1000 ft. may bring one into a essentially new avifauna.")

In Africa, as elsewhere, the oustanding fact is that the montane forest avifauna has no close affinity with that of the lowland forest or of any other biotope. For example, Chapin (1932) after citing 90 species of birds "typical of the Mountain Forest Zone of the Eastern Congo" but not occurring together anywhere, adds: "One might expect that in and about the mountain forests there would be highland races of many of the tropical lowland birds, but these are relatively few"; he cites only a dozen. In fact the difference in avifauna between the montane and lowland forest is nearly as great as that between the lowland forest and the savanna, which is almost total (Moreau 1954). It may be added that the montane avifauna derives practically nothing from the palaearctic or from that of extra-tropical South Africa. This is the more surprising since great numbers of palaearctic birds annually come to tropical Africa to winter, and since the African montane flora outside the forest shows much influence from both temperate zones.

The individuality of the montane forest bird community is emphasized by the extent to which it reappears on widely separated mountains. For example, of the 33 species around the southern end of Lake Nyasa, 29 occur also in the Usambara Mountains, 700 miles to the north; and the isolated Namuli Mountain of Portuguese East Africa shares 23 of its 25 montane birds with Usambara. More striking still, of 41 species on Cameroon Mountain that do not occur in lowland forest, 20 also inhabit mountains 1300 miles away on the eastern side of the Congo basin. Three of the 20 are not even distinguished subspecifically.

Further, in addition to these 20 species, seven others are members of superspecies.[2] The outstanding example is *Francolinus camerunensis*, which has apparently close relatives on mountains in Angola, French Somaliland, and Kenya. Even in the vestigial montane forest of southwestern Angola, which counts about 20 species (Hall 1960a), half of them appear in mountain forests some 1500 miles to the east.[3]

It may be emphasized that, as the foregoing facts suggest, the typical montane forest bird community does not merely occupy the discontinuous remnants of former vast montane areas, but is found on both ancient crystalline and recent volcanic mountains. It is, at the same time, remarkable that this community is very poorly represented south of the Zambesi or on the Abyssinian highlands, which are so extensive and in parts so apparently suitable for these birds. The Abyssinian montane forests, unaccountably, support only 16 passerines, including endemics, compared with 41 in the Kenya montane and 35 in the far smaller area of Usambara. (By contrast the Abyssinian montane, unlike any other south of the Sahara, supports populations of an otherwise exclusively palaearctic species, the red-billed chough, *P. pyrrhocorax.*)

Abyssinia and southern Africa apart, the size of the common element in the bird communities shows that these ecological islands, however widely separated, were colonized by an already integrated community rather than by a process of random dispersal, as in oceanic islands. As already noted, the whole problem of how, when, or where this community evolved has been discussd before, quite inconclusively; the fact that it is hardly represented either on the Abyssinian plateau or in subtropical South Africa argues against the probability that either of these presumably eligible areas was the site of its development. The immediate problem is how the community as a whole achieved its present dispersal. Evidence of subspeciation of birds on neighboring mountains underlines the reality of the existing ecological barriers. It seems necessary to suppose

2. Groups of allopatric species that appear to be particularly closely related but which are morphologically as distinct as sympatric species usually are (cf. Mayr 1942).

3. No general discussion of mammal distribution in tropical Africa has been published but it is curious that on the available information it seems to differ from that of both birds and plants. I am indebted to Mr. R. W. Hayman, of the British Museum (Nat. Hist.) for discussing this problem with me. He writes: "I cannot find that the montane forest mammal communities are particularly specialized as a whole, though a few species are restricted, or almost entirely restricted, to such habitats. In general there is no clear-cut division between the montane forest and the lowland forest mammals. . . . Among the smaller mammals I can think of no comparable examples [to the birds] at all. Where mountain races are recognized for some species they are generally closely related to widespread lowland forms." Some species, such as *Colobus polykomos*, and some closely-knit groups such as the squirrels of the subgenus *Aethosciurus*, occupy and have differentiated in isolated montane forests, but also occupy considerable areas of lowland forest. Hayman adds that Cameroon Mt. "seems to have no special mammalian affinities with East Africa except for an *Otomys* above the tree-line" (cf. Eisentraut 1957). Perhaps further taxonomic work on the mammals and exploration of other montane areas in the Cameroons may eventually reduce the apparent incongruity in distribution of mammals and birds.

a great reduction, however temporary on the geological time-scale, in the efficacy of these barriers, so as to facilitate the wholesale invasion of each montane island. The process need not have been simultaneous over the whole present extension of the montane bird community, and over any given part of Africa it may have taken place more than once. If the age of most bird species is accepted as post-Pliocene, as suggested above, then the dispersal of the montane community has probably taken place since the middle of the Pleistocene. It is indicative that the montane birds of Cameroon Mountain, one of the most remote of the ecological islands, include on the one hand several species that are endemics but definitely members of superspecies occurring also on East African mountains, and at the other extreme three species that are not differentiated even subspecifically from their East African relatives. Unless this is the result of a great difference in specific variability it means that a corridor of montane conditions was established more than once during the Pleistocene.

One difficulty in seeing how the dispersal came about lies in the variety and often the severity of present ecological barriers. For example, the montane birds of Marsabit Mountain in northern Kenya are islanded by subdesert steppe (in the terminology of the 1959 "Vegetation Map of Africa"), lying at barely 2000 ft. above the sea although 300 miles inland. To place such a montane island in contact with others would require a great reduction in temperature and increase in humidity. As an intermediate example, for Namuli Mountain at about 15° S. in Portuguese East Africa, which is surrounded by low-altitude savanna carrying much *Brachystegia* woodland, climatic modifications similar to those for Marsabit would be needed, but in lesser degree. Much the same applies to the connection that must be postulated for the vestigial montane forest of south-western Angola, probably by way of the southern rim of the Congo basin, where *Brachystegia* predominates today. But the land lies above 3000 ft. and the the higher altitude means lower minimum temperatures, so that a greater modification of rainfall than of temperature would be needed. (About half a dozen typically montane species of birds still linger in evergreen patches "as much as 100 yards across, dependent on ground water" [Benson and White 1957] the *mushitu* of Northern Rhodesia.) Finally, Cameroon Mountain, though geographically remote from the mountains on the east of the Congo basin, is separated on the most direct line by the lowland forest of the Congo basin. On this line a severe reduction in temperature but no increase in rainfall would be needed to put the montane forests in contact. A little further north the northern rim of the Congo basin offers a somewhat longer means of contact but at a higher altitude, though still too low for montane conditions. It is at present under savanna, so that a marked change in rainfall amount and distribution as well as a reduction of some 5° C. in temperature would be needed to enable forest to occupy it. To put the montane forest islands of Kenya and northern Tanganyika in contact would require not only a reduction in temperature greater than 5° C. but also an increase of rain (and/or "occult precipitation") up to 50 inches over the present, with (most important) concomitant reduction of dry season.[4]

Whatever the past climatic amelioration in East Africa, the fact that the typical montane bird community does not occur in the apparently suitable forests of the Abyssinian plateau suggests that the present ecological isolation of this area has persisted for a long time. In a direct line from the main Kenya Highlands the barrier is exceptionally severe—over three hundred miles, which is at best *Acacia-Commiphora* steppe and to a considerable extent the most severe biotope met with in tropical Africa, namely, subdesert. However, the extreme northern reaches of the East African highlands, the Imatong-Didinga group of mountains on the Uganda-Sudan border, which carry much of the typical montane avifauna, are separated from the Abyssinian highlands by only about two hundred miles, mostly of *Acacia-Commiphora* and with no subdesert.

The fact that Abyssinia lacks so many of the typical African birds, while the Cameroons do not, is the more remarkable because some of the species which do not appear in Abyssinia are more closely related to the birds in India than to any elsewhere in Africa. This applies certainly to three species confined to montane forest on the eastern edge of the Congo Basin, and, subject to more

4. I am indebted to Dr. L. S. B. Leakey for pointing out, after seeing this paper in draft, that the volcanic Chyulu Range, in southeastern Kenya, north of Kilimanjaro, is of special interest in connection with the questions discussed in this paper. It is a line of cinder cones, some fifty miles long and four miles wide, with a sheet of basaltic lava extending for some five miles on either side. Temperley (1960) after geological examination thinks that the entire range is of Upper Pleistocene age, i.e., less than 40,000 years old, during much of which it must have been active. He also found evidence that in some form the range was "in existence during the last Pluvial period." In fact the effect he describes could, it seems, have been due to either the Gamblian Pluvial, or the more recent Makalian or Nakuran humid phases that Leakey has described elsewhere in East Africa.

A detailed account of the Chyulu birds has been given by Van Someren (1939), together with subsidiary references to the plants. It is clear that much montane forest is present and a rich and typical montane avifauna, many components of which show either marked or incipient subspecific differences from that of Kilimanjaro. The climate is evidently peculiar, there is a great deal of mist, and such typical montane trees as olive (*Olea sp.*) and "cedar" (*Juniperus procera*) are reported on lava flows as low as 4000 ft. above sea level. In view of this, and since in the 45 miles between the main Chyulu forests and the forest girdle of Kilimanjaro only one stretch of 15 miles is below 4000 ft., it appears that no more than a slightly cooler and wetter climate than the present might suffice markedly to reduce the climatic barrier between the two montane areas. Complete elimination would, however, probably require a considerable change in climate since towards its southern end the gap is within the rain-shadow of Kilimanjaro.

If during the "pluvial" referred to by Temperley the climate changed enough for the Kilimanjaro montane forest to expand across to the Chyulus, then the observed subspecific differentiation has all taken place since then. On the other hand, if the climatic barrier between the two montane areas was at no time eliminated it was nevertheless so narrow that colonization by individual movement and transport of seeds might over the years conceivably have been on a scale sufficient to build up a typical forest biome on the Chyulus. Such a conclusion can, in the exceptional circumstances, be drawn without invalidating the basis of the preceding discussion, especially as regards the Cameroon montanes, which are isolated by a gap fifty times as wide as that in the case of the Chyulus.

information on the life history, to two species on East African mountains and one on Cameroon Mountain (Hall & Moreau, in press). Of the two East African species, *Artisornis metopias* and *Apalis moreaui*, each occupies isolated mountains as far apart as the Kenya-Tanganyika border and the neighborhood of Lake Nyasa. There are of course many dry-country affinities between African and Indian animals and plants, but these montane forest birds raise special difficulties. It is hard to accept a postulate of a forest connection via southern Arabia, with the implication that it somehow bypassed Abyssinia. Moreover, on paleoclimatic grounds it is difficult to envisage such a connection as existing much later than the Miocene (Moreau 1952). (Yet one of the species in question, the owl *Phodilus prigoginei*, looks so like the Indian *P. badius* that it would be treated as conspecific if it were not so widely separated.) On the other hand it is extremely difficult to imagine how individual colonization from one continent to the other could have occurred from the parent stock of the species concerned.

Before leaving the question of the montane birds it may be mentioned that there are two species characteristic of the moorland above timberline, namely, a sunbird, *Nectarinia johnstoni*, and a bird of the thrush family, *Pinarochroa sordida*. The locations of the moorland biome are of course separated from each other not only by the same lowland biotopes as separate the montane forests but also by the forests themselves. The distribution of the sunbird is similar to that of many forest species—the Kenya Highlands, Kilimanjaro, Ruwenzori and mountains further south along the Albertine Rift, and a remote area at the uniquely low altitude of 7000 ft. on the Nyika Plateau at the north end of Lake Nyasa. The *Pinarochroa*, on the other hand, occupies the highest parts of the Abyssinian plateau as well as the moorlands of Kenya and northern Tanganyika. It seems impossible to envisage anything but random dispersal for these birds of the open country above timber line, but in this connection it is necessary to recall that floristically the moorlands with their giant *Senecio* spp. and giant *Lobelia* spp. have affinities and differences of the greatest interest (cf. Bruce 1934, Hedberg 1955).

THE LOWLAND FOREST AVIFAUNA

West Africa

The lowland forest of West Africa is divided into two blocks, west (Upper Guinea) and east (Lower Guinea) of the "Dahomey gap" which runs approximately between Accra and Lagos and is almost entirely savanna. The Lower Guinea forest (now subject to much attrition by agriculture) stretches across southern (now "Western" and "Eastern") Nigeria to the southern Cameroons, Gabon, and the Congo. At the head of the Gulf of Guinea it is constricted by Cameroon Mountain and the Bamenda-Banso highlands to a lowland corridor little more than fifty miles wide (see Fig. 2). Between the avifauna of the

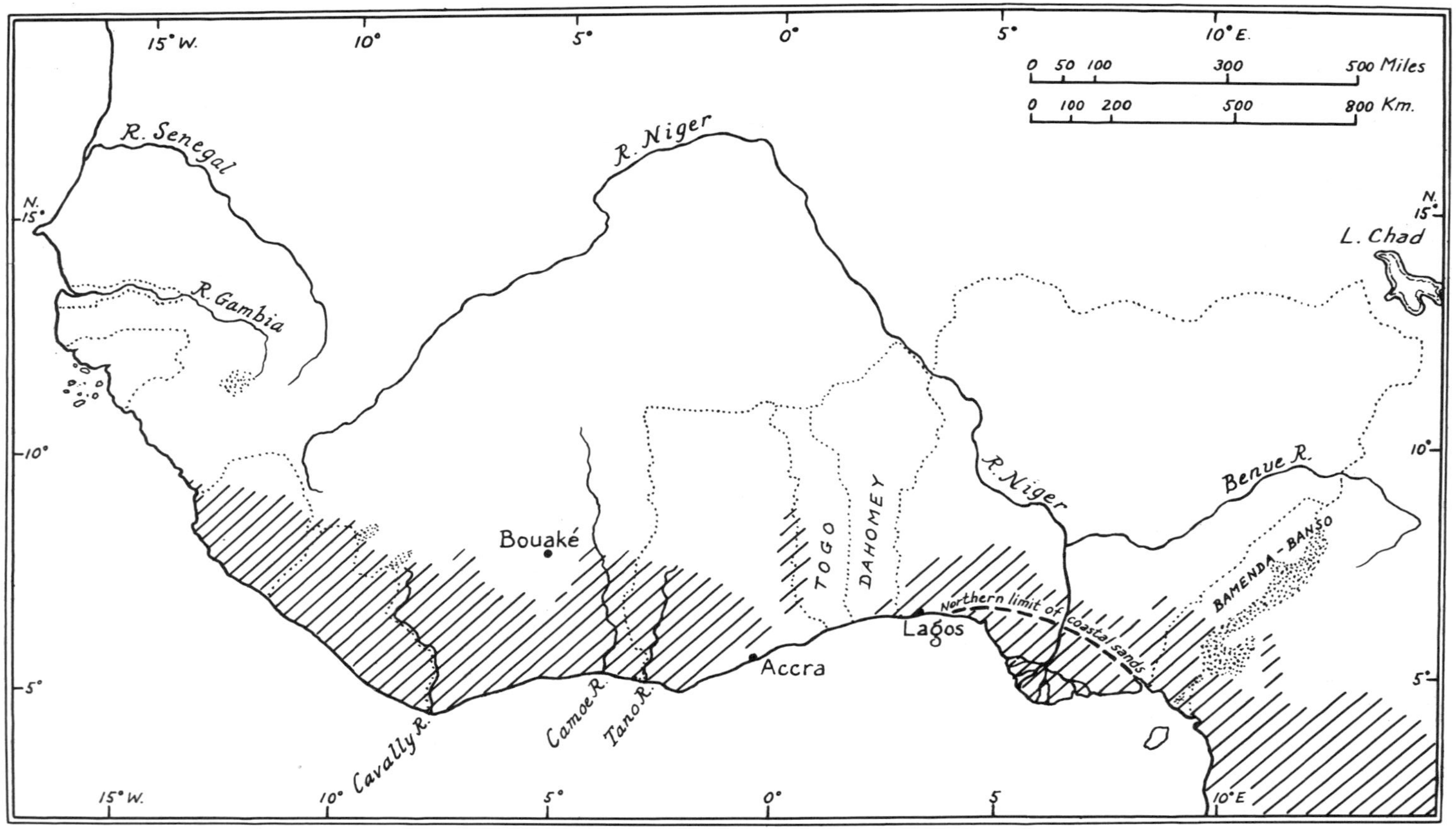

FIGURE 2
Tropical West Africa, to show lowland forest areas (hatched). Montane areas are dotted.

Upper Guinea forests and that of the southern Cameroons portion of the Lower Guinea forest, immediately to the east of the Bamenda constriction, there are many differences, both specific and subspecific, but most of them do not occur at the Dahomey gap or at the Bamenda constriction, as might have been expected on present ecological grounds. In fact, as Marchant (1954) has shown, only about one quarter of the total avifaunal change takes place near these present ecological barriers, while half occurs within the Nigerian lobe of the Lower Guinea forest where there is no barrier other than the Niger River, which must surely be negligible to birds. Thus the avifaunal evidence calls for a gap, considerable in both size and duration, in the neighborhood of the Lower Niger, in substitution for the present gap further to the west.

It is difficult to fit any climatic postulate to such premises. However, Marchant in his discussion points out that prior to the deposition of the Coastal Plain Sands of Nigeria, which "have usually been regarded as Pleistocene or Recent" and cannot be older than late Pliocene, the bulge formed by the Niger Delta would have been absent. The coast would have run about due east for over three hundred miles from Lagos, so that the Niger entered the sea almost at the present northern limit of the forest. A southward recession of the forest line by no more than fifty miles could have produced a savanna gap nearly three hundred miles wide. Marchant suggests that these conditions could most easily have been fulfilled during the high sea level of an interglacial. There is, however, no apparent reason why the production of such a gap in the forest within Nigeria should have been accompanied, as the premises require, by the closing of the Dahomey gap. Perhaps some shift in the Benguela Current (and intensity of the cool upswelling), such as Davies (1957) has invoked in somewhat vague terms in connection with changes in the Accra area, could be partly responsible.

A study of the forest mammals of West Africa by Booth (1958) leads to somewhat similar conclusions. For example, the chimpanzee and the tree-hyrax of Western Nigeria belong to Ghana forms, not to those found east of the Niger. This suggests some past closing of the Dahomey gap but not necessarily an important gap in the neighborhood of the lower Niger, since for many of the species involved even the river itself could be a serious barrier, as it would not be for birds. However, a complete analysis of the specific and subspecific ranges of all the forest primates led Booth to conclude that the forests west of the Niger had been greatly fragmented at one or more stages in the Pleistocene, indeed, reduced to three "refuges": (1) west of the Cavally River in Liberia; (2) in the eastern Côte d'Ivoire and Western Ghana, between the Camoe River and the Tano River; (3) west of the Niger "either in riparian forest or in the Benin area." The first two form part of the present Upper Guinea block, and, as might be expected from the greater individual mobility of birds and consequent gene-flow where the habitat is continuous, there is no corresponding avian evidence. Booth has pointed out that in Côte d'Ivoire just west of the Camoe River there is a constriction in the forest belt at the present day (cf. "Vegetation Map of

Africa" 1959) formed by the "Bouaké V," which could produce a gap in the forest given any deterioration in the climate affecting Upper Guinea.

East Africa

In the forests of the coastal zone of northern Tanganyika (including the foothills of Usambara) there are subspecies of three birds, *Neocossyphus rufus, Illadopsis (Malacocincla) rufipennis,* and *Spermophaga ruficapilla,* which are typically West African, extending from the Atlantic to the vicinity of Lake Victoria. The species are not represented further south, which would suggest a possible connection by way of Rhodesia and Portuguese East Africa. *Ploceus weynsi,* nearest relative of and possibly even conspecific with *P. golandi,* another bird of the relict evergreen forests on the Kenya coast, is found in the same western area. And I am indebted to Dr. L.S.B. Leakey for pointing out that this is true of several mammals, including *Cercocebus galeritus, Anomalurus orientalis,* and *Atherurus turneri.* All these coastal populations appear to be separated from their relatives by five hundred miles or more of country that today is in part very hot and dry and in part montane, equally unsuitable conditions for the animals concerned.

The total number of species is small but taken together they seem to indicate that at one time the country that now isolates them had more and better distributed rain, with somewhat higher minimum temperatures than now (only one of the birds is found as high as 3000 ft., in Usambara). The increase in humidity fits the requirement for intercommunication of the montane forest islands but the temperature argues against it. Hence the climatic change that would permit west-to-east communication of these lowland East African animals could not coincide with the change required for the montane birds. The west-to-east connection may indeed have been earlier than the other, for it is more difficult to envisage a general rise in temperature over the intervening country at any time since the build-up of the highlands along the Rift Valley in Kenya and northern Tanganyika to their present extent and dimensions. On the basis of the rather vague generalizations quoted by Moreau (1952), the East African mountains involved probably reached their present significant dimensions by early Pleistocene, and if this is correct the west-to-east connection must be pushed back accordingly.

It may be added that floristically the evergreen forest of the East African coastal belt shows strong links with West Africa at the generic level, but consists of very distinct species (expert botanical opinion quoted by Moreau 1952). The same generalization apply to Mollusca (Dr. B. Verdcourt, *in litt.*). Subject to the rate of speciation, which is unknown, this reinforces the suggestion of a comparatively early date for the ecological connection between the East African coastal strip and West Africa, via the Victoria basin. Verdcourt has also commented on the Mollusca from Dr. Leakey's sites on Rusinga Island

(Kavirondo Gulf, Lake Victoria), which are attributed to the Miocene, i.e., before the elevation of the Kenya Highlands. He finds that some of these fossils "undoubtedly are more closely related to those which occur only in the coastal areas at the present time than they are to species at present in the Kavirondo area."

OTHER AVIFAUNAL PROBLEMS

Angola

Hall (1960b), discussing the avifauna of Angola, put most of the country in the "Brachystegia Zone," except for narrow belts parallel to the coast—the "Acacia Zone" ("desert" and "semi-desert") and "Escarpment Zone." The latter is a narrow triangle tapering to only a few miles in the south and for the most part ranges through only about 3000 ft. of elevation. But it harbors a notable number of endemic species, and the avifaunal differences to east and west of it are remarkable. Hall has postulated that at some period "Kalahari-like conditions" invaded Angola (presumably from the southeast) and temporarily replaced the Brachystegia, with its associated birds, by Acacia. *Ex hypothesi,* the Brachystegia biome eventually returned from the north or northeast. Meanwhile the Escarpment Zone, owing to its topography and its proximity to the Atlantic, remained comparatively unaffected by the aridity of its hinterland. Hence it could provide a refuge for species requiring humid conditions and at the same time maintain a barrier between the coastal belt and the interior.

Dry-country Birds

The larks (Alaudidae), with their main representation (about 50 species) in Africa, are perhaps more than any other family characteristic of dry and very open country. The species are particularly numerous, with over a dozen endemics, in southwestern Africa, from the Karroo to the coastal "desert" of Angola and the highveld of the Transvaal. The family is also exceptionally well represented, though to a lesser extent, in the northeast of tropical Africa, between the Kenya Highlands and the Gulf of Aden. It may be inferred that these two areas, and especially the southwestern, have continuously provided a variety of dry open habitats for a long time; this is supported by the very high specialization of xerophytic flora particularly in the Karroo. At the same time, however, Dr. Verdcourt tells me he was greatly struck by the similarities of the flora of South West Africa distinct from the Karroo with that of the Ogaden ("Abyssinian Somaliland") (cf. other botanical opinion quoted by Moreau 1952).

Study of the larks or other birds gives no indication that these southwestern and northeastern dry areas of tropical Africa have had a direct connection in the Pleistocene; but one monotypic lark genus, *Heteromirafra* (*ruddi*), is comprised of one montane grassland population near the Drakensburg and one on the Somaliland plateau; while one weaver bird, *Ploceus rubiginosus*, typical of Acacia country in northeastern tropical Africa from the Sudan to northern Tanganyika, reappears in South West Africa; and *Francolinus levaillantoides* (=*gariepensis*), centred on the Kalahari, reappears in the area Lake Rudolf-Somaliland (Hall, unpubl.).

SUMMARY AND CONCLUSION

Peculiarities of bird distribution in tropical Africa give rise to the following postulates. It is unfortunate that too little is known of the evolution of birds for this avian evidence to suggest, by itself, any chronological framework.

1. At some period or periods, presumably during the Pleistocene, an improvement in amount and distribution of rainfall and a reduction of temperature permitted extension of montane forest throughout East Africa, from the mountains on the southern border of the Sudan west to the Cameroons, south at least to the mountains south of Lake Nyasa and thence west to western Angola. The changes need not have affected the whole of tropical Africa simultaneously.

This postulate is to some extent satisfied if the local geological evidence for Pleistocene pluvials is accepted as applying to sufficiently extensive areas (see discussion and references in Clark 1960). As for temperature, most evidence from former glacial extensions is open to question, since increased precipitation is as likely to be responsible as lowered temperature; and the 1000-1500 ft., which is inferred from pollen data as the extent to which the "montane-loving flora" descended (*ibid.*), is far too little for our needs.

2. This climatic change did not extend over the northeast of tropical Africa sufficiently to put the Abyssinian plateau in communication with the East African montane areas.[5]

3. A period of greater and more regular rainfall, associated with higher rather than lower temperature, linked the lowland forest of the Congo and Lake Victoria basins to the East African coast. This pluvial cannot have overlapped one of those mentioned in (1) above, which presupposes a considerably cooler climate.

4. In West Africa west of the Cameroons the lowland forest was interrupted for a longer period or more decisively in southern Nigeria than it has been in recent times between Accra and Lagos. This might be due in part to a regional or subregional climatic change and also to different coastal topography and modified ocean currents. Mammal data, however, suggest greater fragmentation of the forest and consequently more general causes.

5. Note, however, that two of the widespread dominants of montane forest, *Juniperus* and *Podocarpus*, reached Abyssinia. This could conceivably have been before the evolution of the typical montane bird community.

5. Most of Angola experienced a period of drier climate sufficient to substitute *Acacia* for *Brachystegia.* This could well be associated with the period of aridity in the latter part of the Middle Pleistocene, which culminated in the extension of sheets of wind-blown sand over "vast portions" (unspecified) of the Congo (Mortelmans 1957).

6. Prolonged existence of a variety of dry, open habitats occurred in southwestern Africa and to a less marked degree in the interior of the Somalilands. There is some evidence for an ecological connection between these two areas.

Postscript, 1966: Since the above was written, evidence from several sources has demonstrated that during the glaciation temperatures in low latitudes were reduced by at least the 5° C. postulated on p. 33 (see review in Moreau, *Proc. Zool. Soc. Lond.* 141 (1963): 395-421).

BIBLIOGRAPHY

Benson, C. W.
1948. "Evergreen Forests near Blantyre: Comparative Variety of Bird Species," *Nyasald. J.*, 1:45–52.
Benson, C. W., and C. M. N. White
1957. *Check List of the Birds of Northern Rhodesia.* Lusaka: Government Printer.
Booth, A. H.
1958. "The Zoogeography of West African Primates: A Review," *Bull. Inst. Franc. Afr. Noire,* 20:587–622.
Bruce, E. A.
1934. "The Giant Lobelias of East Africa," *Kew Bull.*, 1934, pp. 61–88, 274.
Chapin, J. P.
1923. "Ecological aspects of bird distribution in Tropical Africa," *Amer. Nat.*, 57:106-125.
1932. "The Birds of the Belgian Congo," *Bull. Amer. Mus. Nat. Hist.*, 65.
Chapman, F. M.
1917. "The Distribution of Bird-life in Columbia," *Bull. Amer. Mus. Nat. Hist.*, 36.
Davies, O.
1957. "Climatic and Cultural Sequences in the Late Pleistocene of the Gold Coast," J. D. Clark (ed.), *Proc. 3rd Pan-Afr. Cong. Prehist.* (Livingstone 1955), pp. 1–5.
Eisentraut, M.
1957. "Beitrag zur Säugetierfauna des Kamerungebirges und Verbreitung der Arten in den verschiendenen Höhenstufen," *Zool. Jb.*, 85:619–72.
Hall, B. P.
1960a. "The Ecology and Taxonomy of Some Angola Birds," *Bull. Brit. Mus. (N. H.) Zool.*, 6:369–453.
1960b. "The Faunistic Importance of the Scarp of Angola," *ibid.*, 102:420–39.
Hall, B. P., and R. E. Moreau
1962. "A Study of the Rare Birds of Africa," *Bull. Brit. Mus.* (*N. H.*) *Zool.*, 8:315-378.

HEDBERG, O.

1955. "Afroalpine Vascular Plants," *Symbol. Bot. Upsal.*, 15(1):1–411.

KEAY, R. W. J.

1959. "[Explanatory notes on] Vegetation Map of Africa." London: Oxford University Press.

MARCHANT, S.

1954. "The Relationship of the Southern Nigerian Avifauna to Those of Upper and Lower Guinea," *Ibis*, 96:371–9.

MAYR, E.

1942. *Systematics and the Origin of Species*. New York: Columbia University Press.

MOREAU, R. E.

1952. "Africa Since the Mesozoic," *Proc. Zool. Soc. Lond.*, 121:869–913.

1954a. "The Main Vicissitudes of the European Avifauna Since the Pliocene," *ibid.*, 96:411–31.

1954b. "The Distribution of African Evergreen-forest Birds," *Proc. Linn. Soc. Lond.*, 165:35–46.

MORTELMANS, G.

1957. "Le Cénozoique du Congo Belge." In J. D. Clark (ed.), *Proc. 3rd Pan-Afr. Cong. Prehist. (Livingstone 1955)*, pp. 23–50.

SOMEREN, V G. L. VAN

1939. "Museum's Expedition to Chyulu Hills," *J. E. Afr. Uganda Nat. Hist. Soc.*, 14:1–151.

STORER, R. W.

1960. "Adaptive Radiation in Birds." In A. J. Marshall (ed.), *Biology and Comparative Physiology of Birds*, Vol. I. New York and London: Academic Press, pp. 15–55.

TEMPERLEY, B. N.

1960. "A Study of the Movement of Ground-water in Lava-covered Country," *Overseas Geol. and Min. Res.*, 8:37–52.

OBSERVATIONS ON THE ECOLOGY OF SOME LARGE AFRICAN MAMMALS

F. BOURLIÈRE

Of all the continents, tropical Africa is remarkable in the richness and abundance of its mammalian fauna. Nowhere else in the world is there to be found so many species of wild ungulates, many of which have extremely large populations. The quantity of animal protein so produced is a considerable food resource for totally or partially carnivorous animals, namely the *Carnivora* and Man. Such a condition is certainly not new in Africa. On the contrary, study of the fauna of various Pleistocene sites tends to suggest that the large ungulates were more varied and numerous during the last million years than they are today. One may ask, therefore, whether this enormous source of *potential food* (which was probably even then, as today, not fully exploited by the felids, hyaenids, and canids) was not a decisive factor in *the rapid increase of partially zoophagous primate populations,* having recently left their native forest habitat for a more or less new environment and having developed good hunting methods permitting them to gather abundant food all seasons of the year. This might explain *the acceleration of the rate of evolution* of a group suddenly able to exploit partially unused food resources.

One can probably never estimate the population density of the various Pleistocene African ungulates. But if they are admitted to have been *at least* as important as those in certain parts of the continent still protected against modern man's intervention, it is possible to attempt a minimal estimation which has every chance of approximating the actual. On this basis, certain hypotheses can be submitted to quasi experimental verifications.

To begin with, the large differences in the biomasses of game animals which appear today not only between the forest, the savanna, and the desert, but also between the different types of grass cover, have no justification peculiar to our era. It must have been the same in the past, which suggests the existence of *preferred zones* constituting *optimal habitats* for Australopithecines where this group's evolution had every chance to be more rapid than elsewhere. Someday it will certainly be possible, thanks to the progress of paleoclimatology and paleoecology, to broadly delimit those zones. We would not be astonished if they closely coincided with those precisely where the anatomical remains or cultural traces of our distant ancestors are found to be the most numerous.

THE ABUNDANCE OF GAME UNGULATES IN DIFFERENT AFRICAN HABITATS

The improvement of various methods of evaluating large wild-animal populations (terrestrial and aerial censuses, studies of sample areas) has permitted the accretion during the past five years of substantial quantitative data which indicate in a particularly striking manner *the large inequality of population density* of big game in the different vegetation zones of tropical Africa. Even though the figures dealt with in the following pages were not all obtained by the same techniques, and even though some correspond to one census while others are averages of counts made for several consecutive months over several years, the differences they show are too large to be laid to chance alone. Indeed, most of them correspond to the impressions of a trained observer visiting successively the different habitats in question.

Along with the population densities, we have tried to estimate the variations of what the cattle-breeders call grazing capacity (or carrying capacity) of the various environments. To determine the number of individuals of different species per area unit (square kilometer) is not sufficient. An elephant and a Grant's gazelle each count as one in a census, but the amount of meat which they represent is very different. Therefore, in each case the standing crop biomass of the different specific populations has been estimated by multiplying the number of individuals by the minimum adult weight of each species. Thus, over-estimation caused by attributing to the young the weight of an adult is more or less compensated for by the underestimation of the weight of the oldest individuals.

We have the most numerous and most precise estimations for the *different types of East African savanna.* It is with them, therefore, that we will begin, particularly since these open habitats shelter the greatest concentrations of wild ungulates known today in the tropics.

Our own observations were carried out from 1957 to 1959 in the Albert National Park, in northeast Kivu, all along the Ugandian frontier. They dealt with the game animal populations which frequent the 600-square-kilometer short-grass savannas of the Rwindi-Rutshuru plain, south of Lake Edward (Bourlière and Verschuren 1960). During this period, in addition to the bimonthly censuses carried out by the conservation officer, Cl. Cornet d'Elzius, counts were made in four linear zones of major types of vegetation of the plain. The mean values obtained for 1959 are given in Table 1. It is noteworthy that in these open savannas at the edge of the Congolese forest, 69.2% of the ungulate biomass (±16,300 out of ±23,550 kg.) is represented by only two species—the hippopotamus and the elephant—both large animals.

TABLE 1
POPULATION DENSITY AND UNGULATE BIOMASSES OF THE SAVANNAS OF THE RWINDI-RUTSHURU PLAIN
(Averages of six censuses in Albert National Park in 1959)

Species	*Adult Weight in Kg.*	*Actual Count for 600 Square Km.*	*Density per Square Km.**	*Biomass per Square Km. in Kg.**
Elephant	3,000	1,026	1.7	5,100
Hippopotamus	1,400	4,800	8	11,200
Buffalo	500	7,402	12.3	6,150
Topi	130	1,199	2.0	260
Waterbuck	150	760	1.2	195
Uganda kob	70	4,976	8.3	581
Reedbuck	40	61	0.1	
Bushbuck	50	53	0.09	
Grey duiker	10	1		
Warthog	70	603	1	70
Forest hog	140	35	0.05	
			Total:	> 23,556

* In this Table, as in the following, the figures should be multiplied by 2.5 to obtain the values per square mile.

TABLE 2
POPULATION DENSITY AND BIOMASSES OF THE PRINCIPAL UNGULATE SPECIES OF QUEEN ELIZABETH NATIONAL PARK
(According to Bere's estimation, 1960)

Species	*Adult Weight in Kg.*	*Actual Numbers Estimated in 1,670 Square Km.*	*Density per Square Km.*	*Biomass per Square Km. in Kg.*
Elephant	3,000	2,500	1.5	4,500
Buffalo	500	12,000	7.2	3,600
Hippopotamus	1,400	12,000	7.2	10,080
Warthog	70	1,500	0.9	63
Waterbuck	150	2,400	1.4	210
Uganda kob	70	3,500	2.1	147
Topi	130	2,600	1.5	195
			Total:	> 18,795 kg.

In Table 2 the same figures are calculated for the whole (1,670 square km.) of Queen Elizabeth National Park, in Uganda, on the eastern shore of Lake Edward. Neighboring Albert Park, it has the same types of vegetation as the plains of Rwindi-Rutshuru just mentioned. We based our calculation on the figures published by R. M. Bere in 1960, based on the results of a certain number of sample counts carried out from 1956 to 1958 by different observers (G. A. Petrides and W. G. Swank, W. M. Longhurst and J. Savidge, etc.). In spite of the different methods used for population count, the concordance of the population densities of most of the species per square kilometer is striking. Again it is noteworthy that the same two large animals by themselves represent 77.5% of the total biomass (±14,600 out of ±18,800 kg.).

TABLE 3
POPULATION DENSITY AND BIOMASSES OF UNGULATES IN NAIROBI NATIONAL PARK
(Averages of S. I. Ellis' twelve censuses, 1960–1961)

Species	*Adult Weight in Kg.*	*Actual Count for 116 Square Km.*	*Density per Square Km.*	*Biomass per Square Km. in Kg.*
Zebra	290	1,929	16.6	4,814
Grant's gazelle	60	367	3.2	192
Thomson gazelle	22	323	2.8	62
Waterbuck	150	128	1.1	165
Impala	60	655	5.6	336
Bushbuck	50	14	0.1	5
Wildebeest	200	2,757	23.8	4,760
Giraffe	1,200	93	0.8	960
Warthog	70	230	2	140
Hippopotamus	1,400	4	0.03	42
Kongoni	140	1,220	10.5	1,470
Eland	300	56	0.5	150
Grey duiker	10	4	0.03	0.3
Rhinoceros	1,300	11	0.09	117
Reedbucks (2 species)	40	6	0.05	2
Dik-dik	5	2	0.01	0.05
Steenbuck	15	1.5	0.01	0.15
			Total:	>13,215

Table 3 summarizes the results of the first twelve censuses carried out under the direction of S. I. Ellis, from September 1960 to August 1961, in Nairobi National Park, Kenya. In that protected zone situated on the edge of the Masai steppe, open *Themeda* savannas alternate with a few *Acacia* stands. The altitude is higher than in the two preceding cases (±1,500 m. instead of ±1,000 m.). The number of ungulate species is larger than in the Albert or Queen Elizabeth National Park, but the total biomass per square kilometer is less (±13,200 kg.). At the same time, unlike the basin of Lake Edward, the small ungulates are in a very large majority, the species weighing more than a ton representing only a minute percentage of the total biomass. The results of the censuses done in 1959 in the thornbush savanna of the Tarangire Game Reserve in Tanganyika are close enough; H. F. Lamprey indicates that the twelve species represent a mean annual biomass of ±12,250 kg. The figure of ±15,750 kg. per square kilometer indicated by L. M. Talbot *et al.* (1961) for an *Acacia* savanna in Kenya is also close to Lamprey's figure.

TABLE 4
POPULATION DENSITY AND BIOMASSES OF UNGULATES OF THE SAVANNAS OF SERENGETI
(According to the aerial censuses of B. and M. Grzimek in January, 1958)

Species	*Adult Weight in Kg.*	*Actual Count for 10,000 Square Km.*	*Density per Square Km.*	*Biomass per Square Km. in Kg.*
Grant's and Thomson gazelle	40	194,654	19.5	780
Wildebeest	200	99,481	9.9	1,980
Zebra	290	57,199	5.7	1,653
Topi	130	5,172	0.5	65
Eland	300	2,452	0.2	60
Impala	60	1,717	0.2	100
Buffalo	500	1,813	0.2	100
Kongoni	140	1,285	0.1	14
Giraffe	1,200	837	0.08	96
Waterbuck	150	284	0.02	3
Oryx	200	115	0.01	2
Elephant	3,000	60	0.006	18
Roan antelope	250	57	0.006	1
Rhinoceros	1,300	55	0.006	8
				Total 4,692 kg.

Table 4 is based on the results of the aerial census of Serengeti National Park, Tanganyika, made in January, 1958, by M. and B. Grzimek. The number of ungulate species is also higher here than in Kivu or in western Uganda, but the biomass per square kilometer is distinctly less than in Nairobi National Park. Perhaps there was underestimation of real numbers—which is frequently the case when the airplane is used to take a census of species of small or average size. The fact that our calculation rests on only a single census, and not on the average of many, should also be taken into account. It is nevertheless striking that the calculated biomass (±4,700 kg. per sq. km.) is very close to that (>4,850) obtained by Dr. R. M. Stewart and L. M. Talbot at the time of their aerial census of the combined plains of Loita, Mara, and Serengeti, in May 1961. Here again the percentage of ungulates weighing more than a ton is, as in Nairobi, very low.

For the different *open environments of the Rhodesias* and of South Africa there is, unfortunately, very little precise information. R. F. Dasmann and A. S. Mossman (1961) have, nevertheless, just published the results of their 1959–60 censuses on 125 square kilometers of a ranch in Southern Rhodesia (Henderson Ranch, West Nicholson) not yet used for raising cattle. The calculation of the biomasses of ungulates present in that open mopane woodland (Table 5) gives a total biomass of ±4,500 kg. per sq. km.

TABLE 5
POPULATION DENSITY AND BIOMASSES OF UNGULATES IN THE STUDY AREA, HENDERSON RANCH, W. NICHOLSON, SOUTHERN RHODESIA, 1959–1960
(According to Dasmann and Mossman, based on road-strip counts)

Species	*Adult Weight in Kg.*	*Actual Numbers Estimated in 125 Square Km.*	*Density per Square Km.*	*Biomass per Square Km. in Kg.*
Impala	60	2,100	16.8	1,008
Zebra	290	730	5.8	1,682
Steenbuck	15	200	1.6	24
Warthog	70	170	1.4	98
Greater kudu	250	160	1.3	325
Wildebeest	200	160	1.3	260
Giraffe	1,200	90	0.7	840
Duiker	10	80	0.6	6
Waterbuck	150	35	0.3	45
Buffalo	500	30	0.2	100
Eland	300	10	0.08	24
Klipspringer	10	10	0.08	1
Bushpig	70	10	0.08	5
Sharpe's grysbuck		?		
Bushbuck		?		
			Total:	>4,418

For *Central and West Africa* there is no true census. However, we can cite the count made by J. Dragesco (*in litt.*) in March 1961, in the subdesert steppe west of Oum-Chalouba (*Chad*): 1,500 dorcas gazelle, 400 oryx, 300 dama gazelle, and 15 addax in an area 30 by 40 kilometers. This would mean an ungulate biomass on the order of only 80 kg. per sq. km.

For the Sahara, the figures are also much lower. During his reconnaissance of December, 1959, and January, 1960, in the Majâbat al Koubra, the immense "empty quarter" situated within the borders of Mauritania and Mali, Th. Monod saw 66 addax from December 21, 1959 to January 15, 1960, between the El Ghallâwija and Arawan, and a minimum of 140 others from January 19, 1960, to February 5, 1960, between Arawan and Aratan. Given that all these animals were observed within one kilometer to each side of the observer's route, this would indicate a density of 4.1 addax per 100 square kilometers on the first journey and of 17 per 100 square kilometers for the second—which is certainly remarkable for such a complete desert. J. Valverde (*in litt.*) has generously sent us some counts which he made in 1955 in Rio del Oro. Dorcas gazelle averaged 36 per 100 square kilometers, with the extremes going from 2 to 710. For dama gazelle Valverde counted densities reaching 147 per 100 square kilometers in *Aizoon* Reg, that is, in the regions which also sheltered the largest populations of dorcas gazelle. In these environments the biomass of gazelles reached a figure on the order of 190 kilograms per square kilometer, which is not negligible.

In dense forest, estimations are even more rare. The only one of any value is published by W. B. Collins (1959). A tree-counting party, marking every tree in a zone of 250 square kilometers in the Tano Nimri Forest Reserve in Ghana, counted all the animals it saw or killed during four months. The results are given in Table 6. It shows that the biomass of the three forest ungulates

TABLE 6
POPULATION DENSITY AND BIOMASSES OF THE PRINCIPAL MAMMALS OBSERVED IN THE TANO NIMRI FOREST RESERVE, GHANA, FROM FEBRUARY TO MAY, 1954
(According to Collins, 1959)

Species	*Adult Weight in Kg.*	*Actual Numbers Observed in 250 Square Km.*	*Density per Square Km.*	*Biomass per Square Km. in Kg.*
Philantomba maxwelli	8	79	0.31	2.48
Cephalophus dorsalis	20	38	0.15	3
Neotragus pygmaeus	4	7	0.03	0.12
				5.60 kg.
Colobus polykomos	10	916	3.6	36
Colobus badius	8	621	2.4	19
Cercopithecus diana	5	144	0.57	2.8
Cercopithecus mona	5	127	0.50	2.5
Cercocebus torquatus	8	83	0.33	2.6
Colobus verus	4	5	0.02	0.08
Pan troglodytes	40	22	0.09	3.6
				66.6 kg.
Sciuridae	?	numerous	?	?
Manis species	2	48	0.02	0.04
Porcupines (*Hystrix cristata?*)	20	38	0.15	3
				>3.4 kg.

present is almost negligible—only 5.6 kilograms per square kilometer—and hardly greater than that of the scaly anteaters and the porcupines (±3.5 kg. per sq. km.). In that forest the most abundant mammals by far are the arboreal rodents (not counted), and the seven species of primates which represent a biomass of about 65 kg. per sq. km.—more than ten times that of the small forest antelopes. This is not astonishing when one remembers that in ancient, dense forests the grass cover is almost absent in the undergrowth, thereby nourishing only a very small number of non-climbing mammals. Only the arboreal animals can reach the crowns of the trees. By contrast, in secondary forest, the situation is, according to all probability, better for the duikers and the dwarf antelopes. In Ituri the biomass of forest ungulates is certainly higher than in Ghana, due to the presence of the okapi, the Bongo, the small Red forest buffalo, the Giant forest hog, and the Elephant.

What conclusions can be made from the figures summed up in Table 7? In the first place, population of wild ungulates in the different African habitats vary considerably; also, it is the savannas (*sensu lato*) which shelter by far the greatest biomasses and the largest variety of species. These figures take on their full significance when compared with those obtained in other latitudes and on other continents. In temperate zones, only the North American prairies sheltered, before the invasion of the white man, herbivorous fauna of comparable richness. It is difficult to formulate a precise idea of this today and only study of certain protected zones (not overgrazed) would allow us to determine the conditions that prevailed two centuries ago. As an indication, here are the actual numbers, as of August 1960, in the 75 square kilometers of the Bison Range National Reserve in Wyoming, numbers apparently in keeping with what the vegetation can support, the prairie showing no signs of overgrazing: bison, 425; elk (Wapiti), 75; mule deer, 325; white-tailed deer, 200; pronghorn, 100; bighorn sheep, 100. Taking the mean weight of the adult bison as 400 kg., of the elk as 300 kg., of the mule deer as 100 kg., of the white-tailed deer as 75 kg., and of the two other species as 60 kg., a total biomass of about 3400 kilograms per square kilometer is obtained—which compares unfavorably with most of the East African savannas.

As for the prehistoric European forests, they sheltered an ever lower number of ungulates capable of providing nourishment for those who first hunted there. In the "primitive forest" of the Pol'ana Mountain in Slovakia, Turcek (1953) found a biomass of herbivores which did not surpass 500 kilograms per square kilometer (of which 350 are represented by a single species, the Red deer *Cervus elaphus*). In the deer forests of Scotland—severely altered by centuries of exploitation—V. P. W. Lowe (1961) estimates an average of 10.7 Red deer, or a biomass of ±1,000 kg. per sq. km. The steppes of southern Russia, with their large herds of saiga antelopes, produce little protein per hectare in game ungulates, gregarious rodents (ground squirrels and voles) being the principal consumers of the vegetation. A. Bannikov (1961) indicates a mean density of 0.8 per square kilometer for the saiga antelope in the entire area of distribution of the species and 10 per square kilometer for the zones of hibernal or estival concentration—which corresponds respectively to biomasses of 28 and 350 kilograms per square kilometer. In the Canadian tundras, A. W. F. Banfield finds, for caribou, 6 per square kilometer, a biomass on the order of ±800 kilograms per square kilometer.

There is then no doubt that *the African savannas represent an ideal type of*

TABLE 7

COMPARISON OF THE NUMBER OF SPECIES AND THE BIOMASSES OF UNGULATES PER SQUARE KILOMETER IN DIFFERENT AFRICAN HABITATS

Place	*Type of Vegetation*	*Number of Ungulate Species*	*Biomasses in Kg. per Sq. Km.*	*Kind of Census T=terrestrial A = air*	*Area Covered by census (in Sq. Km.)*	*Year*	*Authors*
Sahara (Rio del Oro)	*Salsola* Reg	1	0.3	T	100	1955	Valverde
Sahara (Rio del Oro)	*Aizoon* Reg (after rain)	2	190	T	100	1955	Valverde
Sahara (Mauritania)	Erg of Majâbat al Koubra	1	4 and 17	T	400 and 1700	1960	Monod
Tchad (West of Oum-Chalouba)	Subdesertic steppe	4	80	T	1,200	1961	Dragesco
Southern Rhodesia	Open mopane woodland	15	4,418	T	125	1959-60	Dasmann & Mossman
Tanganyika (Serengeti)	Thornbush steppe and savanna	15	4,692	A	±10,000	1958	Grzimek
Tanganyika-Kenya (Serengeti-Mara)	Thornbush steppe and savanna	>8	>4,865	A	38,700	1961	Stewart & Talbot
Sud Kivu (Luama)	Forest-savanna mosaic	>5	>5,800	T	75	1953-54	Pirlot
Tanganyika (Tarangire Game Reserve)	Thornbush savanna	22	>12,261	T	1583	1959	Lamprey
Kenya (Nairobi National Park)	Masai steppe (over-grazed)	18	13,215	T	116	1960-61	Ellis
Kenya	*Acacia* savanna	?	15,760	T	?	1960	Talbot
Uganda (Queen Elizabeth National Park)							
Entire Park	Open savannas and thickets	11	18,795	T	1,670	1960	Bere
Undergrazed Area	Open savannas and thickets	7	11,100	T	30	1956-57	Petrides
Moderately overgrazed Area	Open savannas and thickets	9	13,360	T	14	1956-57	Petrides
Severely overgrazed Area	Open savannas and thickets	8	31,028	T	23	1956-57	Petrides
Nord Kivu (Albert Park)	Open savanna at forest edge (partially over-grazed)	11	23,556	T	600	1959	Bourlière & Verschuren
Ghana (Tano Nimri Forest Reserve)	Dense rain, forest	3	5	T	250	1954	Collins

habitat for any large-sized carnivore or omnivore with mainly carnivorous habits.

The second conclusion to be reached from Table 7 is that *the East African savannas* themselves *fall naturally into two types* with regard to their ungulate fauna: 1. those bordering the Congolese forest area with a greater density per square kilometer, but where 70% or more of the biomass is represented by two large species, the elephant and the hippopotamus; 2. the lower savannas of East Africa and of the Rhodesias which shelter smaller biomasses of game animals, but where the number of species is larger and the size of the animals smaller, which makes them more vulnerable to primitive hunting methods. It is possible that the present emptiness of the West African savannas is due to recent human factors. But the fact that their grass covers are very different from those which dominate the east of the continent must not be ignored. A glance at J. M. Rattray's (1960) recent map clearly shows this difference. In the Sudanese and the sahelian zones of the west, the most frequent grasses are the *Andropogon* and *Cenchrus*, while the *Themeda* and *Hyparrhenia* grasslands predominate in the east. The former, then, support, very generally, a pastoral load much weaker (12 to 14 hectares per cow) than the latter (0.4 to 7.7 hectares). This suggests that a floristic factor is equally pertinent in explaining the relative poverty of West Africa in wild ungulates.

SOME OTHER PARTICULARITIES OF THE UNGULATE FAUNA IN THE EAST AFRICAN SAVANNAS

Other than the abundance of ungulate fauna, what other aspects of the East African savannas favor a terrestrial primate capable of hunting his favorite prey in groups?

First of all, the openness of the region facilitates pursuit by sight and the attraction of animals around certain temporary zones: water holes (especially in the dry season), places where grass grows back after a brush fire (perhaps started by lightning), and salt licks. This is particularly obvious in the open plains of Kenya, Tanganyika, and the Rhodesias, where the visibility is good in all seasons, while the height of grass after the rains impedes hunting by sight in many of the savannas closer to the Congo rain forest.

A second peculiarity of the ungulate fauna of the African savanna is the pronounced gregariousness of many species, which live more or less permanently in herds that may reach and surpass thousands of head. Such behavior certainly facilitates game drives and allows the cornering of a group of animals in places where they can be killed by simple methods.

Another ecological characteristic of potential interest to the first human hunter-collectors is the tendency of many African ungulates to drop their young in any month of the year. In our work in Albert Park (F. Bourlière and J. Verschuren 1960), we found that all the species had this tendency, even the Topi, whose young are mainly born during a well-defined season. For his part,

W. F. H. Ansell (1960) showed, in Northern Rhodesia, that the zebra drop their young any time from March to October, and that the duikers, bushbuck, the greater kudu, the hippopotamus, the rhinoceros, and the elephant generally have their young in any month of the year. Farther from the equator, notably in South Africa, the cycle of reproduction seems to be much more distinct. The young of many species of ungulates are very vulnerable in the days following their births, and capturing them offers few difficulties—which means an easy source of food for all predators familiar with the behavior of their prey. Even a newborn buffalo can, in these circumstances, be captured without danger, if his mother has been put to flight with the rest of the herd by a game drive. He may even decide to follow someone who passes close to him—which makes it possible to bring him back to camp, and even to keep him alive a day or two before killing him.

The figures and observations given in this report show that if the East African savannas at the beginning of the Pleistocene resembled those of today, they must have constituted an ideal hunting terrain for early man, even if they had only rudimentary means at their disposal. It was certainly easier for them to find abundant food there than to defend themselves against the large carnivores. But if our distant ancestors were in the habit of spending the night in the trees (or of taking refuge there in times of danger), this peril would have been considerably reduced. Moreover, in proportion as their populations became larger under the influence of a more abundant and more regular nourishment, the felids and the canids must rapidly have learned to avoid this enterprising biped who showed himself more and more capable to compete with them (at least partially) for food.

BIBLIOGRAPHY

Ansell, W. F. H.
1960. "The Breeding of Some Larger Mammals in Rhodesia," *Proc. Zool. Soc. London*, 134:251–74.

Bannikov, A. G.
1961. "L'écologie de *Saiga tatarica* L. en Eurasie, sa distribution et son exploitation rationnelle," *La Terre et la Vie*, 108:77–85.

Bere, R. M.
1960. "An Outline Survey of the Status of the Major Mammal Species (and the Crocodile) in the Uganda National Parks in 1960." *Report and Accounts of the Trustees of the Uganda National Parks, for the year ended 30th June, 1960*, pp. 13–49.

BOURLIÈRE, F., and J. VERSCHUREN
1960. *Introduction à l'ecologie des Ongules du Parc National Albert.* Bruxelles: Institut des Parcs nationaux du Congo et du Ruanda-Urundi.

COLLINS, W. B.
1959. *The Perpetual Forest.* Philadelphia, New York: Lippincott.

DASMANN, R. F., and A. S. MOSSMANN
1961. "Commercial Utilization of Game Mammals on a Rhodesian Ranch." Paper presented at the annual meeting of the Wildlife Society, California Section, Davis, California, January 1961. Mimeographed.

GRZIMEK, M. and B.
1960. "Census of Plains Animals in the Serengeti National Park, Tanganyika," *J. Wildlife Management,* 24:27–37.

LAMPREY, H. F.
1960. In: *Annual Report of the (Tanganyika) Game Department.* Dar es Salaam, pp. 1–17.

LOWE, V. P. W.
1961. "A Discussion on the History, Present Status, and Future Conservation of Red Deer (*Cervus elaphus L.*) in Scotland," *La Terre et la Vie,* 108:9–14.

MONOD, TH.
1960. *Patrouille Majabat 1959–60. Rapport preliminaire* (mimeographed). Dakar: Institut français d'Afrique noire.

PETRIDES, G. A.
1961. "Ecological Research as a Basis for Wildlife Management in Africa." Paper presented at the CCTA/IUCN Symposium on the Conservation of Nature and Natural Resources in modern African States, Arusha. Mimeographed.

PIRLOT, P.
1956. "Recensement de grands Mammifères dans la plaine de Luama," *Bulletin agricole du Congo Belge,* 47:341–66.

RATTRAY, J. M.
1960. "Tapis graminéens d'Afrique." *Etudes Agricoles,* No. 49. Rome: United Nations Food and Agriculture Organization.

STEWART, D. R. M., and L. M. TALBOT
1961. "Loita-Mara-Serengeti Aerial Survey." Departmental Report. Mimeographed.

TALBOT, L. M., H. P. LEDGER, and W. J. A. PAYNE
1961. "The Possibility of Using Wild Animals for Animal Production on East African Rangeland, Based on a Comparison of Ecological Requirements and Efficiency of Range Utilization by Domestic Livestock and Wild Animals." Report presented at the Lake Manyara Conference. Mimeographed.

TURCEK, F.
1953. "Ecological Analysis of the Bird and Mammalian Population of a Primeval Forest on the Pol'ana Mountain (Slovakia)," *Bull. Intern. Acad. Tchèque Sci.,* 53:81–105.

CONTINENTAL VERTEBRATE FAUNAS OF THE TERTIARY OF NORTH AFRICA

C. ARAMBOURG

Our knowledge of the successive faunas of the continental vertebrates of North Africa during the Tertiary is far from being as advanced as that concerning Europe or America. This is a result of the scarcity of fossil strata, a direct consequence of the geological history of this region of Africa.

For the most part, this area is made up of marine formations which largely overlapped the edge of the African continental shelf until relatively recently, and which, in consequence, created few conditions favorable to the formation of continental vertebrate fossil sites.

We shall see, however, that sites do occur in two kinds of areas under very different conditions.

The first, which I shall call the *Atlas Zone*, corresponds to the broad zone of Pyrenean-Alpine folds, which make up the largest part of the Maghreb, extending from the north of Mauritania to the Tunisian edge of the Gulf of Syrte. Con-

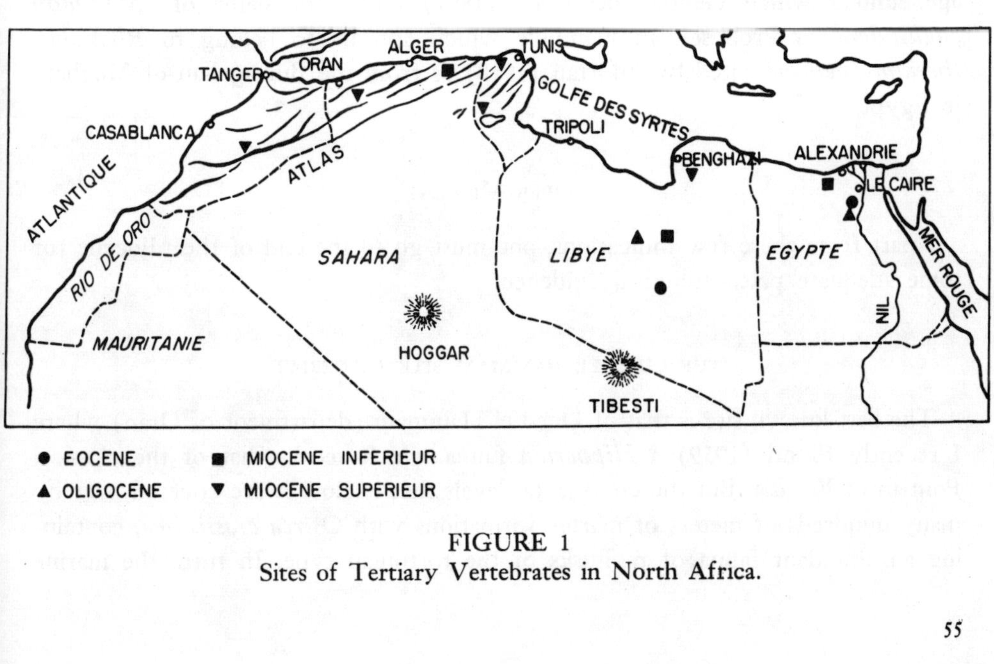

FIGURE 1
Sites of Tertiary Vertebrates in North Africa.

tinental vertebrate sites appear only sporadically and are limited to a few rare lacustrian formations which are relatively recent and hardly extensive.

The second, the *Nilotic-Saharian Zone*, corresponds to the flat region extending from the south end of the Atlas Mountains to the Red Sea and includes all of Libya, Cyrenaica, and the northern Nile Valley. In this region the continental formations are spread over a considerable expanse, and they have supplied the major part of the known African continental vertebrate fossils.

I. THE ATLAS ZONE

No site of Paleogene age is known in this region, where the entire stratigraphic series is marine.

Lower and Middle Miocene

For a long time, certain continental "red" formations in Algeria, subsequent to the Eocene were attributed to the Aquitanian, but with no traces of corresponding fossils. Recently, at least a part of these "Aquitanian" deposits have been recognized as belonging to the Upper Miocene, a much more recent era. Nevertheless, it would seem, according to certain scattered finds, that many of these "red beds" are a little older and go back to the beginning of the Miocene. This at least is the case for those of the Isserville region (Kabylie). Deperet (1897) describes a *Mastodon (Trilophodon) pygmaeus* from there which was recently found in the Burdigalian of Libya.

Finally, in the south of Tunisia, the Cherichera region long ago provided some fossil remains of mammals coming from a "Miocene" level of indeterminate age, among which Gaudry described (1891) under the name of "*Mastodon angustidens*" a Proboscidian mandible which appears to belong to *Rhynchotherium Spenceri*, cited by Fourtau (1918) as from the Burdigalian of Moghara in Egypt.

Upper Miocene

Apart from these few indications, one must go to the end of the Miocene for some adequate paleontological evidence.

THE OUED EL HAMMAM SITE (ALGERIA)

The best known site is that of Oued el Hammam (department of Oran) where I recently found (1959) a *Hipparion* fauna which recalls that of the classical Pontian of Eurasia. But the continental levels that enclose it are covered over by many hundreds of meters of marine formations with *Ostrea crassissima*, containing an abundant fauna of mollusks of the tortonian type. In turn, the marine

Pliocene of Astien facies, with its characteristic molluscan fauna, covers this formation. That is to say, the fauna of Oued el Hammam belongs to the last part of the Miocene and clearly prior to the classical Pontian fauna. In fact, all the elements of the former, although close to those that characterize the Pontian fauna, are nevertheless distinct.

Another site, that of Marceau in the region of Algiers, is contemporary with that of Oued el Hammam and complements its fauna.

The following is a list of the elements coming from these two sites:

Proboscideans: *Turicius* sp., *Mastodon* sp.
Perissodactyles: *Dicerorhinus primaevus*, *Hipparion africanum*.
Artiodactyles: *Palaeotragus Germaini*, *Samotherium* sp., *Damalavus Boroccoi*, *Gazella praegaudryi*, *Tragocerus* sp., *Cephalophus* sp.
Carnivores: *Hyaena algeriensis*.
Rodents: *Hystrix* sp.
Tubulidentes: *Orycteropus mauritanicus*.
Primates: *Macaca Flandrini*.
Birds: *Struthio* sp.

The genus *Hipparion* is the oldest fossil equid known in Africa, because no known earlier sites, Burdigalian or Paleogene, have provided fossils of this group. We are therefore dealing with an eastern migration, but one which reached the African continent during the Upper Miocene and hence was earlier than in Europe. This fact considerably modifies—as I have already had occasion to stress —the ideas maintained by certain geologists and paleontologists about the correlation between the continental formations of the Old World and the New, and about the definition and the limit of Neogene stages of these two hemispheres.

THE BENI MELLAL SITE (MOROCCO)

Here we have a travertine deposit, located east of the *meseta* and near the junction of the Middle and High Atlas mountains.

Its fauna is mostly composed of micromammalia, the relations of which are difficult to establish. However, the presence of certain genera, such as the *Cricetodon* and a Hyracoid, militates in favor of an Upper Miocene or Pontian Age.

THE PONTIAN STAGE (?)

The presence of a Pontian fauna has been noted in a few rare parts of North Africa: by Dalloni (1915) near Nemours in the department of Oran; in Morocco, in the argil of Camp Berteaux, by Bourcart (1937), and Choubert and Ennouchi (1946), following the discovery of *Hipparion* teeth. But, as we have just seen, other data will be necessary to determine the age of these layers with certainty.

Also, Gobert in 1907, found in the Tozeur region of Tunisia, a few elements of a fauna that included the genus *Merycopotamus* with various antelope, which P. Thomas and Solignac attributed to the Upper Miocene. Boule (1910), then

Solignac (1931), placed these *Merycopotamus* beds in the Pontian stage. But it must be noted that this Indian genus appeared in India well before the Pontian, and its presence is in no way characteristic of this stage. The same is true of the genus *Hipparion* recovered later by Passemard (1928) from the Tozeur beds, a genus which appeared in Africa, as we know (see page 57), before the end of the Miocene.

More recently (1934) Roman and Solignac noted the existence, near Douaria (Tunisia), of a Pontian fauna site, including, with *Merycopotamus*, a rhinoceros (*Rh. pachygnathus*) and a sivatherine (*Helladotherium Duvernoyi*); but these records have never been subjected to a detailed description.

Pliocene

The Pliocene period is, for all North Africa, an almost complete paleontological gap.

Only a few lacustrian formations of the Constantine-setiferous plateaus can reasonably be attributed to it: limestone of the St. Arnaud Cemetery and from Aïn el Bey (department of Constantine), as well as that from around Mascara (department of Oran). The only important fossil coming from these sites is a small *Hipparion*, well defined by Pomel (1897) under the name *Hipparion sitifense*.

II. THE NILOTIC-SAHARIAN ZONE

Paleogene Faunas

The first discoveries made in this zone were those from the Fayum, in the Nile Valley, which brought to light the oldest continental fauna thus far recognized in Africa. But the levels containing this fauna correspond to those already elevated in the Paleogene, and, in consequence, all the African fauna from the beginning of the Tertiary remains unknown to us. Until recent years, the Fayum sites appeared to be unique in Africa; but recently (1960-61) the search for oil of the T. O. T. A. L. Company has uncovered new sites of the same age in the Libyan desert.

A. THE FAYUM SITES

Numerous classic studies have been dedicated to these sites, so that only a summary of the data will be given. It is a question of concordant fluvio-marine and fluvial-lacustrine formations, but in which the succession of vertebrate faunas at various levels permits the determination of a certain number of subdivisions.

1. Eocene. The base of the sites is formed by marine levels which appear to correspond to a part of the Eocene prior to the Lutetian. Above, there are fluvio-marine beds showing:

a. a lower zone rich in cetaceans (*Prozeuglodon isis*, *Protocetus atavus*, *Eocetus Schweinfurthi*), in sirenians (*Protosiren Fraasi*, *Eotheroides aegyptiacum*, *libycum*, *abeli*, *majus*, *Trichechus Coulombi*), in crocodiles (*Tomistoma kerunense*), and fish;

b. an upper zone which corresponds to the end of the Lutetian and constitutes the horizon of Qasr-el-Sagha. It contains, with certain of the preceding species, other cetaceans (*Zeuglodon Osiris*) and various terrestrial mammals (*Moeritherium Lyonsi* [1]), primitive ungulates allied both to proboscideans and to sireneans, and *Barytherium grave* (another ungulate with proboscidean tendencies), and reptiles and fish.

2. Oligocene. The rest of the Fayum series belongs to the Oligocene and includes a fauna essentially characterized by the appearance of true proboscideans, by the abundance and the differentiation of hyracoids of which some are giant forms, by the abundance of the anthracotheres, and by the presence of catarrhine primates. The essential elements are the following: [2]

Insectivores: *Metolbodotes Stromeri.*

Bats: *Vampyravus orientalis*, *Ptolemaia Lyonsi.*

Rodents: *Phiomys Andrewsi*, *Metaphiomys Beadnelli.*

Carnivores: *Sinopa aethiopica*, *Metasinopa Fraasi*, *Apterodon altidens*, *macrognathus*, *minutus*, *Pterodon africanus*, *leptognathus*, *phiomensis*, *Hyaenodon brachycephalus.*

Primates: *Moeripithecus Markgrafi*, *Apidium phiomense*, *Parapithecus Fraasi*, *Propliopithecus Haeckeli.*

Proboscideans: *Moeritherium Andrewsi*, *Palaeomastodon Beadnelli*, *Barroisi*, *intermedius*, *parvus*, *Phiomia serridens*, *minor*, *Osborni*, *Wintoni.*

Embrithopodes: *Arsinoitherium Zitteli.*

Hyracoids: *Pachyhyrax crassidentatus*, *Saghatherium antiquum*, *annectens*, *euryodon*, *macrodon*, *sobrina*, *Geniohyus mirus*, *diphycus*, *gigas*, *magnus*, *micrognathus*, *subgigas*, *Bunohyrax fajumensis*, *affinis*, *major*, *Megalohyrax eocaenus*, *minor*, *niloticus*, *pygmaeus*, *suillus*, *Titanohyrax palaeotherioides*, *Andrewsi*, *Schlosseri*, *ultimus.*

Artiodactyles: *Mixtotherium mezi*, *Rhagatherium aegyptiacum*, *Brachyodus Andrewsi*, *Fraasi*, *Gorringei*, *parvus*, *rugulosus.*

To this mammalian fauna must be added a bird (*Eremopezus eocaenus*), some reptiles (*Crocodilus articeps*, *megarhinus*, *Tomistoma gavialoides*, *Testudo ammon*, *Beadnelli*, *Isis*, *Podocnemis fajumensis*, *Stereogenys libyca*), and fish.

There is no need to stress the uniqueness of the fauna of Fayum, which distinguishes it from its contemporaries in other parts of the world.

1. Other species of this genus have been described from the same sites, but the taxonomic value of these forms is disputable.
2. This list corresponds to what has been described. Taxonomic value of species attributed to a particular genus is debatable.

B. THE LIBYAN DESERT SITES

For a long while, the Fayum fauna was known only in the Nile Valley. The first indication of its extension toward the west came about in 1951 following the discovery by Kikoine of a few *Moeritherium* teeth (Arambourg, Kikoine, Lavocat 1951) in an Eocene layer near Gao in the Sudan. But recently (1959–60) the geologic research carried on by T. O. T. A. L. Company, uncovered important sites in the Libyan desert with Eocene and Oligocene vertebrates (Arambourg and Magnier 1961).

1. Eocene. The Eocene sites are about 520 kilometers south of Syrte Major. They belong to a thick subhorizontal formation much cut by erosion, running between 18° and 19° longitude and 26° latitude for nearly 100 kilometers. This outcrop, unnamed on the maps, has been designated by T. O. T. A. L. geologists as "Djebel Coquin." [3]

This series rests on Paleocene marine beds. Its lower part, marine, includes only fish remains (Shark teeth) and a sea serpent of the genus *Pterosphenus.* This level is overlain by a thick clay series, sometimes gypsiferous, the upper part of which is rich in vertebrate remains. The whole is covered by sandstone attributed to the Oligocene containing only plant fossils, with a profusion of silicified wood.

The fauna of Djebel Coquin corresponds to that of the horizon of Qasr-el-Sagha of Fayum. It is a fluvio-marine fauna which also includes the typical association *Barytherium grave*, in relative abundance, and *Moeritherium Lyonsi.* The other fossils are fish of the genera *Pristis, Fajumia, Lates,* etc., crocodiles of various forms, a short-snouted Nilotic type, the others of the long-snouted type (*Tomistoma*) and *Dyrosaurus,* as well as palustral tortoises and one sea serpent, *Gigantophis.*

2. Oligocene. The presence of terrestrial vertebrates in an Oligocene level was discovered by Magnier, in the environs of the Zella oasis, south of Syrte Major (Arambourg and Magnier *loc. cit.*). This involved an estuarine deposit the outcrop of which, unfortunately limited, did not permit more extensive investigation. Nevertheless, a few characteristic forms were found: *Palaeomastodon, Phiomia Wintoni, Megalohyrax palaeotherioides, Brachyodus* cfr. *Gorringei,* carnivore, paludal turtle, and crocodile.

This typical fauna is sufficient to indicate the extension of the Oligocene beds of Fayum into Libya.

3. It is probably the same layer which was noted by Bellair, Freulon, and Lefranc (1954) under the name of Dor-et-Talha.

The Neogene Fauna

A. THE BURDIGALIAN SITE OF MOGHARA

The first elements of Miocene fauna of North Africa were discovered (Fourtau 1920) in Egypt, in the Moghara oasis 150 kilometers southwest of Alexandria. The Burdigalian marine beds there are associated with continental levels where Fourtau collected numerous vertebrate remains. The list follows:

Mammals: *Schizodelphis* cfr. *sulcatus, Delphinus van Zelleri, Teleoceras Snowi, Aceratherium* sp., *Brachyodus africanus, Moneyi, Moneyi* var. *strictidentata, Masritherium Depereti, Mastodon angustidens* var. *libyca, Mastodon (Rhynchotherium) Spenceri, Hyaena* sp., *Dryopithecus mogharensis, Pliopithecus Tandyi.*

Reptiles: *Crocodilus Lloydi, Tomistoma Dowsoni, Gavialis* sp., *Podocnemis aegyptiaca, Bramlyi, Trionyx Senckenbergianus.*

Fish: *Pristis* sp., *Myliobatis* aff. *angustidens, M.* aff. *meridionalis, M.* sp., *Sphyrna prisca, Synodontis* sp., *Lates* sp., Silurids indet.

B. THE BURDIGALIAN SITE OF ZELTEN (LIBYA)

This site, discovered in 1960 by Magnier, is in Cyrenaica, 150 kilometers south of Syrte Major. It extends over a considerable area to the east from Djebel Zelten, where it rests on Oligocene marine beds and is overlain by Helvetian marine beds.

It is remarkable for its abundant silicified wood, whole tree trunks, sometimes complete with branches, and which, lying in the same direction in certain places, suggest the idea of a flood deposition. The vertebrates collected are as follows:

Mammals: *Mastodon pygmaeus, Mastodon (Trilophodon)* sp., *Brachypotherium Snowi* ?, *Brachyodus* sp., *Hyoboops africanus, Prolibytherium Magnieri, Bunolistriodon Massai, Afrocyon Burolleti,* etc.

Reptiles: *Tomistoma africanum, Euthecodon* ?, *Crocodilus* cfr. *niloticus* ?

Fish: Silurides, *Lates*

Birds: Aepyornithide (*Psammornis* ?).

This fauna complements and adds significantly to that of Moghara. *Prolibytherium*, notably, is a generalized Sivathere, the existence of which is related to that of another giraffid of the Burdigalian of eastern Africa, *Climacoceras africanus*, each of them apparently found at the base of two giraffid phyla—on the one hand *Sivatherinae*, on the other *Giraffidae*—indicating the African origin of the giraffids.

Moreover, the genera *Afrocyon* and *Bunolistriodon* are related to the fossil forms of Asia rather than with those of other regions.

C. THE PONTIAN (?) SITE OF SAHABI

To the north of the above, in the neighborhood of Benghazi in Cyrenaica, this fluvial-marine site probably belongs to the end of the Miocene. Its fauna, although abundant, has unfortunately been only very incompletely described; it is noteworthy that it includes a long snouted mastodon, *Stegotetrabelodon syrticus* and *St. libycus*, along with *Stegolophodon sahabianus*, *Stegodon syrticus* *Pentalophodon sivalensis*, *Merycopotamus*, some bovids, some crocodiles, and fish typical of the Upper Miocene.

The Pliocene

The only fossiliferous place of this age currently known is Ouadi Natroun, in the Nile Valley; it has supplied only rare elements, among which a hippopotamus (*H. protamphibius Andrewsi*) and an antelope (*Hippotragus* (?) *Cordieri*).

CONCLUSIONS

In spite of the considerable gaps which yet remain in our knowledge, it is nevertheless clear that the Tertiary fossil fauna of North Africa—just as that of the rest of the continent—are characteristically endemic, which clearly distinguishes them from those of Europe, and even more so from those of America. The presence in the Upper Eocene of large proboscidiform ungulates, and, since the Oligocene, of true proboscideans, highly diversified hyracoids, various suiformes and catarrhine primates—some already oriented in the anthropomorphic direction—are indications of the role Africa probably played as the center of evolution for numerous groups, particularly for the hominids. What we know of Neogene faunas only corroborates these first data and demonstrates, in the absence of the late-immigrating equids, the importance in Africa of the artiodactyl group.

Furthermore, if the mesogean rifts were an obstacle to the exchange of fauna between Africa and Eurasia during the major part of the Tertiary, it is evident that territorial connections between Africa and the Indian regions brought about the establishment of an Ethopian-Indian biogeographic block which, until a relatively recent epoch, dominated the faunal relations between these two continents.

Finally, it should be noted that since the Eocene the successive faunal associations of the North African regions present a remarkable ecological similarity. The mammals in general belong to tropical savanna types of Sudanese character. The aquatic animals, fish and reptiles, are of Nilotic character. But among the crocodiles must be noted the association, since the Eocene, of African types with broad, short snouts (*niloticus type*) or with narrow snouts (*cataphractus* type), with the true long snouted gavialoids (*Tomistoma* or *Euthecodon* genera) which persisted in Africa up to the Villafranchian.

Thus it is known, and I have so indicated in another symposium, on "Early Man and Pleistocene Stratigraphy in the Circum-Mediterranean Regions," that this endemic African characteristic is preserved, in its essential aspects, until the time of the Quaternary, and that it is only toward the end of that period that certain Eurasian elements penetrated to North Africa, probably by way of the Near East and Suez.

BIBLIOGRAPHY

Arambourg, C.

1959. "Vertébrés continentaux du Miocène supérieur de l'Afrique du Nord," *Publ. Serv. Carte Géol. Algérie* (Alger), N.S. Paléontologie Mémoire No. 4.

1963. "Le genre *Bunolistriodon* Arambourg 1933. Bull. Soc. Géol. Fr. (Paris), Sér. 7, V; p. 903–911, 3 fig., 3 Pl.

Arambourg, C., J. Kikoine, and R. Lavocat

1951. "Découverte du genre *Moeritherium* Andrews dans le Tertiaire continental du Soudan," *C. R. Acad. Sci.* (Paris), 233:68–70.

Arambourg, C., and P. Magnier

1961. "Gisements de Vertébrés dans le bassin tertiaire de Syrte (Libye)." *C. R. Acad. Sci.* (Paris), 252:1181–3.

Bellair, P., J. Freulon, and J. Lefranc

1954. "Découverte d'une formation à Vertébrés et Végétaux d'âge tertiaire au bord occidental du désert libyque (Sahara occidental)," *C. R. Acad. Sci.* (Paris), 239:1822–4.

Boule, M.

1910. "Sur quelques Vertébrés fossiles du Sud de la Tunisie," *C. R. Acad. Sci.* (Paris), 150:812–13.

Bourcart, J.

1937. "Sur la découverte de molaires d' *Hipparion* à la base de la série lacustre à argiles smectiques de Camp-Berteaux (Maroc Oriental)," *C. R. S. Soc. Géol. France* (Paris), No. 8, pp. 79–80.

Choubert, G., and E. Ennouchi

1946. "Premières preuves paléontologiques de la présence de Pontien au Maroc," *C. R. S. Soc. Géol. Fr.* (Paris), No. 11, pp. 207–8.

Dalloni, M.

1915. "Le Miocène supérieur dans l'Ouest de l'Algérie: couches à *Hipparion* de la Tafna," *C. R. Acad. Sci.* (Paris), 161:638–41.

Depéret, C.

1897. "Découverte de *Mastodon angustidens* dans l'étage cartennien de Kabylie," *Bull. Soc. Géol. Fr.* (Paris), Sér. 3, 24:518–21.

Fourtau, R.
1918 and 1920. "Contribution à l'étude des Vertébrés miocènes de l'Egypte" (1st and 2nd eds.), *Geol. Surv. Egypt* (Le Caire), 2nd ed.

Gaudry, A.
1891. "Quelques remarques sur des Mastodontes à propos de l'animal du Cherichera," *Mem. Soc. Géol. France* (Paris), Mém. 8.

Pomel, A.
1897. "Les Equidés," *Publ. Serv. Carte Géol. Algérie* (Alger), Monogr. Paléont.

Roman, F., and M. Solignac
1934. "Découverte d'un gisement de Mammifêres pontiens à Douaria (Tunisie septentrionale)," *C. R. Acad. Sci.* (Paris), 199:1649–50.

Solignac, M.
1931. "Le Pontien dans le Sud Tunisien," *Annales Univ. Lyon (Sc. et Médecine)*, n. s. 1, Fasc. 48.

PLEISTOCENE MAMMAL FAUNAS OF AFRICA, WITH PARTICULAR REFERENCE TO SOUTHERN AFRICA

H. B. S. COOKE

INTRODUCTION

THE FOSSIL MAMMALS of Africa were virtually unknown until the latter half of the nineteenth century when a number of French scientists, notably Thomas and Pomel, began to describe material from the northwest African coastal region. At this time the equatorial and southern parts of the continent were paleontologically unexplored and it was not until the first two decades of the present century that systematic descriptions of mammal fossils from this region started to appear. The valuable bibliography of the fossil mammals of Africa by Hopwood and Hollyfield (1954) lists almost 450 living and extinct species and subspecies of Pleistocene mammals recorded up to 1950; the writer has records of a further 59 names to the end of 1960. The dates on which each variety was first mentioned have been analyzed in the histogram given in Figure 1, which thus presents some idea of the activity of paleontological description, decade by decade.

Of these 500 records, about 180 are not separated from living types and the remainder are ascribed to forms supposedly extinct. Although Hopwood and Hollyfield effected some revisions in their lists there are many obvious synonyms amongst the Proboscidea, Equidae, and Suidae, some of which have been dealt with since their account appeared. Omitting these and also the listed subspecies, the number of reasonable records to 1960 is reduced to 350, of which 130 are not separated from living species.

It is clear that the criteria employed for the evaluation of the fossil species varies greatly from authority to authority and from region to region so that a good deal of co-ordinated revision is necessary before effective interregional comparisons can be made. It is also apparent that each of the various major sites has furnished only a partial fauna whose character has been controlled by the local environment of preservation and is not representative of the whole spectrum of contemporary life. Families or even orders which are common at

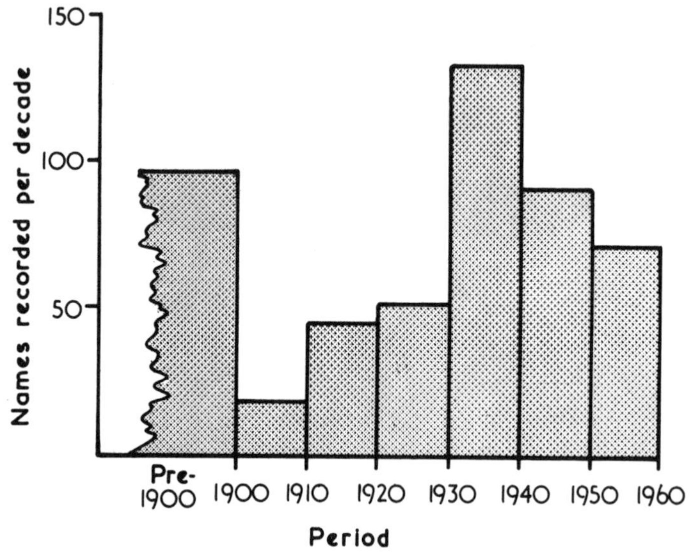

FIGURE 1
Histogram showing rate of recording of Pleistocene fossil mammals in Africa.

one site are virtually unrepresented at others. This renders very difficult the determination of the true age relationships between the families even within a small area and still more difficult the task of effecting correlations between regions. Yet unless the time sequence can be established it is virtually impossible to reconstruct the successive contemporary faunas in the different areas or to attempt any sort of ecological interpretations.

Although Pleistocene fossil mammals have been recorded from many places in Africa, very few areas have furnished reasonable assemblages. The more important sites are plotted on the accompanying map (Fig. 2) and it will be seen that they are very widely scattered over the continent. In the vast Sahara five localities have furnished a few specimens of some interest. West Africa is singularly blank. In Egypt and the Sudan there are a few minor fossil localities but only two have provided significant collections. Cyrenaica is almost devoid of important sites so that in the whole of north Africa it is only in the coastal belt of Morocco, Algeria, and Tunisia that the Pleistocene mammals have been well studied. A large area of little-known territory separates this region from the rich and well-studied area of the East African plateau. Relatively small collections of Pleistocene mammals have come from Angola, the Congo Republic (formerly Belgian), and Northern Rhodesia, but Southern Rhodesia and South Africa have yielded important assemblages from a number of sites. In the whole vast continent, there are thus only three major regions in which the Pleistocene mammals are reasonably well known and each of these regions possesses today an ecological environment which is quite distinctive.

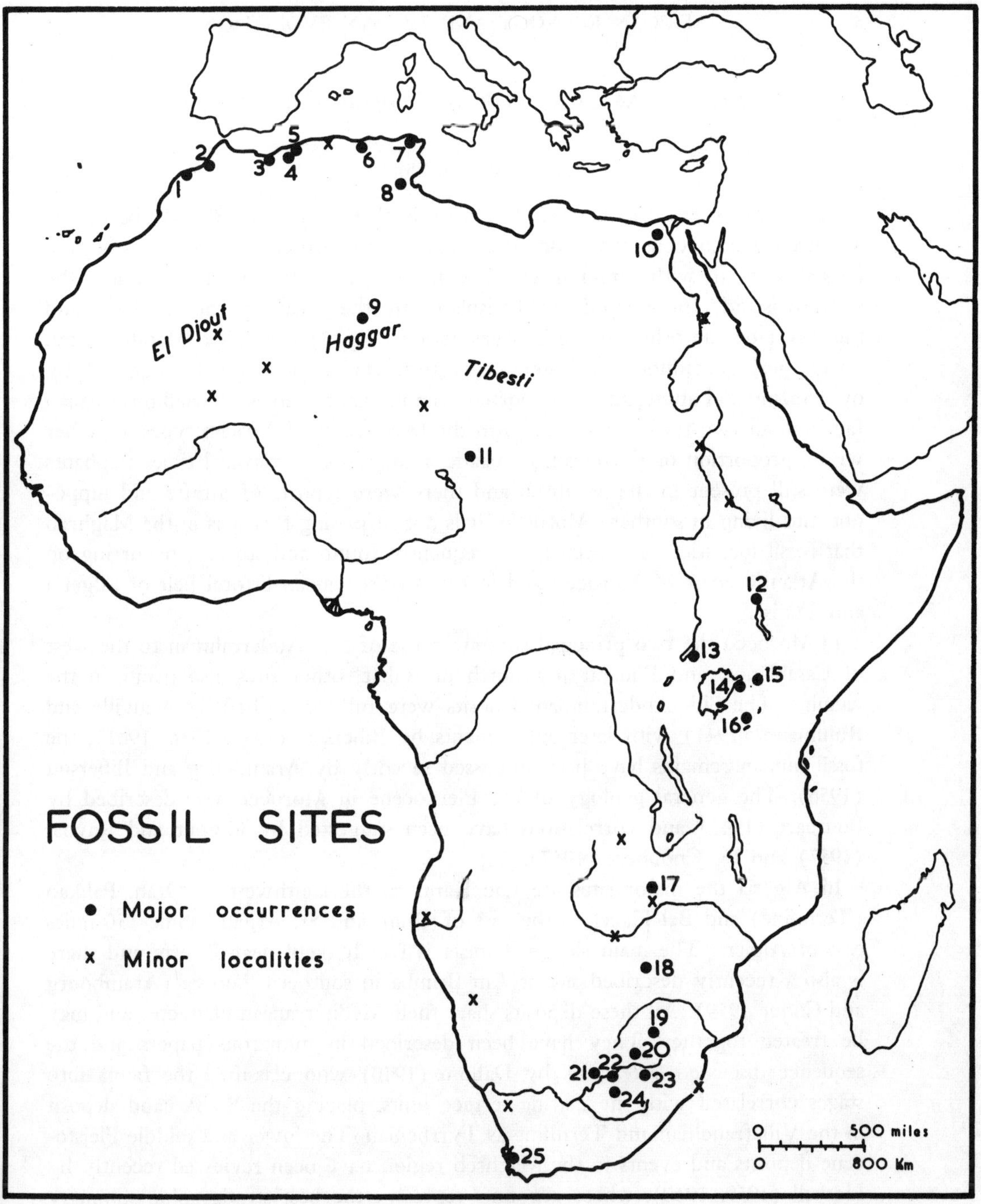

FIGURE 2

(1) Sidi Abderrahman (2) Fouarat (3) Lac Karar (4) Ternifine (5) Bel Hacel (6) St. Arnaud (7) Lac Ichkeul (8) Aïn Brimba (9) Tihoudaïne (10) Wadi Natrun (11) Koro Toro (12) Omo (13) Kaiso (14) Kanam, Rawi, Kanjera (15) Nakuru-Naivasha basin (16) Laetolil, Olduvai (17) Broken Hill (18) Chelmer (19) Makapansgat (20) Sterkfontein area (21) Taung (22) Vaal River gravels (23) Cornelia (24) Florisbad, Vlakkraal (25) Hopefield.

NORTH AFRICA: THE MAGHREB AND SAHARA

The Maghreb

In North Africa the best studied area is that known to the Arabs as the Maghreb, which constitutes the Atlas area of the northwest and the associated coastal strip, all with a rainfall above 8 inches a year. Its full name means "the western island" for it is effectively isolated by the ocean on the one hand and the desert on the other. Its natural vegetation (Fig. 3) comprised Mediterranean scrub, bush, deciduous, and some evergreen forest, now much degraded by overgrazing and abuse, and it supported even in historic times a varied mammalian fauna about two-thirds of which form the basic stock of African types, together with a proportion of European or Asiatic immigrants. In classical times elephants were still present in the Maghreb and there were reports of giraffe and hippopotamus living in southern Morocco. It is not surprising that it is in the Maghreb that fossil localities have been most frequently found and studied, occurring on the Atlantic coast of Morocco and in the Mediterranean littoral belt of Algeria and Tunisia.

In Morocco the two principal sites are those of Sidi Abderrahman to the west of Casablanca, and Fouarat just north of Rabat; other sites also occur in the vicinity. The Sidi Abderrahman deposits were fully described by Neuville and Ruhlmann (1941) with later amendments by Biberson (1955, 1956, 1961); the fossil human remains have been discussed recently by Arambourg and Biberson (1956). The general geology of the Pleistocene in Morocco was described by Bourcart (1943) and correlations have been suggested by Gigout and Raynal (1957) and by Choubert (1957).

In Algeria the major sites are Lac Karar to the southwest of Oran, Palikao (Ternifine) and Bel Hacel to the east of Oran, and St. Arnaud some 150 miles east of Algiers. The main site in Tunisia is Lac Ichkeul near Bizerte and there is also a recently described site at Aïn Brimba in southern Tunisia (Arambourg and Coque 1959). All these deposits share their Mediterranean character and may be treated together. They have been described in numerous papers and the sequence discussed, *inter alia*, by Dalloni (1940) who classified the fauna into stages correlated with the marine terrace units, placing the St. Arnaud deposit in the Villafranchian and Ternifine as Tyrrhenian. The lower and middle Pleistocene deposits and events in the Maghreb region have been reviewed recently by Howell (1959, 1960), whose accounts represent the best stratigraphic summary available. The data given in Howell's two papers have been employed freely in the construction of Table 1 but modifications have been made to accord more closely with the latest views of Arambourg (*in litt.*) and Biberson (1961). Some stage names have been added and the position of the new Tunisian site of Aïn Brimba is shown.

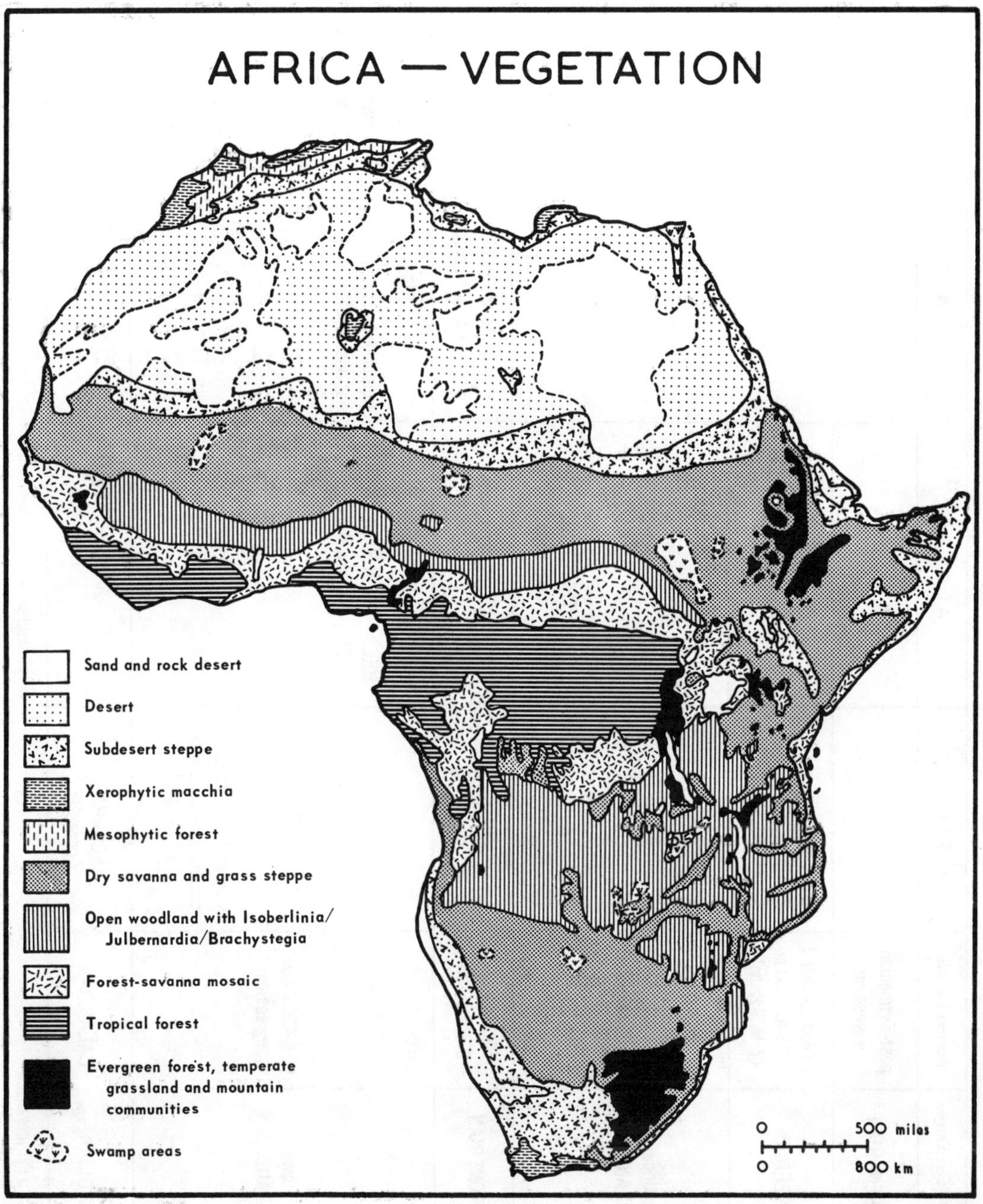
AFRICA — VEGETATION
Sand and rock desert
Desert
Subdesert steppe
Xerophytic macchia
Mesophytic forest
Dry savanna and grass steppe
Open woodland with Isoberlinia/Julbernardia/Brachystegia
Forest-savanna mosaic
Tropical forest
Evergreen forest, temperate grassland and mountain communities
Swamp areas
0 500 miles
0 800 km

FIGURE 3

TABLE 1
Outline Correlation of Lower and Middle Pleistocene Deposits in North Africa

Age	*North Sea Stages*	*Marine Levels*	MOROCCO	ALGERIA	TUNISIA	STONE CULTURE
MIDDLE PLEISTOCENE	DRENTHIAN	post-Tyrrhenian regression				VI, V (MIDDLE ACHEULIAN)
MIDDLE PLEISTOCENE	NEEDIAN	TYRRHENIAN I + 28 to + 30 m (? Milazzian)	SIDI ABDERRAHMAN: karstic caves in Great Dune	marine conglomerates		IV (MIDDLE ACHEULIAN)
MIDDLE PLEISTOCENE	Upper TAXANDRIAN	Romanian regression; SICILIAN II +55 to +60 m	SIDI ABDERRAHMAN: Great Dune; conglomerate; –	LAC KARAR; –; TERNIFINE lake beds		III, II, I (LOWER ACHEULIAN)
	CROMERIAN	SICILIAN I + 80 to + 100 m	marine beaches and marine deposits	marine beaches and marine deposits	high level (+ 100 m) terrestrial beds	
VILLAFRANCHIAN	Lower TAXANDRIAN	post-Calabrian regression	calcareous lacustrine deposits; (erosion); uplift; reddened loams (Marmora) conglomerates (Rharb, Arbaoua*); uplift	tilting; BEL HACEL; AÏN HANECH lake beds*; SAINT ARNAUD LAKE BEDS	tilting; (upper beds not observed); LAC ICHKEUL BEDS	*pebble tools; *pebble tools
VILLAFRANCHIAN	TIGLIAN	CALABRIAN	FOUARAT deposits	AÏN BOUCHERIT lake beds; SAINT ARNAUD LAKE BEDS	LAC ICHKEUL BEDS; AÏN BRIMBA deposits	
	Pliocene sea		Plaisancian – Astian sediments			

Arambourg (1949, 1952) has summarized the general succession of the faunas and their characteristic elements, recognising four stages as follows: (1) *Villafranchian*, separated into a *lower* division represented at Fouarat in Morocco, at Lac Ichkeul and at Aïn Boucherit (lower St. Arnaud), and an *upper* division at Bel Hacel and Aïn Hanech (upper St. Arnaud); (2) *Middle Pleistocene*, represented at Ternifine, Lac Karar, and in the Rabat-Casablanca beds; (3) *Upper Pleistocene* from many sites of post-Tyrrhenian date, including cave and surface deposits, often associated with stone implements of Levalloisian-Aterian affinities; and (4) "*Recent*" with upper Paleolithic and Neolithic associations, lacking the extinct faunal elements. Arambourg's lists of species are highly selective and it has been necessary to search the available literature in order to prepare the stratigraphic distribution chart given in Table 2. This includes the most important elements and an attempt has been made to revise the nomenclature in the light of the latest available information. A large number of Bovidae have been omitted as the criteria for their determination seem to be unsatisfactory.

It is clear from the tabulated data that the "Villafranchian" fauna consists almost exclusively of extinct species, few of which persist into the "Middle Pleistocene" assemblages. This may well be accounted for by the long stratigraphic interval indicated in Table 1. It is also obvious that the post-Pleistocene species form essentially the living fauna and that these elements make their appearance progressively during the Middle and Upper Pleistocene. The nature of the Middle Pleistocene assemblages, however, is characteristic of the savanna country of central Africa and the bovid forms (which are not shown in the table) add considerably to this general impression. It is tempting to suggest that the presence of a tropical fauna, including such moisture-loving species as the hippopotamus, demands a very different climatic setting from that of the present day and, indeed, many authorities have suggested that the Sahara region was not a desert during at least portions of Pleistocene and post-Pleistocene time. The extinctions of the faunas are then ascribed to phases of severe climatic deterioration.

The Sahara

That there is evidence for climatic change in this region is undeniable, for the Sahara is dotted with upper Paleolithic to Neolithic rock engravings and fossil remains of elephant, rhinoceros, hippopotamus, and giraffe (see Mauny 1957), and even of domestic animals (Esperandieu 1955), whose present range lies well outside the areas of the petroglyphs. Fossil remains of greater antiquity have been found at other sites, the most important of which are Koro-Toro and Tihodaïne. Koro-Toro lies to the northeast of Lake Chad and has been only briefly described (Abadie *et al.* 1959; Coppens 1960) but the fauna includes a Villafranchian assemblage which is suggestive of a link between that of St. Arnaud and Lac Ichkeul in North Africa on the one hand and that of Olduvai

TABLE 2

STRATIGRAPHIC DISTRIBUTION OF PRINCIPAL PLEISTOCENE MAMMALS IN NORTH AFRICA

VILLAFRANCHIAN	MIDDLE PLEISTOCENE	UPPER PLEISTOCENE	POST-PLEISTOCENE
pebble tools	Acheulian	Micoquian Aterian	Capsian Neolithic
Fouarat Lac Ichkeul Bel Hacel Ain Boucherit Ain Hanech Ain Brimba Guyot Ain Jourdel	Sidi Abderrahman Lac Karar Temifine Maison Carée (Palikao) Sidi Zin	surface deposits cave deposits	surface deposits cave deposits
	Atlanthropus mauritanicus Simopithecus cf major	Homo neanderthalensis	Homo sapiens →
Macaca cf trarensis			Macaca inuus →
		Canis aureus —	→
			Vulpes vulpes →
		Lycaon picta —	→
	Ursus arctos lateti		Ursus arctos —?
			Genetta genetta —
	Crocuta crocuta —	→	
Hyaena cf striata —	— Hyaena striata —	—	→
			Felis libyca
Machairodus sp.	Machairodus sp.		
	Panthera leo —	—	→
Anancus osiris →			
Loxodonta africanava			
Archidiskodon cf meridionalis or recki			Loxodonta africana
	Loxodonta atlantica —	—	→
	Palaeoloxodon jolensis		
Stylohipparion libycum →			
Equus numidicus	Equus mauritanicus —	→	
Ceratotherium cf simum —	—	—	→
		Rhinoceros mercki	
	Sus scrofa —	—	→
Omochoerus maroccanus →	Afrochoerus sp.	Phacochoerus africanus —	→
Hippopotamus hipponensis	Hippopotamus amphibius —	—	→
Camelus sp.	Camelus thomasi —	→	
		Megaceroides algericus —	→
Giraffa Giraffa sp.	Giraffa cf camelopardalis —	—	→
Libytherium maurusium →			
Bos sp.	Bos primigenius —	—	→
	Homoioceras antiquus —	—	→
	Redunca maupasi —?		
	Alcelaphus buselaphus —	—	→
	Gorgon cf taurinus —	→	
Antidorcas sp.			
Gazella cf cuvieri —	— Gazella cuvieri —	—	→
Gazella setifensis →	Gazella dorcas —	—	→
	Gazella rufina —	—	→
	Gazella atlantica —	→	
		Gszella tingitana	
		Ammotragus lervia —	→
Numidocapra crassicornis			
Capra sp.			

in East Africa on the other. A skull of Australopithecine character has also been found (Coppens 1961), so that the site has great scientific possibilities. At Tihoudaïne, deposits have furnished stone implements of Acheulian, Aterian, and Neolithic type (Arambourg and Balout 1955). The sites lie in the region of the elevated and mountainous Hoggar massif and the Acheulian hand axes occur abundantly in a stratified argillaceous layer with intercalated diatomites. The associated fauna includes an elephant (cf. *Archidiskodon recki*), *Equus mauritanicus*, *Bos primigenius*, *Homoioceras antiquus*, white rhinoceros, hippopotamus, a suid, carnivores, and a number of antelopes, the whole assemblage suggesting equivalence to that of Lac Karar and other sites of later middle Pleistocene age. The Aterian deposit lies about 12 kilometers east of the hand-axe site, resting on a deposit of alluvial material which is covered by diatomites carrying Neolithic tools on its surface. Arambourg's interpretation suggests that the Acheulian deposit was laid down toward the end of the Middle Pleistocene and that the rainfall was high enough to form a small lake in a depression in the crystalline rocks. The predominance of sandy material suggests the presence of aeolian formations in the vicinity, belonging possibly to an earlier period. Fresh water deposition took place in Aterian times, followed by complete desiccation. This was succeeded by a return to humid conditions and lake formation during the Neolithic, when a substantial human population is indicated by the abundant tools. Burnt bones of white rhinoceros occur and these are taken to imply an extensive vegetation since vanished and replaced by desert dunes. In another paper, Balout (1952) has put forward interesting arguments favoring amelioration of climate in the Sahara during interglacial times and suggests that the Neolithic wet period may correspond to the Climatic Optimum (Hypsithermal) and the Acheulian deposit to the Riss-Würm interglacial. In Egypt, Huzzayin (1941) has also suggested equating the Neolithic wet phase with the postglacial Climatic Optimum.

Alimen (1955) in common with a great many other workers has erected a structure of correlation based on the supposed equivalence between pluvial episodes and glaciation and, accordingly, with regressions of the sea. As far as the Neolithic is concerned, this should place the humid period as equivalent at latest to the Dryas (about 10,000–11,000 years ago) whereas Balout's interpretation would make it younger (about 5,000–7,000 years ago). Radiocarbon determinations may serve to settle the matter and at present they suggest a better fit with the younger chronology. It seems highly probable that the relationship is not one of simple equivalence and it is not difficult to visualize conditions which would involve precipitation changes in the present belts of arid or semiarid climate which were in the opposite sense to those taking place in the periglacial areas and in the equatorial rainfall belt. McBurney (1960) has discussed the interpretation very objectively and, while recognizing the reality of climatic fluctuation in North Africa, sounds a cautionary note regarding their interpretation for correlation purposes.

The belt of the Saharan and Libyan deserts stretches right across the continent from the Atlantic to the Red Sea and for about four degrees on either side of the Tropic of Cancer it offers a formidable obstacle to the passage of animals. The boundary coincides approximately with the 4-inch isohyet in the north and with the 6-inch isohyet in the south. Almost half of this area is completely covered by shifting sand or by bare rock surfaces which support no life at all (Fig. 3). In the remainder, scanty desert shrub and bunch grass exist but their distribution is discontinuous and often confined to drainage lines. Surface water is rare and occurs only at scattered oases where the underground water table intersects depressions in the surface. Man can cross the desert because of his ability to find his way from oasis to oasis and to carry with him reserve supplies to last through the barren stretches, but animals cannot do this and are effectively barred from crossing the desert. It is true that after exceptional rains animals move a long way into the desert fringe to enjoy the unusual, if temporary, vegetation which springs up from dormant seeds, but this is a long way from allowing passage across the whole barrier. A general and continued increase in rainfall would, of course, extend the range of such movement and under exceptionally favorable conditions certain routes might allow the passage of animals right across the belt.

As far as man is concerned, one of the most important primitive routes certainly lay through the central Hoggar and Tibesti massifs (Fig. 2) which have peaks rising to 9,000 feet or more with consequent reduction in local temperatures. Although the rainfall in these areas is not significantly higher than that of the rest of the desert belt, the mountains do maintain a slightly better vegetation, including some xerophytic evergreens and mountain grassland. Old drainage lines, leading away from the mountain masses, particularly to the west of the Hoggar, testify to a runoff in Neolithic times much greater than that of the present; the same is probably true also for Acheulian times. The valley of the Nile offers another route for man across the desolation but long stretches of the river run through pure desert devoid of any source of food for animals. Near the west coast man has established the "Mauritania road" and this might well have offered a route to animals under a moister climate if permanent water were available. Rock engravings of elephant and giraffe are plentiful along this route, as they are also along the Hoggar and Tibesti routes. Interest also attaches to the Djouf or Juf depression in the western Sahara as there is clear evidence that the headwaters of the Niger formerly drained into it and that it is only in geologically recent times that the river was diverted past Timbuktu by capture from the lower Niger. A number of Neolithic sites testify to the reality of this drainage, for typical West African species are found as fossils at sites like Arouane and Taoudeni, the latter almost on the Tropic. Fossil remains of hippopotamus and the swamp rat *Thryonomys* occur at both sites, indicating abundant water and at least marginal vegetation. Amelioration of climate might well have opened a link between the Djouf lake area and the habitable region of the Maghreb, perhaps along the course of the Wadi Saoura which drained from the Atlas southwards towards the same depression.

On the other side of the picture there is also evidence for the former equatorward extension of the desert by about five degrees of latitude from its present southern limit. There exists in this zone a belt of dunes which resemble in structure the shifting sands of the Saharan Erg but which are now firmly fixed and "fossilized" by a cover of vegetation (Grove 1958). The rainfall in this belt is about 10 to 20 inches a year and it is clear that to restore them to their former devegetated state would involve a decrease of about 50 percent in the present precipitation. It would seem, therefore, that the present-day climate of this region lies about midway between the drier and the wetter phases of Pleistocene fluctuations. There is plentiful geological evidence for the existence of dry conditions in this belt also in Tertiary and Cretaceous times and it is clear that the existence of the arid zone in and about these latitudes is a direct result of fundamental features of the atmospheric circulation. It is probable that, during the Pleistocene, aridity was the normal condition, offering a barrier to free migration across the desert, and that the periods of higher rainfall did not eliminate the barrier but merely opened up a small number of possible routes across it.

EAST AFRICA

The region of East Africa is part of the elevated African plateau and most of it lies 4000 feet or more above sea level. Considerable areas are covered by Tertiary and Pleistocene lavas and the cones of many great volcanoes rise to more than 10,000 feet. Several are above 14,000 feet and the two highest are Mount Kenya (17,040 feet) and Kilimanjaro (19,565 feet), the latter being the highest point in the continent. Mount Ruwenzori, which rises to 16,795 feet above sea level, is not a volcano but a mass of ancient rocks which was uptilted during the late Tertiary and Pleistocene fracturing which cut the plateau to form the rift valley systems. The Eastern Rift runs from southern Abyssinia through the center of Kenya into Tanganyika and, except for Lake Rudolf, is marked by a line of small and shallow lakes. The Western Rift curves from the southern Sudan along the western border of Uganda and Tanganyika and contains Lakes Albert, Edward, and Tanganyika, as well as the volcanically dammed Lake Kivu; it curves on into Lake Rukwa and meets the southward extension of the line of the Eastern Rift to form the trough which contains Lake Nyasa. Lake Victoria is not a rift valley lake but lies in a downwarped shallow basin between the two arms of the rift system. The whole area is tectonically disturbed and both warping and faulting continued from the late Tertiary well into the Pleistocene.

The rainfall in East Africa is very variable but through most of the region is between 20 and 40 inches a year, rising to over 60 inches in the mountains and over Lake Victoria. In eastern and northern Kenya, however, it drops well below 20 inches and this region is semidesert comparable with the southern margin of the Sahara or the drier parts of the South African Karoo. Elsewhere the vegetation is typically wooded steppe with thorn bush, or savanna with deciduous trees (Fig. 3). The highlands are covered by montane grassland or moist woodland and in the mountains this grades upwards through forest to alpine vegetation.

In most parts of Africa the Pleistocene deposits are usually thin and disconnected, but in East Africa there are some thick fluviatile or lacustrine sediments developed in depressions associated with the Rift Valley troughs and downwarps. The most important sites lie in five widely separated areas: (1) in the vicinity of Lake Albert on the Uganda-Congo border; (2) around Homa mountain in the Kavirondo Gulf area of Lake Victoria; (3) on the Omo river in southern Abyssinia at the north end of Lake Rudolf; (4) in the Eastern Rift around lakes Nakuru, Naivasha, and Magadi; and (5) in the Olduvai Gorge area of the Serengeti plains north of Lake Eyasi in northern Tanganyika.

The Lake Albert depression contains a series of sediments which have been explored for oil from time to time since 1923. A lower series of unfossiliferous, arenaceous sediments of possibly Pliocene age is overlain unconformably by the Kaiso Beds (Wayland 1926, Harris *et al.* 1956) near the top of which ferruginous bands occur carrying some vertebrate and invertebrate fossils; stone implements have not been found in them. A younger formation overlies these beds and is known as the epi-Kaiso or the Semliki Beds (de Heinzelin 1955) from which some stone implements and fossils have been recovered.

In the Kavirondo Gulf area there are three main groups of deposits (Kent 1942) comprising an older series, the Kanam Beds, an intermediate group known as the Rawi Beds, and the younger Kanjera Beds. The upper part of the Rawi sequence has been termed the Rawi Fish Beds. In addition to animal fossils, human remains were found at Kanam and at Kanjera in 1932 (Leakey 1935).

The northern end of Lake Rudolf, which is today a basin of internal drainage but was formerly linked to the Nile system, is fed by the Omo river which drains from the highlands of Abyssinia. In its southern portion this river is flanked by extensive sediments known as the Omo Beds. They have yielded abundant fossil mammals but no stone implements, the principal sites being in the vicinity of Bourillé (Arambourg 1935, 1947) and Shungura (Leakey 1943), both in Abyssinia itself, and also from Todenyang a few miles south of the Kenya frontier on the west side of the lake.

In the Eastern Rift between Lake Baringo and Lake Magadi there are a number of fossil-bearing sites belonging broadly to two age groups. The older deposits were at one time regarded as belonging to a great "Lake Kamasia" (Leakey 1934) but it seems probable that this is fictitious. The most important sites are at Kariandusi in the Nakura-Naivasha basin, near Lake Elmenteita, where beautiful Acheulian hand axes occur in gravels and sands overlying diatomites, and at Olorgesailie, northeast of Lake Magadi where literally tens of thousands of hand axes cover the ground, having been eroded from a number of temporary land surfaces within a lacustrine sequence that includes diatomite layers; some fossil remains also occur. The younger group of sediments lies unconformably on the Kariandusi beds in the Nakuru-Naivasha basin and represents a complex sequence of deposits resulting from fluctuating water levels in the major dischargeless basin. They have been distinguished as representing three groups—Gamblian, Makalian, and Nakuran—each associated with humid conditions in the basin.

By far the most important deposits are those exposed in the Olduvai Gorge area of northern Tanganyika, where a thick sequence of stratified beds provides a clear and complete succession of stone age cultures and fossil material (Leakey 1951). In his original accounts Reck (1914, 1926) distinguished four conformable series which he termed, in order of age, Beds I, II, III, and IV, together with a younger, markedly unconformable deposit designated Bed V. To the south of the Gorge proper lies another sequence cut through by the Vogel River and comprising a thick series of yellow and grey tuffs, known as the Laetolil Beds, which have yielded fossil mammals and primitive pebble tools. The geology of this area has been described by Kent (1941) who regards the Laetolil beds as older, in the main, than the Olduvai beds, but Leakey (1951) in his extensive account of the Gorge appears to regard them as the probable equivalent of Bed I. Near the shores of Lake Eyasi is still another deposit, known as the Eyasi Beds, which furnished a human skull in 1935 and which is probably of an age between that of Beds IV and V at Olduvai. A brief description of the Olduvai beds has been given recently by Pickering (1960) who is engaged on a full scale survey the importance of which is very considerable in view of the recent discovery of Australopithecine and possible human fossils in Beds I and II by Dr. and Mrs. Leakey (1958a, 1959a, 1961).

In the case of most of the deposits in this area, attempts have been made to reconstruct the climatic setting, on foundations of varying degrees of security and with varying degrees of success. The stratigraphic-climatic scheme was initiated by Wayland in Uganda and pursued chiefly by Leakey in Kenya and Tanganyika.[1] Four East African "Pluvials" are recognized on this scheme and are termed Kageran, Kamasian, Kanjeran, and Gamblian, followed by two postpluvial wet phases termed Makalian and Nakuran. The main features of this interpretation are summarized in Table 3, which is based on the data given by Leakey (1950, 1951) and Cole (1954). Arambourg (1951) tacitly accepted the climatic inferences but makes a slightly different grouping of the deposits in the lower part of the sequence, as shown in the last column of the table.

The validity of the climatic interpretation was disputed by Solomon (1939) and, although some of his arguments cannot be upheld, his criticisms were disregarded by other workers rather than disproved by them. The present writer

1. For summaries of the numerous papers see Cole 1954. The most important original digests are Wayland 1934, 1952; Leakey 1931, 1950, 1951; and Nilsson 1940.

TABLE 3

The Stratigraphic Correlation of Pleistocene Events in East Africa According to Leakey's Climatic Scheme

AGE	CLIMATIC STAGE	TANGANYIKA	UGANDA	EASTERN RIFT AND KAVIRONDO GULF	STONE CULTURE			CORRELATION Arambourg 1951
Epi-Pleisto-cene Recent	NAKURAN wet phase dry MAKALIAN 2 wet phase 1			L. Nakuru +145 feet L. Nakuru +335 feet L. Nakuru +375 feet	NEOLITHIC AND MESOLITHIC			
UPPER	dry	Olduvai V		reddened surfaces and aeolian sands	Magosian			
UPPER	GAMBLIAN pluvial 3 2 1	cutting of mature valley at Olduvai Eyasi beds	(epi-Kaiso and Semliki beds) 30 ft terrace of Kagera river	L. Nakuru +375 ft Gamble's Cave, +510 ft Enderit Drift, +720 ft Malewa Gorge, etc.	Upper Kenya Capsian C B A Lower Kenya Capsian	Final Stillbay Stillbay Proto-Stillbay Developed Levallois	Upper Sangoan Middle Sangoan Lower Sangoan	Olduvai V, Gamble's Cave, etc.
UPPER	3rd interpluvial	earth movements and volcanic activity; unconformities; drying of lakes			Fauresmith	Early Levallois	Lower Sangoan	
MIDDLE	KANJERAN pluvial	Olduvai IV	100 ft terrace of Kagera river	Kariandusi diatomites Olorgesailie beds Kanjera Beds	Early Fauresmith 6 5 4 3 2 1 Acheulian stages	Pseudo-Stillbay early flake culture (Kenya Hopefountain)	Early Sangoan Proto-Sangoan	Olduvai IV
MIDDLE	2nd interpluvial	Olduvai III (red)		Upper Rawi (fish) Beds	Early Acheulian I			Olduvai III
MIDDLE	KAMASIAN pluvial	Olduvai II Olduvai I and Laetolil		Lower Rawi Beds	5 4 3 2 1 Chellean stages of the Chelles-Acheul culture Oldowan			Olduvai I & II
LOWER	1st interpluvial	–	Kaiso bone bed	Upper Omo Beds	Oldowan			
LOWER	KAGERAN pluvial	–	270 ft terrace of Kagera river	Main Omo Beds Kanam	Early Oldowan Advanced Kafuan Early Kafuan			Laetolil Omo Kanam Kaiso

(Cooke 1958) reviewed the situation at some length and concluded that the only well documented evidence for climatic changes in East Africa is that for the Gamblian Pluvial, the two later wet phases (Makalian and Nakuran) and the intervening dry period, as shown in the Nakuru-Naivasha basin. The inferences relating to deposits of pre-Gamblian date are unsatisfactory, though there is evidence favoring a general climate somewhat wetter than that of the present day in the same areas. It does not necessarily follow that the climatic sequence already deduced is false, but that there is good reason to question the method of its derivation and the soundness of the correlation procedure. Flint (1959a, 1959b) has reached similar conclusions as a result of independent studies and states:

> I conclude that climate as a primary basis of any stratigraphic scheme is undesirable. This conclusion . . . does not constitute an attack on the reality of climatic change in East Africa, for such change seems truly to have occurred. It seeks only to re-establish climate in the secondary position in which it properly belongs in any stratigraphic scheme. . . . The three entities discussed—rock units, faunal zones, and cultural materials—constitute, as a group, a reasonable basis from which can emerge broad correlation that will inspire confidence, and it is suggested that this basis be adopted for East Africa. . . . A locality admirably adapted as a starting point is the Olduvai Gorge, where distinctive rock units, fossils, and cultures exist in magnificent exposures.

Pickering (1960) has also voiced doubts concerning the climatic interpretation at the Olduvai Gorge and suggested that correlations may best be achieved by working on orthodox geological lines. Bishop (1960), too, has shown that much of the archaeological and climatic interpretation in Uganda is questionable and in need of revision. The Uganda sequence is further discussed by Bishop and Posnansky (1950), who conclude that although climatic changes have undoubtedly taken place it is seldom possible to evaluate the climates in Uganda and that the environmental changes are largely controlled by tectonic movements and volcanic activity. At the moment, therefore, the well-publicized and widely known climatic-stratigraphic interpretation of the East African deposits must be regarded as suspect but it may still be some years before a new and satisfactory scheme is proposed on the basis of the fresh field studies which are at present under way in the territories concerned. The phenomenal discoveries by the Leakeys in the Olduvai Gorge during the past three years will add enormously to the faunal picture, but the new data are not yet available for interpretation.

Hopwood, in Leakey's book *Olduvai Gorge* (1951), discussed and listed the Olduvai fauna. He deplored the tendency to use the terms Kageran fauna and Kamasian fauna and proposed to recognise three faunal stages in East Africa as the Kaiso fauna, the Olduvai fauna, and the Gamble's Cave or Gamblian fauna. Leakey, in the next chapter, dissents from this view and proposes to recognize four faunal stages, corresponding to his Kageran, Kamasian, Kanjeran, and Gamblian pluvial episodes. MacInnes, in an appendix to Leakey's chapter, attempted two mathematical analyses of the faunas of Olduvai and other sites, accepting Leakey's correlations. In his first analysis he found the following results for the 125 species examined:

Confined to Stage I (Kaiso, Omo, Kanam)	22.4%
Common to Stages 1 and 2	4.8%
Confined to Stage II (Laetolil, Olduvai I & II, Lower Rawi)	19.2%
Common to Stages 2 and 3	24.8%
Confined to Stage III (Olduvai IV, Kanjera, Upper Rawi, and Olorgesailie)	16.8%
Common to Stages I–III	12%

His second analysis concerns the percentage of extinct forms in each stage per se, with the following results:

	Stage I	*Stage II*	*Stage III*
Making first appearance	100% *	72.3%	31.3%
Becoming extinct	57	40	41
Persisting to next stage	43	60	59

* MacInnes leaves this blank for some unaccountable reason.

As a large number of the 125 forms considered by MacInnes are recorded from only one locality, such occurrences are not of much use in comparing different deposits. In order to minimize this difficulty, the writer (Cooke 1958) compared only the extinct forms known to occur in more than one deposit, this category embracing little more than one quarter of the species then recorded. Analysis of these forms led to the conclusion that the faunal trends of the stages proposed by Leakey have a considerable reality but that the boundaries are by no means clear cut or easy to define. The correlation proposed by the writer is given in diagrammatic form in Figure 4 and should be compared with the more rigid divisions set out in Table 3. The Kaiso Beds are regarded as being older than the Omo beds and the latter are considered to have a closer affinity with the Laetolil and Bed I deposits than has been suggested hitherto. Leakey (1959a, 1961) now reports an unconformity at the top of Bed I and states that the fossil material discovered recently has affinities with the Omo fauna. Details are still not available and are awaited with great interest.

Little new published information has appeared since the lists given by Hopwood, Leakey, and MacInnes in 1951. Dietrich (1950) has made additions to the bovid fauna and Leakey (1958b) has given a full account of the East African Suidae. Leakey and Whitworth (1958) have reviewed the species of *Simopithecus* and MacInnes (1953) has described a new lagomorph. The writer has examined a great deal of the original material from the East African sites and has tried to prepare a reasonably critical list, which is presented in Table 4. The names given in the body of the table are almost exclusively forms occurring in more than one deposit, numbering 71 in all, of which 40 are extinct; 59 additional forms from single records are added at the end of the table.

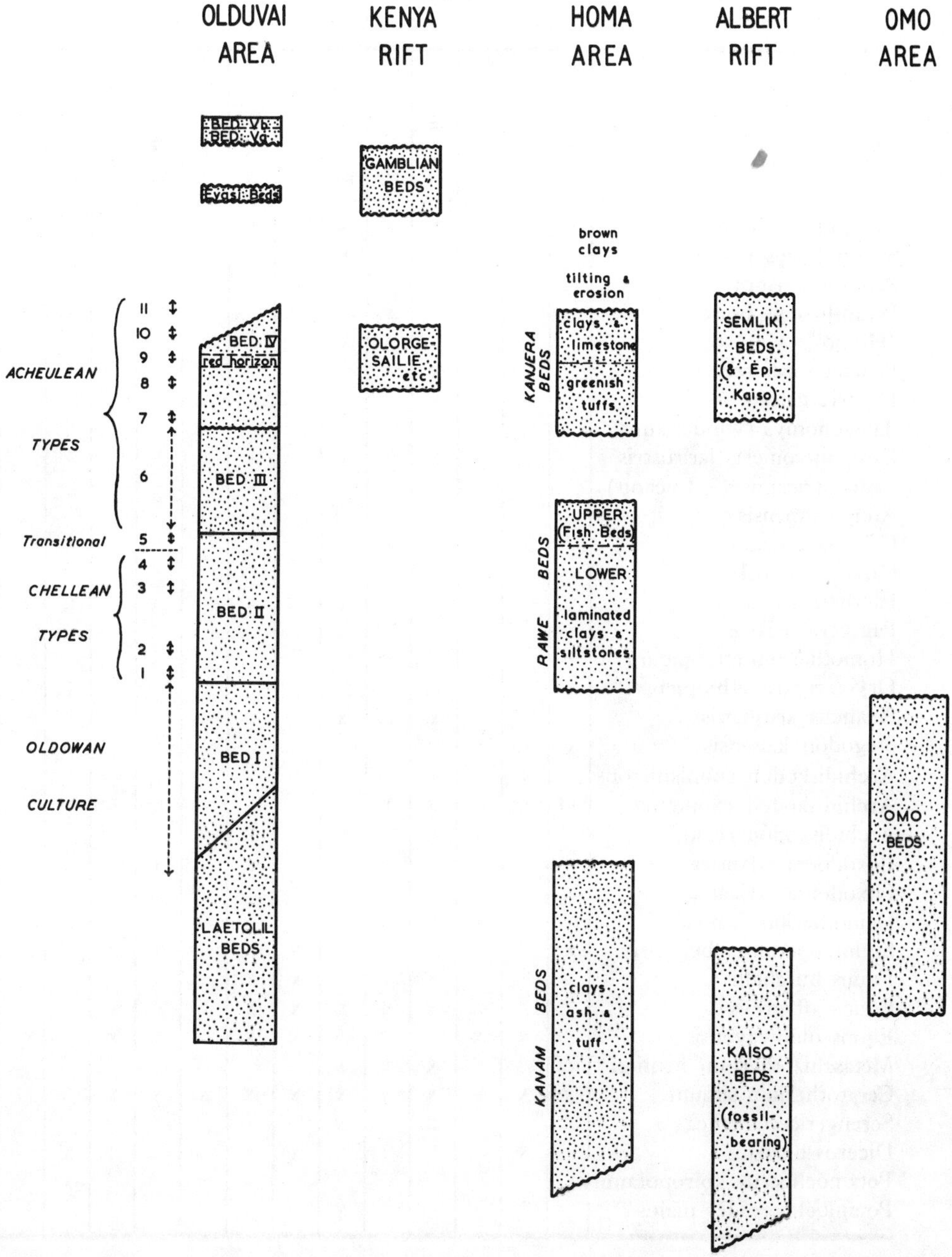

FIGURE 4

Possible correlation of the major fossil-bearing and implement-bearing deposits, based on interpretation of the mammalian fauna with additional data from the stone age cultures; stippled areas contain datable material. The vertical scale is based on stratal thickness rather than time.

TABLE 4

Distribution of Fossil Mammals in East African Pleistocene Deposits

	Kaiso	*Kanam*	*Omo*	*Laetolil*	*Bed I*	*Bed II*	*Bed III*	*Bed IV*	*Rawi*	*Kanjera*	*Olorgesailie*	*Eyasi*	*"Gamblian"*
Simopithecus oswaldi			cf	x	x	x	x	x		x			
Simopithecus jonathani						x		?					
Papio neumanni													x
Paranthropus boisei					x								
"Homo"		x				x						x	x
Pedetes surdaster				x								x	
Hystrix galeata				x									x
Thryonomys swinderianus												x	x
Canis mesomelas latirostris				x		x		x					
Canis africanus (? Lycaon)				x		x		x					
Aonyx capensis						cf							x
Crocuta crocuta		x		x	x							x	
Caracal caracal				cf								x	
Panthera leo						x		x				x	x
Panthera pardus				?	?							x	
Homotherium ethiopicum	?		x										
Orycteropus aethiopicus				cf						cf		cf	x
Anancus kenyensis		x		x	cf	x							
Stegodon kaisensis	x	x											
Archidiskodon subplanifrons		x		x									
Archidiskodon exoptatus	cf	x	cf	x	x								
Archidiskodon recki			x	x	x	x	x	x		x	x		
Loxodonta atlantica										x			
Loxodonta africana													x
Deinotherium bozasi		x	x	x	x	x							
Stylohipparion albertensis	x	x	x	x	x	x	x	x		x	x		
Equus burchelli				x		x	x	x				x	x
Equus aff grevyi			?		x	x	x	x			x		
Equus oldowayensis		x	x		x	x	x	x		x	x		
Metaschizotherium hennigi	cf			x	x	x							
Ceratotherium simum	x	x	x	x	x	x	x	x	x	x	x		
Serengeticeros efficax				x		x							
Diceros bicornis		x			x	x	x	x				x	
Potamochoerus koiropotamus												x	x
Potamochoeroides majus		?			x	x		x					

TABLE 4 (continued)

	Kaiso	*Kanam*	*Omo*	*Laetolil*	*Bed I*	*Bed II*	*Bed III*	*Bed IV*	*Rawi*	*Kanjera*	*Olorgesailie*	*Eyasi*	*"Gamblian"*
Mesochoerus limnetes	x		x	x	?								
Mesochoerus olduvaiensis					x	x	x			x			
Metridiochoerus andrewsi	?		x						x	x			
Notochoerus euilus	x		x	x	x								
Notochoerus (?) hopwoodi						x	x						
"Tapinochoerus" meadowsi			?	?	x	x	x	x			x		
Orthostonyx brachyops						x							
Stylochoerus nicoli					?	x	x	x		x			
Phacochoerus altidens					x	x	x	x					
Hippopotamus gorgops					?	x	x	x	x	x	x		
Hippopotamus imaguncula	x	x							x				
Hippopotamus protamphibius			x										
Hippopotamus amphibius	x		x	x						x		x	x
Okapia stillei				x	?								
Giraffa camelopardalis		x	x	x		x		x	x	x	x	x	
Giraffa gracilis			x			x							
Libytherium olduvaiensis		x	x	x	x	x	x	x	x		x		
Strepsiceros strepsiceros					?	x		x				x	
Strepsiceros imberbis			x		?	x		x					
Tragelaphus buxtoni				x		x							
Taurotragus oryx				x	x	x		x				x	
Syncerus caffer			cf	x								x	
Homoioceras nilssoni								x				x	x
Bularchus arok						x	x	x		x			
Kobus ellipsiprymnus			?	?								x	
Adenota kob					x	x	x					x	
Redunca redunca				x								x	
Hippotragus equinus				x			x	x		x		x	
Hippotragus niro						x		x					
Oryx beisa					x							x	
Damaliscus angusticornis				x		x		x					
Damaliscus teste					x	x		x					
Alcelaphus kattwinkeli						x	?	x		x			
Pultiphagonides africanus					x	x							
Beatragus hunteri (?)					x	x		x					

TABLE 4 (continued)

	Kaiso	*Kanam*	*Omo*	*Laetolil*	*Bed I*	*Bed II*	*Bed III*	*Bed IV*	*Rawi*	*Kanjera*	*Olorgesailie*	*Eyasi*	*"Gamblian"*
Gorgon taurinus semiticus					x	x	x	x					
Aepyceros melampus			x	x								x	
Phenacotragus recki								x		x			
Gazella gazella precursor					x	x		x					
Gazella granti					x	x		x				x	
Pelorovis oldowayensis						x		x					

The following single records are not included in the table:

KANAM: Nyanzachoerus kanamensis

OMO: Tragelaphus nakuae, Menelikia lyrocera, Taurotragus derbianus procanna, Syncerus aff brachyceros, Kobus sigmoidalis, Redunca ancystrocera, Oryx gazella, Gazella praethomsoni

LAETOLIL: Cercocebus ado, Serengetilagus praecapensis, Xerus janenschi, Tatera nigricauda, Tachyoryctes splendens, Saccostomys sp, Dendromus sp, Heterocephalus quenstedti, Herpestes palaeogracilis, Herpestes palaeoserengetensis, Hyaena hyaena, Simatherium kohllarseni, Aeotragus garussi, Praedamalis deturi, Gorgon gadjingeri, Praemadoqua avifluminalis, Gazella hennigi, Gazella janenschi, Gazella kohllarseni

BED I: Acinonyx jubatus, Parmularius altidens

BED II: Metridiochoerus compactus, Hippotragus niger

BED IV: Otocyon recki, Hyaena aff brunnea, Felis libyca cafra, "Tapinochoerus" minutus, Tragelaphus scriptus, Tragelaphus spekei stromeri, Philantomba monticola, Hippotragus leucophaeus (?), Nesotragus moschatus, Thaleroceros radiciformis

KANJERA: Lepus veter, Hylochoerus antiquus

EYASI: Herpestes sanguineus, Gorgon taurinus major, Gazella thomsoni

"GAMBLIAN": Pan naivashae, Canis adustus, Lycaon picta, Lutra maculicollis, Atilax paludinosus, Hylochoerus meinertzhageni, Phacochoerus africanus, Syncerus aberrans

The amended list does not alter the correlation suggested in Figure 4. The faunal evidence favors temporal overlap between the deposits and suggests progressive changes in the faunal assemblages without catastrophic interruptions. Some of the faunas are still too scantily known to be placed with assurance and it is clear that much additional taxonomic work is needed even on the richer faunas before good "zone fossils" can be recognized and employed satisfactorily. The Proboscidea and the Suidae offer particular prospects in this regard.

It might be hoped that the faunal material would throw a good deal of light upon the environment and climate of the deposits themselves but the indications are disappointingly vague. Although some of the living species recorded do not now occur at the fossil sites, nearly all of them exist in the vicinity and are not sensitive indicators of environmental change. Extinct species must be compared with their living counterparts and it is not possible to estimate their former range of adaptability. In broad terms all the faunas are typical of the African steppe and savanna assemblages.

The few Kaiso species, of which all except the white rhinoceros and the hippopotamus are extinct, do not suggest an environment different from that of Lake Albert at the present day. The same is true for the Kanam deposit in relation to the Kavirondo Gulf area. At Omo, however, there are some indications of moister conditions as there occur several water-loving forms, such as *Kobus*, which no longer live in the vicinity; on the other hand the presence of *Oryx* suggests that a dry environment was not far away. This is consistent with a more extensive lake swamp but with dry steppe in the surrounding area. It may be remarked that the occurrence of *Aepyceros* is a long way north of the present limits of this species in the East African region.

The Laetolil fauna, which contains the best studied assemblage of bovid species, includes fossil remains of an okapi which might be inferred as indicating the existence of forest areas in the neighborhood, probably on the slopes of nearby volcanoes. The varied gazelles and other forms otherwise suggest that there was no serious departure from the general savanna environment. The Bed I fauna is essentially similar and that of Bed II likewise; an unexpected element is *Beatragus hunteri*, which is now restricted to the Jubaland area of southern Somalia under semiarid conditions. This species is also reported from Bed IV, which contains *Nesotragus moschatus* whose present range is restricted to moist woodland in the coastal belt or in mountain foothills on the plateau. The occurrence of *Aonyx* in Bed II shows the existence of swamps in the vicinity and it may be suspected that the identification of Hunter's Hartebeest in the fauna is incorrect. The Sable Antelope of Bed II is a long way from its nearest present occurrence to the north in the valley of the White Nile or to the south in the Zambezi region. The Bed IV assemblage is very like that of Bed II, though it is richer in variety and includes the earliest occurrence of the giant buffalo *Homoioceras* so far noted in this region; the inferred environment is like that of Beds I and II. The fauna from Bed III is too scanty to indicate any specific setting but it contains no species not already known from Beds II or IV (usually from both) and there is no faunal evidence for a different environment. The assemblages from Rawi, Kanjera, Olorgesailie, and Eyasi are also typically of open or moist savanna aspect but the "Gamblian" deposits embrace a mixture of savanna forms and forest or swamp species. The assemblages as a whole are very consistent and do not present any evidence for catastrophic events or startling ecological changes. The fossil faunas favor climatic conditions very similar to those of the present day in the same areas with a marked bias toward somewhat moister conditions in most cases. There is no faunal evidence for drier conditions.

SOUTHERN AFRICA

Present Environment

As a geographical entity, Southern Africa is defined by Wellington (1955) as the region lying south of the South Equatorial Divide, which forms the watershed between the drainage of the Congo and Zambezi river systems and continues eastward around the northern end of Lake Nyasa to meet the coast in southern Tanganyika near the Mozambique border. Very approximately this is about 11°S. latitude and includes the southern half of Angola, nearly the whole of the Federation of Rhodesia and Nyasaland, Mozambique, South Africa, South-West Africa and the British protectorates of Bechuanaland, Swaziland, and Basutoland. The structure is that of an elevated plateau, depressed in the central Kalahari basin and raised at the margins, which are cut off from the coastal belt by an almost continuous escarpment. The peripheral highlands commonly reach about 8,000 feet above sea level but in the Basuto highlands a considerable area lies above 8,000 feet and the greatest height is a little over 11,000 feet.

The main westward drainage is that of the Vaal-Orange system, rising in the Basuto highlands and the high border of the escarpment in the eastern Transvaal. To the north the Limpopo and Zambezi flow eatward into the Indian Ocean.

A high proportion of the plateau lies above 4000 feet but the Kalahari basin is well below this level and contains within it two important basins of internal drainage—the Etosha pan in the west at 3550 feet and the Okovango swamp and Makarikari depression at 3000 feet in the east. The latter was probably linked in former times to the Zambezi system. The Kalahari region is largely featureless and sand-covered, the sand being primarily the upper member of the Kalahari System, of Tertiary age, which extends from the Orange river at 20°S. to 1°N. latitude (Fig. 5). Only in the southwestern corner of Bechuanaland is the Kalahari a true sand desert. In most areas it is covered by a mantle of vegetation that ranges from scanty grass in the southwest, through grassland and thornveld in most of Bechuanaland, to a fairly dense woodland in the north. Because of loss through seepage into the unconsolidated sand there is little surface water in the whole of this region and the vegetation relies directly on the rainfall, which ranges from below ten inches in the southwest to 50 or 60 inches on the South Equatorial Divide.

In South Africa the precipitation is highest in the summer rainfall region of the eastern highlands and on the east coast but is also high in the narrow southern coast with its winter rainfall; by way of contrast, the west coast to the north of the Orange River is pure sand desert. The vegetation covers a wide range of types from desert through semiarid Karoo shrub and steppe to grassland, thorn

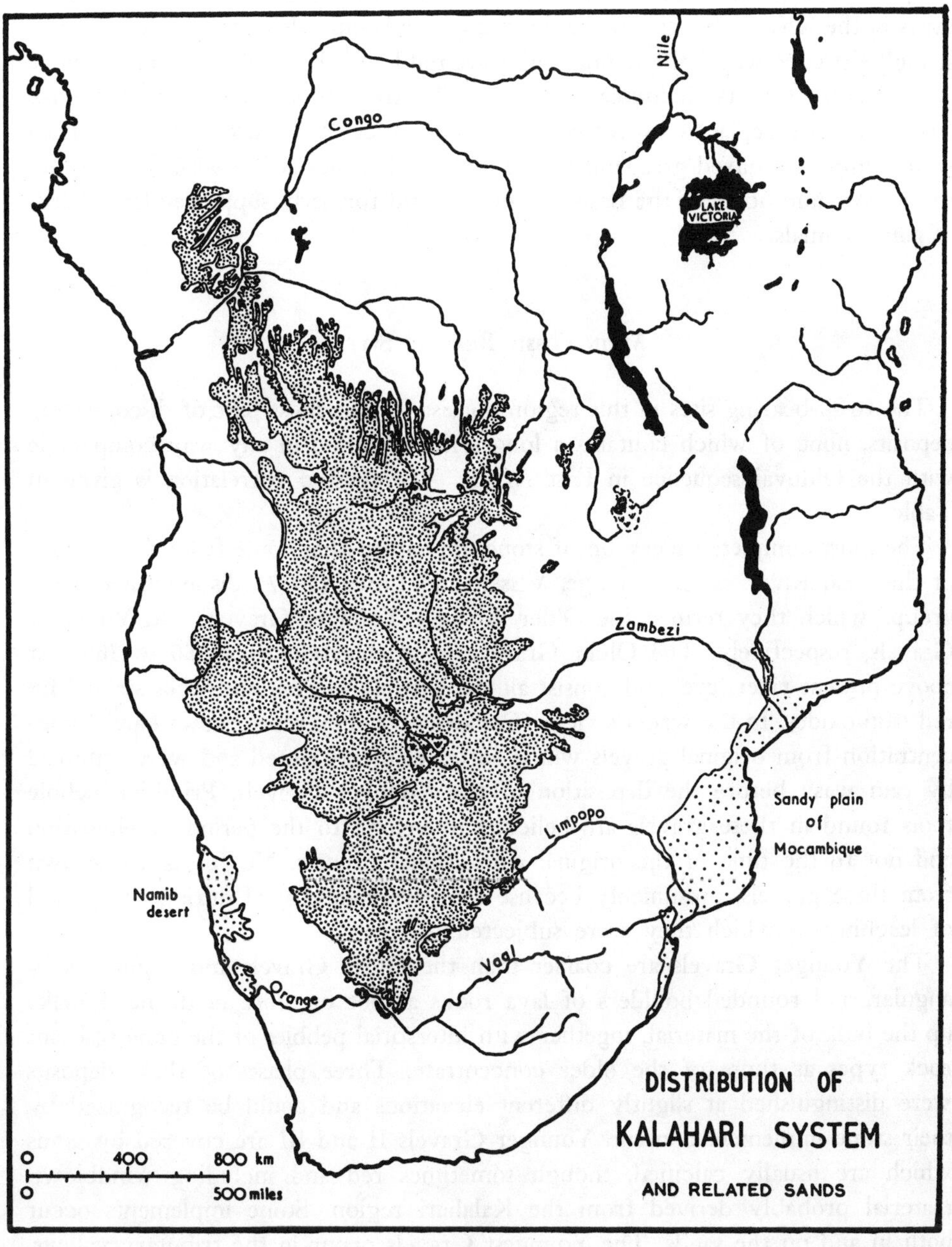

FIGURE 5
Distribution of the Kalahari System, more particularly its upper sand member of presumed Pliocene age.

bush savanna, and deciduous woodland (Fig. 3). Forest-savanna mosaic occurs along the coast and in warm high rainfall areas in the interior. Evergreen forest exists in the Knysna forests on the south coast and somewhat similar forest occurs at high altitudes where the temperatures are mild and the rainfall normally above 50 inches a year. With increased altitude the trees become smaller and above 10,000 feet are replaced by semialpine communities. A feature of South Africa is the temperate mixed grassland of the high veld; it lies above 4000 feet, mainly to the west and north of the Basuto highlands and formerly supported large herds of game animals.

Main Fossil-Bearing Sites

The fossil-bearing sites in this region consist for the most part of disconnected deposits, none of which contains a long series of strata in any way comparable with the Olduvai sequence in East Africa. A suggested correlation is given in Table 5.

The most complete succession of stone implements has come from the deposits of the Vaal River, where Söhnge, Visser, and Lowe (1937) distinguished three groups which they termed the Older Gravels, Younger Gravels, and Youngest Gravels, respectively. The Older Gravels lie at elevations from 80 to 300 feet above present river level and consist almost entirely of resistant rocks stained by red iron oxides. In the writer's view (Cooke 1947) they represent an eluvial concentration from original gravels whose softer elements rotted and were removed by rain wash before the deposition of the Younger Gravels. Primitive pebble tools found in these gravels are believed to belong to the period of eluviation and not to the time of the original depositional process. No fossils are known from these gravels, presumably because they were destroyed during the period of leaching to which they were subjected.

The Younger Gravels are coarser than the Older Gravels and angular, subangular, and rounded boulders of lava rocks a foot or more in diameter make up the bulk of the material, together with interstitial pebbles of the same resistant rock types as those of the older concentrate. Three phases of these deposits were distinguished at slightly different elevations and could be recognized by their stone implement content. Younger Gravels II and III are covered by sands which are usually calcified, though sometimes red and including windblown material probably derived from the Kalahari region. Stone implements occur both in and on the sands. The Youngest Gravels occur in the tributary valleys. The fossil mammals of the Vaal River deposits have been described by the writer (Cooke 1949) but the synonymy has been revised here (Table 7). The material is nearly all derived from Younger Gravels II and III, which are sometimes inseparable. Below Barkly West there are very thick gravels forming a "Middle Terrace" from which a number of rather primitive fossils have come which sug-

TABLE 5

Suggested Correlation of Main Pleistocene Deposits in Southern Africa

PROVISIONAL STAGE TERM	VAAL RIVER BASIN	CAVE DEPOSITS	OPEN SITES	RHODESIA	STONE CULTURE
"Recent"	minor erosion; recent soil formation	numerous caves	surface	Nachikufu, etc.	WILTON SMITHFIELD
FLORISBAD-VLAKKRAAL STAGE	gleying of subsoil and calcification or ferruginization alluvium in minor river valleys terrace gravels ("Youngest Gravels") erosion of minor river valleys	Wonderwerk, etc. ↑ Cave of Hearths ↓	Vlakkraal Florisbad	Mumbwa cave Broken Hill cave Twin Rivers breccia Chelmer	MAGOSIAN MIDDLE STONE AGE COMPLEX SANGOAN & FAURESMITH
VAAL-CORNELIA STAGE	wind action and minor calcification alluvium in valleys Low-level terrace gravels ("Younger Gravels") erosion in major river valleys		Hopefield Cornelia ? Zululand clays	younger terrace gravels of Zambezi river	CHELLES-ACHEUL: LATER
SWARTKRANS STAGE	local calcification; re-distribution of wind-blown sands re-working of high-level gravels ("Older Gravels")	Kromdraai Swartkrans Sterkfontein extension Bolt's farm		older terrace gravels of Zambezi river	CHELLES-ACHEUL: MIDDLE EARLY pebble tools
STERKFONTEIN STAGE	redistribution of sands of Kalahari type original high-level gravels ("Basal Older Gravels")	Makapansgat Sterkfontein Taung		oldest terrace gravels of Zambezi river	
?	erosion of major river valleys	formation of sub-surface caves	? Langebaan phosphate	erosion	
KALAHARI STAGE	formation of primary Kalahari sands, partial consolidation and reddening				

gest that this terrace may belong to phase I of the normal Younger Gravels sequence.

In the original account of the deposits and cultures (Söhnge, Visser, and Lowe 1937) a climatic interpretation was placed on the deposits and this was subsequently amended (Cooke 1947, Lowe 1952). Flint (1959a) has cast some doubts upon the validity of these interpretations and they require review; but the inferences drawn seem broadly correct and it is likely that at least two major wet periods are involved.

Almost certainly older than the Vaal River fossil assemblage are the cave breccias associated with the Australopithecinae. Of these there are two types which may be distinguished as the Taung travertine deposits and the Transvaal dolomite cave deposits. At Taung, in the Cape about 80 miles north of Kimberley, the first ape man from the region was recovered in 1924 and named *Australopithecus africanus* by Dart. The deposit was studied closely by Peabody (1954) who showed that the regional dolomitic cliffs had given rise to four separate phases of growth of great aprons of travertine, each separated from the next by a phase of erosion. Cavities in the aprons were filled by red sand (of Kalahari type) and *Australopithecus* occurred in the oldest of these fillings. Peabody inferred that the travertine deposition took place during the waning stages of a relatively humid period and drew attention to the fact that the climatic cycle demanded by his observations agreed remarkably with an interpretation of the Vaal River sequence. The fauna associated with *Australopithecus* is unfortunately very scanty but serves to date the deposit in relation to the other ape-man breccias.

The Transvaal dolomite cave deposits occur in pre-Cambrian dolomitic limestones in which numerous caves have been cut by meteoric waters. A few of these contain bone-bearing breccia deposits and those at Sterkfontein, Kromdraai, and Swartkrans near Krugersdorp, and at Makapansgat near Potgietersrus (some 200 miles away to the north) have furnished ape-man material. In the vicinity of Sterkfontein many animal fossils have come from the area of Bolt's farm but this has not yielded Australopithecine remains to date. Many of these well-cemented cave breccias are now exposed directly on the surface as a result of stripping away of the former cave roof by subsequent erosion. The breccia fillings have been systematically studied by Brain (1958) whose mineralogical analyses show that the cave deposit was derived essentially from the soil overlaying the regional dolomitic limestone. By evaluating the characteristics of similar modern soils under rainfall conditions ranging from under 2 inches to 60 inches per annum, he has been able to make deductions regarding climatic changes during the periods of infilling. The Sterkfontein deposit, which is probably the oldest, belongs to a period of rainfall below that of the present day, as does most of the deposit at Makapansgat. The Swartkrans deposit seems to correspond to a climate much like that of the present but the Kromdraai breccia is essentially the product of a much wetter environment. A composite curve, based on Brain's work but

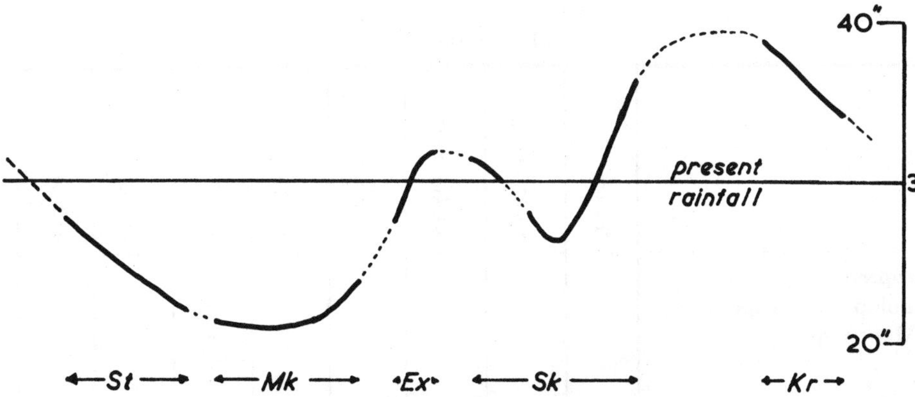

FIGURE 6

Climatic curve deduced for Australopithecine deposits (after Brain, modified by Robinson). St—Sterkfontein type site; Mk—Makapansgat; Ex—Sterkfontein extension; Sk—Swartkrans; Kr—Kromdraai.

incorporating later data from Robinson (1961), is given in Figure 6. This is the most objective and realistic climatic interpretation so far achieved in this region. Primitive stone implements occur at the Sterkfontein extension site (Mason 1961).

A complete faunal list for the cave breccias is given in Table 6.

TABLE 6

Distribution of Fossil Mammals in Cave Breccias in South Africa

	Taung	*Sterkfontein*	*Makapansgat*	*Swartkrans*	*Kromdraai*	*Bolt's Farm*	*Gladysvale*	*Other*
Insectivora								
Proamblysomus antiquus					x	x		
Chrysotricha hamiltoni			x					
Chlorotalpa spelea		x		x				
Atelerix major						x		
Elephantulus langi *		x	x	x	x	x		x
Elephantulus antiquus						x		
Elephantulus cf brachyrhynchus †	x			x		x		
Mylomygale spiersi	x	cf						
Crocidura taungsensis	x							
Crocidura cf bicolor * †		x			x		x	
Suncus cf etruscus †		x	x	x	x	x	x	
Myosorex robinsoni		x	x	x	?	x	x	

Note: * indicates a form also listed in Table VII
† after a specific name denotes a living species

TABLE 6 (continued)

	Taung	*Sterkfontein*	*Makapansgat*	*Swartkrans*	*Kromdraai*	*Bolt's Farm*	*Gladysvale*	*Other*
Chiroptera								
Rhinolophus cf capensis †	x		x			x		
cf Myotis sp						x		
Primates								
Gorgopithecus major					x			
Simopithecus darti			x					
Simopithecus danieli				x				
Parapapio antiquus	x							
Parapapio jonesi	x	x	x	x	x			
Parapapio broomi		x	x			?		
Parapapio whitei	x	x	x			?		
Papio izodi	x							
Papio wellsi	x							
Papio angusticeps					x			x
Papio robinsoni				x	x	x	x	x
Dinopithecus ingens				x				x
Cercopithecoides williamsi	x	x	x	x		x		
Australopithecus africanus	x	x	x					
Paranthropus robustus				x	x			
Telanthropus capensis				x				
Lagomorpha								
Pronolagus randensis * †			cf					
Lepus capensis †	cf			cf		cf		
Rodentia								
Pedetes gracile	x							
Pedetes cf caffer * †						x		
Mystromys antiquus	x							
Mystromys hausleitneri		x	x		x	x	x	
Mystromys darti			x					
Tatera cf brantsi †	x	x	?		x	x		
Desmodillus auricularis †	x				?			
Grammomys cf dolichurus †			x		x			
Dasymys bolti				x		x		
Pelomys cf fallax †		?	x					

Note: * indicates a form also listed in Table VII
† after a specific name denotes a living species

TABLE 6 (continued)

	Taung	*Sterkfontein*	*Makapansgat*	*Swartkrans*	*Kromdraai*	*Bolt's Farm*	*Gladysvale*	*Other*
Rhabdomys cf pumilio	x	x	x		x	x		
Aethomys cf namaquensis †		x	x					
Thallomys debruyni	x					cf	x	
Mastomys cf natalensis †	x	x	x		x			
Leggada cf minutoides †		x	x			x		
Leggada cf major †		x				x		
Dendromus antiquus	x							
Dendromus cf mesomelas * †		x	x					
Malacothrix cf typica †	x				x	x		
? Malacothrix makapani			x					
Steatomys cf pratensis * †			x		x			
Palaeotomys gracilis *	x	x	x	x	x	x	x	?
Prototomys campbelli	x							
Xenohystrix crassidens			x					
Hystrix major			x					
Hystrix cf cristata †		x						
Hystrix africaeaustralis * †			cf			cf	cf	cf
Petromus minor	x							
Gypsorhychus darti	x							
Gypsorhychus minor	x							
Gypsorhychus makapani			x					
Cryptomys robertsi	x	x	x	x	x	x		x
Carnivora								
Canis mesomelas * †	cf					cf		
Canis mesomelas pappos		x	?	x	x	x		
Canis atrox					x			
Canis brevirostris		x						
Canis terblanchei					x			
Vulpes pulcher				?	x			
Aonyx cf capensis †						x		
Suricata suricatta * †						cf		
Herpestes mesotes					x			
Crossarchus transvaalensis					x	x		
Cynictis penicillata †				x				
Cynictis penicillata brachyodon			x					
Lycyaena silberbergi		x		x				
Lycyaena nitidula		?		x				
Leecyaena forfex				x				

Note: * indicates a form also listed in Table VII
† after a specific name denotes a living species

TABLE 6 (continued)

	Taung	*Sterkfontein*	*Makapansgat*	*Swartkrans*	*Kromdraai*	*Bolt's Farm*	*Gladysvale*	*Other*
Crocuta crocuta angella				x				
Crocuta spelea *					x			
Crocuta venustula				x				
Crocuta ultra					x			
Crocuta cf brevirostris			x					
Hyaena brunnea * †				x				
Hyaena brunnea dispar				x				
Hyaena makapani			x					
Hyaena bellax					x	x	x	
Leptailurus spelaeus						x		
Felis crassidens					x			
Therailurus barlowi		x	x	?		x	?	
Therailurus piveteaui					x			
Panthera whitei				?	?			x
Panthera shawi				?	x			
Panthera aff leo				x	x	x	?	
Panthera pardus incurva				x				
Megantereon gracile		x						
Megantereon eurynodon				x	x			
Megantereon *sp nov.*			x					
"Machairodus" transvaalensis						x		
Proboscidea								
cf Loxodonta atlantica						x		
Hyracoidea								
Procavia capensis * †						cf		
Procavia antiqua	x	x	x	x	x			
Procavia transvaalensis	x	x	x	x	x	x	x	
Procavia *sp. nov.*			x					
Perissodactyla								
Stylohipparion steytleri *			?		x	x		
Equus burchelli * †					?	cf	cf	
Equus plicatus *				?	x	x		
Equus helmei *			x		x			x
Metaschizotherium (?) transvaalensis			x					
Ceratotherium simum * †			cf					
Diceros bicornis * †			cf				?	

Note: * indicates a form also listed in Table VII
† after a specific name denotes a living species

TABLE 6 (continued)

	Taung	*Sterkfontein*	*Makapansgat*	*Swartkrans*	*Kromdraai*	*Bolt's Farm*	*Gladysvale*	*Other*
Artiodactyla								
Potamochoeroides hypsodon			x					
Potamochoeroides shawi	cf		x			x	x	
"Tapinochoerus" meadowsi *		x		x		x		x
Notochoerus euilus			x					
Mesochoerus paiceae *							x	
Potamochoerops antiquus		?		x	x	x		
Hippopotamus amphibius * †			cf					
Giraffa camelopardalis *			cf					
Libytherium cf olduvaiensis *			x					
Strepsiceros strepsiceros * †			cf					
Tragelaphus scriptus †						cf		
Tragelaphus angasi * †	cf		cf				cf	
Taurotragus oryx * †			cf			cf		
Syncerus caffer * †			cf			cf		
Syncerus (?) makapaani								x
Cephalophus parvus	x							
Cephalophus pricei			x					
Cephalophus caerulus * †			cf					
Onotragus leche * †							cf	
Redunca darti			x				cf	
Redunca arundinum * †			cf					
Redunca fulvorufula * †			cf					
Hippotragus broomi		x					x	
Oryx gazella * †			cf					
Damaliscus cf lunatus * †						?	?	
Damaliscus cf pygargus * †						x	x	
Alcelaphus caama * †							cf	
Alcelaphus robustus *			cf			cf	cf	
Alcelaphus helmei *			cf			cf	cf	
Connochaetes taurinus * †			cf			cf	cf	cf
Makapania broomi		x	x			x	x	
Oreotragus major			x					x
Oreotragus longiceps	x							
Raphicerus campestris *						?		
Aepyceros melampus * †			cf				cf	
Antidorcas marsupialis * †		?				?	cf	
Phenacotragus vanhoepeni			x			?		
Gazella gracilior			x					
Gazella wellsi *		x				x	x	

Note: * indicates a form also listed in Table VII
† after a specific name denotes a living species

Apart from the cave breccias and the Vaal River gravels there are several other fossil-bearing deposits of note in open sites. The most important of these is at Cornelia in the northeastern part of the Orange Free State, where a thin deposit of marly clays and sands represents an old "pan" filling. The site was described by Van Hoepen (1930) who considered the deposit as largely Pliocene, but the fauna has been reinvestigated by Wells and the writer [2] who regard it as comparable with the Vaal river gravels fauna.

An important site occurs at Hopefield, some 60 miles north of Cape Town, where fortuitous circumstances have led to local stripping of the vegetation and the removal of a thick sand cover to expose a surface littered with bones and late hand-axe tools. Associated with this material, and shown by fluorine measurements to have the same age, are parts of a human cranium and lower jaw, known as "Saldanha Man," which has affinities with Broken Hill man from Northern Rhodesia. The deposit seems to represent an ancient coastal swamp area cut off from the sea by dune ridges. Of the extensive fauna only the human, some carnivore, giraffid, and pig remains have been described (see Singer and Boné 1960 for references) and the Bovidae are being discussed by Wells.[2]

Another interesting human skull has been recovered at Florisbad in the Orange Free State some 50 miles from Bloemfontein. The source of the material is a hot spring whose "eye" occasionally erupted bones and teeth of animals, but the skull was found during the course of systematic excavations by Dreyer in 1932. The human skull and the bulk of the fossil material apparently comes from the lowest of four peaty layers from which fossil pollens have also been recovered and studied (Bakker 1957). The stone implements (Meiring 1956) seem to belong to an early phase of the so-called Middle Stone Age Complex (MSA) of this region, which succeeds the typical hand-axe cultures. The fauna has been restudied by Wells and the writer.[2] Very similar in character is the site of Vlakkraal, five miles away, where several "eyes" of hot springs have furnished abundant material. The stone implements constitute a very homogeneous assemblage belonging to the later part of the MSA Complex and the fauna is clearly related to that of Florisbad (Wells, Cooke, and Malan 1942).

There are a number of significant cave deposits in the area, the oldest of which is that known as the Cave of Hearths in the Makapansgat valley. It has yielded a fine assemblage of late hand axe tools and also a sequence of younger industries embracing the MSA and Later Stone Age, culminating in iron-age deposits. The faunal material is very fragmentary and specific diagnosis is difficult so that conservative nomenclature fails to disclose the possible existence of many extinct forms such as occur at Hopefield, which must be roughly contemporary. Of the many caves which have furnished Middle Stone Age fauna, that at Wonderwerk in the Kuruman district, west of Taung, may be regarded as typical. It

2. As yet unpublished. The writer is indebted to Professor Wells for allowing the information to be used here.

includes only a very few extinct species and these are characteristic of all sites, both cave and open, which are associated with a well-developed MSA industry,

In the coastal area there are two sites which may be mentioned, though they have provided only a very few specimens. The one is on the coast of Zululand in northern Natal, near a point optimistically marked on some maps as Port Durnford. Here in 1907 Anderson discovered bone-bearing clays at mean tide level in a series of marine sediments, and Scott (1947) named four supposedly extinct forms *Bubalus andersoni*, *Elephas* (*Loxodon*) *zulu*, *Hippopotamus ponderosus*, and *Opsiceros simplicidens;* the last two are certainly the living hippopotamus and black rhinoceros but the other two are indeed extinct forms, the buffalo being at present unique and the elephant regarded by the writer (Cooke 1961) as close to or identical with *Loxodonta atlantica.* The other site is at Langebaan, some 20 miles northwest of Hopefield in the southwestern Cape, from which a *Stegolophodon* is reported by Singer and Hooijer (1958) and from which more material is at present being recovered.

In the Rhodesias there occur a number of cave sites associated with Middle Stone Age or Later Stone Age cultures, of which the most important are respectively Mumbwa and Nachikufu (Clark 1950). The faunas are little different from those of corresponding caves in South Africa. Of somewhat uncertain age is the famous Broken Hill cave which yielded the remains of *Homo rhodesiensis.* The skull supposedly belongs to an early stage of the Middle Stone Age Complex but it may be older. The fauna, originally described by Hopwood (1928), has been partially revised by Leakey (1959b) [3] who records the Gerenuk for the first time in a fossil deposit. Not far from Broken Hill is a peculiar fissure-breccia deposit at Twin Rivers containing a similar industry but a very fragmentary fauna.

An important site in Southern Rhodesia is that of Chelmer, some 13 miles from Bulawayo (Bond and Summers 1951) where an alluvial deposit has furnished a good fauna (Cooke and Wells 1951). The fossils come from a layer of alluvium which has not yet yielded any stone implements but rolled hand axes and Middle Stone Age tools occur in a band which rests upon the eroded surface of the fossil-bearing layer. The fauna suggests an affinity with those of Florisbad and Vlakkraal.

The faunal material from the major sites other than the cave breccias is listed in Table 7. There are relatively few specimens which occur in the breccias and also in the other deposits and such forms are marked with an asterisk both in Table 6 and Table 7; most of them are living species from which the fossil forms cannot be separated, usually not because of established identity but because the fossil material is inadequate for better diagnosis.

3. Leakey overlooked an amendment by Cooke (1950) showing that Hopwood's *Diceros whitei* was a composite of a living black rhinoceros and a large buffalo, probably *Homoioceras.*

TABLE 7

FOSSIL MAMMALS OF MAIN MIDDLE AND UPPER PLEISTOCENE DEPOSITS IN SOUTH AFRICA AND RHODESIA

								Rhodesia		
	Vaal River	*Cornelia*	*Hopefield*	*Florisbad*	*Vlakkraal*	*Cave of Hearths*	*Wonderwerk*	*Chelmer*	*Broken Hill*	*Mumbwa*
Insectivora										
Elephantulus langi *						cf				
Crocidura cf bicolor * †						x			?	
Chiroptera										
Rhinolophus cf geoffroyi †						x				
Miniopteris cf schreibersi †						x				
Primates										
Simopithecus sp.			?						?	
Papio ursinus †			x			x	x			
Homo rhodesiensis			cf						x	
Homo cf sapiens				x		x				
Lagomorpha										
Pronolagus randensis * †						x				
Lepus saxatalis †						x				
Rodentia										
Xerus capensis †				x						
Pedetes hagenstadi				x						
Aethomys cf chrysophilus †						x				
Thallomys cf paedulcus †						x				
Saccostomus campestris †						x			?	
Dendromus cf mesomelas * †						x				
Steatomys cf pratensis * †						x				
Palaeotomys gracilis *						?				
Myotomys cf turneri †				x						
Myotomys cf unisulcatus †				x						
Otomys cf irroratus †				x					?	
Hystrix africaeaustralis * †						x	x		cf	cf
Thryonomys swinderianus †						x				x
Carnivora										
Canis mesomelas * †			x			x				
Canis adustus †									cf	cf

Note: * indicates a form also listed in Table 6
† after a species indicates a living form

TABLE 7 (continued)

	Vaal River	*Cornelia*	*Hopefield*	*Florisbad*	*Vlakkraal*	*Cave of Hearths*	*Wonderwerk*	*Rhodesia* *Chelmer*	*Rhodesia* *Broken Hill*	*Rhodesia* *Mumbwa*
Vulpes chama †					x	x				
Lycaon picta †					x	x				
Lycaon picta magnus			x							
Otocyon megalotis †						x				
Mellivora capensis †			x				x			
Aonyx capensis †				x						
Suricata suricatta * †						x				
Herpestes cauui †				x						
Herpestes ichneumon †			cf			x			x	
Atilax paludinosus				x						
Crocuta crocuta †	x				x	x				x
Crocuta spelea *			cf							
Hyaena brunnea * †			cf	x			cf		cf	x
Caracal caracal †						x				
Leptailurus hintoni									x	
Leptailurus serval †			cf							
Panthera leo †			cf						x	
Panthera pardus †						x			x	
Machairodus sp.			?						?	
Tubulidentata										
Orycteropus sp.										x
Proboscidea										
Gomphotherium sp.	x									
Archidiskodon subplanifrons	x									
Archidiskodon broomi	x									
Archidiskodon transvaalensis	x									
Loxodonta atlantica	x		cf							
Loxodonta africana									x	
Hyracoidea										
Procavia capensis *						x				x
Perissodactyla										
Stylohipparion steytleri *	x	x								
Equus quagga †	?			x			x			

Note: * indicates a form also listed in Table 6
† after a species indicates a living form

TABLE 7 (continued)

								Rhodesia		
	Vaal River	*Cornelia*	*Hopefield*	*Florisbad*	*Vlakkraal*	*Cave of Hearths*	*Wonderwerk*	*Chelmer*	*Broken Hill*	*Mumbwa*
Equus burchelli * †	x	x		x	x	x	x	x	x	
Equus plicatus *	x	x	cf	cf	cf		x	cf		cf
Equus helmei *	x			x	x	x	x	x		
Eurygnathohippus cornelianus		x								
Ceratotherium simum * †			cf				x		?	
Diceros bicornis * †	x		cf	x		x	x		x	
Artiodactyla										
Potamochoerus koiropotamus †										x
Tapinochoerus modestus	x	x								
"Tapinochoerus" meadowsi *	x		x							
"Kolpochoerus sinuosus"		x								
cf Orthostonyx sp.		x								
Notochoerus capensis	x									
Mesochoerus paiceae *	x		cf							
Mesochoerus lategani			x							
Stylochoerus compactus	x	x		cf	cf					
Stylochoerus altidens	x									
Phacochoerus aethiopicus †	x			x	x	x	x	cf	x	
Phacochoerus africanus †	x	cf		cf			x	cf		x
Hippopotamus amphibius * †	x		x	x	x			cf		
Hippopotamus gorgops		x								
Giraffa camelopardalis * †		x		x		x			cf	
Libytherium olduvaiensis *	x	x							cf	
Strepsiceros strepsiceros * †	?	cf	cf						cf	x
Tragelaphus angasi * †							x		?	
Taurotragus oryx * †	x	x	x	x	x	x	x	cf	x	x
Syncerus caffer * †	x					x	x			x
"Homoioceras" baini	x	x	cf	x	x	x	x	x	x	
Cephalophus caerulus * †										?
Sylvicapra grimmia †	x			cf	cf		cf			cf
Kobus ellipsiprymnus †						cf	x			
Kobus venterae		cf		x	x					
Onotragus leche * †						x				
Redunca arundinum * †			x			x				cf

Note: * indicates a form also listed in Table 6
† after a species indicates a living form

TABLE 7 (continued)

								Rhodesia		
	Vaal River	*Cornelia*	*Hopefield*	*Florisbad*	*Vlakkraal*	*Cave of Hearths*	*Wonderwerk*	*Chelmer*	*Broken Hill*	*Mumbwa*
Redunca fulvorufula * †										?
Pelea capreolus †						x				
Hippotragus leucophaeus †			cf							
Hippotragus equinus †							cf			x
Hippotragus niger †	x									
Hippotragus *sp. nov.*			x							
Oryx gazella *†							x		cf	
Damaliscus cf albifrons * †	?	?		x	x				x	
Damaliscus cf lunatus * †	x					x			cf	x
Damaliscus cf pygargus * †	?		cf				x			
Damaliscus *sp. nov.*		cf		x	x	cf				
Damaliscus *sp. nov.*			x							
Alcelaphus caama * †	x	cf				cf	x	x		x
Alcelaphus lichtensteini †										?
Alcelaphus robustus *	x					cf				
Alcelaphus helmei *	cf	?		x	cf	cf	cf	cf		
Connochaetes antiquus										
Connochaetes gnou †	x									
Connochaetes laticornutus		x		?						
Connochaetes grandis								x		
Connochaetes taurinus *	x					x	cf	x	cf	x
Lunatoceras mirum			cf	?						
Megalotragus eucornutus		x								
Gen. nov.			x							
Gen. nov.			x							
Oreotragus oreotragus †						x				
Raphicerus campestris * †			cf			x	cf			
Aepyceros melampus * †	x					x				
Antidorcas marsupialis * †	cf	cf		cf	cf	x	x			
Antidorcas (?) *sp. nov.*			x							
Litocranius walleri †									cf	
Gazella wellsi *	x	x	cf							
Gazella *sp. nov.*			x							
Gazella *sp. nov.*		x								
Gazella bondi				x	x	x		x		
"Gazella" helmoedi			x							

Note: * indicates a form also listed in Table 6
† after a species indicates a living form

None of the faunal assemblages presents very clear indications of climatic conditions. If, as Brain's work on the cave breccias suggests, the rainfall varied in the Sterkfontein and Makapansgat areas from its present mean of around 30 inches a year to 20 inches in dry periods and 40 inches in the wetter ones (Fig. 6) the regional climate remained within the range of most of the characteristic steppe and savanna fauna. Nearly all the forms represented should, by analogy with their living counterparts, have tolerated such changes without difficulty. Forest and desert species are equally lacking. Groups like Canidae and Equidae, which avoid forest areas and like open plains, are found side by side with forms which dwell typically in moist thickets. This is entirely consistent with the terrain. At Makapansgat the presence of *Oryx gazella* indicates dry conditions at no great distance from a well-watered bushy valley. The Taungs area is a good deal drier than the Sterkfontein and Makapansgat areas and this is reflected to some extent in the fauna but water was certainly available in fair quantity as the few Bovidae and the occurrence of crabs would suggest. The faunas younger than the cave breccias are again typical of the thornbush and grassland assemblages and do not suggest drastic changes of climate or environment.

Correlation and Dating of Faunas

The relative dating of the Australopithecine breccias has been discussed effectively on geological grounds by Brain (1958) and on faunal evidence by Ewer (1957). Their actual age has been considered by a number of authorities, the most recent of whom are Howell (1955, 1959) and Wells and Cooke (1957), who accept an upper Villafranchian date, and Kurtén (1957, 1960), who rejects this dating as fallacious and proposes a Cromerian age. It has now become very clear as a result of unpublished work by Wells and the writer that even in late middle and upper Pleistocene deposits, many bovid species once thought to be living forms now prove to be largely extinct and even include a number of extinct genera. Of the 14 bovids at Cornelia 9 are undoubtedly extinct, at Hopefield the proportion is 9 out of 15, at Florisbad is 7 out of 12, and at Vlakkraal 5 out of 10. This surprising discovery suggests very strongly that in the older faunas a still higher proportion of the Bovidae will prove to be extinct types. Even before these new figures were available Wells (1957) showed that extensive pre-Pleistocene differentiation must have taken place in Africa and that the Cephalophini, Neotraginae, and Alcelaphinae must be essentially African in origin. Other groups which are supposed to have developed in Eurasia certainly were present in Africa at an early date. The same is apparently true also for the elephants (Cooke 1961) and it is very probable that as much or more migration took place from Africa to Eurasia than the reverse.

It is a mistake to ignore the endemic nature of most of the African fauna and to exaggerate the Asiatic connections. Kurtén's attitude is the traditional one that "Africa is a refuge for scores of forms which have long since vanished from the Palaearctic scene, and . . . that the southern continents have played this role for ages past." It must be remembered that many of the elements in the European Pleistocene appeared and disappeared as a result of favorable and adverse climatic changes of great magnitude. Africa was habitable at all times and was not so much a refuge as a source. It may be expected that African mammals diversified within the continent *before* they spread to Eurasia and their first appearance there would not serve to date their occurrence in African deposits, nor would their disappearance from Europe have valid chronological implications in Africa. Many of Kurtén's arguments rest on an assumed Asiatic origin for his guide fossils and it is most unfortunate that virtually nothing is known of the Pliocene in Africa so that it is not possible to settle the matter by fossil evidence. Kurtén's figures (1960) are out of date—through no fault of his—and from the revised list given in Table 6 the following amended results are obtained:

		Total	*Extinct*	*Extinct forms*
A:	Taungs, Sterkfontein and Makapansgat (combined)	98	68	69%
B:	Swartkrans and Kromdraai	59	47	79%

In group A, nearly all the forms which are not separated from living species are small rodents and bovids and the identifications are based on scanty material. It is significant that better material almost invariably results in recognition of the fact that the species differ from the living forms, often very considerably. It is quite probable that further studies will reduce the numbers of supposedly living species at Taung, Sterkfontein, and Makapansgat to not more than 10%. The Bovidae of Swartkrans and Kromdraai are unstudied and the figures are thus not affected by the incorporation of a number of supposedly recent specific names so that the percentage quoted is probably realistic. Using Kurtén's own figures for the relative frequencies of extinct species in Europe (1960: Table 1), the older Australopithecine breccias would be no later than Günz and the Swartkrans and Kromdraai deposits late Günz or early Cromerian. Although the present writer has been at some pains to dispute some of Kurtén's dating criteria, it must be admitted that some of his arguments are cogent to the issue. It would seem realistic to suggest that the older Australopithecine deposits (with *Australopithecus*) are upper Villafranchian and that the latter ones (with *Paranthropus*) are close to the Villafranchian-Cromerian boundary, but it is still not certain on which side they lie.

Owing to their particular environment of occurrence, the cave breccias contain a rather different cross section of the mammals than occur in the open sites, both in South Africa and in other parts of the continent. Paleontological correlation is thus rendered extremely difficult. No species are common to the faunas of the cave breccias and those of the Lower Pleistocene of North Africa. Compared with East Africa, there are a few common or related species. *Simopithecus* occurs at Makapansgat and Swartkrans and also at Omo, Laetolil, and Olduvai, though the species are different. *Paranthropus boisei* of Bed I, Olduvai, is certainly distinct from *P. robustus* of Swartkrans and Kromdraai, but there is considerable doubt regarding the validity of generic separation. Sabre-toothed cats are found at Kaiso and Omo as well as in the cave breccias of South Africa but the species are quite distinct, and machairodonts are also known from later deposits. Very fragmentary elephant remains occur at Makapansgat but cannot be specifically identified so that particular importance attaches to an undescribed incomplete lower molar from Bolt's farm which was in the collection of the University of California. It seems to belong to *Loxodonta atlantica* but it is perhaps also allied to Arambourg's *A. africanavus* from North Africa. The Bolt's farm material probably corresponds in age to the Swartkrans and Kromdraai deposits. *Stylohipparion* occurs in deposits of widely differing ages and is not diagnostic, though in North Africa it seems to be confined to the Villafranchian. Chalicotheres occur at Makapansgat and at Olduvai but the relationship is not clear. The Suidae are important for correlation and dating: *Notochoerus euilus* occurs at Makapansgat and also at Kaiso, Omo, Laetolil, and Bed I in East Africa; "*Tapinochoerus*" *meadowsi* occurs in an early form at Sterkfontein, Swartkrans, and Bolt's farm and in a more advanced form in the Vaal river gravels and at Hopefield, a similar early form existing in East Africa in Bed I and a more advanced form in the later horizons at Olduvai. The primitive *Potamochoeroides shawi* of Makapansgat is allied to *P. majus* from East Africa (though clearly not specifically identical with it) and is also close to the diverging line of *Omochoerus-Mesochoerus limnetes; Omochoerus maroccanus* from North Africa is another allied form. The antlered giraffid *Libytherium* is confined to the Villafranchian in North Africa but extends into the Middle Pleistocene in East and South Africa, where it has been referred to *Sivatherium* (Singer and Boné 1960) although there is good reason to believe that the latter genus does not, in fact, occur in Africa at all. The time range of *Libytherium* makes it of little use in correlation. The Bovidae do not at present provide data of value for correlation and the diversity of forms is remarkable.

Although ecological differences prevent too close a comparison, the faunas suggest strongly that the sequence in East Africa from the Kaiso and Kanam levels through Omo and Laetolil to Bed I corresponds fairly closely in time to the sequence Sterkfontein and Makapansgat to Swartkrans and Kromdraai. Although the evidence is extremely tenuous, the Villafranchian fauna of North Africa could well be contemporary with these East African deposits and the ape-man breccias. The occurrence of pebble tools in similar relationships in all three areas may be significant and if the North African beds are truly pre-Cromerian as has been suggsted by several authorities, this would provide additional grounds for keeping at least the major part of the ape-man deposits within the Villafranchian.

As far as the younger faunas in South Africa are concerned, the Vaal River and Cornelia assemblages bear comparison with those of Olduvai Beds II to IV, though very few species are actually shared between the two areas. Among the suids the strange "*Afrochoerus*" of Olduvai and Kanjera has been found in the Vaal River and at Cornelia,[4] perhaps also with *Orthostonyx* at the latter site. The long-range species "*Tapinochoerus*" *meadowsi*, *Libytherium olduvaiensis*, and *Stylohipparion* are shared. *Hippopotamus gorgops*, which is so typical of the Olduvai beds, has been identified at Cornelia by Hooijer (1958). The elephants of the Vaal River gravels include archaic forms which may well be derived through erosion from earlier deposits; the commoner forms which may belong to the gravels are probably *Archidiskodon transvaalensis* and *Loxodonta atlantica*. The Hopefield fauna ties in well with Bed IV at Olduvai, including a *Mesochoerus* which is very close to *M. olduvaiensis*.

The faunas definitely associated with well-developed Middle Stone Age cultures in South Africa include only a few extinct species, most of which are larger than their living counterparts (*Equus helmei*, "*Homoioceras*" *baini*, and *Alcelaphus* cf *helmei*) and which survive from the earlier faunas; in East Africa the "Gamblian" deposits contain *Homoioceras nilssoni* but other extinct forms are very rare or absent and it is suspected that there is a time gap between the typical Gamblian and the faunas of Olduvai Bed IV and its equivalents, the gap being bridged in South Africa by the Florisbad-type fauna. An outline correlation is suggested in Table 8.

SOME FEATURES OF THE FOSSIL FAUNAS

As has been mentioned above in discussing the age of the Australopithecine deposits, there is a great deal of evidence which favors the existence of a considerable endemic pre-Pleistocene mammalian fauna which included practically all the elements that compose the typical spectrum of African wild life. Despite the lack of Pliocene deposits there is increasing testimony that the Pleistocene species have evolved primarily within the continent itself, exporting emigrants rather than admitting immigrants. Parallel development has undoubtedly produced forms which have superficial resemblances to their European or Asiatic near-contemporaries but the discovery of better preserved and more complete material usually serves to provide evidence that the African fossils are *not* identical with the Eurasiatic ones. Had Africa been open to free—or even limited —immigration from Eurasia after a late stage in the Pliocene it would be expected that more of the characteristic Asiatic animals would have been found in this continent than is actually the case. The African environment would certainly have been suitable for deer and camels, for example, yet they are represented only by rather scanty elements in the northern littoral belt, and even those are late immigrants.

The distinctive mammal faunas of Africa in Oligocene and Miocene deposits

4. Van Hoepen's genus *Stylochoerus* pre-dates Leakey's *Afrochoerus*.

TABLE 8

PROVISIONAL CORRELATION OF MAJOR PLEISTOCENE DEPOSITS IN AFRICA

Possible Age	*South Africa*	*East Africa*	*North Africa*
"Recent"	Deposits with Later Stone Age cultures and no extinct faunal elements	Deposits with Neolithic and Mesolithic cultures and no extinct faunal elements	Deposits with Neolithic and Capsian associations
"Upper Pleistocene"	Vlakkraal and other deposits	"Later Gamblian" beds and Eyasi Beds	Deposits with Micoquian and Aterian cultures
	Florisbad and Chelmer	"Early Gamblian" beds	
"Middle to Upper Pleistocene"	Hopefield	Olduvai Bed IV +	
	Cornelia	Olduvai Beds III to IV Semliki Beds	Lac Karar etc.
"Middle Pleistocene"	Vaal "Younger Gravels"	Olduvai Beds II and III	Ternifine
"Cromerian or uppermost Villafranchian"	Kromdraai Swartkrans Sterkfontein extension	Olduvai Bed II (in part)	
"Villafranchian"	Makapansgat, Taung and Sterkfontein	Olduvai Bed I Laetolil Omo Kaiso Kanam	Aïn Hanech Bel Hacel etc. Koro-Toro Lac Ichkeul, Aïn Brimba, Aïn Boucherit, Fouarat etc.

favor a considerable measure of isolation from Eurasia even at those times, when the present barrier of the Red Sea was only a tectonic trough intermittently occupied by freshwater lakes or a temporary sea. The violent Pliocene rifting must have created an effective barrier and, in particular, the artiodactyls which came from Asia must have evolved in Africa for some considerable period before they reached the highly diversified state in which they are found in the earliest Pleistocene deposits. The equine *Stylohipparion* seems to be an independent development in Africa from a Pliocene invader and the same may be true of the chalicothere, of *Libytherium*, and some of the carnivores. The origin of the zebrine horses is obscure but they must represent one of the rare early Pleistocene invasions whose existence must be admitted even in the face of contrary evidence for other groups.

There are two marked trends shown by the Pliocene-Pleistocene mammals—conservatism on the one hand and dynamic expansion on the other. An extreme example of the first is the white rhinoceros, which seems to have persisted almost unchanged from the final Pliocene to the present day. The fossil faunas show that it existed formerly all over the continent, including North Africa, but today it is restricted to a limited area from Lake Albert into the southern Sudan and in historic times existed between the Zambezi and the Orange rivers and also in Natal, where it still survives in the Hluhluwe Game Reserve. The forms in these two widely separated areas are distinguished on no more than a subspecific level and their parting from the basic stock is probably not a matter of great antiquity. The white rhinoceros is a grasseater and the territory which intervenes between the northern and southern occurrences seems basically suitable for the species at the present day and would not exclude it under drier conditions so that it may well be an overdevelopment of woodland and forest in the past which was responsible for the separation. Both the white and the black rhinoceros lack incisor teeth and are not met as fossils outside Africa.

Another apparently conservative form is the giraffe, which seems to have survived through the whole of the Pleistocene with little variation, though an extinct species is known in East Africa. It occurs throughout Africa south of the Sahara except in wet, heavily forested areas and it existed in the Sahara and North Africa both as a fossil in the Middle Pleistocene and in Neolithic times. The reticulated giraffe survives in the dry area east of Lake Rudolph and the Tana River where the vegetation is decidedly scanty and the giraffe must be highly adaptable. Its relative, the okapi, is by contrast a forest-dweller which has probably evolved in Africa over a long period and is now confined to the eastern Congo; a fossil form at Laetolil suggests the former existence of forest areas in this region though not to the extent of obliterating the steppe and savanna.

The white rhinoceros and the okapi are by no means the only herbivores with very restricted or peculiar distribution. Among the Bovidae, the great eland is now limited to much the same area as the white rhinoceros in the region west of the White Nile, but some of the fossil forms in southern Africa may be allied to it rather than to the more typically southern *Taurotragus oryx*. The rare Bongo, unrecorded as a fossil, exists in the coastal and upland forested areas of West Africa and in the forest of the Kenya highlands, though not in the intervening 2000 miles of low-lying Congo forest or in the forested areas of the Western Rift. Again there is a suggestion of former more continuous forest country and probably also of a cooler environment. More extensive swampy conditions are suggested by the present discontinuous distribution of the Sitatunga and of the Lechwe. The former occurs in West Africa, in the White Nile and southwards into Uganda and northern Tanganyika and, after a gap across Lake Tanganyika, occurs again in Northern Rhodesia. The Lechwes are absent in West Africa and do not occur in Uganda or Tanganyika. The Nile species is confined to the marshy areas west of the White Nile and the southern species lives in the vast swampy areas of Rhodesia and northeastern South West Africa; the latter or an allied form occurs as a fossil in the Gladysvale cave breccia and in the deposits of the Cave of Hearths, suggesting very wet conditions locally at these periods.

Apart from the few forms whose abnormal distribution has been mentioned above, the Bovidae are in general very widespread and are associated with the large carnivores and scavengers which prey upon them. Half of the surface of Africa is covered by dry savanna and grass steppe or by open woodland and mixed savanna (Fig. 3). Within this vast area the mammalian fauna has the same essential composition, though local features control the precise relationships and balance of the various species. The Pleistocene deposits demonstrate that the fauna had the same general character in the past, even though the species have changed with progressive extinction of archaic forms and steady development of the present-day varieties. Many of the Pleistocene species were larger than their modern counterparts and some deserve to be termed giants; they were probably overspecialized and failed to survive a changing environment but the reason for their gigantism is an unsolved problem.

Although the archaic forms provoke most attention in the fossil faunas, they do not represent the most interesting features of the assemblages. The extensive pre-Pleistocene diversification of the Bovidae has already been mentioned briefly. The Cephalophini and Neotragini are clearly indigenous and the Alcelaphini are probably also an African group (Wells 1957). The living Alcelaphines comprise only three genera with twelve species between them but the fossil material provides nine new species of these genera and five undeniably new genera from South Africa and at least five additional species of the living genera and three new genera from East Africa. Some of the genera, such as *Makapania* from Makapansgat, are somewhat aberrant, as also are the forms from Hopefield, and it is clear that the Alcelaphini possess a long history in the continent. The sole representative outside Africa is the Villafranchian *Damalops* of India, which Wells suggests may be a stray emigrant.

Another feature of the bovid fauna of the Pleistocene deposits is the variety of gazelles which occur as fossils. The unspecialized genus *Gazella* has four living species in North Africa and there are at least three extinct species in this region which have not been reported elsewhere; they inhabit very dry areas at the present time and one species, *G.rufina*, may be extinct. West Africa has only one species, the sub-Saharan steppe another, and there are two in Somalia. East Africa has three living species but there are three additional fossil species in this area. South Africa has no living species of *Gazella* but there are at least five fossil members of the genus ranging from the ape-man breccias to the Vlakkraal upper Pleistocene. Two of these gazelles have some resemblance to the living Springbuck *Antidorcas marsupialis* but they are not ancestral to it for an *Antidorcas* close to the living species occurs in the fossil state in the later cave breccias and persists through all the Pleistocene deposits; in North Africa it occurs at Aïn Brimba and Aïn Boucherit. An undescribed new species from Hopefield is a close relative. *Antidorcas* must have evolved from a gazelle in the Pliocene of Africa and there can be little doubt that the Impala *Aepyceros* was also differentiated by the beginning of the Pleistocene for it is recognisable in the cave breccias, perhaps by a more robust species which occurs at Omo and Laetolil in East Africa as well. The Impala shows a preference for open woodland and does not move far from water. Springbuck are predominantly dwellers in open fairly dry grassland or grassland with very open bush and they range into semi-desert scrub areas where they can exist for long periods with little or no water. They migrate over considerable distances to new grazing lands and their restriction to Africa south of the Cunene and Limpopo rivers is surprising. The true gazelles have habits very like those of the Springbuck. The affinities of the South African fossils seem, rather strangely, to lie with the North African gazelles rather than with those of East Africa, and one may suspect a Pliocene spread of ancestral rather generalized gazelles about Africa, followed by separate diversification in South, North, and East Africa without much intercommunication. The peculiar gazelline fossil antelope *Phenacotragus* is another specialized African development, represented by one species at Makapansgat and by another in the upper Olduvai beds. The peculiar Gerenuk also represents a specialized member of this family and it is unfortunate that it is known as a fossil only by fragments at Broken Hill, a long way south of its present habitat.

It would be tedious to discuss the features of the Hippotragini and Strepsicerotini, but these groups must also have lived in Africa in the Pliocene for they appear in the early Pleistocene deposits where they are already well differentiated and distinct from their Asiatic fossil relatives. The Reduncini, as Wells points out, have a curious distribution and it is possible that migration was into Asia rather than the reverse. The Boselaphini are unrepresented in Africa and the history of the Bovini is obscure. *Syncerus* is present in the earliest cave breccias and, although *Homoioceras* is superficially like an Asiatic buffalo, it has unique African features; furthermore there is probably more than one genus of giant buffalo at present placed in *Homoioceras*, and the East African *Bularchus* is another purely

African variant within this group. Extensive evolution of African Bovidae through much of the lamentably unknown Pliocene is clearly indicated in sub-Saharan Africa with limited transfer taking place from time to time across the Sahara barrier. Within sub-Saharan Africa climatic changes may have caused temporary migration but there is no evidence of wholesale population transfer or destruction.

Another group which has evolved almost explosively in Africa is the pig family. Leakey (1958) lists 11 genera (including the three living ones) and 23 species from East Africa, to which must be added at least eight species from South Africa. Although there are a number of synonyms which may reduce the number of species by a half, there are undoubtedly at least ten genera. Two of these, *Stylochoerus* (= *Afrochoerus*) and *Orthostonyx* have most unusual canine teeth whose evolution is obscure. The others range from types with brachyodont somewhat *Sus*-like molars to hypsodont forms following three different evolutionary trends. In the earliest Pleistocene deposits in East Africa and in South Africa, a giant specialized pig with hypsodont molars (*Notochoerus euilus*) occurs with diverse brachyodont and small hypsodont forms which show the main trends of the evolutionary divergence. The giant pig must be the product of a fairly lengthy evolutionary history separate from that of the Asiatic suids. It is impossible to guess the habitats of these creatures but it may be inferred from their associations that the brachyodont forms were thicket-dwellers and the more hypsodont types were grass-eaters.

The fossil elephants of Africa have long been referred to Eurasiatic forms, though Arambourg (1947) demonstrated clearly that "*Palaeoloxodon antiquus recki*" bore only a superficial and unreal resemblance to *Palaeoloxodon;* in skull structure it is clearly an *Archidiskodon.* The present writer (Cooke 1961) has recently reviewed the position and suggested a possible phylogeny for *Archidiskodon* and *Loxodonta* in this continent. Arambourg (1945) has also shown clearly that *Anancus* and perhaps other basic forms evolved in Africa and it would again seem likely that this continent has been a greater center of dispersal than has been imagined hitherto. The living African elephant most probably arose from an *atlantica*-like ancestor in the West African forest region and its dispersal into other parts of the continent only in the upper Pleistocene may very well be due to greater extension of forest areas, which would have provided ready passage over the normal steppe areas where elephants can survive only with difficulty.

The carnivores present some serious problems. The hyenas are compared with Asiatic forms, and if the inferred relationships are correct there must have been a late Pliocene or early Villafranchian invasion. *Hyaena* is undoubtedly African and although *Crocuta* is reported from the lower Pleistocene there is some doubt concerning the accuracy of this generic reference.[5] The Villafranchian assemblages include sabre-toothed cats which in the various regions are referred to different Eurasiatic genera but whose mutual affinities suggest that such reference may well be unjustified and that they all represent African differentiates.

5. L. S. B. Leakey, verbal communication.

The normal Canidae and Felidae appear well established in the early Pleistocene and some fossil forms of the latter are tiger-like so that one may wonder whether *Panthera* did not perhaps originate in Africa.

There is thus a great deal of evidence which favors the diversification of the larger African mammals in Africa itself during Pliocene times, out of contact with Eurasia after a certain amount of interchange of stocks had taken place (perhaps in the Pontian). The Villafranchian faunas are fundamentally African in character and more and more of the Eurasiatic resemblances are proving to be convergent rather than genetic. Obviously, also, the older faunas were less completely differentiated and possessed closer affinities with extra-African groups derived from common basic stocks. During the Pleistocene there has been a progressive decline in the archaic forms and a rise of modern species. The geographic distribution of the fossil herbivores demonstrates that the broad ecological conditions did not vary greatly but there is evidence in most areas which favors former extension of forests and of swampy conditions to the present savanna areas, though this need not have occurred synchronously. In a few cases lower temperatures are suggested but the evidence is scanty. Although some interchange of species took place, the faunas of the different regions maintained during Pleistocene times the same order of distinctiveness which is characteristic today. In some respects the southern part of the continent shows signs of isolation from the north but there is no evidence for the existence of major barriers. Despite the geological data suggestive of drier conditions at times during the Pleistocene, there is little faunal testimony for conditions too dry to sustain a normal steppe assemblage; it may well be that fossil-bearing deposits belonging to these intervals are rare or unstudied. It also seems clear that most of the African fauna developed south of the almost permanent barrier of the Sahara but that migration across this barrier was possible at times, probably during the final Pliocene or early Villafranchian, during the late middle Pleistocene and for the last time in the final Pleistocene. Mammals of Eurasiatic character penetrated the North African coastal belt during the late Pleistocene but did not move southwards across the Sahara, though it is conceivable that it was by this route that the horses reached sub-Saharan Africa in the early Villafranchian. Ecological interpretations are severely handicapped by inadequate data on the habits and limitations of the living species and it is very possible that further studies of the insectivores and rodents in particular would be highly profitable.

BIBLIOGRAPHY

ABADIE, J., J. BARBEAU, and Y. COPPENS

1959. "Une faune de Vertébres villafranchiens au Tchad.," *C. R. Acad. Sci.* (Paris), 248:3328–30.

ALIMEN, [MARIE]-HENRIETTE

1955. *Préhistoire de l'Afrique.* Paris: N. Boubée et Cie.

ARAMBOURG, C.

1935. "Esquisse géologique de la bordure orientale du lac Rodolphe." Mission scientifique de l'Omo, 1932–33. *Bull. Mus. Hist. Nat. Paris,* No. 1, Fasc. 1.

1938. "Mammifères fossiles du Maroc," *Mem. Soc. Sci. Nat. Maroc,* 46:1–72.

1945. "Anancus osiris, un mastodonte nouveau du Pliocène inférieur d'Egypte," *Bull. Soc. Géol. France,* Sér. 5, 15:479–95.

1947. "Contribution a l'étude géologique et paléontologique du bassin du Lac Rodolphe et de la basse vallée de l'Omo." *Mission scientifique de l'Omo,* 1932–33, Part II, pp. 1:232–562.

1949. "Les gisements de vertébrés villafranchiens de l'Afrique du Nord," *Bull. Soc. Géol. France, Sér.* 5, 19:195–203.

1950. "Les limites et les corrélations du Quaternaire Africain," *C. R. Int. Geol. Cong.* (London 1948), Part XI, pp. 49–54.

1952. "La paléontologie des vertébrés en Afrique du Nord française." *19e Cong. Geol. Int. (Algiers 1952). Regional Monograph.*

1955a. "L'Atlanthropus de Ternifine," *Libyca* (Algiers), 2:425–39.

1955b. "Récentes découvertes de Paléontologie humaine réalisées en Afrique du Nord française (L'*Atlanthropus* de Ternifine-L'Hominien de Casablanca)," *Actes 3e Pan-Afr. Cong. Préhist.* (Livingstone 1955), pp. 186–94.

ARAMBOURG, C., and L. BALOUT

1955. "L'Ancien Lac de Tihodaïne et ses gisements préhistorique," *Actes 2e Pan-Afr. Cong. Préhist.* (Algiers 1952), pp. 281–92.

ARAMBOURG, C., and P. BIBERSON

1956. "The Fossil Human Remains from the Paleolithic Site of Sidi Abderrahman (Morocco)," *Amer. J. Phys. Anthrop.,* 14:467–89.

ARAMBOURG, C., and R. COQUE

1959. "Le gisement villafranchien de l'Aïn Brimba (Sud-Tunisien) et sa faune," *Bull. Soc. Géol. France,* Sér. 6, 8:607–14.

BAKKER, E. M. VAN Z.

1957. "A Pollen Analytical Investigation of the Florisbad Deposits (South Africa)," *Proc. 3d Pan-Afr. Cong. Prehist.* (Livingstone 1955), pp. 56–67.

BALOUT, L.

1952. "Pluviaux interglaciaires et Préhistoire Saharienne," *Trav. Inst. Rech. Sahar.* (Algiers), 8:9–21.

Biberson, P.

1955. "Nouvelle observations sur le Quaternaire côtiers de la région de Casablanca," *Quaternaria,* 2:109–47.

1961. "Le cadre paléogéographique de la préhistoire du Maroc atlantique." *Publ. Serv. Antiquités Maroc,* Fasc. 16.

Bishop, W. W.

1960. "A Review of the Pleistocene Stratigraphy of the Uganda Protectorate." Commission de cooperation technique en Afrique au Sud du Sahara (CCTA), Regional Committees for Geology, Leopoldville 1958 (1960), pp. 91–105.

Bishop, W. W., and M. Posnansky

1960. "Pleistocene Environments and Early Man in Uganda." *Uganda J.,* 24:44–61.

Bond, G., and R. Summers

1951. "The Quaternary Succession and Archaeology at Chelmer, near Bulawayo, Southern Rhodesia," *S. Afr. J. Sci.,* 47:200–204.

Bourcart, J.

1943. "La géologie du Quaternaire au Maroc." *Rev. Scient.,* 81:311–36.

Brain, C. K.

1958. "The Transvaal Ape-Man-Bearing Cave Deposits," *Transvaal Mus. Mem.,* No. 11.

Choubert, G.

1957. "Essai de la corrélation entre les cycles marins et continentaux du Pléistocène au Maroc," *C. R. Acad. Sci.* (Paris), 245:1066–9.

Clark, J. D.

1959. *The Prehistory of Southern Africa.* London: Penguin Books.

Cole, S. M.

1954. *The Prehistory of East Africa.* London: Penguin Books.

Cooke, H. B. S.

1947. "The Development of the Vaal River and its Deposits," *Trans. Geol. Soc. S. Afr.,* 49:243–59.

1949. "Fossil Mammals of the Vaal River Deposits," *Union S. Afr. Geol. Surv.,* Mem. No. 35 (III).

1950. "A Critical Revision of the Quaternary Perissodactyla of Southern Africa," *Ann. S. Afr. Mus.,* 31(4):393–479.

1958. "Observations Relating to Quaternary Environments in East and Southern Africa," *Trans. Geol. Soc. S. Afr.,* Annexure to vol. 61.

1961. "Further Revision of the Fossil Elephantidae of Southern Africa," *Palaeont. Afr.,* 7:59–63.

Cooke, H. B. S., and L. H. Wells

1951. "Fossil Remains from Chelmer, near Bulawayo, Southern Rhodesia," *S. Afr. J. Sci.,* 47:205–9.

Coppens, Y.

1960. "Le Quaternaire fossilifère de Koro-Toro (Tchad). Résultats d'une première mission," *C. R. Acad. Sci.* (Paris), 251:2385–6.

1961. "Découverte d'un Australopithecine dans le Villafranchien du Tchad," *C. R. Acad. Sci. (Paris),* 252:3851–2.

Dalloni, M.

1940. "Notes sur la classification du pliocène supérieur et du quaternaire de l'Algerie," *Bull. Soc. Géog. et Archaeol. Oran,* 61:8–43.

Dietrich, W. O.

1950. "Fossile Antilopen und Rinder Äquatorial afrikas (Material der Kohl-Larsen'schen Expeditionen)," *Paläontographica,* 99:1–62.

Ennouchi, E.
1950. "Les mammifères du quaternaire de Rabat," *Bull. Soc. Sci. Nat. Maroc.*, 28:34–6.

Esperandieu, G.
1955. "Domestication et élevage dans le Nord de L'Afrique au Néolithique et dans la protohistoire d'après les figurations rupestres," *Actes 2e Cong. Pan-Afr. Préhist.* (Algiers 1952), pp. 551–73.

Ewer, R. F.
1957. "Faunal evidence on the dating of the Australopithecinae," *Proc. 3d Pan-Afr. Cong. Prehist.* (Livingstone 1955), pp. 135–42.

Flint, R. F.
1959a. "Pleistocene Climates in Eastern and Southern Africa," *Bull. Geol. Soc. Amer.*, 70:343–74.
1959b. "On the basis of Pleistocene correlation in East Africa," *Geol. Mag.* (London), 96:265–84.

Gigout, M., and R. Raynal
1957. "Corrélation des phénomènes marins et continentaux dans le Quaternaire marocain," *C. R. Acad. Sci.* (Paris), 244:2528–31.

Grove, A. T.
1958. "The Ancient Erg of Hausaland, and Similar Formations on the South Side of the Sahara," *Geog. J.*, 124:528–33.

Harris, N., J. W. Pallister, and J. M. Brown
1956. "Oil in Uganda." *Geol. Surv. Uganda*, Mem. No. 9.

Heinzelin, J. de
1955. "Géologie régionale du fosse tectonique sous le parallèle d'Ishango," *Inst. Parcs Nat. Congo Belge*, Mission J. de Heinzelin de Braucourt (1950), Fasc. 1.

Hooijer, D. A.
1958. "Pleistocene remains of Hippopotamus from the Orange Free State," *Navorsinge Nat. Mus.*, (Bloemfontein), 1(2):259–66.

Hopwood, A. T., and J. Hollyfield
1954. "An annotated bibliography of the fossil mammals of Africa (1742–1950)," *Fossil Mammals of Africa*, No. 8, Brit. Mus. (Nat. Hist.).

Howell, F. C.
1955. "The Age of the Australopithecines of Southern Africa," *Amer. J. Phys. Anthrop.* (n.s.), 13:635–62.
1959. "The Villafranchian and Human Origins," *Science*, 130:831–44.
1960. "European and Northwest African Middle Pleistocene Hominids," *Curr. Anthrop.*, 1(3):195–232.

Huzayyin, S. A.
1941. "The Place of Egypt in Prehistory. A Correlated Study of Climates and Cultures in the Old World," *Mem. Inst. Egypt*, vol. 43.

Kent, P. E.
1941. "The Recent History and Pleistocene Deposits of the Plateau North of Lake Eyasi, Tanganyika," *Geol. Mag.* (London), 78:173–84.
1942. "The Pleistocene Beds of Kanam and Kanjera, Kavirondo, Kenya," *Geol. Mag.* (London), 79:117–32.

Kurtén, B.
1957. "Mammal Migrations, Cenozoic Stratigraphy, and the Age of Peking Man and the Australopithecines," *J. Paleont.*, 31:215–27.
1960. "The Age of the Australopithecines," *Stockholm Contrib. Geol.*, 6(2):9–22.

LEAKEY, L. S. B.

1931. *The Stone Age Cultures of Kenya Colony*. Cambridge: Cambridge University Press.

1934: "Changes in the Physical Geography of East Africa in Human Times," *Geog. J.*, 84:296–305.

1935. *The Stone-age Races of Kenya*. Cambridge: Cambridge University Press.

1943. "New Fossil Suidae from Shungura, Omo," *J. E. Afr. Uganda Nat. Hist. Soc.*, 17:45–61.

1950. "The Lower Limit of the Pleistocene in Africa," *C. R. Int. Geol. Cong.* (London 1948), Part IX, pp. 62–5.

1951. *Olduvai Gorge*. Cambridge: Cambridge University Press:

1958a. "Recent Discoveries at Olduvai Gorge, Tanganyika," *Nature* (London), 181:1099–1103.

1958b. "Some East African Pleistocene Suidae," *Fossil Mammals of Africa*, No. 14, Brit. Mus. (Nat. Hist.).

1959a. "A New Fossil Skull from Olduvai," *Nature* (London), 184:491–3.

1959b. "A Preliminary Re-assessment of the Fossil Fauna from Broken Hill, N. Rhodesia." In Clark, J. D. "Further Excavations at Broken Hill, Northern Rhodesia," *J. Roy. Anthrop. Inst.*, 89:201–32.

1961. "New Finds at Olduvai Gorge," *Nature* (London), 189:649–50.

LEAKEY, L. S. B., and T. WHITWORTH

1958. "Notes on the Genus Simopithecus, with a Description of a New Species from Olduvai," *Occ. Pap. Coryndon Mem. Mus.* (Nairobi), No. 6, pp. 1–26.

LOWE, C. VAN RIET

1952. "The Vaal River Chronology: An Up to Date Summary," *S. Afr. Arch. Bull.*, 7:1–15.

MASON, R. J.

1961. "The Earliest Tool-makers in South Africa," *S. Afr. J. Sci.*, 57:13–16.

MAUNY, R. A.

1957. "Répartition de la grande 'faune ethiopienne' du Nord-Ouest africain du Paléolithique à nos jours," *Actes 3e Pan-Afr. Cong. Préhist.* (Livingstone 1955), pp. 102–5.

MEIRING, A. J. D.

1956. "The Macrolithic Culture of Florisbad," *Researches Nat. Mus.* (Bloemfontein), 1(9):205–37.

MACINNES, D. G.

1953. "The Miocene and Pleistocene Lagomorpha of East Africa," *Fossil Mammals of Africa*, No. 6. Brit. Mus. (Nat. Hist.).

MCBURNEY, C. B. M.

1960. *The Stone Age of North Africa*. London: Penguin Books.

NILSSON, E.

1940. "Ancient Changes of Climate in British East Africa and Abyssinia," *Geog. Annaler*, 22:1–79.

NEUVILLE, R., and A. RUHLMAN

1941. "La Place du paléolithique anciens dans le Quaternaire marocain." *Publ. Inst. Hautes Etudes Maroc.* (Collection Hesperis), No. 8.

PEABODY, F. E.

1954. "Travertines and Cave Deposits of the Kaap Escarpment of South Africa, and the Type Locality of *Australopithecus africanus* Dart," *Bull. Geol. Soc. Amer.*, 65:671–706.

Pickering, R.
1960. "A Preliminary Note on the Quaternary Geology of Tanganyika," Commission de Cooperation technique en Afrique au Sud du Sahara (CCTA), *Regional Committees for Geology*, Leopoldville 1958 (1960), pp. 77–89.

Reck, H.
1914. "Erste vorläufige Mitteilung über der Fund eines fossilen Menschenskelets aus Zentral-afrika," *Sitzungsber. Gesell. Natur. Freunde*, Berlin, pp. 81–95.
1926. "Prähistorische Grab- und Menschenfunde und ihre Beziehungen zur Pluvialzeit in Afrika," *Mitteil. aus den deutschen Schutzgebieten*, 34.

Robinson, J. T.
1961. "The Australopithecines and Their Bearing on the Origin of Man and of Stone Tool-making," *S. Afr. J. Sci.*, 57:3–13.

Scott, W. B.
1907. "A Collection of Fossil Mammals from the Coast of Zululand," *Geol. Surv. Natal and Zululand*, 3rd and Final Rep., pp. 253–62.

Singer, R., and E. L. Boné
1960. "Modern Giraffes and the Fossil Giraffids of Africa." *Ann. S. Afr. Mus.*, 45(4):375–548.

Singer, R., and D. A. Hooijer
1958. "A Stegolophodon from South Africa." *Nature* (London), 182:101–2.

Söhnge, P. G., D. J. L. Visser, and C. van Riet Lowe
1937. "The Geology and Archaeology of the Vaal River Basin," *Union S. Afr. Geol. Surv. Mem.*, No. 39.

Solomon, J. D.
1939. "The Pleistocene Succession in Uganda." In T. P. O'Brien, *The Prehistory of Uganda Protectorate*, Cambridge: Cambridge University Press.

Van Hoepen, E. C. N.
1930. "Fossile perde van Cornelia, O.V.S.," *Paleont. Navorsinge Nat. Mus.* (Bloemfontein), 2:13–24.

Wayland, E. J.
1926. "The Geology and Palaeontology of the Kaiso Bone Beds," *Geol. Surv. Uganda Occ. Pap.*, 2:5–11.
1934. "Rifts, Rivers, Rain and Early Man in Uganda," *J. Roy. Anthrop. Inst.*, 64:333–52.
1952. "The Study of Past Climates in Tropical Africa," *Proc. 1st Pan-Afr. Cong. Prehist.* (Nairobi 1947), pp. 56–66.

Wellington, J. H.
1955, *Southern Africa*, vol. 1. Cambridge: Cambridge University Press.

Wells, L. H.
1957. "Speculations on the Palaeogeographic Distribution of Antelopes," *S. Afr. J. Sci.*, 53:423–4.

Wells, L. H., and H. B. S. Cooke
1957. "Fossil Bovidae from the Limeworks Quarry, Makapansgat, Potgietersrus," *Palaeont. Afr.*, 4:1–55.

Wells, L. H., H. B. S. Cooke, and B. D. Malan
1942. "The associated fauna and culture of the Vlakkraal thermal springs, O.F.S.," *Trans. R. Soc. S. Afr.*, 29:203–33.

THE LATE TERTIARY AND PLEISTOCENE IN THE SAHARA

THEODORE MONOD

"I do not know what is worse—an archaeologist with unrestricted imagination or one with a complete lack of that quality so necessary in life, art, and science."
ALFONSO CASO

"Je cherche autant à détruire mon hypothèse qu'à la vérifier." CLAUDE BERNARD

"I fear you are on rather ticklish ground with the Pluvials. We have yet much to learn. . . . *We need more facts.*" E. J. WAYLAND (21-IX-1933)

CONTENTS

INTRODUCTION

IN AGREEING TO PREPARE this paper, I recognized the editorial difficulties and other delays which would prevent me from spending more than few short weeks on it, thus producing a work of singularly deceiving character.* First, the subject touches simultaneously on a whole series of disciplines, requiring the collaboration of a team of specialists rather than the adventurous effort of a single researcher whose good will can never make up for his lack of information and competence. Moreover, it became quickly apparent that any attempt at a real synthesis is still premature. The literature, however considerable, deals more with theories, opinions, and suppositions than with observed facts. These themselves are often subject to interpretations which are hazardous, ambivalent, or even impossible—for example, the paleoclimatic inferences to be drawn from the nature of deposits, or the long-term correlations, the ethnic bearers of this or that cultural unit (industries, rock-art representations, etc.), or prehistoric chronology, and so forth.

In all these areas, dependable observations and precise data are sadly lacking. How many industries can be found *in situ* or paleontologically noted throughout the Sahara, when fewer than half a dozen prehistoric sites, all of which are recent (Neolithic), have been dated by C-14?

An added difficulty in which the researcher so often finds himself lies in the impossibility of judging, at a distance, the value of a publication. Thus, one sees the most serious and, in their own geographic field, the most competent authors at times, citing works or reports of debatable merit that, though taken from good sources, do not always have the value of the distant reader may be tempted to give them.

* I want to draw attention to the fact that the original text was in French and that, in spite of a very careful correction of the translation, I cannot feel certain that all the niceties of the manuscript have been adequately rendered. In addition, I must apologize to the various authors of quotations in French if any misinterpretation should have crept into the translation of their words. – Th. M.

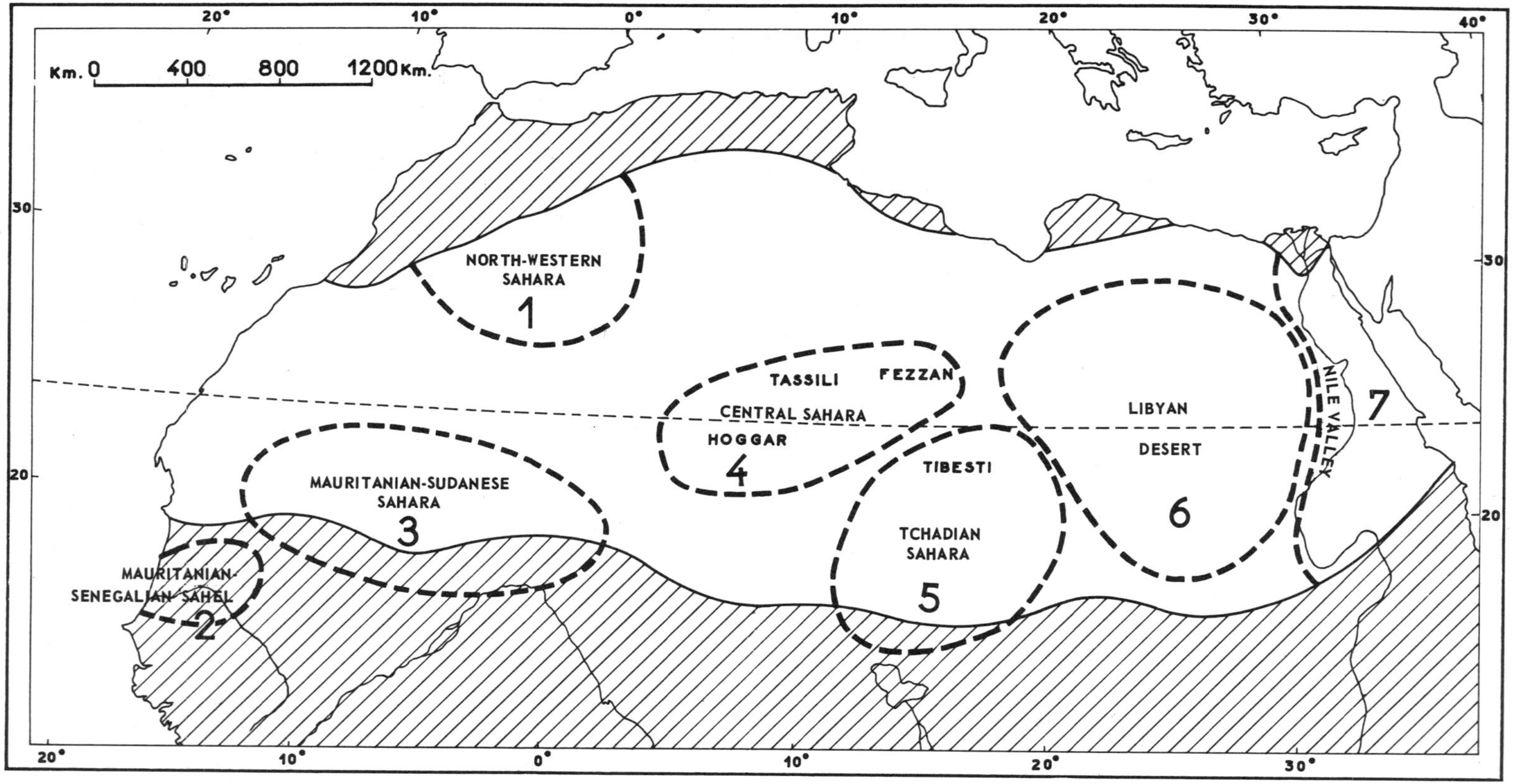

FIGURE 1
Approximate location of major regions as used in this paper.

Given these conditions, with such a basic and serious lack of documentation the only reasonable and possible proposal here was a rearrangement of certain bits of evidence, and a careful attempt at confronting and interpreting them, with the hope of placing them in an historico-geographical framework large enough that future syntheses may find the ground to some extent cleared.

As for the plan of the paper, it will understandably fail to satisfy the reader. In a geographical area as extensive as this, it was necessary to introduce regional divisions (Fig. 1), which in a very general way gives rise to certain important themes. This runs the risk, if not of actual duplication then at least of an inevitable dispersion of data, which only the inclusion of a detailed index, itself out of the question, could excuse. I am not unaware of the problems of the adopted plan; but after trying at first to give more room to general chapters and to limit the regional sections to geological data, I found it preferable, on the whole, to give more importance to the latter. One of the major perils of archeology, of paleoclimatology, and of Saharan Pleistocene geology, has too often been the imprudent use of generalizations and the failure to recognize the primary necessity of detailed studies on a modest and efficiently regional scale.

The Nile Valley, although crossing a considerable stretch of the Sahara, belongs only partially to it, and represents a foreign element. The problems of the Nile Valley remain largely *sui generis* and will receive in this work somewhat marginal attention.

I will sum up my proposed task by saying, with an author who came across similar difficulties in the prehistory of North Africa: "This is why you will find here a valuation of analyses, rather than one of syntheses." [1]

Finally, a few words concerning the bibliography, whose composition must seem, at first, surprising. There was no question, obviously, of assembling anything like a complete bibliography on a subject dealing simultaneously with such a wide area and so many fields; I had to limit myself. Outside of a certain number of general sources, I have listed mostly those works whose omission would have been held against me, and local works—of special interest, perhaps, to non-French researchers—which may be less familiar with the continuously growing Saharan literature in French. So, for example, I have tried to help the reader by giving as complete a list as possible of General P. Huard's publications on the Chad area.

I. THE GEOGRAPHIC FRAMEWORK

The Sahara is unquestionably the largest [2] and most beautiful desert in the world. While its eastern border is rather arbitrary, since the Red Sea could

1. Christian Courtois, in C. André Julien, *Histoire de l'Afrique du Nord* (2nd ed.), 1951, p. 31 (translated).

2. With an area of 7,000,000 to 9,000,000 square km., depending on whether or not the adjacent semideserts are included (H. Schiffers, *Die Sahara*, 1950, p. 14).

easily be considered as an intra-Saharan accident,[3] I will confine myself to the usual African concept of the Sahara without stressing the problems raised by its definition or the drawing of its limits, problems put forward and discussed many times in the past few years.

However, it would be helpful to recall the position the Sahara holds in the Oikumene. Forming the southwest extremity of the great eremian zone that cut diagonally across the ancient world from Mauritania to Mongolia, the Sahara, increasingly from east to west, shares the "marginal" character of the African continent, with regard to Asia, and is, consequently, marginal to the regions which appear to have seen the origin of domestication and agriculture. These regions, to go back even further into the past, are responsible for at least a part of the animal and plant life of the Sahara, and, more generally, all nonforested Africa.

On the other hand, the Sahara, dividing the Mediterranean from the "true" Africa, will always present a contradictory appearance of barrier and connective. With all the possible variations in time, space, and various aspects of human activity, the Sahara, though all too often considered an obstacle, is actually a device for sorting and filtering. Although the "Lehmkomplex" of paleomediterranean culture was established in the Sudan, and the wheel reached the southern edge of the desert, the swing-plow, apparently finding no reason to penetrate it, never crossed it.

We spoke of the Sahara. The singular is inaccurate. We should say the Saharas, for there are many. The climatologist correctly lists, within a general Saharan climate, a series of subclimates, functions of altitude (the Saharan mountain climate) or of geography, as the northern, southern, eastern (Libyan), western (Mauritanian). This last has many variants itself (J. Dubief 1953), and presents, perhaps, at least in the coastal area, a kind of "residual Sudanese climate" (*ibid.*, p. 17 and note 1). Taken altogether, the Sahara constitutes "a quite complex climatic system, with characteristics varying according to the regions" (*ibid.*, p. 8).

Biogeographically, and especially botanically, the Sahara is divided into two areas, a Mediterranean (and palaearctic) and an African (and palaeotropic), by the major phytogeographical separation of two floristic "empires," Palaearctis and Palaeotropis, or Holarctis and Pantropis (Monod 1957, pp. 76–81). That a double and symmetrical deterioration of two original populations, by converging in the heart of the desert under the effect of a centripetal gradient of increasing aridity, could show an apparent identity, would not change anything. There is indeed an extratropical (subtropical) Mediterranean Sahara on this side of the effective "Frostgrenze"—and a tropical (thermotropical) African Sahara beyond that (*ibid.*, p. 79). From the human point of view, one might compare this last

3. In this regard, the idea of a "Saharo-Arabian Belt" (S. Huzayyin 1941, pp. 3, 399) is much more satisfying than that of a purely African Sahara.

to a roof with a double slope, one of the north and one south, the wheat-barley and the Guinea-corn—bulrush-millet side.

Many other factors contribute to and indeed require the subdivision of the Great Desert. Relief, morphology, climate, vegetation, and human activity can be invoked, but this is not the place to go into details set forth by the geographers (cf., for example, H. Schiffers 1951). However, it would be useful to draw attention to one important feature. There was value, in certain periods of prehistory, in recognizing meridian compartments of circulation and habitat, separated by zones devoid of human habitation. The existence of these north-south belts appears in the plot of the great trans-Saharan caravan trails. One rediscovers them in the distribution of types of dromedary harness, finding in sequence from west to east, areas of Moorish saddle, of the Tuareg saddle (and a similar one with pommel *en palette*, the *Berbushiyya* or *Kuntiyya*), and finally of the pack saddle, beginning with the Teda and extending to the northwest of India as far east as the dromedary itself. Meridian circulations are easy in the "practicable" corridors. These have few lateral contacts, being separated by the severe no man's lands of Tanezrouft between corridors I and II, Tenere between II and III, the Libyan desert between III and the Nile Valley, which plays the part of a fourth corridor, thread-shaped and linear.

II. RESULTS AND UNCERTAINTIES

It is always surprising to discover the paucity of precise, attested, correctly observed and interpreted facts from which wholesale synthesis of Pleistocene geology and prehistoric archeology are constructed. Much has been published, but not enough is known to permit our going beyond the modest framework of regional observations and yielding to the dangerous fascination of premature generalizations. What Reeve (1959) tells us of the state of our knowledge of the origin of domestic animals in the Near East is true also for other subjects: valid data are sadly lacking.

Imprudent interpretations, hasty conclusions, excessively precise schemes and hazardous correlations should inspire much caution. As Karl Butzer recently put it so well (1960, p. 626), "With the great mass of field results, good or bad, building up information on the Pleistocene, it will become increasingly necessary to pause and reflect on the validity of evidence proposed. Uncritical employment of data would eventually be disastrous for a coherent understanding of the recent geological past."

Since R. Capot-Rey stated (1953, p. 83) that we do not know "how many Pluvials there were in the Sahara," we should not hasten to set up too remote correlations and, especially, we should not use in the Sahara chronological terminology devised for East Africa; though locally valid, its application to the whole of the continent seems too hazardous to be recommended yet. In utilizing European Glacial terminology in Africa, care should be exercised: ". . . correlations based on an assumed universal equivalence of glacial and pluvial episodes must be made with caution" (H. B. S. Cooke, 1957, p. 53).

It is evident that chronology and its terminology must be first specified within regional units. G. Mortelmans (1950) [4] has proposed such a scheme for East and South Africa. Comparable division will be defined for the rest of Africa, including the Sahara, where paleoclimates showed regional variation within a general framework whose principal traits are beginning to appear.

If it is generally admitted that for the zone which interests us, the Glacial-Interglacial fluctuations parallel the Pluvials and Interpluvials (more or less arid) [5] and that the Interglacials coincide with the Interpluvials,[6] it seems still difficult or impossible to compare the diverse local chronologies. This is especially true since much that is known is still only hypothetical, and at times even modified considerably from one paper to another by the same author.

Acknowledgment must be given to C. O. Sauer for his critiques (1956, pp. 52–53) of "such naively nominal climatic constructions" and to W. Meckelein for having stressed (1959, p. 92) both the frailty of our knowledge (the number of the changes in climate acknowledged in the Sahara varies according to M. Schwarzbach [1953, p. 165] from 1 to 10) and some salutary reminders: sand is not necessarily an indication of an increase of aridity; an increase of alluviation does not imply a local change of climate; river erosion probably does not coincide with either of the two peaks, extreme humidity or aridity, but with the transitional periods (principally with the phase H→A), as G. Choubert (1956, p. 589) also has suggested: "erosion would be more likely at the limit of the pluvial and interpluvial (cataglacial phase)".[7]

Although there are uncertainties, the whole tableau of the African Pleistocene becomes constantly more complete, and general syntheses endeavoring to cover the entire continent are proposed, for example, those of H. Alimen (1955) and R. Furon (1960). Leakey's (1953), while pertaining essentially to East Africa, should not be ignored as classic chronology, having often served as a frame of reference even, at times, for the Sahara.

A certain number of works directly concerning the Sahara must be cited here.

4. With two large divisions: I. North of the Zambezi (with two subdivisions: a) East Africa-Great lakes and b) Katanga-Kasai-north of Angola-Northern Rhodesia). II. South of the Zambezi.

5. For the Levant, however, N. Shalem (1950) did not deem it necessary to accept pluvial repercussions of glaciations and attributes the "changes of climate" to local geomorphological causes.

6. R. Capot-Rey, following L. Balout (cf. *infra*, p. 126) speaks however of an ancient Pluvial (Acheulian) corresponding, in the Sahara not to a glaciation but to an Interglacial, here warm and humid. S. Huzayyin (1956, p. 314) noted that in the middle latitudes, an Interglacial could coincide either with an Arid (if the Interglacial is cold, e.g., Riss-Würm) or with a Pluvial (e.g., Mindel-Riss, if it is hot).

7. "Anaglacial" and "cataglacial" are hybrid words of debatable origin. In reference to comparable periods of the Pluvial-Interpluvial cycles, "anapluvial" and "catapluvial" would not be more correct; with the prefixes *ad*- and *ab*- one would have "adpluvial" and "appluvial" (from a theoretical "abpluvial"), like the "adaxial" and "abaxial" of the botanists.

Those of L. Joleaud (1933b, 1936c, 1936d, 1936e, 1938, etc.) shows a very wide knowledge, although their chronological tables must be used with a certain caution. Three stages are defined in the evolution of mammals during the present Dry Period (post-Neolithic): —2000 to —1500, —1500 to —500, —500 to +300. The great work of S. A. Huzayyin (1941) deals with the climatic and archeological evolution of the entire Saharo-Arabian belt and gives for the Sahara the following sequence: [8]

1. 1st Pluvial, 2 phases (Pebble Culture to Acheulian)
2. Interpluvial, very dry (Acheulian)
3. 2nd Pluvial, 2–3 submaxima (Acheulian, Levalloisian, Upper Paleolithic)
4. Postpluvial, arid, with maximum in Upper Paleolithic (to Mesolithic)
5. Neolithic: "fairly wet phase" (c. —5500 to —2500; figures very similar to those of K. W. Butzer (1957): c. —5000 to —2400).

According to S. Huzayyin (*op. cit.*, p. 212), the Sahara should have played an important role in the Lower Paleolithic in the assimilation and the diffusion of industries. The system proposed by F. E. Zeuner (1952) rests upon meteorological considerations establishing that the Saharan Pluvials were not only the "secondary effects of the glacial phases," but imply a conjunctive influence of a Mediterranean Pluvial watering the northern Sahara while, in the south, a displacement toward the north of the thermal Equator provoked a corresponding advance from the northern limit of the monsoon and consequently a Saharan vegetation" (*ibid.*, pp. 68–69). Chronologically, the result is (p. 69):

1. Two short pluvials dating back to about 666,00 years ago
2. A long dry period (with short wet interruptions) from 430,000 to 230,000 years ago
3. A very long Saharan pluvial from 230,000 to 70,000 years ago
4. A possible drier phase about 150,000 years ago

Numbers 3 and 4 would be contemporary with the Penultimate and Last Glaciations plus the intervening Last Interglacial.

It remains to be seen to what degree this "test case for or against the applicability of the astronomical theory to the Saharan belt" will be verified by the facts. We cannot yet decide the question.

The contribution of meteorology to the study of Saharan paleoclimates has noticeably increased in the last two decades. The remarks of J. Dubief (1951, 1952, 1956, etc.) are important in this regard. The role of the "Saharan, or Sudano-Saharan depressions," running east to west (Sudan), then north and northeast (Sahara) and finally west to east (Mediterranean), seems very important. If these become rare, as they seem to have done from east to west (thus the extreme dryness of the Libyan Sahara, "the senescent desert") along with an

8. I have transcribed it with several additions from a more recent version by the same author (1956, p. 313).

increase in the strength of the etesian winds [9]: "it follows . . . that the aridity of the Sahara is due, to a small degree, to the northward retreat of polar-front lows or of the Sudanese monsoon to the south, and, to a greater extent, to a diminution in the frequency of Saharan and Sudano-Saharan depressions" (Dubief 1952, p. 8). It must be noted that the play of these Saharan lows gives a certain climatic autonomy to the area involved. A humid Sahara could be formed "independently of an increase in the rainfall of neighboring areas to the north and south" (*ibid.*); this is an idea to which we will return. Pushing the hypothesis a little further, the author adds: "One can think that at the time of an interglaciary the rainfall decreased in the temperate and equatorial zones, while it increased in the present desert zone. In other words, the terrestrial climate tended to balance itself. In the glacial periods, on the other hand, the contrasts were emphasized; the rainfall increasing both towards the poles and towards the equator, but diminishing or even disappearing in the Sahara" (*ibid.*).

The climatic optimum, around —6000 to —5000, the "cattle period," would have been "humid or semiarid." Afterwards, there was not a very desertic phase (since the Sudanese aquatic elements did not disappear) and the "garamantic period," in the first centuries of our era, ought to have been even a little less arid than at present (*ibid.*, p. 2).

With L. Balout (1952; and 1955, pp. 76–82, Fig. 12) we see reappearing one of those "Pendulations-Theorien" to which biogeographers have so often had recourse.[10] The Sahara cannot be treated as a homogeneous climatic entity. The fluctuation of zones of more or less opposite characteristics obliges us to consider as distinct the climatic, and consequently the human, history of a northern and a southern Sahara, one becoming increasingly arid while the other experiences an increase of precipitation, and vice versa. I have summarized in a chart borrowed from the author the significance of these conclusions (Table 1).

It should be noted that Balout (1955) describes a "Great Pluvial of the Last Interglacial," this latter being for the Sahara "the last great period of humidity, much larger than that of the Neolithic" (pp. 80–81).

Many geographers seem to have adopted Balout's theory, for example R. Capot-Rey (1953, p. 88) and J. Tricart and M. Brochu (1955, p. 171; J. Tricart 1956). For the latter there are two types of pluvials, those of the middle latitudes (e.g., North Africa) which coincide with the glaciations and those of the lower latitudes (south of the Sahara) occurring in the interglacial periods. The pluvial periods of North and West Africa are not contemporary. There is a reversal of phases in the cycles (Tricart 1956, pp. 164, 167).[11]

9. In the optimum European climate, the weakening of the Trade Winds and the etesian winds can explain the northward movement of the Monsoons. (Dubief 1951, p. 189).

10. Cf. Th. Monod 1957, pp. 26–27.

11. The author states (1956, p. 164) that, on the southern edge of the Sahara, "the last pluvial is Neolithic, and thus postglacial." But is this really a true Pluvial, in the usual and exact sense of the word?

TABLE 1
L. BALOUT'S CHRONOLOGICAL TABLE (1955)

CHRONOLOGIE GLACIAIRE	PULSATIONS CLIMATIQUES	INDUSTRIES PREHISTORIQUES
dernier interglaciaire Riss-Würm	"Sahara des Tchads" Faune tropicale (à hippopotames) des cuvettes lacustres (ex : Tihodaïne)	Paléolithique inférieur final (Acheuléen de faciès africain à hachereaux) (ex : Tihodaïne) progression probable Sud-Nord?
dernière glaciation Würm	Dessèchement du Sahara méridional et extension max. du désert ves le Sud. Hernie saharienne du climat tempéré chaud. Montagnes humides. Réseau de vallées du Mzab. Flore méditerranéenne du Hoggar. Faune de savane (ex : Tiouririne)	Levalloiso-Moustérien? Paléolithique supérieur Atérien progression Nord-Sud
stade de retrait des glaciers würmiens optimum climatique post-glaciare	Dessèchement du Sahara septentrional (sauf pluies d'orage au pied de l'Atlas). Légère pulsation humide du Sahara méridional due à l'extension de la mousson d'été. Fossilisation des dunes sahéliennes. "Sahara des Egyptes" Faune de savane passant à la steppe.	Arrivée d'H. sapiens Substratum capsien? Néolithique Art rupestre Homme d'Asselar
actuel	Recul des influences de mousson. Mort des "Egyptes" Aggravation des conditions désertiques par l'homme.	Période historique

D'après L. BALOUT, 1955, p. 80-81 et 120

J. Büdel (1953) notes simply that the climate of the Sahara during the Glaciations was "im ganzen etwas feuchter als heute" (p. 261) but H. Flohn (1953, Fig. 1) suggests that during the Würm the Sahara was somewhat less warm and dry, especially in the northern half; its southern limit was roughly what it is today.

Concerning climate antagonism between the northern and southern Saharas,

W. Meckelein (1959, p. 142) states that there is an intermediate "nucleus," a "Vollwüstenraum," which itself only gives evidence of a single Pleistocene pluvial.

K. W. Butzer, to the contrary, reacts strongly against the theory of a climatic antagonism; for him the phenomena are symmetrical on the two Saharan borders, north and south, Mediterranean and Sudanese, those of the etesian and of the monsoon [12]: "Das Neolithikum war mindestens ebenso feucht im Bereich der Winterregen wie im Bereich der Sommerregen, d. h. gleichzeitige Südverlagerung und lokale Verstärkung des Westdrifts sowie allgemeine Intensivierung und grössere Ausdehnung der Monsunregen. Es war also das Subpluvial von c. 5000–2400 v. Chr. eine Zeit verstärkter südmeridionaler allgemeiner Zirkulationsform" (1958, p. 39). The climatic pulsations of the desert manifest themselves simultaneously on its many frontiers (*ibid.*, pp. 38–39). It is necessary, then, once and for all to have done "mit dem Gespenst einer 'wandernden Sahara,' das heute nach 30 Jahren in den Arbeiten einiger Forscher aus der Westsahara weiterlebt" (*ibid.*, p. 39).

The terms *pluvial* and *glaciation*, *interpluvial* and *interglacial* are not necessarily always synonymous. Moreover, there may exist "another form of non-glacial circulation favoring greater precipitation in the trade-wind belts" (1957, p. 113) which does not contradict the role attributed by J. Dubief to his "Saharan Depressions." K. W. Butzer (1957b, p. 110) thought he could reject, "as certainly not tenable" for explicit meteorological reasons, the Balout-Dubief hypothesis. However, in the Neolithic, the northern edge of the Saharo-Arabic zone underwent an increase in humidity, not aridity, as Balout points out. On the other hand, the humid character of this subpluvial diminished from north to south, which makes the intervention of a south-north increase in humidity provoked by a revival of the monsoon less plausible (*ibid.*, p. 110).

Another especially important work by K. W. Butzer (1957c) is entitled "Late Glacial and Postglacial Climatic Variation in the Near East." A chronological and paleoclimatic synthesis, it should certainly be of use to the geologist and prehistorian as a reference system for a zone that largely covers the Sahara.[13] From this chart we can set up what the petroleum geologist would call a "log," a "standard series."

Thus it seems necessary to reproduce here the essentials of that chronology.[14]

0. Last Pluvial

1. Postpluvial I, with probable maximum in 11th millenium; drier and cooler; Aeolian action (loess of Gurgan); Fayum (±dry); Upper Paleolithic.

2. Subpluvial I, probably about the 9th millenium; pre-Neolithic; equivalent to

12. S. Huzayyin (1956, p. 311) pointed out that "during pluvial phases the desert belt was invaded by moisture and rainfall from both sides" and sketched a meteorological explanation of this fact.

13. Although set up in principle for a territory extending from Egypt to Iran.

14. The numbering system is my own.

the phase of Alleröd (and of Makalian?); Capsian (?); Natufian p.p.; Mesolithic (with seals) of Mazanderan.

3. Postpluvial IIa, ca. −8000 to −6500; drier.
4. Postpluvial IIb, ca. −6500 to −5500/5000; climate like today's; Mesolithic.
5. Subpluvial II, ca. −5000 to −2400; duration variously estimated (S. A. Huzayyin 1941, −5500 to −2500; G. W. Murray 1951, −8000 to −4000; H. Bobek (arid): −9000 to −4000; C-14 age from Neo A of Fayum: 4144±250 and 4400±180, of Neo of the cave of Hana Fteah, Cyr.: 4850±350); "fairly insignificant" period geologically but humanly "of great importance" (p. 29).
6. Postpluvial III, ca. −2400 to −850; recent arid phase; ceramic of the C group, Sahara towards Nile (ca. −2500), VI Dynasty.
7. Postpluvial IV a, ca. −850 to +700; climate like today's (a little warmer).
8. Postpluvial IV b, from ca. + 700; climate like today's (a little less warm).

P. Quézel and C. Martinez (1962, p. 318) gave a general chronology for the Central Sahara, which is worth quoting. It rests mainly on the study of pollens as paleoclimatological indicators:

1. To −10,000. Industry: ? Steppes, with possible elements of Mediterranean vegetation. Alluvia. Arid and steppic climate.
2. To −6,000. Moustero-Aterian (in the Sahara). Mixed forest of cedars, oaks, etc., with mountains; of Aleppo pine in the plains. Diatomites, fossil soils of a forest type and swamp soils. Climate temperate and humid with summer drought already marked (Mediterranean humid to subhumid).
3. To −2,800. Saharan Neolithic (to −5,000) then bovidian rock pictures (to −2,800). Aleppo pine, cypress, juniper, etc., with (in the mountains): cedar, oaks, walnut trees, heaths, etc. Diatomites, fossil soils. Climate hot and dry (Mediterranean semiarid), less so in the mountains.
4. To −500. a) Libyco-Berber rock pictures; nearly total disappearance of Mediterranean flora, invasion of the Sahara by the *Acacia* forest. No fossil soils. Mediterranean climate replaced by a Sahelian climate; b) progressive dessication of the Sahara. No fossil soils. Desert climate.

E. A. Bernard's report (1959) is theoretical, basically astronomical, but whose relation with our subject is obvious from the following quick analysis.

The theory postulates almost twenty Pluvials and Interpluvials during the Pleistocene in the area which concerns us, many more than are presently admitted by geologists even in regions studied in great detail.[15]

In the northern intertropical zone a desertification corresponds to an Interglacial, in the north tropical zone to a Glaciation, one which simultaneously caused Polar Front rains on extratropical northern Africa and a desertification of the tropical latitudes.[16] It is important to note that there is no continental synchronization: ". . . the great zonal regions of Africa have undergone their

15. ". . . the complexity of our theoretical sequences may be surprising to many specialists used to the more simplified and less complete sequences found in the literature" (Bernard 1959, p. 28).

16. A regression would coincide with a humid period, a transgression with an arid one (*ibid.*, p. 25).

own paleoclimatic evolution; a sequence established for an area of the continent cannot be extrapolated to any other zone too far removed" (p. 26)—an observation to temper the ardor and daring of future makers of correlation tables.

Another important idea (pp. 29–30): "There are no slow and continuous pulsations," but rather "a series of numerous levels produced . . . in brusque manner, almost disconected if one thinks of them in the total extent of Pleistocene period." This is, after all, a sort of "quantification," with a rapid transition from one level to another. E. A. Bernard's schema, instead of representing a curve of rounded undulations, presents "blocs" with quadrangular outlines, without transition from one to the other, and representing each one of the four major recognized types: polar front Pluvial (glaciation), Displuvial (heavy summer rains, dry winters), Isopluvial (regular rains throughout the year), and Interpluvial and aridification.

Bernard (p. 20) tends to consider the notion of a direct connection between Glaciation and Pluvials to be simplistic: "It is incorrect to speak of a direct cause-and-effect correlation between the climates of polar latitudes and those of the tropics *because each great climatic zone operates from a causality of its own*" (italics mine).

There are at the most "concomitances in the occurrence of extreme situations of regional paleoclimates." As for "the frequently debated question of correlations between glaciations and pluvials," it is "clearly solved by the theory: the intertropical pluvials with convective rains correspond to interglacials; extratropical pluvials with cyclonic rains, as well as the phases of desertification in intertropical regions, are correlative to the glaciations" (p. 30).

If "each great zonal region of Africa experienced its own paleoclimatic sequence during the Pleistocene," [17] the climatic regions "have glided ceaselessly over the surface of the continent, undergoing continual variations in form and position in the course of the last million years. These variations obey, on the whole, a balance from one side to the other of the equator" (p. 34).

Certain authors, to the contrary, seemed, up to now, disposed to accept the possibility of a paleoclimatic history common not only to the entire Sahara but to all of Africa: "The Pluvial-Arid oscillations in Africa are not, therefore, local variations, and can constitute, along with the prehistoric industries which date them, a valid chronological base for the continent" (H. Alimen and J. Chavaillon 1959, p. 2896).

It might be helpful to recall here a subject which has not always received the attention it deserves: the role of fluctuations or *Klimaschwankungen* (as distinct from changes of climate properly called *Klimaänderungen*) in the distribution of species and the changing of living conditions.

In fact, since the end of the last subpluvial, the Neolithic, we have been in

17. A function both of the basic zonality implied by the astronomic theory, and of various climatic functions superimposed on it, e.g., continentality, ocean currents and "upwellings," relief, etc.

an arid climate, no different from the present except for relatively slight fluctuations. The "pluvial" character of the Neolithic, and therefore the nature of the "threshold" which separated it from the subsequent dry period (Postpluvials III-IV of K. W. Butzer) should certainly not be exaggerated. The existence of a large number of Sahelo-Saharan water-habitats containing fish and sometimes crocodiles prevents us from admitting at any rate since the Neolithic, a deterioration grave enough to have obliterated these aquatic biotopes. Although a considerable number have disappeared, strewn today with many fish vertebrae and crocodile scales, others have persisted to the present—even if we cannot usually deduce the reason for their perpetuation.[18]

That the ecological equilibrium of these fauna is precarious is evident. The permanence of the habitat is fragile. Today, each flood, in carrying off great numbers of specimens, threatens their extinction. The pools are at times so small as to seem on the verge of drying up, and as if with the next unfavorable fluctuation they would disappear.

In the present state of the desert, a few millimeters of rain, whether more or less than normal, if accompanying a trend of either dry or wet years, are decisive. One has the impression that a slight increase in rainfall would return the giraffe to Adrar or Tibesti, make possible again the raising of cattle in the Saharan zone, the repopulation of the Gilf Kebir, the Gebel Aouenat, and even certain of those desolate and uninhabitable expanses of the western Sahara where admirable pasturage of *had (Cornulaca monacantha)* dots the red sands in compact blue-green clumps.

Even today, however, have not elephants reached Mauritania (17° 15′ N.) (G. Grandidier 1932, G. Duchemin 1949) or tortoises (*Testudo sulcata* Gmelin) the Sudanese Sahara (20° N.)? Did not L. E. Almasy (1942, p. 131) discover in the Gilf Kebir the cadaver, only a few years old, of a . . . cow? Certainly thousands of years have passsed since the lakes of the Azaouad with their crocodiles and Nile perch disappeared. But if the Neolithic fisherman appeared early, herdsmen with their cattle seem elsewhere only to have disappeared recently, or near to returning. The tradition appears in the Libyan Desert where H. Rhotert judiciously noted (1952, p. 118) that "solche mageren Weiden, wie sie in Kargur Talh (Aouenat) und das Wadi Abd el Melik [Gilf Kebir] heute sind, besucht man nur wenn dies ein alter Brauch, aus früheren, noch fruchtbareren Tagen ist."

18. We simply do not know when aquatic species arrived where we find them, whether Neolithic or modern. It is doubtful that Neolithic hydrology would have allowed the distribution of this fauna by direct contact between basins of former origin and establishment. In other words, it is difficult to see in the Neolithic a crocodile traveling from Niger to the central Sahara, a silurid fish from Niger to Biskra or from Senegal to Atar, a water tortoise (*Clemmys leprosa*) from Morocco to Mauritania. Perhaps this distribution is more ancient, dating, for example, from the Middle Pleistocene, or, even earlier, from the Pliocene lakes. Could the Interpluvials, like the Postpluvials, however, have been so desertic as to obliterate all vertebrate aquatic fauna? From the situation today, one may doubt this.

K. S. Sandford for his part (1936) insisted on the often very important effects on flora and fauna of slight, or very slight, increases in rainfall. These occasional rains ("small and purely temporary changes in precipitation," p. 201), with nothing in common with Pluvials or Subpluvials, can be very effective and explain the recent existence of *Limicolaria* at Wadi Hawa and even at Mourdi, of *Pila*, *Lanistes*, and *Aspatharia* at Wadi Hawa, etc. These meridional species lived here, at the extreme edge of their area, "a rather ghostly existence of complete disappearance and reappearance at suitable times" (p. 210; cf. also A. J. Arkell 1949, pp. 109–110).

Concerning the Libyan desert, P. Graziosi (1952, p. 111) wisely insisted on the fact that the presence of great African fauna in the heart of the desert during the Neolithic Subpluvial does not imply major climatic changes: a slight increase in the rainfall, "even restricted to mountainous zones," could suffice to sustain an adequate vegetation in the great wadis. The countryside in general must have been a steppe, offering increased possibilities of contact between those of the south and north.

Correspondingly, even a slight diminishing of the rains could have "disastrous and far-reaching effects upon vegetation and animal life and would thus lead to large scale migrations of human groups" (S. Huzayyin 1956, p. 311), or at least to retreats and concentrations.

The image of a sponge used many times by Huzayyin (*ibid.*, pp. 311, 320) is excellent: "During the whole of the Pleistocene times, the Saharan belt was like a big sponge, which during phases of rainfall attracted groups of migrating people both from Eurasia and from Sudanese and eastern Africa. During phases of aridity the big sponge squeezed its population out in both directions" (p. 320).

One must not forget, however, that a climatic worsening could also provoke a segregating and partitioning of groups and cultures subject to an ecological strain of this type, rather than initiating a migratory process; S. A. Huzayyin noted it himself (1941, p. 332) for the desiccation of the "Late Upper and Final Paleolithic times."

We will return later to the essay by K. W. Butzer (1958) on Neolithic isohyetes in the eastern Sahara (Fig. 11) obtained in correlation with the ecological exigencies of some large mammals. The maps given by R. Mauny (1955, 1956) illustrate the evolution of the distribution of the hippopotamus, the elephant, the giraffe, and the rhinoceros from the end of the Paleolithic (Figs. 9–10) and show, at the same time, the great former extension of these species in the Sahara and, for the first three, the relatively slight distance from the first ancient sites (rock pictures or bone remains) and the northern limits of the present area. A slight increase in the rainfall would certainly suffice to extend this limit considerably in the direction of the desert.

We must not forget, moreover, the role Man possibly played, directly or indirectly, in the destruction of the plant cover, the herding of flocks, the occupation of watering places, etc.

Here we must deplore the poverty of research on Saharan Pleistocene paleontology, in particular concerning the vertebrates. According to R. Mauny the number of sites furnishing bone deposits was, in 1955, 12 for hippopotamus, 6 for the elephant (Fig. 9), and 1 for the rhinoceros and the giraffe (Fig. 10). The work of C. Arambourg (1952), which is limited to the Sahara of North Africa, cites only the *Elephas recki* of Tihodaine.[19]

III. REGIONAL FACTS

This section of the report is devoted to the examination of known facts, or of proposed hypotheses, for certain distinct geographical regions. Considering the difficulties, pointed out above, in the elaboration of distant correlations, particular importance will be attached to the establishment of local sequences. It should not be forgotten in this chapter that the outlined program implies the consideration, in space, of the border regions of the southern Sahara, and, in time, of the recent Tertiary. Notice that this last seems only represented in the northwest with the various types of hamada, and in the south with the "Continental terminal"; it is always almost azoique, and in any case without vertebrates.[19a]

Our review will go from west to east, through a series of regions for which there is no evidence to indicate they went through an identical evolution, and for which, understandably, the chronologies—most often the broad outlines of chronologies—are not directly comparable. It could be, moreover, that they are never comparable except roughly, the noted differences exceeding those which could be ascribed to errors in observation or differences in interpretation. Would it, then, be surprising to discover different histories for a valley (Saoura) fed by the southern overflow of the Atlas Range and rich in detrital alluvia; for a lake (Tihodaine) situated in a central massif and dependent on local precipitation; and finally, for a lake in a plain (Guir-Araouan) probably situated in the basin of a proto-Niger with distant sources?

A. The Northwestern Sahara

The northwestern Sahara is especially interesting in several ways: first, because the important work of Henriette Alimen make it henceforth a *locus classicus* for the study of the Saharan Quaternary; second, because the area touches

19. The indicated chronology was at that time as follows: Villafranchian not attested in the Sahara; Middle Pleistocene (Mindel to Riss & Riss-Würm), "Chado-Zambesian" fauna; Late Pleistocene (Würm), "last great Pluvial" with a "hot, moist" climate (when, Balout says [1955, p. 119], "conditions were fresher, more humid, and more forested").

19a. We are aware of the lengthy debate concerning the lower limit of the Pleistocene and the abundant literature it has prompted (cf., for example, H. B. S. Cooke 1950, F. E. Zeuner 1952, G. Choubert 1950, etc.). Choubert noted (p. 85) that if the decision of the 1948 Congress of London (Villafranchian = early Pleistocene) is not accepted, and if the marine fauna of Morocco are relied upon more than the mammals "a good part of the Villafranchian of northern Africa would laterally correspond to the Pliocene."

Morocco, and in recent years the study of the Quaternary has been actively pushed in that country, where the great problem of correlations between marine and continental formations is particularly insistent.

We must return for a while to the question of the end-Tertiary. The dates in this case still disputed, in spite of numerous recent works, and the fauna is as rare as it is deceiving (molluscs).

The northwest Sahara, in its vast monotonous plains, abounds in a series of beds of varying ages, but generally horizontal and still more often, unfortunately, azoique: the hamadas. According to F. Joly (1952), three groups can be distinguished, of which two—the early hamadas (cretacean) and the intercalary hamadas (Oligocene, with *Clavator* and *Limicolaria;* cf. P. Jodot 1951, 1956)—do not concern us here. The third group, the upper hamadas (e.g., the "Hamada du Dra"), comprises sites topped with a limestone slab (which can, especially toward the south, exist alone) with frequent silifications. These beds appear to be of lacustrine origin (fresh water mollusks, gyrogonites of Characeae).[20] G. Choubert (1946, p. 737) speaks of a Pliocene slab of limestone corresponding to a lake with a surface area comparable to that of the Caspian, roughly 2000 by 250 km. Fossile here are very rare: *Limnaea, Planorbis, Physa, Melania*[21] (P. Jodot and R. Lavocat 1954). The age of this formation remains doubtful. It is probably Pliocene rather than Villafranchian, according to R. Lavocat (1954), or "Plio-Villafranchian" (F. Joly 1954).[22] In the Saoura, H. Alimen thought she could attribute the upper Hamadian slab—by continuity and in spite of the apparent absence of fossils—to the Early Pliocene, probably to the Plaisancian, while Jodot suggests (1955, p. 68), "Early Pliocene, probably toward the top of the middle zone." The Villafranchian of the Saoura rests on the Hamadian slab which must be Pliocene; consequently, so must be all the identical beds of the northwestern and western Sahara. Only the discovery of mammalian fauna would solve the question once and for all.

In the South Oranais, the limestone cap of the "Terrain des Gour," which A. Cornet (1952, p. 34) prefers to call "Continental Tertiary" rather than "Mio-Pliocene," could be "Pliocene or early Quaternary" (*ibid.*, p. 35). The existence of lacustrine Villafranchian limestones must not be excluded; we will see that it has already been noted in the Sudanese Sahara (P. Jodot and S. Rouaix 1957).

20. A hypothesis of an origin not sublacustrine but pedological and subaerian has been formulated (M. Auzel and A. Cailleux, 1950).

21. Of the group *tuberculata;* I believe I recognized the *Melanoides tuberculata* in an Hamadian flint (1951); G. Rocci noticed the same species in the Hamada of Hammami, and that of Aftassa (*Bull. Dir. Féd. Mines and Géol. A. O. F.*, No. 21, I (1957), p. 49), along with *Linnaea* and *Planorbis*. But one must not forget how delicate the distinction can be between the Hamada and certain lacustrine levels dating from the Pleistocene.

22. Formerly, Joly (1952) prudently judged the upper hamadas as "Pontico-Pliocene" or "Mio-Villafranchian."

Let us go on to the Pleistocene, the Quaternary about which H. Alimen could say (1958, p. 9): "A study of such a delicate chronology as the Quaternary is difficult anywhere, and even more so in the desert. . . ."

We will begin with the Guir-Saoura basin, with the mountains of Ougarta, regions where the Pleistocene has been the object of numerous publications, e.g., H. Alimen 1952, 1954, 1955a, 1955b, 1956a, 1956b, 1957a, 1957b, 1957c, 1958, 1960; H. Alimen and J. Chavaillon 1959, 1962; H. Alimen, J. Chavaillon, and

TABLE 2

CHRONOLOGY AND CORRELATIONS PROPOSED BY H. ALIMEN, J. CHAVAILLON, AND J. MARGAT FOR MOROCCO, TAFILELT, AND GUIR-SAOURA (1959)

MAROC		TAFILALT (Ziz-Rheris)		GUIR – SAOURA – OUGARTA			AFRIQUE ORIENT^LE^ (corrélation hypothétique)	
Classification marocaine	Industries en place au Maroc	Indices	Niveaux et industries de surface	Classification saharienne	Indices	Industries en place	Classification	Industries en place
Actuel et Sub-actuel		a et a′		Actuel	Q^{2b}			Fer Njoroen Gumbien
Rharbien	Néolithique	a″	Néolithique et ibéro-maurusien	Guirien	Q^{2a}	Néolithique	Makalien	Wiltonien (−5.000)
				Aride post-Saourien	Q^{1b-c}	Néolithique		Elmenteitien
Soltanien ou Grimaldien	Atérien	q_5	Néolithique Ibéro-maurusien et Atérien final	Saourien (3e Pluvial)	Q^{1a}	Atérien	Gamblien (3e Pluvial)	Stillbayen Capsien (−11.000) Moustérien
				Aride post-Ougartien		Acheuléen final	Aride post-Kamasien	Fauresmithien
Tensiftien	Micoquien Acheuléen évolué	q_4	Néolithique à Moustéro-Atérien	Ougartien (2° Pluvial): Ougartien II (couches V de Kerzaz)	Q'_b	Acheuléen sup. " moy. " inf.	Kamasien (2e Pluvial) à 2 phases	Acheuléen Chelléen Oldowayen
Amirien	Acheuléen archaïque "clacto-abbevillien" de Casablanca	q_3	Calcaires lacustres de Salah et conglomérat de Sigilmassa Atérien prédominant	Ougartien (2° Pluvial): Ougartien I (couches de Taourirt)	Q'_b	Pebble-culture très évoluée confinant au Clacto-Chelléen	Kamasien (2e Pluvial) à 2 phases	Acheuléen Chelléen Oldowayen
				Aride post-Mazzérien		Kafouen évolué	Aride post-Kaguerien	
Salétien	Chelléen Pebble-culture	q_2	Atérien prédominant	Mazzérien (1er Pluvial)	Q'_a	Kafouen récent	Kaguérien (1er Pluvial) à 2 phases	Kafouen { évolué récent ancien }; Kafouen primitif
				Aride post-Villafranchien		Kafouen ancien	Kaguérien (1er Pluvial) à 2 phases	Kafouen { évolué récent ancien }; Kafouen primitif
Moulouyen	Pebble-culture	q_1	Conglomérat du J. Erfoud	Pluvial Villafranchien	Vil.	Kafouen primitif	Kaguérien (1er Pluvial) à 2 phases	Kafouen { évolué récent ancien }; Kafouen primitif

ésenté au Congrès Préhistorique de France-Monaco août 1959 — H. ALIMEN, J. CHAVAILLON, J. MARGAT

J. Margat, 1959; H. Alimen, J. Chavaillon, and G. Conrad, 1959; J. Chavaillon, 1956a, 1956b, 1958, 1960; J. and N. Chavaillon, 1957a, 1957b.

The work by Alimen, Chavaillon, and Margat (1959): "Contribution à la chronologie préhistorique africaine. Essai de corrélation entre les dépôts quaternaires du bassin Guir-Saoura (Sahara) et bassin du Tafilalt (Maroc)" is the best report of results so far obtained, and furnishes the latest word on a question of chronology which is continually improved in successive publications. One could complete it, first, by the schema of climatic cycles presented by H. Alimen (1958, Fig. 89) (Fig. 2), previous to her adoption (in collaboration with J. Chavaillon and G. Conrad) of the terms Mazzerian, Ougartian, Saourian, and

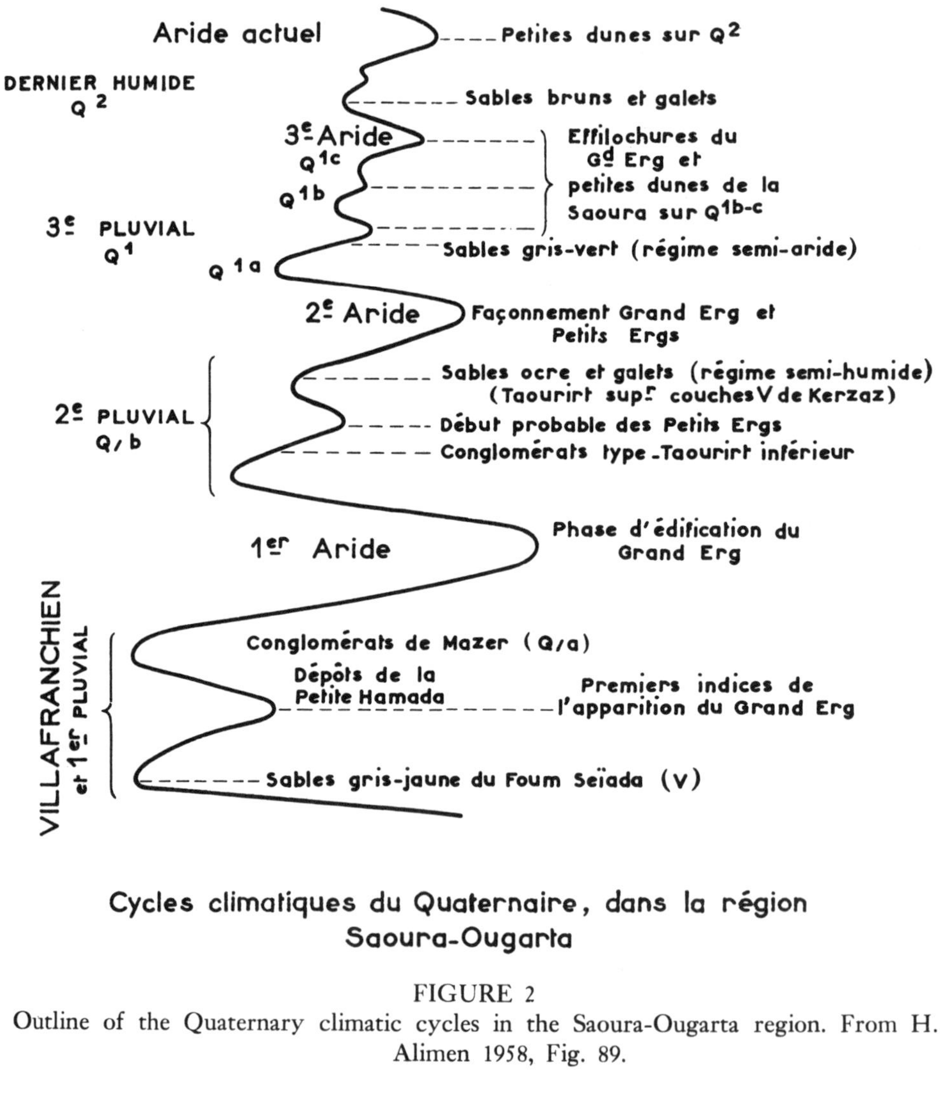

FIGURE 2
Outline of the Quaternary climatic cycles in the Saoura-Ougarta region. From H. Alimen 1958, Fig. 89.

Guirian, which replace the graphic notations used until then;[23] secondly by the accuracy brought to the interpretation of the second by J. Chavaillon (1960).

Here is the present stratigraphy adopted (Figs. 3–4)[24] from the *Essai* of H. Alimen, J. Chavaillon, and J. Margat (1959):

1. *Villafranchian Pluvial* [or "Aïdian," J. Chavaillon 1961]. Vast alluvial sheets (sands, gravel, conglomerates) of red or pink color, resting directly on the Hamadian Pliocene (or Paleozoic).
2. *Arid post-Villafranchian.* Deposits of the same color (talus, brecchias, sandy-loess strata, sandstones), topped with a brownish-red fossil soil. The pluvial-arid cycle of the Villafranchian contains the most ancient pebble tools: earliest Kafuan.
3. *First Mazzerian Pluvial.* Conglomerates and sands—early Kafuan, then later Kafuan.[25]
4. *Post-Mazzerian dry phase.* Sandy clay deposits, aeolian sands, screes, fossilized by a reddish-brown fossil soil or a "meulièriforme" sandstone—Developed Kafuan.
5. *Second Taourirtian Pluvial* [ex-Ougartian I]. Conglomerates—very evolved "pebble culture approaching Clacto-Chellean" (*Essai,* p. 11) or perhaps belonging to the middle Acheulian (J. Chavaillon 1960, p. 183).
6. *Post-Taourirtian dry period*—Erosion.
7. *Third Pluvial* [or Second Pluvian, second phase]. Ougartian s. str. [ex-Ougartian II]. Varicolored pebbles and ochre sand on a brown-red fossil soil—middle to upper Acheulian (*Essai,* p. 11 says "lower to upper").

J. Chavaillon's note (1960) distinguishing a Taourirtian from an Ougartian posed an important problem: Is it a matter of two pluvials (2nd and 3rd) separated by an interpluvial, whose deposits would be (*Essai,* p. 11) "nonexistent or thin," or of a single Pluvial (the 2nd) with 2 maxima?

8. *Post-Ougartian dry period.* Aeolian sands (under sandstone crusts or white sandy clays) or talus—final Acheulian.
9. *Fourth Saourian Pluvial* [ex-3rd]. Gray-green sands, rubble, Limnaeas, black fossil soils, etc.—Aterian.
10. *Post-Saourian dry phase.* Sandstone crust—talus, etc. Neolithic.
11. *Guirian wet episode.* "Small humid recurrence, marking the transition from the post-Saourian dry period to the present-day period" (*Essai,* p. 12)—Neolithic.

23. Having experienced certain difficulties myself in using these, it may be helpful to show the following correspondences:

Mazzerian	= Q/a (or Q.a)
Ougartian	= Q/b (or Q.b)
Saourian	= Q^{1} (subdivided: $Q^{1a} + Q^{1b-c}$
Guirian	= Q^{2a}
Modern	= Q^{2b}

24. These authors seem to have applied it to a very extended area since they termed it "Saharan" and made it parallel, term by term, with the East African sequence.

25. The text (*Essai,* p. 10) says: Villafranchian and post-Villafranchian arid phase/earliest Kafuan and Mazzerian/early plus later Kafuan, and the attached table: Villafranchian and arid post-Villafranchian/primitive Kafuan plus ancient Mazzerian/later Kafuan. In reality the later Kafuan should originate (H. Alimen and J. Chavaillon 1959, p. 2894) at the same time from the beginning of the Mazzerian and from the post-Villafranchian dry phase. It is useful to note that the number of pieces discovered was small up to 1959; a dozen for primitive Kafuan, 19 for early Kafuan, 27 for later Kafuan, and 2 for evolved Kafuan; in 1962 "environ 110 pièces" were known.

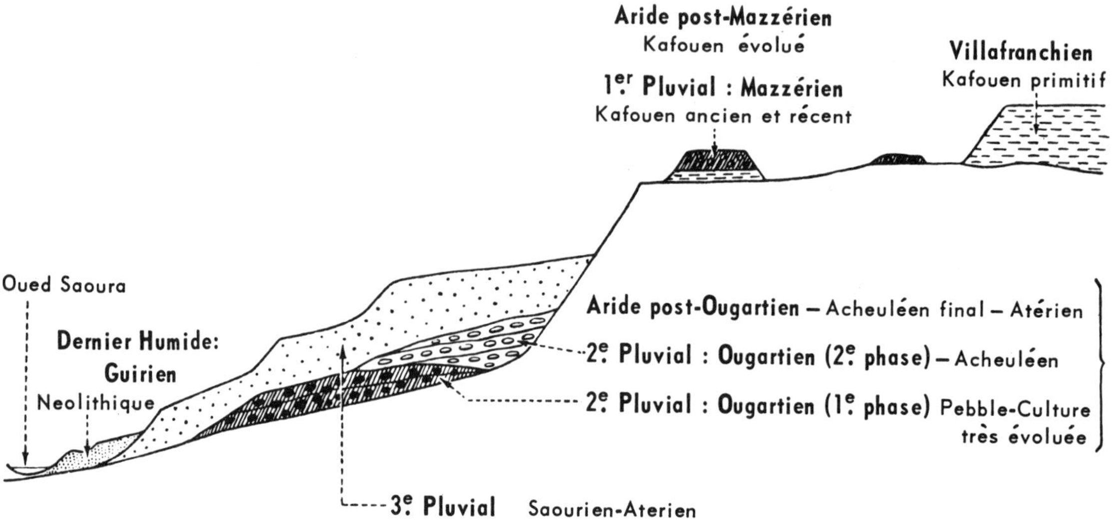

FIGURE 3
Outline of the Quaternary stratigraphy in the Saoura-Ougarta zone. From H. Alimen and J. Chavaillon 1959, Fig. 1.

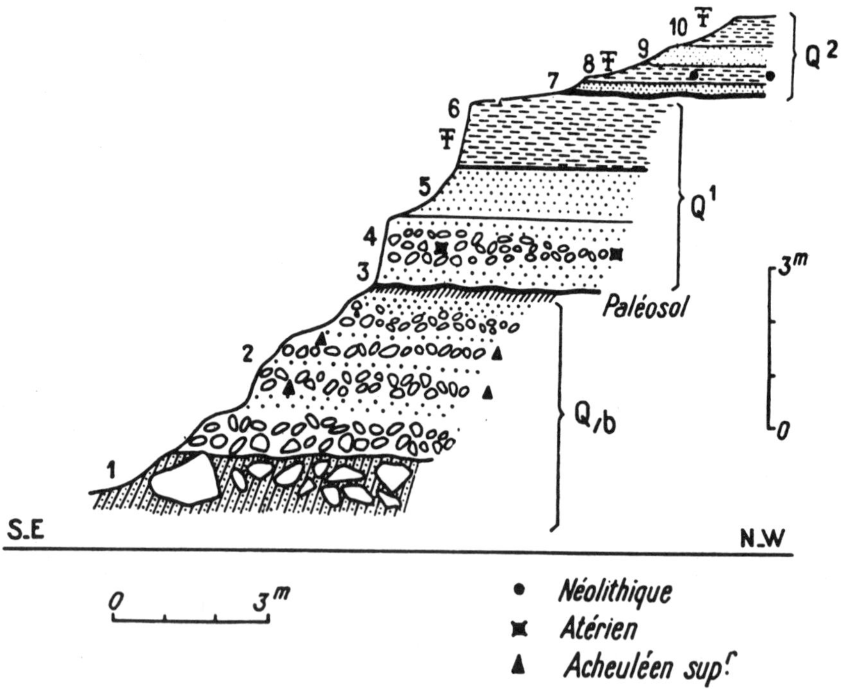

FIGURE 4
Outline of the Quaternary stratigraphy in the Ougarta Mountains (confluence of the Wadi Lacba and the Khenig et-Tlaïa). From J. and N. Chavaillon 1957b, Fig. 2; cf. also *idem* 1957a p. 3.

This is the Saouran sequence, henceforth locally considered classic. The Wadi Guir presents a very comparable succession, "but is characterized by the considerable development of Mazzerian conglomerate alluvium, by the extension of sandy formations of the last Guirian Humid, and by several morphological traits, being somewhat independent from the Guir basin in proportion to that of Saoura, and of a less markedly Saharan character" (*Essai*, p. 15).

The *Essai*, while recognizing the absence of precise paleontological and archeological criteria (no mammals, industries *in situ* only in Guir-Saouran basin), and the necessity of resorting to morphological and lithological arguments, proposes a comparison with continental Morocco where a sequence of five pluvials has been identified (+ the Neolithic Subpluvial) (Moulouyian, Saletian, Amirian, Tensiftian, Soltanian (Grimaldian), Rharbian), and in particular with the Ziz-Rheris (Tafilalt) basin. Despite the absence of industries *in situ* in the latter area, the proposed correspondences appear, on the whole, satisfactory (*Essai*, pp. 6–8, 16–18). It is possible, then, to speak of correlations rather clearly established for the northwest Sahara and of a "common Pleistocene evolution" (*Essai*, p. 19). It could be tempting to equate the various terms of the Guir-Sahara sequence with that of the European glaciations on the one hand and the Pluvial-Interpluvials of East Africa on the other. It is not certain we could safely try this, and the creation of local stratigraphical vocabularies would be wise. The Saourian is perhaps the Northwest Sahara equivalent of the Kenya Gamblian, but we hesitate to accept this as definitely established. Arambourg (1962, p. 262), however, adopts the following correlations:

Mazzerian	Kaguerian (Olduvai I)	Pebble Culture
Ougartian I	Kamasian (Olduvai II)	Early Acheulian
Ougartian II	Kanjerian (Olduvai IV).	Acheulian
Saourian	Gamblian	Moustero-Aterian
Last humid stages	Post-Gamblian humid stages	Neolithic

We have seen above that in Morocco the continental Quaternary had its regional terminology. This has been the object of many studies dealing both with littoral deposits (which do not concern us here) and with continental deposits. An essential point is evidently the correlation of the marine and continental cycles (G. Choubert 1953, 1956, 1957, etc; M. Gigout 1957; M. Gigout and R. Raynal 1957, etc.).[26]

In 1956, G. Choubert, F. Joly, M. Gigout, J. Marçais, J. Margat, and R. Raynal published an *Essai de classification du Quaternaire du Maroc*" [27] which proposed the terms Moulouyian (Villafranchian), Saletian (Günz?), Amirian (Mindel), Tensiftian (Riss), Soltanian or Grimaldian (Würm), and Rharbian (post-Pluvial). The authors see an alternation of the marine transgressions and the

26. Cf. also G. Choubert 1955, etc.

27. Completed by Gigout and Raynal 1957, and Choubert 1957.

"pluvial" continental formations. The pluvial terraces, of climatic nature, correspond to the regressions. Thus erosion occurred during the transgression and alluviation during the regression (M. Gigout and R. Raynal 1957). G. Choubert remarked as early as 1953 (p. 2) that "the filling-in of the wadis should have taken place during the Pluvials and could be explained by the greatness of the "charge" (load) transported. On the other hand, erosion should occur during the Interpluvials (cf. Choubert 1956, pp. 582-85). It goes without saying that this description, and the paleoclimatic interpretation of the Moroccan Pleistocene, including the continental and even the pre-Saharan, could only be applied *grossissimo modo* to remote Saharan areas and to less and less a degree as the distance increases. May one hope to find valid parallelisms between the piedmont Pleistocene where there are abundant rubble and conglomerates, and that of the great flat basins where a very thin Pleistocene (a few dozen meters at the most) is entirely sandy or calcareo-orgilaceous? Very detailed work on the problems concerning the Pleistocene of Atlantic Morocco can be found in the recent works of P. Biberson (1961a, 1961b, 1963).

B. The Mauritano-Senegalese Sahel

There are few things to be said about the late Tertiary: "no fossils, neither microfauna or macrofauna, have been found in the Terminal Continental." (P. Elouard 1959, Fig. 188); its age is therefore unknown: "it is post-Lutetian and pre-Tyrrhenian" (*ibid.*, F. 189).[28] Geologists considered it Pliocene, mid-Pliocene, or Oligo-Pliocene, but admittedly without solid proofs. It could perfectly well have begun in various ages, according to the region.

The lower Senegal valley and the adjacent regions of Mauritania (e.g., Trarza) have been closely studied in the last few years. In spite of the difficulties in interpretation in the stratigraphical study of a flat and lithologically monotonous area (mostly sand and clay), useful attempts at Pleistocene chronology have been proposed for several different regions.

For the delta and the lower Senegal valley (J. Dubois and J. Tricart 1954; J. Tricart 1955a, 1955b, etc.; P. Michel 1960) one finds:

A. Pliocene substratum: Continental Terminal; subarid

B. Lower Pleistocene
- (1) Lateritic carapace; humid
- (2) Polygenetic glacis with lateritic debris; lacustrine limestone; alternating: humid-subarid

C. Upper Pleistocene
- (3) Gravel; subarid
- (4) Red dune, very arid; pre-Ouljian regressions
- (5) Beach with *Arca* (+4 to 6m), humid; transgression
- (6) Yellow dunes, arid; pre-Flandrian regression

28. Why not rather pre-Villafranchian if the basal Pleistocene really belongs to this stage?

(7) Dunkirkian Delta, Flandrian transgression; climate a bit more humid than now
(8) Premodern to present

It should be noted that arid periods correspond to regressions, thus to glaciations; humid periods to transgressions, thus to interglacials.[29] J. Tricart (1955, p. 14) saw there the proof of an alternation of climatic oscillations between one part of the Sahara and another "since the North African pluvials are contemporary with glacial periods and the Senegalese pluvials, on the contrary, with interglacials."

In an area a little farther north (Lake Rkiz and Aftout de Boutilimit), P. Elouard and P. Michel (1958) found a very comparable series (the figures correspond to the Senegalese series above);

(1) Lateritic carapace
(2) Fans of gravel or lateritic conglomerates
(2b) Lacustrine limestones—humid
(3) Fans of gravel
(4) Red dune
(5) Terrace, *Arca senilis*—Ouljian
(6) Lacustrine limestone of the "Goud"-Flandrian

In 1959, P. Elouard proposed a regional terminology which is valid for Lower Mauritania and the lower Senegal valley. The future will show how far from its point of origin the terminology can be utilized. If the situation of the Guir-Saoura basin conferred upon the latter certain characteristics, the Atlantic Mauritano-Senegalese coast could also present its peculiar traits, in particular the proximity of the ocean, with the participation of eustatic cycles of transgression and regression. There is nothing to prove that the central zones of the west Sahara had an identical history.

P. Elouard's sequence follows. Upon a clayey-sandstone Terminal Continental corresponding to "sheet-floods spreading elements reworked from the continent submitted to an important erosion" in an arid climate (P. Elouard, 1959, p. 190 and Fig. 31), there were:

A. Lower Pleistocene
(1) Carapace; sudanian climate
(2a) Sands and lateritic gravels; humid } Ajouerian
(2b) Lateritic conglomerate (rocks "à raisins secs"); contrasting seasons; } Ajouerian
(3) Calcareo-sandstones and shelly limestones; Inchirian (= Tyrrhenian I); humid.
B. Upper Pleistocene
(4) Red dune; Ogolian; very arid
(5) Beach with *Arca* (+4 to 6m.); Ouljian; humid; lagoonal transgression
(6) Yellow dunes; Aftoutian; arid; pre-Flandrian regression

29. J. Tricart and M. Brochu (1955) correlated 4 with Riss, 5 with Riss-Würm, and 6 with Würm.

(7) Recent beach and limestone of "Goud";[30] Flandrian (Dunkirkian); humid; transgression

(8) Premodern littoral chain; Sbarian

The sequence may vary from region to region and P. Elouard admits four types for South Mauritania: Aftout es-Saheli, Tafolli-Inchiri, Aftout ech-Chergui, Senegal Valley, and Trarza (1955, *passim* and Fig. 31).

It is difficult to make a useful comparison of this sequence, which has few fossils (no vertebrates) and in which prehistory played no role, with that of the northwest Sahara.[31] P. Elouard has attempted a comparison of his south Mauritanian sequence with those of Morocco, the Spanish Sahara, north Mauritania, and Senegal (1959, pp. 249–52). A hybrid scale employing side by side terms relating to glacio-marine eustatism (e.g., Ouljian, Flandrian) and designating continental levels (e.g., Ogolian, Aftoutian) is *a priori* a delicate enterprise for useful comparison of the areas belonging to one or the other of these categories. Besides, if the Pluvials and Interpluvials of north and south Sahara are not similar in relation to marine transgressions and regressions, one can hardly see how a valid chart of correlation could be presented. I, myself, tried to do this for a long time and finally admitted defeat.

Even in a hydrographical basin with marine base-level (Senegal in the area of Kayes, Kolimbine, and Karakoro), one must recognize that there is a slight time lag between the curve of the variations of sea level and that of climatic oscillations (P. Michel 1960, p. 22).

One last remark: in stratigraphies, one tries to give importance to the Villafranchian, to which the ferruginous carapace is at times attributed. Do we consider seriously enough the length of that period, of which F. Clark Howell recently wrote, "Half a million years is perhaps a modest estimate" (1959, p. 831). Should we not abandon the practice, in tables, of giving equal compartments to the Villafranchian and a postglacial Subpluvial? Since 500,000 years on one side and 3,000 on the other are graphically and thus visually given as identical, we may lose a sense of proportion.

C. The Mauritano-Sudanese Sahara and Sahel

1. Atlantic Deposits

The Pleistocene of the coast between Cape Timris and Cape Blanc has been the subject of several studies (e.g., Th. Monod 1945, A. Blanchot 1952, 1957).

30. The "Turritelles" of the Limestone of "Goud" (P. Elouard 1959, p. 230) are evidently *Melanoides tuberculata*, a fresh-water species.

31. For the Pleistocene of the Cape Verde peninsula and the paleoclimatic conclusions to be drawn from it, see especially F. Tessier 1954, and 1959.

The problems should be reconsidered by a Pleistocene specialist.[32] In the present modest state of our knowledge, we can probably accept, at least provisionally, the following sequence:

1 Continental Terminal—streaked sandstone of the Tirersioume, etc.
2 Ferruginous crusts and carapace
3 Group of marine calcareo-sandstones and shelly-limestones (El Aioudj, Sbeyat) plus calcareo-sandstone bearing *Helix* and fossil dune with *Helix* (Aguerguer)
4 Deposition of major dunes (Akchar-Azeffal)
5 Beach with *Arca*
6 Present beaches, coastal chain, muddy alluvium, etc.

For comparison with the Mauritanian south, one may suggest these equivalences: 2, Villafranchian; 3, Inchirian;[33] 4, Ogolian; 5, Ouljian; 6, Flandrian and present.

Lacking mammals and industries *in situ*, all this remains very hypothetical.

The Inchirian calcareo-standstone poses a great problem. I had thought it identical with the hamadian slab, typical more of the east and northeast (Th. Monod 1945, p. 31). But this "Inchirian" could not be Tertiary, and it seems difficult to "age" it. Should one, on the contrary, "rejuvenate" the Hamada (or at least its terminal slab)? This is an eventuality to keep in sight. These formations unfortunately have not yielded any useful fossils.[34] It could be that it is a matter of identical and recurring facies, of differing ages according to region, Pliocene (Lower) at Dra (P. Jodot 1955), Pleistocene farther south.

The Pleistocene of the Spanish Sahara has been the object of recent studies (M. Alia Medina 1949 and 1955) whose results are difficult to interpret because they depend in part on works which are themselves, in some respects, outmoded (e.g., Th. Monod 1945).

2. ADRAR, MAJÂBAT, ETC.

In this immense area, which stretches from the meridian of Atar to the Adrar of Iforas and from Hank to the Sahel, the only sign of the dated Tertiary seems to be the discovery in Aregchach (Erg Chech), at 23°15′N and 7°30′W, of a shelly limestone containing *Limicolaria kem-kemensis* Jodot. Its age is presumed to be Aquitanian and would indicate (this does not seem evident to me at first) a "wooded savanna with Sudanese flora, bordering the subtropical rain forest" (P. Jodot and S. Rouaix 1957, p. 375).

32. The work now in progress of G. Lecointre, who has been able to visit the southernmost part of Morocco and Villa Cisneros, will certainly provide valuable new information.

33. Here, it is actually only a question of equivalence: for the Inchirian it is more the south which is aligned with the north.

34. Ostracodes have been cited on the coast (R. Dars 1957), aside from the Characea which I recognized there.

On the other hand, Hamadian veneers abound, especially in the north, with the calcareous, lacustrine matrices stuffed with wind-shaped grains of their calcareo-sandstones. Their age is unknown (Th. Monod 1958, pp. 117–18); 50 km. northeast of Tajnout Haggeret an interesting small fauna has been found in a lacustrine limestone (probably Hamadian) (P. Jodot and S. Rouaix 1957) with a Villafranchian species (*Hydrobia acerosa* Bourgt.) and three Pliocene (*Melania* (*Brotia*) *verrii* de Stef. var. *acuta* Foresti, *Hydrobia* cf. *iberica* Jodot and *Cardium* (*Cerastoderma*) *glaucum* Brug. var. *crassa* Defr. The authors considered the fauna Villafranchian.

As early as 1935 (*ibid.*, p. 296–7) I noted the possibility of distinguishing, in the western Sahara, in the extension of Pleistocene lakes, "two maxima of diminishing amplitude, separated by a severe drying out phase in the course of which the major system of dunes could have been installed."

In 1953 (p. 1233), J. Richard-Molard confirmed the existence of two lacustrine deposits in the Adrar: "the wet periods transformed the tayerets (of the Maqteir) into sheets of fresh water at least twice, as is proved by the calcareous crusts which today occupy the depressions, superposed. . . ." One of the levels would be Paleolithic (Lower), the other Neolithic. The presence of these two lacustrine levels has since been confirmed (F. Joly in Th. Monod 1956, p. 650; Th. Monod 1958, etc.).

Resuming with R. Mauny in 1958 the study of the great Acheulian site around El Beyyed (Mauny 1962), we found a certain number of hand axes *in situ,* sealed in a white calcareous level of lacustrine origin surmounted with red gravel of indeterminate age (Monod 1962). By comparison with the Pleistocene of the Majâbat, it seems that this evolved Acheulian, where the wooden technique predominated, is found in GB1, that is, in the first of the two lacustrine beds. In 1963 a number of new sites with Acheulian implements *in situ* have been discovered by P. Biberson, mostly in the Aghmakou-Tazazmout-El Beyyed region.

The study of the Pleistocene of the Majâbat and adjacent regions presents difficulties which could explain uncertainty of the results obtained, or more correctly, glimpsed. In effect, it is a matter of thin, discontinuous outcroppings, appearing separately at random in slits in the sandy covering of reduced thickness, monotonous lithology, and without "useful" fauna, and where the connection of deposits to industry is rarely determinable.

In my work of 1958 (pp. 118–139 and 165–8; cf. Fig. 11) can be found a report on the Pleistocene of the Majâbat. The observed sequence appears to be the following: [35]

1. SA: varicolored sands and clays of the Terminal Continental; only in the southeast.
2. SB: extended sheet of white sand, alluvial.

35. Explanation of abbreviations: SA, clayey sands; SB, white sands; GB, grey-blue; CL, milk-chocolate; TH, *trâb hamrâ* (red land); TB, *trâb beydhâ* (white land).

3. GBI: first lacustrine clayey level (calcaro-clayey–sandy).
4. Period of dune edification, emanating from SA material, making up the major systems which are presently red (TH).
5. CL: clayey brown sand.
6. GB2: second clayey lacustrine level (partly diatomites).
7. TB: presently existing white shifting dunes.

The fauna and flora of GB1 and 2 (Th. Monod 1958, pp. 130–9), though still abundant, do not furnish elements of chronological discrimination. They denote, on the contrary, at any rate, the presence in GB1 of extremely extended sheets of fresh water.[36] A sampling of diatomite from north of the Akle Aouana is judged by E. Manguin (in Th. Monod (1961, pp. 615–9) "abounding in euplanctonic diatoms, accompanied by a certain number of current species of plankton from the great African lakes."

Even if the observed series clearly implies a number of climatic oscillations and several wet periods, it is still impossible to extract from it a coherent picture capable of comparison with other Saharan or Sahelian sequences. Even the number of wet periods to distinguish remains doubtful, since one does not know the age of the SA (Villafranchian or Pliocene?), the validity of the combination SB plus GB1 (a pluvial with transition from a pluvio-lacustrine to a lacustro-palustrian system, or two distinct Pluvials), or the exact significance of CL, interpreted by me as a fossil soil, "a deposit born of the filling-in of depressions by the results of the rain-wash of dunes under the influence of a climate at most semiarid with accentuated seasonal oscillations."

If the wet GB2 phase seems easy to correlate with the Neolithic subpluvial (Guirian, etc.) it is useless, in my opinion, to set up a table, even very hypothetical, of conceivable correlations with the northwestern Sahara or the lower Senegal. Such tables, always very pleasing to the eye, tend to crystallize simple suppositions without emphasizing their largely hypothetical nature.

The deposit-industry correlation here is too uncertain to permit using it in an attempt at correlations. The Neolithic seems bound to GB2, the evolved Acheulian to GB1 (with the possibility of an anterior stage in the SB). There is no indication of the Aterian, which appear only toward the north; a relation with the CL would not be surprising.

We may wonder why the Majâbat sequence seems to be only roughly comparable to the others.[37] Our ignorance is probably affected by the very nature of the deposits—their thinness, the absence of sections, etc.; but there is perhaps something else. Is one part of the series (the base) really lacking, either because it never existed or was later removed? Is the second hypothesis absurd in a

36. I myself spoke of "veritable inland seas" some of which reached several hundred kilometers in diameter (1935, p. 296), in an article in which I assumed the existence of two lacustrine periods (Lower Paleolithic and Neolithic) (pp. 296–7).

37. Which do not seem to be mutually comparable without reference to an extra-African chronology.

region where the sand represents the most coarse sediment? Or indeed should one hold certain elements, the SB for example, as largely comprehensive and able to represent by themselves what constitutes elsewhere a series of distinct levels? One must not forget that towards the southwest, the SB may attain 60 meters (wells sections).

With the Azaouad [38] and its Sahelian edges, we come to a region rich in GB2 deposits akin to a Neolithic industry with harpoons and fishhooks of bone (Th. Monod and R. Mauny 1957) and an abundant aquatic fauna.

H. Radier (1955, p. 7) notes: "In the Paleolithic—very wet monsoon climate, erosion of hydrographic network which is now fossil; in the Würm—very arid climate, breaking up of the hydrographic network, sand invasion, formation of ergs by mobilization of the alluvia left by the great rivers; in the Neolithic—wet monsoon climate, development of a vegetation stabilizing the dune reliefs, fossilization of *ergs;* finally, return of the monsoon to the south." The lacustrine calcareous-clay deposits of interdune valleys are encased in the dune system and, obviously, H. Radier (p. 6) endorses only GB2, Neolithic, without taking pre-dune GB1 into account.[39]

The wetness of the Neolithic Subpluvial should not be exaggerated. I know that fish of 1.5 meters and more (*Lates niloticus*) and hippopotami require definite amounts of water. But here we are on the edge of the Niger basin and we have known for a long time the part that river must have played in maintaining the Neolithic lakes of Azaouad, as it still does for Faguibine and several others. Aerial photography recently showed a very sinuous effluent of the Niger, leaving it at Timbuctu to go straight north, towards Araouan.[40] Perhaps Neolithic lakes depended more on the contribution of rivers than on local rainfall, as does present-day Chad. H. Jacques-Felix has made the pertinent statement (1947, p. 66) that "a few herbivore fossils argue more for a wet paleoclimate than the mass of aquatic fossils which testify only to the presence of water. . . . The residual fauna, today isolated in pools in the middle of the desert, has no more significance for the previous climate than goldfish have for the climate of an apartment." This is true, but if edaphism can explain aberrant distributions, they are still limited and strictly defined. Climate itself is regional—and besides, the Neolithic Azaouad also had herbivores.

For the middle Niger and neighboring regions, Y. Urvoy (1942) has distinguished a series of "dry" and "wet" periods, essentially on morphological bases: H1, S1, H2, S2, H3, and the present ("less wet than the H's and wetter than the S's") which would perhaps correspond—"a risky and entirely provisional

38. At least as it is customarily and conveniently called. A recent article by F. Poussibet explains the true sense of the word (1961).

39. Just as he treats the sandy mass as a unit, without distinguishing the SB underlying the dunes, which is genetically if not morphologically very defendable if, as I believe, the dune is only "constructed" SB.

40. A. Clos-Arceduc 1955, G. Palausi 1955; cf. Y. Urvoy 1942, Figs. 23–24.

indication" (p. 86)—to: Riss (H1), Chellean (S1), Würm (H2), post-Glacial Optimum (S2), post-Glacial Recurrence (H3).

To the east of the Azaouad, the Pleistocene of the Tilemsi valley, bordered on the left by the Precambrian of the Iforas massif and on the right by the Cretaceous Eocene, still awaits a detailed study. Practically nothing is known, and even the age of Asselar Man and the exact conditions of its situation is being questioned. Relevant bibliography can be found in L. Balout in H. V. Vallois and Hallam M. Movius, Jr. 1952, p. 273, and L. Balout 1955, pp. 387–8.

D. The Central Sahara (Hoggar, Tassili, Fezzan)

Arbitrary though it may seem, this grouping seems useful to me.

For this area I will draw attention to three subjects: two sites, Tihodaine and Tejerhi, which have been the object of serious study, and the recent progress of the paleo-palynology of the central Sahara.

The Tihodaine site is important not so much for the presence of the Acheulian *in situ* as for a datable fauna of the middle Pleistocene (L. Joleaud 1936c, etc.; Ch. Devillers and J. M. Peres 1939; Ch. Devillers 1948; C. Arambourg 1948; and C. Arambourg and L. Balout 1955; photos (site and tools) in Anon. (1956).[41]

The Tihodaine basin comprises (Arambourg 1955) two superposed lacustrine levels, the first clayey-sandy, the second clayey (diatomites). Level 2 contained scarcely more than mollusks, level 1 yielded Acheulian with cleavers and a vertebrate fauna: [42] a large non-loxodont elephant "which seems identical" with *Elephas recki* Dietrich [43] (C. Arambourg 1955, p. 283), the white rhinoceros (*Ceratotherium simum*), a zebra (*Equus mauritanicus* Pomel, the amphibious hippopotamus, some large Bovines (*Bos primigenius* and *Homoioceras antiquus*), diverse antelopes, among which a gnu (*Gorgon taurinus prognu* Pomel), an eland (*Taurotragus* sp.) a strepsicere (*S.* aff. *imberbis*), a Bubalis (*Alcelaphus bubalis*), a gazelle,[44] a pig,[45] a dog.[46] It is, for C. Arambourg, the fauna of Kanjerian (ex-upper Kamasian of East Africa.

For C. Arambourg (1955, p. 286) "the end of the filling up of the basin corresponds to a more lacustrine period which appears to have begun at the beginning of the upper Pleistocene"; according to him, then, the upper diatomites would have been contemporaneous with the adjacent Aterian industries. For Ch. Devillers (1948, p. 191) these levels, with mollusks and "the remains of fish

41. The existence of bones at Tihodaine seems to have been known through native information to H. Duveyrier (*The Tuareg of the North*, 1864, p. 85), who called the location Tehodayt-tan-Tamzerdja.

42. Cited according to Arambourg 1955.

43. See, on this species, C. Arambourg, *Mission Scientifique de l'Omo, 1932–33*, Part I, Fasc. III (1947), p. 253 [97]–269 [113] and *passim;* a molar from Tihodaine is illustrated (Fig. 11A).

44. Ch. Devillers (1948, p. 190) adds: *Oryx*.

45. *Ibid.: Phacochoerus.*

46. Ch. Devillers (1948, p. 190) adds: "Dog (2 species); some remains of birds and crocodiles."

(*Clarias*), birds and mammals (of which a *Bos* represented by several massacres)" would be Neolithic.

J. Y. Thébaut (1961, p. 33) gave the following section of the Tihodaine basin, with reference to the Saoura chronology of H. Alimen:

4. Upper diatomites (3rd Pluvial)
3. Upper sands (2nd Arid)
2b. Lower diatomites } (2nd Pluvial)
2a. Clayey green sands } (2nd Pluvial)
1. Lower sands (1st Arid)

Several other sites of the central Sahara could be cited here, in particular those of In Guezzam (*Hippopotamus amphibius*, and *Kobus sigmoidalis* of the lower Pleistocene of East Africa), *Crocodylus niloticus*, tortoises, fishes—Silet (*Thryonomys logani* [Romer and Nesbitt 1930], buffalo, gazelle, ostrich, etc.)—Titerine Crater with possibly 2 molluscan fauna, one of the "middle Pleistocene," the other of Neolithic (P. Jodot 1958).

In the extreme south of Fezzan, at Tejerhi, P. Bellair (1949, 1953) studied an interesting site where he noted:

1. Clays—2nd Pluvial [Kamasian]
2. Ancient white erg [47]
3. Calcareous slab—3rd Pluvial [Kanjerian]
4. Ancient erg, Acheulian
5. Breaking up of the calcareous crust; 4th Pluvial [Gamblian]; Moustero-Levalloisian
6. Clays from the dhaya—Neolithic subpluvial
7. Present-day erg

The mollusks have been studied by E. Fischer-Piette (1948) and by P. Jodot (1953) who consider the first level as "Upper Paleolithic."

We might compare Tejerhi to the Majâbat, as, for example, 2—SB, 3—GB1, 4—TH, 5—CL, and 6—GB2. But Araouan and the Fezzan are far enough apart to forego a similar comparison.

So, we are happy to see, there are new methods of exploring the past and dating deposits, recently used in the Sahara, which will place chronologies, and the comparison of chronologies, on a firmer footing. Neither palynology nor C-14 furnishes automatic and infallible answers.[48] Causes of error and uncertainty exist, which we must acknowledge. But palynology and C-14 dating give to the other methods an assistance whose value cannot be doubted.

One series of answers is already known (A. Pons and P. Quézel 1956, 1957, 1958; P. Quézel and A. Pons 1958; P. Quézel and C. Martinez 1958; G. Delibrias,

47. Dune sands?

48. H. von Wissman (1956, pp. 282–3) notes two possible causes of error with pollens: (1) the possibility of "aclimatic" diagrams in the case of samples taken "in the subtropical and temperate belt" from shaded areas, and (2) the carrying of grains over a great distance.

H. Hugot, and P. Quézel 1959; P. Quézel 1958, 1960). I will summarize quickly.

Some terraces (Oued Ahor, Tin Tessandjelt, In Eker), some fossil soils (Meniet), a subfossil guano of *Procavia ruficeps,* etc., have furnished pollens. Tree pollen came from the following genera of species: Aleppo pine (*Pinus halepensis*), Atlas cedar (*Cedrus libanitica atlantica*), Barbary Thuya (*Callitris articulata*), Duprez cypress (*Cupressus dupreziana*), various junipers (*Juniperus*), green oak (*Quercus ilex*) and others, alder (*Alnus glutinosa*), "micocoulier" (*Celtis australis*), maple (*Acer* spp., e.g. aff. *monspeliensis*), daphne (*Daphne* aff. *gnidium*), mastik trees (*Pistacia lentiscus* and *terebinthus*), the tree heath (*Erica arborea*), *dimorphous* ash (*Fraxinus xanthoxyloides*), elder (*Sambucus* sp.), phillyrea (*Phillyrea* aff. *media*), willow (*Salix* sp.), jujube tree (*Ziziphus* sp.), linden (*Tilia* sp. and *T. platyphyllos*), walnut (*Juglans regia*).

It was, thus, a "typically Mediterranean forest vegetation, quite similar to that presently occupying the semiarid Saharan Atlas mountains or Aurès mountains" (P. Quézel 1960, p. 357) and which contains a number of species no longer extant in the Sahara. One species (alder) is extremely localized in North Africa (Rif-Numidian littoral), another is even unknown there (linden).

In time, the studied flora stretches from ca. —10,000 (Moustero-Aterian) to the end of the postglacial Optimum (—2800). Dating by C-14 gives —2730±300 for the guano and —3450±300 for a Neolithic fossil soil at Meniet (with *Homoioceras antiquus, Bos ibericus* (?), *Thryonomys* sp., *Varanus* aff. *niloticus, Testudo* sp., *Limicolaria*).

A time evolution appears certain, as the elements with the most noticeable northern affinities are found only in the most ancient strata (Moustero-Aterian) while the most thermo-xerophile species (*Olea, Ziziphus, Cupressus,* etc.) appear in the Neolithic and, particularly, in a recent phase of the Neolithic.

We could have, then:

1. Moustero-Aterian: mesophile flora rich in deciduous elements, oaks, maples, lindens, alders, but also cedars.
2. Neolithic: tends to a predominance in xerophiles, green oak, Aleppo pine, junipers.
3. Terminal Neolithic: cypress, olive, "micocouliers," still some Aleppo pines.

A recent passage by P. Quézel (1960, p. 358) is worth quoting:

"The Würmian has been characterized in the central Sahara by a relatively cold, dry climate, with important variations of temperature. There was a steppe-flora, dominated by Salsolaceae and grasses.

"The debut of the last interpluvial,[49] stretching approximately from 10,000 to 6000 B.C. is, on the contrary, characterized by a temperate humid climate, probably of Mediterranean humid or sub-humid type and which saw the whole of the Central Sahara invaded by Mediterranean forest species (cedar, Aleppo

49. *Sensu lato:* it is, rather, a postpluvial preceding the Neolithic subpluvial.

pine, green oak) and even by elements of the oak forest (lindens, maples, alders, etc.).

"The middle part of this same interpluvial up to a very recent period (—2800), that is to say, up to the end of the climatic Optimum, saw a progressive deterioration in the climate, which became hotter and dryer, remaining, all the while, of Mediterranean type, as is proven by the flora corresponding to the Neolithic Sahara, where one finds the Aleppo pine and the green oak side by side with clearly xerophile elements (olive, cypress, junipers, etc.)"

A general chronological scale for the Central Sahara, with tentative correlations to Europe, is given in P. Quézel and C. Martinez in a recent paper (1962) which sums up all previous work for that particular region.

At Adrar Bous (eastern border of Aïr massif), the following species have recently been reported from Aterian diatomites: *Pinus* sp., *Juniperus* sp., *Juniperus phoenicea, Cupressus* sp., *Cupressus dupreziana, Callitris quadrivalvis, Phillyrea* sp., *Alnus glutinosa, Myrtus nivellei, Pistacia lentiscus, Tilia* sp., etc., From Neolithic fossil soils: *Cupressus* sp., *Juniperus* sp., *Myrtus nivellei,* etc.

Similar research is in progress for other regions of the Sahara. It is unnecessary to emphasize its importance, not only for the history of vegetation and for recent paleoclimatology, but also for archeology (e.g., the existence of agriculture as proven by the pollen of cultivated plants).

E. The Chadian Sahara

Let us examine the morphological basin of the Chad, of which the Ténéré is part.[50]

For the latter, a recent note by H. Faure (1959) furnishes a useful outline. The author lists (p. 2809) the following terms for a section at Agadem, situated close to a cretaceous cliff, which will explain certain of the terms:

1. Ancient dune I, orange sand and ancient debris; dry
2. Lacustrine with diatomite, reeds; humid
3. Ancient dune II; dry
4. Ancient reg and alluvia; humid
5. Recent dune; now dry

One of the two humid periods may be Paleolithic, the other, Neolithic. Note that the observations are generally in agreement with those in the Majâbat, rather similar physiographically. We might thus say: 1—SA, 2—GB1, 3—CL, 4—GB2, 5—TB, with the two following difficulties: first, the SB appears to me to be genetically alluvial and "dune" only in a later stage of its history, and second, if 3 = CL, this latter, it seems to me, by no means represents a "dune" element.

50. Cf. the maps of G. Ruhle, *Physiogeographie des saharisch-sudanischen abflusslosen Gebietes* (Tschadseebecken), Leipzig 1929, and those by J. Dubief in *Essai sur l'hydrologie superficielle au Sahara,* Algiers, 1953.

Dealing with the Chadian Sahara, the two well-known works, by H. Freydenberg (1908) and by G. Garde (1911), both very much "catalogues of observations," scarcely yield a useful sequence.[51]

As for the chapter on the evolution of the Chad in the Pleistocene by H. Jacques-Félix (1947, pp. 66–75, Figs. 50–52), it must remain, without an adequate stratigraphical, paleontological, and archeological base, in the realm of hypothesis.

Various recent works show a number of new facts and lay the foundations for a stratigraphy of the Chadian Pleistocene (J. Barbeau 1960; A. T. Grove and R. A. Pullan 1961; J. Pias 1958, 1960; J. Pias and E. Guichard 1957); it appears, from the work of Barbeau and Pias, particularly, that we must envisage a sequence consisting of four principal lacustrine or fluvio-lacustrine transgressions (or invasions), corresponding to the Pluvials and separated by dry phases.

For the Logone-Chari region, J. Pias thinks he can distinguish four transgressions (1958):

1. Kelo sand (on the continental shore of the paleo-Chadian "sea")
2. Clayey sand series with calcareous nodules
3. Recent sandy series (e.g., in the north and to the east of the lake) and which, in the course of a dry phase, will furnish the material for the modern dunes
4. Sandy clay, muddy clay, etc. (with the invasion of the interdunes by the lake, at various points)

Further east, a comparable sequence occurs (1960):

1. Ancient sandy series
2. A clayey sand series, with gravel. Reg.
3. Recent sandy series
4. Alluvial muddy-clayey to clayey-muddy, recent to present.

For his part, J. Barbeau (1961) acknowledges four "invasions":

1st stage: Manga—Goz Terki (+30 m.)
2nd: Goa—Her Sayal (+20 m.)
3rd: Bahr el Ghazal major (Massaguett-Am Djemena) (c. +10 m.)
4th: Bahr el Ghazal minor (+4–5 m.)
Similar dune groups correspond to each dry period separating wet periods.

Thus a Pleistocene series, with four transgressions corresponding to four Pluvials, has been described. The ground is clearly covered from a geological

51. Freydenberg's work (1908, p. 63) was criticized by Garde (1911, pp. 173, 181) concerning the postdunar situation of lacustrine beds in boat-shaped hollows, judged to be infradunar by Freydenberg. K. W. Butzer (1957, p. 112) cites only Freydenberg. He, along with Garde, considers the white sand at the bottom of the shaft as dune ". . . for it is found everywhere, and only a dune regime would be capable of creating a strictly sandy deposit in such a vast region" (p. 63). We know today that the white sands of the Majâbat seem to question the argument.

point of view, but the picture must be completed by paleontological observations and the archeological correlations necessary to enhance its value. Undoubtedly, the years to come will bring a number of improvements.

Already discoveries are increasing, particularly in northeast Chad where its drainage, the Bahr el Ghazal stretches in the direction of the Sahara and the paleo-Chads of the Bas-Pays. There, important discoveries have given evidence to the presence of many Pleistocene fauna, of which one is Villafranchian (J. Abadie, J. Barbeau, and Y. Coppens 1959; Y. Coppens 1960a, 1960b, 1961a, 1961b, 1962a, 1962b, 1962c).

There are 8 known main sites:

1. Goz Terki (1955)
2. Toungour (1955)
3. Koula (1957) and Koula-Ri-Katir (1960)
4. Bochianga (1959)
5. Wadi Derdemy (1960)
6. Yayo (1960)
7. Yekia (1960)
8. Dama Rezou (1960)

Of these sites, Bochianga and Yekia should be the oldest (presence of mastodon and elephant), since all the others only yielded the elephant.

It is a matter of fluviatile sands, of sandstone, of diatomites, etc.

The fauna, not yet studied, comprise: *Loxodonta africanava* Arambourg (according to Y. Coppens, the term *Archidiskodon* must be abandoned), *Elephas recki*, *Anancus*, *Stegodon*, *Hyaena* cf. *striata*, a Felid, *Ceratotherium simum*, *Stylohipparion*, *Hippopotamus* (*Hexaprotodon*) sp.,[52] *Notochoerus* nov. sp., *Potamochoerus* (?), *Phacochoerus*, *Libytherium maurusium*, *Giraffa* cf. *camelopardalis*, *Menelikia*, *Redunca*, *Hippotragus*, *Kobus*, an Alcelaphine, a Camelide, *Crocodylus niloticus*, *Cr. cataphractus*, *Cr. barbeaui* nov. sp., a Gavial, *Trionyx*, fishes (Siluriformes, *Lates*), finally an Australopithecine.

This is an early Villafranchian fauna, more or less contemporary with that of Kanam, Kaiso *pro parte*, Olduvai Bed I *pro parte* and older than Kaiso (type), Omo, Laetolil, Olduvai Bed I *pro parte*.

Recent information from Y. Coppens (*in litt.*, 23 April, 1963) indicates that the elephant from the Australopithecine site could be not *africanavus* but rather an early mutation of *atlanticus;* if so the site may be upper Villafranchian, the lower Villafranchian represented by the sites with mastodon, *Elephas recki*, *Elephas africanavus*.

The region yielded a second fauna of indeterminate age, having furnished, in a fluviatile sand, mollusks, fish (*Bagrus*, *Synodontis*, *Heterobranchus*, *Clarias*, *Arius*,

52. Cf. *omaguncula* and not *H.* (*Tetraprotodon*) *protamphibius* Arambourg (oral info., Y. Coppens).

Lates), a tortoise (*Trionyx*), a crocodile, some antelopes; this fauna should be equally represented at Yayo.

Lastly, a third Neolithic level contained *Loxodonta africana*, *Hippopotamus amphibious*, *Giraffa camelopardalis*, and *Phacochoerus africanus*.

The Australopithecine found by Y. Coppens at Yayo was presented at the Symposium on the Evolution of Vertebrates held at the Muséum National d'Histoire Naturelle (Paris) from May 29th through June 3rd, 1961. It was a fragment of the cranium, rather eroded, whose status cannot be determined until after a detailed comparative study, in particular with the new Australopithecines from Olduvai.[53]

Y. Coppens is preparing to publish on these fauna and some new research on the terrain is in progress. We would like to see these obviously important finds completed by the discovery of associated industries.

Farther northeast there is the site at Ounianga, in its unforgettable saharo-lacustrine landscape. There is a fauna there about which we are still badly informed,[54] although we have known it for a long time (L. Joleaud and J. Lombard 1933a, 1933b; L. Joleaud 1936c, etc.): an elephant, some hippopotami, a giant wild boar (*Sus scrofa*), some fishes (*Lates*), etc., were found there. L. Joleaud considers the site Kamasian; more precise information is needed for a decision. The recent discovery of *Loxodonta africana* by Y. Coppens (*in litt.*, 10 March, 1961 and 23 April, 1963) might make it necessary to date this fauna somewhat later.

The palynological study of two diatomites of Borkou (P. Quézel and C. Martinez 1958, 1962) has revealed the existence of "at least two wet periods during which northern influences greatly affected that region" (p. 243). In the Neolithic, one witnesses the "brutal replacing of Mediterranean flora with Sahelian flora, very probably caused by a very sudden, important variation in climatic conditions" (p. 242). The species noted are: *Cupressus dupreziana*, *Callitris* sp., *Juniperus oxycedrus*, *Cedrus libanitica atlantica*, *Pinus halepensis*, *Pinus* sp., *Ephedra tilhoana*, *Quercus* aff. *coccifera* or *ilex*, *Ficus* sp., *Acacia flava*, *A. raddiana*, *Myrtus nivellei*, *Olea europaea*,[55] *Artemisia tilhoana*, *Erica arborea*.[56]

K. W. Butzer (1957a, p. 112) attributes to M. Dalloni (*Mission to Tibesti*, 2(1934): pp. 137–41) the mention of lacustrine deposits associated with Mousterian, in Borkou. I was not able to find any indication of this in Volume I of the *Mission to Tibesti*. Besides, if comparison is still permitted, according to what we know—or guess—about the western Sahara, the two major lacustrine periods should belong to the early Paleolithic and the Neolithic.

53. L. S. B. Leakey, "New Finds at Olduvai Gorge," *Nature*, 189:649–50.

54. But the research done in 1957 by A. J. Arkell and Y. Coppens in 1961 will certainly provide more information.

55. One expected rather *O. laperrinei*.

56. This species was found alive in the Tibesti area (Ph. Bruneau de Miré and P. Quézel, 1959).

F. The Libyan Desert

Though the Libyan desert has yielded much archeological material (industries, rock engravings and paintings), the geology of its Pleistocene is still little known. Where A. Knetsch (1950) acknowledges at least five Pluvials since the end of the Tertiary, W. Meckelein (1955) recognizes only one, belonging to the *Altquartär*, a second should be attributed to the Tertiary (pp. 93–6).[57] There was not, in the Pleistocene, a true "Feuchtzeit" (in the sense of wet climate), but rather the permanent dryness of a "Wüstensteppenklima," with at most some fluctuations, one of which would be "lediglich eine relativ regenreichere Zeit, eben ein Pluvial" (p. 96). Zonal distinctions would have to be made, with a central "Kernraum" of the "Extremwüste" type remaining unchanged since a very long time, since there are no definite traces of a morphological evolution implying a Pleistocene Pluvial. It is here that a desert climate has remained from the Tertiary to the present day. This idea of a "nucleus" that escaped the wet oscillations is important, especially since there seem to be several of these "Kernwüsten," suggesting a "Kernzone" equally divided between Ténéré and Tanezrouft.

It will be helpful to know, and Meckelein's future publications will probably tell us, if the distribution of industries corroborate the conclusions of the morphologist and geologist.

Meckelein (p. 142) seems to lean toward the theory of a climatic alternation on the northern and southern edges of the Sahara: "Einer Glazialzeit in Europa steht in der Sahara im Nordabschnitt zwar auch ein Pluvial, in Süden dagegen gleichzeitig eine Trockenzeit gegenüber. . . . Während eines europäischen Interglazials aber herrschte im Saharischen Trockenheit, in seinem Süden jedoch mit vorrückender Monsunfront ein Südpluvial." These climatic oscillations remained marginal on both sides of the "Vollwüstenraum" in a single Pleistocene Pluvial, including an "Extremwüstenraum" apparently without any Pluvial at all.

The affirmation of F. Machatschek (*Das Relief der Erde*, Vol. 2 (1955), p. 185) that "die innersten Teile der Sahara, namentlich die weniger hoch gelegenen, und der Kern der Libyschen Wüste an dem Quartären Pluvialperioden überhaupt kaum teilgenommen haben," cited by W. Meckelein (p. 134) is criticized by him. The sheets of sand in the Libyan desert are too vast, signs of fluvial erosion too manifest in the reliefs (G. Aouenat, Gilf Kébir, etc.) for Pleistocene climatic oscillations to be considered foreign to that region. The Pleistocene deposits of the Libyan desert, especially perhaps of its contours and their piedmont plains (G. Aouenat, G. Arkenu, G. Kissu, Gilf Kébir, etc.) where their relations to the industries could be studied, deserves careful research.

Here should be noted the interesting attempt of K. W. Butzer (1958) in starting

57. W. Meckelein's observations apply principally to the Fezzan (eastern) and might have been recalled à propos of the central Sahara; but the expedition passed 18° to the east.

with modern ecological limits expressed in pluviometric minima, 50 mm. for the giraffe, 100 mm. for the elephant, 150 mm. for the rhinoceros and hippopotamus[58] compared to the distribution of their rock pictures in the eastern Sahara,—to trace the isohyetes in the Neolithic (ca. —5000 to —3000) (Fig. 14). Compared to those of the present, they show a simple shift, never more than 250–300 km. and very often less (100, for example). If the Tibesti (>150 mm.) could have presented a marked Sahelian character,[59] the quasi-totality of the eastern Sahara remains below the isohyete of 100 mm., and a very great part still more arid (<50 mm.). It is important to note that two bridges of 50–100 mm. unite Ennedi to Tibesti and the latter to a Djado-Fezzan-Tassili group; also that the phenomena are symmetrical, and consequently synchronous, on the two edges of the desert. One must acknowledge "ein gleichzeitiges Vordringen oder Rückweichen dieses riesigen Wüstenkomplexs auf allen Fronten hin" (pp. 39–40).

A provisional outline of the paleoclimatic sequence for the Libyan desert will be found in K. S. Sandford (1933; see below, p. 179). He returned in 1936 to the same question with respect to the distribution of mollusks, but without attempting any correlations, even with Egypt: "All such correlations are considered to be, at the present stage of our knowledge, unprofitable and misleading" (p. 217).

G. The Nile Valley

Leaving aside the voluminous literature on the recent geological story of the Nile Valley, I shall confine myself to an account of the recent reports of K. W. Butzer (1959a, 1959b, 1959c, 1961, 1962) which seem to offer the best resumé of the question.

Here is the first part of the series (1959a, pp. 65–6) (Table 3).

Miocene to Lower Pliocene. Cutting of present Nile Valley and tributaries to at least 125 m. below mean sea level, probably in response to a very pronounced marine regression. Tributaries could apparently not keep apace with rapid incision of main valley.

Upper Pliocene (Astian to Plaisancian). Deposition of marine and freshwater beds in gulf flooding Nile Valley to over 180 m. above m. s. l.

Pleistocene. Calabrian. Return to fluviatile conditions. Various gravels between 115 and 230 m. over alluvium in Lower Egypt, probably eustatic. Large scale erosion in Red Sea Hills and transport of masses of gravel to Nile.

Sicilian. Nile course to +98 m. on surface of Western Desert north of Mallawi in response to Mediterranean sea level of 103 m.

Shorter period of vertical incision on lower Nile.

58. This last is an aquatic animal, therefore in a certain measure independent of local climatic factors.

59. Of which there are still ample traces today, especially on the southwest side of the massif.

TABLE 3
PRE-NEOLITHIC SEQUENCE IN EGYPT (FROM K. W. BUTZER 1959A, P. 65)

	Nile and Wadis	Climate	Vegetation
5000 BC			
Helwan Culture	No wadi activity, renewed aggradation	Dry, warmer	Desert
Epi-Levallois III	No wadi activity, Nile downcutting	Dry, cool	Desert
Epi-Levallois II	Wadi activity, Nile aggradation	Damper, cooler	Dry-steppe
Epi-Levallois I } Upper Levallois }	First silt aggradation by floods	Dry	Semi-desert
30000 BC			
Levallois	Torrential aggradation Little wadi activity No wadi activity	Damp, cooler Damper, warm Dry, warm	Dry-steppe Thorn-savanna Desert
Acheulio-Levallois	Renewed aggradation Little wadi activity	Damp, cooler Dry	Etesian-steppe Semi-desert
Acheul	Torrential aggradation Some wadi activity	Damp, cooler Damp, warm	Etesian-steppe Thorn – or grass-savanna
Acheul	No wadi activity	Dry, warm	Desert
Lower Acheul } Abbevillian }	Torrential wadi – and Nile aggradation	Damp, cooler	Etesian-steppe
300,000 BC			

Nile course to +78 m. on Western Desert from Mallawi to the sea in response to m. s. l. of +80–85 m.

Post-Sicilian, Pre-Mindel. Long period of intermittent vertical incision in entire valley. During part of this time longer phase of rubefication of calcareous sands (red earth) (with some precipitation).

Mindel (?) ("Abbevillian to Lower Acheulian"). Climatic aggradation on upper Nile and tributaries with gravels deposited to 25–30 m., gradually falling off towards the north.

Brown soil formation with strong calcareous precipitation below surface (strong precipitation and cooler).

Tyrrhenian I. Lower to Middle Acheulian. Vertical incision upstream, eustatic aggradation and redeposition of older material by Nile downstream following a course in 25–30 m. on Western Desert north of Minya, bed rising towards north in response to sea level of +35 m. Locally dry.

Longer period of red earth development (considerable precipitation.)

Riss ("Middle and evolved Acheulian, pre-Micoquian"). Climatic aggradation of Nile and wadis in Upper Egypt to 15 m., apparently falling off rapidly north of Sohag to 8 m., at Mallawi. Vertical incision downstream.

Red earth formation (?).

("Acheulio-Levalloisian"). Renewed aggradation of Nile and wadis in upper Egypt to 9 m., falling off towards north. Vertical erosion downstream?

Period of brown soil formation with discontinuous calcareous horizon (considerable precipitation and cooler).

Tyrrhenian II. (Monastirian). Aggradation of Lower Nile to 10–13 m. rising towards north in response to sea level of +15–20 m. Beds are merely redeposited older materials from further upstream showing no traces of local wadi activity. Pronounced local aridity.

Moderate period of red earth development (some precipitation).

Würm ("Late Lower Levallois"). Aggradation of 4 m. torrential gravels in Upper Egyptian wadis.

Moderate period of brown soil formation with precipitation of calcareous gypsum (some precipitation and cooler).

("Upper Levallois to Epi-Levallois I"). Inauguration of modern hydrographical regime with surge of exceptionally high, silt-bearing annual floods. Sea level low or falling. Vertical erosion of wadis. Very short phase of slight reddish weathering with some moisture. Epi-Levallois I phase with little local wadi activity.

("Epi-Levallois II"). Short period of wadi aggradation and wadi activity. Incision of Nile begins.

("Epi-Levallois III"). Vertical incision of Nile with gradual cessation of all wadi activity. Some tendency to formation of arid, light brown soil (slightly moister, cooler?). Ultimately intense local aridity with greater aeolian activity.

To sum up, after the appearance of the "low desert" during the Pliocene in what is now the Nile Valley, there occurs in the Pleistocene series of terraces, built up and fitted together, some of which contain industries:

25–30 m.: Abbevillian and Lower Acheulian (south), Lower and Middle Acheulian (north);

12–15 m. and 8–9 m.: Middle or Late Acheulian and Acheulian-Levallois;

3–4 m. and 1–2 m.: early Levallois to Epi-Levallois II.

An important event at the end of the mid-Paleolithic was the appearance of Sebilian silts, marking the beginning of modern hydrographical rhythm: summer floods, receding winter waters, and the transport of Ethiopian silts. Until then—and this fact is important—the Nile sediment was entirely detrital, with floods of torrential character.

During the early Paleolithic, three principal types of climate may be distinguished:

1. Humid and temperate. Vegetation of a valley, which is not then an inundation plain: etesian steppes (cf. Cyrenaica) with groves of sycamores, acacias, willows, tamarisks. The living conditions were propitious for the establishing of gatherers, hunters, and fishermen. This type of climate was that of the early Abbevillian-Acheulian, but also of a later Acheulian, and of an Acheulian-Levallois.

2. Dry and warm. Comparable to the present; vegetation more xerophile (acacias). The climatic deterioration lessened the possibility of human habitation; Acheulian, Acheulian-Levallois.

3. Warm, but wetter. Vegetation more or less Sahelian ("Dornsavanne" with "sudanese" fauna); Acheulian.

It goes without saying that these are recurring climatic types which can be found in several periods within the ancient Paleolithic.

There are several types of climate and thus of vegetation in the middle Paleolithic, but what is now the desert remained, on the whole, habitable and inhabited.

The upper Paleolithic finds, about 30,000 years ago, in the Sebilian, the hydrological system still functional today with the primarily silty sediment that characterizes it. The climate tends toward desertification: Sebilian and Epi-Levallois are no longer found on the plateau, but in the wadis or the Nile Valley. Towards —18,000 to —16,000 the climate is desert-like, with floods of silt and aeolian deposits. Human life found, in or near the water, a very abundant exploitable fauna: mollusks, (e.g., kjökkenmödiger of the Epi-Levallois II), fish, reptiles, mammals, denoting an amphibious fauna or a wooded gallery on the edge of the desert. The juxtaposition of the hippopotamus or the crocodile with *Bubalis, Bubalus, Bos brachycerus* and *primigenius* and *Gazella dorcas isabellina* is significant.

The reduction of the floodable zone, with the progress of the erosion of the valley in the Epi-Levallois III again limited the possibility of occupation, with the minimum towards —10,000, which a slight climatic amelioration around

—9,000 will not perceptibly modify; also the appearance, with a Neolithic subpluvial (after —5000) of a new population which found the terrain not inhabited, or only slightly so.

It is important to know the origin and the evolution, both vertical and horizontal, of the modern zone of occupation, the alluvium properly so-called.

Between —8,750 and —3,500 the level of the Mediterranean probably rose more than 50 m. This caused a shift from erosion to upbuilding in the Nile, first in the delta, then farther and farther up-stream. The process, of course, was not regular, and the author estimates that 60% of the Nile silt existed before the Old Kingdom (ca. —2,850) and that a final sedimentation, since the Ptolemaic times, deposited 20–25% of present silt. Basic fertile material probably appeared in the Neolithic.

What was the Nile Valley like when Paleolithic man—hunter, gatherer, and fisherman—came, before the technical influence of Neolithic and protohistoric agricultural civilizations was felt? It has often been accepted that the valley, covered with papyrus swamps and semiaquatic jungles, constituted such an obstacle for the human population that it could only be overcome by recent techniques (drainage, irrigation) which farmers brought from the adjacent desert to colonize the valley. E. Dechambre wrote, for example (1951a, p. 111): "This development [canals and dams] dates from the prehistoric era and has been constantly improved. But previously Egypt presented only a minimum of favorable conditions for the establishing of man. The river rolled its powerful tides down the valley, and its floods destroyed all attempts at settling by men. In the neighboring swamps and in the delta, a stagnant and fetid water barely covered a blackish slime; islets of sand, wooded knolls, plateaus of good earth appeared in a sea of mud. Food resources were laughable. As for the Nile, it is true that its waters offered numerous varieties of fish, but not very succulent ones. . . ." Does a river that abounds in *Lates* and *Tilapia,* to mention only two genera, merit such a severe judgment?

But even subjected to high and low waters, the valley remained accessible to man, with a raised minor bed, bordered with banks, and lateral floodable plains, rather like the Bas Chari-Logone or the middle Niger. The inundated plains of Macina or of Yaérés, with their patches of trees indicating island villages, resemble probably very closely flooded Egypt, where only certain deeper basins had a dense palustrine vegetation throughout the year. The notion of limitless impenetrable swamps as a major obstacle to human occupation undoubtedly is a myth.

It is nevertheless evident that the palustrine habitat, for example the *Cyperus papyrus* swamp, will play a prominent role in Egyptian civilization. One recalls the work of E. von Rosen (1929)—"Did prehistoric Egyptian culture spring from a marsh-dwelling people?"—where he proposes interesting comparisons with the Batwa of Lake Bangweolo.

In the delta itself, however, the picture could be different; but at the height of the Flandrian transgression the open sea did not pass beyond the region of the present Lakes of Idku, Burullus, and Manzala, and the area of permanent swamps and lakes remained relatively small. Along the channels of the delta, the vestiges of ancient sediment (turtle backs), the mounds which only human occupation could raise, permitted establishment of numerous little villages since before —4000.

The Neolithic and its Subpluvial (ca. —5000 to —2400) would be, climatically and culturally, a favorable period for the development of human technical and economic activities. With the extension of vegetation (acacias, tamarisks, sycamores, willows near the water; steppe on the low desert), the ecology would favor animal life; the elephant lived at Fayoum.

Zoological data (bones or representations) for the Predynastic period have multiplied. These the author analyzes in detail, emphasizing the uncertainties attached to certain identifications (*Bubalis* and *Addax*, *Gazella dorcas* and *leptoceros*, lion and panther, etc.) The absence of the camel is mentioned, but the author, with reason, ignores a problem given so much recent attention. Only the discovery of new facts could prompt its reconsideration.

I shall reproduce here, for its own interest, the list of mammals of the "altägyptische Grossfauna" (1959a, pp. 42–3), showing clearly the expected juxtaposition of palaearctic, Saharan, and Sudanese elements:

[*Camelus dromedarius*], *Dama schaeferi*, *Giraffa camelopardalis; Oryx algazel*, *Addax nasomaculatus*, *Bubalis busephalus*, *Bubalis lelwel*, *Gazella dorcas* and ssp. *isabella*, *Gazella leptoceros loderi*, *Lithocranius walleri*, *Cephalophus sylvicultrix*, *Ammotragus lervia*, *Capra ibex nubiana*, *Bos primigenius*, *Bubalis* aff. *caffer*, *Loxodonta africana*, *Equus asinus africanus*, *Diceros bicornis*, *Ceratotherium simum*, *Hippopotamus amphibius*, *Sus scrofa*, *Felis leo*, *Felis pardus*, *Hyaena striata*, *Hyaena crocuta*, *Canis aureus*, *Canis lupaster*, *Canis vulpes*.

An important fact to note is the existence of a zoo-ecological discontinuity around —3600, between Nagada I and II, a date at which one finds the elephant and the giraffe becoming rare, if not disappearing, in the Nile Valley and in the eastern desert.

In the Predynastic and during the Old Kingdom, the low desert was not a "Vollwüste" as it is today, but rather a desert-like savanna, with bushes, tufts of grasses, some trees—a tableau which is confirmed by paintings of desert hunting scenes from the 5th and 6th Dynasties. Concerning the arable part of the valley, the author cites several autochtonous species (e.g., trees) [60] and borrows from S. Passarge a description of the ancient spontaneous vegetation of the delta. It shows that the permanent papyrus swamp, of "Sudd" type, would only be found at certain points, and that elsewhere wooded banks and low-water pasturage would form on the contrary a landscape not at all hostile to human life.

What about precipitation prior to the Dynastic period? The existence of a

60. Among which the *Mimusops schimperi*—at first a little surprising, the species seeming especially known in South Arabia and Ethiopia and not represented in R. Muschler.

Neolithic Subpluvial (c. —5000 to —2400), with a maximum previous to Nagada II, is acknowledged. Indeed this wetter "Neolithic" stage extends (diminishing, however) into the Predynastic and even beyond, since it is only in the 6th Dynasty that one can consider settled the dry period that has remained until the present.

A particularly interesting diagram might pinpoint the extension of vegetation in Egypt and the Sudan for the Subpluvial (—5000 to —3000). In the diagram, the zones are limited by isohyetes: >150 mm. (good pasturage), 100–150 mm. (fair pasturage), <50 mm. (hardly any pasturage). We see that an important part of Egypt—not only the presently arable strip—is accessible to man, at least in the form of pastoral economy, and probably with small centers of culture in favorable spots, even outside the valley, as are still evident in the various Saharan massifs.

At the end of the Pre-Neolithic (Epi-Levallois III, Helwan), the desert was probably not truly inhabited, nor habitable, but with the Neolithic, surface sites and also the petroglyphs appear on the desert plateau.

The character of the country in the Subpluvial naturally facilitated the diffusion of the new grain economy into western Asia, the Middle East, and North Africa. K. W. Butzer reverts here to H. Helbaek (1958) on the subject of the origin of the culture of wheat (*Triticum dicoccum*) and barley in the Fertile Crescent between 650 and 1400 m.

A pertinent observation of the author is that if the Subpluvial favored the diffusion of agriculture, it could equally have allowed the northward advance of Sudanese fauna. Perhaps it is more a matter of a resumption of this advance, because for the Maghreb at any rate, there are some elements of great Tropico-African fauna in the lower Paleolithic, although the true African elephant only appeared with the Capsian.

Departing somewhat from the plan (Pleistocene and prehistory) of this report a little, I think it useful to give a résumé of what K. W. Butzer tells us of the evolution of fauna in Dynastic times.

There is a second zoological discontinuity, that around —2850 to —2600: elephant, giraffe, and rhinoceros disappear from Egypt for good;[61] lion, panther, and Barbary sheep diminish noticeably. The phenomenon would have primarily climatic causes, provoked by a dry interlude during a subpluvial which was nearing its termination.

The author gives us (1959a, pp. 96–109) a copious and precise inventory of the fauna represented in the bas-reliefs in tombs, for a series of Dynasties and for each of these by locality, with the necessary iconographical references. This inventory is summarized below. The number of representations is by species and by dynasty; percentage of that number by era.

The following conclusions can be drawn from these facts:

61. The gerenuk gazelle (*Lithocranius walleri*) too, perhaps.

1. Ca. —3600 (Nagada I to Nagada II); first discontinuity of fauna: diminution of (Neolithic) fauna from the tropical savanna, increase in antelopes.

2. Between —2800 and 2600; second discontinuity of fauna: rhinoceros, elephant, giraffe, perhaps also the wild camel disappeared; diminution of lion and Barbary sheep, increase of antelopes and gazelles.

3. Beginning of the 5th Dynasty (ca. —2480): representations of desert gazelles increase, the Oryx diminishes; larger animals withdrew, both to the steppe and to the eastern mountains; the desert loses its population. At the beginning of the 6th Dynasty big hunts were held in walled parks, and the age of pyramids can be said to mark a third stage in the evolution of fauna. Ca. —2000 a condition was reached that continued to the Middle Ages.

4. In the New Kingdom, only swamp and water fauna remain significant: hippopotamus, crocodile, urus, birds, fishes; there is no longer big game in the valley (outside of the parks), but some survived in the wadis of the eastern desert and the steppes to the west of the delta.

Of this stock, there remain only dorcas gazelles and ibex. Lion and oryx disappeared from the eastern desert long ago; bubalis, Barbary sheep, wild ass since the 19th century; lion, ostrich, addax, and oryx disappeared from the Mediterranean steppe or the oases at the same time. Here, man was the agent of destruction. The last hippopotami died in 1658 in the delta and in 1850 in Upper Egypt.

K. W. Butzer acknowledges that animals shown in bas-reliefs are there by virtue of choice: as sporting game, or for symbolic or cultural reasons. It could also be that they represent domestic or semidomestic animals. Certain articles are pertinent: C. Gaillard, "Les tâtonnements des Egyptiens de l'Ancien Empire à la recherche des animaux à domestiquer" (*Rev. Anthrop.*, 1912); Ed. Dechambre, "Le Sahara, centre primitif de domestication" (1950 [1951], pp. 147–51), and "Discussion del l'interprétation de figurations animales anciennes" (1951a, pp. 105–15). In the latter, the author asks if Saharan desiccation is not responsible for the beginnings of domestication, by a more and more efficient control of primitive wild herds, and the occupation of the Nile Valley, considered originally uninhabitable (which is precisely what K. W. Butzer considers unacceptable).

There remains a consideration of historic times. One can acknowledge that between —2480 and —200 the rainfall gradually descended to its present level, and K. W. Butzer believes that he can place the end of the Subpluvial at the beginning of the 6th Dynasty. With the diminution of the rains, aeolian actions increased.

Four principal dune deposits can be distinguished:

1. Ancient dunes: —2350 to —500
2. Lower recent dunes I: +300 to +800
3. Lower recent dunes II: +1,200 to +1,450
4. Upper recent dunes: since +1700

Summing up the whole of his paper, the author gives the following table (1958a, p. 115–16):

−16,000 (?) to −5,000—dry, with intense Aeolian deposition and probably a short interruption ca. −9,000

−5000 to −2350—humid interval (Neolithic Subpluvial) with consolidation of marginal dunes and great advance of alluvium

−2350 to −500—dry, with intense Aeolian depositions; ancient dunes in Middle Egypt.

−500 to −300—silty sedimentation accrued by stronger inundations with extended recovery of marginal dunes by substantial covering of Nile silt; local climate dry.

+300 to +800—generally deeper floods, more intense Aeolian deposition; lower recent dunes I; locally dry

+800 to +1200—heavy floods and some increase in precipitation locally

+1200 to +1450—dry; lower recent dunes II

+1450 to +1700—heavier floods; locally dry

Since +1700—modern dunes (upper recent) in Middle Egypt, in present conditions; dry.

IV. DISTRIBUTION OF PRIMATES[62]

This section will be short, since monkeys no longer exist in the Sahara except at Tibesti and are rarely found portrayed in rock drawings and are unknown in a fossil state.[63] A very useful source is P. Huard's article, "Archéologie et préhistoire: contribution à l'étude des singes du Sahara oriental et central" (1962a).

The only possible question is, in effect, that of the present northern limit of African primates. The case of *Macacus sylvanus* (L.) of Barbary being completely distinct from that of the monkeys of intertropical Africa, let us consider those of baboons, *Papio papio* (Desmarest) and *P. doguera* (Pucheran and Schimper), of Patas, *Erythrocebus patas* (Schreber), and somewhat indirectly, of *Callitriche* or green monkey, *Cercopithecus aethiops* (L.)

There appear to be two species of baboons, one more western (*Papio papio*), the other more eastern (*P. doguera*) whose zone of contact cannot be exactly determined. Toward the west, P. L. Dekeyser found *P. doguera* as far as Dabola (Guinea).

The northern limit of *Papio papio* (Fig. 6) seems to pass close to Gambia, between the sea and Tambacounda, then climbs to Tagant (18° N. L. Decloitre 1954). The species exists near Bamako and in the rocks of Bandiagara. More to the east (Chad), the limit of the *Papio* genus is around 13° N. The northern limit of *P. doguera* to the east of Chad is not precisely known. One finds the species in the Sudan, for example at Jebel Marra (Setzer 1956, p. 468) and in Ethiopia where *Comopithecus hamadryas* and *Theropithecus gelada* also exist. Protected by the sandstone massifs north of Oudai, the doguera Baboon extends to Ennedi.

62. I am grateful here to P. L. Dekeyser, Chargé du Département de Zoologie at I.F.A.N. for the useful information he generously furnished me.

63. One *Papio atlanticus* (P. Thomas) of the late Pliocene in Algeria (Constantine) is known.

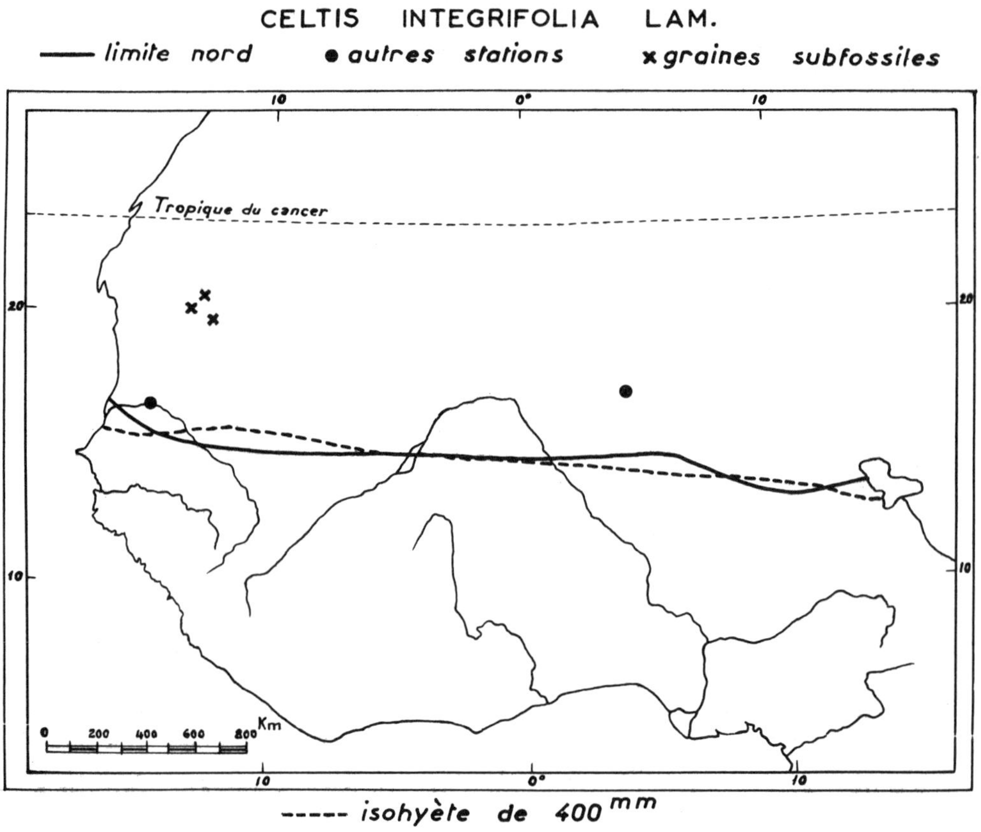

FIGURE 5
Northern limit of the tree, *Celtis integrifolia.*

It is found as well in Tibesti, particularly to the southwest, where it is represented by the subspecies *Papio doguera tibestianus* Dekeyser and Derivot 1960 (see J. Bigourdan 1950, P. L. Dekeyser 1952, and P. L. Dekeyser and J. Derivot 1960, on the subject of the baboons of Tibesti).

The first skull studied showed, in comparison with extra-Saharan *Papio doguera,* a reduction in size, with severe dental wear associated with phenomena of destruction. Reports of five new specimens in 1960 confirmed the previous observations: "Small size and poor physical state are probably the common lot. In particular, dental caries are frequent; with the exception of one young specimen, the teeth are worn, apparently lacking hardness; many are broken or totally missing and traces of violent combat are evident: jaw fracture (in the type itself), premaxillary and mandibulary mutilations having caused the total loss of the incisors and the formation of a scar pad in one of the specimens. In this same individual the entire tibia-fibula is extremely deformed; one can suppose that this deformation is the result of poor knitting of multiple fractures or more

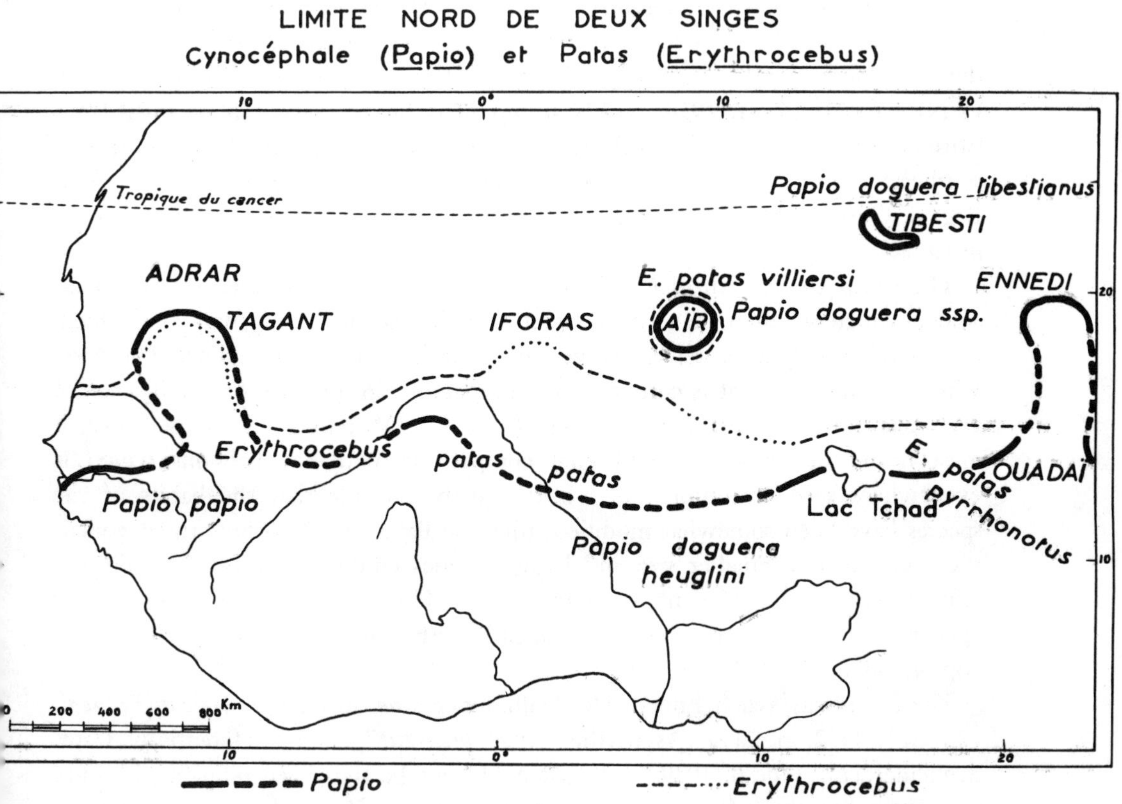

FIGURE 6
Northern limit of two monkeys, baboon and red monkey (Patas).
Sketch with the assistance of P.-L. Dekeyser.

probably a manifestation of an ancient osteomalacy" (pp. 1455–6).[64] Can one think that the population of baboons at Tibesti were submitted to the influence of an isolated phenomenon? Monkeys seem to inhabit only a small part of Tibesti (the southwest slopes) and can live only in certain valleys reminiscent in appearance and in certain respects of the flora of the Sahel. Thus, baboons are hardly less isolated than aquatic relicts, for example the crocodile of the Archei guelta in Ennedi.

Papio doguera exists also in the Aïr Massif, and hardly seems less isolated than in Tibesti. It is, perhaps, a particular subspecies.

The northern limit of Patas (Fig. 9) has been only vaguely determined. The subspecies *Erythrocebus patas patas* has been found from the south of Mauritania through Chad—Lake Rkiz, for instance, north of the Niger bend, etc. The Tagant and the Iforas were cited by L. Joleaud (1933a, p. 852, note 1), but also the Patas in Tibesti. The presence of the species in Tagant and in the Adrar

64. A photo of the Baboon of Tibesti appeared in the *Journal Brossette,* No. 18, 1953, p. 13.

of Iforas is not impossible *a priori*, especially in the more Sahelian southern areas of these massifs. It exists in Aïr, in an apparently isolated situation. A. Villiers did not find it between Zinder and Agadès (P. L. Dekeyser 1950, p. 423, but F. Nicolas (Tamesna, 1950, p. 17) found it as far as the south of Ingal, in a latitude not much more southerly than Agadès: "Dense thickets of the south and southeast [from Azawar]; goes north again in winter to the Tadarast as far as Bagam, G'irmawar and I-n-Wagger, in the *tiggart* (*Acacia arabica*) of the pool's margins."

The subspecies of Aïr is *Erythrocebus patas villiersi* (Dekeyser 1950): "That animal which seems to have no contact with its Sahelian congeners and which, most likely, will never have, evolves in an entirely different milieu. The food which it can dispose of is certainly not comparable, in quantity or quality, to the customary diet of the Sahelo-Sudanese Patas. . . . Moreover, in a closed area, it must find itself in serious and vital competition with the baboons which it everywhere seems to avoid. . . . It is equally possible that the habits of the species have been somewhat modified; thus the Patas of Aïr voluntarily frequent the rocky places. Smaller size and larger canines could, under these conditions, only reflect actions of a physiological order. Thus a body change could only appear since no cross breeding seems possible with individuals in a normal habitat" (pp. 424–5).

The Patas may reach Ennedi (R. Malbrant, *Faune du Centre African Français* (2d ed.), 1952, p. 219). As to the "*singe pleureur*" cited in Tibesti by Capt. Schneider (*Le Tibesti*, 1939, p. 39), it could not be the green monkey; it could be a Patas, but the information requires confirmation.[65] The latter exists at Jebel Marra (ssp. *pyrrhonotus* Hemprich and Ehrenberg; cf. Setzer 1956, pp. 470–1).

Let us add finally that L. Joleaud (1936a, p. 4) cited the vervet, or green monkey (*Cercopithecus aethiops*), in Aïr. The fact has not been confirmed and it is probably an error (confusion with Patas, specimen in captivity?). On the other hand, there is a Jebel Marra *Cercopithecus æthiops marrensis* Thomas and Hinton (*Proc. Zool. Soc. London* [1923], p. 248; cf. Setzer 1956, p. 470).

It is hardly necessary to remind the reader that Sahelian monkeys (and so far as there are any, Saharan monkeys) are neither forest creatures or tree dwellers. They are ground animals of the savanna. Baboons frequent rocky sites (slope debris, cliffs, etc.), although that preference is not exclusive. We will note, however, that at the northern (Saharan) limit of their area we find only mountainous massifs (Aïr, Tibesti, Ennedi), the only habitats in these latitudes able to offer them accessible water (springs and gueltas).

65. Doesn't Schneider also mention the "marmot" in Tibesti?

V. THE HUMAN POPULATION

A. The Skeletal Remains

Again we are dealing with a subject on which there is little to say, in view of the rarity of ancient human skeletal remains. However, three excellent sources allow me to dispense with a discussion of details.

First, we have, by L. Balout (in the *Catalogue des Hommes fossiles* of H. V. Vallois and Hallam L. Movius, Jr., 1952) the section "Afrique du Nord et Sahara" (pp. 255–73 [Asselar, p. 273]), and in the same work, "Egypt and Sudan" by A. J. Arkell (pp. 275–8).[66]

Again, by L. Balout, "Les hommes préhistoriques du Maghreb et du Sahara" (1955 [Asselar, pp. 387–8]).

Finally, a recent article by R. Mauny (1961), "Catalogue des restes osseux humains préhistoriques trouvés dan l'Ouest-Africain," lists 28 Neolithic sites,[67] of which 26 are Saharan or Sahelian. Most of the material has not yet been studied.

Further, "The Prehistoric Background" of "Living Races of Africa" by L. Cabot Briggs (1958) should be consulted.

On the whole, we know literally nothing for the Sahara, neither of the men of the hand axes, nor of the men of the Aterian tanged points. North Africa is a little more favored but the Sahara has yet neither its Olduvai nor its Ternifine.

Despite the rarity of prehistoric human skeletal remains, some hypotheses may be formulated of the history of the populating of the Sahara. The essential references on this question are those of L. Cabot-Briggs (1955, 1958), where all pertinent details are to be found.

The proposed schema, which the author himself qualifies with frankness and humor as "speculative and perhaps seemingly fanciful" (1958, p. 19) is, even if the hypothesis part is necessarily high, a very clear essay. The following are, for the Post-Paleolithic, more or less the main points:

I. Mesolithic
   A. Three distinct groups (of which the first two are probably of Oriental origin) [68]

66. Remains attributable first to the Khartoum Mesolithic (D. E. Derry, pp. 31–3 in A. J. Arkell, *Early Khartoum*, 1959); second, to the Upper Pleistocene (Singa: A. S. Woodward, "A fossil skull of an Ancestral Bushman from the Anglo-Egyptian Sudan," *Antiquity* 1938, 12: pp. 190–5); third, to the Pleistocene (Qau el-Kebir, N . . . , "Fossil Human bones, possibly of Pleistocene Age, found in Egypt," *Nature* 1923, 112: p. 250).

67. The age of Asselar man is still disputed. It had been attributed to the Upper Paleolithic.

68. For the Afro-Alpines, L. Cabot Briggs (1955, p. 89) asserted that their origin could be sought among the brachycephals of southwest Asia: "from or through Palestine."

1. Paleomediterranean (A)[69]—Iberomaurusian; in the Sahara: Aterian?—Possibility of Negroid influences; date of arrival, −10,000 or more.
2. African Mediterraneans (B) Capsian; same possibilities of Negroid influences; arrival, before −8,000.
3. African Alpines (C); arrival same as the African Mediterraneans or a little later; not archaeologically discernible.

B. A mixed group resulting from hybridization of preceding groups.

4. Mechta-Afalou type (A x B x C) (industry: Ibero-Maurusian and Upper Capsian.)

II. Neolithic

5. Northern group: the Mesolithic population evolved, toward the south (Sahara) with possibilities of contact with the Negro or at least Negroid peoples.
6. Southern group: "highly probable if not certain" that Neolithic culture had been introduced to the South Sahara "by half-Hamite Negroes" (*ibid.*, p. 13).

If in this current, first east-west, then north-west and north, the industry included "armature" tools, they would have appeared in the Sahara at the same time from the north (i.e., Capsian tradition) and from the Nilotic Sudan. We shall return to the role of the Nile-Chad or Nile-Tibesti transversal as great axes of population in the Sahara.

In summary, the populating of the Sahara would then be the result of an occupation converging from the north (Mesolithic Capsian, then Neolithic of a "Mediterranean" source) and from the south (Neolithic *lato sensu* of a Negro or at least Negroid origin).

The North African Mediterraneans were the source of the Proto-Berbers and the Berbers who, until the coming of the Arabs, formed a relatively homogeneous group (Teda, Tuareg, Proto-Moors), subject, at contact with Saharan Negroids, to a cross-breeding which was more or less intense, sometimes dominant (Haratines), but always perceptible.[70]

Also, it is not surprising to find the present-day Saharan groups made up of disparate elements such as Berbers *lato sensu* (B), Negroids (and Negro as slave property) (N), Arabs (A). For example: Haratines (N x B), Teda (B x N), Tuareg (B[x N]), Moors (B[x N]), Mozabites (B[x N x A]), Chambas (A[x N]), etc.

The history of Saharan populations, very crudely schematized, is a series of major stages.

First, establishment of prehistoric stocks in the north and south which, having made contact, hybridized more or less in a savannan Sahara with large Sudanese animals, traversed by small clans of hunters and gatherers. If the lacus-

69. The Boskopoid traits of this element could have been acquired on the upper Nile and influenced by other Negro or Negroid influences "on its way west before turning northward into Northwest Africa" (1955, p. 89).

70. More in morphology than in blood: Teda; more in the blood than in the morphology; Moors (L. Cabot-Briggs).

trine harpoon-using aspect is as old here as at Shaheinab, we must admit the existence of sedentary ripicole communities of hunter-fishers. Perhaps on the southern edge of the desert there was a cereal agriculture, or the beginnings of one (millet in flooded areas, floating rice?) and a sedentary peasantry, while bovidian pastoralism seemed to stretch over the whole of the Sahara.

After the Neolithic wet phase, with the worsening of the climate, and the decrease of vegetation (in which men and animals played a part), an insular segregation developed little by little. The big game began to decrease in number. Life "à la peule" became impossible and gave way to the life of the oases and Bedouin and "open-sea" nomadic life. Culture was more and more restricted to small areas and finally to "points"—the oasis, peopled by the Haratine gardeners, islets of sedentary life in the middle of a no man's land now of oceanic extent, where pastoral nomadism and commercial contact demanded new modes of transportation, the horse and cart first, and probably for 2,000 years, the dromedary. Supreme good fortune of the shepherd and the warrior, the camel, beast of burden and mount, permitted, in the heart of a desert about to close up on definitely secluded punctiform oases, the permanence of human life in a Sahara that became "le plus beau désert du monde." Saharans of all description, Berbers of all kinds, future Teda, Tuareg or Moors, etc., settled into the type of economy that has typified them to the present day, resting equally on the herd, the caravan, and the ghazzi.

L. Cabot-Briggs (1957, p. 198), while admitting that it is "little more than thoughtfully reasoned guesswork," gave a good résumé of the little we can surmise concerning the four principle groups of post-Pleistocene immigrants:

1. The Teda "who might be considered negrified descendants of stragglers from the earliest westward migrations of African-Mediterranean proto-Berbers that took place probably as long ago as Middle Mesolithic times";

2. The Haratines which we may doubtfully consider as "having roots in the extreme westward and north-westward extensions of a racially half-Hamitic Negroid migrational flow, coming possibly from somewhere around the headwaters of the Nile, that probably introduced Neolithic cultures into Northwest Africa";

3. The proto-Berbers of the north who "once settled in the high plateaus that separate the east-west ridges of the Atlas Mountain System, began to make contact during the Neolithic with Negro or heavily Negroid peoples to the south, and it may be supposed that in the process they established southern outposts of Berber-speaking mixed Africans-Mediterraneans scattered progressively across the western Desert."

4. Finally, "a series of general northward tidal movements of Negroes from south of the Sahara, which we may suppose gradually submerged both the earlier negroid proto-Harratin and the ancestors of the modern Teda, but left the ancestral Tuareg almost untouched."

After several thousands years of relative Saharan stability, as much ethnic as climatic since the major groups were established and the desiccation, at an end, could get no worse, the irruption of the Arabs in the Middle Ages modified language and religion more than blood or way of life. It provoked, however, that enormous development of trans-Saharan commerce carrying across the desert a gold indispensable to a Mediterranean economy importing Asian luxuries, and black slaves which did not all stop in the Maghreb.

B. Human Habitats in Time and Space

1. Sites and Industries: Kind and Distribution

Obviously, we will not retrace here the archeological history of the Sahara, from the Pebble Culture to the Neolithic—first because I am not capable of it, but also because our purpose is not to go into the details of industrial and typological history.

Thus I shall limit myself, after an attempt to sketch in the general lines, to try to analyze the very modest knowledge which we possess of the habitats, considered in their rapport with their environment, and to whatever can be glimpsed, at the same time, of the geography of the industries, and, if necessary, of technological areas.

One turns to the "Chronologie Préhistorique du Sahara" by H. Alimen (1957a) for a résumé of all Saharan industries.

Very sketchily and considering only the Sahara properly so called—i.e., without its northern steppe borders, the Nile Valley, etc.—we would have the following great "industry families."

First, Pebble Culture, known from the South Morocco, Saoura-Ougarta, Tassili, Fezzan, and Tibesti; not to be confused with pebble tools which can accompany the various Acheulians and even the Aterian.

Then the various hand-axe industries, accompanied by Clactonian flakes and cleavers. The question of the Abbevillian (Chellean) in the Sahara does not seem quite clear; "little or not at all represented in the Sahara," writes H. Alimen (1957, p. 80); later, R. Vaufrey says: "It is possible that among the innumerable hand axes found in the Sahara some were Chellean, but we do not have stratigraphic proof" (1958, p. 108). True, and the generally very dispersed character of the hand-ax industries does not facilitate regroupings by homogeneous groups. On the other hand, almost instinctively, the collector makes a choice, and for obvious reasons when he is riding a camel and especially with "heavy" industries, tends to neglect what is "ugly" (or doubtful) to profit by the beautiful wood-worked hand axe of the evolved Acheulian. Naturally the ages remain unknown because of lack of stratigraphy, and the quotation marks around the word "Chellean" must be kept provisionally; but it seems to me certain that hand axes with Chellean facies abound across the Western Sahara.

While these subdivisions may seem a little theoretical and often difficult to apply to the concrete, and further, since pebble tools and stone-worked hand axes survive from the ages for which they are considered characteristic, it is certain that the Acheulian touched vast regions of the Sahara. It is locally associated with a fauna, including *Elephas* (*Archidiskodon*) *recki* from the Middle Pleistocene (Kamasian, *sensu lato*, in Kenya).

It is difficult to determine what, in the surface collections of the Sahara, dates from the Middle Paleolithic. Objects of Levalloiso-Mousterian technique are certainly noted here and there; even locally "mousteroid" groups can be described without, however, determining their age.

With the Upper Paleolithic, we see the northern (and central) Sahara developing late, from north to south, an Aterian which could extend to the Neolithic. L. Balout (1955) is very positive in regard to this: "Still further south, in the central Sahara, the Aterian attains an ultimate stage: here nothing hinders it until the Neolithic. . . . It lasted in Morocco and the Sahara until, or almost until, the Neolithic. Würmian in its origins, the Aterian seemed to follow the entire retreat of the Würm glaciation, that is to say, the Flandrian transgression, and subsisted at Tanger as well as in Tidikelt or at the foot of Tassili-n-Ajjer, until several thousand years before our era. . . . In the Sahara, the Aterian is as much 'Upper Paleolithic' as 'Mesolithic' " (pp. 334–5).

Let us add that this final Aterian—and the fact is a bit disturbing—which the Neolithic succeeded everywhere,[71] could admit "Neolithizing forms" (Moroccan and pseudo-Saharan points) (L. Balout 1958, p. 312). We know, too, that C. B. M. McBurney and R. W. Hey (1955, p. 240) asked in what measure the tanged point and the pressure flaking from the "Maghreb Neolithic" could find their origin in "a belated Aterian," an important problem since it draws attention to the possible extent of regional evolutions and consequently to that of eventual convergings: "With regard to the predominance of pressure-flaking and bifacial work generally, it should not be forgotten that these are extremely widespread characteristics of primitive agricultural societies of many different ages and regions" (p. 251).

There does not seem to have been either an Ibero-Maurusian or Capsian[72] in the Sahara, and the Mesolithic is still rarely recognized.

The Neolithic is still far from being adequately studied and understood, in spite of the enormous mass of surface collections. R. Vaufrey has insisted, in numerous works, (e.g., 1936, 1938, 1939, 1946, 1953, etc.) on the importance of a "Neolithic of Capsian tradition" on the Sahara:[73] "The Saharan Neolithic

71. The 5th Aterian or "Tingitan" of Morocco is "rightly pre-Neolithic" (L. Balout 1955, p. 311).

72. I am not ignoring the reserves which various industries at Tanezrouft, Tidikelt, Tademait, etc., recall to this subject (cf. L. Balout 1955, pp. 443–4).

73. And one might add, in Africa, since the author meets it again in Egypt, at Khartoum, etc., and considers the South African Wilton as "a neighboring form of the Neolithic of Capsian tradition" (1953, p. 138).

(Mauretanian), a Neolithic of Capsian tradition whose microlithic elements derive from the Capsian [74] was created by successively borrowing actual Neolithic elements fom Egypt" (1936, p. 634). In this layer, which covers the Sahara and makes its influence felt even beyond the southern limits of the desert, one distinguishes three aspects, "typical," "maugrébin" and "Saharan" (R. Delacroix and R. Vaufrey, 1939, p. 311).[75] The latter can allow variants, e.g., "Lacustrine Neolithic" or "Sudano-Mauritanian Neolithic" (1953, p. 129). If there really are (and I will return to this) Neolithic "provinces," born of the local physiography and lithology, one could probably solve the difficulties of admitting the superposition of these industrial groups. Take the *baten* of Tichit, for example. According to R. Vaufrey it is an area of the "Sudano-Mauritanian type," probably the "final stage of the Neolithic" (*ibid.*, p. 129) with perhaps a few "Paratumbian" elements, even ancient. But there is also some "lacustrine" Neolithic, with objects of bone. Undoubtedly the remaining elements of the industry are different here from around the lakes of the Araouan region, but we cannot affirm that the Tichit lakes are necessarily younger. Moreover, they rest on a schisto-phtanitic infra-Cambrian and wash the foot of a sandstone-quartzite cliff, while there exists no rocky out-cropping near Araouan, which could, by itself, account for the differing content of the respective "tool kits."

If it is possible that the Neolithic carriers of a Capsian tradition could propagate certain elements (e.g., microliths and microburins) while moving towards the south and meeting other Neolithics who were meridional and probably Negroid, is it then certain that the various waves of southern Neolithics of the south did not themselves possess, outside of their own diagnostic tools, the usual series of backed blades, orange or lunate quarters, trapezoids, etc.? To put it differently, is it certain that the "microlithism" of Saharan "armatures" is monophyletic and totally derived from a Constantino-Tunisian industry?

The problem of the origin of the bifacial Saharan arrow remains unsolved. The presence of Egyptian influences in the Saharan Neolithic (the pistiliform arrow and the south-Algerian knives of the Fayum type published by L. R. Nougier [1955] have been cited) [76] does not imply that the reappearance of the bifacial retouch necessarily has its point of origin on the Nile. We will touch on this again further on. It would be imprudent to disregard the possible extent

74. Let us remember that J. Bobo (1955) distinguishes, in the Souf, a Neolithic "of Capsian tradition with some Saharan influences" and a Neolithic "of Saharan tradition with some Capsian influences," this last implying "the arrival of people bringing with them perfected tools and armament." This would actually mean two Neolithics.

75. The three modes of Neolithic installation imagined by L. Balout (1955, pp. 453–4) could perhaps correspond to these: *evolution* in place by a "neolithization" of the Capsian; *colonization* in the Ibero-Maurusian territory; finally, by *invasion* in the Sahara: "The new arrivals brought (into Aterian country) a Neolithic already loaded with tradition."

76. One of them comes from "Haei el Hameida"; it should read Hassi (or Haçi); the place is in Algeria and it is difficult to see why it should be "south-Moroccan," except as a historic echo of a situation before the French conquest.

of convergences and re-inventions. The arrowhead is not even peculiar to the Ancient World.

We must be careful to give credit to regional diversities that show periods of relative isolation. For example, C. B. M. McBurney and R. W. Hey (1955) believed they were able to settle on the existence of two North African Neolithic provinces, having as a boundary the meridian of the Syrtic Gulf (pp. 237–69, 272–3).

The eastern province occupies the north of the Libyan desert from Djebel Akhdar (Cyrenaica) to Fayum and Kharga; the much larger western province, comprising Tripolitania and the Maghreb, is found closely bound to the Neolithics of the Sahara (Hoggar-Tibesti, etc.), Nigeria, the Sudan, and the Upper Nile. It should be noted that this hypothesis agrees with that of a SE-NW axis of population, directed from the Sudanese Nile toward Barbary and crossing the Sahara diagonally. Not only could Maghrebian Neolithic elements have come from the southeast (more than from the eastern Mediterranean coast), but others could have had Saharan regions (perhaps certain mountainous areas) as centers of diffusion. This agrees with hypotheses formulated on the Saharan origin of elements belonging to Egyptian or Nile industries.

For H. Alimen (1957) a series of "hiatuses," corresponding to dry periods, were wedged into the Saharan chronology (without implying a general uninhabitability of the desert): during the Pebble Culture, then during the Acheulian, between the Acheulian and the Aterian, the Aterian and the Neolithic, and finally in the post-Neolithic (present).

With the Neolithic we have a period where the provinces are differentiated. This is a novelty for the archeologist because neither the Lower Paleolithic nor the Aterian (in the Sahara) permits, at least for the present, similar distinctions. There was an Acheulian; there are Neolithics. The character of these could have been conditioned one by one or simultaneously, by physiography (intradune, perilacustrine, "tassilian" [cliff villages], etc., habitats), by the local lithology, and probably by idiosyncrasies of the regional group, its own cultural heritage, the influence of its past migrations, the evolutionary potential which characterize it, etc.

Thus it should not be surprising to see prehistorians trying to distinguish, with varying luck, industrial facies more or less characteristic of a region, Tidikeltian with arrows with truncated ailerons, Ounanian with warped augers,[77] Tenerean with necked hatchets, Asselarian, Eglabian and Enjian with arrows of polished schiste.[78]

77. First considered Upper Paleolithic, in fact probably Neolithic.

78. Points of schiste retouched after polishing have been attributed, at times, to the Eglabian. In reality, the Eglabian technique (this name is badly chosen because it deals with a little-known locality in Aouker, and not with the well-known massif of the same name more than 1000 km. away) is of "wholly polished" points (H. Hubert 1921, p. 394). Points polished and then denticulated by retouching, "contrary to the polished points of the Eglabian type" (p. 395) are Enjian (p. 396).

Let us return to a few regional remarks.

Northwestern Sahara. In the region of the Saoura-Ougarta Mountains the Pebble Culture, the Acheulian, the Aterian, and the Neolithic (H. Alimen and J. Chavaillon 1956, 1959; J. and N. Chavaillon 1957a, 1957b; N. Chavaillon-Dutrievoz (1956) are now known *in situ.* Men of the evolved Acheulian seem to have lived on banks or islets, in a relatively wooded environment, with "varied and plentiful fauna" (J. and N. Chavaillon 1957b, p. 630). In the Aterian, there was a complete change in the landscape; the dune invaded the Acheulian areas, man inhabited "little hard-bottom passages of the Quaternary or Pliocene crust surrounded by dunes." There are still little lakes, "narrow dead-arms" constituting the watering spots of the population (N. Chavaillon-Dutrievoz 1956, p. 640). Finally in the Neolithic, human habitation "in the absence of the small lakes, already dried up, was reduced to temporary camping of more or less nomadic tribes"; "less episodically, however," than the present-day long-range nomads (*ibid.*). This idea of a Neolithic already adapted to apparently severe desert-like conditions is interesting. It does not seem to apply to the entire Sahara, where the Neolithic Subpluvial seems often, on the contrary, to be sedentary, steppe-like, and bovidian.

Southwestern Sahara. Generally, we can state that there were two principal industries, one of which (Acheulian) was connected with rocky plateaus, the other (Neolithic) to dune sands. Undoubtedly, a heavy tool assemblage is connected, of necessity, to the outcroppings which furnish its main materials (in this case, quartzites) but physiography must also be considered. One wonders if the extreme abundance of hand axes in Adrar does not correspond to a habitat more or less limited to the southeast by the extension of lakes or swamps from the Majâbat. If the white sands (SB) of the latter are really fluvio-lacustrine, it is necessary to admit the existence of a lake as large, perhaps, as the Caspian Sea, and which, after its drying out, could not be refilled by the last pre-Neolithic pluvial.[79]

The distribution of hand-axe industries in the Adrar suggests that two types can be distinguished. While the plateaus are literally strewn with hand axes, giving evidence at least of an active circulation [80] and being perhaps the oldest examples,[81] we find elsewhere, in segregations connected perhaps with a more concentrated habitat and linked with water sources (river banks), an evolved Acheulian.[82] Habitation can therefore be very dense (e.g., at El Beyyed, and over 200 sq. m.

79. If this is the case, we would like to see here an argument to prove that the Neolithic lakes of the Araouan type were principally fed by fluvial supplies whose origin was then less distant, extra-Saharan (Haut-Niger, Chari-Chad) or not (Tilemsi, Azaouak).

80. L. Balout (translated from: 1952, p. 107): "All the evidence points to a human population not dense but having at least explored all that immense territory, through constant roaming over the millennia."

81. As much for morphological and technological reasons (absence [or only rarity?] of cleavers) as because of physical state (from corroded to very corroded).

82. Even very evolved, with quasi-Micoquian elements, and sometimes already on the dried out lacustrine beds (Ijâfen).

of a site extending over several acres: 30 hand-axes, 2 spheroids, 227 various flake-tools.[83] Are these concentrations the sign of a climatic evolution? Do they reflect a change in social behavior? We do not know for sure.[84] Similarly, we cannot imagine what the landscape looked like—for example, the type of vegetation for a period which is sometimes said to be "warm and humid"—for the lack of indispensable paleontological data.

If the Aterian is practically unknown in the southwestern Sahara, the Neolithic abounds there, probably ubiquitous although particularly connected with the dunes [85] on the whole. Archeological provinces can certainly be distinguished there, corresponding to regional facies. It cannot be proved that these are more a matter of successive types than of local responses to different environmental conditions (physiography plus lithology).

On the part of the Sahara which interests us here, one is tempted to distinguish from this point a principal type (with variations) of very great distribution—the Sudano-Mauritanian of R. Vaufrey (1953), and two more localized subtypes.

The generalized type (Adrar to Azaouad) [86] is, finally, the common Neolithic of dunes with its usual assemblage of scrapers (very abundant), blades, microliths, arrowheads, polished celts, lower grindstones and grinders, potsherds, etc. The influence of regional lithology makes itself felt by the development, especially perhaps in the west, of an industry of phtanite plates (hatchets, adzes, chisels, gouges, etc.).

Aouker and the Baten of Tichit (with the dependent reliefs—Rkiz, cliffs of the Dhar) are rich in grindstones, in shafts of grooved sandstone. Particular types of polished points (of Eglabian or Enjian workmanship) appear there, with very numerous axes with polished cutting edges; there are a few bone objects (harpoons and fish hooks).

The half-worked, half-polished axes would probably be placed by R. Vaufrey in the "para-Tumbian" whose presence he indicates in that region (1947).

In fact, for that author, this "para-Tumbian"—common in Guinea, Mali, etc., and representing "a whole tool assemblage of cultivators and carpenters, including grindstones, pestles, and grinders," (*ibid.*, p. 228)—would extend to Hodh and Aouker. It suggests primitive farmers "spread out in elementary families who are forced by the poor quality of the soil to frequently shift the place of their exploitations" (p. 231), while the "hunters and breeders of the Saharan and Maghrebian Neolithic who had merely to pick their vegetable foodstuffs, live in groups" (*ibid.*); these Saharan Neolithics should have

83. R. Mauny, 1962, fig. 6.

84. Do we even know if the accumulations of hand axes like that of El Beyyed indicate a habitat in the true sense of the word, or a secondary concentration—unlikely, considering the extreme newness of the pieces—or even a gigantic workshop?

85. There are other types of habitat: in still humid valleys (Amder, with *Celtis-Limicolaria*), on lacustrine shores, at springs (Ksar Torchane), on cliffs (Dhar Tichit-Walata), etc.

86. These two toponyms are used solely to "situate" in a general way an area which, in fact, begins farther west and goes farther east.

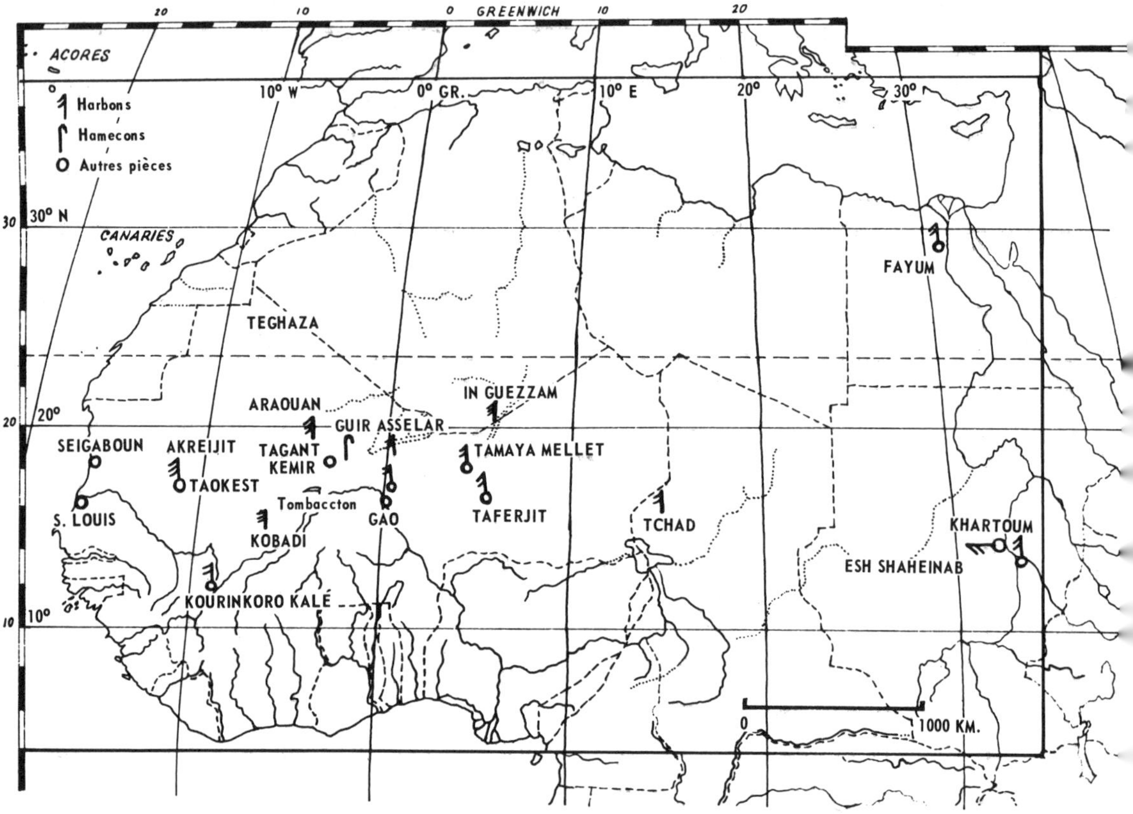

FIGURE 7

Sites yielding bone implements. From Th. Monod and R. Mauny 1957, Fig. 1. Since that date three other sites have been discovered; see R. Mauny and F. Poussibet 1962, Fig. 1, map.

succeeded the "para-Tumbian" farmers. The hypothesis is interesting, but locally the distinction between "para-Tumbians" and Neolithics seems quite difficult. Were they distinct, moreover, how would one know if they succeeded each other, and in what order? It is more probable that the Neolithic of Aouker and Baten, itself with Negroid racial support, in contact with agricultural populations in the south, was marked by their technological influence.

A "lacustrine" facies, characterized by bone harpoons and fish hooks, closely reminiscent of the harpoons of Shaheinab (Figs. 7–8), developed timidly in Aouker but more forcefully in Azaouad (and farther to the east) (Th. Monod and R. Mauny 1957). If it were contemporary with the latter, it should be considered as a relatively archaic Neolithic.[87] Is it this, or is it "an evolved but not final Neolithic"? (*ibid.*, p. 246). We still do not know, just as we do not

87. L. Joleaud (1935, p. 13) considered this "Nigritian" Neolithic of the Azaouad as previous to the "Saharan" Neolithic.

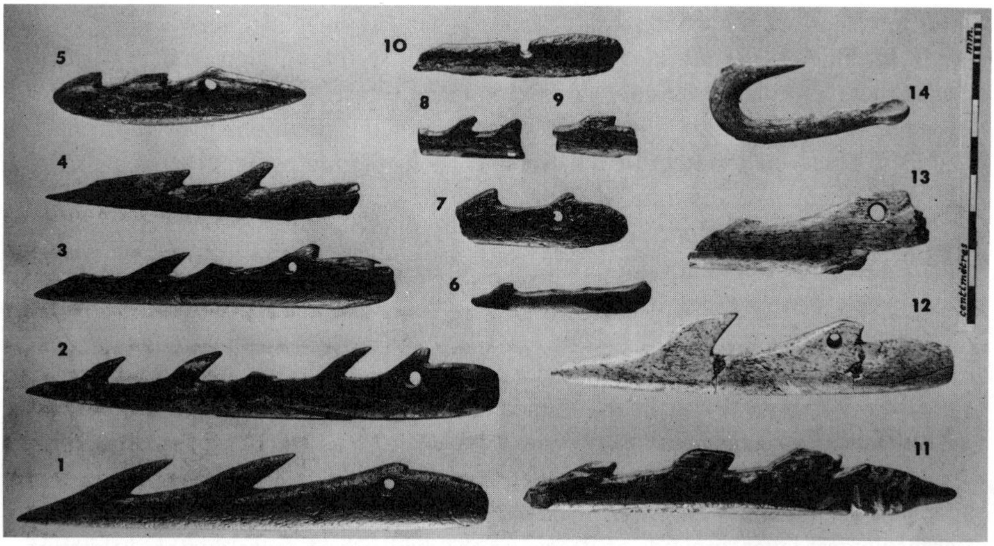

FIGURE 8
Bone implements of the South-Saharan lacustrine Neolithic: Araouan region (1–10), Chad (11), Kobadi (12–13), Karkarichinkat south (14). From Th. Monod and R. Mauny 1957, Fig. 2.

know of its chronological relations with that of Aouker. Only direct datings will clear it up.

In visiting the Neolithic villages of the Tichit cliff one cannot avoid the impression of their relative youth, for in places well protected by rock shelters there is still a soil of fine dust with potsherds sticking up through it, and bits of charcoal.

The dune settlements raise a series of problems. While vast sandy surfaces, the undulating plains of the Le-Mreyye for example, are relatively poor in Neolithic remains and were probably only traversed or visited (perhaps by groups of hunters) interdune settlements in systems of heavy bands, or intra-dune with funnel systems are of a truly incredible richness. Northeast of Ijafen, for example, where I had to cross 40 such bands, I do not believe I crossed a single interdunary corridor without discovering signs of Neolithic.[88] On the other hand, in the "funnel" regions (Adafer, Aklé, Aouana, etc.) each funnel encompasses at least some Neolithic. I counted in the course of a one-day camel stage on a linear itinerary 156 sites of this type. Admitting that an observable surface during a day's march is 25 $\times$ 250 sq. km. and that the area of funnel formations is 100,000 sq. km., with 50 and 100 funnels per 50 sq. km., one gets a total of

88. Which means that if one had followed the corridor instead of cutting across it, he would have found it to be a veritable Neolithic highway which one could follow for perhaps 200 km. or more.

1,000,000 and 2,000,000 Neolithic sites. The evaluation is very crude and perhaps too low, but such astronomical figures stagger the imagination.

It is difficult to imagine the nature of this Neolithic intradunary habitat. Was there a "village" or only an encampment (more or less temporary) in the funnels? Or a Neolithic countryside previous to the partitioning of the funnels and still based on a system of transverse cordons permitting long habitable strips between them? At any rate, it is necessary to accept a large-scale grid pattern of an occupation, if not locally dense and thick [89] at least in a thin cover, perhaps diffuse but ubiquitous. Was it a question of: few people in one place, but people everywhere? In any case, one sees nothing comparable today where the groups with a semiarid habitat are either cultivators grouped in villages, or more or less nomadic hunters-gatherers.

But is it certain that these Neolithics, in spite of their grindstones and pottery, were sedentary and practiced agriculture? We will return to this question. I am ready to state, however, that in spite of the incredible abundance of remains habitation was sparse. One does not find—or no longer finds—a true archeological layer, mounds of debris, or the evidence of more durable habitation sites. The general impression is that of a country actively traversed more than a territory occupied by sedentary peoples. Were these hunter-gatherers or pastoral people? The abundance of heavy objects (grindstones and grinders) as well as of pottery implies rather, in the case of non-sedentary populations, domestic animals and even pack animals. It would be important in the future to collect the ungulate teeth of these sites systematically.

The Chadian Sahara. Before turning to the paleoecological conclusions to be drawn from the rock paintings, let us consider the technological aspect of the facts. The little there is on this subject will be substantially illuminated by the future publications of G. Bailloud and A. J. Arkell, especially for Ennedi.

The importance of the Chad-Nile "hiatus" applies equally to the direction of the east-west shifts, in particular to the southern Sahelian fringe of the desert, and to the possibility of currents from the opposite direction assuring the arrival from the Sahara of certain elements as far as the Nile Valley which appeared foreign (e.g., Wavy Line pottery of Sudanese Meso- and Neolithic) to that region.

G. Bailloud (1958, pp. 16–18) cites, for Ennedi, various Paleolithic industries (Acheulian, rare; Levalloisian, well represented; no Aterian; tools of the leptolithic type); one Mesolithic industry (in caves and rock shelters, Wavy Line pottery of Khartoum type); several Neolithic industries distinguished by their pottery: Early (in caves or rock shelters, Dotted Wavy Line pottery), Middle (in caves or rock shelters; Hohou type pottery), Late (hut floors; increase of polished celts—perhaps in relation to the increase in wooden dwellings—and stone bracelets; "fine red" pottery) and the Terminal, passing

89. If kitchen debris, mounds of rubbish, etc., existed, little remains today: a ruminant's tooth or a piece of bone.

into the Iron Age. A certain correlation is established between the archeological sequence and that of the rupestrine styles: Iron Age—Camelin, Late Neolithic–Late Bovidian, Middle Neolithic–Middle Bovidian, Early Neolithic–Early Bovidian(?).

A. J. Arkell (1959) indicates the evolved Acheulian near Sarra and between Tekro and Ounianga, the Aterian at this last point, Dotted Wavy Line pottery from the Nile to Ennedi. The Simple Wavy Line could have remained Nilotic, but we have seen above that G. Bailloud indicates it in Ennedi. The question is far from clear, and in a more recent paper (1962) Arkell, contrary to his previous suggestions, adopts the view that both types of Wavy Line pottery originated in the Nile valley and spread west from there possibly even as far as Ahaggar.

Libyan Desert and Nilotic Sudan. K. S. Sandford on the one hand (1933, p. 219) and R. F. Peel and R. A. Bagnold on the other (1939, p. 292) have insisted on the relation between Lower and Middle Paleolithic distribution [90] and elevated land in the Libyan desert. It has not been observed on the surface of sandy sheets. For K. S. Sandford (1933, p. 219) this is the result not so much of the total destruction of stones (a hypothesis which may be thought a bit surprising) as of the fact that the sands were "inattractive or uninhabitable, even for hunting." In the case of heavy tools (hand axes) should not one think first of the proximity of the primary material?

One observation worthy of attention—and pathetic—concerns a case of human over-crowding, reported by K. S. Sandford (*Geog. J.*, 82 (1933): 127): "One of the most remarkable things that I saw in the whole expedition was the extraordinary concentration of late Middle Paleolithic implements in the Laqiya depression. Man seems to have been packed into that small area rather like sheep. He could not live outside at that particular time. His implements are amazing in their abundance, which exceeds any concentration that I have ever seen in the far more habitable Nile Valley. One of the disasters of the human race in this part of the world is seen in the concentration in an area and, finally, in complete withdrawal."

The attempt at a Paleoclimatic sequence for the south of the Libyan Desert merits reproduction here (*ibid.*, p. 222):

1. Lower Paleolithic. Virtually whole of present desert area available, at least for hunting, except perhaps inner parts of present sand sheets. "Laterite"-forming climate still persisting between W. Hawa, central oases, Sarra Triangle, and western mountains (?).
2. Middle Paleolithic. Major deterioration of climate, "laterite"-forming climate ended (?).

90. "Neither Myers nor ourselves noticed a single Upper Paleolithic implement throughout the trip, and it seems to be generally agreed that the Libyan desert was uninhabitable during this period"(K. S. Sandford 1933, p. 292). Is this remark still valid?

3. Lower Sebilian. At close, aridity complete, all oases abandoned, desert virtually uninhabited except along borders, arid area at its maximum (?).

4. Later Paleolithic. The sandy *goz* south of W. Hawa probably recolonized by plants, but probably desert at outset. Oldest rock pictures (?).

5. Mesolithic, Neolithic, Recent. Some water established across *goz* from mountains towards Nile. Alternations of rainy and desert periods in country between W. Hawa, central oases, Sarra Triangle, and western mountains.

W. B. K. Shaw remarks, however (1933, pp. 223–4), in attributing to cultivators the lower grindstones found between Tekro and Uweinat, that "given a slight increase in rainfall this country would probably grow crops": in north Kordofan cultivation is carried on with around 250 mm.

This was not the last climatic oscillation causing human displacements and especially movements from the desert to the Nile Valley. Around —2500 according to O. H. Myers (1939, p. 289) a deterioration followed the Saharan Neolithic subpluvial which perhaps caused the west-east movement, or movements, before the introduction in the Nile of the "Saharan" pottery of the Nubian C-group, "and the last pulse action of the Sahara was thereby caused, affecting cultures as remote as southern Europe and Nubia, and perhaps even farther. In the origin of this people we can perhaps look to the lake areas explored by M. Théodore Monod." [91]

In the Nile Valley, one could imagine Saharan influence at a much earlier date. At Khartoum (A. J. Arkell 1947, 1949a, 1950a, 1955) they were "Mesolithic negroid fishers and hunters." Pottery (Wavy Line) was found, but there was no proof of domestication nor of agriculture. The lower grindstones and grinders found there could have been used for grinding fragments or clay for pottery. The presence of hygrophilous mammals (*Onotragus, Atilax, Thryonomys*), of mollusks of the genus *Limicolaria* and seeds of *Celtis* imply a more humid climate (probably around 500 mm. as against 164 at present). This "Khartoum Mesolithic," sometimes regarded as already Neolithic, has been related, because of its harpoons, to the lacustrine south-Saharan Neolithic. For A. J. Arkell (1949, p. 115) there existed "a mesolithic fishing and hunting culture common to negroid peoples between 19° and 15° N. all the way from near Timbuctoo to Kassala."

Not far from Khartoum, at Shaheinab (A. J. Arkell 1949b, 1953, 1955), the Khartoum Neolithic or "gouge Culture" seems a little younger than the preceding site, with polished pottery, polished celts and gouges, bone celts and harpoons, fish hooks of shell, amazonite beads, etc. There are domestic animals (sheep, goats, perhaps dogs) but still no proof of agriculture. There are so many resemblances to the Fayum Neolithic that A. J. Arkell thinks Fayum A and Shaheinab could have been contemporary.[92] The climate is dryer than at Early Khartoum,

91. Lacustrine Sudan-Saharan Neolithic.

92. E.g., circa —3900, in spite of the deviation in known C-14 dates, —4200±250 and —4440±180 for Fayum A, 63110±450 and —3490±280 for Shaheinab (cf. A. J. Arkell 1955, p. 345).

and the three more or less paludal mammals of the latter site were no longer to be found at Shaheinab.

A. J. Arkell (1950, pp. 125–6; 1955) wonders if certain elements did not come from the west, for example, the Tibesti region, from where they would have reached the Sudanese Nile (harpoons), then Fayum (harpoons and arrowheads). He even supposes that the Shaheinab amazonite could have come from the site which I discovered at Egueï Zoumma in the Libyan Desert (1948). That seems a bit far, in any case.

2. THE EVIDENCE OF ROCK ART

L. Cabot-Briggs in 1958 (pp. 15–17, 38) gave a pertinent critique of the archaeological value of rock engravings and paintings, the study of which may sometimes do harm to more commonplace, more difficult, and more efficacious archeology. Why undertake laborious digging "when one can catch the public eye and public funds so much more easily with pictures of rock paintings that allow of romantically farflung and fascinating interpretations?" (p. 38). This sally, which expresses the opinion of some North African prehistorians, contains a healthy warning against certain temptations to do the easier task. But the interest and importance of rock paintings remain nevertheless great as long as one never loses sight of the many handicaps that effect their study: rare possibility of archaeological dating, drawing "from memory," successive copies, uncertainties about patinas, styles, the individual element and "schools," errors or jokes, etc. All this is true, but in representations from memory, when they are abundant (the case of an ocean liner [93] would be different), one surely finds an element of the local surroundings. When the rock artist draws giraffes, cattle, or chariots, he has, if not before his eyes at least nearby, his big game, his herd, or his cart.[94] The author concludes that the rock paintings give us "little if anything" that

93. Or that of marine animals, of which representations have occasionally been described in the interior of South Africa (e.g., Walter W. Battiss, "Prehistoric Fishing Scenes," *S. Afr. J. Sci.*, 41:356–360; H. Breuil, "Sea Animals Amongst the Prehistoric Rock Paintings of Ladybrand," *ibid.*, pp. 353–6; H. L. Wells, "Marine Animals in a Rock Painting near Fouriesburg, O.F.S.," *ibid.*, 42:236–9; C. Van Riet Lowe, "Rock Paintings of Marine Animals in the Interior of South Africa," *S. Afr. Arch. Bull.*, vol. 2 (1947), Part VI; A. J. H. Goodwin, "A Fishing Scene from East Griqualand, *S. Afr. Arch. Bull.*, 4:51–3; Walter W. Battiss, The Artist of the Rocks, 1949, pl. XI. Representations of boats have been recognized right in the middle of the Sahara (e.g., P. Graziosi 1942, pl. 69 and 103, and J. M. Freulon, *Trav. Inst. Tech. Sahar.*, 11:125, pl. I).

94. G. Charles-Picard, in his article "Civilisations antiques au Sahara" (*La Revue de Paris*, 65:132–7) defends the peculiar hypothesis that the Sahara cart drawings are only "memories": "The camel drivers relaxed some time between trips in the great cities of the province. While there, they succumbed to that folly [the hippodrome]. On their return to the country they illustrated with naive images their stories of the marvelous festivals which they attended." This strange supposition does not withstand examination. We would have to acknowledge that the proto-Libyans lent to the urban "jockeys" of the "great cities of the province" their shield, their javelin, their plumed headresses . . .

ordinary archaeology could not reveal to us. Is this true? How many giraffe bones have been discovered in the Sahara, where its representations are abundant? How many remains of horses, carts, javelins, or "Libyan" shields, etc.? And is one not surprised at how much a work such as P. Huard's *Préhistoire et Archéologie au Tchad* (1959b) owes to rock pictures?

For the last 25 years, Saharan rock art has been the object of a great number of studies. The number increases each year. Although many are only summary reports of discoveries, some are serious contributions to the paleoethnological exploitation of the material or to a chronology of the figures. We will not attempt a general, even rather succinct, summary covering the entire problem. I will limit myself to some analyses and regional comments, while stressing that which can shed light on the environment, the ways of life, or the ethnic character of the groups under discussion.

The naturalistic art of the hunters, with its large incisions, is dated in the Southern Maghreb where it is associated with the Neolithic of Capsian tradition (R. Vaufrey 1939, *ubi. litt.;* L. Balout 1955, pp. 474–9). Should one associate with it the very similar art of the Tassili-Fezzan region? R. Vaufrey (1936, p. 625) is convinced of this: "It is very likely that the two arts are the work of the same men and go back to the same archeological epoch." L. Balout (1955, p. 479), however, seems inclined to separate them: "If one can follow its extension [South Oranais art, etc.] to the south, at least as far as the Beni-Abbés region, there could be no question of identifying it with the art galleries of the central Sahara. These belong to another world which I consider more recent, at least in part." This remark is plainly justified if one thinks of the Tassilian paintings, but are the great archaic naturalistic drawings of *Bubalus antiquus* in Fezzan so different from those in Barbary that represent the same species?

If the eastern limit of the big game engraved in naturalistic style seems to be situated near the meridian of Tibesti (P. Huard 1960, p. 566, note 2), the Marhouma station, studied in detail by H. Alimen (1954), although clearly "marginal" as to style, is connected to the South Oranian complex in certain elements ("intellectual and cultural relationships"). A paleoethnological analysis of Group II or "principal" (pp. 55–81) provides useful information about the life of Neolithic men in the Sahara—searching for food (neither domestic animals nor agriculture), rites and cults (hunting, fertility, orants, and gods). As to environment (p. 98) there must have been ". . . an important vegetation, and large areas of water, at least a high humidity. Flamingos, shoe-bill,[95] marsh birds and fish-eating birds evoke (as do wart-hogs, buffalo, and rhinoceros) vast muddy areas[96] probably established on the present location of fossil sebkhas. Outside

95. A zoologist would probably hesitate to identify these two birds, and further, flamingos are strictly microphagous.

96. Here, a zoologist would hesitate to connect the Phacochere, which abounds in the dry Sahelian steppe, to a normally muddy environment.

of the sebkhas, on the ancient terraces of the Saoura and in the ravines which cut the flanks of the Djebel, groves of trees grew, probably more or less related to talhas,[97] present day tamarisk and palms. . . ."

The Spanish Sahara and Mauritania, with several archaic engravings, have furnished a particularly large number of Libyco-Berber graffiti, of little interest except for those portraying chariots, most often two-wheeled biges, some four-wheeled. An inventory of rupestral Saharan chariots found up to 1950 has been given by Monod and Cauneille (1951) and since completed by H. Lhote (1953, 1957b) and R. Mauny (1955). At the latter date more than 250 chariots were known, only for the former French West Africa. These representations pose various problems—e.g., origin and geographical distribution. There is a tendency to seek the origin of all Saharan chariots rather automatically in the Eastern Mediterranean. A work which John Paget is preparing may bring new views on this. As to the routes that chariots could take across the Sahara, the evident distribution is, for the regions under discussion, the same as that of the representations and, obviously, that of the rocks on which they were executed, when canyons and plateaus probably did not especially favor these vehicles. At most, it is probable that the most likely zones were chosen to bypass the rocks, and that vast surfaces (of reg) could be crossed regularly, aside from any possibility of rock-art representation.

One finds in the central Sahara (Fezzan included), the domain where abundance and variety of representations are the richest. For example, think of the astonishing harvest of Tassilian paintings recently gathered by H. Lhote (1957–1958), in addition to what Tassili, like Fezzan, had already yielded to E. F. Gautier, M. Reygasse, R. Perret, L. Frobenius, P. Graziosi, H. Breuil, etc., and the complexity of material that H. Lhote reports, for the Tassilian paintings alone—some 16 stages and 30 styles.

There is no doubt that these prodigious archives on stone constitute a sort of historical epitome from the Neolithic (perhaps early) to present day, capable of giving us a great deal of information on the various populations involved, their way of life, their arms, their domestic animals, their game, etc.

As for the general chronological framework, the one that K. W. Butzer proposed in 1958 (pp. 43–4) largely corroborates long accepted divisions (bubaline, bovidian, equidian, and camel groups; precamel-camel group; groups of hunters (Sudanese fauna), shepherds with large cattle, horsemen, camel nomads, etc.).[98]

1. Neolithic hunters (ca. —4500 to —3600). Big game: hippopotamus, crocodile, rhinoceros, elephant, ancient buffalo (*Bubalus*), giraffe, ostrich, antelopes.

2. Shepherds and Neolithic nomad hunters (from ca. —4000). Herds of bovids and big game; retreat of the most significant species, hippopotamus, rhinoceros,

97. *Acacia raddiana.*

98. It is useless to insist, I think, on the ultra-schematic character of these divisions and on the great overlapping which they admit, as I have tried to emphasize (1951, p. 201, Fig. 2).

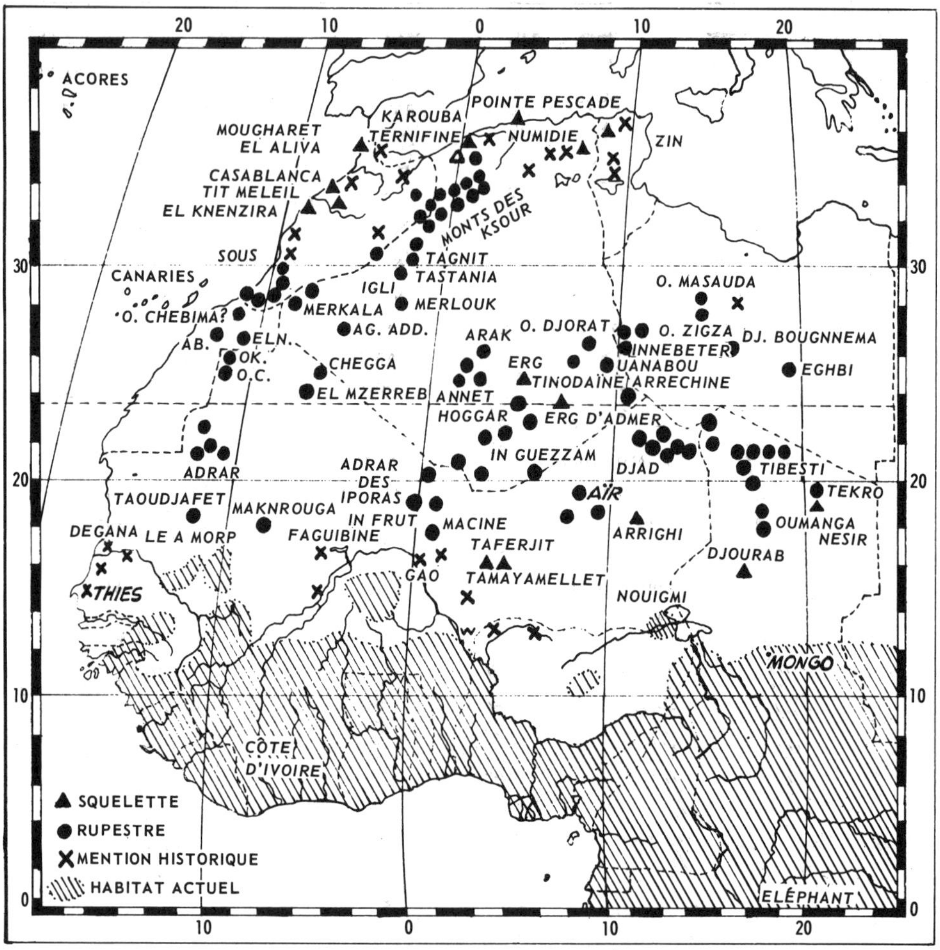

FIGURE 9
Ancient distribution of elephant. From R. Mauny 1957, Fig. 3.

and elephant, which disappeared in the east after 2750 and later (around 2000 B.C.) in the central Sahara.

3. Warriors and war chariots (after —1500 in the east, towards —1200 in the center). Horse, giraffe, ostrich, antelopes.

4. Nomadic camel drivers (from our era). Camel, ostrich, Barbary sheep. In 1956 (p. 270) R. Mauny adopted the following groups for rupestral Saharan sites:

1. Naturalistic group, with large Ethiopian fauna (—5000 to 2000)
2. Neolithic group with bovidian shepherds (—2500 to 1000)
3. Horse group (—1200 to our era)
4. Libyco-Berber group (—200 to +300)
5. Arab-Berber and modern group (after +700)

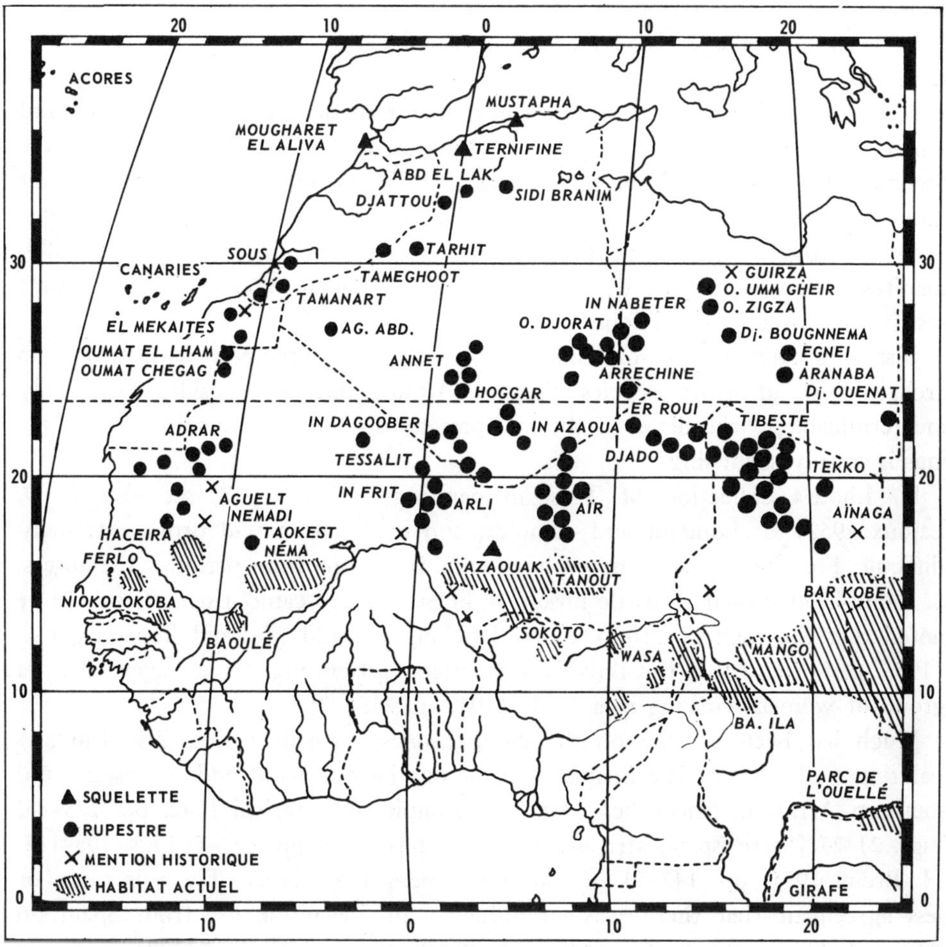

FIGURE 10
Ancient distribution of giraffe. From R. Mauny 1957, Fig. 4.

All of the numerous proposed chronologies are variations on this general theme.

In 1953 P. Graziosi established three large groups for Libyan representations:

1. Hunters
2. Cattle herders
3. Camel drivers.[99]

J. Tschudi (1955) distinguishes an equidian stage (caballin): "1. Epoche der Jäger und der alten tropischen Fauna. 2. Pastorale-Epoche" [with agriculture]. 3. "Garamantes" Zuchtpferd-Epoche. 4. "Cameline" Epoche. In a meritorious

99. There were, perhaps, cattle in 1., and the horse stage with bitriangular people (javelins and round shields) is not individualized; it constitutes a 3rd subgroup of 2.

endeavor the author attempts a division of psycho-social traits of various groups: 1. a "stark entwickelter Krieger, und Jagdsinn"; 2. a "hervortretende Kastenunterschiede, warscheinlich bedingt durch rassische Verschiedenheiten (Rassenwiechungen und dorgestellt durch Ünterschiede in den Kopfformen (lang und schmal, breit und rund"; 3. a "starke Tradition in Familienleben und wichtige Stellung der Frau. Die Frau tritt neben dem Mann auf, ja sie reist selbst mit ihm und halt wie er die Züge der Pferde. Sie ist meistens mit einem starken Fettsteiss dargestellt, was ebenfalls auf matriarkale Formen hinweist"; 4. finally, manifests "Freude an Fortbewegung und Reise." One wonders whether the data justify such precise deductions.

Just as a zoologist is sometimes surprised at the precision with which an archaeologist identifies species that he would himself consider reasonably indeterminable,[100] the detail of the interpretations made on the subject of paintings is equally surprising.[101]

H. Lhote's collection of Tassilian paintings (1956, 1957; see also J. D. Lajoux 1962) is abundant and complex, and exploitation will prove extremely difficult. For the moment the author seems to accept four periods: 1. hunters; 2. round-headed men; 3. cattle raisers; 4. horsemen and camel-riders. It should be noted that the men in group 2 could belong to "old Negroid human stock" (1956, p. 54) and that the Bovidians practiced agriculture "as is suggested by a group of women working in a field" (1957, p. 235).[102]

Much has been written on the comparison of Tassilian (and also Libyan) painting with that of the Spanish Levant on the one hand and of eastern and southern Africa on the other (see, for example, P. Graziosi 1942, pp. 273–85, Figs. 21–25 ["arte ispano-africana"]; H. Rhotert 1952, pp. 121–5, Figs. 108–114; H. Breuil 1955, pp. 147–51; P. Bosch-Gimpera 1955; etc.). There is more or less agreement that there was an influence of Levantine art from Spain on Africa: "How can we fail to recognize, in this great bovidian [Tassilian] art, the moving reincarnation of the artistic spirit of the Spanish Levant, prolonged into the Egyptian art of the New Kingdom (1500 to 1300 before our era)?" (H. Breuil 1957, p. 108.)

The similarities are not doubtful—they are sometimes identical—but do they

100. The identification of two engravings at Wadi Zirmei (NE of Tibesti) as *Megaceroides algiricus* (A. J. Arkell 1959) appears debatable to me; the jugular pendeloque (of the specimen in Fig. 8) is rarely found, to my knowledge, except on domestic bovids.

101. In H. Breuil's 1955 work on the paintings of Tassili, there are many captions of this type: "Woman cooking near a shelter," "Marriage celebrated in the family," "Plumed archer with trousers, walking on hide rug," "Scene of the sale of one woman by three others (Josephine!)," "Tavern scene," "Footrace," "Market of Women." This latter is the sale of young white girls; to the right and left two salesmen "praise their goods." None of these subjects is unlikely, but how can one be so certain?

102. We admit the argument is debatable: how can an agricultural scene be distinguished graphically from one of gathering or digging?

necessarily imply diffusion and in the indicated sense? [103] In all cases the intervals are enormous and regional groups remarkably insular. And to the northwest of Tassili, towards the Spanish Levante, we find scarcely anything but the engravings of the bubaline group, without known landmarks for the paintings, bovidian or prebovidian.

Tibesti, thanks to the numerous publications of P. Huard, has become one of the regions whose rupestral representations—engravings and paintings—are the best known and studied. Here again we meet that general sequence which seems most applicable to the whole of the Sahara, at least in its broad outlines.[104]

A. Neolithic hunters
   Hunters-engravers (big Ethiopian fauna: Gonoa); [105] after ca.–4000
B. The contribution of eastern Hamites—pastoral phase, bovidian stage.
   1. Archaic; still quite near A, but with attempts to domesticate indigenous cattle, *Bos ibericus* (Gonoa, Gira Gira); ca. –4000
   2. Early; in the northwest: hunter-shepherds (Oudingueur, 1st layer); on the Libyan slope: shepherds-painters
   3. Middle; (Bardai; Oudingueur, 2nd layer; Areun, 1st layer); ca. –2000
C. Horse-camel group
   4. Late (phase of the pastoral series); Iron Age—beginning of the Christian era
   5. Modern phase

For Ennedi, the major part of known rupestral representations (reported by G. Bailloud) is still unpublished, but a preliminary description (G. Bailloud 1958) indicates a typically Saharan chronology:

1. Prebovidian period (drawings) (Early Neolithic?)
2. Bovidian period
   a. Early (drawings and paintings) (Early Neolithic?)
   b. Middle (drawings and paintings) (Middle Neolithic)
   c. Late (drawings and paintings) (Late Neolithic)
   d. Final (drawings and paintings) (coming of iron)
3. Camel period (Iron Age)
   a. Early
   b. Late
   c. Modern

The Libyan Desert has yielded, in turn, a considerable mass of representations (cf. the classic works of H. A. Winkler 1938, 1939a, 1939b; P. Graziosi 1942; H. Rhotert 1952 (*ubi. litt.*); K. W. Butzer 1958; etc.). If the classification principles adopted by H. A. Winkler do not seem directly applicable to the

103. L. Kohl-Larsen (1960) seems to acknowledge a displacement from South Africa to East Africa, and thence to the Sahara.

104. See, for example, 1957, p. 216; 1960d, pp. 179–201, with "Aire des Chasseurs néolithiques du Tibesti nord-ouest" (Fig. 1) and "L'apport hamitique oriental au Tibesti nord-ouest" (Fig. 2).

105. What P. Huard wrote of the hunting techniques (1959c, p. 40) is most interesting.

entire Sahara, it is because the author had to take also specifically Nilotic elements into account. The schema follows:

1. Pre-Cattle period
   a. Earliest hunters
2. Cattle period
   b. Early Oasis dwellers (Dakhla); sedentary, cultivators, probably in contact with archaic hunters: "A hunting people joins a cultivating people. Probably the hunter shared the porridge of the Oasis man, and the Oasis man got his giraffe joint when the hunter came back from a lucky expedition." (1939, pp. 34–5).
   c. Autochtonous mountains dwellers; ". . . a race of cattle breeders, wearing the Libyan sheath, probably the forefathers of those peoples who appear in historic times as speaking Hamitic languages" (1939, p. 308).

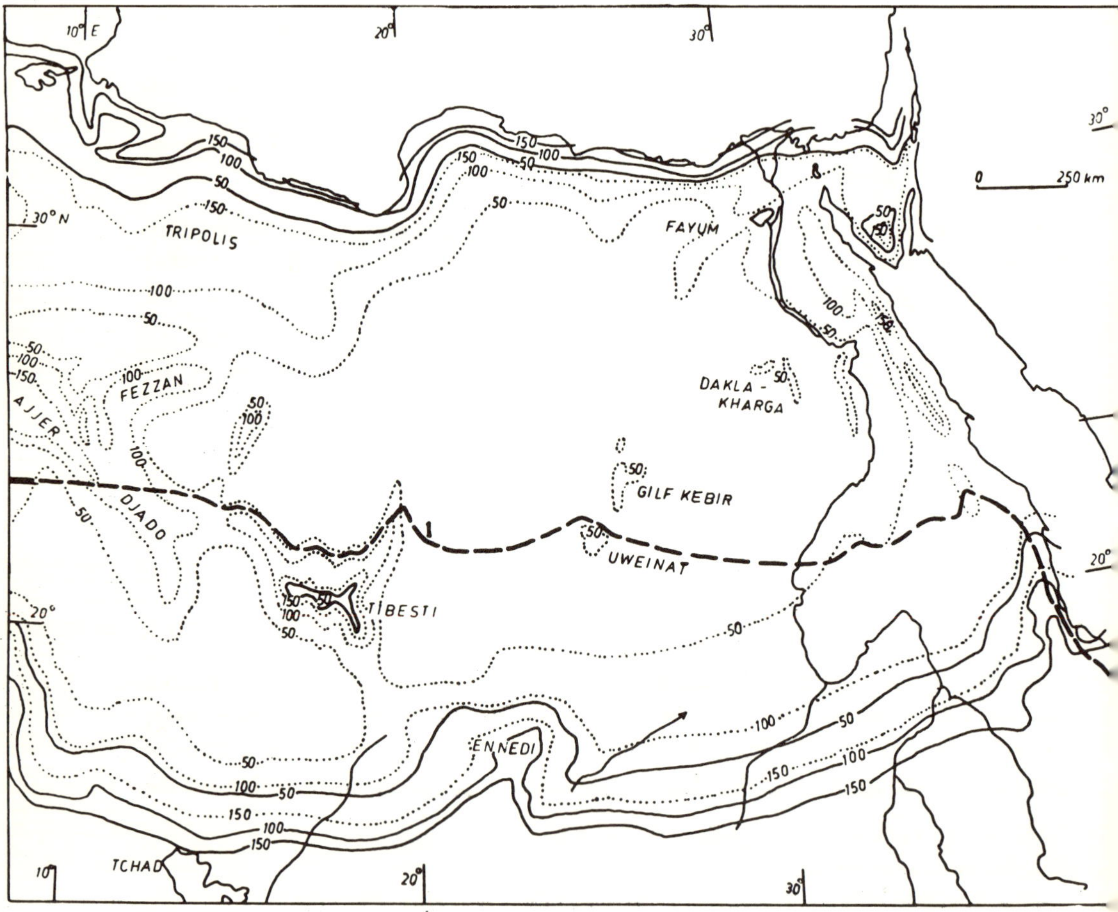

FIGURE 11
Present-day and Neolithic isohyets in the eastern Sahara. From K. W. Butzer 1958, Fig 2; see modified version of the same map in 1961, Fig. 3.

d. Eastern invaders; arrived from and via the Red Sea (with "straight wooden boats, probably dugouts, with high vertical prows and sterns") (*ibid.*).
e. Early Nile Valley dwellers; characterized by a type of papyrus boat; more or less contemporary with the Gerzeans.

The hypothesis which attempts to attribute locally (Gebel Aouenat, for example) the engravings to men and paintings to women must be considered very debatable (cf. H. Rhotert 1952, p. 113).

The importance of Rhotert's work, *Libysche Felsbilder* (1952), cannot be emphasized enough. When H. Breuil (1957, p. 109) wrote that "Saharan art, starting with the great hunt, continues for a long time in the pastoral life (unmixed with agriculture) of the Saharan cattle herders," and when he speaks of a fusion of two currents "connecting the still hunter-like pastoral life to the great art of the West," he concurs with the hypotheses of H. Rhotert. Rhotert says that, as far as engravings are concerned, the rupestral bovidian Saharan art is born of the contact of immigrant Hamitic shepherds with the hunter-engravers already there. The paintings, more recent on the whole, show the penetration from northwest to southeast of distant influences of Spanish-Levant origin. This is a possible explanation; time will tell how much of it to retain. The total absence of landmarks between Spain and the Central Sahara is annoying, even if the rock medium is not obligatory and the paintings, often small or very small in size, could have been done on other surfaces: clay, ceramic, wood, skin, etc.

While H. Rhotert closely analyzed the content of the representations and the paleoethnological information to be drawn from them, he expresses himself with a happy sobriety: "Was wermögen uns diese Bilder über das Leben der Viehzüchter zu sagen? Es ist wenig genug: dass sie Herden hatten, dass sie die Kühe wahrscheinlich von hinten gemolken, und dass sie in Hütten gelebt haben" (p. 103). One cannot go much further, once the objects or present details have been carefully catalogued (the bow, for example), without risking a frivolous interpretation, as tempting as it is dangerous.

Has the landscape changed very much since the Neolithic? It does not seem so at all. Already the pastureland of the massifs are becoming insular, one no longer goes—especially with cattle—from Aouenat to the Nile,[106] and at the end of the Subpluvial the Libyan desert plain is "kaum noch bewohnt" (p. 116). Concerning the valleys which still disclose a strip of vegetation in the heart of inhospitable mountains, H. Rhotert experienced the impression that one has on the very edge of habitability: "hätten diese Täler (Karkur Talh, Aouenat) ein wenig üppigere Vegetation, die sie ja sicher einmal gehabt haben, und einige Wasserstellen, dann waren sie das denkbar günztigste Weideland . . ." (p. 12),

106. O. H. Myers (1939, p. 289) supposed that the pottery of C-Group (ca. —2500, Third to Sixth Dynasties) could have reached the Nile Valley, whether from Aouenat, via Bir Tirfawi and Bir Sleb, or chiefly via Erdi, Merga, Laqiya, Selima, and Wadi Halfa, or even by a more southerly route, well frequented in the Meroitic era, via Wadi Hawa, Wadi el Malik and Ed Debba.

and the landscape remained pleasant: "Karkur Talh macht . . . einen romantischen, freundlichen und wohnlichen Eindruck. . . . Das Grün der Pflanzen ergibt mit dem Gelb des Landes und dem Rot des Gesteins einen wohltuenden, lebhaften Farbeindruck" (p. 11). However, how many thousands of years has it been since the bovidian archers and herdsmen disappeared?

Finally, let me cite a singularly evocative page from N. Heseltine (1960, p. 146), discovering at Mourdi a spectacle which took him right back to the middle of the Bovidian period: "In the last clear light at sunset, when the dust and the heated air seemed suddenly to clear, so that every distant hill stands out in sharp outline, we came on a sight which was worth much fatigue to see. Among scattered dunes, tall grass and shrubs and solitary trees a huge herd of cattle moved, lowing as they do in the evening, lit up by the last rays of the sun. A grassy plain stretched as far as I could see, broken only by jagged blue hills on the eastern horizon; the Saharan sky was lifted up in an immense dome, vaster than any sky in the world, dwarfing the trees and hills on the plain, and under it streamed the level line of beasts. Round their flanks ran naked boys with spears, followed by the greyhound-like local dogs which have slightly webbed paws for running on the sand. So this was the Mourdi, and there before me was the Sahara of ten thousand years ago, preserved in this remote corner on the edge of the blank spaces on the map. This was how the cattle people had looked, moving slowly over Tanezrouft, and Ténéré, reg and hamada, when the grass flourished and the wadis were rivers. The boys with their spears, and even the shape of some of the cattle could be from a frieze painted by other cattle people fifty centuries before."

3. WAYS OF LIFE AS FUNCTIONS OF ENVIRONMENT

We must recall here the excellent articles by J. D. Clark (1955 and 1960). The characteristics of a given culture depend on a series of factors: physiography, climate, natural resources (their nature and quantity), finally the "culture-pattern" itself, partially dependent on environment of course but also charged with "internal elements," traditional and racial heritage, facts of acculturation, its own evolutionary potential, etc. Cultures are modified under the triple influence of changes in the environment, borrowings from exterior contact, finally from immigration of new racial stocks.

However, it should not be forgotten that an external factor, for example a change of climate, can by itself only constitute a stimulus provoking a response from the part of the organism involved, an autonomous, spontaneous, unforeseen response, capable of going from resigned surrender to successful invention. S. Huzayyin (1956, p. 312) put it well: faced with a serious deterioration of environment, man could accept the lethal character of the aggression and disappear, or adapt himself to the new conditions by perfecting his tools, for example, or seek through emigration the promise of a new biological equilibrium.

The environment sets the scene, but man remains the actor. Ph. V. Tobias (1961, p. 34) showed this clearly with regard to the Bushmen: "A people whose curious, genetically controlled anatomy plays little part in adapting them to desert life, whose physiological responses acclimatize them to hot, dry conditions, and whose cultural pattern finally makes it possible for them to bear the extremes of heat and cold that characterize their harsh environment. In the Bushman it is not too much to say that we see epitomized the stages of man's evolution—from slow, structural, genetic adjustment, through smooth and reversible functional accommodation, to swift, intelligent, cultural adaptation and, with this, the final mastery of the world around him." [107]

Note that the environment would not always remain what it still is for a primitive man, a simple element, *inter pares*, of the biocenosis. It would not be long before man (from the hunt by fire, probably, and at least from forest clearance by the axe) [108] became himself a factor in the alteration of environment,[109] in areas to an unexpected degree, as indicated by the titles of several articles by W. Knoche (1937, 1938a, 1938b, 1939) of which one is called "Über die Möglichkeit anthropogener Auslosung von Seismen."

The physical framework. If human activity had a major dependence on the milieu for such a long time, this itself would be conditioned through a series of factors whose global aspect does not prevent research into the particular, operating on the human population either directly (natural resources, e.g. game or comestible vegetable products), or indirectly (climatic fluctuations, seasonal or not, influencing wild animals or plants).

I will only call to mind two groups of factors here: physiography and natural resources.

It is presently almost impossible to propose a coherent table for the physiography of the Pleistocene and the post-Pleistocene. Indeed, the large morphological units, mountainous massifs, plateaus, basins, etc., were there when man appeared. But it is impossible to imagine, other than in quite general terms, the aspects of the various Saharan regions for these remote periods. We do not even know if, in periods of habitability, abundantly witnessed by, say, the dispersion of hand axes, it is necessary to imagine, on the slabs of present day structural surfaces scarcely covered with sand, a soil capable of supporting forest vegetation. It is not certain, and one can think that the Adrar of the hand-axe man, for example, resembled areas physiographically and lithologically comparable, situated today in the Sudanese zone, with wooded savannas and open forests. In fact,

107. See, also, P. V. Tobias 1961a.

108. One should recall the experience reported by S. Jorgensen, "Forest Clearance with Flint Axes," *Fra Nationalmuseets Arbejdsmark*, 1953, pp. 109–110 (cited by S. Cole, *The Neolithic Revolution*, 1959, p. 35, Fig. 25, pl. X): three men, with silex polished axes, cleared 500 sq. km. of birch forest in four hours.

109. In K. J. Narr (1956) one can find pertinent remarks on the respective roles, for various cultural levels, of the adaptation to environment and the modifying activity of environment.

we know nothing, except that the few elements of known fauna seem more savanna-like.

Natural resources. There is no doubt, and archeology sometimes bears this out, that hunting as well as gathering (and locally fishing) played an important role in prehistory.

We are reduced to pure supposition for the Paleolithic, but with the Neolithic (in the "cultural level" sense more than the chronological sense of the word), the evidence is growing, based simultaneously on industry, the content of rupestral art, and the inventory of kitchen middens.

We may admit that, on several occasions, the Sahara has experienced more humid phases which spread out either Sahelian or Sudanese vegetation (wooded savannas, thorn savannas, etc.) or Mediterranean vegetation (Aleppo pine and juniper forests, etc.) across the surface of the Sahara.

It is, thus, allowable to visualize what game man had at his disposal. In addition to a few palaearctic (e.g., wild boar and Barbary sheep) or extinct types (ancient buffalo = *Bubalus*), there were species which still populate the tropical African savannas: [110] elephant, rhinoceros, giraffe, various gazelles, zebras, wild asses, wart hogs, etc., and, more locally, hippopotamus.[111]

In the Sudanese Sahara and the Sudan, the Neolithic kitchen middens yielded numerous remains of mammals (elephant, hippopotamus, rhinoceros [Esh Shaheinab, Tiouririne], buffalo, antelope, gazelles, wart hogs, carnivores, rodents, etc.), birds, reptiles (in particular a fresh water tortoise, *Trionyx*), fishes (particularly Nile perch [*Lates*] and various Siluriforms) and mollusks.

It goes without saying that the habitat will offer quite a few more animal resources, birds and their eggs (particularly ostriches, bustards, guinea fowls, water birds), reptiles (lizards and tortoises), various fishes, locally crustaceans,[112] various insects [113] (especially locusts, but also larvae and caterpillars, gnats,[114] etc.), mollusks (*Limicolaria* and perhaps aquatic Lamellibranches).

Semiarid environments are in reality much less unproductive for hunter-gatherers than a superficial analysis would lead us to believe. The alimentary

110. We refer here obviously to large animals.

111. H. Lhote (1951, 189–90) speaks of Acheulian "camps" at Tihodaine and Admer with bones of big game (antique elephant, giraffe, hippopotamus, zebra); C. Arambourg (C. Arambourg and L. Balout 1955, p. 283) cites *Elephas* cf. *recki*, white rhinoceros, a zebra (*Equus mauritanicus*), hippopotamus, *Bos primigenius*, a wildebeest, an eland, a strepsicere, a bubalis, etc., but without mentioning any direct relation between the fauna and industry.

112. Crabs (*Telphusa fluviatilis* to the north, *Potamonautes* to the south) or Branchiopodes of temporary pools. The *Artemia salina* is still eaten in the Fezzan, (cf. R. Bellair, La Ramla des Daouada (Fezzan), *Trav. Inst. Rech. Sahar.*, VII, 1951, pp. 69–85, 3 fig., III pl.).

113. The eating of insects, very common among monkeys, "belongs to his [prehistoric man's] phylogenetical tradition" (F. S. Bodenheimer 1951, p. 19).

114. The fact is known among the Ounia of Ounianga in the Chadian Sahara, and has even a mythological connotation with the story of Mide, the "goddess" who eats flies (L. Carl and J. Petit, *La ville de sel*, 1954, pp. 121–4).

lists given for the Australians and the Bushmen are in this regard very instructive. For example, the following list on the western Australians, published by Sir George Gray (cited by F. S. Bodenheimer 1951, p. 75), includes: "six kinds of kangaroo, twenty-nine of fish, one whale, two species of seals, wild dogs, three kinds of turtle, emus, wild turkeys, two opossums, eleven species of frogs, four kinds of fresh water shellfish, every sort of sea shellfish except oysters, four kinds of edible grubs, eggs of birds and lizards, five animals of the rabbit class, eight snakes, seven iguanas, nine species of mice and rats, twenty-nine different roots, seven mushrooms, four species of gum, two kinds of manna, two species of by-yu (nut of the Jamia palm), two sorts of mesembryanthemum, two of small nut, four of wild fruits, beside the seeds of several plants."

Along with the hunt, collecting is practiced, its intensity varying according to regions and seasons. The collected vegetable may be a delicacy, a condiment, sometimes a useful complement to an insufficient diet; at times, if famine threatens, it may be a means of survival, pounding hearts of palm or date pits into powder.

In certain regions of the Sahel and the Sahara, gathering may play an important role. J. Chapelle, writing on the Toubou (1957, pp. 191–5), gives an excellent idea of the role of the wild plant in human nourishment. For wild cereals, "their harvest calls for veritable expeditions of women using asses and sometimes camels."

Regarding the work of R. Story, "Some Plants Used by the Bushmen in Obtaining Food and Water,[115] J. G. Adam and I have noted (*Bull. IFAN*, 22, 1960A, No. 3, p. 114) how interesting it was to "find with the Bushmen, in a very similar biological setting, a population whose alimentary customs give an idea of the dietary habits of the Sahelo-Sudanese gatherers, prehistoric or not."

However, in just the Sahara and its Sahelian fringe, the number of wild plants capable of providing nourishment is relatively high. We may classify the principal plants as follows: [116]

1. Mushrooms: *Terfezia* spp.
2. Rhizomas and pulpy stems: *Cistanche Phelypaea*, *Cynomorium coccineum*, *Typha elephantina*.
3. *Brèdes: Gynandropsis pentaphylla*, *Portulaca oleracea*, *Rumex* spp., *Eruca sativa*, *Schouwia purpurea*, *Solanum nigrum*, etc.
4. Seeds of herbaceous plants: *Boerhaavia* sp., *Chenopodium* spp., *Aizoon canariense*, *Cassia italica*, *Colocynthis vulgaris*, *Limeum* sp., *Glinus lotoides*, *Glossonema boveanum*, *Rogeria adenophylla* (S), *Blepharis* spp., etc.
5. Caryopses of graminaceous plants: *Panicum turgidum*, *Panicum* spp., *Aristida pungens*, *Cenchrus prieurii* (S), *C. biflorus* (S), *C. ciliaris* (S), *Sporobolus spicatus*, *Sorghum* sp., *Echinochloa* spp. (S), *Latipes senegalensis* (S), etc.
6. Fruits: *Salvadora persica*, *Ziziphus* spp., *Balanites aegyptiaca*, *Cordia gharaf*, *Cocculus pendulus*, *Boscia senegalensis*, *Maerua crassifolia*, *Capparis aphylla*, *Celtis*

115. *Bot. Surv. of S. Afr.*, Mem. No. 30, 1958.
116. S designates a more particularly Sahelian species.

integrifolia, Rhus oxyacantha, Grewia tenax, Grewia spp. (S), *Sclerocarya birrea* (S), *Hyphaene thebaica* (exocarpe and corozo), etc.

7. Aromatic plants *à infusion* (often more or less medicinal): *Solenostemma argel, Artemisia* spp., *Brocchia cinerea, Paronychia* spp., *Cymbopogon schoenanthus, Myrtus nivellei, Mentha* spp., *Salvia chudaei, Ocimum* spp.

8. Gums and nourishing mannas: gum of *Acacia* spp., manna of *Tamarix.*

In the unfortunately little known work of Créac'h, *Aliments et Alimentation des indigènes du Moyen-Tchad* (Marseille 1941), is a chapter of great interest, with lists of species,[117] concerned with the vegetal foods used as substitutes by the Chad Arabs. For these Arabs, living in a Sahelian environment, the author notes (pp. 41-55) among the substitute vegetal species a group of 30 wild grasses, including the "kreb" on one hand, various rices on the other, several plants *à infusion,* one oleaginous plant (seeds of *Balanites*), 4 species with edible subterranean parts, 4 with edible stalks, 12 with edible leaves, 7 with edible fruits, 10 with edible grains, and finally, 3 with edible gum (Acacias). One may with profit consult the chapter of A. Chevalier, "Plantes spontanées du Sahara et de ses confins utilisées dans l'alimentation" (pp. 803-14 in A. Chevalier, 1932) and the recent work of P. Simonneau (1960).

On the whole, collected vegetal foods seem important first in certain massifs, especially Tibesti,[118] and secondly in the Sahel with the "Kreb" grasses, harvested most often with baskets.[119] Even though collected foods may be appreciable locally and seasonally, they are still only a supplementary source of nourishment.

The seeds, or more exactly the cores of *Celtis,* merit some attention (Fig. 5). These small, hard, wrinkled cores have often been found both in prehistoric stations and in fossil soils and are frequently associated with a terrestrial mollusk (*Limicolaria chudeaui, flammata* and *kambeul*). We know a) a *Celtis* sp. at Kharga (E. W. Gardner 1935, p. 504, pl. XXXII, Fig. 6); b) *Celtis integrifolia* at Khartoum and Shaheinab (A. J. Arkell 1947, 1948, 1949, 1949a, and 1953), with *Limicolaria;* c) the same species in the Adrar in Mauritania (Iriji, Toungad, Amder (Neol.), Amatil), with *Limicolaria chudeaui;* d) *Celtis australis* at Meniet in the "Upper Saharan Neolithic" (P. Quézel and A. Hugot 1957); e) some seeds of *Celtis* in a Neolithic pottery in the eastern Sahara (Ehi Dohar, 20°5′ N–14°45′ E, unpublished; R. Mauny, oral communication). Some impressions of *Celtis* fruits have been found in the Paleolithic from Hadhramaut (G. Caton-Thompson and E. W. Gardner, *Geog. J.,* 43:28). A. J. Arkell has remarked on several occasions that the presence at Early Khartoum or Shaheinab of *Celtis* as well as *Limicolaria* implies a rainfall of at least 450 mm., more than three times that of today. The *Limicolaria chudeaui* which abounds in a subfossil state in the southwestern Sahara (cf. E. Fischer-Piette 1949, p. 234) has been considered by P. Jodot (1951,

117. Where the botanical names are often to be taken with caution.

118. Where the industry of the Coloquinte seed is especially developed.

119. But sometimes with sickles (J. Chapelle 1957, p. 193), a fact that is very interesting for the functional interpretation of certain prehistoric sickles (Natufian, for example).

p. 168) as indicating "an abundant vegetation of underbrush[120] or of humid forests."

The same author holds that present-day Limicolaria require "the great humidity of rain forests, of wooded savannas or of pastures along rivers. . . ." (p. 176). Actually, the *Limicolaria Kambeul* (which could well be identical to *L. chudeaui*) lives in Senegal as far as the St. Louis region, in a very Sahelian and notably semiarid environment. It is no less certain that the subfossil *Limicolaria* of the Sahara bear witness to a less desert-like climate, perhaps Senegalese but no more than that. As for *Celtis integrifolia*, it is a bank-dwelling species in the Sahelian zone (edges of pools and intermittent streams) but it descends in the south to the forest borders (Fig. 5). Its northern limit is clearly marked by an isohyete of 450 mm. which allows us to think that in the Neolithic the Mauritanian Adrar had a Senegalese climate.

Hunting[121] *and Fishing*. For the various levels of the Paleolithic,[122] while in other parts of Africa certain interesting deductions can be made (J. D. Clark 1960), we know practically nothing for the Sahara. It is useless, therefore, to guess the hunting methods which hand-axe men, for example, might have used.[123]

We are a little better informed about the Neolithic. The rock paintings and drawings constitute, as indicated above, an astonishing gallery of animal art where wild fauna, alone at the beginning, continues up to the singularly impoverished form of Libyco-Berber graffiti where ostrich, oryx, or Barbary sheep hunts are still often shown. Early Neolithic drawings are of hunters armed with bows and perhaps throwing-sticks. Shields, javelins, and lances seem to have appeared later, along with iron. The chapters by Lhote: "La Chasse au Sahara d'après les gravures et peintures rupestres" (H. Lhote 1951, pp. 201–30) are very helpful on this.

If Neolithic groups of hunters ignorant of domestic animals existed in the Sahara—and petroglyphs bear witness to this—it is likely that some hunting methods still in use today were used then—for example, various kinds of traps (one with radial points), while the bow is no longer used. Hunters of the bovid period, since exploitation of big game and breeding, had at that time, as well as today, to be associated, remained essentially archers; but later hunter-herdsmen acquired javelins with metallic points and shields. Horse and cart are not long in appearing and one has only to add camel and gun to the picture to reach the Sahara of today.

In this, the hunt of the Tuaregs (G. Brouin 1950; H. Lhote 1951), of the

120. Probably: "treelike."

121. See H. Lhote, "La Chasse au Sahara dans les temps préhistoriques in H. Lhote," 1951, pp. 187–99.

122. In *Wildbeuterkulturen*, it is still necessary to determine what, in the Paleolithic, belongs to *Urkultur*, to the *Grosswild-Nahjäger* and *Fernajäger* (cf. K. Dittmer 1954).

123. To judge the importance of fire (Omer C. Stewart 1956, p. 120), would we not have to know when its use appeared: in the Acheulian?

Tedas (J. Chapelle 1957), and of the Nemadi (D. Brosset 1932), for example, can certainly give an idea of a Neolithic hunt. The snares, the running hunt, hunting with dogs, with nets,[124] the production of dried meat, etc.—all this is certainly very old [125] and there have been exceptional cases here and there, of individuals or groups, of survivals of a very archaic type of life. The hunter from Takolokouzet (Aïr), dressed in hides, an "example of what men in ancient times were" and a "true bush-animal himself" encountered in Aïr (G. Brouin 1950, pp. 447–9) is, in this regard, symbolic.

The tradition of *Steppenjäger* is still there, close to the surface, and perpetuates itself in a certain number of Sahelian or Saharan groups of professional hunters—Nemadi,[126] Azza,[127] Haddad Cherrek et Nichab,[128] Gow.[129]

Fishing is known to us only through the fishhooks and harpoons of the lacustrine Neolithic as already mentioned.

From Gathering to Agriculture. The excavations at the Neolithic site of Esh Shaheinab have not furnished one sign of agriculture, but emmer (*Triticum dicoccum*) and barley are known in Egypt from various Neolithic sites (Fayum, Merimde, etc.) from about –4000 (V. and G. Tackholm 1941, pp. 242–3, 288–9).

Did the Saharan Neolithic have some kind of agriculture or not? It is generally admitted, at least implicitly, that the heavy tools (querns, pounders, grinders, pestles) are to be taken as "agricultural."

This impression, which I myself have had for a long time, is often expressed. H. Hubert (1920, p. 435) suggests Neolithic agriculture for northwest Mauritania. The grinding stones imply the pounding of some grain, whether imported or, more probably, home grown and the worked axes were probably hoes. For N. Heseltine (1959, p. 26) in Ennedi "the millstones show that some settled agriculture was practiced—or at least trade in grains."

According to A. Chevalier (1938, 1947, 1949a, 1949b), a Saharan protoculture may date from the end of the Paleolithic, and the Sahara (with its southern frontiers) may have been a true "center of origin for cultivated plants" (e.g., dates, certain jujubes, Guinea corn, bulrush millet, digitaria, rices (*Oryza glaberrima*), various common culinary plants, etc.). To an "ancient" (sorghum and millet) culture, new cultures of eastern origin may have been added (wheat, barley,

124. Very interesting schemas of the Addax hunting with nets in G. Brouin, 1950, fig. 5, and in J. Chapelle, 1957, fig. 1, p. 203.

125. Unequally old, to be sure, certain methods being anterior to domestic animals, others not.

126. Cf. Th. Monod, "Chasseurs sahariens et sahéliens," *Bull. Intern. Com. on Urgent Anthrop. and Ethnol. Res.*, No. 2 (1959), pp. 99–100; *Inst. für Völkerkunde* (Wien), Bull. No. 2, 1959, p. 24.

127. Cf. G. Brouin (1950), pp. 449–55; and J. Chapelle 1957, pp. 201–7.

128. H. Carbou, "La région du Tchad et du Ouadaï, I, 1912, pp. 49–72. Publications de la Faculté des Lettres d'Alger. Bull. de correspondance africaine, vol. 47/48.

129. A. V. Dupuis-Yakouba, *Les Gow ou chasseurs du Niger, Paris,* 1911.

pulses, etc.).[129a] "The querns that one encounters in rather large numbers in the Sahara could have served as well for grinding wheat and barley as for indigenous grains" (1938, p. 321).

R. Schnell, for his part, does not hesitate to write (1957, p. 91; cf. also pp. 105–133, 130): "Neolithic sites often rich in querns abound in the savannas and even up to the Sahara, where this cereal agriculture must have been quite developed."

We have no material proof for the Saharan Neolithic as we do for *palafittes* or Egypt, proof in the form of cultivated grains, of the existence of agriculture, since the "cake of petrified meal" indicated by G. le Rumeur in a kitchen-midden at Tamaya Mellet (1934, pp. 303–4, pl. I, Fig. 5) has never been studied.[130] A. Pons and P. Quézel mention in the Neolithic fossil soils of Meniet some "traces of the cultural activity of man" in the form of "cereal pollens" (1957, p. 35).

As for the Neolithic of the Kharga "peasants," G. Caton-Thompson, concludes: "Although, unlike Fayum, no concrete evidence for agriculture was forthcoming, the artifacts and grinding stones proclaim it . . . ," [131] while at Armant, S. Huzayyin states that, "as to the grinders and the quern there can be no mistake about their use:" these objects are "connected primarily with agricultural food." [132]

Concerning Neolithic Saharan populations who used heavy grinding material, (e.g., Ouaran, Aklé, Aouker, Azaouad, etc.), we might suggest the following in support of the agriculture hypothesis:

1. the existence of an identical material related to an evolved agriculture (Fayum, etc.);
2. the presence of permanent settlements, sometimes of actual villages;
3. the relation to the grindstones, etc., of an enormous quantity of ceramics: [133]
4. the possibility that many of the "polished axes," in particular those with asymmetrical edge, were in reality hoes;
5. the difficulty of imagining an agglomeration of sedentary people or even of seminomadic cattle breeders supplying a large part of their diet and all of their vegetable foods from gathering;
6. the apparently constant association of the quern with agriculture in West Africa (where it exists, that is) (cf. R. Schnell 1957, *passim;* A. Prost 1954, p. 80; L. Pales 1955, pl. L, LI, LXIV).

129a. The hypothesis of a Saharan evolutionary center (perhaps "Sahelian" then) of plant cultivation and of successive "layers" of food species, the first local, the others introduced, fits into the C. O. Sauer-H. von Wissmann sequence which admits both a central western hearth, Africano-South-Arabic in origin, of cultivated grains of *Hirsenbau,* and the arrival from elsewhere of a *Halmgetreidebau* of wheat and barley.

130. The photograph alone does not establish the nature of the object, with any certainty.

131. "Kharga Oasis in Prehistory," *London,* 1952, p. 38.

132. In Sir R. Mond and Oliver Myers, "Cemeteries of Armant 1," *London,* 1937, p. 225.

133. Practically absent in the culture of steppe-hunters (e.g., Bushmen).

7. the fact that among the Tuareg of the central Sahara the wooden mortar seems to be used generally for millet, dates, cheese, and grasshoppers, while wheat is ground on the quern (J. Nicolaisen 1954, Fig. 12).

Recently A. Balachowsky, in a work devoted to insects harmful to Saharan cultures affirmed that there appeared in the Saharan Neolithic "not one trace of agriculture." [134] He evidently believes, although not saying so, that the grinding tools were used for wild grains. A quern could, of course, serve this purpose (A. Prost 1954, p. 80) and many other things (grinding earth for pottery, condiments, pigments, perfumes, etc.). Among the Paiute of the Great Basin, who are in a sense the Bushmen of North America, the gathered grains are crushed with a pounder on a flat grindstone (C. Daryll Forde, *Habitat, Economy and Society*, Methuen & Co., London, 1939, p. 36, Fig. 13); in A. Maurizio (1932) there is similar mention of the use of grinding stones for the grains *Elymus giganteus* and *Arundo villosa* among the Mongols (p. 69), as well as for the "wild rice" (*Zizania aquatica*) of North America among the Indians (p. 72).

Let us add that the sickle itself does not seem necessarily associated with agriculture.[135] In discussing Natufian sickles, K. I. Narr (1959, p. 121) recently concluded that they imply neither agriculture nor the gathering of wild grains,[136] and could have been used on other grasses (reeds, straw, etc.) destined for other uses (basket woık, thatch, etc.).

For Tibesti, P. Huard (1960, p. 198) concludes that "toward the end of the first millennium, the material life of the population of Bardaï-Aozou, dressed in skins, was based on animal breeding and the gathering of edible wild seeds. . . . There is no certain proof that they (the Teda) were acquainted with cereal or date culture, practiced in Herodotus' time at Augila and Fezzan."

The question of Neolithic agriculture in the Sahara deserves some attention, and anyone interested in the question should not fail to read G. Camp's chapter, "Un Problème non résolu: l'agriculture préhistorique," with a whole section on sickles (1960). The list of arguments given above in its favor does not take into account certain distinctions. One should proceed by regions, and not speak of the "Sahara" generally; also only the most general meaning can be attributed to the word "quern," or "lower grindstone." Finally, we should not lose sight of the fact that in principle, sedentary agriculture implies, in all probability, the village.

What do we know for certain? On one hand we have many remains scattered

134. Mission Scient. Tassili des Ajjer, III, Zool. pure et apliquée, *Inst. Rech. Sahar. Univ. Alger.*, 1958, p. 9.

135. H. Von Wissmann (1957, p. 87) and others (*ibid.*, note 30) seem to admit the conclusion sickle = agriculture; the cereal could be a "millet."

136. Which are gathered in baskets or with rods. Note however that J. Chapelle (1957, p. 193) mentions the use of a sickle for wild grasses among the Toubous. This should be substantiated.

so extensively that we can hardly think (despite the pottery and grinding stones, generally of slight thickness) they represent more than encampments of hunter-gatherers. On the other hand, there are concentrated settlements (with cinders and cooking debris) representing either cliff villages (Rkiz, Dahar Tichit-Oualata, Ennedi, etc.) or fishing villages (Baten de Tichit, Azaouad, Azaouak). In the villages in sandstone areas, grindstones are plentiful, but they are here of the thick, heavy, often even funnel-mouthed variety (and perhaps used rather as mortars).

In the southern Sahara, then, one must distinguish between: first, a Neolithic "of encampments," inter- or intradunary, with flat grindstones,[137] and second, a "village" Neolithic which has various aspects: a) a "lacustrine" type with fish-hooks and harpoons; b) some types less dependent—or not dependent at all—upon open water, sometimes *planitiaires* (with huts, probably of wood and straw), sometimes of rock (with partitioned shelters, low walls, etc.) and at least in the latter case, characterized by the abundance as well as the volume of grindstones.

Might the sequence be, very roughly: hunter-gatherers, hunter-fishers, and finally farmers? Did the last also practice animal husbandry, too? We do not know. Just as we cannot at present know if the various modes of life were contemporary or not, and if so in which order they came. Even if the harpoons of the Azaouad and those of Shaheinab had an undeniable common ancestry, this does not at all mean that they were of the same age.

Basically, economic patterns seem to match what we might imagine the nature of the phytoclimatic regions to have been. In this view, we may see a more or less Sahelian Sahara accommodating on the one hand steppe-dwelling hunter-gatherers (with pottery, however) and on the other cattle-breeding herders. At the same time, on the southern fringe where the rainfall was sufficient (probably 500 mm. or more) sedentary peasants were able to establish the cultivation of Guinea corn and bulrush millet. Meanwhile, hunter-fishers occupied the edges of a series of lakes, of which the modern Faguibine gives us an excellent picture. We may quite easily imagine that the Neolithic in the middle of the southern Sahara saw the establishment of a mixed economy comparable to that of the Middle Niger zone of inundation, with cereal agriculture, very active fishing, migratory herds, and "supplementary" hunting. It is of course evident that these various activities may have had different racial or ethnic foundations.

Domestication. If one is increasingly ready to admit that cattle herding, and consequently nomadism, followed the beginning of agriculture rather than originating with the hunter-gatherers, one should take this fact into account as much in the evaluation of contacts between hunter-gatherers and hunter-herders as of Saharan Neolithic chronology. On the one hand, this "Neolithic" complex may be at least "diphyletic," if it represents the results of a more or less effective

137. For which the often used name, "Saharan Neolithic," will be kept despite its questionable geographic merit.

mixture of two "basic" cultural layers.[138] On the other hand agriculture and pastoralism, far from representing two successive levels (of which herding would be the older), may express, then, two different but synchronous responses to different environmental conditions on the part of a "common," already mixed, "substratum." We should have, therefore, more of a functional adaptation, an "environmental specialization" as C. O. Sauer would call it (1956, p. 58), than a fundamental heterogeneity.[139]

The ideas of C. O. Sauer (1952, 1956) dealing with the priority of agriculture over domestication,[140] its origin in humid tropical Asia and the hypothetical sequence tubercles with dogs, pigs, and poultry—small-grain cereals, sorghums, millets, etc.—barley with goat and sheep—wheat and bovines, with their successive geographic "creative centers," arrayed east to west, have been reworked and developed by H. Von Wissmann alone (1957) or in collaboration with F. Kussmaul (1958) and H. Poch, G. Smolla, and F. Kussmaul (1956). A comparable view is found in K. Dittmer (1954, pp. 161 *et seq.*, pp. 245 *et seq.*, *et passim*).

Whether or not one can share views based largely on hypothesis (as their defenders would be the first to admit),[141] one cannot deny the importance for Saharan paleohistory of notions such as a cultural level of *Hirsenbau* (with a possible secondary Yemen-Sudan hearth), of an appearance of cattle pastoralism following the *Hirsenbau,* and a general southeast-northwest orientation of hearths of the agricultural-pastoral sequence, etc.

The Mesolithic site of Early Khartoum contained no trace of a domestic animal (D. M. A. Bate 1949, p. 27), but Esh Shaheinab, a Neolithic site, yielded, goat (or sheep?) in small quantity. These are not indigenous; D. M. A. Bate suggests (1953, p. 13) that the dwarf goat of Neolithic Esh Shaheinab could have reached the Upper Nile from Barbary by way of Hoggar and Tibesti.[142]

In fact, in the Sahara, the origin of the domestic animals becomes as conjectural as that of agriculture. When we find Nile Valley itself full of uncertainties and so many confusions in zoological identification,[143] we can imagine the extent of problems concerning the Sahara, where osteological material is so little known but

138. Which it would be tempting to search for at their source, under autonomous forms if they exist, anterior to the intra-Saharan contamination.

139. Which does not prevent one from thinking, however, that racial bases of two modes of life, pastoralism and peasantry, could have been identical.

140. ". . . all the domestication of the herd animals (except for reindeer) was effected by sedentary agriculturists living between India and the Mediterranean . . . nomadic cultures depend on agricultural people for some of their needs and, thus lacking a self-contained economy, can hardly have originated independently" (1956, p. 58).

141. Can the first "phase," tropico-humid (Gulf of Bengal), with a protoculture of tubercles and trees, fishing and intertidal gathering, dog, pig and poultry, for example, find at present any archaeological support?

142. But would not the *Capra promaza* Pomel be a Neolithic domestic goat (cf. C. A. Reed 1959, p. 1638)?

143. The same representation has been identified by specialists: *Bos primigenius* (Urus), *Syncerus caffer* (African buffalo) and *Bos indicus* (Zebu) (cf. J. W. Jackson in R. Mond and O. H. Myers, Cemeteries of Armant, 1, 1937, p. 254. H. Milford, Oxford University Press.)

where the rock-art furnishes evidence of considerable interest, at least in reference to the Neolithic (G. Esperandieu 1954, 1955; H. Lhote 1955).

As we have seen, there was cattle-raising in the Neolithic Sahara. Rock art has preserved abundant examples of it for us, sometimes in the form of actual frescoes (Mertoutek, Tassili of the Ajjers, Aouenat, etc.). There seem to have been multiple breeds, variously horned (cf. the table of H. Rhotert 1952, pp. 102–3), of the *Bos africanus* and the *Bos ibericus* (G. Esperandieu 1955; H. Lhote 1955). Zebus were introduced from the outside.

The history of sheep, goats, asses, horses, dogs, and pigs in the Sahara is no clearer than that of cattle (G. Esperandieu 1955; for the horse as well, H. Lhote 1953).

One ingenious hypothesis (E. Dechambre 1951, 1951a) considers the Sahara a "primitive center of domestication," this having been forced on man by the evolution of the environment—in this case, desiccation. The hunter, attached to a particular herd whose geographic isolation is increased by lack of waterholes, finally spares, controls, and manages his game which has become livestock. Milk allows less killing for meat. This new civilization, with its domestic Saharan animals perhaps fell back upon the Nile Valley where, later,[144] the representations or oryx and addax would be those not of domesticated menagerie animals but "of scenes related to the ancient life of Saharan peoples," depicting animals maintained in captivity especially to furnish victims for sacrifice (1951, p. 114).

E. Dechambre's hypothesis raises again the possible influence of the Sahara on Egypt, already suggested by Saharan Neolithic pottery.[145] Note that this applies only to species whose domestication has proved only temporary, probably because more "profitable" species had been introduced, that is, species whose domestication, besides, is by no means proven to have always preceded the practice of agriculture. If goats and sheep were among the first domestic animals, as has been supposed, and if this association with man was made at the same time and in the same regions as the beginnings of cereal agriculture, the hypothesis of hunters→breeders does not apply to them (cf. J. Nicolaisen 1954, p. 105; H. von Wissmann 1957; H. von Wissmann and F. Kussmaul 1958, etc.).

The question of the existence of African centers of domestication, in particular where progressive desiccation is thought to have facilitated such attempts, remains open. G. Espérandieu (1955, p. 544), asking if "areas of spontaneous pastoral culture" could have existed in Africa, lists among the conditions "favorable to domestication": "desiccation, keeping away big game while the savanna favored the increase of the number of herbivores, the paucity of waterholes forc-

144. Cf. L. Joleaud, *Bull. Soc. Géogr. Archéol. Oran,* vol. 38 (Fasc. 150, 1918), p. 105, notes 4–5. This refers rather to the addax than to the waterbuck and the kudu. The author cites as domestic animals in the Old Kingdom the hartbeest nubian ibex, and the Barbary sheep. We may add that many rock engravings seem to represent giraffes held by halters. Were these captured or domestic animals?

145. R. Mond and O. H. Myers, *Cemeteries of Armant,* 1, *London,* 1937, pp. 267–76. H. Milford, Oxford University Press.

ing close association of domesticable animals with their masters, and finally the necessity for prolonged rest due to the process of rumination inducing the ovines and bovines to seek human protection."

In opposition to the Dechambre hypothesis, C. A. Reed (1959, p. 1637) after H. Lhote (1955) criticized these "propinquity" and "riverine-oasis" theories. The last periods of drying were either pre-Neolithic (and pre-domestication) or post-Neolithic (with the domestic fauna already in place). As far as the Near East is concerned, the theory would not make sense in any case since, in fact, except in the desertic basins, "there was relatively little climatic change" from upper Pleistocene to "early Recent."

The same author, after careful consideration of available evidence, concluded that agriculture and domestication (with the possible exception of the dog) originated in the Near East, "in the hilly, grassy and open-forested flanks of the Zagros, Lebanese and Palestinian mountains" (p. 1632).[146] Even if the *Bos ibericus* has been domesticated in Barbary, it is hard to see what species could have been in the Sahara.

In fact, theories requiring a climatic deterioration to account for technological progress or a cultural acquisition raise an often discussed problem: do evolutionary human accelerations coincide with "crises," or, on the contrary, with periods of climatic stability?

A. J. Arkell (1945, p. 2) noted that two reactions to increased aridity were possible: "One was to look after the herd that they used to hunt, and domesticate it, and live on its products," the other "to drop grass seed on the mud flats left by the falling river, to tend the crop, to water it when necessary." But is the transition from hunting to domestication likely, and, on the other hand, if a climatic optimum appeared simultaneously with agriculture, could drying up explain the origin of the latter?

It is true that if the climate of Early Khartoum, considered Mesolithic, was more humid than today, the presumed absence of agriculture could be explained by a greater spontaneous productivity of the habitat.

For E. W. Haury (1958, p. 71), "Shifting ecological determinants forced man to reorganize his life if he was to survive." Thus the rhythm of climatic evolution determines that of human behavior: "adapt or perish"[147] . . . The increase in aridity, by forcing concentration of population,[148] would probably facilitate "the beginning of settled life and the cultivation of plants" (p. 71). Moreover, "the early possession of grinding tools made the transition from the gathering of food plants to the cultivation of plants, such as corn, an easy one."

146. The goat between —7000 and —6000; cattle, sheep and hogs "some time thereafter."

147. Col. R. Meinertzhagen, *Birds of Arabia*, 1954, p. 69, n. Edinburgh: Oliver and Boyd.

148. Already acknowledged by S. Huzayyin for the "late Upper and Final Paleolithic times," followed by an "intense localization of culture-groups in semisecluded areas" (1941, p. 334). A dramatic example of concentrations has been quoted above (p. 179).

Here (beginning of the Neolithic) the author recognized the role of forces "operating in an optimum cultural climate" (p. 72).

There is no doubt that the Neolithic Subpluvial not only facilitated certain important cultural acquisitions but their dispersion as well. "This relaxation of the 'climatic crisis,'" writes K. W. Butzer, (1957, p. 27), "was intimately associated with the rapid spread of the earliest food-producing economy and the establishment, or rather, the resumption of widespread cultural contacts and intercommunications during the Neolithic."

In the Near East, the "food-producing revolution" may not have been in any way "stimulated by the challenge of a post-Pleistocene climatic change" (C. A. Reed 1959, p. 1638). This is understandable if C. O. Sauer (1956, p. 66) is correct: "planting and domestication did not start from hunger but from surplus and leisure. Famine-haunted folk lack the opportunity and incentive for the slow and continuing selection of domesticated forms."

Perhaps that dangerous optical illusion, "telescoping" of stages of the past, and which increases with the distance, risks making us lose sight of the necessarily gradual character of many discoveries and innovations. Probably millennia passed while gropings toward various agricultures and domestications were made. Consequently, these do not have, and cannot have, specific "dates of birth." Any credit must be given to survivals, to industrial lagging. Probably Africa continued to manufacture in iron what it had in stone—namely the hoe and axe, which are not "new things" but "the same thing" in a different material.[149]

The gradual character of the environmental transformations and the concomitant human adaptations is especially evident in the problem of Saharan wells. Some attain considerable depths; for example, in the Azaouad, Hassi Touil and Erakchiwen measure 84 m., El Mamoun and Tazelaft 92 m., and El Khatt, . . . 110 m. Since it is out of the question that present-day nomads chose such sites and sank wells *ab initio*, we are tempted to think that digging has taken place over the centuries and millennia in the search for a water table. What was only a shallow hole in the Neolithic has become a 100 m. well (cf. F. Hernandez-Pacheco 1945).

C. Axes of Migrations

How was the Sahara populated? Whence came, successively, the peoples who made up the various ethnic groups identified with certain tools and modes of

149. One finds along with this survivals of technique independent of material: in the Eneolithic, flint, copying "sometimes servilely the new forms of metal" (L. R. Nougier 1950, p. 643); some stone axes of Agordat (Erythrea) and Sesibi (Wadi Halfa District) were considered by A. J. Arkell as "stone copies of the Egyptian two-lugged metal axe" (Mary D. Leakey, *J. E. Afr. Uganda Nat. Hist. Soc.*, vol. 17, Nos. 3–4 [77–8], 1943, p. 192); in the Bronze Age, metallic models were copied in stone (Saint-Just Péquart, *Bull. Soc. Sci. Nat. Nancy* [n.s.], vol. 3, no. 1 [1938], p. 12 and Fig. 1/14–5.)

life that archeology permits us to name: Pebble Culture, hand-axe culture, Aterian, the various Neolithic cultures? Reference has already been made in this report to problems concerning human remains, industries, and rupestral representations, but we must return to them.

First, for the longest part of history (Lower Paleolithic), our knowledge is nil. We know no more of the origin of the pebble-tool men than of the hand-axe men. Obviously suppositions can be made, and one can imagine that hand-axe industries in the interior of Africa moved from south to north (African character of the flake cleaver, abundant in the Sahara, but rare in Barbary). But in fact, our ignorance is complete.

We know little more for the Aterian, except that prehistorians of North Africa seem to consider it an element that was brought to the Sahara from the north by some unknown race.[150]

As we saw above, it is possible that during the Mesolithic, Afro-Mediterraneans (L. Cabot Briggs 1958, p. 11) and their industry (Capsian) arrived in Barbary via the Sahara (Fezzan?), coming from the Nile via the Sudan. This is, obviously, pure supposition.

The situation scarcely improves with the Neolithic; however, a few rare osteological documents date from it, and the evidence from rupestral representations is added to that of the industries. On the whole, available materials seem to support the idea, well explained by L. Cabot Briggs a propos the Saharan races in the Neolithic (1958, pp. 12-14), of a peopling of the desert from two principal "points of departure" on the northern and souther borders. The descendants of the Capsians, established on the northern border of the Sahara, penetrated more or less deeply into the desert.[151] Contacts at first probably more or less accidental were established across the Sahara by the random travels of small groups of "explorers." But these contacts have technological importance and favored the diffusion of Neolithic tool assemblages. A general advance of Negroids from the south across the Sahara, however, must be envisaged: "At some time about 6000 years ago, perhaps, Neolithic, negroid peoples from the Sudan began to spread northward through the entire length of the then relatively fertile Sahara" (L. Cabot Briggs, 1958, p. 19). It might have been "a strongly negroid population with a visibly substantial Hamitic component" (p. 13). Although we are ignorant of its origin, this element seemed to introduce Neolithic culture via the south probably more than 5000 years ago, or perhaps rather Neolithic cultures, as one should perhaps distinguish that of the steppe hunter-herdsmen from that of cultivators.

In any case, the mountainous massifs were colonized and were to become in their turn both secondary centers of diffusion and "bastions," which, as havens

150. Is there some connection between the Aterian and the mesolithic Paleomediterraneans (L. Cabot Briggs 1958, p. 10)?

151. This movement was perhaps started by the Capsians themselves.

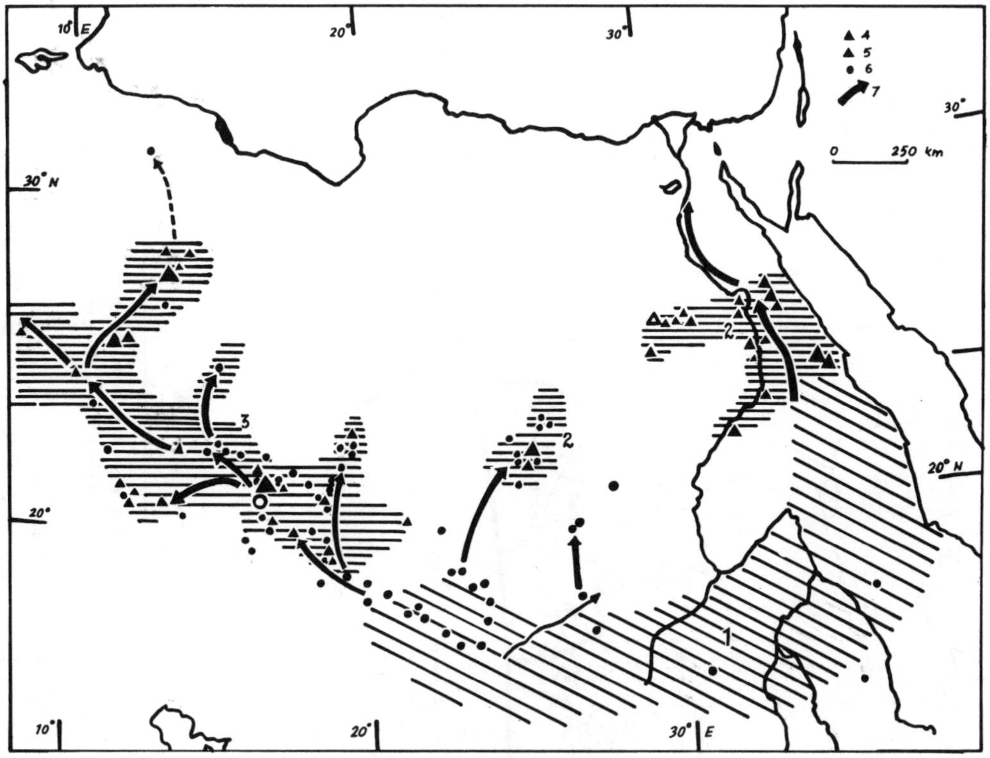

FIGURE 12
Neolithic hunters and herders in the eastern Sahara. From K. W. Butzer 1958, Fig. 3.

for cattle after the last deteriorations of climate, were prolonged probably longer than other areas. Locally perhaps "bovidian" life lasted until the arrival not only of horses but of camels.

We are indebted to H. Rhotert (1952, p. 119) for a schema which summarizes his basic hypothesis of the origin and development of rupestral art in the Libyan desert (Fig. 13). According to him, it can be explained by the meeting of three currents, two from the northwest, the third from the southeast. In effect the author suggests a vast migration of herdsmen, entering Africa via the Eritrean region and moving toward the west across Sahelian pastureland, toward Chad and the Niger.[152] From this major axis—which is the east-west arm of a right angle (like a carpenter's square) whose other side is directed toward eastern and southern Africa (cf. Th. Monod, 1944, Fig. 23)—lateral branches, perpendicular to the

152. A. J. Arkell (1944, 1950a) has already insisted on the role of the Meroe-Niger axis, and indicated for example the discovery of copies of Coptic lamps from the fifth to seventh centuries in West Africa (Ghana).

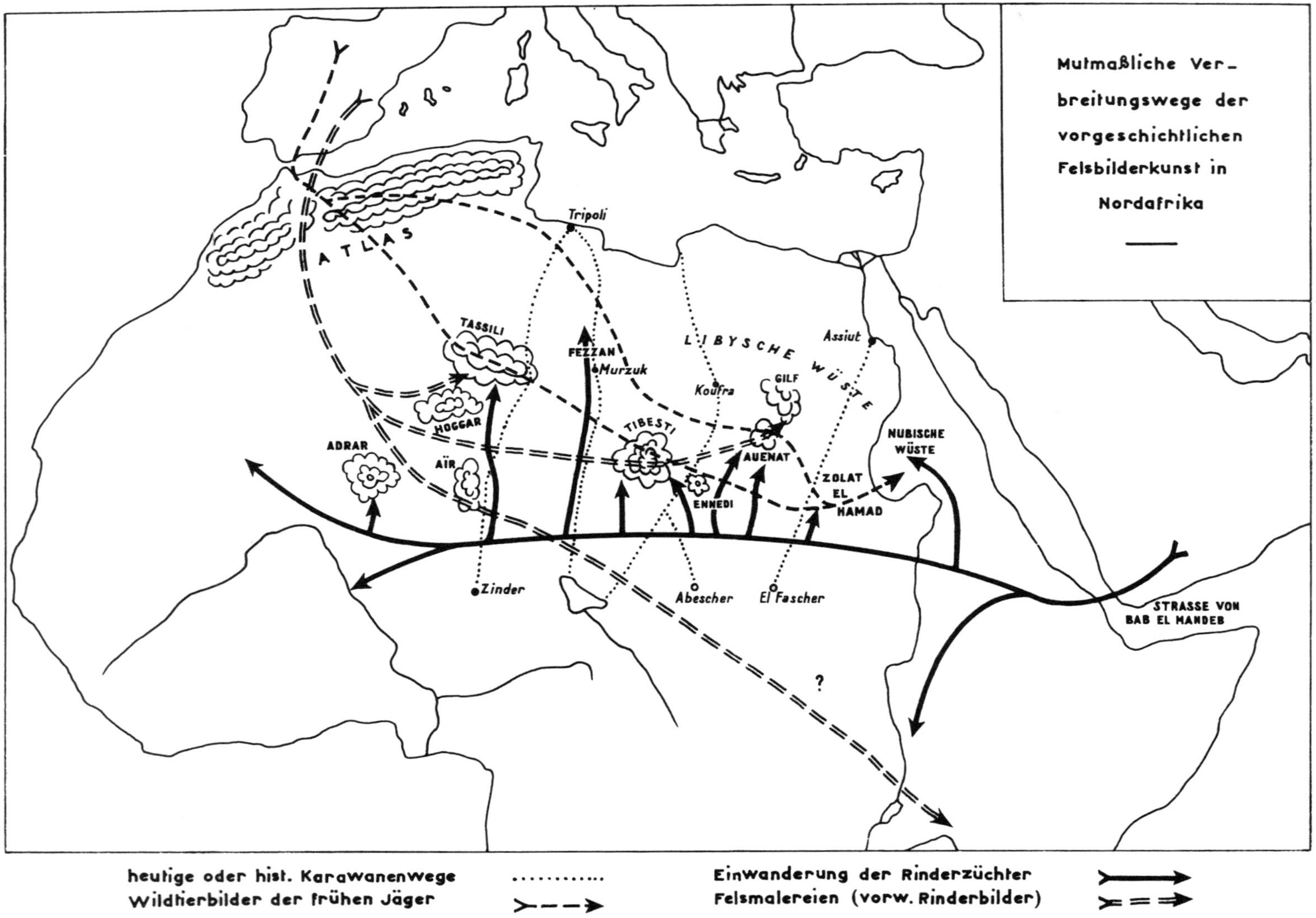

FIGURE 13

Possible migration axes of naturalistic art (engravings), paintings, and bovidian pastoralism. Map by H. Rhotert 1952, Map II.

main trunk detached themselves to the north to establish themselves in the Sahara, toward Aouenat and Gilf Kebir, Ennechi, Tibesti, Fezzan, Tassili, etc. There, the herdsmen came in contact with an already settled group whose art appears to come from Maghreb, and through this latter, from the Franco-Cantabrian cradle. These hunter-engravers initiated the herdsmen in the practice of petroglyph art, and the rocky walls were soon covered with pictures of bovids. Later another process, painting, came in its turn to be adopted by the people of the Cattle Period. It can be considered to have come from the west also, since Gilf Kebir and Aouenat mark its eastern limit (p. 121), and the close resemblances of style between it and the art of the Spanish Levant [153] suggest that possible origin.

Such is, if correctly understood, I think, the conclusion of H. Rhotert, who would himself willingly recognize its hypothetical character as he is not unaware "dass wir nur tastend wie Blinde langsam in die Vorzeit einzudringen vermögen" (p. 125).

Many aspects of it are obviously criticizable. We note, for example, that the fact that the paintings in the Sahara seem to have an eastern limit does not imply their western origin, because, farther toward the southeast one finds great quantities of paintings in analogous style, to the point that some were inclined to find their origin in Africa more than in Spain. The enormous distances between regions with paintings (Spain-Hoggar/Tassili,[154] Aouénat-East Africa, etc.) and our complete ignorance of the racial characters of the artists obscure the problem. But if the bovidians of the Sahara are "Negroids," "semi-Hamites," etc., could the East Spanish Neolithics have belonged to the same stock? Could different races have produced styles so strikingly similar? To put it another way, could one admit that artistic styles, in an archaic and thoroughly isolated epoch, be propagated independently of a definite ethnic support? Should one not attribute a part and perhaps a large part to the phenomena of reinventions and convergences in spite of the most troubling Levant-Sahara-South African resemblances? Could not a similar mode of life—here that of the steppe hunters or of the hunter-herdsmen—"spontaneously" provoke the development of a wall art if not completely identical, at least evidently parallel? While we can easily admit that the arrival of hunters-herdsmen in contact with hunters-engravers soon caused the

153. The chronological problems raised by this complex are not clarified. One should read with great interest the recent work of E. Anati (1961) where H. Kuhn's classification is improved by the distinction, in the interior of Group II, of two phases: Alfera-Volltorta ("classic") and Villar del Humo ("seminaturalistic"). The author thinks he can recognize (p. 700) in the Levant a population "which never passed through the Mesolithic stage, nor through the Neolithic in the true sense of the word, and which, from hunting typical of Paleolithic life, passed slowly to a way of life based on breeding and hunting, without ever making agriculture an important basis of its economy": in fact, a kind of "epipaleolithic." Notice the interest of these remarks for Saharan steppe cultures.

154. H. Lhote (1953, p. 1734) wisely emphasized that the hypothesis of a progress of paintings from the northwest to the southeast "does not rest on any concrete fact."

appearance of bovidian petroglyphs,[155] is it as permissible to think that these same hunter-herdsmen might have waited, perhaps for a long time, for an outside stimulus from a singularly distant source, to begin grinding ochre and daubing silhouettes of cows or archers? There is nothing to prove that one would be correct in considering the cattle stage of Libyan paintings a homogeneous entity. What H. Lhote revealed to us of those of Tassili [156] must cause extreme prudence, in the face of material probabily still more complex that we think, and to which we are always tempted to apply an excess of schematization which is certainly useful, but perhaps rather artificial.

Who can prove for example that Saharan humans obediently traversed the technical and cultural stages that our manuals recommend for them? Nothing, in fact, is less certain, and it is absurd to imagine a gradual passage from one mode of life [157] still "paleolithic," that of predominantly hunting (with naturalistic rock engravings) [158] to another of hunter-herdsmen who seek in the vegetable kingdom only those things that grow wild.

However it may be—and in our present stage of knowledge, these discussions remain dangerously "in the air" and rather in vain—we saw that H. Rhotert imagines a more or less east-west route for the herdsmen, from the Nile towards Chad and the Niger, with successive south-north branchings off toward the Sahara.[159] This conception has been criticized. H. Lhote on one hand (1953, p. 1733; and in H. Breuil 1955, p. 80) and K. W. Butzer (1958, p. 46 and Fig. 3) reject this "Sahelian" hypothesis. For them the major axis of the Saharan immigration of bovidian eastern Hamites must be directed from southeast to northwest (Fig. 12) via Ouadai, Ennedi, Tibesti, Tassili, etc. For H. Lhote it is the rarity of bovidian representations in Aïr that should indicate that the principal east-west flux divided long before it could have reached these massifs from the south. If the fact is definite—and H. Lhote presents it as if "confirmed"— it is very important. For K. W. Butzer (1958, p. 46): "Es fehlen zwischen Ennedi und Aïr am Südrand der Sahara auf einer Strecke von 1500 km alle Spuren dieser Rinderzüchterkultur obwohl sie im ostlichen Sahel doch vereinzelt belegt ist." It is true, but we must take into account that on the 14th parallel, we are, from Ouadai to Tegama (on the meridian of Aïr) in a sandy Pleistocene where even a Roman legion could scarcely find a place to engrave an inscription.

Besides, rocks in themselves are not enough to attract an artist; they must be the kind that lend themselves readily to carving. While sandstone (Ennedi, Tibesti, Tassili, etc.) constitutes a choice material, eruptive rocks seem least

155. While nothing obliges us to conclude that these cattle men were ignorant of all graphic activity.

156. Cf. also H. Lhote, in H. Breuil 1955, p. 79.

157. And not necessarily from one age, obviously.

158. It is astonishing that the oldest rock art chronologically is also the most "beautiful." Such consistent workmanship obviously implies attempts, gropings, an apprenticeship. Why have these left no traces on the rocks? Were there not other media even more favorable to "scrawls" (ostrich eggs, wood, plates of schist, skin, bone, etc.)?

159. This agrees with the existing divisions of the Sahara into meridian belts.

favorable.[160] As to metamorphic volcanic rocks, they are nearly all worthless: now, Aïr, like the Iforas, is essentially Pre-Cambrian.

Thus we should not hasten to consider the distribution of representations as an adequate indication for the migration of the bovidians. Nevertheless, let us add that Ennedi, as shown in the works of G. Bailloud, is rich in paintings, and if the bovidians could have passed only through there, they did certainly pass there. This is an important revelation.

It is noteworthy that an anthropologist, L. Cabot Briggs (1955, p. 90), seems to admit the possibility of a double channel of Nilotic or East African Negroid contributions, the main one following the southern edge of the desert as far as the Niger "and thence northward" (to Morocco), the secondary one, after leaving Egypt, using a more northern route to find a new center of dispersion "somewhere in the neighborhood of the Tidikelt." These views remain very hypothetical, and the author is the first to acknowledge it. They suggest, however, a more complex history that one would have thought possible until the distribution of paintings was known. One day, however, C-14 datings in sufficient number will inform us not only about the chronology but even about the direction itself of the movements. The east-west migration of lacustrine culture can only be established if the harpoons of Araouan are found to be notably more recent than those of Shaheinab.

VI. BY WAY OF CONCLUSION

At the end of a long analysis whose apparent richness of details should not give the illusion of completeness, and which is above all a copious record of our ignorance, there are not and should not be any "conclusions," properly speaking. Actually, is an attempt like this one not premature at this point? Would it perhaps be more valuable just to keep still, and wait 25, 50, or 100 years—given the condition of hard work from now until then—when it might be possible, finally, to speak and really say something?

Lacking conclusions—or if one prefers, by way of conclusion—perhaps I can formulate a few remarks pertinent to the general theme proposed, involving the role of the Sahara in the life of prehistoric men.

In a larger geographical framework, the Sahara is the southwestern extremity of a belt of more or less arid regions extending diagonally across the Ancient World from Mauritania to Mongolia. In Africa it separates two worlds, or more exactly, through it passes the frontier between two worlds, the subtropical Mediterranean and Blad-Es-Sudan, the "Pays des Noirs." It remained in a very marginal situation with respect to Asia which seems to have witnessed, at least

160. On the opposition of two neighboring localities, one granitic, the other of sandstone, see H. Rhotert (1952, pp. 10–12, 28): the sandstone locality is the richest and the only one to have paintings and engravings simultaneously: "In dem ganzen vons uns bereisten Gebiet wurden niergends Gravierung auf Granit oder einem ähnlich harten Stein [. . .] gefunden" (p. 12).

since the end of the Paleolithic, the appearance of at least most of the "key" elements that were to revolutionize human economy: grain, cattle, pottery (?), and later, iron. Thus, it was, in many respects a "lands-end" situation, a cul-de-sac, making the Sahara an enormous, deep gulf, on whose beaches many great swells of distant origin [161] come to die.

But the picture would not be complete if we dared to forget the great duration of pre-Neolithic civilizations and the cyclical character of environmental changes. The desert phenomenon may be incredibly old—Saharan Paleozoic sandstones contain aeolian grains—but it does not signify the general theme around which climatic oscillations will, in the course of an unknown number of successive cycles, change the land from the most frightful desert (it might even have been more "beautiful" than the present day desert) to wooded savanna, even to open forest.

Clearly, we must suppose wet cycles during which the Sahara was covered with steppes, savannas, and lakes, to appreciate its role in human evolution. In such periods it is not the Sahara itself as such which was benefitted, but the habitat. One supposes this type of environment was extended over most of Africa aside from the dense forest.[162]

We have often stressed the advantages which a Primate in the process of hominization would find in a wooded savanna, while acquiring the erect posture necessary to the development of the hand as well as the vision. It had neither sufficient speed for the steppe nor a quadrumanal adaptation to tree life (see, for example, M. Nestourkh 1960, pp. 114, 213, et *passim*). With respect to this, Henno Martin's chapter (1957, pp. 134–42) "Man's Evolution," has a special flavor from the fact that the author survived a long time, with Hermann Korn, deep in the Namib, a life in many respects very like that of the Bushmen, which is to say, like Neolithic men.

Savanna of the Sudanese type or, especially in the rainy, puddle season the Sahelian savanna, facilitates crossing, approach, and maneuver that distinguish it from both the true desert and the great forest. This benefitted the herdsman, horsemen, and wagon-driver, all of which oriented the migrations toward finally polarizing economic-political creations such as the black empires of the Middle Ages.

Another factor favoring the beginning of Man was the abundance of natural products, from the herds of ungulates to the wild grain, seed grain, seeds of coloquinte. And after the steppe-hunter, bovid husbandry, exploiting the same habitat, under another form, prospered in its turn.

One should beware of considering the Sahara, when it was like present-day savannas, as an homogeneous entity. Along with the advantages, there have always been obstacles and of various orders—physiographic (mountains, large lakes), climatic (former desert "nuclei" and, of course, arid periods), and Man

161. Direct or indirect; courses of a Near East→(Barbary or Sahel)→Sahara type extend the itinerary without displacing the focus of diffusion.

162. Cf. Th. Monod 1957, *passim* (*ubi litt.*).

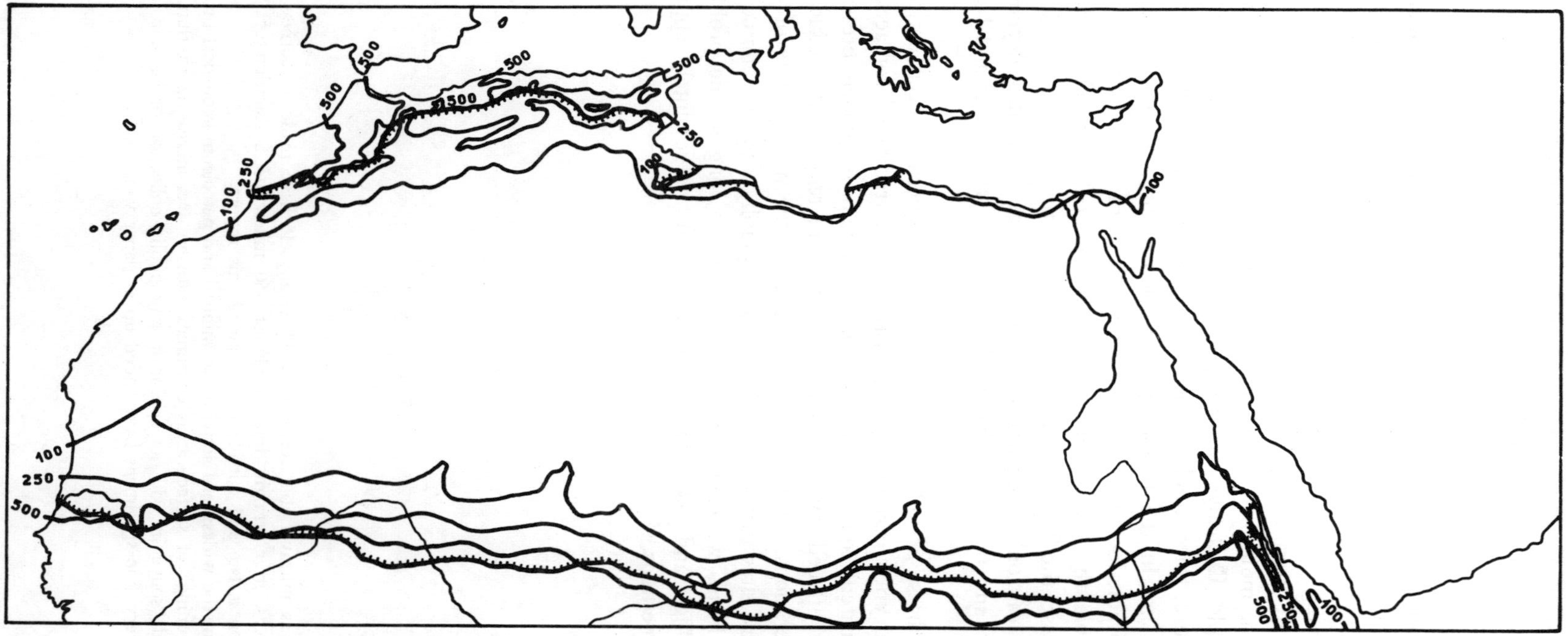

FIGURE 14
Comparison between the isohyets and the limits of rain agriculture (potential). Simplified from F. R. Falkner 1939, Map 1; even during the Neolithic, this type of agriculture does not seem possible for the Sahara.

himself, as many reasons for partition and appearance of regional diversities, adequately explaining the necessity of avoiding all adventurous generalization and of maintaining a wisely limited geographic framework in research.

Beside these obstacles, there are incidentals. In the most ancient phases, the autonomy of a small band of hunter-gatherers was limited only by the environment itself, which dominated their way of life, its tools, etc. Later, certainly from the beginning of the Neolithic, the intra-Saharan populations were not completely cut off from the "outside." Contacts, slow and sporadic though they might have been, multiplied. Consider the transfer of sea shells or amazonite beads, for example. Exchanges were negotiated, products of the hunt for those of the new culture, for example, or useful objects, tools, bones, magic or medicinal substances, etc. In the periods of increased aridity, if some small groups persisted in favorable locations—the insular condition of oasis or altitude pastures—it is improbable that their economic autonomy could have been complete. Today, in any case, the Saharans, though they live *in* the desert, do not live *from* the desert, or at least only live partially from the desert, and remain dependent on the Guinea-corn and bulrush millet of the south and the wheat and barley of the north.

Even in the driest periods, the Sahara has never perhaps been entirely uninhabited. The dryness certainly caused disruption of traditional contacts, closed routes, "encapsulated" some groups; but it is not certain that human presence was ever completely effaced, even though restricted and temporarily reduced.[163]

These considerations will seem a little vague. In the present state of Saharan prehistory, where sites that have yielded a stratigraphy can be counted on the fingers of one hand, we cannot be more precise so long as the problems of Pleistocene geology and associated industries cannot be resolutely approached with adequate methods.[164]

163. Nothing was untraversable for the camel (I myself have made a journey between two watering places, in a straight line of almost 560 miles and a duration of 22 days), but its introduction is late and outside the scope of this report.

164. Note that in the widespread activities of mineral and petroleum geological prospecting, a deplorable "skimming" of surface areas continues with an ever-increasing rhythm, and too often the collected "souvenirs" do not appear in any public collection or become known to specialists: some very fine specimens have ended up as paper weights. . . .

BIBLIOGRAPHY

ABADIE, J., J. BARBEAU, and Y. COPPENS

1959. "Une faune de Vertébrés villafranchienne au Tchad," *C. R. Acad. Sci.* (Paris), 248:3328–30.

ALIA MEDINA, M.

1949. "Contribucion al conocimiento geomorfologico de las zonas centrales del Sahara Español," *Inst. Et. Afr.*, Madrid.

1955. "Sobre las variaciones climaticas durante el Cuaternario en el Sahara Español," *Africa* (Madrid), 12:544–6.

ALIMEN, [MARIE]-HENRIETTE

1952. "Formations récentes." In "Les Chaines d'Ougarta et la Saoura," *19e Cong. Géol. Int.* (Algiers 1952): 83–118.

1954. "La station rupestre de Marhouma (Sahara occidental)," *Trav. Inst. Rech. Sahar.* (Algier), Mém. No. 1.

1955a. "Découverte de nouveaux dépôts et d'industries préhistoriques en couche dans les alluvions quaternaires de la Saoura (Sahara occidental)," *C. R. Acad. Sci.* (Paris), 240:1652–4.

1955b. "Présence d'Acheuléen dans les alluvions de l'Oued Saoura. Chronologie quaternaire du Sahara nord-occidental," *Bull. Soc. Préhist. Fr.*, 52:480–92.

1955c. *Préhistoire de l'Afrique.* Paris: N. Boubée et Cie.

1956a. "Chronologie du Paléolithique ancien au Sahara nord-occidental," *C. R. Acad. Sci.* (Paris), 242:2023–5.

1956b. "Fouilles dans les alluvions à Paléolithique ancien dans la région de Kerzaz (Sahara occidental)," *Bull. Soc. Préhist. Fr.*, 53:648–55.

1957a. "Chronologie préhistorique du Sahara," *Actes 3e Pan-Afr. Cong. Préhist.* (Livingstone 1955), pp. 80–85.

1957b. "Terrasses climatiques de type pluvial aride du Sahara nord-occidental," *Proc. 5th Int. Cong. INQUA*; 7–8 (Abstract).

1957c. "Tertiaire et Villafranchien au Sahara nord-occidental (Saoura-Ougarta)," *C. R. Somm. Soc. Géol. Fr.*, No. 12: 238–41.

1958. "Sables quaternaires du Sahara nord-occidental (Saoura-Ougarta)," *Publ. Serv. Carte Geol. Algérie* (n.s.), No. 15.

1960. "Découverte d'un atelier de l'Acheuléen supérieur, en place, à la limite du 2e Pluvial et du 3e Pluvial, dans les Monts d'Ougarta (Sahara occidental)," *Bull. Soc. Préhist. Fr.*, 57:421–3.

1961. ["Tableau chronologique Sahara nord-occidental et Maghreb"], *6th Int. Cong. INQUA* (Varsonie).

Alimen, [Marie]-Henriette, and J. Chavaillon

1956. "Industrie acheuléenne in situ de l'Oued Fares, dans les Monts d'Ougarta (Sahara occidental)," *Bull. Soc. Préhist. Fr.*, 53:202–14.

1959. "Découverte de la Pebble-Culture in situ au Sahara nord-occidental. Son age et son evolution," *C. R. Acad. Sci.* (Paris), 248:2894–96.

Alimen, [Marie]-Henriette, J. Chavaillon, and G. Conrad

1962. "Position stratigraphique et évolution de la Pebble Culture au Sahara nord-occidental," *Actes 4e Pan-Afr. Cong. Préhist.* (Leopoldville 1959), Sér. 8, No. 40: 3–26.

1959. "Formations arides et paléosols quaternaires au Sahara nord-occidental," *C. R. Somm. Soc. Géol. Fr.*, No. 5: 104–105.

Alimen, [Marie] - Henriette, J. Chavaillon, and J. Margat

1959. "Contribution a la chronologie préhistorique africaine. Essai de corrélation entre les dépôts quaternaires du bassin Guir-Saoura (Sahara) et du bassin du Tafilalt (Maroc)," *Cong. Préhist. Fr.* (Monaco). In press.

Almasy, L. E.

1942. *Unbekannte Sahara* (3rd ed.), Leipzig: F. A. Brockhaus.

Anati, E.

1961. "Quelques réflexions sur l'art rupestre d'Europe," *Bull. Soc. Préhist. Fr.*, 57:692–712.

[Anonyme].

1956. "Collections préhistoriques," Musée du Bardo (Planches), Album No. 1.

Arambourg, C.

1948. "Observations sur le Quaternaire de la région du Hoggar," *Trav. Inst. Rech. Sahar.* (Algiers), 5:7–18.

1951. "Les limites et les corrélations du Quaternaire africain," *18th Int. Geol. Cong.* (London 1948); 49–54.

1952. "La paléontologie des Vertébrés en Afrique du Nord française," *19th Cong. Geol. Int.* (Algiers 1952). Regional Monograph.

1962. "Etat actuel des recherches sur le Quaternaire en Afrique do Nord," *Actes 4e Pan-Afr. Cong. Préhist.* (Léopoldville 1959), Sér. 8, No. 40: 255–277.

Arambourg, C., and L. Balout

1955. "L'ancien lac de Tihodaïne et ses gisements préhistoriques," *Actes 2e Pan-Afr. Cong. Préhist.* (Algiers 1952); 281–292.

Arkell, A. J.

1944. "Archeological Research in West Africa," *Antiquity*, 18:147–50.

1945. "An Outline History of the Sudan," *Inform. Service, Khartoum.*

1947. "Early Khartoum," *Antiquity*, 21:172–81.

1948. "The Historical Background of Sudan Agriculture." In *Agriculture in the Sudan* (J. D. Tothill, ed.), London: Oxford University Press, pp. 9–18.

1949a. *Early Khartoum*, London: Oxford University Press.

1949b. "Excavations at Esh Shaheinab, Sudan (1949)," *Proc. Prehist. Soc.*, No. 4: 42–49.

1950a. "The Sudan Archaeology and Excavations," *Archaeol. News Letter*, 2:124–8.

1950b. "Gold Coast Copies of 5th–7th Century Bronze Lamps," *Antiquity*, 24:38–40.

1953. *Shaheinab. An Account of the Excavation of a Neolithic Occupation Site.* London: Oxford University Press.

1955. "The Relations of the Nile Valley with the Southern Sahara in Neolithic Times," *Actes 2e Pan-Afr. Cong. Préhist.* (Algiers 1952); 345–346.

1959a. "Preliminary Report on the Archaeological Results of the British Ennedi Expedition," *Kush*, 8:15–26.

1959b. "Elks in the Sahara: Unique Rock Drawings from Tibesti, which Throw Light on the Early Saharan Climate," *Illust. London News*, November 21, 1959: 690–691.

1962. "The Distribution in Central Africa of One Early Neolithic Ware (Dotted Wavy Line Pottery) and Its Possible Connection with the Beginning of Pottery," *Actes 4e Pan-Afr. Cong. Préhist.* (Léopoldville 1959), Sér. 8, No. 40: 283–287.

AUZEL, M., and A. CAILLEUX

1950. "Silicifications nord-sahariennes," *Bull. Soc. Géol. France*, Sér. 5, 19:553–9.

BAILLOUD, G.

1958. "Mission des confins du Tchad. I. Recherches préhistoriques et archéologiques (1956–1957)," *Avant-projet de rapport*, 1958. Mimeographed.

BALOUT, L.

1952a. "Le peuplement préhistorique de l'Algérie," *Cong. Préhist. Fr.* (Paris 1950): 106–114.

1952b. "Pluviaux interglaciaires et Préhistoire saharienne," *Trav. Inst. Rech. Sahar.* (Algiers), 8:9–21 (reprinted in L. Balout, *Préhistoire de l'Afrique du Nord*, 1955, pp. 76–82).

1955a. *Préhistoire de l'Afrique du Nord*. Paris: Arts et Métiers Graphiques.

1955b. "Les hommes préhistoriques du Maghreb et du Sahara," *Libyca* (Algiers), 2:215–422.

BARBEAU, J.

1961. "Morphologie du Quaternaire des abords orientaux du Tchad," *Bull. Inst. Equat. Rech. et Et. Géol. et Minières*, No. 14: 73–82.

BATE, DOROTHEA M. A.

1947. "An Extinct Reed-Rat (*Thryonomys arkelli*) from the Sudan," *Ann. Mag. Nat. Hist.*, 14:65–71.

1949. "The Vertebrate Fauna." In A. J. Arkell, *Early Khartoum*, London: Oxford University Press, pp. 16–28.

1950. "The fauna of Esh Shaheinab," *Archaeol. News Letter*, 2:128–9.

1953. "The vertebrate fauna." In A. J. Arkell, *Shaheinab*, London: Oxford University Press, pp. 11–19.

BELLAIR, P.

1949. "Le Quaternaire de Tejerhi (Fezzan), *C. R. Somm. Soc. Géol. Fr.*, No. 9: 161–162.

1953. "Le Quaternaire de Tejerhi," *Inst. Hautes Et. Tunis, Publ. Scient.*, 1:9–16.

BERNARD, E. A.

1956. "La paléoclimatologie, spécialement du Quaternaire africain, vue sous la lumiére moderne des lois de la climatologie physique et de la géophysique," *I.N.E.A.C.*, 7e *Réunion du Cons. Scient. Afr.* (Yangambi). Mimeographed.

1959a. "Théorie astronomique des pluviaux et des interpluviaux du Quaternaire africain," *Actes 4e Pan-Afr. Cong. Préhist.* (Léopoldville 1959): 67–95.

1959b. "Les climats d'insolation des latitudes tropicales au Quaternaire. Théorie astronomique des pluviaux et interpluviaux africains," *Bull. Acad. Roy. Sci. Colon.* (n.s.), :344–64.

BIBERSON, P.

1961a. "Le cadre paléogeographique de la préhistoire du Maroc atlantique," *Publ. Serv. Antiquités Maroc*, Mem. 16.

1961b. "Le Paléolithique inférieur du Maroc Atlantique," *ibid.*, Mem. 17.

1963. "Human evolution in Morocco, in the framework of the paleoclimatic variations of the Atlantic Pleistocene," this volume.

BIGOURDAN, J.
1950. "Présentation de deux pièces zoologiques intéressantes de la collection de l'I.F.A.N.," *C. R. 1ere Conf. Int. des Africanistes de l'Ouest* (Dakar 1945), 1:189.

BLANCHOT, A.
1952. "Notice explicative sur la feuille Port-Etienne-Est," Carte Géol. A.O.F. au 500. 000
1957. "Les formations récentes de Mauritanie occidentale," *Bull. Dir. Féd. Mines et Géol. A.O.F.*, No. 20: 9–91.

BOBO, J.
1955. "Un faciès mésolithique saharien: le "faciès d'El-Oued," sa place dans l'ensemble des industries du Souf," *Actes 2e Cong. Pan-Afr. Préhist.* (Algiers 1952): 493–505.

BODENHEIMER, F. S.
1951. *Insects as Human Food: A Chapter of the Ecology of Man.* The Hague: W. Junk.

BOSCH-GIMPERA, P.
1955. "Le problème de la chronologie de l'art rupestre de l'Est de l'Espagne et de l'Afrique," *Actes 2e Cong. Pan-Afr. Préhist.* (Algiers 1952): 695–9.

BREUIL, H.
1955. "Les roches peintes du Tassil-n-Ajjer," *Actes 2d Cong. Pan-Afr. Préhist.* (Algiers 1952): 65–219.
1957. "L'Occident, patrie du grand art rupestre," *Mélanges Pittard:* 101–113. Paris: A. G. Nizet.

BRIGGS, L. C.
1955. "The Stone Age Races of Northwest Africa," *Amer. School of Prehist. Res.*, Peabody Museum, Bull. 18.
1957. "Living Tribes of the Sahara and the Problem of their Prehistoric Origin," *Proc. 3rd Pan-Afr. Cong. Prehist.* (Livingstone, 1955), pp. 195–99.
1958. "The living Races of the Sahara Desert," *Papers Peabody Mus. Amer. Archaeol. and Ethnol.*, Vol. 28, No. 2.

BROUIN, G.
1950. "Notes sur les Ongulés du Cercle d'Agadez et leur chasse," in "Contribution a l'étude de l'Aïr," *Institut Français d'Afrique noire*, Mém. No. 10: 425–55.

BRUECKNER, N. D.
1958. "Le rôle de la latérite dans l'éstablissement de la chronologie quaternaire dans l'Ouest africain," *West Afr. Sci. Assn.* (2d Conf. inter-territ.). Mimeographed.

BRUNEAU DE MIRÉ, P., and P. QUÉZEL
1959. "Sur la présence de la Bruyère en arbre (*Erica arborea* L.) sur les sommets de l'Emi Koussi (Massif du Tibesti)," *C. R. Séances Soc. Biogéog.*, Nos. 314, 315, 316: 66–70.

BÜDEL, J.
1953. "Die 'Periglazial'-morphologischen Wirkungen des Eiszeitklimas auf der ganzen Erde," *Erdkunde*, 7:249–66.

BUTZER, KARL W.
1957a. "Mediterranean Pluvials and the General Circulation of the Pleistocene," *Geog. Annaler*, 39:48–53.
1957b. "The recent climatic fluctuation in lower latitudes and the general circulation of the Pleistocene," *ibid.*, 39:105–13.
1957c. "Late glacial and postglacial climatic variation in the Near East," *Erdkunde*, 9:21–35.
1958. "Studien zum vor- und frühgeschichtlichen Landschaftswandels der Sahara und Levante seit dem klassischen Altertum. II. Das ökologische Problem der Neo-

litischen Felsbilder der östlichen Sahara," *Abhandl. Akad. Wiss. u. Liter. Mainz, Math.-Naturwiss. Kl.*, No. 1: 1–49.

1959a. *Idem.* III. "Die Naturlandschaft Ägyptens während der Vorgeschichte und der Dynastischen Zeit," *ibid.*, No. 2, pp. 43–122.

1959b. "Contribution to the Pleistocene Geology of the Nile Valley," *Erdkunde*, 13:46–7.

1959c. "Some Recent Geological Deposits in the Egyptian Nile Valley," *Geogr. J.*, 125:75–9.

1960. "On the Pleistocene Shore-lines of Arabs' Gulf, Egypt," *Jour. Geol.*, 68:626–37.

1961. "Les changements climatiques dans les régions arides depuis le Pliocène." In *Histoire de l'utilisation des terres arides*, UNESCO: pp. 35–64.

1962. "The Pleistocene Sequence in Egypt and its Implication for Pluvial-Glacial Correlation in the Sahara," *Actes 4e Pan-Afr. Cong. Préhist.* (Léopoldville 1959), Sér. 8, No. 40: 133–139.

CAMPS, G.

1960. "Aux origines de la Berbérie-Massinissa ou les débuts de l'histoire," *Libyca*, Archéol.-Epigr., Vol. 8.

CAPOT-REY, R.

1950. "A propos de l'age de l'Homme d'Asselar," *Bull. Institut Français d'Afrique noire*, 12, pp. 1128–31.

1953. *Le Sahara français.* Paris: Presses Universitaires de France.

CHAPELLE, J.

1957. *Nomades noirs du Sahara.* Paris: Plon.

CHAVAILLON, J.

1956. "Quaternaire de la vallée du Guir (Sahara nord-occidental)," *C. R. Somm. Soc. Géol. Fr.*, No. 13: 231–33.

1958. "Industrie archaïque du Paléolithique ancien, en place dans les alluvions de l'Oued Guir (Sahara Nord-Occidental)," *Bull. Soc. Préhist. Fr.*, 55:431–43.

1960. "Précisions apportées à la chronologie quaternaire du Sahara nord-occidental (subdivisions de l'Ougartien)," *C. R. Somm. Soc. Géol. Fr.*, No. 7: 182–83.

1961. "Aperçu du Quaternaire de la vallée de l'Oued Messaoud (Sahara occid.)," *C. R. Acad. Sci.* (Paris), 253:1714–16.

CHAVAILLON, J., and N. CHAVAILLON

1957a. "Chronologie du Quaternaire saharien, depuis les derniers dépôts du Paléolithique ancien jusqu'au Néolithique, *C. R. Acad. Sci.* (Paris), 244:1663–66.

1957b. "Présence d'industries acheuléenne, atérienne et néolithique dans les alluvions du Kheneg et Tlaia (Sahara nord-occidental)," *Bull. Soc. Préhist. Fr.*, 54:636–44.

CHAVAILLON-DUTRIEVOZ, N.

1956. "L'Atérien d'Anchal (Monts d'Ougarta, Sahara nord-occidental)," *Bull. Soc. Préhist. Fr.*, 53:637–47.

CHEVALIER, A.

1938. "Le Sahara, centre d'origine de plantes cultivées," in "La vie dans la région désertique nord-tropicale de l'Ancien Monde," *Mém. Soc. Biogéog.*, 6:307–22.

1947. "Les Jujubiers ou *Ziziphus* de l'Ancien Monde," *Rev. Bot. Appl.*, 27:470–83.

1949a. "L'origine des plantes cultivées dans l'Afrique du Nord et le Sahara," *Trav. bot. dédiés a René Maire* (Algiers 1949), pp. 51–56.

1949b. "A quelle époque a pris naissance la protoculture au Sahara?," *Rev. Bot. Appl.*, 29:418–19.

CHOUBERT, G.

1946. "Note préliminaire sur le Pontien au Maroc," *Bull. Soc. Géol. Fr.*, 15:677–764, and *Notes et Mém. Serv. Géol. Maroc*, No. 85:677–764.

1950. "Réflexions au sujet du Pliocène continental," *Notes et Mém. Serv. Géol. Maroc,* No. 76: 13–91.

1953. "Sur les rapports entre les formations marines et continentales quaternaires," *C. R. Acad. Sci.* (Paris), 237:826–28.

1955. "Note sur la géologie des terrains rècents des Doukkala," *Notes et Mém. Serv. Géol. Maroc,* No. 13.

1956. "Les rapports entre les formations marines et continentales quaternaires," *Proc. 4the Int. Cong. INQUA* (Rome-Pisa, 1953), 576–90.

1957a. "Essai de corrélation entre les cycles marins et continentaux du Pléistocène au Maroc," *C. R. Acad. Sci.* (Paris), 245:1066–9.

1957b. "Essai de corrélation des formations continentales et marines du Pleistocene au Maroc," *Proc. 5th Int. Cong. INQUA* (Madrid-Barcelona). Mimeographed.

1959. "Compléments à la note intitulée 'Essai de corrélation des formations continentales et marines du Pléistocène au Maroc,'" *Com. Géomorph. Périglaciaire Maroc.*

CHOUBERT, G., F. JOLY, M. GIGOUT, J. MARCAIS, J. MARGAT, and R. RAYNAL

1956. "Essai de classification du Quaternaire continental du Maroc," *C. R. Acad. Sci.* (Paris), 243:504–6.

CLARK, J. D.

1955. "Environment and Culture-Contact in Prehistoric Africa South of the Sahara," *Actes. 2e Pan-Afr. Cong. Préhist.* (Algiers 1952): 359–65.

1960. "Human Ecology during Pleistocene and Later Times in Africa South of the Sahara," *Cur. Anthrop.,* 1:307–24.

CLOS-ARCEDUC, A.

1955. "Le fleuve fossile de Tombouctou," *Tropiques* (Paris), No. 371: 36–40.

COOKE, H. B. S.

1948. "The Plio-Pleistocene Boundary and Mammalian Correlations," *Geol. Mag.* (London), 85:41–7.

1957. "The Problem of Quaternary Glacio-Pluvial Correlation in East and South Africa," *Proc. 3d Pan-Afr. Cong. Prehist.* (Livingstone 1955): 51–55.

1958. "Observations Relating to Quaternary Environments in East and Southern Africa," *Trans. Geol. Soc. S. Afr.,* Annexure to vol. 61.

COPPENS, Y.

1960a. "Mission paléontologique dans le Nord de la République du Tchad." Mimeographed.

1960b. "Le Quaternaire fossilifère de Koro-Toro (Tchad). Résultats d'une premiere mission," *C. R. Acad. Sci.* (Paris), 251:2385–6.

1961a. "Un Australopithèque au Sahara (Nord-Tchad)," *Bull. Soc. Préhist. Fr.,* 58: 756–7.

1961b. "Découverte d'un Australopithéciné dans le Villafranchien du Tchad," *C. R. Acad. Sci.* (Paris), 252: 3851–2.

1962a. "Deux gisements de Vertébrés villafranchiens du Tchad," *Actes 4e Pan-Afr. Cong. Préhist.* (Léopoldville 1959), Sér. 8, No. 40: 299–315.

1962b. "Prises de date pour le gisements paléontologiques quaternaires et archéologiques découverts au cours d'une mission de deux mois dans le Nord du Tchad," *Bull. Soc. Préhist. Fr.,* 59:260–7.

1962c. "Découverte d'un Australopithéciné dans le Villafrancien du Tchad. Note préliminaire," *Colloques Internat. C. N. R. S.,* 104, Problèmes actuels de paléontologie (évolution des Vertébrés): 455–9.

CORNET, A.

1952. "L'Atlas saharien sud-oranais," *19e Cong. Géol. Int.* (Algiers 1952).

Créac'h, P.
1941. "Aliments et Alimentation des Indigènes du Moyen Tchad (Afrique Equatoriale Française)," Marseille.

Dars, R.
1957. "Observations pétrographiques sur des échantillons du calcaire gréseux mauritanien," *Bull. Dir. Féd. Mines et Géol. A.O.F.*, No. 20: 107–110.

Dechambre, E.
1951. "Le Sahara, centre primitif de domestication," *C. R. Séances Soc. Biogéog.*, No. 238: 147–51.
1951a. "Discussion de l'interprétation de figurations animales anciennes," *La Terre et la Vie*, No. 2: 105–15.

Dekeyser, Pierre-Louis
1950. "Mammiféres," in "Contribution à l'étude de l'Aïr, *Institut français d'Afrique noire*, Mém. No. 10.
1952. "A propos de la tête osseuse d'un Cynocéphale du Tibesti," *Bull. Institut Français d'Afrique noire*, 14, pp. 537–44.

Dekeyser, Pierre-Louis, and J. Derivot
1960. "Sur de nouveaux spécimens de Cynocéphales du Tibesti (Mission Carl et Petit 1957), *ibid.*, No. 22, pp. 1453–6.

Delcroix, R., and R. Vaufrey
1939. "Le Toumbien de Guinée Française," *L'Anthrop.*, 49:265–312.

Delibrias, G., and H. J. Hugot
1962. "Datation par la méthode dite 'du C-14' du néolithique de l'Adrar Bous (Ténéréen)." In *Missions Berliet Ténéré-Tchad*, Paris: Arts et Métiers Graphiques, pp. 71–72.

Delibrias, G., H. Hugot, and P. Quézel
1959. "Trois datations de sédiments sahariens récents par le radio-carbone," *Libyca* (série Anthrop. Prehist. Ethnogr.), 5:267–70.

Devillers, C.
1948. "Les dépôts quaternaires de l'Erg Tihodaïne (Sahara central)," *C. R. Somm. Soc. Géol. Fr.*, No. 10: 189–91.

Devillers, C., and Jean-Marie Peres
1938. "Notes sur quelques gisements de coquilles fluviatiles du Sahara central," *Bull. Mus. d'Hist. Nat.*, Paris. 11:473–8.

Dittmer, J.
1954. *Allgemeine Völkerkunde, Formen und Entwicklung der Kultur*. Braunschweig: F. Vieweg.

Dresch, J., and G. Rougerie
1960. "Observations morphologiques dans le Sahel du Niger," *Rev. Géomorph. Dyn.*, 11 (Nos. 4, 5, 6):49–58.

Dubief, J.
1950. "Evaporation et coefficients climatiques au Sahara," *Trav. Inst. Rech. Sahar.*, 6:13–44.
1951. "Alizés, harmattan et vents étésiens," *ibid.*, 7:187–90.
1952. "Rapport sur l'évolution des régions arides dans la passé et à l'époque actuelle," UNESCO/NS/AZ/112. Mimeographed.
1953. "Le climat saharien," *Maroc-Medical*, No. 342.
1955. "Note sur l'évolution du climat saharien au cours des derniers millénaires," *Proc. 4th Int. Cong. INQUA* (Rome-Pisa), 2:848–51.

Dubois, J., and J. Tricart
1954. "Esquisse de stratigraphie du Quaternaire du Sénégal et de la Mauritanie du Sud," *C. R. Acad. Sci.* (Paris), No. 22: 2183–5.

Duchemin, G.
1949. "Les Eléphants de Mauritanie," *Notes Afr.*, No. 44: 127–9.

Elouard, P.
1959. "Etude géologique et hydrogéologique des formations sédimentaires du Guebla mauritanien et de la vallée du Senegal."

Elouard, P., and P. Michel
1958. "Le Quaternaire du Lac Rkiz et de l'Aftout de Boutilimit (Mauritanie)," *C. R. Somm. Soc. Géol. Fr.*, No. 12: 245–8.

Esperandieu, G.
1954. "Les animaux domestiques du Nord de l'Afrique d'après les figurations rupestres au cours des périodes préhistoriques et protohistoriques," *Bull. Soc. Zootechnie d'Algérie*, Fasc. 2: 23–68.
1955. "Domestication et élevage dans le Nord de l'Afrique au Néolithique et dans la préhistoire d'après les figurations rupestres," *Actes 2e Pan-Afr. Cong. Préhist.* (Algiers 1952): 551–73.

Falkner, F.
1938. "Die Trockengrenze des Regenfeldbaues in Africa," *Pet. Geog. Mitt.*: 211–14.
1939. "Beiträge zur Agrargeographie der afrikanischen Trockengebiete," *Geog. Abhandl.*, vol. 3.

Faure, H.
1959. "Sur quelques dépôts du Quaternaire du Ténéré (Niger), *C. R. Acad. Sci.* (Paris), 249:2807–9.

Fischer-Piette, E.
1948. "Sur quelques Mollusques fluviatiles du Sahara (Aïr, Itchouma, Fezzan)," *Bull. Mus. d'Hist. Nat.*, Paris, 20:180–1.
1949. "Mollusques terrestres et fluviatiles subfossiles récoltés par Th. Monod dans le Sahara occidental," *Jour. Conchyl.*, 89:231–9.

Flöhn, H.
1953. "Studien über die atmosphärische Zirkulation in der letzten Eiszeit," *Erdkunde*, 7:266–75.

Freydenberg, H.
1908. *Etude sur le Tchad et le bassin du Chari.* Thèse. Paris.

Furon, R.
1960. *Géologie de l'Afrique* (2d ed.). Paris: Payot.

Garde, G.
1911. "Description géologique des Régions situées entre le Niger et le Tchad et à l'Est et au Nord-Est du Tchad." Paris: A. Mermann.

Gardner, Elinor W.
1935. "The Pleistocene Fauna and Flora of Kharga Oasis, Egypt," *Quart. Jour. Geol. Soc. London*, 91 (Part 4):479–518.

Gigout, M.
1957. "Chronologie du Quaternaire récent marocain. Principes de la corrélation fluvio-marine," *C. R. Acad. Sci.* (Paris), 244:2404–7.

Gigout, M., and R. Raynal
1957. "Corrélations des phénomènes marins et continentaux dans le Quaternaire marocain," *C. R. Acad. Sci.* (Paris), 244:2528–31.

Grandidier, G.
1932. "Les eléphants de Mauritanie," *La Terre et la Vie*, 2:130–4.

Graziosi, P.
1942. *L'arte rupestre delle Libia.* Naples: Edizioni della Mostre d'Oltremare.
1952. "Les problèmes de l'art rupestre libyque en relation à l'ambiance saharienne," *Bull. Inst. Fouad 1er du Désert*, 2:107–13.

Grove, A. T., and R. A. Pullan

1961. "Some Aspects of the Pleistocene Paleogeography of the Chad Basin," in this volume.

Hernandez-Pacheco, F.

1945. "Los pozos del Sahara español e hipotesis de su construccion," *Investigacion y Progreso*, 16:1–13.

Heseltine, N.

1959. "Toubou and Gorane, Nomads of the Chad Territory. Notes on Their Origins," *S. Afr. Arch. Bull.*, 14:21–7.

1960. *From Libyan Sands to Chad*, London: Museum Press.

Howell, F. C.

1959. "The Villafranchian and Human Origins," *Science*, 130:831–44.

Huard, Paul

1952a. "Etat des recherches rupestres au Tchad," *Tropiques*, No. 345: 40–45.

1952b. "Les gravures rupestres de Gonoa (Tibesti)," I, *ibid.*, No. 346: 38–46.

1953a. "Les gravures rupestres de Gonoa (Tibesti)," II, *ibid.*, No. 349: 35–48.

1953b. "Les gravures rupestres du Tibesti," *Magazine de l'Afr. du Nord* (Algiers), June 1953.

1953c. "La faune disparue du Tibesti," *Vétérinaires*, September 1953.

1953d. "Gravure rupestres des Confins nigéro-tchadiens," *Bull. Institut Français d'Afrique noire*, 15 (No. 4): 1569–81.

1953e. "Art rupestre au Tchad," *Encycl. Mens. Outre-mer*, 3:313–17.

1953f. "Gravures rupestres de la lisière nord-occidentale du Tibesti," *Trav. Inst. Rech. Sahar.* (Algiers), 10:75–106.

1953g. "Répertoire analytique des stations rupestres du Sahara oriental français (confins nigéro-tchadiens-Tibesti-Borkou-Ennedi)," *J. Soc. Afr.*, 23:43–76.

1954a. "Les gravures rupestres d'Oudingueur (Tibesti)," *Tropiques*, No. 360: 35–45.

1954b. "Le dessèchement du Sahara tchadien," *Tropiques*, No. 365: 52–56.

1955. "Populations anciennes du Tibesti," *Encycl. Mens. Outre-mer:* 366–70.

1956. "Les peintures rupestres du Tchad," *ibid.*, Nos. 71–72:317–20.

1957. "Nouvelles gravures rupestres du Djado, de l'Afafi et du Tibesti," *Bull. Institut Français d'Afrique noire*, 19 (Nos. 1–2): 184–223.

1959a. "Les cornes déformées sur les gravures rupestres du Sahara sud-oriental," *Trav. Inst. Rech. Sahar.* (Algiers), 17:109–31.

1959b. "Préhistoire et Archéologie du Tchad," *Bull. Inst. Et. Centrafr.* (n.s.), Nos. 17–18: 5–20.

1959c. "Le Tibesti des chasseurs," *Notre Sahara*, No. 7: 33–48.

1959d. "Aspects géographiques du Tibesti," *ibid.*, No. 12: 34–47.

1960a. "L'âge pastoral au Tibesti," I, *ibid.*, No. 10: 17–28.

1960b. "L'âge pastoral au Tibesti," II, *ibid.*, No. 14: 13–24.

1960c. "Contribution à l'étude du cheval, du fer et du chameau au Sahara oriental, I. Le fer," *Bull. Institut Français d'Afrique noire*, 22 (Nos. 1–2):134–78.

1960d. "Données paléthnologiques." In P. Huard and M. Charpin, Contribution a l'étude anthropologique des Teda du Tibesti," *Bull. Institut Français d'Afrique noire*, 22: 179–201.

1960e. "Gravures rupestres de la région d'Edjeleh," *Bull. Soc. Préhist. Fr.*, 57:564–72.

1961a. "A propos des gravures rupestres de la région d'Edjeleh," *ibid.* (Fasc. 11–12): 657–8.

1961b. "Les groupes sanguins des Téda du Tibesti (résumé d'une publication du Dr. M. Charpin)," *Bull. Institut Français d'Afrique noire*, 23 (Nos. 1–2):328–9.

1961c. "Les figurations d'animaux a disques frontaux et attributs rituels au Sahara oriental," *Bull. Institut Français d'Afrique noire*, 23 (Sér. B):476–517.

1962a. "Archéologie et zoologie: contribution a l'étude des singes au Sahara oriental et central," *Bull. Institut Français d'Afrique noire*, 24 (Sér. B):86–104.

1962b. "Art Rupestre." In *Missions Berliet Ténéré-Tchad*, Paris: Acts et Métiers Graphiques, pp. 123–48.

1962c. "La préhistoire et l'archéologie en ex-A.E.F.-Cameroun; description du Tibesti et itinéraires archéologiques (Gonoa, etc.)," in *Guide Bleu Afrique Centrale* (Paris), pp. 365–76.

1962d. "Figurations sahariennes de boeufs porteurs, montés et attelés," *Riv. di Storia dell'Agricoltura*, 4: (23 pp.).

1963a. "Gravures rupestres de l'Ennedi et des Erdis," *Bull. Inst. Rech. Scient. Congo*, (24 pp., to appear).

1963b. "A propos des bucranes à cornes déformées de Faras (Nubie)," *Kush*, 11: (18 pp., to appear).

1963c. "Nouvelles données séro-anthropologiques sur les Teda du Tibesti (résultats du Pr. J. Ruffié, mission Prohuza)," *Bull. Institut Français d'Afrique Noire*, 25 (Sér. B) : (2 pp., to appear).

1963d. "Figurations sahariennes de bovins à pendeloques jugulaires," *Riv. di Storia dell'Agricoltura*, (20 pp., to appear).

1964a. "Nouvelle contribution à l'étude du fer au Sahara et au Tchad," *Bull. Institut Français d'Afrique Noire*, 26 (Sér. B) : (90 pp., to appear).

1964b. "Influences culturelles du 'Groupe C' au Sahara tchadien," (18 pp., to appear).

1964c. "Pentiures rupestres du Tibesti oriental," (24 pp., to appear).

In preparation:

"Contribution à l'étude du cheval, du fer et du chameau au Sahara oriental. II. Le chameau."

"Ramassage et premiers travaux agraires au Sahara tchadien."

"Contribution à l'étude des 'haches à gorge' au Sahara tchadien."

Huard, Paul, and Cnes Bacquié

1963. "Matériaux pour l'étude de l'âge du fer au Djourab (Tchad)," *I. Tongour*, (8 pp., to appear).

1964a. "Matériaux pour l'étude de l'âge du fer au Djourab. II. Maledeinga," (8 pp., to appear).

1964b. "Un établissement islamique dans le désert du Tchad: Ouogayi," (15 pp., to appear).

Huard, Paul, and Max Charpin

1962. "Aspect sociologique d'une enquête anthropologique sur les Teda du Tibesti," *Bull. Institut Français d'Afrique Noire*, 24 (Sér. B):575–83.

Huard, Paul, and Jean Marie Massip

1963a. "Gravures rupestres du Tibesti méridional et du Borkou," *Bull. Soc. Préhist. Fr.*, 60: (10 pp., to appear).

1963b. "Sites à porterie au décor 'en vague' du Tibesti," *Bull. Soc. Préhist. Fr.*, 60: (10 pp., to appear).

1964. "Nouveaux centres de peintures rupestres au Sahara nigéro-tchadien," *Bull. Soc. Préhist. Fr.* (10 pp., to appear).

Hubert, H.

1920. "Le dessèchement progressif en Afrique Occidentale," *Bull. Com. Et. Hist. Scient. A.O.F.* [III], No. 4: 401–67.

Hubert, H., P. LaForge, and G. Vanelsche

1921. "Objects anciens de l'Aouker," *Bull. Com. Et. Hist. Scient. A.O.F.* [IV], No. 3: 371–444.

Huzayyin, S.
1941. "The Place of Egypt in Prehistory. A Correlated Study of Climates and Cultures in the Old World," *Mém. Inst. Egypte,* Vol. 43.
1956. "Changes in Climate, Vegetation and Human Adjustment in the Saharo-Arabian Belt with Special Reference to Africa." In *Man's Role in Changing the Face of the Earth* (William L. Thomas, Jr., ed.), Chicago: University of Chicago Press.
Jacques-Felix, H.
1947. "La vie et la mort du lac Tchad. Rapports avec l'agriculture et l'élevage," *Bull. Agron., Sect. Tech. Agric. Trop.,* No. 3.
Jodot, P.
1951. "L'étrange lumachelle néogène à *Limicolaria kem-kemensis* n. sp. des Hammadas du Sud Marocain présaharien," *Notes et Mém. Serv. Géol. Maroc,* No. 85:157–61.
1953. "Gastéropodes lacustres du Quaternaire de Tejerhi (Fezzan)," *Inst. Hautes Et. Tunis, Publ. Scient.,* 1:21–69.
1956. "Les subdivisions du Pliocène dans le Nord de l'Afrique (Algérie-Maroc) d'après les faunes de Mollusques continentaux," *Notes et Mém. Serv. Géol. Maroc,* No. 126.
1958. "Classification et climatologie des Mollusques du Quaternaire dans l'Ahaggar (Sahara)," *C. R. Somm. Soc. Géol. Fr.:* 367–70.
Jodot, P., and R. Lavocat
1951. "Sur la faune de Gastéropodes des terrains de couverture de la Hammada du Dra," *C. R. Somm. Soc. Géol. Fr.,* No. 15: 297–9.
Jodot, P., and S. Rouaix
1957. "Découverte de Mollusques continentaux de l'Aquitanien et du Villafranchien au Sahara mauritanien et soudanais," *C. R. Somm. Soc. Géol. Fr.,* No. 16: 375–7.
Joleaud, L.
1933a. "Etudes de géographie zoologique sur la Berbérie. Les Primates: le Magot," *C. R. Cong. Int. Géog.* (Paris 1931), 2:851–63.
1933b. "Chronologie des Phénomènes Quaternaires, des Faunes de Mammifères et des Civilisations," *5e Congr. Int. Archéol.* (Algiers 1930): 13–46.
1934. "Etudes de géographie zoologique sur la Berbérie. Les Reptiles. Les Crocodiles," *Bull. Soc. Zool. Fr.,* 58 (No. 6):397–404.
1935. *idem.,* "Les Poissons," *Rev. Géog. Maroc.,* Vol. 19, No. 1.
1936a. "Les débuts de la domestication, d'après la chronologie des gravures rupestres," *16e Congr. Int. Anthrop. Archéolo. Préhist.* (Bruxelles 1935): 924–38.
1936b. "Les Mammifères de la Libye et du Sahara central au temps de l'Antiquité classique," *Rev. Afr.,* Nos. 368–9.
1936c. "Gisements de Vertébrés quaternaires du Sahara," *Bull. Soc. Hist. Nat. Afr. Nord,* 26:23–39.
1936d. Evolution géographique de la faune des Mammifères du Sahara central pendant la période actuelle de dessèchement, *C. R. Séances Soc. Biogéog.,* 13:21–3.
1936e. "Essai stratigraphique sur les faunes de Mammifères quaternaires et leurs relations avec les hommes fossiles du Sahara," *Rep. 16th Int. Geol. Cong.* (Washington 1933).
1938. "Histoire de la formation d'un désert: paléogeographie du Sahara." In "La vie dans la région désertique nord-tropicale de l'Ancien Monde," *Mém. Soc. Biogéog.,* 6:21–47.
Joleaud, L., and J. Lombard
1933a. "Mammifères quaternaires d'Ounianga Kebir," *C. R. Acad. Sci.* (Paris), 196:497–9.

1933b. "Conditions de fossilisation et de gisement des Mammifères quaternaires d'Ounianga Kebir," *Bull. Soc. Géol. Fr.* (5) 3:239–43.

JOLY, F.

1952. "Le Tafilalt. Le problème des hamada." In "Aspects de la géomorphologie du Maroc," *Notes et Mém. Serv. Géol. Maroc*, No. 96: 83–91.

1954. "Les Terrains de couverture." In "Les Hamada sud-marocaines," *Trav. Inst. Scient. Chérif.* (série générale), No. 2: 45–71.

KNETSCH, A.

1950. "Beobachtungen in der libyschen Sahara," *Geol. Rundschau*, 39:40–59.

KNOCHE, W.

1937. "Der Einfluss von Vegetationsbränden auf die Witterung," *Meteor. Zeitschr.*, Heft 7: 243–54.

1938a. "Algunos apuntes sobre la produccion del anhidrido carbonico antropogeno," *An Soc. Cient. Argentina*, 126:41–6.

1938b. "Una posible influencia antropogena sobre el aumente de la nubosidad," *ibid.*: 471–3.

1939. "Uber die Möglichkeit anthropogener Auslösung von Seismen," *Mitt. Geogr. Ges. Wien*, Vol. 82.

KOHL-LARSEN, L.

1960. *Die Bilderstrasse Ostafrikas.*

LAVOCAT, R.

1954. "Reconnaissance géologique dans les Hammadas des Confins algéro-marocains du Sud," *Notes et Mém. Serv. Géol. Maroc.*, No. 116.

LE RUMEUR, G.

1934. "Les témoins d'une civilisation ancienne dans le Cercle de Tahoua," *Bull. Com. Et. Hist. Scient. A.O.F.*, 16:299–318.

LAJOUX, J. D.

1962. *Merveilles du Tassili N'Ajjer.* Paris.

LEAKEY, L. S. B.

1953. *Adam's Ancestors* (4th ed.). London: Methuen and Co.

LHOTE, HENRI

1951. "Le chasse au Sahara dans les temps préhistoriques" (pp. 187–99); and "Le chasse au Sahara d'après les gravures et les peintures rupestres" (pp. 201–30), in H. Lhote, *La chasse chez les Touaregs*, Paris.

1953. "Le Cheval et le Chameau dans les peintures et gravures rupestres du Sahara," *Bull. Institut Français d'Afrique noire*, 15 (No. 3): 1138–1228.

1953a. C. R. de H. Rhotert. "Libysche Felsbilder, 1952," *Institut française de l'Afrique noire*, Bull. 15 (No. 4):1731–34.

1955a. "Le Tassili-n-Ajjir. Description géographique et principaux groupes de roches peintes." In H. Breuil 1955 (this bibliography), 67–72.

1955b. "Comparaison avec d'autres sites peints d'Afrique septentrionale," *ibid.*, pp. 125–46.

1955c. Chapters: "L'Ahaggar pré- et protohistorique," (pp. 53–89), and "Le peuplement du Sahara central et l'origine des Touaregs," (pp. 90–153), in *Les Touaregs du Hoggar* (2e éd.), Paris: Payot.

1957a. "Peintures préhistoriques du Sahara. Mission H. Lhote au Tassili," Paris.

1957b. "Les gravures rupestres d'Aouineght (Sahara occidental). Nouvelle contribution à l'étude des chars rupestres du Sahara," *Institut française de l'Afrique noire*, Bull. 19 (Nos. 3–4):617–58.

1958. *A la découverte des fresques du Tassili.* Grenoble: Arthaud.

MARTIN, H.
1957. *The Sheltering Desert.* London.
MCBURNEY, C. B. M.
1960. *The Stone Age of Northern Africa.* London: Pelican Books.
MCBURNEY, C. B. M., and R. W. HEY
1955. *Prehistory and Pleistocene Geology in Cyrenaican Libya,* Cambridge: Cambridge University Press.
MAUNY, R.
1955. "Autour de la répartition des chars rupestres du Nord-Ouest africain," *Proc. 2d Pan-Afr. Cong. Prehist.* (Algiers 1952): 741–6.
1956. "Préhistoire et zoologie: la grande 'faune éthiopienne' du Nord-Ouest africain du Paléolithique à nos jours," *Bull. Institut Français d'Afrique noire,* 18 (No. 1): 246–79.
1957. "Répartition de la grande 'faune éthiopienne' du Nord-Ouest africain du Paléolithique à nos jours," *Proc. 3d Pan-Afr. Cong. Prehist.* (Livingstone 1955): 102–5.
1961. "Catalogue des restes osseux humains préhistoriques trouvés dans l'Ouest africain," *Bull. Institut Français d'Afrique noire,* 23 (Nos. 3, 4):388–410.
1962a. "Les industries paléolithiques de la region El Beyyed-Tazamount (Adrar de Mauritanie)," *Actes 4e Pan-Afr. Cong. Préhist.* (Leopoldville 1959), Sér. 8, No. 40: 179–93.
1962b. "Plages soulevées de la région de Nouakchott-Sebkha de Ndghamcha," *ibid.*: 279–87.
MAUNY, R., and F. POUSSIBET
1962. "Nouveaux sites à harpons et faune subfossile de l'Azaouad (Sahara malien)," *Notes Africaines,* No. 93: 1–5.
MAURIZIO, A.
1932. *Histoire de l'alimentation végétale depuis la préhistoire jusqu'a nos jours.* Paris:
MECKELEIN, W.
1959. "Forschungen in der zentralen Sahara. 1. Klimageomorphologie," Braunschweig: G. Westermann.
MICHEL, P.
1960a. "L'evolution géomorphologique des bassins du Sénégal et de la Haute Gambie, ses rapports avec la prospection minière," *Rev. Géomorph.* Dyn., 10 (Nos. 5–6 to 11–12):117–43.
1960b. "Note sur l'evolution morphologique des vallées de la Kolimbine, du Karalkoro et du Sénégal dans la region de Kayes," *Bur. Rech. Géol. Min.,* Dakar.
MONOD, TH.
1935. "Compte-rendue sommaire de la première partie d'une mission au Sahara occidental," *Bull. Mus. d'Hist. Nat.,* Paris, 7 (No. 5):293–9.
1941. "Le Sahara barrière ou trait d'union?" *Dakar, Sénégal,* 1941. Reprinted as "Le Sahara: 'étanche' ou 'perméable?,' " *Cahiers Charles de Fouchauld,* No. 10 (1948): 96–103.
1944. *Au bord de l'Océan Ténébreux: Atlantique et Afrique,* St. Louis, Sénégal.
1945. "La structure du Sahara atlantique," *Trav. Inst. Rech. Sahar.* (Algiers):3:27–55.
1948. "Sur un gisement libyen d'amazonite et 'l'émeraude' dite 'des Garamantes.' " In Th. Monod, "Reconnaissance au Dohone," *Inst. Rech. Sahar.* (Algiers), *Miss. Scient. Fezzan,* Vol. 6: 151–4.
1951a. "Peintures rupestres du Zemmour français (Sahara occidental)," *Bull. Institut Français d'Afrique noire,* 13 (No. 1): 198–213.

1951b. "Un fossile hamadien dans un silex taillé," *Notes Afr.*, No. 48: 17–18.

1956. "Notes sur l'excursion D 10 (Adrar mauritanien) du XIXe Congrès Géologique international (4–28 octobre 1952)," *Bull. Institut Français d'Afrique noire,* 18 (No. 2): 633–51.

1957. "Les grandes divisions chorologiques de l'Afrique. Rapport présenté à la Réunion de Spécialistes sur la phytogéographie" (Yangambi, 29 juillet–8 août 1946), Publ. Commission de Cooperation Technique en Afrique au Sud du Sahara (CCTA/CSA) No. 24, London.

1958. "Majâbat al-Koubrâ. Contribution à l'étude de l' 'Empty Quarter' ouest-saharien," *Institut Français d'Afrique noire,* Mém. No. 52.

1961. "Majâbat al-Koubrâ" (supplément), *Bull. Institut Français d'Afrique noire,* 23 (No. 3):591–637.

1962a. "Notes sur le Quaternaire de la région Tazazmout-El Beyyed (Adrar de Mauritanie), *Actes. 4e Pan-Afr. Cong. Préhist.* (Léopoldville 1959), Sér. 8, No. 40: 177–88.

1962b. "Majâbat al-Koubrâ: un 'Empty Quarter' ouest-saharien." In Hermann von Wissmann-Festschrift, Tübingen, Germany.

Monod, T., and Capitaine Cauneille

1951. "Nouvelles figurations rupestres de chars du Sahara occidental," *Bull. Institut Français d'Afrique noire,* 13 (No. 1):181–97.

Monod, Th., and R. Mauny

1957. "Découverte de nouveaux instruments en os de l'Ouest africain," *Actes 3e Pan-Afr. Cong. Préhist.* (Livingstone 1955): 242–7.

Monod, Th., and C. Toupet

1961. "Sahara-Sahel." In "Histoire de l'utilisation des terres dans la zone aride," Paris: *UNESCO:* 263–77.

Mortelmans, G.

1950. "Le Quaternaire de l'Afrique Sud-Equatoriale: Essai de Corrélation," *3d Cong. Nat. Sci.* (Bruxelles): 62–64.

Mortelmans, G., G. Choubert, and H. Hollard

1952. "Découverte d'industries du groupe de la 'Pebble Culture' sur le reg ancien des plaines du Dra (Sud marocain)," *C. R. Acad. Sci.* (Paris), 235:1680–2.

Myers, Oliver H.

1939. "The Sir Robert Mond Expedition of the Egypt Exploration Society," *Geog. J.*, 93 (No. 4):287–91.

Narr, Karl J.

1956. "Early Food-producing Populations." In *Man's Role in Changing the Face of the Earth,* (William L. Thomas, Jr., ed.), Chicago: University of Chicago Press, pp. 134–51.

1959. "Frühe Hundevorkommen und ihr Kulturgeschichtlicher Ort," *Berlin, Beitr. Zur Vor- und Frühgesch.*, 2:119–25.

Nestourkh, M.

1960. *L'origine de l'homme.* Moscow: Academy of Sciences of the U.S.S.R.

Nicolaisen, J.

1954. "Some Aspects of the Problem of Nomadic Cattle Breeding Among the Tuareg of the Central Sahara," *Geogr. Tidsskr.*, 35:62–105.

Nougier, L. R.

1955. "Influence égyptienne dans le Néo-énéolithique saharien," *Actes. 2e Pan-Afr. Cong. Préhist.* (Algiers 1952):641–5.

PALAUSI, G.
1955. "Au sujet du Niger fossile dans la région de Tombouctou," *Rev. Géomorph. Dyn.*, 6:217–8 (also, *Bull. Dir. Féd. Mines et Géol. A.O.F.*, 1957, No. 20: 143–7).
PEEL, R. F., and R. A. BAGNOLD
1939. "Archeology, Additional Notes," *Geog. J.*, 93 (No. 4):291–5.
PIAS, J.
1958. "Transgressions et régressions du lac Tchad à la fin du Tertiaire et au Quaternaire," *C. R. Acad. Sci.* (Paris), 246:800–803.
1960. "Sédimentation au Quaternaire dans l'Est de la cuvette tchadienne (massifs du Ouaddai et de l'Ennedi. Plaines de piedmonts), *ibid.*, 250:1514–16.
PIAS, J., and E. GUICHARD
1957. "Origine et conséquences de l'existence d'un cordon sableux dans la partie Sud-Ouest de la cuvette tchadienne," *C. R. Acad. Sci.* (Paris), 244:791–3.
PONS, A., and P. QUÉZEL
1956. "Premiers résultats de l'analyse palynologique de quelques paléosols sahariens," *C. R. Acad. Sci.* (Paris), 243:1656–8.
1957. "Première étude palynologique de quelques paléosols sahariens," *Trav. Inst. Rech. Sahar.* (Algiers), 15:15–40.
1958. "Premières remarques sur l'étude palynologique d'un guano fossile du Hoggar," *C. R. Acad. Sci.* (Paris), 246:290–2.
POUSSIBET, F.
1961. "Notes sur l'Azaouad," *Bull. Institut Français d'Afrique noire*, 23 (Nos. 3–4): 573–95.
PROST, A.
1954. "Pierres à moudre en pays songay," *Notes Africaines*, No. 63: 80.
QUÉZEL, P.
1958. "Quelques aspects de la dégradation du paysage végétal au Sahara et en Afrique du Nord," *Union Int. Conserv. Nature* (7e Réun. Techn. Athens 1958), preprint No. RT7/Ie/5.
1960. "Flore et palynologie sahariennes," *Journées Inf. Médico-Soc. Sahar.* (Paris 1960): 41–45; and *Bull. Institut Français d'Afrique noire*, 22A (No. 2):353–60.
QUÉZEL, P., and H. HUGOT
1957. "A propos de quelques graines fossiles du gisement préhistorique de Meniet," *Bull. Soc. Hist. Nat. Afr. Nord*, 48:370–2.
QUÉZEL, P., and C. MARTINEZ
1958. "Etude palynologique de deux diatomites du Borkou (Territoire du Tchad, A.E.F.)," *Bull. Soc. Hist. Nat. Afr. Nord*, 49:230–44.
1962. "Premiers résultats de l'analyse palynologique de sédiments recueillis au Sahara méridional à l'occasion de la Mission Berliet-Tchad." In *Missions Berliet-Ténéré-Tchad*, Paris: Arts et Métiers Graphiques.
QUÉZEL, P., and A. PONS
1958. "A propos de l'étude palynologique de quelques sédiments sahariens récents," *Bull. Liaison Sahar.*, No. 29: 77–80.
QUÉZEL, P., and J. Y. THEBAULT
1959. "Palynologie et datation du volcanisme récent de l'Ahaggar," *Bull. Scient. Econ. Bureau Rech. Min. Algérie*, No. 6: 59–64.
RADIER, H.
1955. "Extension de la série d'Ydouban au Nord-Est de Tombouctou (Soudan)," *Bull. Dir. Féd. Mines et Géol. A.O.F.*, No. 11 (mimeographed).

Reed, Charles A.

1959. "Animal Domestication in the Prehistoric Near East," *Science*, 130:1629–39.

1960. "A Review of the Archeological Evidence on Animal Domestication in the Prehistoric Near East." In Robert J. Braidwood and Bruce Howe, *Prehistoric Investigations in Iraqi Kurdistan* ("Studies in Ancient Oriental Civilization," No. 31), Chicago: University of Chicago Oriental Institute.

Richard-Molard, J., and R. Mauny

1953. "Contribution à la préhistoire de l'Adrar mauritanien septentrional et du Makteir," *Bull. Institut Français d'Afrique noire*, 15 (No. 3): 1229–41.

Rhotert, H.

1952. *Libysche Felsbilder.* Darmstadt: L. C. Wittich.

Romer, A. S., and P. H. Nesbitt

1930. "An Extinct Cane-Rat (*Thryonomys Logani* n. sp.) from the Central Sahara," *Ann. Mag. Nat. Hist.*, 6:687–90.

Rosen, Eric von

1929. "Did Prehistoric Egyptian Culture Spring from a Marsh-dwelling People?" *Riksmuseets Etnogr. Avd., smärre meddel.* No. 28.

Sandford, K. S.

1933. "Past Climate and Early Man in the Southern Libyan Desert," *Geog. J.*, 82 (Nos. 2–3):219–22.

Setzer, Henry W.

1956. "Mammals of the Anglo-Egyptian Sudan," *Proc. U.S. Nat. Mus.*, 106:447–587.

Täckholm, V. and G. (collab. M. Drar)

1941. *Flora of Egypt*, Vol. 1, Cairo:

Tappen, N. C.

1960. "Problems of Distribution and Adaptation of the African Monkeys," *Cur. Anthrop.*, 1:91–120.

Tessier, F.

1954. "Notice explicative sur la feuille Dakar-Est," *Carte Géol. A.O.F.*

1959. "Termitiéres fossiles dans la latérite de Dakar (Sénégal). Remarques sur les structures latéritiques," *Ann. Fac. Sc. Univ. Dakar*, 4:92–132.

Thebaut, J. Y.

1961. "Caractères de façonnement continental de certaines formations détritiques du Hoggar," *Bull. Bur. Rech. Géol. et Minières*, No. 4:17–47.

Tobias, P. V.

1961a. "Physique of a Desert Folk. Genes, Not Habitat Shaped the Bushmen," *Nat. Hist.*, 70:16–25.

1961b. "New Evidence and New Views on the Evolution of Man in Africa," *S. Afr. J. Sci.*, 57:25–38.

Tricart, J.

1955a. "Carte géo-morphologique du delta du Sénégal," *Bull. Assn. Géog. Fr.*, Nos. 251–252: 98–117.

1955b. "Aspects sédimentologiques du delta du Sénégal," *Geol. Rundschau*, 43:384–97.

1956. "Tentative de corrélation des périodes pluviales africaines et des périodes glaciaires," *C. R. Somm. Soc. Géol. Fr.*, Nos. 9–10: 164–7.

1960. "Présentation d'une feuille de la carte géomorphologique du delta du Sénégal au 1/50.000," *Rev. Géomorph. Dyn.*, 10:106–16.

Tricart, J., and M. Brochu

1955. "Le Grand Erg ancien du Trarza et du Cayor (S. W. de la Mauritanie et N. du Sénégal)," *Rev. Géomorph. Dyn.*, 6:145–76.

Tschudi, J.
1955. "Die Felsmalereien in Edjeri, Tamrit, Assakao, Meddak (Tassili-n-Ajjer)," *Proc. 2d. Pan-Afr. Cong. Prehist.* (Algiers 1952): 761–7.
Urvoy, Yves
1942. "Les bassins du Niger," *Mém. Institut Français d'Afrique noire,* No. 4.
Vallois, H. V., and H. L. Movius, Jr. (eds.)
1953. "Catalogue des Hommes fossiles," *19e Cong. Géol. Int.* (Algiers 1952), Fasc. 5: 61–376.
Van Campo, Madeleine, and R. Coque
1960. "Palynologie et géomorphologie dans le Sud tunisien," *Pollen et spores,* 2 (No. 2):275–84.
Vaufrey, R.
1936. "L'âge de l'art rupestre nord-africain," *Bull. Soc. Préhist. Fr.,* 33:624–39.
1938. "L'âge de l'art rupestre nord-africain," *IPEK,* 12:10–29.
1939. "L'art rupestre nord-africain," *Arch. Inst. Paléont. Humaine, Mém.* No. 20.
1946. "L'art rupestre nord-africain," *La Nature,* No. 3112: 149–51; No. 3113: 165–8.
1947. "Le Néolithique para-Toumbien. Une civilisation agricole primitive du Soudan," *Rev. Scient.,* 85:205–32.
1953. "L'âge de la pierre en Afrique," *J. Soc. Afr.,* 23:103–38.
1955. *Préhistoire de l'Afrique.* I. *Maghreb.* Paris: Masson et Cie.
Winkler, Hans A.
1938. "Rock-Drawings of Southern Upper Egypt," I. London: Oxford University Press.
1939a. *idem,* II. London: Oxford University Press.
1939b. "Rock-Pictures at Uweinat," *Geog. J.,* 93:307–10.
Wissmann, H. von
1957. "Ursprung und Ausbreitungswege von Pflanzen und Tierzucht und ihre Abhängigkeit von der Klimageschichte," *Erdkunde,* 11 (Heft 2):81–94; 11 (Heft 3):175–93.
Wissmann, H. von, and F. Kussmaul
1958. Article: BADW, D: "Histoire des origines du nomadisme et ses aspects géographiques," in *Encycl. Islam.,* nouv. éd. (I. livr. 14): 899–906.
Wissmann, H. von, H. Pöch, G. Smolla, and F. Kussmaul
1956. "On the Role of Nature and Man in Changing the Face of the Dry Belt of Asia." In *Man's Role in Changing the Face of the Earth* (William L. Thomas, Jr., ed.), Chicago: University of Chicago Press, pp. 278–303.
Wright, H. E., Jr.
1960. "Climate and Prehistoric Man in the Eastern Mediterranean." In Robert J. Braidwood and Bruce Howe, *Prehistoric Investigations in Iraqi Kurdistan* (Studies in Ancient Oriental Civilization No. 31), Chicago: University of Chicago, pp. 71–97.
Zeuner, F. E.
1952a. "The Plio-Pleistocene Boundary," *Proc. 1st Pan-Afr. Cong. Prehist.* (Nairobi 1947): 55–59.
1952b. "Mediterranean and Tropical Pluvials," *ibid.:* 66–69.
Zohrer, Ludwig G. A.
1952–53. "La population du Sahara antérieure à l'apparition du Chameau," *Bull. Soc. Neuchat. Géog.,* 51:3–133.

SOME ASPECTS OF THE PLEISTOCENE PALEOGEOGRAPHY OF THE CHAD BASIN

A. T. GROVE AND R. A. PULLAN

THE CHAD BASIN, occupying some three-quarters of a million square miles of west-central Africa, stretches from the central Sahara to the Congo watershed. At the rim of the main, saucer-shaped depression rise several mountain massifs: Aïr, the Hoggar, Tibesti, and Ennedi in the Sahara; Jebel Marra in the Sudan; the Jos Plateau in Nigeria; and Adamawa in the Cameroun Republic. The lowest parts of the basin, comprising the Djourab and Bodélé depressions are no more than 500 feet above sea level, whereas Lake Chad, 300 miles to the southwest, stands at 920 feet (Fig. 1).

Because it lies on the fringes of the desert, annual evaporation losses from the surface of Lake Chad are very great and are believed to exceed 80 inches (Bouchardeau 1956). The level of the lake is maintained in part by 10 to 20 inches of rain falling on its surface between June and September, but its continued existence depends mainly on the combined flow of the Logone and Shari rivers entering from the south. Nigerian rivers seem to contribute relatively little to the upkeep of the lake level. If the Logone and Shari were to cease to flow the lake would probably dry up within a couple of years, for it is very shallow.

Although the waters of Chad are fresh near the Logone-Shari delta, salinity increases towards the north and there is a slow drift of water from the southern delta region towards the northern lake shore. Lake and delta together can be compared in some respects with the Okovango swamps, Lake Bangweulu, the Nile Sudd, and the Inland Niger Delta.

The roughly circular shape of the Chad basin is broken by the catchment of the Benue in the southwest where the upper courses of the Gongola, Faro, and certain other rivers of the Benue system run northeast toward Chad before bending sharply away from it. The Benue valley seems to have encroached northeast into the Chad basin, abstracting rivers that formerly flowed toward the lake. None of these rivers has a discharge comparable with that of the combined flow of the Logone and Shari, and the loss of their supplies of water is not enough to explain the diminished extent of Chad today as compared with its great size at some stages in the Pleistocene.

At the present time, the Logone in flood overflows its banks in various places

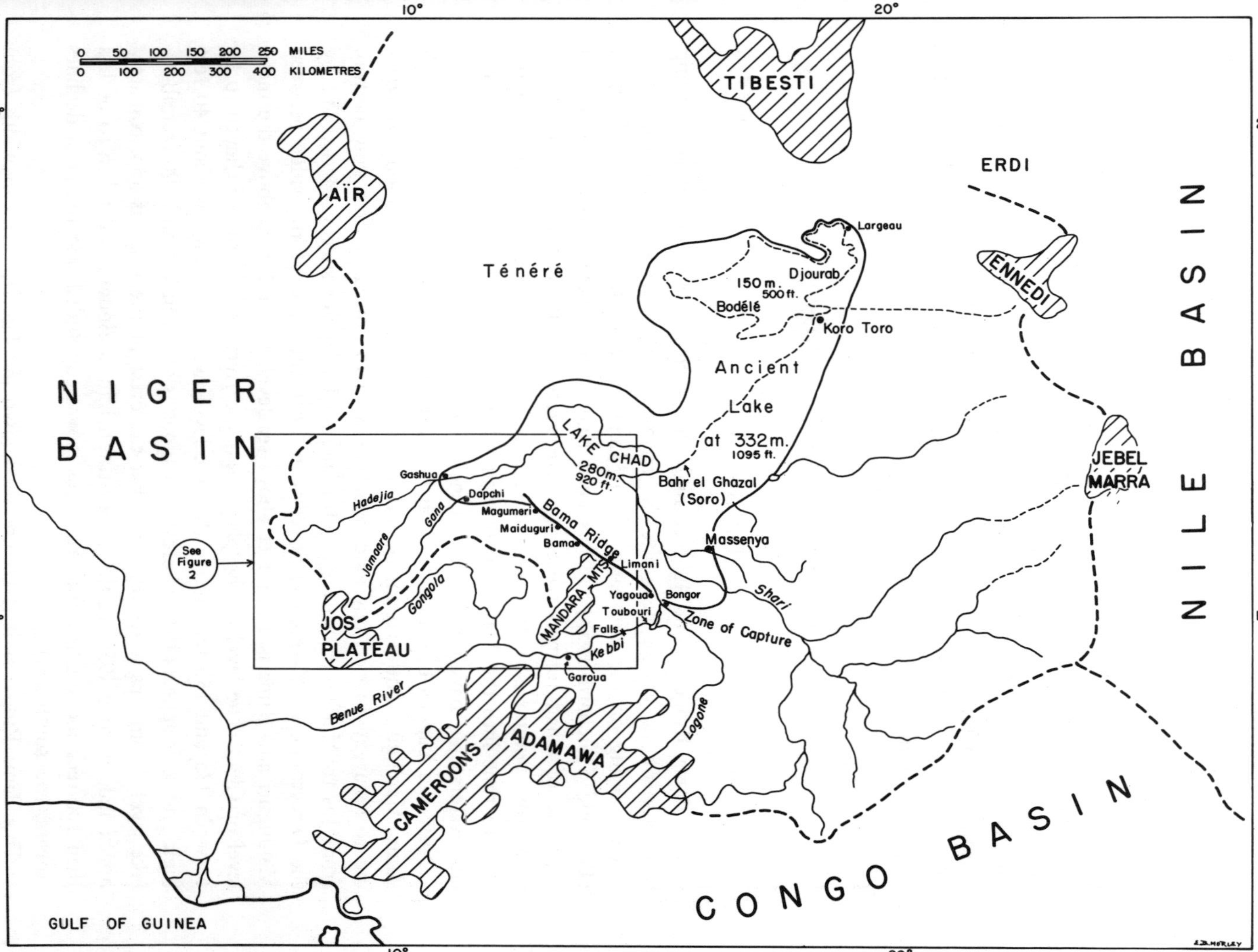

FIGURE 1

The Chad Basin showing its bordering highlands, present-day lake, and the outline of a Pleistocene lake that is believed to have overflowed to the Atlantic via the Benue River.

above Bongor, and some of the water escapes westward through the lakes and swamps of the Toubouri depression, and over the Gauthiot Falls on the Mayo Kebbi to reach the Benue and the Atlantic (Tilho 1939). This has often been regarded in the past as an instance of river capture in progress, the victim throwing herself into the pirate's arms, as it were (Fig. 1). However, buried channels in the valley above Garoua and the laterized remains of a gravel fan in the zone of capture indicate that the volume of water finding its way from the Logone into the Benue was once much greater than now (Roch 1952). The current tendency, Roch suspects, is for the Logone to erode its channel deeper into the basin fill; its capture is not imminent.

Evidence of former conditions more humid than those of the present is afforded in that part of the Chad basin lying within northeast Nigeria by the strandlines of old lakes much larger than the present one, and by still earlier Pleistocene sediments (Fig. 2).

ANCIENT LAKE STRANDS

The Bama Ridge

The Bama Ridge, about 40 feet high, has recently been described running from Limani, a village near the Nigerian-Cameroun frontier, to Yagoua, a town standing on the left bank of the Logone across the river from Bongor. (Pias and Guichard 1957). It has been suggested that this represents an old shoreline of Chad formed when the lake stretched 80 miles farther southwest than now. Bongor is in the zone of capture where the Logone overflows from time to time into the Mayo Kebbi, and when Chad stood at this shoreline the lake waters must have overflowed to the Atlantic via the Benue.

Now the Yagoua-Limani ridge can be traced northwest into Nigeria through Bama, Maiduguri, and Magumeri, and altogether is some 250 miles long. According to a survey carried out in connection with the building of a new railway from Jos to Maiduguri, and a careful aneroid barometer survey carried out by the Directorate of Geological Surveys in 1959, the crest of the ridge between Maiduguri and Bama is 1,095 feet above sea level, i.e., 175 feet above the mean level of Chad. The level of the plains near the river at Bongor is believed to be between 135 and 150 feet above the lake level, while the ridge is about 40 feet high, and so it appears that its crest at Yagoua is between 175 and 190 feet above lake level. From this it would seem that the crest of the ridge is approximately horizontal. (Grove 1959; in this paper the height at Bongor was taken to be less than 100 feet above Chad and from this it was concluded that the crest declines toward the southeast.)

This Bama Ridge, as it is commonly known in Nigeria, varies in width from a few hundreds of yards to over a mile, and runs more or less straight across

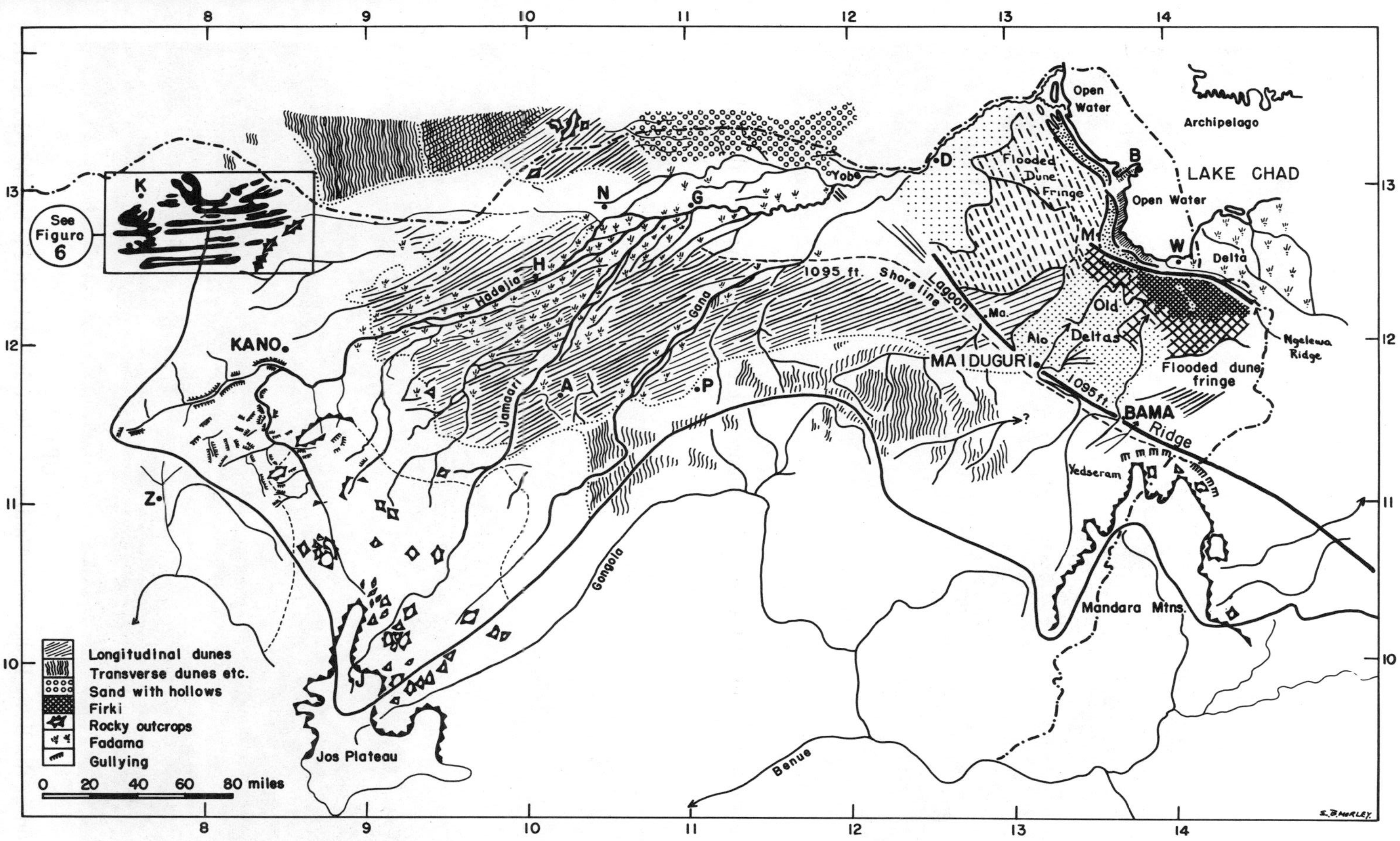

FIGURE 2

The southwestern sector of the Chad Basin in Nigeria. The hill masses, old dunes, and shorelines are shown diagrammatically. Firki is a black, cracking, clayey soil; fadama is land waterlogged or flooded in the rainy season. The letters stand for the following: A—Azare; B—Baga; D—Damasak; G—Gashua; H—Hadejia; J—Jos; K—Katsina; M—Mongonu; Ma—Magumeri; N—Nguru; P—Potiskum; W—Wulgo; Z—Zaria.

FIGURE 3
Section in the Bama Ridge near Maiduguri where the Alo River cuts through it.

FIGURE 4
Section in the Bama Ridge near Maiduguri where it has been cut by the Alo River.

country from Yagoua to Magumeri and beyond to Daberi. The sediments building it vary in calibre from fine silt to subangular gravel more than an inch in diameter. The sands include some well rounded grains that may be of aeolian origin, but the coarseness of certain beds, the presence of flakes of biotite mica, and the angularity of most of the sand grains indicates that water, not wind, has been the main formative agent. The variable character of the sedimentary layers, their cross-bedding and the general form of the ridge in plan and cross section persuade me that it is a single barrier beach or, locally, a series of such beaches (sometimes called offshore bars or lake-bars) formed when Chad stood at a level of 1,095 feet above present sea level.

The source of the sediments in the ridge is believed to be at least in part the crystalline rocks of the Mandara Mountains which project northwards into the basin almost as far as the Bama Ridge (Fig. 5). Pias noted that the calibre of the material in the ridge increases from Yagoua toward the northwest. The valley floors between the northern spurs of the mountains are covered with alluvial spreads of varied material of the kind found in the ridge sediments. Such debris was presumably carried down to the shore of the ancient lake and then drifted to the northwest and the southeast by winds from a northeasterly quarter, thereby building the barrier beaches comparable to those characteristic of the southern Baltic, the Gulf of Lions from the Rhone Delta to the Pyrenees, and the Gulf of Mexico between the Tuxpan River and Cape San Blas.

FIGURE 5
The northern end of the Mandara Mountains, with plains stretching north (to the left) towards the Bama Ridge.

Straight beaches of this kind can develop in less than 6,000 years along coasts where relatively unconsolidated rock is exposed to wave attack. (Russell 1958). At the southern end of Chad, northeasterly winds (which blow for most of the dry season) would have had a long fetch over open water, possibly of some 600 miles, and would have been capable of giving very large waves approaching the shore parallel to the Bama Ridge, and breaking onto it.

On the southwest side of the ridge lies a long narrow depression which probably held a lagoon, separating the barrier beach from an inner shoreline. The ridge and parallel depression continue to influence the present drainage pattern, rivers flowing toward Chad from the southwest turning at right angles to run along the floor of the old lagoon and breaking through the ridge in only a few places. Although some streams have cut into it, the ridge is remarkably well preserved. Northwest of Magumeri it is a less obvious feature, and the 1,095-foot contour swings away from it westwards toward Dapchi and Gashua, suggesting that an embayment of the ancient lake 30 or 40 feet deep stretched towards the present positions of the lower courses of the Hadejia, Jamaare, and Gana rivers (see Fig. 2).

The shoreline of the ancient lake outside Nigeria can be traced at least in part. A few miles east of the Shari River, close to Chad's southern shore, rounded waterworn pebbles have been found a hundred feet above lake level on the five rocky pinnacles known as the Hadjer el Hamis, and farther east, old water levels up to 130 feet above the present level of Chad are indicated by red staining on inselbergs (Barbeau 1957). From north of Massenya the old shoreline has been traced for 400 miles to a position east of Koro Toro (Pias 1960). Air photographs and 1:1,000,000 maps (I.G.N., 1961) show features, which are believed to be the same shoreline at about 1,100 feet, continuing as the Taimanga which runs northwest within a few miles of Largeau and then west, a little to the south of the 18th parallel. There are signs of old lagoons in this region as well. The northwest shore cut across Ténéré, while higher ground probably formed a broad peninsula north of the site of the present lake.

This ancient Chad occupied some 120,000 square miles and was comparable in shape and extent to the largest lake in the world at the present day, namely the Caspian Sea. Lake Victoria, only one quarter the size, considerably modifies the climate within several miles of its shores, and so the effects of this greater Chad may well have been felt over a zone some tens of miles broad. Consequently, large savanna animals might have flourished in Borkou and southern Tibesti because of the lake nearby and its influence on local climates and vegetation.

The age of this lake, standing at about 1,095 feet in the Maiduguri-Yagoua region, remains uncertain. The shore features are well developed at this height presumably because the level of the lake, controlled by the Bongor overflow to the Mayo Kebbi, either remained steady for several centuries, perhaps thousands of years, or repeatedly rose to that level. Near Bama the barrier beach is a complex feature made up of several bars lying close together, but there is no clear indication that the bars differ in age very widely, and so the old lake standing at 1,095 feet may be taken to belong to a single period. The good state of preservation of the shoreline forms in Nigeria suggests that this period was during a part of the Upper Pleistocene.

There are some indications that the former existence of a lake of such great size necessarily implies that not only the local but also the regional distribution of rainfall differed considerably from that of the present day. Evaporation losses from the surface of a lake in these low latitudes is high, and even if it is supposed that temperatures were 5° C. lower than now (Flint 1957), it is unlikely that the figures would have been much lower than those estimated for Lake Victoria, i.e., about 65 inches annually (Walker 1956), the mean annual temperatures at Entebbe being 6°C. lower than at Fort Lamy. Since the area of the ancient lake was some 20 times greater than that of Chad today, it would appear that the total annual supply of water from rivers and rainfall together required to balance evaporation losses must have been about 16 times greater, some 400 million acre-feet. This does not take into account an unknown volume discharged from the lake into the Benue system. In other words, Chad at this stage was receiving annually a volume of water equal to one-third the annual discharge of the Congo.

The Ngelewa Ridge

A second much lower ridge rises about 20 feet above the level of Chad within a few miles of the present lake shores. It has been recognized on the southeast side of the lake running north of Ngouma to the southeast of Tourba (Pias 1958), and can also be traced from a point about 25 miles west of Fort Lamy along the left bank of the Ebeji river, through Gambaru to Mongonu. Sometimes referred to as the Ngelewa Ridge, it is a few hunded yards wide with twin crests in places, composed of rounded white sand and, like the Bama Ridge, probably a barrier beach. It enclosed a large lagoon on the south side where various sediments accumulated that will be described in a later section.

When Chad stood at the level of this low ridge it must have reached up to the outfall of the Soro channel or Bahr el Ghazal at the southeast corner of the lake, thereby allowing water to overflow along the channel towards the Bodélé depression. When the ridge was formed is not clear, and the degree to which it indicates a decrease in the aridity of the region is uncertain, because it is not known how much water escaped at the time to the Bodélé depression. The lake is known to have flooded the Soro channel for a distance of 60 miles in 1867, and some water moved along the depression in 1956; but the ridge is probably several centuries older than this.

PLEISTOCENE SEDIMENTATION IN NORTHEAST NIGERIA

The central parts of the Chad basin both within and outside the limits of the ancient lake at 1,095 feet are floored by continental sediments of Quaternary age, called the Chad formation in Nigeria. Thickest in the southwestern parts of the basin, they consist at Maiduguri of some 1,800 feet of lacustrine clays with impersistent arenaceous beds. Below Maiduguri are three zones containing sandy horizons, at 130 to 170 feet, 700 to 900 feet, and 1,300 to 1,800 feet below the surface (Barber and Jones 1960). These beds are thought to have been deposited as alluvial fans at the mouths of rivers flowing into a lake. Diatomaceous earth and diatomite occur at various levels; the main species involved is said to be a form of *Melosira* not older than the Pleistocene.

Amongst the few vertebrate remains that have been found in these sediments in Nigeria is a jaw of *Hippopotamus imaguncula*, found in a well at a depth of 190 feet and attributed to the Villafranchian (Lower Pleistocene). It seems possible that these formations are comparable in age to the oldest fossiliferous sediments at Koro Toro (Abadie, Barbeau, and Coppens, 1959; Barbeau 1961), but unfortunately there are practically no exposures in Nigeria and the chances of finding fossil material in wells is slight. Pollen analysis may give some indication of the climates at the time these clays accumulated.

ANCIENT DUNES

Systems of dunes probably of differing ages give some indication of the extension of arid conditions south of the present limits of the Sahara at different stages in the Pleistocene. Transport of sand by the wind in the Sudan zone of the present day appears to be limited to areas with a mean annual rainfall of less than 6 inches. There may be some drifting of sand in more humid regions (Prescott and White 1960), but much of this can be attributed to accelerated erosion caused by clearing of the vegetation accompanying cultivation and grazing. In general with more than 6 inches of rain the vegetation is dense enough to reduce the wind velocity near the ground below the critical value for sand blowing, and the soil structure presents some resistance to wind action.

Wind-blown Sands of the Kano-Katsina Area

Much of the country around Kano is masked by aeolian sands which lack any clear dune pattern except for one or two very small east-west fixed dunes. On minor watersheds the old land surface emerges, much of it being lateritic ironstone giving poor grazing land, whereas most of the sandy drift is under permanent cultivation. (The rainfall is about 34 inches, ninety per cent falling in the four months from June to September.) Alongside the main watercourses, wide-spreading systems of deep gullies cut back into the break of slope above the flood plains of the seasonal streams, stripping off the mantle of alluvial sand and depositing it in widening river channels.

About 50 miles east of Kano, sand has accumulated against the eastern, i.e. windward, slopes of large rocky hills, such as those as Dutse, and has streamed downwind on the north and south sides. Immediately to leeward of these hill masses are marshy hollows where no sand was deposited.

This arrangement helps to explain a puzzling striped pattern of soils and land use between Katsina and Kazaure, where east-west strips of cultivated sandy soils are separated by uncultivated lateritic ironstone (Fig. 6). The strips run across country for more than 50 miles. The local relief is of the order of 100 feet and the strips were evidently formed after the main features of the existing drainage system and relief of the Chad-Niger watershed had evolved. The clue to the situation is provided by the positions of several groups of quartzose hills rising several hundreds of feet above the general level of the surrounding gneiss and granite plains at the eastern limit of the strips. Each hill mass stands at the eastern end of an uncultivated strip, and it may be surmised that under arid conditions in the past, sand moved westward under the influence of dominantly easterly winds, and while strips to leeward of the hill masses remained bare for long distances downwind, those to leeward of the main gaps between the hills

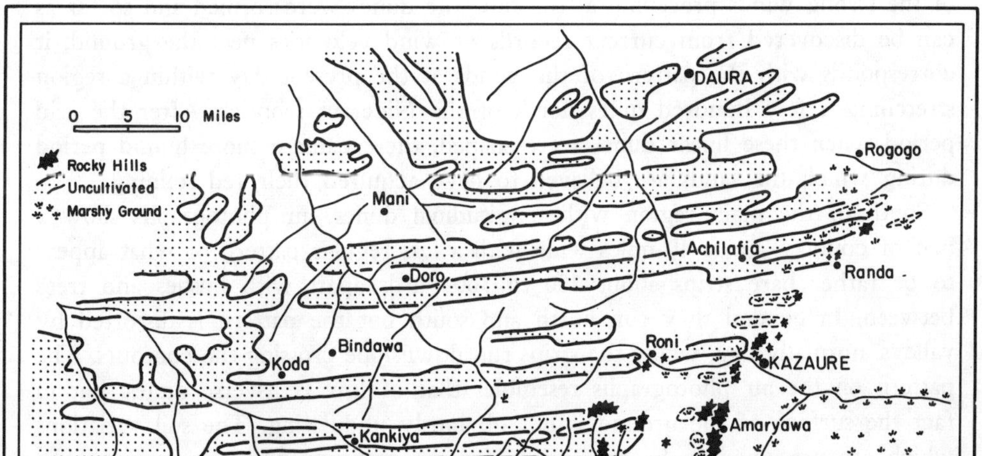

FIGURE 6

The Land-use pattern of northern Katsina and northwest Kano Provinces in Nigeria, showing alternating strips of cultivated and uncultivated land. The cultivated soils are derived from sands which are believed to have drifted westwards through the gaps between the hills lying at the eastern end of the uncultivated strips. The soils of the uncultivated strips consist largely of lateritic ironstone.

were covered with sand. The sandy strips now under cultivation (annual rainfall 30 inches) do not retain any obvious dune forms and the exact mode of their formation remains uncertain. Vegetation need not necessarily have been entirely lacking at the time when the sand was on the move, but it was probably very scanty.

The Ancient Erg of Hausaland

In the northern Nigerian provinces of Kano, Bauchi, and Bornu, linear dunes less than 5 to more than 50 feet high occupy a region of more than 12,000 square miles, stretching west-southwest from a line joining Nguru and Maiduguri to within about 40 miles of Kano. This ancient erg of Hausaland (Falconer 1911) stretches south of the 30 inch isohyet and is in many areas today well settled and productive, but it is reasonable to suppose that when the dunes were shaped, at no very remote period, the climatic zones lay some 300 miles south of their position at the present day with the 6 inch isohyet where the 30 inch isohyet now lies (Grove 1958).

The sands of the linear dunes, as of the long sandy strips in Katsina are generally red-brown in color, the grains mainly subrounded and frosted, with 60 per cent between 0.1 and 0.2 mm. in diameter. The dunes themselves run mainly ENE to WSW, but their direction swings round gradually from NE-SW in the north to E-W in the west. This trend presumably represents the direction of the strong winds prevailing at the time the dunes were formed and so far as can be discovered from current records of wind velocities near the ground, it corresponds with the pattern of the winds at the present day within a region stretching a few hundred miles north of the Nigerian frontier. After the arid period when these linear dunes were shaped, there came a more humid period during which the sands are believed to have acquired their red staining.

South of the main region with longitudinal dunes, air photographs show a belt of country about 40 miles wide with a less regular pattern of what appear to be rather bare strips about 200 to 300 yards apart with bushes and trees between. In general they run north and south but the pattern is distorted by valleys, normally dry, where the strips run downslope crossing the contours. The pattern on the air photographs resembles that of low transverse dunes, but in fact the surface of the ground is not obviously undulating. The soil variations which are presumed to be responsible for the patterns in the vegetation have yet to be described and their paleoclimatic significance is not understood.

The present river systems of the ancient erg of Hausaland postdate the dune ridges. The Hadejia River and many smaller streams rising southeast of Kano conform with the dune pattern and occupy interdunal hollows. The Jamaare and Gana, on the other hand, rising on the Jos Plateau, run from southwest to northeast across the dune lineation. They cut across sand ridges 20 or 30 feet high and appear to have accomplished this by eroding laterally until a gap was cut and then overflowing to the next interdunal hollow at a slightly lower level. The size of abandoned meanders in comparison with meanders of the present day suggests that the rivers acquired their present positions in a somewhat more humid period than now, perhaps at the time when Chad stood at the level of the Ngelewa Ridge.

INTERPRETATION OF LANDFORMS BETWEEN THE BAMA RIDGE AND LAKE CHAD

The area described was studied by field reconnaissance teams of the Soil Survey Section, Research and Specialist Services, Ministry of Agriculture, Northern Nigeria, and the description of these geomorphological units is condensed from the more detailed description given in Higgins *et al.* (1960).

Most of the area between the Bama Ridge and Lake Chad lies below 1,000 feet above sea level, and the general slope of the surface is less than one foot in a mile (1:5,000). Within this area several geomorphological units have been distinguished (Higgins, Ramsay, Pullan, and de Leeuw 1960).

The significance and interpretation of these units, while clearly understood in the southern part of the area where the stratigraphical succession obtained is almost complete, is confused in the northern part where the landforms appear to be older, are poorly preserved, and where the flood plain of the Yobe River is a recent feature crossing some of the units.

South of the flood plain of the Yobe River is a wide spread of compacted sands which have a slightly hummocky surface, irregularly orientated. It is believed that these sands may be old deposits of the river which have been modified at the surface by later aeolian action. The present flood plain contains many outliers of these sands.

To the southeast of these compacted sands and merging into them is a large area of clearly defined sand ridges which are up to 50 feet high, have a length of up to one and a half miles and are spaced two miles apart. These ridges have an assymetrical cross section with the steeper side facing to the east, and they are orientated in a NNW-SSE direction which is the same direction as the sand ridges on the northeast side of the present lake where many are islands for at least a part of the year. The origin of these sand ridges cannot be determined at present.

Linear dunes about 10 feet high, spaced about three-quarters of a mile apart and running from NE to SW extend along the southern periphery of the high ridges described above and they are also found further east and south, to the north of the Mandara Mountains. They are similar to the linear dunes of the fossil erg west of the Bama Ridge and there is only a narrow break occupied by the Bama Ridge between the two fields, which are suspected to be of the same age.

The lakeward parts of the areas occupied by both the NNW-SSE and the

NE-SW sand ridges were inundated by the lake at a time when it seems to have risen 40 feet or more above its present level. Fine-grained grey sediments were deposited in the interdune depressions and become more widespread and deeper toward the lake.

During this extension of the lake, which nearly reached the Bama Ridge in several places, a delta some 20 miles in length was built into the lake, presumably by the rivers Alo and Yedseram. It is now completely isolated from the courses of these rivers and is situated midway between the Bama and Ngelewa ridges (Fig. 2). The original deltaic landforms have been modified considerably by subsequent rainwash and wind action to give a complicated arrangement of circular and linear depressions with irregular ridges and islands of sand. Surface drainage is no longer integrated, but the green appearance of the delta during the dry season indicates that it is still an area with considerable subsurface draining.

The Ngelewa Ridge was probably formed at the same time as or very soon after this delta was built, and it extends from the micaceous-rich sands of the Ebeji River to the eastern edge of the NNE-SSW sand ridges. This ridge is 20 feet above the present average lake level, and sediments accumulated in an extensive lagoon formed on its landward side. In the deeper parts a light gray, greasy clay with a platey structure preserves numerous psuedomorphs of plant roots, suggesting that the lagoon was vegetated either during or immediately after the deposition of the clay, which is rarely more than two feet thick. Resting on this clay, at a depth of between three and six feet beneath the surface, is a thin layer of white sand never more than three inches thick. The sand is invariably very fine and varies in texture throughout the lagoon. It has also been found to show current bedding in one exposure.

The surface deposit of the lagoon is a very dark grey clay which grades into a light loamy clay on the landward side. This clay is highly montmorillonitic in character as shown by the very wide and deep polygonal macrostructure found on the surface during the dry season. The clay is no more than six feet deep and may rest on the lower sand and clay or directly on the yellow sand which formed an irregular floor to the lagoon and which rises as small islands through the clay at the present time. These upper clays are known locally as *firki.*

It is not easy to interpret the significance of the sand horizon between the two clays, but it may well represent a minor period of increased fluvial erosion in the highlands to the south of the lake.

The *firki* clays do not appear to be accumulating under present conditions, as evidenced by the soil profile development, and the small deltas of the Alo and Yedseram rivers are extending across the clays at the present time near the towns of Maiduguri and Dikwa respectively.

These features within a hundred miles of the southwestern shores of Lake Chad thus indicate that after the humid period associated with the Bama Ridge the climate became more arid and linear dunefields were formed, possibly including those west of the ridge and covering much of Hausaland. The lake then rose at least 40 feet above its present average level, flooding the edges of the dune system in the south and west. This was another humid period during which a large delta was built in the lake, but no large barrier beach evolved. Some time during the fall of the lake from this higher level, the Ngelewa Ridge was built and this enclosed a large lagoon in which montmorillonitic clays were deposited. Further falls in the lake level are associated with the formation of a sandy lacustrine fringe on the western and southwestern shore of the lake inside the Ngelewa Ridge, and where the lake margin is free from hydrophytic vegetation a series of high parallel beach dunes have been formed running NNW to SSE. In the south, a sand ridge has formed along the north fringe of the Ebedji delta and is continuing to grow westward into the lake.

HISTORICAL CHANGES IN LAKE LEVEL

Tilho has summarized what is known about the fluctuations of Chad in the last century (Tilho 1911). From the accounts of the early explorers it appears that the lake reached higher levels in some years of the nineteenth century (1823, 1853–4, and 1866–70) than have generally been reached in this century. In the years preceding Tilho's mission of 1907 the level of the lake fell progressively and by the end of the year all that part lying north of the Komadugu Yobe was dry. The lowest levels on record were reached in 1913–15, a period which was markedly dry throughout much of the Sudan zone. This drying up of Chad naturally gave rise to speculations about progressive desiccation and the idea already in people's minds that the Sahara was advancing appeared to be confirmed. But heavier rains caused the lake to recover its normal level within a few years, and in 1954 and later years wider areas were flooded than ever before in this century. In 1963 the lake was still rising.

THE IMPLICATIONS

In time it should be possible to correlate the climatic sequence indicated by geomorphological and other evidence in the Nigerian sector of the Chad basin with the sequences described from Ténéré in the Niger Republic (Faure 1959), the northern part of the Cameroun Republic, and Chad Republic (Pias 1958 and 1959, Barbeau 1961). The fossil material from the Koro-Toro area is likely to be important for helping to date the humid periods so providing a framework to which archaeological and other finds in the highlands at the margins of the basin can be attached.

The results of such correlation may be of even wider significance. The Chad basin when compared with East Africa seems to have been free from violent tectonic activity during the Pleistocene period and may have been scarcely affected at all by earth movements in the upper Pleistocene. Furthermore, the regularity of the latitudinal pattern of climatic belts stretching across the basin and beyond west to the Atlantic and east to the Abyssinian highlands, gives some promise that movements of those belts north-south across the Chad basin were accompanied by shifts of the same magnitude through the whole Sudan region. It may well be possible to link up climatic sequences throughout the region in Senegal, the inland Niger basin and the middle Nile. And it would not be unreasonable to suppose that long-term oscillations in the depth of penetration of the southwest "monsoon" into West Africa were accompanied by climatic conditions in Guinea alternately wetter and drier than those of today.

BIBLIOGRAPHY

ABADIE, J., J. BARBEAU, and Y. COPPENS

1959. "Une faune de Vertébres villafranchienne au Tchad," *C. R. Acad. Sci.* (Paris), 248:3328–30.

BARBER, W., and D. C. JONES

1960. "Geology and Hydrology of Maiduguri," *Records of Geol. Surv. Nigeria*, 1958, pp. 5–20.

BARBEAU, J.

1957. *Bull. de la Direction des Mines et de la Geologie*, No. 8, E.A.F., p. 133.

1961. "Existence d'un dôme villafranchienne dans la région de Koro-Toro (République du Tchad)," *C. R. Acad. Sci.* (Paris), 253:881–3.

BOUCHARDEAU, A.

1956. "Le Lac Tchad," *Annuaire Hydrologique de la France d'Outre-mer.*

FALCONER, J. D.

1911. *The Geology and Geography of Northern Nigeria.* London.

FAURE, H.

1959. "Sur quelques dépôts du Quaternaire du Ténéré (Niger)," *C. R. Acad. Sci.* (Paris), 249:2807–9.

FLINT, R. F.

1957. *Glacial and Pleistocene Geology.* New York: Wiley.

GROVE, A. T.

1958. "The Ancient Erg of Hausaland, and Similar Formations on the South Side of the Sahara," *Geog. J.*, 124:528–33.

1959. "A Note on the Former Extent of Lake Chad," *Geog. J.*, 125:465–7.

HIGGINS, G. M., D. M. RAMSAY, R. A. PULLAN, and P. N. DE LEEUW

1960. "Report on the reconnaissance and semi-detailed soil surveys undertaken in north-east Bornu." *Bulletin No. 14*, Soil Survey Section, Ministry of Agriculture, Northern Nigeria.

INSTITUT GÉOGRAPHIQUE NATIONAL

1961. "Carte de l'Afrique – 1:1,000,000." Paris: Largeau.

PIAS, J.
1958. "Transgressions et régressions du Lac Tchad à la fin du Tertiaire et au Quaternaire," *C. R. Acad. Sci.* (Paris), 246:800–803.
PIAS, J.
1960. "Sedimentation dans l'Est de la cuvette tchadienne," *C. R. Acad. Sci.* (Paris), 250:1514–16.
PIAS, J., and E. GUICHARD
1957. "Origines et conséquences de l'existence d'un cordon sableux dans la partie sud-ouest de la cuvette tchadienne," *C. R. Acad. Sci.* (Paris), 244:791–3.
PRESCOTT, J. R. V., and H. P. WHITE
1960. "Sand formations in the Niger valley between Niamey and Bourem," *Geog. J.*, 126:200–203.
ROCH, E.
1952. "Itinérairies géologiques dans le nord du Cameroun et le sud-ouest du Territoire du Tchad." Office de la Recherche Scientifique d'Outre-mer, Paris.
RUSSELL, R. J.
1958. "Long Straight Beaches." *Technical Report* No. 13. Baton Rouge: Coastal Studies Institute, Louisiana State University.
TILHO, J.
1910. *Documents scientifiques de la mission Tilho* (3 vols.), Paris: Imprimerie Nationale.
1939. "Au sujet de la capture du Logone par la Bénoué," *Rev. Scient.*, No. 3, pp. 159–71.
WALKER, H. O.
1956. "Evaporation from Lake Victoria," *Weather*, 11:384.

THE LATER TERTIARY AND PLEISTOCENE IN EASTERN EQUATORIAL AFRICA

W. W. BISHOP

INTRODUCTION

In such a broad topic it seems desirable to commence by considering briefly the nature of available later Tertiary and Pleistocene stratigraphical evidence from East Africa and the main implications arising from it. Selected field localities representative of the major subdivisions of the period under discussion will then be considered in detail.

A. NATURE OF AVAILABLE EVIDENCE

Direct Evidence

Sedimentary. The deposits from which fossil evidence has been derived may be grouped as:

1. Volcanic. The majority of the important sites are in areas of calcareous volcanic deposits.
2. Lacustrine. The still-water deposits frequently contain a large percentage of primary or derived, fine-grained calcareous volcanic detritus.
3. Fluviatile. The fossils from riverine environments are frequently derived and rolled, or their preservation is poor because of the coarse nature of the deposits.
4. Cave. The material often has almost perfect preservation but is obtained from a rather specialized, local environmental setting.

Morphological. The evidence often allows paleogeographical reconstructions to be made of former landscapes. Of particular importance in East Africa has been the control exercised by tectonic movement (faulting, tilting, or warping) in delimiting lake basins and areas of deposition.

Faunal. The nature of vertebrate and invertebrate fossil assemblages.

Floral. Evidence of former vegetation is either microscopic in the form of pollen (this is rare because of the rapidity of oxidation in tropical areas) or macroscopic, with seeds and fruits occasionally fossilized. Wood and charcoal, although also infrequently preserved, are becoming increasingly important in radiocarbon dating of the last 60,000± years.

Archaeological. Stone artifacts as "second-grade" paleontological indications of hominids, have proved useful in large, stratified and underived assemblages. Their abundance in East Africa and their durability make them particularly useful in areas where bone is unlikely to be preserved.

Inferred Evidence

Ecological. Conclusions as to geographical setting, including vegetation, the part played by tectonic or volcanic activity, and, more rarely, climatic conditions.

Age. As "absolute" chronology of the Tertiary-Quaternary is still in its infancy or limited in application to the later Pleistocene, much of the dating is purely relative and in continental East Africa depends upon that most difficult of subjects, nonmarine vertebrate paleontology.

B. IMPLICATIONS FOR PRIMATE AND HUMAN DISTRIBUTIONS

Distribution in Time

The relative dating of individual localities is of particular importance in controlling reconstructions of distribution patterns for any geological time-horizon. The establishment of a firm time base is essential before discussion of rates of dispersal of particular groups, or of speeds of evolutionary change, can have any real meaning. It will be suggested that at present only five broad time divisions can be recognized for the period under consideration in East Africa, as a framework within which primate distributions and their associated paleoecologies can be discussed.

Distribution in Space

The principal factors affecting distribution patterns based upon fossil evidence falling within particular time zones are suitable environments for life, and of more importance, suitable environments for death. Many of the patterns discernible in the distribution of fossil groups, and particularly the primates, reflect the existence of conditions ideal for the preservation of the creatures after death. Only infrequently can optimum environmental conditions for life be inferred.

Suitable Environments for Life

1. Lacustrine. It may be reasonably argued that the prevalence of artifacts and hominid fossils in deposits of former lakes suggests dense concentrations in lake-edge environments in view of the need of a water supply for hunter and hunted. However, it must be added that unless the lake sediments are of a particular character (see below), fossilization on death would not automatically follow.

2. Fluviatile. The same water-supply argument may be applied to rivers but, in view of the likelihood of less-perfect fossil preservation, the influence of rivers upon former distributions can only be speculated upon in the case of rolled and possibly derived material.

3. Volcanic effects. It seems likely that volcanic regions may at times have inhibited life in the past, for instance in the release of detrimental gas, or as outlined by Du Bois (1959) in causing scorched earth, famine, and possibly death in the form of acid rain or the fatal glass fibres of Pelé's Hair.

However, in East Africa where heavy rainfall results in typical lateritic soils, often with a ferruginous duricrust, on the widespread siliceous Basement Complex rocks, one wonders whether the alkaline volcanic areas may not have played a considerable part in the past as today in supplying calcium, salts, and trace elements. By analogy with deficiency diseases arising from "over-pure" town water supplies, or the need to supply cobalt to sheep in otherwise perfect pastures, it may be that the carbonatitic volcanic centers (Fig. 1) had considerable, if subtle, effects in the past upon primate and general mammalian distributions.

4. Caves. Caves and rock shelters were undoubtedly attractive dwelling sites for the more advanced hominids. In the record of the later Pleistocene the distribution of such suitable shelters may have had some influence upon human distribution patterns.

Suitable Environments for Death, Resulting in the Selective Preservation of Fossils in Certain Areas

1. Lacustrine. Many of the regions of still-water lacustrine deposition in East Africa were and are in rift-valley rain shadow areas, or in other localities with a rainfall of less than thirty inches per annum plus *poor reliability*. Thus, not only would lake basins of internal drainage exercise a centripetal influence on fauna in a semiarid environment, but also the lake levels themselves would fluctuate considerably with variations in the total precipitation received within their catchment areas. The rift-valley lakes in Kenya are still illustrating this feature (e.g., Elmenteita, Naivasha, Natron). To these minor oscillations can be added evidence of changes in level of greater amplitude, reflecting more widespread later-Pleistocene changes in precipitation (Leakey 1931, Nilsson 1931, Flint 1959a, b).

The minor fluctuations have only a local climatic, tectonic, or volcanic connotation, but are nonetheless important because small changes in lake level expose large areas of lake flat. It is these that are recorded as minor non-sequences or land surfaces at Olorgesailie, Olduvai, and in the Kagera Valley (Fig. 3). Similar conditions probably existed earlier at Rusinga and Mfwanganu, etc. (Fig. 1). Any debris in the form of bone or stone left during the life or after the death of man or animal would thus be stratified and entombed by these gentle transgressions.

2. Fluviatile. Death on or near the banks of a river may result in fossilization, but again the problems are those of rolling, or of derivation from some very different and possibly much older environment.

3. Volcanic. The importance of the volcanic centers in suppying fine-grained detritus to the basins of deposition and in providing stratifying materials which are similar to bone, tooth, and tusk in their chemical composition, cannot be overstated. The manner in which volcanic centers dominate many of the areas with well preserved fossil material will be described below for individual localities. In some cases stratification is by the gentle process of smothering by subaerial, wind-borne volcanic ash. The important chemical aspect of the environment of fossilization is illustrated in Table 1. The analyses illustrate the virtually unaltered calcium percentage in the bony structure compared with modern bone and the comparable calcium content of fossil and matrix in a series of Tertiary volcanic deposits from Uganda. This should be compared with Table 2 showing

TABLE 1

Analyst: J. B. Pollock

	UGANDA MID-TERTIARY				*MODERN*	
	Fossil Mammalian Bone Napak	*Fossil Proboscidean Tusk Napak*	*Flaggy Tuffaceous Limestone Napak*	*Tuff from Interstices of Fossils Napak*	*Modern Elephant Tusk*	*Modern Elephant Bone*
SiO_2	1.10	0.21	13.0	14.18	0.11	0.52
Fe_2O_3	0.13	0.07	3.25	4.12	0.26	0.71
FeO				2.08	0.00	0.00
Al_2O_3	3.18	4.84	5.54	4.10	0.89	0.00
TiO_2	0.00	0.00	0.24	0.99	0.00	0.07
MnO	0.29	0.04	0.23	0.31	0.15	0.00
P_2O_5	25.32	31.13	0.14	0.78	46.12	48.07
CaO	48.78	48.53	38.90	36.62	45.23	49.07
MgO	0.47	0.47	1.39	3.27	5.08	0.35
Na_2O	0.26	0.25	2.17	0.33	1.78	1.27
K_2O	0.08	0.09	1.21	0.83	0.36	0.00
$H_2O^{-105°C}$	2.14	2.84	0.65	1.51	23.06	14.88
$H_2O^{+105°C}$	5.31	1.60	5.89	5.42	29.93	46.61
CO_2	13.03	10.02	27.48	24.80	0.00	0.00
TOTAL	100.09	100.09	100.09	99.34	99.98	100.00

approximately 50% loss of calcium in earlier Pleistocene fossils from Uganda with replacement by iron and aluminum and a similar change in later Pleistocene material from Tanganyika. In the majority of environments where calcium is lacking and where replacement by iron or silica does not occur after death, decalcification and the break down of bony structure follow rapidly even if stratification has taken place.

4. Cave. Although cave excavations often provide rich fossil assemblages they are perhaps misleading in yielding typical cave-dwelling forms which are not representative of the total fauna living in an area. In addition, although cave-dwelling forms frequently bring back remains of other animals, this collecting for food is usually highly selective and fossils from a midden or a lair are generally an atypical section of the total animal assemblage in the surrounding area.

TABLE 2

Analyst: J. B. Pollock

	UGANDA, EARLIER-PLEISTOCENE			*MODERN*		*TANGANYIKA LATER-PLEISTOCENE*
	Mammalian Bone Kaiso	*Mammalian Bone Kazinga*	*Mammalian Bone Mweya*	*Modern Elephant Tusk*	*Modern Elephant Bone*	*Mammalian Bone Nyabusora*
SiO_2	9.94	0.90	3.70	0.11	0.52	0.80
Fe_2O_3	19.08	6.97	9.55	0.26	0.71	2.93
FeO	0.15	1.10	0.18	0.00	0.00	0.00
Al_2O_3	13.44	5.00	3.45	0.89	0.00	18.66
MnO	1.66	1.14	2.16	0.15	0.00	0.01
P_2O_5	20.31	28.38	28.94	46.12	48.07	33.44
CaO	21.39	31.65	31.34	45.23	49.07	31.22
MgO	0.69	1.53	0.83	5.08	0.35	0.20
Na_2O	0.05	1.04	0.59	1.78	1.27	3.77
K_2O	0.01	0.06	0.10	0.36	0.00	0.01
$H_2O^{-105°C}$	2.43	0.15	0.27	23.06	14.88	2.04
$H_2O^{+105°C}$	6.19	13.08	9.34	29.93	46.61	1.60
CO_2	3.36	9.05	9.03	0.00	0.00	2.89
TOTAL	100.62	100.05	99.48	99.98	99.93	97.57 ‡

‡ Plus for Nyabusora 1.88% Fluorine.

C. CONSIDERATION OF SPECIFIC LOCALITIES

Age Groupings

Lack of space dictates that certain localities must be passed over and in the light of the above discussion the following will not be considered: riverine deposits, cave deposits, and in general sites falling within the last 60,000 years. During this period the problems become more exclusively archaeological rather than geological, and the detail available for East Africa is too abundant to cover adequately in this paper. In addition the region possesses a wealth of older sites and it is these that the present writer is most qualified to discuss. The localities fall into the following time zones:

1. *Mid-Tertiary*. Fifteen major localities in Western Kenya and Eastern Uganda (Fig. 1 and Table 3), from ten of which primate material has been recovered. All the assemblages, which are in similar geological settings in volcanic areas and sealed beneath lavas or agglomerates, have been labelled Lower Miocene. Although this may well prove to be correct, Le Gros Clark and Leakey (1951) and Whitworth (1958) have suggested that a more recent date is possible, at least for some sites. Savage (*in litt.*) records carnivores from some of the "Miocene" sites which are identical to earlier Tertiary forms. Although the possibility of different ages for some individual assemblages must be borne in mind, all the sites are here considered as broadly contemporary and are included in the term mid-Tertiary.

2. *Time Gap*. It is necessary to commence with discussion of the Mid Tertiary sites as for the period following, having a length of approximately 15 million years, mammalian evidence from East Africa is virtually nonexistent. Holmes (1960) quotes 25 million years for the base of the Miocene and 11 million years for the base of the Pliocene. Depending upon the age finally established for individual sites the time gap may be as long as 15 million or as short as 10 million years. The exceptions are a thin deposit on Maboko island which may be as late as Pontian (Lower Pliocene, Whitworth 1958) and an important new site, probably of Lower Pliocene age, at Fort Ternan near Tinderet Mountain in Kenya which is at present being investigated by Leakey (1960).

3. *The Earlier, Middle, and Later Pleistocene*. These names were applied to East Africa by Wayland (1933) and are preferred here because they are purely relative terms which avoid confusion or direct association with the European Lower, Middle, and Upper Pleistocene faunal subdivisions. They are also in keeping with the title chosen for this paper. In addition they are broad time divisions which avoid the climatic inferences which have been criticized by Cooke (1958) and Flint (1959a, b). Both the writer in Uganda and Pickering in Tanganyika have found it impossible to relate with certainty the theoretical climatic divisions to the evidence observed in rock units in the field.

As used by the writer, the *Earlier Pleistocene* broadly corresponds to the oldest faunal zone of Hopwood (1951) and to the Omo-Kanam faunal zone (Clark 1957). The onset of the *Middle Pleistocene* approximates to the incoming of stone artifacts of recognizable consistent design.

The *Later Pleistocene* as used here spans the effective range at present of radiocarbon dating, that is, approximately the last 60,000 years. It represents the most ideal and constant Pleistocene time zone within which correlation can be applied with some confidence over long distances. Its commencement almost coincides with the incoming of the group of cultures referred to as First Intermediate, including Fauresmith, Sangoan, Acheulo-Levallois, and comparable material (Clark 1957, and Flint and Deevey 1959, in which Lamont Dates V include an age of 43,000± years for the Sangoan at Kalambo Falls). It is also the period in which widespread climatic fluctuations can be recorded with certainty within enclosed basins of tectonic origin (i.e., Naivasha, Elmenteita, and Nakuru). Prior to the *Later Pleistocene*, tectonic movements have frequently altered the form of earlier basins and made extremely difficult the differentiation of purely climatic features from those arising from tectonic causes.

The *Earlier Pleistocene* in East Africa is thus the period for which our knowledge of primates and hominids either from artifacts or fossils still requires to be established. The *Middle Pleistocene* is characterized by hominids associated with the development of the Chelles-Acheul hand-axe culture and its lower boundary is slightly predated by the Oldowan culture. The whole range from Oldowan through Chelles-Acheul corresponds to the Earlier Stone Age in Africa (Clark 1957, p. xxxiii). The Later Pleistocene spans 60,000± years and marks in East Africa the beginning of diversification in stone artifacts. Although an abundance of local archaeological detail is available, only the beginning of this last period will be touched upon in this paper.

1. MID-TERTIARY

The distribution of faunal sites is shown in Figure 1 and published details of the primate specimens recovered are given in Table 3. It is suggested in the Table that fossils may be obtained from three types of lithology.

A. Fully lacustrine conditions predominant, with fish fossils but with some nonsequences. Fine-grained clays, silts, sands, and water-deposited tuffs.

B. Ponding established from time to time by "flash" floods but with numerous breaks in sedimentation during which subaerial deposition was dominant (e.g., Mfwanganu pisolitic tuffs). Some of the localities in this group seem to consist largely of subaerial deposits with only occasional flash flood deposits (e.g., Napak I). Subaerial tuffs predominate with fossil wood sometimes in position of growth and with land gastropods in addition to the mammalian assemblages.

C. Coarse quartzose grits, gravels, and conglomerates with a variable proportion of fine volcanic ash as matrix. These deposits are usually at the bottom of the

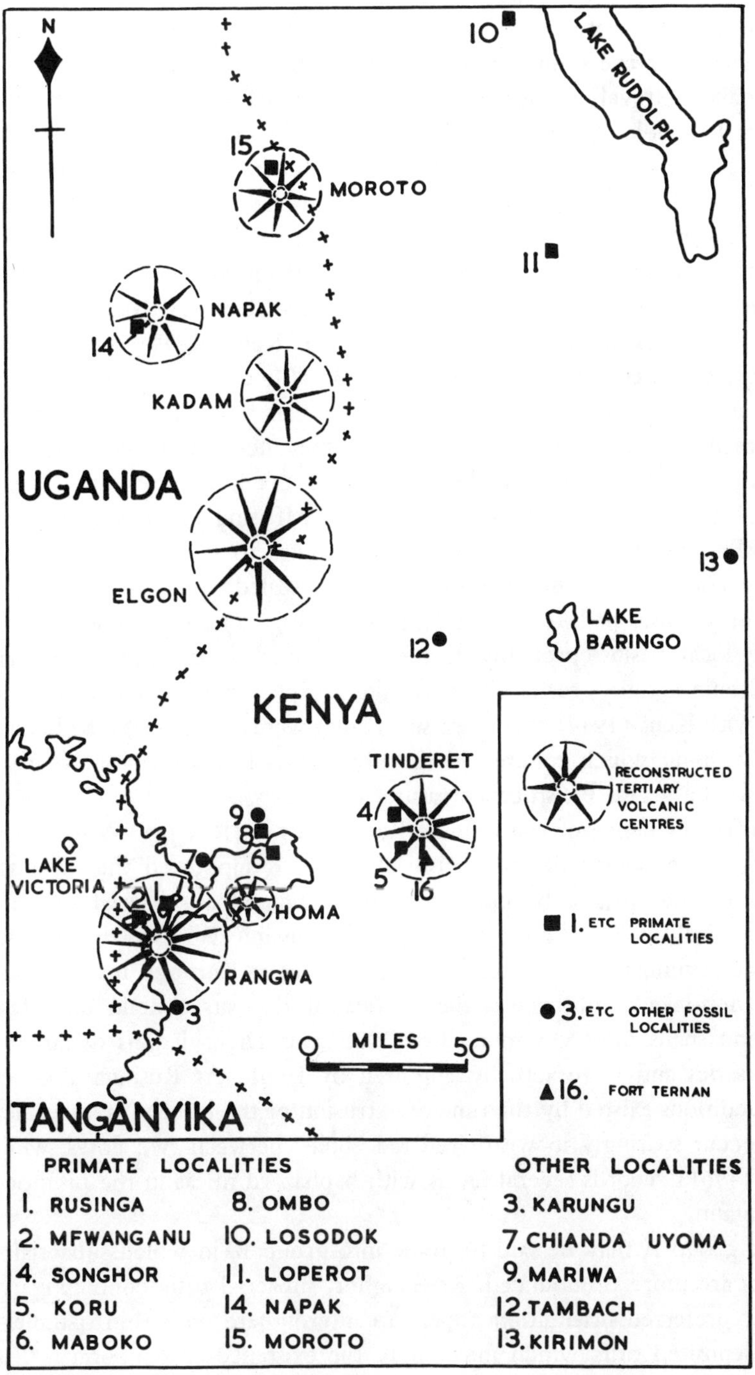

FIGURE 1
Distribution of Tertiary mammalian fossiliferous localities.

volcanic sequence resting directly on Basement Complex rocks. In this group may be included the coarse agglomerates which are very occasionally fossiliferous.

It will be appreciated that these three groups grade into each other and Table 3 indicates only the dominant lithology or lithologies in which fossils occur at each locality. Gravels and grits of group C have yielded fossils at Karungu, Ombo, Mariwa, Losodok, Napak II, and Moroto while some finds are recorded from coarse agglomerate at Rusinga and Mfwanganu. In general, however, this group contains fewer fossils and their preservation is less perfect than elsewhere.

The deposits which contain, in addition to mammals, abundant fish, crocodile, and chelonian remains (e.g., Kathwanga Beds, Whitworth 1954) represent a truly lacustrine environment (group A), as is the case at several Rusinga levels and in the middle series at Karungu (Oswald 1914). Kent (1944) suggested that one large lake could encompass all the Kavirondo "Miocene" sites but it seems doubtful whether even Rusinga and Mfwanganu can be linked with Karungu into one spread of water, despite the fact that they occur now just above the level of Lake Victoria. Several faults have to be taken into account as shown in Figure 2. Similarly, the fact that Rusinga and Mfwanganu lie only 10 miles north of the Rangwa vent while Karungu is 21 miles to the south, may imply that Rusinga and Mfwanganu were completely "sealed" beneath lavas before any depth of deposit had formed at Karungu (Fig. 2). The existence of several shallow local basins, probably formed by impeded drainage consequent upon the volcanic activity, seems the most likely explanation of the facts. The writer agrees with Kent (1944) that these small lakes were not ancestral to Lake Victoria as the warping which resulted in the present shallow saucer of Nyanza can be shown to be a late Pleistocene feature.

From the evidence of the Gumba Red Earths on Rusinga, Whitworth (1954) suggests the existence of ephemeral lakes which supported the savannah fauna implied by the fossils. Shackleton (1951) suggests a subaerial origin for the Rusinga pisolitic Red Earths, while at Mfwanganu Whitworth (1961) records a limestone similar to calcrete. The upper series at Karungu may perhaps imply similar non-lake conditions as the 70 feet of deposits contain only few fossils with land snails most common. Tree trunks of Dicotyledons occur at the top of the series and were actually engulfed by basalt. At Rusinga also permanent land conditions existed by the time of extrusion of the Lunene lavas as mammalian fossils occur sparingly in weathered red "bole" between two flows, while Whitworth (1961) records several levels with haphazard micas in the pisolitic tuffs of Mfwanganu.

Thus group A may be said to grade into group B, in which subaerial volcanic deposits are more pronounced. At Songhor, subaerial tuffs containing fragments without preferred orientation appear to interdigitate with thin, parallel-bedded, water-deposited tuffs which may imply the existence of temporary "flashes" or ponds. At Napak it was originally suggested (Bishop 1958) that ponding had occurred in a shallow valley on the flank of the volcano. However, the fossili-

TABLE 3

Distribution of Tertiary Primates in East Africa Numbers of Separate Specimens

(Based on Whitworth 1958, 1961; Le Gros Clark and Leakey 1951; Le Gros Clark 1952; Hopwood 1934; Le Gros Clark and Thomas 1951, 1952; and Bishop *in litt.*)

	KENYA													UGANDA			K
Associated Volcanic Centres	Rangwa			Tinderet		Not Known								Napak	Moroto	TOTALS	Tinderet
Locality on Fig. 1.	1	2	3	4	5	6	7	8	9	10	11	12	13	14	15	–	16
Fossil Locality	Rusinga	Mfwanganu	Karungu	Songhor	Koru	Maboko	Chianda Uyoma	Ombo	Mariwa	Losodok	Loperot	Tambach	Kirimon	Napak	Moroto	TOTALS	Fort Teman
Miles from Volcanic Centre	9-13	9-13	20-22	10	10	?	?	?	?	?	?	?	?	6	10-12	–	10
Type of Lithology	A & B	A & B	A (B)	B	B	B	?	C (B)	C	B	B & C	B & C	B & C	B & C	C	–	B
Primates Present	Yes	Yes		Yes	Yes	Yes		Yes		Yes	Yes			Yes	Yes	YES 10	Yes
Hominoidea: Proconsul	124	–		37	8	2		–		X	1			2	–	174+	–
Hominoidea: Sivapithecus	1	–		–	–	1		–		–	–			1	–	3	–
Hominoidea: Limnopithecus	63	–		52	5	–		–		–	–			1	–	121	–
Hominoidea: Unspecified	1	9		27	–	1		1		–	–			9	3	51	–
Hominoidea: TOTAL	189	9		116	13	4		1		3	1			13	3	352	X
Cercopithecidae	9	–		–	–	3		1		–	–			X	–	13+	–
Galaginae	8	–		10	–	–		–		–	–			1	–	19	–
TOTAL PRIMATES	206	9		125	13	7		2		3	1			18	3	388+	X

ferous horizon, with persistent, almost flat-lying strata, has now been mapped over an area of 1.7 miles by 1.3 miles, and two further fossil sites have been located. The relations of the beds make it doubtful if a lake could hold water at this point, and it may be significant that *Dinotherium* has not yet been found at Napak.

The bulk of the 100 feet of sediment at Napak I is undoubtedly subaerial with the bedding marking individual outbursts, together with the effects of brief flash floods, and the secondary concentration of layers of calcium carbonate (Table 1). These beds are underlain by lavas and coarse agglomerate and overlain by further agglomerate and so must represent a comparatively quiet phase of activity. The fossils are always fragmentary and frequently show signs of rodent gnawing. With the abundant fossil wood, they represent scattered remains of the flora or fauna which inhabited the temporary land surfaces and which were later sealed beneath wind-borne ash.

At Koru also, Hopwood (1934, p. 437) recorded, "All the bones had been broken before burial and gnawed by rodents as they lay on the surface," while Le Gros Clark and Leakey (1951) say of primate material from all the sites, "The specimens themselves are in every case broken and incomplete." They go on to suggest that this implies that the fragments were "washed in" to the lacustrine sediments. However, the specimens are usually mint fresh and there is a lack of coarse sediment which might be expected to have washed in with them. Kent (1944) notes the unabraded nature of the Rusinga mammalian fossils and the intimate association of bones joined in life by soft tissue while Whitworth (1954) states the Gumba material from Rusinga to be unworn. It seems likely that the fossils largely represent sparsely distributed bony debris which accumulated naturally on temporarily dry lake flats, as occurs in similar situations at the present time. Having been dismembered by carnivores and gnawed by rodents, they were gently submerged and stratified by later transgressions.

Whitworth (1954, p. 91) records from Rusinga the discovery of fossils, including primates, in a small circular pipelike body containing fine green tuff which lacked bedding. The pipe was cut in the well-bedded Flaggy Series and it seems probable that it represents the infilling of a hole developed on a local land surface cut across the flaggy beds. Only detailed excavations at the various localities can settle the exact relationship of fossils to sediments but in general the beds are too sparsely fossiliferous for excavation to yield much material.

It is difficult to discuss faunal differences between the ten localities which have yielded primates, as the bulk of the 388+ pieces come from Rusinga (206) and Songhor (126), and only 56+ from the other eight sites. Of these 18 are from Napak. It can only be commented that the geological setting is broadly the same in each case with coarse agglomerates or lavas overlying fine-grained tuffs or other sediments at Ombo, Koru, Losodok, and Moroto as at Rusinga, Mfwanganu, Songhor, and Napak. Contemporaneity of the sites depends upon the volcanic episodes occurring at the same time from each center. At Maboko

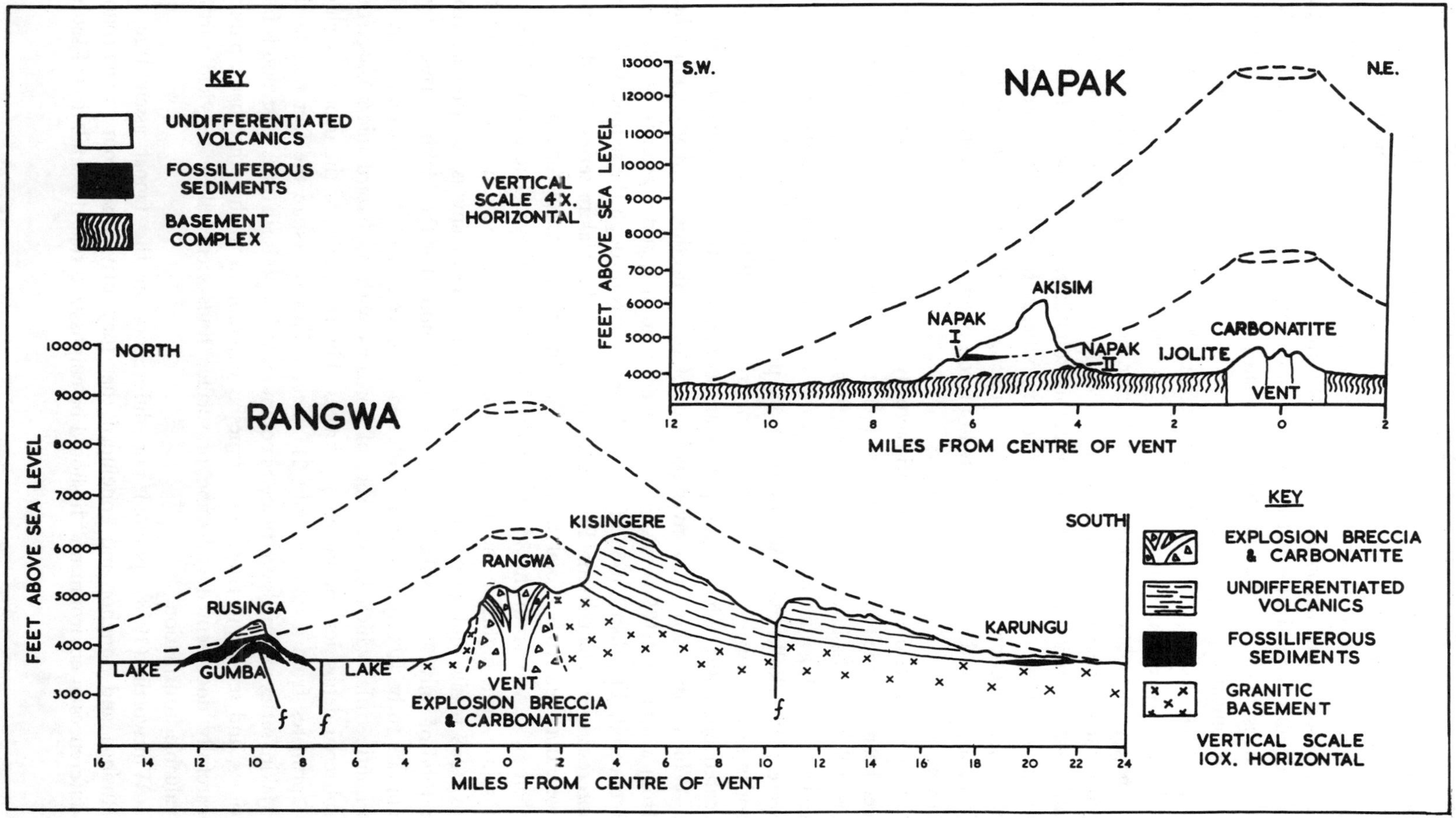

FIGURE 2
Sketch cross sections Napak and Rangwa. Rangwa after McCall 1958.

any former cover seems to have been stripped off by Pontian times (Whitworth 1958) but unfortunately at this site the beds are not exposed naturally and are only known from small artificial sections. Very little material has been obtained from Losodok and Ombo and unfortunately the Koru site seems to have changed considerably since Hopwood's collections were made in 1931 (Shackleton 1951). The beds yielding the 13 Koru primates have not been fully described stratigraphically.

However, Rusinga, Songhor, Mfwanganu, Koru, and Napak seem to imply only slightly different paleogeographical settings. Despite the small number of primate pieces obtained to date from Napak, at least five genera (six species) are represented. It seems probable that the large numbers of pieces from Rusinga and Songhor may arise partly from a more suitable environment for fossilization, but also because of more intense collecting where erosion yields large areas suitable for collection. The exposures at Mfwanganu (9 primate pieces) have been less collected and are of more limited outcrop owing to an extensive agglomerate cover.

It is not possible to account satisfactorily for the fact that primates have not yet been recorded at the other five localities although this may be linked with conditions of fossilization. Primates often seem to occur where remains of small rodents are also numerous. At Napak, out of 1,114 pieces collected in 1958, 148 were teeth, skulls, and jaws; of these, 39 pieces were of *Trilophodon*, 35 were rodents, and only 5 were fragments of primate dentition. Only 9 pieces represented ungulates, which is in quite sharp contrast to the main Kenya sites.

Whitworth (1961) also gives statistics of the collecting at Mfwanganu where from a total of 6,105 fossils only 303 were mammals and 9 primate pieces were recovered. The distribution of these 303 specimens through the sediments showed interesting concentrations. From the 150 feet of the Makira series 211 Mammals were obtained from a total of 4,622 fossils. The greatest fossil concentration, more than 50%, was from the 34 feet of the Upper Red Clays. The 60 to 65 feet of Pisolitic tuffs, higher in the sequence, yielded a further 36 mammals from a total of 1,268 fossils, and the remainder of the sediments as many as 56 mammals from a total of only 215 fossils. It is interesting to note that the Makira series, which appears to be largely lacustrine with bedded micas, yielded 2 primates, while the Pisolitic tuffs, which include several horizons with haphazard micas suggesting subaerial deposition, produced 4 primates. The remainder of the deposits yielded 3 primates from a total of only 215 pieces. This is despite the fact that in the Makira series mammals are twice as common, arthropods six times, gastropods four times, and seeds and fruits three times as common as in the other beds. Similar, carefully documented collecting of surface finds seems likely to yield further valuable information.

Whitworth (1958) pointed out differences in the faunal assemblages of Rusinga and Songhor and concluded that they may arise from "synchronous differences in environment." It should be emphasized that the similarity of setting arises partly from the fact that the various volcanic centers were developed upon a Basement surface sufficiently lacking in features to be referred to as a peneplain (Shackleton 1951, Dixey 1946). However, this surface has considerable relief, at least in places, with large residual hills of gneiss rising over 1,000 feet. Songhor hill itself, on to which the fossiliferous sediments overlap, must have dominated the main site at Songhor during the mid-Tertiary even more than at the present time. This proximity of steep hill slopes to "flats" covered by calcareous volcanic ash may account for the fact that over 100 primate specimens have been found concentrated in this single small gully in the Songhor district.

However, such speculations on local detail cannot detract from the broadly similar picture at the five sites. They were all on the lower flanks or at the foot of volcanic cones which were in process of being built up and probably already of considerable proportions as outlined in Figure 2. Whitworth (1961) suggests 5,000 to 6,000 feet above present lake level for the maximum height of Rangwa volcano while Napak probably stood from 8,000 to 10,000 feet above the Karamoja plain. It is of interest to note that although the fossiliferous sites range from 6 to 22 miles from the volcanic centers with which they are associated (Table 3), there is a pronounced grouping of the Primate localities in the 10- to 12-mile range.

The slopes of these volcanoes when dormant must have been the habitat of at least some of the fossil fauna. One is tempted to draw analogies with the present habitat of the mountain gorilla in Kigezi, southwest Uganda (Donisthorpe 1959). This is between 7,500 and 9,500 feet above sea level (1,500–3,500 feet above the surrounding country) on the slopes of volcanoes which have been active until very recently and which are in areas where lakes impounded by volcanic activity abound.

The analogy cannot be taken too far as the nature of the terrain and its vegetation are dependent upon rainfall. However, the growing mid-Tertiary volcanoes were probably of sufficient height to induce their own local climates and to support dense forests on their upper slopes, even if their general setting was a more arid savannah with ephemeral lakes by contrast with present-day Kigezi. It certainly seems probable that some major rivers with gallery forests would be supplied by sufficient rainfall on the higher hills to flow permanently through the more arid areas. The general setting is perhaps close to the Olduvai-Ngorongoro area at the present time, again with major volcanoes dominating the sedimentation in a small local lake as shown in Figure 4 and as outlined later. This analogy also holds dangers and the rainfall implied in Kavirondo may have been rather heavier than that at present in the local environment of the eastern Serengeti Plain.

The fruits and seeds studied by Chesters (1957, p. 67) suggest "a gallery type forest in which trees festooned with climbers overhang the watercourses" with the presence of numerous lianas. This recalls the present-day vegetation of Maboko Island. Such an environment would be found as a fringe along semipermanent lakes and along water courses. It is also similar in character to the present habitat of *Gorilla gorilla beringei* described by Donisthorpe (1959, p. 4) as of medium sized trees "festooned with lianas which, added to the thick bush and undergrowth, give the forest an appearance of density. A tangle of creepers and ramblers covers fallen trees, bushes and old stumps." This lower forest ranges from 7,500 feet upwards (1,500 feet above the surrounding country) on the Kigezi volcanoes, above the level of cultivation.

Leakey (1952) draws attention to the fact that at Rusinga, open grassland, marshland, and woodland types of fossils are all preserved together and suggests a habitat of grassland country with forest galleries. It seems probable that this is a true picture for the immediate vicinity of the fossil sites, but Primates may also have lived a little way up the volcano slopes where more extensive forest must have existed.

Napier and Davies (1959) from a study of the limbs of *Proconsul* conclude that the foot suggests an arboreal habitat and the upper limb "foreshadows the extreme degree of mobility necessary for brachiation of the pongid type." They state that "tropical rain forests abounded and contained great trees and an abundance of climbing plants." This is, however, at variance with the stratigraphical evidence at the faunal sites except for the local forest galleries already mentioned. Chesters notes that thorny wood is common on Rusinga, as it is also at Napak, while Chaney (1933) writes of the flora from the Bugishu series at the foot of Mount Elgon, which is from a setting similar to the mammalian sites, that the forest "had the general characters of the savannah or woodland vegetation in Central Africa today."

It seems probable, therefore, that although savannah, swamp, and forest forms are all found as fossils, *having died together* upon lake flats or in the vicinity of ephemeral lakes, only the swamp dwellers were strictly in their home environment, and that only at certain times. The savannah forms came in search of water from nearby grasslands at the foot of the volcanoes. At Rusinga these grasslands possibly extended westwards over the present site of Lake Victoria.

The smaller Primates like *Progalago* may well have lived permanently in the local gallery forests, but it is worthy of consideration that the most attractive environments for life, and containing a source of food, for the larger apes were probably on the intermediate forested slopes of the volcanoes. Steep, wooded cliffs and eminences may have had their attractions, as at Songhor, and expeditions into the more open savannah as quadrupedal creatures may also have played their part. At Napak, death occurred on the volcano slopes in company with a more homogeneous faunal assemblage. However, at Rusinga and Mfwanganu it was only on excursions along gallery forests to the vicinity of lake flats that they joined permanently the other members of their deathbed assemblage.

2. TIME GAP

The length of this period which lacks faunal evidence may be established more accurately by means of potassium-argon dating of micas and felspars from fossiliferous volcanic sediments, associated with the fossiliferous horizons. However, one difficulty to be overcome, in addition to the fact that this technique as applied to Tertiary-Quaternary rocks is still in its infancy, is that of ensuring that the micas are primary and not derived from some earlier volcanic source. The new site at present being investigated by Leakey at Fort Ternan and probably of Lower Pliocene age (Leakey 1960) underlines the fact that in East Africa the vulcanicity probably spans a considerable time range in the mid-, and later Tertiary.

3. EARLIER PLEISTOCENE

This is a continuation of the "time gap" in that it is largely a period of negative evidence in hominid development in East Africa. However, the three localities shown in Table 4 are worthy of mention.

Kanam has been considered by most writers as representing the oldest East African Pleistocene fauna at present known. The fossils occur in a series of calcareous clays and fine tuffs outcropping locally at the foot of the dissected volcanic center of Homa Mountain. Assemblages of undoubted artifacts from particular horizons have not been recorded from Kanam. The Kanam mandible (*Homo kanamensis*, Leakey 1936), was obtained from this site in 1932, but could not be proved with absolute certainty to be directly associated with the Earlier-Pleistocene fauna. As a great deal has been written concerning the find, it is not proposed to discuss it further but merely to note that Tobias in a paper to the Pan-African Congress on Prehistory in 1959 removed certain anomalies concerning the morphology of the mandible and stressed that the last word had not yet been said concerning its age.

A single fragment of maxilla with right, second, and third molars referred to *Simopithecus* by Hopwood (1939) is the only primate material so far obtained from the Kaiso Series. Ferruginous sands and gravels in bands up to 2 feet in thickness within lacustrine clays and silts, contain abundant vertebrate and invertebrate fossils and are widely distributed in both the Lake Edward and Lake Albert basins of the Western Rift (see Kaiso and Kazinga, Fig. 3). Their age is generally agreed as Earlier Pleistocene, probably somewhat older than the lithologically similar Omo beds (Arambourg 1943–1947) from which *Dinopithecus* (Arambourg 1947; referred to *Simopithecus* by Leakey and Whitworth 1958) has been recorded. Proboscidean material from the type locality at present being studied by Arambourg suggests an assemblage similar to that of Kanam and broadly equivalent to the Lower Villafranchian. Despite detailed searches and

FIGURE 3
Principal Pleistocene artifact and Primate localities.

the suitability of the beds for the preservation of fossils, no undoubted stone tools have been recorded from the Kaiso series of Uganda. De Heinzelin (1955) records a few possible artifacts from the Kaiso series of the Congo section of the Lake Edward basin. Wayland (1926) and Fuchs (1934) suggested that the fossil ironstones were to be correlated with a period of arid climate, but Cooke (1958) and Flint (1959b) have shown that this has not been proved conclusively. Bishop (1960) suggests that their origin is to be sought in tectonic instability prior to a major movement of the rift valley faults. Despite the existence of attractive lakeside environments with an abundant fauna in a rift valley bounded by less severe escarpments than at present, it must be concluded until positive evidence is found to the contrary, that these deposits either predate the development of recognizable stone artifacts or that tool-making hominids and techniques had not penetrated into western Uganda at this time. However, artifacts of Oldowan type were collected recently from Omo by Clark Howell (personal communication).

Finally mention must be made of the Kafuan culture. In 1919, "a series of extremely primitive pebble tools . . . Pre-Chellean in facies," were found in the gravels of the Kafu Valley in western Uganda (Wayland 1934). In 1939, Lowe extended the Kafuan culture to the Kagera Valley (Lowe 1952). Recent investigation of the Kafu deposits by Bishop (1959) has resulted in a reorientation of the terrace sequence, which no longer supports the suggested Kafuan culture stages. In addition, statistical studies of the gravels revealed prolific natural fracturing which prevents the certain recognition of the presence of pebble artifacts. Similar investigation of the Kagera River 270-foot and 200-foot terraces at the one locality where they are known to exist, in Tanganyika 16 miles downstream from Nsongezi (Fig. 3), has again shown water-rolled quartz pebble gravels with numerous natural fractures. In such a setting it is impossible to recognize rolled-pebble tools with any confidence. As these sites are in riverine gravels they do not warrant further discussion here. In addition, the deposits ascribed to the base of the lacustrine 100-foot terrace sequence at Nsongezi from which Lowe (1952) records derived Developed Kafuan, are assigned by Bishop (in press) to a river terrace, post-MN horizon (Acheuleo-Sangoan) in age.

This lack of an established Kafuan culture in Uganda confirms Oakley's (1957) suggestion that the Oldowan choppers are the oldest recognizable stone tools in East Africa. The argument of Lowe (1952) that the more developed, easily recognizable cultures must have had a gradual origin through a series of more simple forms may not be strictly valid, as once hominids had sufficient intelligence to realize the value of deliberately making stone tools, the fashioning of them after a recognizable pattern may have followed quite rapidly.

4. MID-PLEISTOCENE

As it is essential to consider the Olduvai Gorge sequence as a whole, the subject will be treated under the Middle Pleistocene heading although it should be noted

TABLE 4

Principal Earlier- and Mid-Pleistocene Faunal and Artifact Localities

	CULTURES	KENYA	TANGANYIKA	UGANDA
Later-Pleistocene 30,000 yrs. Circa 60,000 yrs. B.P.	etc. First Intermediate Sangoan Fauresmith, etc.	Not discussed	Not discussed Nyabusora	Not discussed Kalambo Sangoan MN Horizon Nsongezi
Mid-Pleistocene	Earlier Stone Age Chelles-Acheul Hand-axe culture	Kariandusi) Olorgesailie) Kanjera[5.*])	Isimila Olduvai IV Olduvai III Olduvai II[4.*]	Sango Bay Mweya ? Paraa ?
Earlier-Pleistocene	Oldowan No Firm Evidence	Omo Kanam[1.*]	Olduvai I[3.*] [2.*] Laetolil	Kaiso

[1.*] Kanam jaw. [2.*] "Pre-zinj" child parietals, etc. [3.*] Zinjanthropus. [4.*] Olduvai Chellean skull. [5.*] Kanjera skulls.

(hatched) Periods for which reliable evidence is at present lacking that Bed I has been shown by Leakey to contain a fauna which places it in the latter part of the Earlier Pleistocene. In Table 4 the shading indicates the time divisions in Kenya, Uganda, and Tanganyika for which data are lacking at present. This emphasizes the lack of evidence in Tanganyika for the beginning of the Earlier Pleistocene and also the manner in which the evidence from Olduvai gorge stands alone for the first part of the Middle Pleistocene.

The importance of the Olduvai section in our knowledge of fossil primates cannot be overstated and its dominance can be seen in Table 5. Stratified hominid fossils at three levels in Beds I and II and good prospects of further finds, together with a continuous series of stratified, sealed, temporary land surfaces with living floors yielding large artifact assemblages, make it without rival elsewhere in the world.

The detailed analysis of the significance of the fauna and archaeology must await further publications by Dr. and Mrs. Leakey, without whose persistent endeavors the new finds might never have been made, while the first detailed geological investigations have only recently been completed by Dr. R. Pickering of the Tanganyika Geological Survey. However, as far as the paleogeography of the site is concerned the writer feels that in the past certain aspects have received undue attention, while others have been barely touched upon.

The time span of the deposits is shown in Table 4. Leakey (1959) suggests that Bed I is to be correlated with the Upper Villafranchian of Europe, which is shown in Table 4 by placing Bed I as nearly the time equivalent of Omo. The reasons for this new correlation will be seen when the Bed I fauna is specifically identified and published. However, it seems probable that Bed I lies immediately below the base of the Middle Pleistocene and that its contained culture may reveal almost the "beginnings of the development of technical aids" (Clark 1960). This phase passes upwards at Olduvai into the Chelles-Acheul culture sequence (Leakey 1951).

The series of potassium-argon dates for Olduvai Beds I and II, recently published (Curtis, Evernden, and Leakey 1961), suggesting an average age of about 1.75 million years for the *Zinjanthropus* horizon, serves to show how useful this technique may become in applying time zones to lithological units similar to those mapped by Pickering and others. However, the more recent age of 1.3 million years obtained at Heidelberg by Gentner and Lippolt for the basalt underlying Bed I (von Koenigswald *et al.*, 1961) underlines the necessity of checking both the technique and the nature of the samples used in dating, before a new, firm time scale can be set up.

The environment of the hominids responsible for the tools seems to have been the shoreline and flats adjacent to a small local lake with a maximum length of 20 miles and width of 7 to 10 miles (Pickering 1958; 1960, p. 84). The lake was probably initiated as a result of ponding of water on an uneven surface of Basement Complex quartzites and gneisses, ignimbrite, agglomerate, and lava (personal communication from Dr. R. Pickering), possibly by the flow of basalt upon which Bed I rests although this is not certain. Deposition in the lake was dominated to the east by the active volcanoes of Ngorongoro (7,000 feet at present but probably higher during the Middle Pleistocene) and Lemagrut (10,300 feet). As shown in Figure 4, these rise to 2,000 feet and 5,000 feet above the plain while, also to the east, Olmoti rises to 10,100 feet or again approximately 5,000 feet above the plain.

Pickering (1958) records that in the lake beds waterlaid primary tuffs predominate in the east, where the majority of the faunal sites are located, while secondary, derived volcanic materials with sandstones, siltstones, and clays become more important to the west. The fact that a continuous supply of volcanic detritus was available throughout the existence of the lake resulted in a sediment filled basin in which the shallow lake was very susceptible to minor changes in rainfall, tectonic movements, or to changes in precipitation if the area was a basin of internal drainage or when lake level fell below any outlet that may have existed. Figure 4 shows the progressive transgression of the lake toward the west as it was infilled by volcanic material from the east. The lake was probably little more than 11 miles wide at the end of Bed I times but extended gradually westward to a maximum of 22 miles at the close of Bed IV times, with the progressive overlap of Beds I, II, and IV on to the Basement Complex rocks.

That Bed III is connected with the supply of material from the hills is shown by its wedging out westward to disappear at a point 10 miles west of the foot of Ngorongoro and about 6 miles north of Lemagrut. Flint (1959b), Cooke (1958), and Pickering (1960) all consider that the bed represents weathered material from the slopes of the volcanic hills. Pickering (1960) suggests that a period of higher rainfall on the volcanic slopes compared with the present time is probably indicated by Bed III. This is in contrast to previous interpretations of this bed as representing more arid conditions than Beds I, II, and IV.

What is certain is that for weathering on the slopes to progress sufficiently far to give a deep lateritic soil profile and for this material to form such a distinctive red bed in the Olduvai succession it must, in addition to any climatic considerations, represent a phase of quiescence in the volcanic history.

The lake was finally eliminated at the close of Bed IV times by the action of the normal faults shown in Figure 4, which initiated the Ol Bal Bal depression as a new base level of erosion for the predecessor of the modern gorge. Kent (1941) shows how this young and virile river system is cutting back into and capturing an older established drainage pattern in the Laetolil area. The Ol Bal Bal depression (10 miles long by 5 miles wide) is smaller than the original "Lake Olduvai" and fault-bounded rather than lava-dammed. Nevertheless, conditions of deposition within the basin must be very similar to those which existed during the formation of the bulk of the Olduvai beds. Pickering (1960, p. 86) points out that at the north end of the Ol Bal Bal a seasonal lake forms annually, which in years of good rainfall persists throughout most of the year.

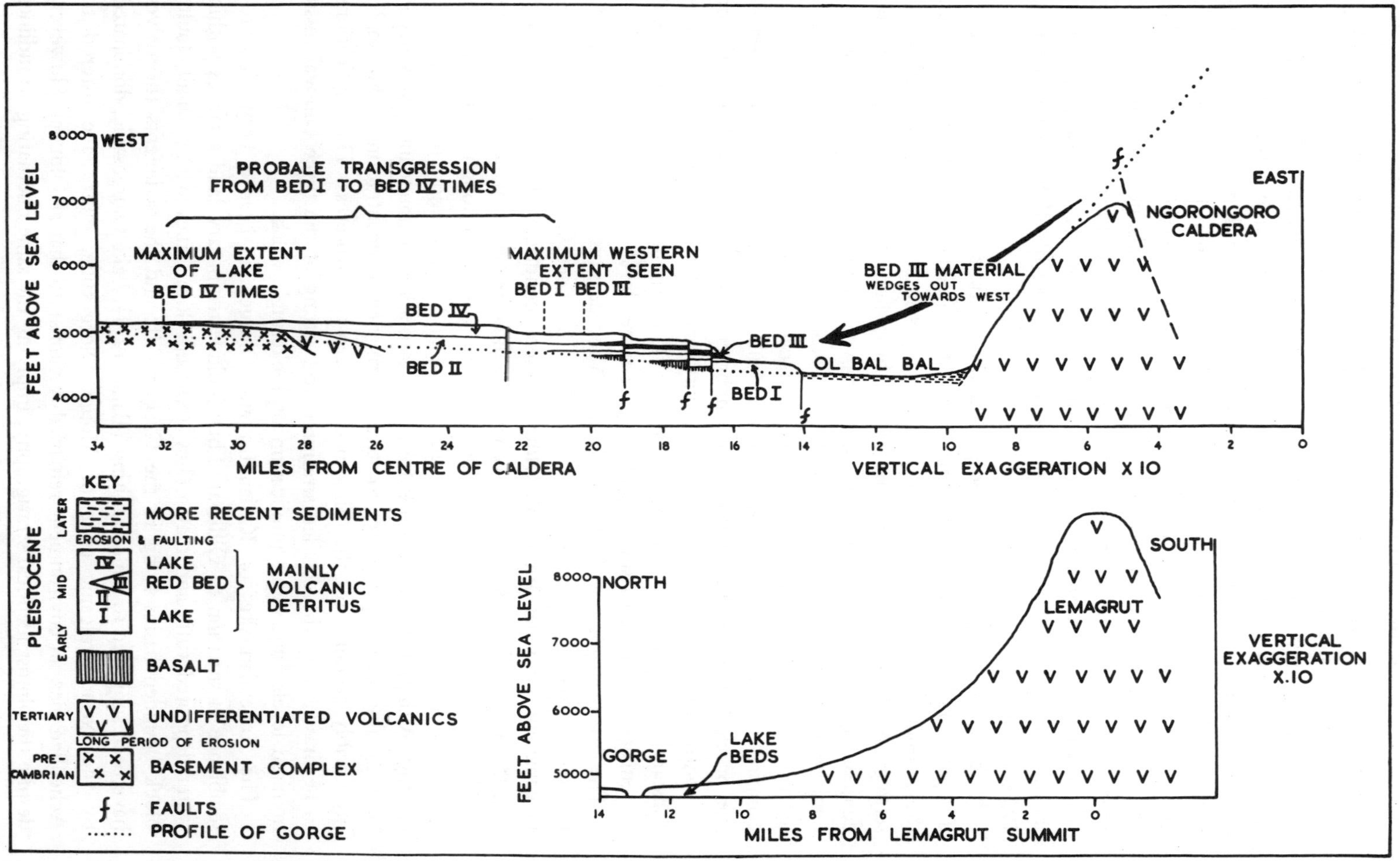

FIGURE 4

Sketch cross sections of the Olduvai region. Western portion of upper section after Pickering 1960.

Water-lain sediments accumulate in the Ol Bal Bal depression and if the active volcano of Lengai erupted violently or the nearby dormant centers again became active, the analogy with Middle Pleistocene conditions would be very close.

The calcareous nature of the volcanic sediment which has played such an important part in the preservation of much of the fossil material at Olduvai is shown by the comparatively recent subaerial tuff which masks the whole area away from the gorge in a layer of "Surface or Steppe Limestone." This grades from a calcareous tuff in the east to a more distinct limestone in the west. The difficulty of identifying undoubted evidence of major climatic changes within the Olduvai sequence is seen in the fact that *dunes* of loose ash exist at the present time within 13 miles of the seasonal lake of Ol Bal Bal (Pickering 1960, p. 86).

In the light of such local variation at the present time in this rain shadow area, it is particularly difficult to isolate major climatic influences in sedimentation from those resulting from tectonic or volcanic changes in the local paleogeographical setting. A detailed analysis of the fauna throughout the sequence would be colored by local conditions but might prove a better guide to widespread climatic fluctuations than lithological variations. The minor oscillation of lake level resulting in the temporary land surfaces which proved such attractive camping grounds for the tool-makers are doubtless to be related to changes in the precipitation received within the catchment area of this shallow basin or to control of the outlet by volcanic activity or minor tectonics. The lake was constantly being infilled with fresh supplies of calcareous volcanic detritus and so remained sensitive to small oscillations of water level.

The broad analogy with the Kavirondo Miocene sites already noted is made the more striking when one also considers the Laetolil area which lies approximately 20 miles south of Olduvai. Here conditions for ponding were absent and fossiliferous subaerial volcanic sediments occur which are possibly equivalent in age to Bed I of Olduvai (Kent 1941). Although some artifacts have been found in the area the abundance and stratification of Olduvai are lacking. Although the conditions for the preservation of abundant fossils occur here the absence of open water to attract permanent concentrations of hominids, presumably accounts for the difference between the two areas. The implications to be drawn from the detailed stratigraphical setting of the three Olduvai fossil hominid levels must await publication of Pickering's work.

The Acheulian sites of Kariandusi and Olorgesailie bear a very close resemblance in situation to Olduvai. They include temporary land surfaces within water-deposited tuffs and clay. They represent local lakes within small, fault-block bounded basins lying in the Gregory rift and the sediments themselves have been subjected to minor faulting (Baker 1958). Lake Olorgesailie, dominated by the extinct volcanic cone of Mount Olorgesailie, must have offered to Acheulian man an environment remarkably similar to that at Olduvai. However, despite the almost identical stratigraphy, implying the same fluctuating strandlines

TABLE 5

DISTRIBUTION OF PRIMATES AND ARTIFACT ASSEMBLAGES BASED ON PUBLISHED DATA (LEAKEY, 1951, ETC.)

		EARLIER			MID								LATER		
		Kanam	Omo	Kaiso	Laetolil	Olduvai I	Olduvai II	Olduvai III	Olduvai IV	Olorgesailie	Kariandusi	Kanjera	Isimila	Nyabusora	MN Nsongezi
Fauna Present		X	X	X	X	X	X	X	X	X	X	X	X	X	–
Preservation		V	F	F	V	V	V	V	V	V	V	V	?	F	–
Nearest Volcanic Center – Miles		3	–	–	12	12-20	12-20	12-20	12-20	8	14	5	–	–	–
Lacustrine Conditions		X	X	X	–	X	X	X	X	X	X	X	X	X	X
Artifacts Present		–	–	–	–	X	X	X	X	X	X	X	X	X	X
Other Primates	Simopithecus	–	–	X	–	X	X	X	X	X	–	X	–	–	–
	Papio	–	–	–	X	–	X	–	X	–	–	–	–	–	–
	Others	–	X	–	X	–	X	–	–	–	–	–	–	X	–
Hominids	Zinjanthropus	–	–	–	–	X	–	–	–	–	–	–	–	–	–
	Others	–	–	–	–	X	X	–	–	–	–	–	–	–	–

V. Volcanic deposits – calcareous F. Ferruginous

and lake flats, with seventeen land surfaces, many of them yielding very large assemblages of Acheulian hand axes, it has not yet proved possible to correlate any of these individually with Olduvai Bed IV horizons.

This perhaps indicates the limits of using artifacts as *fossil assemblages* for purposes of correlation. However, with increased knowledge of typology, "in due course it will be necessary to decide which groups of minor stages of evolution to be seen at Olorgesailie together represent any given major stage from Bed IV of Olduvai" (Leakey 1951).

At Kanjera in lacustrine clays, limestones, and tuffs at the foot of Homa Mountain, fragments of three human skulls were found from beds which also yielded Acheulian hand axes (Leakey 1953). Unfortunately, the skulls are rather fragmentary while the lake deposits have suffered earth movements since their deposition and now only outcrop in a limited fault-bounded area. It is thus impossible to reconstruct more than the broad setting of a lake, again at the foot of an eroded carbonatite volcanic centre. The geological details are recorded in Kent (1942) who also states that the local domestic animals eat the medium grained tuff for the sake of the contained salt. This may have been a feature in all these volcanic areas in the past.

In Uganda there is no faunal evidence for this period. Possibly because of the absence of calcareous volcanic deposits of this age in the Western Rift. However, artifacts from the Semliki series beds at Paraa, Lake Albert, and forms probably derived from near the top of the same series of deposits at Mweya, Lake Edward (Fig. 3), suggest that from Acheulian times man was also living along the margins of the Western Rift lakes.

5. LATER PLEISTOCENE

In the Kagera Valley at Nsongezi (Fig. 3) occurs a complex artifact horizon from which the artifacts have been correlated by Lowe (1952) with those from several levels in Bed IV at Olduvai. Recent excavations by Bishop and Posnansky have shown only one assemblage to exist on the MN horizon which marks a major and probably climatic break in lacustrine and estuarine sediments. These occur at the head of a former embayment of Lake Victoria which extended some 70 miles westward from the present margin of the lake. Since their deposition the sediments have been uptilted to the west and at Nsongezi are now 400 feet above the present lake level. This Later Pleistocene movement was responsible for causing or intensifying the back-ponding in the Lake Victoria-Kyoga area, and gave to the lake its present form. Although the lake recession marked by the MN horizon has led to the stratification of literally millions of artifacts, no fossils have been preserved in this non-calcareous setting, although it seems probable that bone was originally stratified.

The horizon probably represents a rather longer break in sedimentation than the majority of non-sequences described above. However, its time span is certainly much shorter, and its age rather less, than Lowe (1952) envisaged in suggesting a range from Olduvai Chelles-Acheul 6 to 9 inclusive.

Posnansky (1960) writes of the MN artifacts, "The assemblage is culturally mixed in that type tools of two different cultures have been found; nevertheless there is no firm typological evidence to suggest the existence of two separate cultures." The assemblage consists of "a well developed Acheulian facies with certain non hand-axe culture features." The latter if taken alone "would be termed Early or Proto Sangoan."

Some 25 miles eastward, or 31 miles downstream along the Kagera valley, in Tanganyika at Nyabusora, Bishop and Posnansky excavated a stratum in a similar situation to the MN horizon. Nyabusora yielded both fauna and artifacts from a gravelly horizon at a height of 77 feet above the Kagera River, within a series of arenaceous to argillaceous lacustrine deposits. The artifact horizon is overlain by a further 40 feet of lacustrine deposits and the pronounced "flat" representing the bottom of this former arm of Lake Victoria has been raised to 200 feet above present lake level at this point. This flat can be traced consistently upstream into the 400-foot level at Nsongezi and downstream to grade into former strandlines of Lake Victoria at about 100 feet above the lake.

The gravel contained patches of grey ashy material, charred bone, fossil mollusca, bones and teeth of reptilia, fish, and mammalia, together with stone artifacts. The predominance of small bovids and remains of *Hippopotamus* with abundant isolated teeth of *Barbus* and jaws of *Protopterus*, suggest a selected "midden-fauna" possibly accumulated subaquatically at the margin of the lake.

The artifact assemblage closely resembles that from the Nsongezi MN horizon, comprising hand axes, cleavers, picks, cores, flake tools, and waste flakes. Posnansky (private communication) describes the assemblage as follows: "The hand-axes compare closely with the assemblages from Isimila in Tanganyika and Olorgesailie in Kenya although the admixture of Early Sangoan elements suggests that the industry should be thought of as Acheulo-Sangoan rather than as simple Acheulian. In this respect the assemblage can be linked with that from Kalambo Falls."

The type site of the Sangoan at Sango Bay is not considered in detail here as the artifacts occur only on the surface or in shallow hilltop soils lacking stratification. The type sites, at which the Acheulian element is lacking and which may have stood as steep sided islands in Lake Victoria until very late in the Pleistocene, are only some 30 to 40 miles east of Nyabusora.

Isimila (Fig. 3) marks the site of a former small lake ringed by hills. Again, non-sequences marking fluctuations in water level as at Olduvai resulted in large artifact assemblages being stratified between layers of barren sediment.

Sango Bay and Isimila are placed tentatively in Table 4 as are Nyabusora and the Nsongezi MN horizon. By reference to the Kalambo dates all four sites are probably clustered near the base of the Later Pleistocene as defined above. As Nyabusora yields evidence of the knowledge of fire by hand-axe users it is correlated provisionally with Kalambo, where evidence of fire has also been obtained, and placed at the beginning of the Later Pleistocene. Isimila and Sango Bay are considered to be only slightly older.

It is realized that this probably places an undue strain upon the Kalambo Sangoan date and upon the significance of the incoming of the Sangoan both at Kalambo and elsewhere. As this brings the study into the realm of radiocarbon dates, it is proposed to conclude here with a picture of Acheulo-Sangoan hunters and fishers having a knowledge of fire and again favoring lake-strand sites. It is hoped that further absolute dates will shortly be available in East Africa and that a reliable Later Pleistocene chronology will be built up.

CONCLUSION

It has been suggested that both mid-Tertiary and Pleistocene fossil primate and human distributions reflect primarily conditions suitable for fossilization. It is essential to remember that the fossils are members of death assemblages which need not necessarily reveal anything of their conditions of life.

Nevertheless, despite the "dependence of biological inferences on stratigraphical premises" (George 1958) it is possible that the "Miocene" apes were concentrated in attractive habitats in the forests of the volcano slopes and that they may have derived subtle physical benefits from the volcanic soils.

Similarly during the Pleistocene it is impossible to ignore the influence exercised by tectonic basins containing shallow lakes. The Eastern and Western Rifts in particular provided *rain shadow savannah* in which a string of small lakes attracted camps of human hunters and possibly facilitated movement between one sedimentary basin and the next.

Finally, I would express my gratitude to Dr. L.S.B. Leakey who kindly introduced me to the fascination of the East African Miocene and Pleistocene. Also to Dr. R. H. Cummings, Dr. R. Pickering, Dr. M. Posnansky, and Dr. F. Whyte for much stimulating discussion and many helpful comments.

BIBLIOGRAPHY

ARAMBOURG, C.
1943–1947. Mission Scientifique de l'Omo. *Mus. Nat. Hist. Paris,* Fascs. 1, 2, 3.
BAKER, B. H.
1958. "The Geology of the Magadi Area." Degree Sheet 51, SW Quarter, *Geol. Surv. Kenya,* Report No. 42.

BISHOP, W. W.
1958. "Miocene Mammalia from the Napak Volcanics, Karamoja, Uganda," *Nature*, 182:1480–82.
1959. "Kafu Stratigraphy and Kafuan Artifacts," *S. Afri. J. Sci.*, 55:117–21.
1960. "A Review of the Pleistocene Stratigraphy of the Uganda Protectorate," Commission de cooperation technique en Afrique au Sud du Sahara (CCTA), Regional Committees for Geology, Leopoldville 1958 (1960): 91–105.
BISHOP, W. W., and M. POSNANSKY
1960. "Pleistocene Environments and Early Man in Uganda," *Uganda J.*, 24:44–61.
CHANEY, R. W.
1933. "A Tertiary flora from Uganda." *J. Geol.*, 41:702–709.
CHESTERS, K. I. M.
1957. "The Miocene Flora of Rusinga Island, Lake Victoria, Kenya." *Paläeontographica, Bd.* 101, Abt. B: 30–67.
CLARK, J. D.
1960. "Human Ecology During Pleistocene and Later Times in Africa South of the Sahara," *Cur. Anthrop.*, 1(4):307–24.
CLARK, J. D. (ed.).
1957. *Proc. 3d Pan-Afr. Cong. Prehist.* (Livingstone 1955).
COOKE, H. B. S.
1958. "Observations Relating to Quaternary Environments in East and Southern Africa," *Trans. Geol. Soc. S. Afr.*, No. 20, Annexure to vol. 61.
DIXEY, F.
1946. "The Relation of the Main Peneplain of Central Africa to Sediments of Lower Miocene Age," *Quart. Jour. Geol. Soc. London*, 101:242–53.
DONISTHORPE, J.
1959. "A Pilot Study of the Mountain Gorilla in S.W. Uganda," *Uganda J.* 23:1–28.
DUBOIS, C. G. B.
1959. "Recent Volcanic Activity in Kivu District, Belgian Congo." *Uganda J.*, 23:118–23.
FLINT, R. F.
1959a. "Pleistocene Climates in Eastern and Southern Africa," *Bull. Geol. Soc. Amer.*, 70:343–74.
1959b. "On the Basis of Pleistocene Correlation in East Africa," *Geol. Mag.* (London), 96:265–84.
FLINT, R. F., and E. S. DEEVEY (eds.)
1959. "Radiocarbon Supplement," *American Jour. Sci.*
FUCHS, V. E.
1934. "The Geological Work of the Cambridge Expedition to the East African Lakes 1930–31," *Geol. Mag.* (London), vol. 71, Nos. 837, 838.
GEORGE, T. N.
1958. "The Ecology of Fossil Animals," *Sci. Progress*, 46(184):677–90.
HEINZELIN, J. DE
1955. "Le fosse tectonique sous le parallele d'Ishango." *Inst. Parcs Nat. Congo Belge*, Mission J. de Heinzelin de Braucourt, Fasc. 1. Brussels:
HOLMES, A.
1960. "A Revised Geological Time-Scale," *Trans. Geol. Soc. Edinburgh*, 17(3): 183–216.
HOPWOOD, A. T.
1934. "Miocene Primates from Kenya," *J. Linn. Soc. Zool. London*, 38:437–64.

1939. "The Mammalian Fossils." In T. P. O'Brien, *Prehistory of Uganda Protectorate*, Cambridge: Cambridge University Press.

1951. "The Olduvai fauna." In L. S. B. Leakey, *Olduvai Gorge*, Cambridge: Cambridge University Press, pp. 20–24.

KENT, P. E.

1941. "The Recent History and Pleistocene Deposits of the Plateau North of Lake Eyasi, Tanganyika," *Geol. Mag.* (London), 78:173–184.

1942. "The Pleistocene Beds of Kanam and Kanjera, Kavirondo, Kenya," *Geol. Mag.* (London), 79:117–32.

1944. "The Miocene Beds of Kavirondo, Kenya," *Quart. Jour. Geol. Soc. London*, 100:85–118.

KOENIGSWALD, G. H. R. VON, W. GENTNER, and H. J. LIPPOLT

1961. "Age of the Basalt Flow at Olduvai, East Africa," *Nature*, 192:720–21.

LEAKEY, L. S. B.

1931. "East African Lakes," *Geog. J.*, 77 (No. 6):497–514.

1936. *Stone Age Africa:* An Outline of Prehistory in Africa. London: Oxford University Press.

1951. *Olduvai Gorge.* Cambridge: Cambridge University Press.

1952. *Proceedings of the First Pan-African Congress on Prehistory* (Nairobi 1947). Oxford: Blackwell.

1953. *Adam's Ancestors* (4th ed.). London: Methuen and Co.

1959. "A New Fossil Skull from Olduvai," *Nature* (London), 184:491–3.

1960. "Palaeontology," *Coryndon Memorial Museum Annual Report*, 1959, Nairobi, pp. 21–2.

LEAKEY, L. S. B., and T. WHITWORTH

1958. "Notes on the Genus *Simopithecus*, with a Description of a New Species from Olduvai," *Occ. Pap. Coryndon Mem. Mus.* (Nairobi), No. 6, pp. 1–26.

LE GROS CLARK, W. E.

1952. "Report on Fossil Hominoid Material Collected by the British-Kenya Miocene Expedition 1949–51," *Proc. Zool. Soc. London*, 122:273–86.

1956. "A Miocene Lemuroid Skull from East Africa," *Fossil Mammals of Africa*, No. 9. London: Brit. Mus. (Nat. Hist.).

LE GROS CLARK, W. E., and L. S. B. LEAKEY

1951. "The Miocene Hominoidea of East Africa," *Fossil Mammals of Africa*, No. 1, Brit. Mus. (Nat. Hist.).

LE GROS CLARK, W. E., and D. P. THOMAS

1951. "Associated Jaws and Limb Bones of Limnopithecus macinnesi," *Fossil Mammals of Africa*, No. 3, Brit. Mus. (Nat. Hist.).

1952. "The Miocene Lemuroids of East Africa," *Fossil Mammals of Africa*, No. 5, Brit. Mus. (Nat. Hist.).

LOWE, C. VAN RIET

1952. "The Pleistocene Geology and Prehistory of Uganda." Part II: Prehistory; Memoir No. VI, *Geol. Surv. Uganda.*

MCCALL, J.

1958. "Geology of the Gwasi Area," *Geol. Surv. Kenya*, Report No. 45.

NAPIER, J. R., and P. R. DAVIS

1959. "The Fore-Limb Skeleton and Associated Remains of *Proconsul africanus*," *Fossil Mammals of Africa*, No. 16, Brit. Mus. (Nat. Hist.).

NILSSON, E.

1931. "Quaternary Glaciations and Pluvial Lakes in British East Africa," *Geog. Annaler*, 13:249–349.

OAKLEY, K. P.

1957. "Tools Makyth Man," *Antiquity*, 31:199–209.

OSWALD, F.

1914. "The Miocene beds of the Victoria Nyanza," *Quart. Jour. Geol. Soc. London*, 70:128–62.

PICKERING, R.

1958. Degree Sheet 12, S. W. Quarter. Geological Survey, Tanganyika.

1960. "A Preliminary Note on the Quaternary Geology of Tanganyika," Commission de cooperation technique en Afrique au Sud du Sahara (CCTA), Regional Committees for Geology, Leopoldville, 1958 (1960): 77–89.

SHACKLETON, R. M.

1951. "A Contribution to the Geology of the Kavirondo Rift Valley," *Quart. Jour. Geol. Soc. London*, 106:345–88.

WAYLAND, E. J.

1926. "The Geology and Palaeontology of the Kaiso Bone-Beds," *Geol. Surv. Uganda Occ. Pap.* 2:5–11.

1933. *Annual Report Geological Survey of Uganda for 1932*, p. 58.

1934. "Rifts, Rivers, Rains and Early Man in Uganda," *J. Roy. Anthrop. Inst.*, 64:333–52.

WHITWORTH, T.

1954. "A contribution to the geology of Rusinga Island, Kenya," *Quart. Jour. Geol. Soc. London* (1953), 109:75–92.

1958. "Miocene Ruminants of East Africa," *Fossil Mammals of Africa*, No. 15, Brit. Mus. (Nat. Hist.).

1961. "The Geology of Mfwanganu Island, Western Kenya," *Overseas Geol. and Min. Res.*, 8:150–90.

PALEOECOLOGICAL CONDITIONS OF THE LAKE ALBERT—LAKE EDWARD RIFT

J. DE HEINZELIN

A. NEW GEOLOGICAL AND PALEONTOLOGICAL DATA

FROM APRIL TO JUNE, 1960, a geological team from the University of Ghent at the Lower Semliki, south of Lake Albert, sought to settle the question of existing relations between peneplains and sedimentary formations of the Albertine Rift. R. Paepe carefully studied the morphology of the region south of Bunia-Bogoro, and A. Gautier explored the paleontology of the deposits of the valleys of the Sinda and the Mohari. Political events unfortunately interrupted our research at a crucial phase, but the data gathered yield interesting conclusions. We brought back to Belgium much paleontological material which is the basis of this discussion.[1]

The geologists Delpierre and Lepersonne explored the region before us, and what they found is reported in the publications of Lepersonne, Hopwood, and Adam. The following sedimentary formations had been distinguished stratigraphically: terraces, Semliki Series of the Middle Pleistocene, Kasio Series of the Lower Pleistocene, and Mohari Series of the Lower Miocene. Our new data are particularly relevant to the following points:

1. extremely complicated tectonic appearance of the sedimentary deposits in the Rift, calling into question local stratigraphic interpretations;
2. condition of the plateaus bordering the Rift, their relations with the tectonic and the relative dating of the peneplains;
3. accessible base of the sedimentary formations and its relation to the peneplains;
4. discovery of a fauna of mollusks in the Mohari Series.
5. determination of the evolution of the molluscan fauna at the basis of the Kaiso Series;
6. analysis of the terraces and of stone industries.

Let us first examine the question of the peneplains in the region of the plateaus and mountains that dominate the western edge of the Albert Rift. In a geomorphological map, J. Lepersonne distinguishes remains of P I (Mesozoic), ex-

1. These documents are in the process of being analyzed and will be described in a series of articles by Paepe and Gautier. Our purpose here is to present the basic plan of the research.

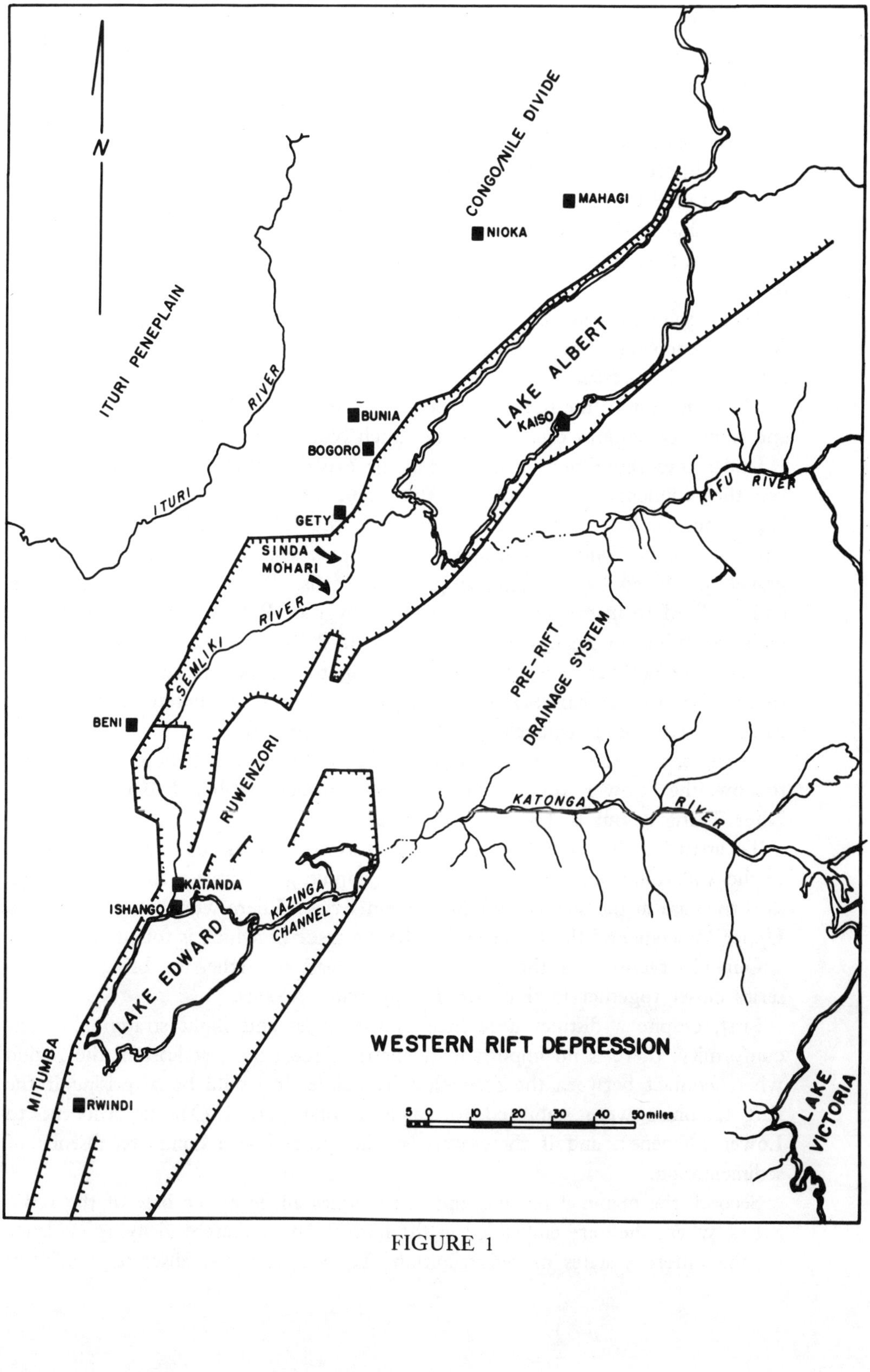

N
CONGO/NILE DIVIDE
MAHAGI
NIOKA
ITURI PENEPLAIN
RIVER
ITURI
LAKE ALBERT
BUNIA
KAISO
BOGORO
GETY
SINDA
MOHARI
KAFU RIVER
RIVER
SEMLIKI
PRE-RIFT
DRAINAGE SYSTEM
BENI
RUWENZORI
KATONGA
RIVER
KATANDA
ISHANGO
KAZINGA
CHANNEL
LAKE EDWARD
WESTERN RIFT DEPRESSION
MITUMBA
RWINDI
5 0 10 20 30 40 50 miles
LAKE VICTORIA

FIGURE 1

tensive surfaces of P II (mid-Tertiary) and P III (Later Tertiary). The following equivalences can be given to this nomenclature: P II = Buganda Peneplain; P III = Ituri = Kyoga Peneplains. Recent evidence suggests that the author had underestimated the effects of the tectonic deformations outside the Rift, and that many of his "escarpements" are faults greatly eroded and not junction lines of different erosion cycles. We accept, rather, the view of Ruhe. that the extension of P II is reduced to a few small insignificant domes, while the Quaternary erosion cycles are everywhere predominant. One must go farther south (Mitumba, from Butembo-Lubero) or farther north to meet the true equivalent of P II.

As a result, the peneplain which was divided by the Rift is P III = Ituri = Kyoga Peneplain. The hydrographic network of the latter extended east to west over the present Rift toward the present Congo basin.

The formation of the Buganda peneplain, characterized by its severe weathering and thick ferruginous crusts, must be much older.

Under these conditions it can hardly be expected that the base upon which rest the sedimentary series of the Rift would be P II rather than P III. This base is accessible in only one known point, the Edo, a small tributary on the left bank of the Sinda where appears a complex assemblage of concretions and gravels produced by changing tropical paleosoils. At first sight these rubefied and oxidized formations may suggest the cover of P II, but a careful analysis indicates rather an analogy with the cover of P III. The results of comparative analyses are not yet available, but, according to the present interpretation, we submit that the Mohari Series (pre-Kaiso, previously attributed to the Lower Miocene) rests on an equivalent of P III and not of P II.

There is, then, an evident contradiction here to the correlations admitted up to now, the "Lower Miocene" resting on a surface attributed to the Pliocene-Later Tertiary. But is this really the case?

To attribute the Mohari Series to the Lower Miocene and the Kaiso Series to the Villafranchian assumes a long stratigraphic gap between these formations, as is the case in the analogous East African series. Where would the Middle and Upper Miocene and the Pliocene then be, no trace of which is found anywhere?

Our observations in the Lower Semliki lead us rather to bring the two series closer together in time for the following reasons.

First, despite a distinct difference in the facies and slight stratigraphic unconformity, there is no important geometric discordance, at least in the region where contact between the formations is visible. It would be surprising if the Rift tectonics were stabilized for such a long period (Middle Miocene to Lower Pliocene), and if there were at the same time a complete absence of sedimentation.

Second, the mammal remains and other bones all lie at the base of the Kaiso Series, where they are embedded in the bottom layer, thereby denying evidence on the different states of redistribution. To postulate two absolutely different fauna under these conditions is unjustified, and there is a good chance that the Miocene species are actually found *in situ* at the base of the Kaiso Series.

Third, our discovery of the molluscan fluvial-lacustrine fauna of the Mohari Series raises new arguments. Up to now, this fauna included twelve species, nine of which persisted at least to the Kaiso Series, and six to the present period. The modern aspect of this fauna is very striking, lacking as it does the rich variety of very special viviparids that characterizes the Kaiso Series. Its state of conservation is also much better, the fossils of the Kaiso Series being almost always greatly limonitic.

The East African "Lower Miocene" sites have also yielded molluscan remains, but they are terrestrial and pulmonate species. Their modern affinities are clearly distinct although Verdcourt's study (Nairobi) is not yet published. I therefore believe that the dating of the East African "Lower Miocene" must be left open to question, and that it will be found to be younger.

As to the Kaiso Series, we shall see that it may be in part older than the Pleistocene.

The locality of Kaiso offers a complex stratigraphy which has not yet been clarified. As a result, the vertebrate fauna can be considered as a mixture of several horizons. Stratigraphically, the site is rather late in comparison to the sequences appearing in the Lower Semliki (Sinda-Mohari) and in the Upper Semliki for the two following reasons:

1. there are no basal facies similar to those of the Lower Semliki, but the terminal and middle facies are certainly found;

2. there are large deposits of *Melania brevissima.* This species appeared in the other two regions only near the top of the series. If the Kaiso site is dated from the Villafranchian, then the base of this series could be older, i.e., Pliocene.

The evolution of facies and fauna of the Kaiso Series is also remarkable. One sees, in broad lines, the following succession of sand facies: silt-clay, clay with limonitic beds, locally with diatomites, and silt-sand. The molluscan fauna clearly shows four groups: (1) pre-Kaiso fauna, relative increase of the bivalves, especially two new species (*Iridina ovata* and *Iridina adami*); (2) majority of spiny carinated viviparids; (3) these vivaparids plus bivalves and introduction of *Platymelania* (*Melania*) *brevissima* and *Pl. bifidicincta;* and (4) same bivalves as in Group I with return of Miocene forms (*Lanistes carinatus*, *Cleopatra dubia*, *Melanoides tuberculata*) and introduction of new ones (*Viviparus alberti* and *unicolor*, *Neothauma dubium*) almost all of which survive today.

Ecological environments appear in the Kaiso Series which are unique in the history of African fauna. There is so far no explanation for their origin in shallow lacustrine expansions and swamps in the process of filling in and drying up. In these, the spiny and carinate viviparids extend to the limits of the series; and beyond, disappear entirely.

Only one other fauna, that of the brackish Pliocene of Slavonia, could be compared to this. Is there a relationship between the two, coming as they do somewhat closer in the stratigraphic scale? Occasional convergence, phylogenetic parents, or migration—no definitive answer can be given; but the point of reference of expected stratigraphic correlations is there.

Several sandy formations known as Semliki beds have been attributed to the Middle Pleistocene in the Lower and Upper Semliki but their paleontological content is poor.

The stone industry deposits of Rwindi appear to be prior to those of Upper Semliki and are, in fact, comparable to a special Abbevilleo-Acheulean, characterized by the abundance of chopping-tools of so-called Rwindi type. The industries of Upper Semliki, such as those of Katanda, are flake industries evolving toward the Levalloisian. The fossil vertebrates of Katanda furnish a landmark in the Middle Pleistocene.

Of the terrace formations cut into the earlier sedimentary series, the oldest group clearly shows the greatest distribution and thickness. This is similarly found in the Lower Semliki south of Lake Edward and doubtless represents a major episode in physiographic and paleoclimatic evolution of the area. This group is the high terraces, with coarse-element fluvial deposits, gravels, sands, red soils, and red, orange, and brown colluvial deposits. It is the last great Pluvial.

In these sediments all the earlier industries are found redistributed as well as more recent flake industries and those of Sangoan affinities. In the non-rubefied soils, colluvial deposits, and the slightly younger gully heads the Upper Sangoan-Lupemban is found.

At one place in the upper soils and red colluvial deposits (Lubilia D 2) appeared the first known quartz-flake industries which become so abundant during the central African Mesolithic and which are particularly evident at Ishango.

The evolution of the subsequent terraces and of the stone industries in the Upper Semliki is already known. The system of terraces is much more complex in the Lower Semliki-Sinda-Mohari, but without typical industries. There, the constitution of the Lower Terrace, geologically very recent, poses a particular problem. Buried peats partly cover a broad area of an old topographic region. This would be inconceiveable today—an altitude of 700 meters in a climate of dry savannas much exposed to sunlight; the formation of peats only begins above 2800 meters at Ruwenzori. What is the answer? Is it a question of recent extensive climatic fluctuations, or were there "warm peats"? Or perhaps certain tectonic movements caused particularly bizarre ecological conditions. The paleobotanic and palynological indications are for the moment lacking.

B. NEW DATA ON THE RUWENZORI

These are two kinds: the cartography of the glacial moraines, and a botanical treatise of the special flora of the high African mountains.

The following glacial stages, from the most recent to the oldest, have been delineated:

1. *Historic, or Lac Gris stage.* The most typical example is the moraine system of Lac Gris. In it can be distinguished moraines from the seventeenth to the eighteenth centuries (based on vegetal development), the main moraine cirque attributed to the nineteenth century and the recent and current recession. The historic moraines descend to lows between 4,300 and 4,250 meters altitude and cover a very small surface.

2. *Lac Noir and Lac Vert stage.* This is a glaciation of Alpine valley type, whose topography is still recent, the moraines scarcely indented by the torrents that run across them. Moraine ramparts and outwashes of this age are responsible for most of the large glacial lakes. The glacial tongues reached down to 3,700 or 3,800 meters, or perhaps even lower in certain valleys.

3. *Butahu stage.* This corresponds to the most extensive glaciation in the interior of the present valley systems. The lowest points reached to 2,900 to 3,000 meters. When a large river cross them, the moraine constructions are cut by the torrents or worn into V-shapes by erosion. Apart from this, the forms are still well preserved, and lateral moraines have sometimes remained intact for a length of more than two kilometers.

4. *Crête Ruamya-Haute-Ruanoli stage.* It is represented by large morainic extensions, the soft, undulating topography of which shows a pedological development and a warping of peats more significant than elsewhere. All of these extensions lack ramparts or other identifiable constructions, and are found at rather high altitudes—from about 3,900 to 4,100 meters. They appear like remnants of older formations conserved between the modern valley cuttings and must have been prior to them. They are no doubt the result of an ice-cap older than the Butahu stage.

Of these four stages, only the historic is suitably dated: it corresponds to the Little Ice Age of the Alps and the United States. In the absence thus far of proper material to be submitted to C-14 analysis, the approximate dating of the stages can only be made morphologically; eventually it can be based on the study of lacustrine bottoms and estuaries. As a result of the large differences in climate, vegetation, and topography, morphological comparisons are difficult.[2]

In relation to the Alps, one can propose an analogy between the Lac Noir-Lac Vert stage and the stages of Alpine recession: Bühl-Gschnitz-Daun; another exists between the Butahu stages and the young end-moraines of the Würm. If, apart from that, one admits the close similarity (Alpine recession = Salpausselka stage of Scandinavia-Finland; young end-Moraines = Brandeburg stage of Germany), one arrives at an age of about 10,000 years to the Lac Noir-Lac Vert and of about 20,000 years to the Butahu.

Likening the Ruwenzorian stages to the Alpine has its justification in the

2. Two more important contributions to the glaciology of Ruwenzori have since been published: H. A. Osmaston, "The Moraines of the Ruwenzori," Map Lands and Surveys, Uganda, 1961: and D. A. Livingstone, "Age of Deglaciation in the Ruwenzori Range, Uganda," *Nature,* 194 (4831):859–60. They do not contradict our statements.

perfect similarity of all aspects of the "Small Historical Glaciation." The latter has been attributed to the modification of the atmosphere by industrial activity, but this does not seem likely.

If these stages do coincide, two important results follow:

1. the paleoclimatic significance of the astronomic curves is doubtful since it is contrary to the parallelisms of the glaciations;
2. high terraces and red soils of the later Pluvial appear related to the moraines of the Ruwenzori and to coincide with the maximum extension of the later European Glacial.

What kind of climatic fluctuations could cause such glacial changes? Only one glacier approaches below 4500 m. elevation and it is in full retreat. The shift of the altitude in relation to the Butahu stage is therefore at least 1,500 m., which corresponds roughly to a 9° C. change of mean temperature the length of the mountain profile.

This rudimentary calculation neglects the incidence of rainfall and cannot be extended to non-mountainous regions, but it suggests possible limits to the size of the mean-temperature variation.

The recent elevation of the mountain by new faults on the western flank must still be taken into account; one can admit an elevation in certain places of 1,000 m. or so since the Middle to Upper Pleistocene. The ice-cap of the Ruamya-Haute-Ruanoli stage may have developed at a lower altitude and have been raised later to the present 3,900 to 4,100 m.; this may indicate a cooling-off greater still than that of the Butahu stage.

The works of Liben on the flora of the high African mountains are in press. With his kind permission, they are summarized here. The dispersion of certain species and the relationships of certain others indicate that important climate fluctuations, principally of temperature, must have extended over all the center of the African continent during recent geological times.

C. REVISION OF HISTORICAL GEOLOGY

We can see the succession of geological events in the region of the western African Rift in the following manner:

Formation of P II = Peneplain of Buganda (Eocene-mid-Tertiary?)
Formation of P III = Kyoga = Ituri Peneplain; drainage toward the Congo
First tectonic depressions, with or without faults; warping of the Mohari Series, Pre-Kaiso (Upper Miocene to Pliocene)
Increase of the Rift tectonic; warping of the Kaiso Series (Pliocene to Villafranchian and Lower Pleistocene)
Increase of Rift tectonic; deposition of the Semliki Series (Middle Pleistocene)
High terraces associated with rubefied soils and various colluvial deposits

Correlations of the latter Pluvial-Glacial in the Ruwenzori
Middle terraces of the Sinda, terraces of Ishango and of the Upper Semliki
Low terraces of the Sinda; extension of a peaty drainage system

The paleoecological interpretation of this succession depends on paleoclimatic indices, many of which, based on the facies of formations, are illusory for many reasons. Among others: (1) the conditions of the genesis of the recent sediments are not well known; (2) in spite of the recent progress of pedology, genetic interpretation of paleosoils is extremely tricky, more so, at least, than has been generally believed; (3) tectonics and climate intervened simultaneously at the time of the deposit of the geological series, and it is particularly difficult to determine their separate roles.

There is, therefore, more inclination than before to use prudence in paleoclimatic interpretations. What other elements have we for interpretation?

Severe modifications and hardening in crusts of P II are the result of a long pedogenesis and of complex variations of climate. These episodes are beyond our subject, since they are found well before the fossiliferous formations of the early Tertiary of East Africa.

Of P III itself, there are few portions left intact; one sees more and more that the Quaternary erosion, while often new, has widely redistributed the upper beds and mutilated the soils. This is another reason for interpretive prudence.

The stratigraphy of the Kaiso Series constitutes perhaps the most important key to African paleoclimatology. On the one hand, it extends over a much longer time than was believed, and on the other, the modifications of fauna and facies are in evidence there. Carinated and spiny viviparid fauna and facies with concretions and with gypsum deposits go together; it is difficult to make the tectonic alone responsible.

The Middle Pleistocene is present but probably very incompletely represented in the Western Rift. With what we know so far, it would be hazardous to build a paleoclimatic sequence on that; there is no trace of extreme climates, except perhaps all the way at the top where the limy and saline concretions abound.

The later Pluvial, with its rubefied soils and extended fluvial terraces, is a reality. It is probably associated with the last major African Glacial and corresponds not only to an increase of precipitations, but also to a lowering of the temperature. Several phases of glacial extension are evident.

Other interpretive factors are found in recent formations, but their objective utilization would necessitate more pedological, sedimentological, and pollinal studies than those available.

The balance sheet of our paleoclimatic conclusions, therefore, must be somewhat restricted.

BIBLIOGRAPHY

ADAM, W.

1957. "Mollusques quaternaires du la région de Lac Edouard." *Inst. Parcs Nat. Congo Belge* (Mission J. de Heinzelin 1950), Brussels, *Fasc. 3.*

HEINZELIN, J. DE

1960. "Carte des extensions glaciaries du Ruwenzori (versant Congolais)," *Union Géog. Intern. Comm. de Géomorph.*, Périglaciarie carte mondiale et Biuletyn perygjalny, No. 11.

1962. "Les formations du Western Rift et de la Cuvette Congolaise," *Actes 4e Pan-Afr. Cong. Préhist.* (Leopoldville 1959): 219–43.

HOPWOOD, A. T., and J. LEPERSONNE

1953. "Présence de formations d'âge Miocène inférieur dans le fossé tectonique du Lac Albert et de la Basse-Semliki (Congo belge)," *Ann. Soc. Géol. Belgique*, 77:B83–113.

LEPERSONNE, J.

1956. "Les aplanissements d'érosion du Nord-Est du Congo-Belge et des régions voisines," *Mém. Acad. Roy. Sci. Colon.* (Cl. Sc. Nat. et méd., n.s.), Vol. 4, Fasc. 7.

RUHE, R. V.

1954. "Erosion Surfaces of Central African Interior High Plateaus," Publ. I.N.E.A.C., sér. sc., 59.

1956. "Landscape Evolution in the High Ituri, Belgian Congo." Publ. I.N.E.A.C., sér. sc., 66.

OBSERVATIONS ON THE ABSOLUTE CHRONOLOGY OF THE UPPER PLEISTOCENE

J. DE HEINZELIN

IT SEEMS USEFUL first to summarize what is known about the chronology of the Upper Pleistocene so as to apply, as far as possible, the geological events and prehistoric sites of Central Africa.

Current methods of absolute dating applicable to the Quaternary can be classified according to:

Radioactivity. Measurement of the residual radioactive nuclide (C-14 method) or measurement of the accumulation of a radiogenic derivative in relation to the parent (K/A method; percentage of equilibrium of radium or ionium; Pa^{231}/Th^{230}).

Chemistry. Atomic or molecular transformation or replacement of a compound over a period of time (fluorine test).

Astronomy. Geometric variation of the earth's orbit over a period of time may perhaps justify a paleoclimatic curve of glacial and interglacial or pluvial and interpluvial periods.

Between the dates thus obtained, or otherwise, interpolations and extrapolations can be made on the relative thickness of deposits, the sediment deposits of the deep bottoms of the sea, the rapidity of evolution of the fauna, and the parallel comparisons of marked climatic fluctuations.

Of the methods mentioned, the astronomical is the only one that is not subject to experimentation nor directly controllable. Therefore, it will be taken up last in order to compare it with the results established by others.

Useful paleoclimatic indices can be deduced from the following observations:

- sedimentary facies, sedimentology, lithology;
- chemical precipitations and alterations;
- paleosoils, periglacial phenomena;
- fauna and flora, pollen analysis;
- foraminifera counts in ocean bottom deep-sea sediment cores;
- isotopic relation 0^{18}–0^{16}.

Three coherent groups of facts can be put together from a critical review of current data: dating of European paleoclimatic sequence up to about —60,000 years (analogous results in the U. S.); stratigraphy of Atlantic deep-sea cores; stratigraphy of continental basin cores.

DATING AND EUROPEAN PALEOCLIMATIC SEQUENCE

The C-14 analyses perfected by de Vries have resulted in dependable dates of around 60,000 years (Fig.1).

The Eemian (later Interglacial) is older and therefore beyond the limits of the C-14 method.

The following is as short a summary as possible of the succession of stages and interstages of the later Glacial, Weichsel-Würm.

After one or more glacial fluctuations, the Amersfoort interstage appears around —60,000 to —62,000 years, followed by a short cold stage and then by the Brørup interstage around —54,000 years to —57,000 years.

The first really extended cold stage, which is called either Würm or Weichsel I, is stage W I, found at between —40,000 and —54,000. It dos not coincide with the maximum extension of the glaciers, which is found much later.

After W I comes an interstage called Göttweig between —31,000 and —40,000, which apparently lies between the deposits of the western European Recent Loess I and Recent Loess II. The climate of this interstage was hardly better than temperate cold and the pedological weathering was not very deep, still a long way from the transformation into the brick earth of the modern soils.

The various groups of the Mousterian appear alternatively from the end of the later Interglacial until the W I–II Interstage. From one of these groups, probably the Mousterian of Acheulean tradition, the Lower Perigordian developed at this moment: the first industry relatable to the Upper Paleolithic in western Europe. In Central Europe the evolution led to the Szeletian.

The glacial advance W II broadly corresponds to the deposit of the Recent Loess II and is found around —28,000 to —31,000 years. This is the period of the typical Aurignacian in both Central and western Europe; there is ample evidence that this industry originated in the Near East and spread westward.

W II is followed by a new interstage called Paudorf, W II–III around —24,000 to —28,000 years, in which the pedological alteration is even less than that of the Göttweig.

The maximum extension of the moraines, Brandebourg stage, between —18,000 and —24,000 years, corresponds to the glacial stage W III. This is the period of the Upper Perigordian, Proto-Magdalenian, and very late Aurignacian in western Europe; the Eastern Gravettian is equivalent to the Upper Perigordian of Central Europe.

Immediately following is an interstage which may be called Laugerie, between —13,500 and —18,000 years, corresponding to different stages of moraine retreat in the northern European plains. This is the period of the Solutrean and the Lower Magdalenian, up to Magdalenian III.

The glacial advance W IV is that of the Götiglacial, around —13,500 to

—12,000. This is the period of the Upper Magdalenian in France and of the Hambourgian in the northern plains.

The final glacial recessions are characterized by close, distinct fluctuations: small interstage of Bölling around —11,000 followed by older Dryas around —10,600 to 10,000; the well-known interstage Allerød of —9,900 to 8,900; younger Dryas, —8,900 to 8,000.

The Holocene boundary has been placed at this level. Pre-Boreal, Boreal, and Atlantic mark a continuous amelioration of the climate until around —4,000, the period called "Optimum climatic" or "Hypsithermal."

The European climate finally arrived at the present stage with at least two short cold periods, one towards —400 and the other from the seventeenth to the nineteenth centuries, called the Little Ice Age.

The last of the Paleolithic is found in the Allerød (Federmesser industries) and the first Mesolithic industries are found in the younger Dryas (Ahrensburgian). The Maglemosian appeared in the Pre-Boreal and developed during the Boreal. The position of the Azilian is uncertain—Allerød, or Pre-Boreal to Boreal.

The Upper Pleistocene of the United States can be described according to the same schematism. There, also, the maximum extension of the moraines is dated from —18,000 to —24,000 years: it is the Tazewell Stage which slightly precedes the Farmdale loess and follows the Tazewell loess. The stages of subsequent recession are well known, and, among others, the Two Creeks Interstage is known to correspond exactly with the European Allerød.

Contrary to the case in Europe, the prior glacial stages are little known and until now little noted. However, no doubt that the glacial extensions of several stages and their collateral formations lie under the Classical Wisconsin, which is the equivalent of European Stages III and IV (de Heinzelin 1958; Frye and coll. 1962).

STRATIGRAPHY OF ATLANTIC DEEP-SEA CORES

B. One of the latest publications on the stratigraphy of ocean floors is that of Ericson, Ewing, Wollin, and Heezen, which reports the results of 221 deep-sea sediment cores of the Atlantic and the Caribbean Sea.

The lithological and sedimentological study of these cores explains some irregularities, such as the effects of slumping or the turbid currents. There remain the "normal cores," many of which provide a continuous profile of deep floor deposits of the Middle and Upper Pleistocene.

The count of the planktonic Foraminifera which have been preserved (not all are, some being subject to dissolution) permitted the location of micropaleontological zones appearing in all the cores. Eighteen species of Foraminifera were counted with relative frequency; one of these is particularly important as a climatic indicator: *Globorotalia menardii* and its subspecies.

The following zones were defined:

TABLE 1

Zone (z), with *Globorotalia menardii* in abundance;
Zone (y), without *Globorotalia menardii* and richer in boreal species than actual sediments;
Zone (x), with *Globorotalia menardii* in abundance, also the variety *Gl. m. flexuosa* (this one now probably extinct), also *Gl. hexagona;*
Zone (w), thin, without *Gl. menardii;*
Zone (v), with *Gl. menardii* and *Gl. m. flexuosa,* thicker than the two zones mentioned above;
Zone (u), without *Gl. menardii,* reached only at certain depths.

Here is a distinct succession of temperate and glacial periods. The presence or absence of Foraminifera is not only due to variation in water temperature but also to oceanic changes such as current, nutritional capacity, etc. *Gl. menardii* is a good index of temperate conditions.

The upper horizons of the cores were dated by the C-14 method and the results extrapolated for lower horizons. The rate of sedimentation is variable from one point to another. For example, for Zone (z) it can vary from 0.5 to 63.5 cm. per 1,000 years; for Zone (y) from 1.0 cm. to more than 20 cm. per 1,000 years. The relative durations of the zones were estimated according to the average length of core parts.

Zone (z) is Holocene, beginning at about 11,000 years. Zone (y) is the later Glacial, which the authors call later Glacial 2–3, beginning about 60,000 years ago. Zone (x) is a short interglacial or a long interstage which has no known equivalent in the continental stratigraphies but is no less real and lasted probably 35,000 years.

Zone (w) is a small glacial or a first phase of the later Glacial, between 95,000 and 115,000 years, which the authors call W I. Zone (v) is later Interglacial, from 115,000 to 235,000 years, and Zone (u) the Riss-Saale Glacial, the duration of which could not be estimated, since no core reached to the base.

To these curves can be compared those Emiliani (1955) obtained by the determination of the isotopic temperature according to the ratio 0^{18}–0^{16} of the Foraminifera tests. In the precipitation of CO_3Ca, the ratio of the oxygen isotopes is a function of the temperature. Consequently, the variation of surface water temperature is around 6° C. The variation of the mean air temperature above oceans would have been approximately the same; variations in high latitudes and especially on the continents must have been greater. The maxima and minima of the different Glacials and Interglacials are almost the same, and the present period can be considered as an interglacial in this respect.

Zones (z), (y), (x), and (v) can be similarly located on Emiliani's curves.

But Emiliani places a different meaning on the oldest ones, compressing into too short a span of the Glacials and Interglacials from Riss-Würm to Günz-Mindel. Ericson, Ewing, Wollin, and Heezen give a much more likely interpretation.

Emiliani's determinations of the isotopic temperature have just lately been integrated with a new method of absolute dating of two deep-sea cores: the Pa^{231}–Th^{230} method.

The following table shows the result.

TABLE 2

Postglacial	0 to 10,000 years, Zone (z) Foraminifera
Upper and Middle Würm	10,000 to 30,000 years, Zone (y)
Würm Interstage	30,000 to 50,000 years, Zone (y)
Lower Würm	50,000 to 65,000 years, Zone (y)
Riss Würm Interglacial	65,000 to 100,000 years, Zone (x), last appearance of *Globorotalia menardii flexuosa*
Riss Glacial	100,000 to 130,000 years, Zone (w)
Mindel-Riss Interglacial	130,000 to 175,000 years, Zone (v)

This agrees with the results of Ericson, Ewing, Wollin, and Heezen, but interpretations differ as to the lower zones.

It appears to me untenable, given the stratigraphic data, to assign only 30,000 years to the later Interglacial and above all to the Riss Glacial, more extensive than any other, when, by contrast, the Würm Interstage would have lasted 20,000 years and the entire Würm 55,000 years. There is hardly any doubt that Zone (w) (the Riss Glacial of Emiliani and colleagues) is a pre-Würm phase, which remained undetected until now because it left no visible moraines (its moraines were buried under the younger ones), loess deposits, or periglacial phenomena in Western Europe.

The older data of Piggot and Urry (1942) and Phleger and colleagues (1951 and later) can equally well be compared to these.

Piggot and Urry dated their cores by the ionium method or by the "percentage of equilibrium" of radium, and the counts of warm and cold species of Foraminifera were done by Cushman. The curve made by the sum of the warm species rises to a maximum of —75,000, probably corresponding to Zone (x), and descends to a minimum of —100,000 to —85,000, probably corresponding to Zone (w). Comparison with the other zones is less valid.

Phleger and his colleagues have confirmed that the temperature fluctuations of the surface water during the Quaternary were of the order of 6° C.

STRATIGRAPHY OF CONTINENTAL BASIN CORES

To my knowledge, there exists one good example of a core from a continental basin, that of the Great Salt Lake, Utah. Eardley and Gvosdetsky used all possible methods of analysis on it. Their results can be summarized as follows:

TABLE 3

Postglacial	0 to 12,000 years
Upper Wisconsin	12,000 to 22,000
Wisconsin Interstage	22,000 to 38,000
Lower Wisconsin	38,000 to 45,000
Interglacial	?45,000 to 115,000
Glacial	?115,000 to 185,000
Sangamon Interglacial	185,000 to 245,000
Illinoian Glacial	245,000 to 308,000
Interglacial	308,000 to 370,000
Kansan Glacial	370,000 to 525,000

For the Pleistocene no correspondence is found between Table 3 and the data in Tables 1 and 2. A glacial or a pluvial here corresponds to Zone (v) (of Tables 1 and 2; later Interglacial of Ericson and colleagues; Mindel-Riss of Emiliani). It is true that in Table 2 the authors linked this glacial with Emiliani's Würm I. This is only one example among the hundreds of jumbled contradictions to which the game of equivalences between Glacials, Pluvials, Stages and Substages has led.

The core chronology of the Great Salt Lake agrees no better with that recently given for the Wisconsin Glacial in the United States.

TABLE 4

Recent (Postglacial?)	0 to 5,000 years
Valderan	5,000 to 11,000
Two Creeks Interstage	11,000 to 12,500
Woodfordian	12,500 to 22,000
Farmdale Interstage	22,000 to 28,000
Altonian	28,000 to 50,000 or 70,000

To summarize, an extraordinary uncertainty emerges from the efforts, conducted with such refined methods as absolute dating, isotopic determinations, sedimentology, and micropaleontology. Nevertheless, there are many traits in common between the several schemas.

UPPER PLEISTOCENE CLIMATIC FLUCTUATIONS

Let us now try to draw a general picture of the Upper Pleistocene climatic fluctuations with the help of the above information.

We will agree to call Interglacial all positive climatic fluctuation reaching a magnitude of 6° C. of mean temperature in temperate regions in relation to the Glacials and a length of more than 10,000 years; we will call Interstadial all fluctuations less well marked. The stadials are found within Glacials and are minor fluctuations.

The Holocene has an Interglacial value; it began with the Pre-Boreal about ten thousand years ago.

The later Glacial, as defined by continental formations of the Boreal hemisphere, began shortly before —60,000. It is divided into four stadials, only one of which is distinctly marked in the deep-sea cores. A careful study of several of Emiliani's (1955) curves (A 179–4) appears to indicate a first maximum of cold around —42,000 (W I), then around —30,000 (W II), an interstadial around —27,000 (Paudorf?), finally a long cold period between two maxima at —21,000 (W III) and —12,000 (W IV); this schema is far from being always distinct and is not found in all the cores. W I is probably calculated by default.

The curve of generalized temperature lately published by Emiliani and collaborators clearly indicates an average rise of 2°5 C. between 50,000 and 30,000 years (Würm Interstadial).

The later Glacial corresponds to Zone (y) of Ericson, Ewing, *et al.*, which they labelled W II–III. Here it can be seen that denoting the stadials and interstadials by numbers can be a source of confusion.

Zone (x) actually has the value of an Interglacial and not of a Würm-Weichsel Interstadial. It is impossible to adapt it either to Amersfoort, Brørup, or Göttweig, and it is better to admit that it has no recognized equivalent among continental formations. It is the same for Zone (w), which has the value of a Glacial-pre-Würm-Weichsel, if one wishes, but not W I. All the stratigraphy of the sediments and the paleosoils should be reviewed in this respect.

Zone (v), of long duration, must be assimilated to the Riss-Würm or Eemian later Glacial of the nordic stratigraphy. It may be that at the peak of the classic Eemian the equivalent of Zones (w) and (x) will be encountered some day.

Thus the beginning of the later Interglacial is brought to around 175,000 or 240,000 years (depending on the methods of dating and extrapolation in the deep-sea cores), which definitively points to a long chronology of the Pleistocene (much more than a million years if the rest of the stratigraphy matches this rhythm).

The indications given by the Ar-K method of absolute dating on the age of the Pliocene should be recalled in this context, especially the two below, chosen from those which are stratigraphically the surest:

Upper Pliocene of Mendocino County, California: 5.2×10^6 years
Upper Pliocene of Kruisschans, Anvers, Belgium: 7×10^6 years

It should be added that the last is not far, stratigraphically or paleontologically, from the Tertiary-Quaternary boundary with the first appearances of the Villafranchian fauna.

There is no reason to hold to the fatidic date of one million years for the beginning of the Quaternary; more probably one should count in several millions of years. In conclusion, the "long chronology" of the Pleistocene seems to be in favor.

AFRICAN PALEOCLIMATE AND PRE-HISTORY

Let us now examine how the known points of African paleoclimate and pre-history are ranged in this schema.

The possibility of coincidence of climatic phases is based on the following arguments: (1) definite parallelism of the Ruwenzorian and Alpine moraines of the "Little Historic Glaciation" and the probable parallelism of the older moraines;[1] (2) extension over the tropical and equatorial Atlantic of climatic fluctuations on the order of 6° to 9° C. corresponding to the fluctuations of the Boreal hemisphere.

The equivalence of later European Glacial = Central African Last Glacial and Pluvial can be generally accepted. In it the Lower and Middle Sangoan and numerous industries of Mousterian affinities take the place of the Mousterian and its variants.

The dry phase from 43,000 to 40,000 B.P. at Kalambo Falls is placed almost at the level of the Göttweig Interstadial and its Interpluvial value is certainly not tenable.

The following pluvial phase dated at Kalambo Falls, Florisbad, Cave of Hearths, between 40,000 to 35,000 B.P., and at Mufo, Angola, at 12,000 B.P. can very well be equated to the maximum glacial extension in Europe and America (our stages W II, III, IV).

The succeeding short dry phase between 12,000 and 9,000 B.P. corresponds with the Allerød to Pre-Boreal, and to the disappearance of the large glaciers.

Eric Higgs (1961) has recently reviewed the results of absolute dating by the C-14 method in the Mediterranean region. It can be summarized in the following manner: early dry period at 50,000 B.P.; colder and more humid phase from 49,000 to 41,550 B.P.; drier and warmer phase from 40,500 to 30,000 B.P.; colder and more humid phase from 30,000 to 10,000 B.P., with short sudden

1. See discussion in my first article (Chap. 9).

increases in temperature around 11,000 B.P. Even though the dates do not correspond exactly, the schema, it appears, is very similar to that which Desmond Clark made for subequatorial Africa, with a dry phase around —40,000. As to the episode at —11,000, it is paralleled by the Allerød, which still supports the idea of the universality of temperature fluctuations.

Quézel's palynology studies (1960) of the Sahara show that from this period there was a constant progress of aridification below the Tropic.

The Rhodesian Lupembian of Kalambo Falls dated from 27,000 to 29,000 B.P. is approximately contemporary to the Upper Perigordian or Gravettian. It can already be placed with the Upper Paleolithic. The upper Pietersburg and the final Lupembian have the same age as the Upper Magdalenian and the Hamburgian.

The Lupembo-Tshitolian matches the height of the Federmesser industries and perhaps of the Azilian. The Smithfield A of the Matjes River Cave almost corresponds to the Ahrensburgian, and the Upper Magosian to the Maglemosian.

It is at about this level that I am now tempted to place the Ishango industry, thereby pushing back the previously published estimations.

The Capsian is found at about the time of the Boreal-Atlantic transition, parallel to a large part of the European Mesolithic. Not far away, undoubtedly, is the Mesolithic of Khartoum, then the sites with harpoons in the Sudan and of the Saharian regions.

The Neolithic, which begins around —7,000 B.C. in the middle of the Boreal in Jericho is found already at —5,000 B.C. in the Fayoum, followed by the Tasian, the Badarian (Chalcolithic), the site of Merimde, and finally the predynastic Egyptian shortly after —4,000 B.C.

ASTRONOMICAL METHODS

It remains to examine whether the curves deduced by astronomical methods are somehow comparable to the objective schema which result from our present knowledge.

It is known that astronomical methods are based on the succession of periodic modifications of certain geometric factors of terrestrial motion. These fluctuations do not, or hardly, influence the total quantity of solar heat received by the earth but do influence the distribution of this quantity during the year and according to climatic zones.

The factors in question are as follows:

1. Eccentricity e of the terrestrial orbit, which can vary from 0 to 0.053 with a mean period of 96,600 years. At the maximum of eccentricity the difference of insolation between the two extreme points of the orbit reaches 20%.
2. Inclination ε of the axis of terrestrial rotation on the orbital plane, or obliquity, which can vary from 24°4″ to 21°8″ with a period of 41,000 years.

3. Angle π: position of the perihelion in relation to the earth-sun direction at the moment when the planet passes the autumnal equinox.
4. Perihelion: point of orbit closest to the sun.
5. Equinox: Point of orbit when the terrestrial axis of rotation is perpendicular to the earth-sun axis. The period is an average of 21,000 years and can vary from 13,000 to 26,000 years.

It is admitted that the best conditions for glaciation and deglaciation in the northern hemisphere are created by the following coincidences:

$$\varepsilon \text{ maximum} \quad \begin{cases} \pi = 90^\circ = \text{small interglacial} \\ \pi = 270^\circ = \text{large interglacial} \end{cases}$$

$$\varepsilon \text{ maximum} \quad \begin{cases} \pi = 90^\circ = \text{large glacial} \\ \pi = 270^\circ = \text{small glacial} \end{cases}$$

The curves presented are generally those of the relative variation of the climatic latitude (for example 65°) calculated according to the intensity of solar radiation during the estival half of the year.

Several authors, and principally Zeuner, have looked for correspondences between the theoretical points of glaciation and the known stratigraphic successions.

Several calculations, with slightly different corrections, have been published: Spitaler, Soergel, Köppen and Wegener, Milankovitch, Brouwer and Van Woerkom.

These last two curves, very similar, lead to the following succession, for temperate boreal latitudes:

Interglacial around	−10,000
Glacial	−20,000 to −25,000
Interglacial	−40,000 to −60,000
Glacial	−70,000
Interglacial	−80,000 to −100,000
Glacial	−110,000
Interglacial	−120,000 to −170,000
Glacial	−180,000

The correspondence with the Würmian and the Holocene is very weak, and, besides, one can hardly see how to extrapolate further.

Very recently Bernard (1959) presented a more complete study especially for the low latitudes. Taking account of the global climatic circulation of the globe, he distinguishes:

Pluvial of polar front = glaciation of the high latitudes
Displuvial = intense summer rain, winter dryness (low latitudes)
Isopluvial = regular rain throughout the year (low latitudes)
Interpluvial and aridification

Accordingly he forecasts the following succession in extratropical northern zones:

Aridification from	−4,000 to −13,000
Glaciation "	−13,000 to −33,000
Glaciation "	−62,000 to −80,000
Aridification "	−80,000 to −87,000
Glaciation "	−97,000 to −120,000
Aridification "	−122,000 to −135,000
Aridification "	−170,000 to −180,000
Glaciation "	−180,000 to −197,000, etc.

Similarly in equatorial zones he forecasts:

Displuvial from	−2,000 to −11,000
Aridification "	−18,000 to −33,000
Isopluvial "	−39,000 to −52,000
Aridification "	−52,000 to −75,000, etc.

It seems to me difficult to draw conclusions on the possible coincidences with the objective stratigraphical data and their absolute dating.

In the calculation of the astronomical curves, a large number of oversimplifications must be admitted that jeopardize accuracy. Among others: that the absorption of radiation is proportional to the thickness of the atmosphere traversed; that the reflection is considered as though the earth were a solid without hydrosphere; that the atmosphere is at rest; that the accumulation of snow and rain depends on climates with a cold summer.

It certainly seems that the fluctuations calculated for the 65° latitude north are too weak to justify such extensive glaciations; the glaciations also affected the tropical and equatorial zones where the fluctuations are weaker. Furthermore, the astronomical theory is opposed to the universality of glaciations, which seems to be more and more probable.

Other factors, more important than the geometric variations of the terrestrial movements and on which they can only be superimposed, must intervene.

Very probably the cycles of solar activity will have to be used to explain the variations of terrestrial climate. Already the small solar cycle of eleven years is manifest in many terrestrial phenomena: growth of plants, varved sedimentation, retreat of glaciers. The eleven-year cycle is equally plainly manifest in the deposits of saline varves of the Sea of Zechstein (Permian) next to perhaps other cycles of from 90 to 100 and from 380 to 400 years.

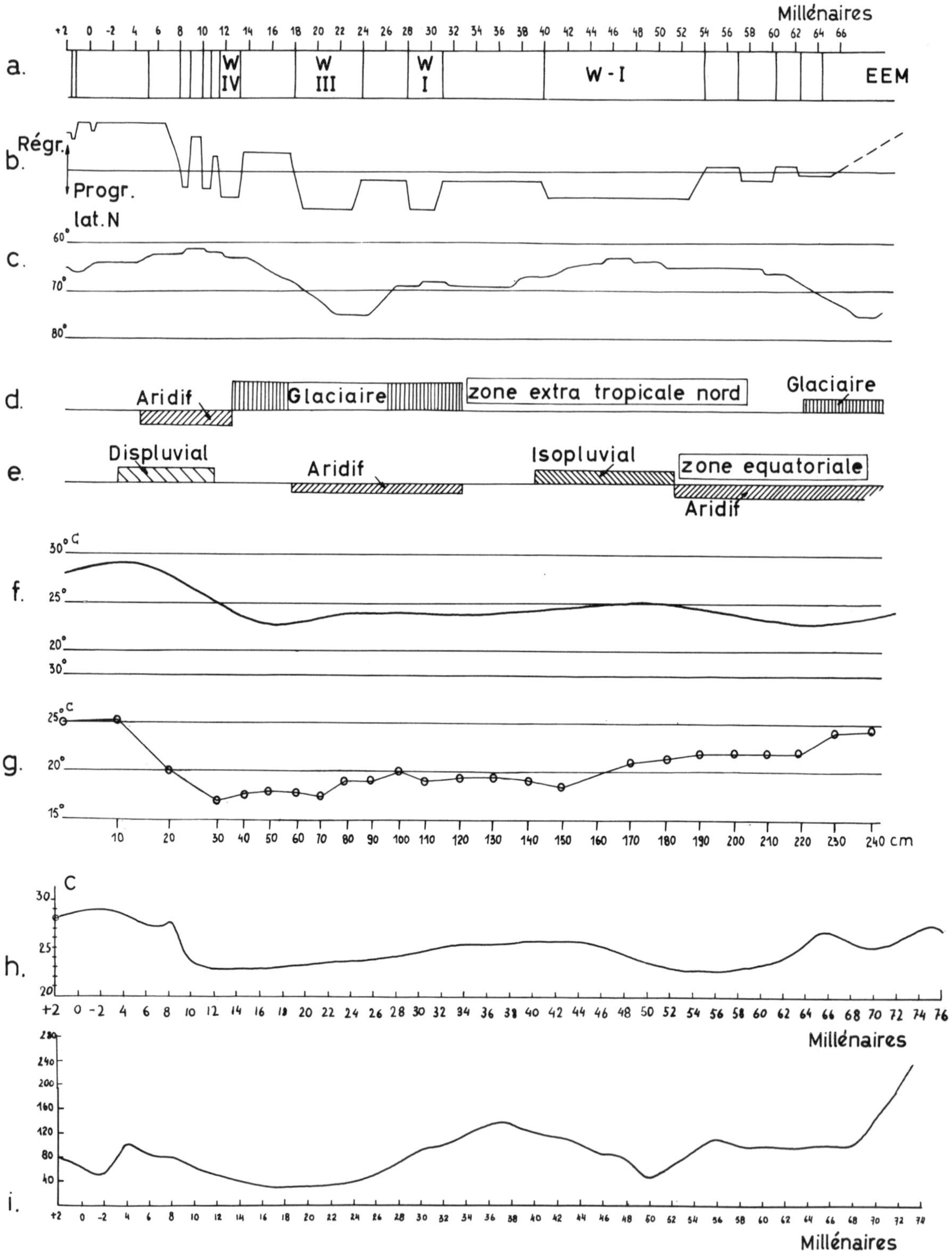

FIGURE 1 *Comparisons of*

(a) Stratigraphy of the Last Glacial (Würm-Weichsel)
Position of the stages and interstages in relation to the scale of the milleniums according to present knowledge.

(b) Schematic presentation of the glacial advances and regressions or of the relative climatic variations in Europe, according to (a).

(c) Astronomical climatic curve according to A. J. J. Von Woerkom (1953).
"The Astronomical Theory of Climate Changes," in *Climatic Change* (H. Shapely, ed.), reprinted in R. F. Flint (1957), *Glacial and Pleistocene Geology*, Figure 27–2, page 508.

(d) Theoretical climatic variation in north extra-tropical zone according to Bernard (1959).

(e) Same as (d). Theoretical climatic variation in equatorial zone.
Displuvial: intense summer rains, dry winter.
Isopluvial: regular rainfall all year.

(f) Generalized curve of the isotopic temperature variation O^{16}–O^{18} in the deep-sea sediment cores of the Caribbean Sea and the equatorial Atlantic, according to Emiliani (1955).

(g) Isotopic temperature curve obtained from one species of Foraminifera:
Globigerina dubia, in the core A-179-4 according to Emiliani (1955). Depth reported in nonlinear manner, apparent acceleration of the sedimentation in glacial period.

(h) Generalized curve of the isotopic temperature variation O^{16}–O^{18} according to Emiliani (in Rosholt, Emiliani, Geiss, Koczy, and Wangersky 1961).

(i) Total of the warm Foraminifera species in the deep sea core P-127 of the Caribbean Sea according to Piggot and Urry (1942). Determinations by J. A. Cushman.

the methods for the last 70 millenia.

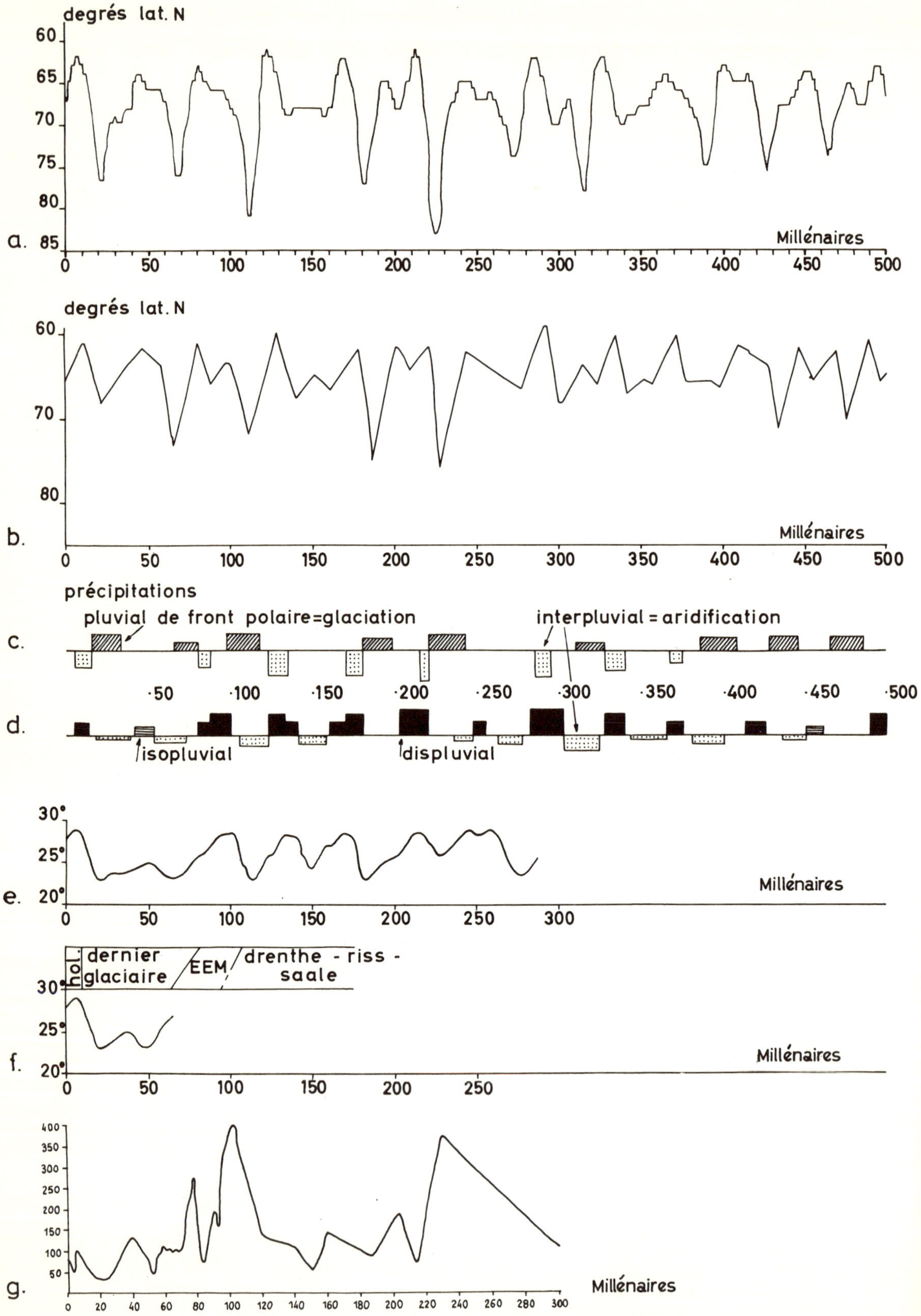

FIGURE 2 *Comparisons of*

(a) Astronomical climatic curve according to Van Woerkom (1953). Relative variation of the climatic latitude at 65° north, calculated according to the intensity of solar radiation during the estival half of the year.

(b) Astronomical climatic curve of Milankowitch (1938), reprinted in J. E. Charlesworth (1957), *The Quaternary Era*, Vol. 2, Figure 325. Same as (a); other corrections.

(c) Theoretical climatic variation in north extra-tropical zone according to Bernard (1959).

(d) Same as (c). Theoretical climatic variation in equatorial zone.
Displuvial: intense summer rains, dry winter.
Isopluvial: regular rains all year.

(e) Generalized curve of the isotopic temperature variation O^{16}–O^{18} in the deep-sea cores of the Caribbean Sea and the equatorial Atlantic according to Emiliani (1955).

(f) The preceding curve corrected by H. De Vries (Radiocarbon dates for Upper Eem and Würm-Interstadial samples. Eiszeitalter und Gegenwart, 9:10–17, 1958.)

(g) Total of warm species of Foraminifera of the deep-sea core P-127 of the Caribbean Sea according to Piggot and Urry (1942). Determinations by J. A. Cushman.

the methods for the last 500 millenia.

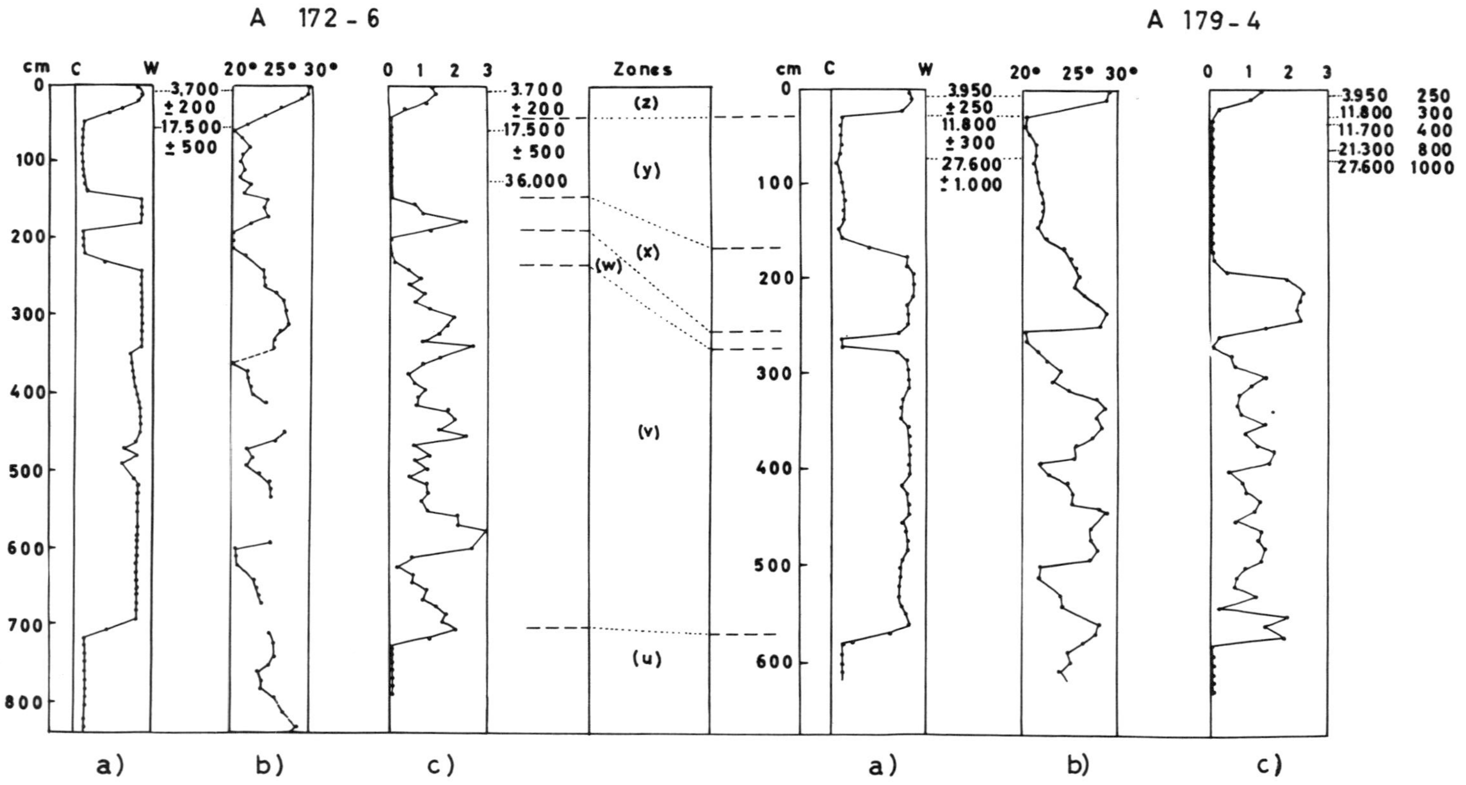

FIGURE 3
(a) Relative proportion of warm-water and cold-water plankton Foraminifera. (b) Isotopic temperatures obtained by Emiliani on *Globigerinoides sacculifera*. (c) Frequency of *Globorotralia menardii*. Documents taken from Ericson, Ewing, Wollin, Heezen (1961).

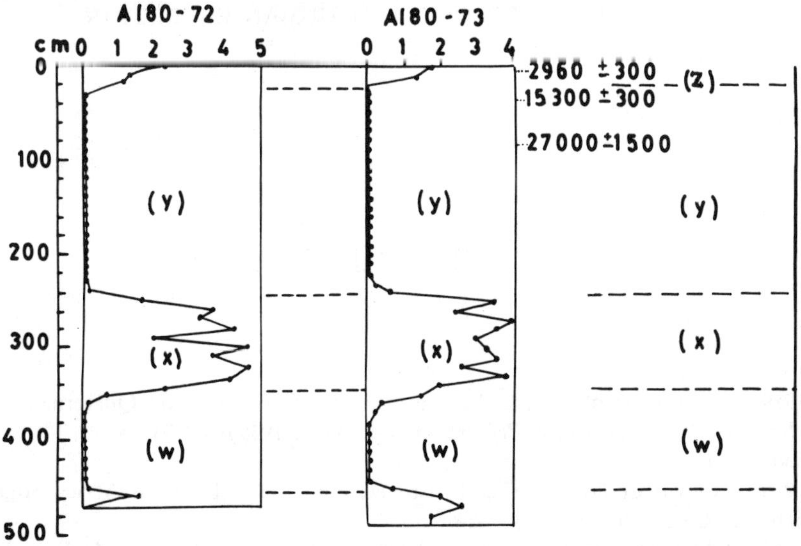

FIGURE 4
Frequency of *Globorotalia menardii* and location of the paleontological zones. Ericson, Ewing, Wollin, Heezen (1961).

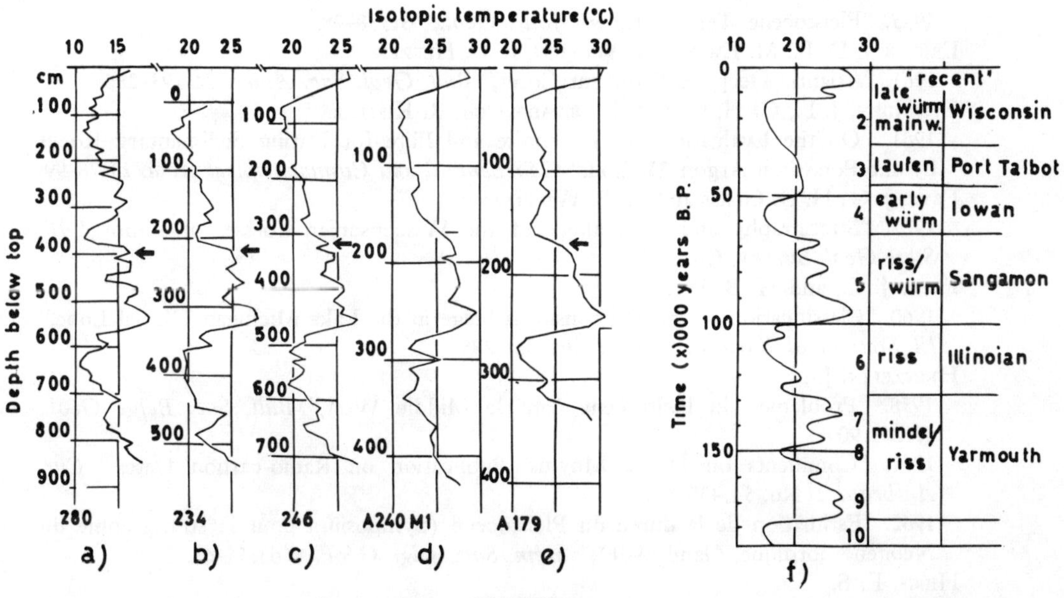

FIGURE 5
(a) to (e) Isotopic temperatures obtained by Emiliani on *Globigerinoides sacculifera.* (a) Core 280, North Atlantic; (b) and (c) Cores 234 and 246, equatorial Atlantic; (d) and (e) Cores A 240-M1 and A 179–4, Caribbean Sea. The trunks indicate the disappearance of *Globorotalia menardii flexuosa.* (f) Generalized curve of the temperatures in the scale of absolute time. Absolute dating by the Pa^{231}–Th^{230} method extrapolated beyond 150,000 years. The interpretation of the Alpine and American nomenclature which is relevant is that given by the authors. Rosholt, Emiliani, Geiss, Koczy, Wangersky (1961).

BIBLIOGRAPHY

BERNARD, E. A.

1959. "Théorie astronomique des pluviaux et interpluviaux du Quaternaire africain." *Actes. 4e. Pan. Afr. Cong. Préhist.* (Leopoldville 1959): 67–95.

CLARK, J. D.

1960. "Human Ecology during Pleistocene and Later Times in Africa South of the Sahara." *Cur. Anthrop.*, 1:307–24.

1962. "C14 Chronology in Africa South of the Sahara." *Actes, 4e Pan-Afr. Cong. Cong. Préhist.* (Leopoldville 1959): 303–13.

EARDLEY, A. J., and V. GVOSDETSKY

1960. "Analysis of Pleistocene Core from Great Salt Lake, Utah." *Bull. Geol. Soc. Amer.*, 7:1321–44.

EMILIANI, C.

1955. "Pleistocene Temperatures." *Jour. Geol.*, 63:538–78.

ERICKSON, D. B., M. EWING, G. WOLLIN, B. C. HEEZEN

1961. "Atlantic Deep-sea Sediment Cores," *Bull. Geol. Soc. Amer.* 72:193–286.

EVERNDEN, J. F., G. H. CURTIS, J. OBRADOVITCH, R. KISTLER

1961. "On the Evaluation of Glauconite and Illite for Dating Sedimentary Rocks by the Potassium-Argon Method," *Geochemical and Cosmochemical Acta*, 23:78–99.

FRYE, J. C., H. D. GLASS and H. B. WILLMANN

1962. "Stratigraphy and Mineralogy of the Wisconsonian Loesses of Illinois," *Ill. State Geol. Survey*, Circ. 334.

FRYE, J. C., and H. B. WILLMAN

1960. "Classification of the Wisconsinian Stage in the Lake Michigan Glacial Lobe," *Ill. State Geol. Surv.* (Urbana), Circular 285.

HEINZELIN, J. DE

1958. "Problèmes du Pleistocene dans le Middle West," *Bull. Soc. Belge Géol.*, 67:265:90.

1961. "Comments on H. L. Movius' Publication on Radio-carbon Dates," *Cur. Anthrop.*, 2(No. 5):434–6.

1962. "Estimation de la durée du Pleistocene (Symposium pour la stratigraphie du Néogène nordique, Gand 1961)," *Mém. Soc. Belge Géol.*, 8(6):55–63.

HIGGS, E. S.

1961. "Some Pleistocene Fauna of the Mediterranean Coastal Areas," *Man* (August 1961), No. 166.

HOLMES, A.

1960. "A Revised Geological Time-Scale." *Trans. Geol. Soc. Edinburgh*, 17(3): 183–216.

HURLEY, P. M., R. F. CORMIER, J. HOWER, H. W. FAIRBAIRN, W. H. PINSON

1960. "Reliability of Glauconite for Age Measurements by K-Ar and Rb-Sr Methods," *Bull. Amer. Assoc. Petr. Geol.*, 44 (No. 11): 1793–1808.

KULP, J. L.
1961. "Geologic Time Scale," *Science*, 133:1105–14.

LIVINGSTONE, D. A.
1962. "Age of Deglaciation in the Ruwenzori Range, Uganda," *Nature*, 194(4831): 859–60.

MOVIUS, H. L.
1960. "Radiocarbon Dates and Upper Paleolithic Archaeology in Central and Western Europe." *Cur. Anthrop.*, 1 (Nos. 5–6): 355–91.

PIGGOT, C. S., and W. D. URRY
1942. "Time Relations in Ocean Sediments." *Bull. Geol. Soc. Amer.*, 53:1178–1210.

QUÉZEL, P.
1960. "Flore et palynologie sahariennes," *Institut française de l'Afrique noire*, Bull. 22A (No. 2): 353–60.

RICHTER-BERNBURG, G.
1957 and 1958. "Isochrome Warven im Anhydrit des Zechstein 2; Die Korrelierung isochromer Warven im Anhydrit des Zechstein 2," *Geol. Jahrbuch*, 75:629–46; 74:601–10.

ROSHOLT, J. N., C. EMILIANI, J. GEISS, F. F. KOCZY, and P. J. WANGERSKY
1961. "Absolute Dating of Deep Sea Cores by the Pa^{231}/Th^{230} Method," *Jour. Geol.*, 69:162–85.

VOLCHOK, H. L., and J. L. KULP
1942. "Age Determination of Deep-sea Cores: The Ionium Method," *Bull. Geol. Soc. Amer.*, 53:1386.

WOERKOM, A. J. J. VAN
1957. "The astronomical theory of climate changes." In H. Shapley (ed.), *Climatic Change*, Cambridge: Harvard University Press, pp. 147–57.

BRIEF REMARKS ON THE VEGETATION OF THE MOUNTAINOUS REGIONS OF EAST CONGO[1]

L. LIBEN

IN STUDYING the mountain flora of Central Africa, one is struck by the similarities found in mountainous massifs often separated by considerable distances. This inevitably implies a common origin, which must be located in time and space. Particularly in the Congolese mountains, the phytogeographic interest revealed by the Ruwenzori—which is, except for the adjacent Virunga, very isolated in relation to the other massifs of comparable altitude—must be stressed.

The following are among the most interesting results of flora analysis.

1. the affinities of the mountain species (with discontinuous range or not) vary with altitude: those of the lower levels are mostly African elements, while those of the upper level belong mostly to extratropical or cosmopolitan genera;

2. the rate of endemism increases noticeably with altitude.

These and other facts lead to the general conclusion that as altitude increases, the flora appears to come from more distant origins, not only in space (for example, the orophytes), but also in time: the increasing rate of endemism argues in favor of a substantially longer isolation.

How could systematic groups presently isolated on high massifs have migrated from one to another? There are only two possibilities: either propagation between neighboring plants, or transportation over long distances of seeds or other propagation devices.

The former is practically excluded at the present time; it assumes either a different climate, or some connection between massifs that are now isolated.

As for the latter, it has never been demonstrated experimentally and supposes a happy combination of unlikely circumstances. Nevertheless, if climatic variations intervened at crucial times, they could have greatly facilitated the process

1. These brief remarks present the general conclusions of a memoir now in press, entitled *Nature et origine du periplement végétal (Spermatophytes) des contrées.*

by reducing the distances between step stones. As Simpson (1952) noted, a slight probability must not be confused with no probability; it should also be remembered that these probabilities increase with time.

In conclusion, discontinuous transports over long distances cannot be excluded, even though they may be improbable; they must be admitted in certain cases. Nevertheless, we suspect that most migrations took place in a continuous manner in a different context from what we know today.

The flora settlement of the high peaks results from horizontal as well as vertical migrations, continuous or discontinuous. Thus, the hypsophilic plants can be divided into two categories: endogenous hypsophilic species when the vertical component is predominant; exogenous hypsophilic species when their settlement is the result of a predominantly horizontal migration.

As for the endogenous species, if one limits oneself to the three most characteristic genera (arborescent *Senecio* and *Lobelia*, *Alchemilla*) one finds that their rate of endemism rises with altitude; that the forms which live lowest are those that exhibit the most primitive nature; that certain traits evolve in a parallel way as a function of altitude on isolated massifs.

These three genera, therefore, present the double phenomena of concentric endemism and convergent evolution.

These facts will be more easily explained if one admits that the present hypsophilic forms are born, independently on each massif, from forms which previously lived at low altitudes in the midst of a flora which has disappeared today and which occupied the entire East African plateau, the existence of which was already suggested by the brothers Fries (1948) and by Hedberg (1959).

While nothing is known concerning the time when this flora appeared in East Africa, we may attempt to guess the time when it disappeared. The paucity of Virunga in endemic forms and the particular affinity of their flora with that of Ruwenzori seems to indicate that these mountains were populated mainly from forms already differentiated on the Ruwenzori before the building of volcanoes, that is, already in the Tertiary; it is around this time that the old East African submountain flora would have sought refuge in altitude.

As to the exogenous hypsophilic species, of extratropical origin, their introduction could occur only under two conditions: if the mountains had attained a sufficient altitude, and if there were vacant habitats ready to receive them. From this point of view the volcanoes must have played an important role.

Once the ascension of the old East African flora had begun, the total territory was left open to the various migratory swarms which probably succeeded each other in the following order:

1. nordic;
2. Afro-austral, including numerous sclerophyllous genera of the Afro-sub-Alpine (given this, the Cape flora would not have been a relic of a hypothetical "Randflora," but a very old nucleus differentiated on the spot, and which, by the later Tertiary, already partially migrated toward the north);

3. the autochthonous low level flora (largely forest), then differentiated in altitude toward the more mesophilic species;

4. finally, a dryer climate would have permitted the introduction of the Sudano-Zambezian flora, formed of diverse elements: the hygrophilic forest flora, and xeric elements from the north and south. At the same time, the hygrophilic forest flora retreated vertically to more favorable places, the zone of greatest pluviosity appearing at the foot of the growing mountains.

The scheme outlined here, therefore, supposes the existence on the East African plateau, in the Tertiary, of a relatively mesophilic climate, permitting the old submontane flora (with *Lobelia*) to coexist with the hygrophilic forest flora. The climate would have progressively dried and warmed while the mountains built up. This outline does not, however, take account of the minor climatic fluctuations which could have occurred one way or another. One cannot exclude, for instance, the possibility of temporary connections between mountain massifs at lesser vegetation levels (transition forest and mountain forest). So far, the vegetation can teach us only about the global effect of climate and relief variations.

If, as the analysis of high mountain flora seems to indicate, the Virunga were populated partly by species differentiated on the Ruwenzori, one must suppose that the latter massif had, quite early, at least the indication of an Alpine stage; on the other hand, one must take account of the recent uplift (Middle Pleistocene) of the order of 1,000 meters mentioned by de Heinzelin (1955). The essential condition which agrees with my hypothesis is that the Virunga did not begin to build up until after the ancient submountain flora had already sought refuge at a level higher than their beginning level.

A remark concerning the "African Orophytes" [2] should be added. There are species of Guinean and Sudano-Zambezian alliance, generally rather plastic from the ecological viewpoint and often linked more or less to the presence of man. It is possible, therefore, that weak thermal fluctuations permitted their passage from east to west along the transversal mountain arcs, as Chapin (1932) suggested.

2. These are species "which, in their range of distribution in intertropical Africa, show a true mountain character and are found in two or more floral regions" (Lebrun 1956).

BIBLIOGRAPHY

Chapin, J. P.
1932. "The Birds of the Belgian Congo, I," *Bull. Amer. Mus. Nat. Hist.*, Vol. 65.

Fries, R. E., and Th. C. E. Fries
1948. "Phytogeographical Researches on Mt. Kenya and Mt. Aberdare, British East Africa," *Kungl. Svenska Vetensk. Akad. Handl.*, Stockholm, Series 3, vol. 25, No. 5.

Hedberg, O.
1959. "The Phytogeographical Position of the Afroalpine Flora," *C. R. 9e Cong. Internat. Bot.* (Montreal 1959), 2:155.

Heinzelin, J. de
1955. "Le fossé tectonique sous le parallèle d'Ishango," *Inst., Parcs Nat. Congo Belge,* Mission J. de Heinzelin de Braucourt (1950), Brussels, Fasc. 1.

Lebrun, J.
1956. "Les 'Orophytes africains,'" *C. R. Conf. Internat. Africanistes de l'Ouest* (6e Session, San Tomé 1956), 2:121–8.

Liben, L.
1962. "Montaynuses du Congo oriental," *Acad. Roy. Belg., Cl. Sc.*, No. 4, 2nd Ser., 15:13, 195 pp.

Simpson, G. G.
1952. "Probabilities of Dispersal in Geologic Time," *Bull. Amer. Nat. Hist.*, 99:163–76.

PLEISTOCENE ENVIRONMENTS IN SOUTHERN AFRICA

GEOFFREY BOND

I. INTRODUCTION

Reviews of the Pleistocene period in Africa by Flint (1959a, b), Clark (1959, 1960), and Cooke (1958) have recently appeared. Each discusses the problem from his own particular angle, and Flint has provided a framework of criteria for assessing the climatic significance of various geological observations, with particular regard to arid and semiarid environments.

Many papers describing individual sites have included good observational data on the geological events, but the interpretations placed upon these data have varied. It has been possible to make the geological data fit almost any paleoclimatological scheme, and while this has been interesting, it has not led to a reasonable picture of paleoclimatology in Southern Africa.

The scope of the present paper is limited geographically to southern Africa, including the territories of Northern Rhodesia, Southern Rhodesia, Nyasaland, Mozambique, Angola, South-West Africa, Bechuanaland, and the Republic of South Africa. Unfortunately the state of knowledge varies within this region, and it is impossible to make a reconstruction of Tertiary or Pleistocene environments having uniform validity for the whole region. It is hoped to use Flint's criteria for the assessment of paleoclimatological environment, and to apply them as uniformly as possible to a selection of sites within a limited part of the region, wherever adequate descriptions or personal knowledge will permit.

The region chosen comprises the Northern Transvaal, a small part of the Northern Cape Province, Bechuanaland, and Southern and Northern Rhodesia. This includes a wide range of present topographic and climatic zones, probably ranging from conditions too arid for early hominids to those in which the rainfall was more than adequate, with a corresponding range of vegetation types. By this means it is hoped that some idea may emerge of the limits between which hominids at various stages could live.

Environment involves a number of factors, including topography, vegetation, and climate. These are all variables, and climate itself includes a number of subheadings of which rainfall is one, and the one that is most usually dealt with in paleoclimatological studies. However, climate also involves temperatures, which are much more difficult to estimate. Within the southern African region only very minor variations in topography have taken place since the Middle Tertiary. These are discussed below and shown to be so small that they can have had very little influence on climate. Vegetation has certainly varied with climate, and can be regarded as a function of it provided climatic changes are of such long duration that the vegetation is able to reach a balanced condition. Climate must also have been responsible for controlling the availability of water supplies, and the lower limit is important. From the point of view of topography and stability, this region has an advantage over East Africa. Against this, under the conditions prevailing throughout southern Africa, deposits are lamentably thin and scattered, whereas in East Africa they are spectacular and thick. When uniformity of interpretation is reached, reliable deductions will be possible, and comparisons can be made with the deposits in East Africa where volcanic activity and tectonics have introduced considerable complications.

The problem in the Pleistocene resolves itself into the selection of a number of suitable sites where investigations have been reasonably complete. It would be ideal if sites could be selected from different toptographic regions equally scattered over the whole region. Unfortunately at this stage it is not possible to do this, as the total number of sites adequately known is limited, and their regional distribution is far from uniform. In Northern Rhodesia the field is limited to the Kalambo Falls site at the extreme north of the territory, and the Victoria Falls area in the extreme South. Neither of these provides any evidence at all of conditions during the Lower Pleistocene, and interpretation must begin in or just before the Older Stone Age. In Southern Rhodesia the middle Zambezi valley provides a certain amount of evidence, but the best sites are Lochard on the plateau near Bulawayo, and Khami waterworks west of Bulawayo. These two sites are sufficiently close together to be combined as one section, which overlaps in the Middle Stone Age. In the Union the Vaal River provides a long section which can be amplified at the base by the evidence from the Australopithecine caves. It may be possible to reassess the Cape Coastal area on the basis of Hopefield, but the stratigraphy there is not entirely satisfactory. Nyasaland at the moment is a complete blank, since no site there has been adequately described. Portuguese East Africa is as yet insufficiently known to contribute significantly. South-West Africa is in almost the same state, but Angola is very much better known since Pleistocene deposits there yield economic minerals.

The procedure adopted in this paper has, therefore, been first to review the Mio-Pliocene background, then the tectonic stability of the region and the extent of Pleistocene erosion. These factors having been discussed and found to be of very minor importance, the main factor of hominid environment must be climate and climatic change. For the selected region, an attempt is made to deduce climatic limits under which hominids existed. By this means some idea of the range of environments tolerable to hominids at various stages of evolution may be deduced. The remainder of the subcontinent is then briefly considered in the light of these tentative conclusions. What seems to emerge is that Australopithecines had a narrow environmental tolerance, but that during Older Stone Age times, man's ancestors became progressively emancipated from environmental control. Now, of course, man is the most cosmopolitan of all mammals, a remarkable advance which must be due to his technical achievements. Australopithecines may have been "facies" fossils; modern man will certainly never be so regarded by our paleontologically-minded descendants.

The problem of correlating Pleistocene deposits in Africa is a very difficult one, yet if comparisons between regions are to be made, it is essential that it should be tackled. One day it may be possible to subdivide the Pleistocene on the basis of human evolutionary types, but until much more material is available, this is only a hope. However, when it becomes possible to link artifact assemblages with hominid types, we may at least be able then to correlate indirectly by human evolutionary stages.

Correlation is also possible broadly on the basis of non-hominid mammalian fossils, but again they are rather rare, and the problem is complicated by the survival in Africa of archaic types. Presumably these archaic types have survived in Africa because whatever the changes in local climate and vegetation, there have always been areas of retreat into which mammalian species could migrate. Thus those mammals which prefer forest conditions may have spread widely during the wetter phases of climate, but have always been able to survive the drier phases by migration into high ground where forest conditions were able to continue throughout the driest phases. Correlation by mammals, therefore, can only be on broad lines, whereas archeological studies of the present kind need correlation to very fine limits. Perhaps the ideal evidence will come from absolute dates, but at present the only method available is Carbon 14, which can generally only cover the last 45,000 years. Physicists may evolve an absolute date method to cover the vital period before this back to the times when K40-A40 dating can take up the story again, but until this is achieved, there is a nasty gap in the absolute record. For this period we are, therefore, forced back on artifacts as a means of correlation. Their use has been discussed at some length by Flint, and I think it can be generally agreed that in a small region, say Southern Rhodesia, correlation by artifacts occurring as reasonably large assemblages can be accepted. Using the broader subdivisions of artifact types, it should be reasonable to correlate over large regions, particularly in the Older Stone Age where the rate of evolution of culture was slow. However this shows a progressive acceleration with time. More care is needed in the Middle Stone Age, and even more in the Later Stone Age. Meanwhile, however, in the absence of anything better, we have very little option but to use this method.

II. THE MIO-PLIOCENE BACKGROUND

Any study of southern Africa which deals with the environment of Early Man and his ancestors should begin as far back in geological time as the Miocene. Early primate fossils of this age from the African continent have only been found in East Africa, and therefore lie outside the region with which this paper is concerned. However, one should examine the rest of southern Africa to see what the environment is likely to have been. The region seems to have been very little changed topographically during this period, but there must have been elevation on the continent as a whole to give rise to the present river system.

The subcontinent had been reduced to a peneplain at the end of the Cretaceous. This was uplifted and a new cycle initiated in the early Tertiary, but by mid-Tertiary times peneplanation was again reaching perfection, with upstanding residuals of the earlier surfaces remaining. However, over very large areas of the interior there was little difference in altitude between the two surfaces, and great areas must have had little relief.

There do not appear to be any inland fossiliferous deposits of Miocene age within the region under review. Much of it was covered by the Kalahari Sands, which are presumed to be of Miocene-Pliocene age. These accumulated under very arid conditions which cannot have been hospitable for animal life of any kind, and it seems unlikely that hominids would have been very much at home under these conditions. The only suitable environment would have been mountainous areas and the coastal strip, in parts of which reasonable rains no doubt fell even during the Pliocene. This long period of desert and semidesert seems to have ended towards the end of the Pliocene. This coincides in time with the onset of the Pleistocene glaciations in higher latitudes, and it is tempting to think that there was some connection between the change of climate there and the amelioration of conditions here.

Pliocene evidence to fill the gap between the Miocene and the lower Pleistocene is needed, but it hardly looks as though the interior of southern Africa is the place to search for it. Coastal regions, particularly on the East, might possibly be worth investigating, but there is a dearth of known Pliocene fossiliferous beds in which to look. The southern African region, therefore, is unlikely to provide the missing link between the Miocene apes and the Australopithecines, though the raised beaches around Lake Nyasa should be investigated.

Cooke (1958) has attempted to reconstruct the main Pliocene drainage pattern in East Africa, but little on these lines has been done south of Lake Tanganyika.

Cooke suggests a late Pliocene-early Pleistocene age for the formation of the Lake Tanganyika trough. This agrees fairly well with the writer's own observations at Kalambo Falls (Bond, MS) and might be applicable to Lake Nyasa as well. Coming further south, a tentative reconstruction of the Mashi-Zambezi-

Shire-Limpopo system can be suggested, but little has been done south of the Limpopo.

The mid-Zambezi is an ancient trough. West of the Gwaai River in the region of Dett in Southern Rhodesia, the Kalahari Sand, of probably Miocene-Pliocene age, rests on a very level surface and now appears on interfluves where the present cycle of erosion has bitten into it. Its base is at about 3000 feet above sea level. In the mid-Zambezi basin, however, typical Kalahari Sands rest on a pronounced surface at about 2000 ft. (Bond, MS) in the vicinity of Makwa and Binga. Only remnants of sand remain, but some of them are many square miles in extent. Further northeast the present cycle of erosion working up the tributaries has destroyed the surface, and no sands remain. This sand-covered bevel is backed on the southeast by scarps which lead up to older erosion surface remnants. The Kalahari Sand is banked against these scarps and runs up old valleys cut in them. In this part of Southern Africa, therefore, a valley-like feature existed in pre-Kalahari Sand times in which a proto-mid-Zambezi flowed northeastward. There is evidence that this ancient river was already entrenched in the 2000' surface before Kalahari Sand times. At two points in the present course of the mid-Zambezi, at Makwa and Binga, Pipe Sandstone appears in the present bed of the Zambezi, 400 to 500 feet below the 2000 ft. surface. In the Sanyati Hills area, between the Sengwa and Bumi rivers, similar deposits occur in a deep valley-like depression south of the present river channel, which had been cut to almost present river level. Before the Kalahari arid period, therefore, a deep river valley existed in this area in which the proto-mid-Zambezi flowed northeastward. Kalahari deposits filled this valley and presumably levelled it off with unconsolidated deposits.

No deposits are known in the Limpopo trough which could be correlated with the Kalahari Sands. However, the Limpopo is now a smaller river than the Zambezi, runs through a lower rainfall area, has a smaller catchment, and yet has cut its valley deeper. The inference is that it is at least as old a drainage line. If the following tentative reconstruction of the main drainage of the region is accepted, this is quite logical.

The Okovango, Mashi, and Upper Zambezi drain south from their headwaters toward the Kalahari, and all of them take a very sudden bend to the east when they reach the Mababe depression. The water then flows eastward past Livingstone and over the Victoria Falls, emerging from the swamp area at the Katambora rapids. This is peculiar behavior in major rivers, and some explanation of it is called for. It seems likely that at some time in the past the Mashi and Upper Zambezi flowed straight through the Kalahari and had direct connections with the Limpopo, perhaps through the present Macloutsi river valley. At this time it is also possible that the Shire, flowing out of Lake Nyasa, flowed straight on south through the flat lands of Mozambique, and that the Shire and Limpopo had a joint outlet to the sea in this area. The Middle Zambezi at this time would have been an entirely separate river system, forming a right bank tributary of the Shire. The vast alluvial plain of Portuguese East Africa may well have been formed by this old river system, the surface having been reworked since by minor stream action. The separate middle Zambezi had a headwater system of tributaries which included the present Matetsi, Lukunguni, and Deka rivers. Thus there would have been a freshwater connection from the headwaters of the upper Zambezi through the Kalahari into the Limpopo and up the Shire which would be available for the distribution of freshwater fish. A connection of this kind is required in reasonably recent geological time to account for the distribution of tiger fish, which are unable to go either up or down waterfalls of more than a few feet in height. Thus, the Murchison Falls on the Shire River cut off the whole of the Lake Nyasa region, which contains no tiger fish. The same applies to the small falls on the Sabi-Lundi system above which there are no tiger fish, yet they occur above and below the Victoria Falls on the present Zambezi, which are much higher than any of these others. However, if the drainage is reconstructed in this way for the comparatively recent geological past, the distribution of tiger fish is quite understandable, the difficulty being that either tiger fish must have evolved extremely slowly, or the change in drainage must have been geologically very recent. This difficulty has not been resolved, and the geological age of the diversion may be too great in relation to the average rate of evolution.

The present drainage pattern must have been evolved from that suggested above by a gentle uplift of the Rhodesian axis in the region of the present Makarikari Lake with a complementary gentle syncline in the Mababe depression. This ponded the waters of the upper Zambezi and Mashi, and broke the connection with the Limpopo system. It can be shown, by measurements on the floor of the Kalahari Sand, that Livingstone lies in a gentle syncline, which is the easterly continuation of the Mababe depression. The ponded waters, therefore, seem to have flowed along the axis of this syncline eastward, and to have spilled over into the old middle Zambezi drainage system just above the Matetsi junction.

The left flank of the Matetsi valley rises about 800 feet above the Zambezi, and it seems that the spilling of the water over this flank of the valley provided the initiation of the Victoria Falls gorge. The gorge emerges suddenly from the flank of the valley, but on each side of it there is a distinct nick on the skyline which may possibly indicate the spillway channels which were abandoned after the central one had cut down preferentially, so that all the water was concentrated in it. Thus it appears as though the upper Zambezi and Mashi were captured by the middle Zambezi and a whole new section added to the Zambezi channel along the axis of the Livingstone syncline. This section, from Katambora to the Matetsi junction, has very few tributaries, and certainly looks like a recent addition to the Zambezi drainage pattern. It is also suggested that about this time the old Shire-Limpopo mouth was abandoned, and the Zambezi broke through to the sea around its present mouth. This has a noticeably small delta for the size of the river, and also suggests that it is a comparatively recent discharge point. It is worth noticing in passing that head-water erosion by the Deka or Matetsi would eventually have succeeded in making this river capture without tectonic diversion. The date of this event cannot be fixed accurately, but it certainly appears to have occurred in pre-Pleistocene times, and therefore the lower Pleistocene drainage pattern in this part of the country was very similar to the present day.

It would be interesting to hear whether attempts to reconstruct Tertiary drainage patterns further south have been made and if so, what conclusions have been reached.

It would appear that during the Mio-Pliocene period much of this region was so arid that hominid life would be impossible. Probably the Southern Cape, part of the eastern coastal area and the mountainous regions along the rim of the subcontinent would have had sufficient rainfall for occupation, but the desert must have stretched virtually from coast to coast, isolating this southern area from East Africa. No suitable Pliocene fossiliferous deposits have so far been discovered, and the chance of their being found within this region seems remote.

III. CRUSTAL STABILITY SINCE THE MID-TERTIARY

An environmental factor which must be considered is the stability of the region in relation to tectonic disturbances. In discussing the Mio-Pliocene history of the Zambezi drainage pattern, it was suggested that the only movements involved are extremely minor ones, indicating considerable stability for the region over a very long geological period. This is further supported by evidence from the middle Zambezi. Profiles for the section of river between the end of the Batoka Gorge and the beginning of the Kariba Gorge have been drawn for the present river and for three Pleistocene terraces. They are all parallel, even the nick points being comparable in all profiles. As might be expected, these nick points have moved upstream during this period. However the essentially parallel nature of the profiles indicates no disturbance by tilting during the time involved. The oldest recognizable terrace contains very rare rolled Acheulean hand axes and cleavers, and has unrolled Sangoan tools on its surface. During this period, therefore, there has been no tilting. Mapping has also disclosed a widespread surface older than these terraces, which is provisionally correlated with a late Tertiary peneplain, and again it parallels the present profile of the river. Throughout the Pleistocene, therefore, there has been no appreciable tilting in this length of the river, which is about 250 miles, and this indicates considerable regional stability.

It has also been possible to plot many older partial erosion surfaces, going back to the surface below the Kalahari Sand, which is presumably mid-Tertiary. The oldest of these surfaces has been tilted about 300 feet in a distance of 250 miles, which is a very minor movement. The evidence from the mid-Zambezi region, therefore, indicates very considerable stability over the whole period of time considered in this paper. Since the Zambezi trough has been a segment of the African shield which has, over a long period of geological time, been subjected to depression and faulting, stability here may indicate stability of the rest of this part of the continent during the same period of time. Further south, in the Vaal River valley, Cooke (1958) has also shown that the region has been stable for a very long time. While there may have been local deformations round the coast, it does not seem that southern Africa has been subject to disturbances on a scale comparable with East Africa since the Miocene. From this aspect, therefore, conditions in South Africa have been much simpler than those in East Africa, and one environmental variable is much reduced in significance.

IV. PLEISTOCENE EROSION

Much of southern Africa consists of erosion surfaces of various ages. As might be anticipated the amount of lowering during Pleistocene has been negligible. Furthermore, on these surfaces incision by rivers is extremely slow, since gradients are very gentle. This is well illustrated by such rivers as the headwaters of the Bembesi system, which include the Lochard site and similar deposits in the Inkwinkwisi valley just to the north. The intermittent wisps of gravel at its base which have yielded nothing but pebble tools are near present stream-bed level. In fact, the lens of gravel in which Dr. Brain (then about 11 years old) found the first pebble tool in 1943, was exposed in the present river channel and has since been completely destroyed by stream erosion. The same can be said of much of the peneplained country in Bechuanaland and Northern Rhodesia. The total amount of erosion since pre-Chellean time has, therefore, been negligible in terms of land form evolution, and hardly even amounts to changes of minor details on plateau surfaces.

In the middleveld topographic region less certainty exists. No Older Stone Age sites occur, and it is difficult to decide whether erosion has destroyed them or whether they never existed, but since Sangoan times very little erosion has occurred. Perhaps they were rare, and the steeper gradients of the rivers have removed them. However, the relationship of the highveld Older Stone Age sites to their lowveld counterparts makes it certain that the three types of topography existed with very little change throughout the Pleistocene.

Turning now to the major river valleys such as the Zambezi, Limpopo, and Vaal, more Pleistocene incision than on the plateau is in evidence, but it is still only a relatively small amount. Indeed, one of the major difficulties in interpreting the Pleistocene succession at the Victoria Falls, lies in the vertical condensation of the succession. Even the older gravels are very near present river level. In the Middle Zambezi there is a greater vertical range, the oldest gravels being generally between 100 and 150 feet above present river level. This, by comparison with the general relief of the region, is a negligible amount, and even in the

Middle Zambezi any wandering Australopithecine would see very little difference today from what he was accustomed to. In the Vaal River valley the oldest terraces have only about the same height above the present channel. Only the details of the topography, therefore, have been modified during this period, in high-, middle-, and lowveld areas.

In the more mountainous regions such as the Drakensberg and the eastern districts of Southern Rhodesia, where the streams have steep gradients and rainfalls are high, more erosion must have occurred during the Pleistocene, but as practically nothing is known of older Pleistocene deposits, there is no yard-stick by which to measure it.

The Kalambo lake basin at the extreme north of Northern Rhodesia is perhaps the nearest approach to a mountainous area in which Older Stone Age deposits are known. It is an area of considerable relief, but the relationship of the oldest lake beds to the modern scenery indicates a negligible amount of topographic change even under these conditions.

In the Coastal area, eustatic changes of sea level have taken place, but the coastal region is narrow and the area involved even at maximum sea level, must have been very small.

By and large, the topography of the whole southern African region has undergone only very minor modifications during the whole of the Pleistocene. There have been no great faulting movements and no vulcanicity. From the point of view of primate environment, therefore, the topographic factor can be taken as having remained substantially constant over the whole period, and the problem is reduced to one of paleoclimatology.

V. PLEISTOCENE CLIMATES

A. Southern Rhodesia

Throughout the Pleistocene, the country has, topographically, looked much as it does now, and has been divisible into three regions: 1) the highveld plateau, 2) the middleveld region, 3) the lowveld.

The highveld plateau averages between 4000 and 5000 feet above sea level, and is the watershed region of the country on which the main development of the present population has taken place, because of its coolness and comparative freedom from diseases.

The middleveld region is one of moderate topographic relief, in which erosion has been dominant over deposition throughout the Pleistocene.

The lowveld region flanks the country on the north and south, forming the Zambezi and Limpopo troughs, and it is in these areas that the late Tertiary surface has been developed, mainly as a valley-floor feature. The main rivers have preserved some record of Pleistocene events, but the country is hot, unhealthy, and inhospitable.

It is interesting to compare these three zones in prehistoric time. Knowledge is to some extent limited by the amount of exploration which has been undertaken, but it seems fair at this stage to say that the highveld plateau was the area in which Older Stone Age man chose to live, and in which sites of this age are comparatively abundant (Fig. 6 in Summers 1960). The middleveld region is poorly known by comparison, but so far practically no Older Stone Age material has been found in it. The lowveld region is perhaps better known, but Older Stone Age material is extraordinarily rare. It does seem that by choice Older Stone Age man avoided the middle and lowveld country, and since presumably the population was small, he was able to exist comfortably on the plateau without having to colonize the less hospitable regions. In Sangoan and Middle Stone Age times (Figs. 8 and 11 in Summers 1960), there is a very much wider spread of archaeological material covering all three regions. Probably this is due to increased population, which forced a wider distribution of hunting parties, some of which had to live permanently in regions which were less suitable from the point of view of climate. Technical advancement may to some extent have made this easier. In Magosian and Later Stone Age times the population seems to have been dispersed over the whole country irrespective of topography and climate (Figs. 13 and 14 in Summers 1960).

THE SOUTHERN RHODESIA WATERSHED REGION

There is an abundance of Older Stone Age sites on the plateau country north of Bulawayo, particularly in the headwater region of the Bembesi drainage system.

The best known sites are at Lochard (Jones 1946, Bond 1946) which were described many years ago, and it is time that the evidence was reassessed. The older part of the succession is as follows:

Lochard (Bond 1946)
Erosion
Ferricrete
Kunkar
Factory sites undisturbed on Surface of Alluvium I
Alluvium I—rare Acheulean
Basal Gravel—with Pebble Tools only
Erosion

This part of the succession lies between two periods of erosion, and includes alluviation and chemical alteration of two kinds. The cultural range is from Pebble Tools to Sangoan, but the main interest is in the Acheulean and its associated Alluvium I.

There are several sites, all within a radius of 3 miles in the Lochard area, but the succession is the same in all of them, even though they occur in separate river valleys. In view of the stability of the region, changes in the geological record can reasonably be ascribed to climatic variations.

In the original paper the basal gravel was said to yield rare Pebble Tools only. In the course of many visits since that date, this statement is still true. Several more good pebble tools have been found since, but no other cultural material has come to light. None of these tools is rolled, and the origin of this gravel, which only occurs as thin wisps and pockets in depressions on the undulating granite floor, needs discussion. The material is poorly sorted for size, and most of it is subangular to subrounded. It does not seem to have been accumulated by stream action, and is much more like a hill creep deposit, formed by the gradual movement of entirely locally derived material down the transverse slope of the valley.

The absence of any cultural material bridging the gap between Pebble Tools and advanced Acheulean at Lochard, has always been puzzling, as there is no appearance of unconformity between the gravel and the overlying alluvium.

The accumulation of the alluvium itself also presents a problem. It was originally said to be sterile, but occasional hand axes have been found from time to time embedded in it. These are weathered but unrolled, and if the alluvium was the product of the stream itself, this is hard to understand. However, a different origin for this deposit can now be suggested. This has occurred to the writer as a result of seeing deposits in the area around Lilongwe in Nyasaland, and in parts of Northern Rhodesia. It is now suggested that Alluvium I at Lochard is a fossil "dambo." This suggestion can be tested in future by laboratory examinations of deposits from typical dambos in Northern Rhodesia and Nyasaland and comparison with the Lochard deposit.

Dambos are long shallow valleys on peneplain surfaces, in which there is no definite drainage channel. The floor of the valley is level as a result of sedimentation, and drainage percolates through the deposit without cutting a proper channel. Stream sedimentation under these conditions is impossible, and the deposit is almost entirely due to lateral transport down the transverse profile into the axis of the valley, a process which at some time in the history of the valley became dominant over longitudinal transport by the stream. It appears that such a condition can only occur when stream gradients are low, in which state axial transport normally only occurs in floods. If for any reason the channel is choked, vegetation will readily stabilize the deposit in the channel. Once this occurs aggradation by colluvial action will begin, and prevent the stream from reopening a proper channel. A swampy area will normally be formed, covered by grass and reeds but devoid of trees, which make a well-defined fringe above the level of waterlogging of roots.

This process seems to have definite climatic limits. Dambos today are typically developed in Northern Rhodesia and parts of Western Nyasaland where the rainfall is over 40 inches per annum. They are very rare in Southern Rhodesia where rainfall seldom exceeds 30 inches in areas of suitable topography, though on the Mashonaland plateau an approach to this form is evident. In just the same way, pans are climatically limited at the present day, with an upper limit of about 25 inches per annum. If future laboratory work can confirm the field evidence that Lochard is a fossil dambo, a valid deduction of climate at the time of formation of Alluvium I can be made. Meanwhile the field similarities can be stated, and a provisional deduction made on this basis.

Apart from the discontinuous gravel at its base, Alluvium I is a remarkably uniform deposit and does not contain pebble bands. The only coarse material is confined to isolated artifacts. The deposit is poorly graded, devoid of visible bedding, and contains much quartz and kaolinized felspar set in a fine grained matrix. Professor Flint and I spent much time discussing this peculiar lithology on the site, and agreed that it was unlike normal stream alluvium. We even considered mud flows down the valley side, which however seemed unlikely in such a locality. It was only after his visit that the general similarity with dambos became apparent. The material is just what would be expected to creep down the lateral valley profile into an axial swamp, and resembles a transported soil. It seems to be all locally derived, and while a sudden mud-flow seems out of the question, the same material transported by soil creep over a long period of time would be expected to produce just such a deposit. The surface of the deposit is flat, and intersects the valley side in just the same way as modern dambos.

The scattered unrolled but weathered hand axes which are occasionally found in the deposits are easily understood under these conditions. Dambos today form swamps with dense reeds and grass, the Savannah trees failing abruptly at their margins, and would have been good hunting areas in Older Stone Age times, as well as permanent watering places.

If this similarity is substantiated by laboratory work, then we can conclude that Alluvium I at Lochard and similar deposits elsewhere in the watershed area of Matabeleland were accumulated under a climate with more than 40 inches per annum of rainfall, in contrast with the present average of 25 for this region.

The factory sites on the undisturbed top of the alluvium may well date from the end of the period of dambo formation, being conveniently near water and sources of raw materials; but the isolated hand axes within the deposit indicate occupation of this region during its formation. At Lochard site C there are three factory sites in a distance of less than 100 yards. It is a pity that they were exposed by natural erosion, but they probably indicate no more occupation than three temporary camps, perhaps belonging to the same hunting party visiting the area within a very short period of time.

The significance of kunkar and ferricrete at Lochard must also be reconsidered.

The age of chemical deposits of this kind is not always easily established. At Lochard the kunkar forms a capping of solid limestone a foot thick on Alluvium I, passing down into isolated concretions of an average diameter of several inches. Acheulean factory site artifacts are heavily encrusted by kunkar, but not by ferricrete. None of the later cultural material is coated by kunkar. Here the evidence seems fairly conclusive that a period of calcification followed soon after the Acheulean but preceded the Sangoan. There is fairly abundant Middle Stone Age cultural material in the lag gravels at the site, but none of this shows calcareous encrustation. This period of calcification seems, therefore, to be fairly well fixed in time. It occurred after the final Acheulean and before the Sangoan and Middle Stone Age.

Ferruginization at Lochard takes the form of a thin sheet of ferricrete, now much broken up by erosion. However, in one area there is a continuous sheet of many square yards reaching a thickness of about two inches. Sangoan tools have been found embedded in this layer, and a fair amount of Middle Stone Age material was discovered lying loose on top of it. The Sangoan material is heavily encrusted by iron oxides, but the Middle Stone Age material is free from encrustation. While perhaps not as strong as the dating evidence for the calcification period, it looks as though this period of ferruginization postdates Sangoan and predates the Middle Stone Age.

Under present climatic conditions the streams at Lochard are entirely seasonal, flowing mainly after rain as flash floods in well-defined channels. The rainfall is between 20 and 25 inches per annum average. It is tentatively concluded that during the formation of Alluvium I it must have resembled the dambo country of Northern Rhodesia and Nyasaland with a rainfall of about 40 inches.

The period of calcification which followed would, following Flint (1959a), indicate much drier conditions, probably less than 20 inches. The succeeding period of ferruginization may indicate a temporary oscillation towards much wetter conditions at or just after Sangoan times, but perhaps too much reliance should not be placed on it.

The remainder of the Lochard succession is less clear. Luckily, however, the succession at Khami waterworks, which is about 45 miles in a direct line southwest, can take up the story from this point.

The Khami site has been described by Cooke (1957) who dealt with the cultural succession, and Bond (1957), in which paper a new approach to the interpretation of such deposits was suggested. This involved an attempt to deduce the transporting power of colluvial action in terms of average annual rainfall. Until it has been more thoroughly tested, the results should be viewed with extreme caution.

The succession at Khami begins with the Sangoan, and correlation with Lochard by this means should be reasonably secure. The climatic curve deduced from three profile studies at Khami, all of which were similar, was held to indicate a wetter period during the Middle Stone Age, during which the rainfall reached a maximum of about 35 inches. This was followed by a drier phase during the Magosian in which the rainfall probably fell to about 20 inches. These limits are supported by the over-all freshness of felspar grains and the absence of any horizon of secondary calcium carbonate. There has been no standing water condition at Khami, so that the state of the felspar (particularly potash felspar) should reflect the general conditions of weathering at the time of accumulation. Above 40 inches per annum chemical decomposition of felspar seems to predominate over mechanical disintegration. At Lochard felspars are completely kaolinized in Alluvium I, but if this was accumulated under dambo conditions, they could have been kaolinized after deposition. The fact that they are still recognizable suggests that they were transported fairly fresh, and that rainfall probably did not much exceed 40 inches even when the dambo was actively forming.

Absence of blown sands at Khami and Lochard does not necessarily indicate that the climate never fell to 15 inches per annum or less. It may have done, but since there is no convenient nearby source of loose Kalahari-type sand, raw material for such sands was lacking.

Combining the results for Lochard and Khami, the following table gives a tentative picture of conditions from early in Acheulean times onwards for the plateau region of the southern part of Southern Rhodesia without recourse to the controversial question of "cut and fill" in river valleys.

TABLE 1
SUMMARY OF RAINFALL EVIDENCE, MATABELELAND PLATEAU

Cultural Stage	*Relative Rainfall*	*Suggested Absolute Rainfall*
Later Stone Age	As now	About 25 in./annum
Magosian	Slightly drier	About 20 in./annum
Middle Stone Age	Wetter	Up to 35 in./annum
Sangoan	Wetter	Up to 40 in./annum
(Calcification)	Drier	Less than 20 in./annum
Late Acheulean	About as now	About 25 in./annum
Alluvium I (Acheulean)	Wetter	About 40 in./annum
Chellean & Early Acheulean (not known)	?	?
Pebble Tools	As now (or a little less)	About 20 to 22 in.*

* By analogy with Transvaal.

Under highveld conditions, therefore, later Acheul people were capable of living under conditions which involved 35 to 40 inches of rain per annum, some 50% higher than present day. The local vegetation would, under such conditions, have resembled the *Brachystegia* woodland of the Mashonaland plateau. The Lydiate 25-foot terrace (Jones & Bond 1948) containing Acheulean comparable with Lochard and therefore roughly contemporary, is at present in *Brachystegia* woodland. If the rainfall at that time was 50% higher, then Acheul people must have been living in a climate with about 50 to 55 in. Under these conditions the Lydiate area and much of the Mashonaland plateau would have been covered in Forest. On the plateau of Southern Rhodesia, therefore, environmental tolerance of Acheul man could have ranged from 35 inches of rainfall and *Brachystegia* savannah, and 50 to 55 inches and Forest country. No lower limit can be suggested in this region. Neither is there any direct evidence for the makers of the pebble tools. By analogy with the Transvaal Australopithecine caves, conditions then must have been very much the same as now. The Middle Stone Age evidence shows a tolerance just as wide as for the later Acheul people, and the same can be said of the Sangoan people.

THE VICTORIA FALLS AREA

Topographically the area around the Victoria Falls is at an altitude of just on 3,000 feet. It cannot therefore by included in the lowveld or the highveld, but it is not typical middleveld, since the relief is so low. The reason for this may be in the possible diversion of the Upper Zambezi as outlined earlier in this paper.

It is an area in which very intensive archaeological studies have been carried out by Clark (various papers, chiefly summarized in 1950, 1958) and others. The stratigraphy is therefore well known, but suffers from vertical condensation due to lack of down-cutting by the river, and "big river" problems. The Zambezi is already a long way from its source region, and changes in river behavior at the Victoria Falls may have been conditioned by changes far upstream. For these reasons it is unwise to place much reliance on periods of erosion and aggradation by the Zambezi, and one must turn to locally derived materials such as wind-blown and other sands, and chemical alterations.

The stratigraphic and cultural succession is summarized below.

TABLE 2
SUMMARY OF RAINFALL EVIDENCE, VICTORIA FALLS

Geological Succession	*Relative Rainfall*	*Suggested Av. Rainfall In./Annum*	*Cultures*
Erosion to present contours	*25 ins./annum at present*	—	—
Sandy calcareous alluvium (Erosion)	*Drier*	18 to 20	Wilton
Sands (Blown)	*Drier*	Less than 15	Magosian
Calcification & pink alluvium (Erosion)	*Drier*	18 to 20	M.S.A.
Younger gravels & Sands (Erosion)	(River deposits	No deduction)	
Younger gravels (Erosion)	(River deposits	No deduction)	
Sands (Blown)	*Drier*	Less than 15	
Ferruginization Land Rubble	*Wetter*	About 40	Sangoan
Sands	?	?	
Older Gravels II	(River deposits	No deduction)	Late Acheulean
Ferruginization	*Wetter*	About 40	
Older Gravels I (Erosion)	(River deposits	No deduction)	E.S.A.

The distance between the Plateau region of Matabeleland and the Victoria Falls area is about 250 miles, but they are today in practically the same rainfall belt. Because of the nature of the evidence at the Victoria Falls, more emphasis is laid on phases of relatively drier climate since there is an abundance locally of loose sand available for redistribution whenever climate became suitable. The evidence from two separate areas can perhaps be combined at least tentatively into a single picture. Neglecting the Zambezi fluviatile evidence, the M.S.A. is a bad blank, but this can to some extent be filled by extrapolation of the watershed succession.

It is not possible to say how wet the wettest periods have been. The evidence does not demand a figure much greater than 40 inches per annum. This has happened several times. In the drier phases the evidence from the watershed does not seem to indicate any period of much below 20 inches, while at the Victoria Falls it may on some occasions have been as low as 15. It is unlikely that the Zambezi ever dried up in these drier periods since it rises on ground with a much higher rainfall average.

In general terms the E.S.A. seems to have been a long period of comparative wetness, which by Late Acheulean times had dried to about the present climate. There was no part of Southern Rhodesia under semidesert, a much reduced Savannah region (*Brachystegia*, *Mopane*, and *Terminalia*) and a greatly increased forest area, (evergreen and *Baikaia*). Water would be abundant and perennial over most of the country, and dambo conditions flourished in the Matabeleland plateau region. No doubt there were oscillations within this generally wetter period.

The First Intermediate (Sangoan) people seem to have existed under a rather similar phase of climate, which however was probably rather short-lived, and sandwiched between a phase of climate like the present in final Acheul times and a phase of climate drier than now, in which wind-blown sands appear at the Victoria Falls. Whether the "Sangoan" wet phase was long enough to allow the vegetation pattern to be re-established is not known, but it has been claimed that the Sangoan was a forest culture.

The post-Sangoan dry phase was fairly intense and the vegetation pattern must have had time to be established. In this there are small areas of semidesert in the Limpopo and western Matabeleland where "Kalahari" conditions spread eastward. Forest was almost eliminated and practically the whole country was Savannah Woodland. Water would be strongly seasonal except in the larger river valleys. Game probably had seasonal migrations, as take place in the eastern Kalahari today, and man the hunter would be forced to do likewise unless he could exist on a purely vegetable diet.

Middle Stone Age times saw a return to much wetter conditions, which must have resembled Older Stone Age times, with reduced Savannah, perennial streams and increased forest. The semidesert must have retreated westward far into the Kalahari region.

The Second Intermediate Stone Age (Magosian) saw a hardening of climate, but again probably only a short one. How much effect on vegetation took place, it is hard to say. Forest probably gave place to Savannah, but it is unlikely that much semidesert developed.

In Southern Rhodesia, therefore, from what we can at present learn about early environments, it does not seem that they limited post-Acheulean man's distribution to any marked degree. The topography of the region has not been greatly changed during the period, the landscape having been modified only in the finer details. The main variable has been climatically controlled vegetation changes, and since early Sangoan times man's *distribution* has been virtually independent of this factor. His *way of life*, however, may well have been constantly adapted to suit the change of environment.

Evidence before the Sangoan is scanty and stratigraphically poor. It does look, however, as though late Acheulean man preferred to live on the watershed, under conditions very like the present, and who can blame him if numbers were small and population pressure had not developed.

Evidence from pebble tools, which presumably take us back to Australopithecine days, is scanty also, but does suggest (it does no more) that the makers, whoever they were, preferred open Savannah country, as they did elsewhere in the Transvaal, East Africa, and North Africa.

There may have been a tendency in the past to overestimate the extent of climatic changes in the African Pleistocene. What quantitative evidence there is suggests that at wet maximum the increase has been no more than about 50% of the present figure. For the more arid phases the decrease does not seem to have been more than about 40%. These figures apply from Acheulean times to recent. It is perhaps significant that for the earlier period of the Australopithecines Brain (1958) found quantitative variations of about the same order of magnitude.

As a working hypothesis, therefore, these figures can be used as a guide even outside the areas where they were obtained.

Distributions of cultures in relation to topographic regions has already been discussed. The distribution maps given by Summers (1960) for the various stages are most helpful in this respect. There seems no evidence that Southern Rhodesia was at any time uninhabitable during the period covered, though the absence of Chellean and earlier Acheulean is puzzling.

Summers (*op. cit.*), with the help of Wild, attempted to reconstruct vegetation patterns at various times in the light of paleoclimatic evidence. From the point of view of early man, vegetation could be divided more simply into Forest, Savannah, and semidesert, but a comparison of Summers' distribution maps and reconstructed vegetation maps shows no recognizable correlation, and this factor seems to have played little part in distribution. So long as there was food and water, he could exist.

B. Bechuanaland

This territory is peculiarly important, as it embraces a desert focus which has probably existed since the mid-Tertiary, but has contracted and expanded as long term climatic changes took place. Throughout the whole period, however, there has probably been some area, either within the present limits of the territory or in adjoining South-West Africa, where conditions were too arid for primate occupation. Therefore distribution maps, like those prepared by Summers (1960) for the Pleistocene in Southern Rhodesia, would be of prime importance and could be used to deduce lower limiting conditions at various stages of hominid evolution.

Bechuanaland, however, is very poorly known territory geologically, and this is particularly true of the Pleistocene. Its later-Tertiary history seems to have been a long period of continuous desert conditions during which the great spread of the main Kalahari Sand accumulated. At this time it must have been an extremely inhospitable region, except in the immediate vicinity of the major rivers draining south through it. Tectonic diversions of all these rivers at some time in the later Tertiary may have occurred, but while they were flowing they must have rather resembled some parts of the present Nile. They might have provided corridors for the migration of animals through the region on north-south lines. No fossiliferous deposits dating from this time are known, and their discovery seems unlikely.

When the Pleistocene began the main rivers were already in their present courses, and much of the region therefore probably uninhabitable by hominids.

The present knowledge of Quaternary events in the whole region is almost entirely due to Wayland (1954). In this paper he summarized all the information he had obtained during his term of office as founder and first Director of the Geological Survey of Bechuanaland Protectorate. Unfortunately he has never published details of his excavations and other stratigraphic information. The over-all picture is, therefore, not very clear. There is no evidence of major warping, and the generally flat topography indicates little change in this aspect of environment throughout the period. There is evidence of climatic changes, which must have been reflected in vegetational shifts of the belts of desert, semidesert and Savannah. The details are hard to interpret from the meager published account.

The oldest deposits falling within the span of Geological time considered here, are the Botletle beds, which Wayland (1954, p. 40) regards as early Pleistocene, but which others have placed earlier, in the Tertiary. Their age has not yet been finally determined as they are largely unfossiliferous, and those few forms found are not diagnostic of age. However they appear to be water-laid sediments, but they might be connected with the mid-Tertiary river system and, therefore, no reflection of climatic conditions within the territory. Pleistocene events and deposits include "fossil" river channels, chemical alterations and blown sands.

In keeping with its geographical position the chemical deposits are mainly calcretes and silcretes. Ferruginous alterations are rare, unless the reddening of blown sands is included. While this is probably a postdepositional effect (it is easily removed if sands are exposed to transport either by wind or water) it should be distinguished from the accumulation of ferricretes. What ferricretes there are seem to be restricted to the eastern part of the country between Lobatsi and Francistown, where conditions are very similar to the Matabeleland plateau.

Wayland (*op. cit.*, p. 30) lists the following kinds of Pleistocene evidence in Bechuanaland:

1. Accumulations
   a) Windblown sands
   b) Bedded calcretes
   c) Silcretes
   d) Ferricretes
   e) Diatomite deposits
   f) Riverine red sands and grits and bouldery horizons
   g) High-level boulder beds
   h) Low-level boulder beds
   i) Buff-coloured arenaceous deposits (Notwani River)
   j) Black earths
   k) Chemical deposits of dripstone in special environments
2. Erosional effects
   a) Fossil rivers
   b) Valley and stream bed features
   c) Wave-cut beaches
   d) Limestone caves

There is, therefore, a good deal of evidence to be obtained in Bechuanaland, but it has yet to be put in good order.

Wayland gives little stratigraphic information. In his section on paleoclimates he shows that Older Stone Age sites are often covered by blown sands of Kalahari type, and from his text a table has been compiled. It is very tentative, and a difficulty is that localities are only in broad terms. Furthermore the region covers present rainfall zones from about 10 up to 25 inches per annum. It is therefore not a conveniently homogeneous one.

However, it seems that during Olden Stone Age times part of the region was reasonably well watered and occupied by makers of the hand-axe cultures. Probably at this time there occurred a considerable extension westward of Savannah, like the present Matabeleland plateau. It is doubtful if the rainfall increased enough to allow *Baikaia* forest to spread very much even on Kalahari Sand country.

Wayland's paper gives no hint of any wetter period in the Middle Stone Age; but the deltaic sediment covering a late Still Bay hunting site north of the Makarikari Lake is now being eroded by the Nata River (Bond & Summers 1954). On the basis of this slender evidence, a wetter phase is included at the end of Middle Stone Age times.

TABLE 3
TENTATIVE PLEISTOCENE SUCCESSION IN BECHUANALAND

Geological Succession	*Relative Rainfall*	*Suggested Absolute Rainfall.*	*Cultures*
	Present about 15 to 20 in./annum		
Black soils and calcrete nodules	Slightly wetter	18 to 20	L.S.A.
Blown sands	Drier	Less than 15	Magosian
Nata Delta	Wetter	?20 to 25	
Blown sands	Drier	Less than 15	M.S.A.
Calcrete	Slightly wetter	18 to 20	
	?	?	Fauresmith/ Sangoan
Silcrete	Very dry	10 to 12	
Calcrete	Slightly wetter	18 to 20	
Blown sands	Drier	Less than 15	
Gravels of fossil rivers	Wetter	? plus 25	O.S.A.
Blown sands	Drier	Less than 15	

Wayland's evidence of blown sands in Magosian times seems to link up with similar evidence of drier climates in the adjoining parts of Southern Rhodesia.

The meager evidence seems to point to the ability of Older Stone Age man to exist in Savannah country, but later people seem to have been sufficiently adaptable to live in even drier climates.

The evidence from Southern Rhodesia shows that they were also capable of living in much wetter climates at the same stage of cultural evolution. Again it seems that Sangoan and later man was environmentally cosmopolitan.

C. Northern Rhodesia

Distribution maps of the various stages of Stone Age cultures have been compiled by Clark (1957). He also included vegetation maps for present day conditions and for various rainfalls above and below the recent figures. The territory is joined to the Southern Rhodesia region, and the two together comprise an area in which present rainfall varies from 50 to 60 inches in the north down to less than 15 in the southeast, with a correspondingly wide range of vegetation types, from Forest, through Savannah, to semidesert.

There is no evidence in Northern Rhodesia of tectonic disturbances during the period, other than very gentle basin and swell movements which give rise to such swampy areas as Lake Bangweolu. Such movements are not accurately dated, but probably occurred in later-Tertiary times. The topographic environment, therefore, has remained substantially constant. As with most of southern Africa the changes have, therefore, been climatic with consequent shifts of vegetation belts.

The two areas where Pleistocene successions are best known are Kalambo in the extreme north, and the Victoria Falls in the extreme south. Unfortunately the Kalambo succession, though known in detail as a result of the beautiful excavation techniques employed by Professor Clark, cannot yet be clearly interpreted in terms of climate. Pollen analyses by van Zinderen Bakker have shown that in late-Acheulean times, temperatures were significantly lower than now, presumably indicating increased cloudiness and hence more rain. Flint's criteria cannot be applied, since no chemical alterations occur and no blown sands could be expected in an area which at present receives about 50 inches per annum. Ferruginization could be expected, but although some of the deposits are distinctly ferruginous, no ferricretes occur. Calcification also seems to be absent, perhaps suggesting that lowest rainfalls were well above 18 to 20 inches. This could also be expected if the maximum changes in rainfall suggested elsewhere in the paper applied in the Kalambo area also.

The Victoria Falls area was dealt with under Southern Rhodesia (perhaps rather unfairly since most of the work was done by Prof. Clark!) and need not be repeated here.

Professor Clark has pointed out how much work remains to be done in Northern Rhodesia, which is more than twice as large in area as Southern Rhodesia, and his distribution maps must be used with suitable caution in consequence. However, for the Older Stone Age the pattern is somewhat similar to Southern Rhodesia. Pebble tools are scarce, but those sites so far discovered lie mainly on the highveld. Since the Sangoan and later cultures have been found so widely in the territory, this can hardly be taken to indicate lack of exploration, but destruction by erosion in middle- and lowveld topography might be the cause. However no site is known in the lowveld of the middle Zambezi, and this suggests that the distribution is a real one. Very little Chellean is known (but more than in S. Rhodesia), and the pattern, such as it is, resembles that for pebble tools. Acheulean and Hope Fountain are only slightly more extensive, again being largely highveld with rare occurrences in the mid-Zambezi valley. The Sangoan is widespread in various topographic regions, and Middle and Later Stone Age are as widely dispersed in Northern as in Southern Rhodesia.

The evidence from Northern Rhodesia, therefore, supports that from Southern Rhodesia. Certainly by the end of Older Stone Age times man had mastered all the local environments, extending into vegetation types not present in Southern Rhodesia. Although vegetation maps for various rainfalls have been prepared (with the help of Dr. Wild), less use can be made of them until further geological successions have been evaluated in terms of climate. Generally speaking it seems that the northern part of Northern Rhodesia must have at all times been relatively wetter than any part of Southern Rhodesia, though as far north as the

Kafue river conditions closely resemble those of part of Mashonaland in Southern Rhodesia. The Kalambo evidence probably can be taken as showing that late Acheulean man was quite at home in rainfall areas of more than 60 inches per annum.

D. North Transvaal

The great interest of this region is the occurrence of cave breccias yielding Australopithecine remains. Thanks to the most elegant work of Dr. C. K. Brain (1958), the climatic sequence for this period is probably more reliably known than for the later deposits anywhere else in southern Africa, and this in spite of their extreme age. It is also the most important period from the point of view of hominid environment, since at this very early stage hominids were probably just as dependent on their environment as any other mammal.

The outstanding importance of Brain's work is his conclusion that while the climate varied over the sequence of cave breccias, the Australopithecines of Sterkfontein, Swartkrans, and Kromdraai all flourished under a rainfall of 20 to 30 inches per annum (present average, 28). This is not significantly different from present day conditions. Thus the overall environment of the Transvaal Australopithecines must have been closely comparable with the present day, apart from man made changes. In view of the tectonic stability, lack of vulcanicity and small degree of topographic change which has taken place since that time over the whole southern Africa region, it is reasonable to conclude that over the whole region environmental conditions at that time were very much as at present, with perhaps a tendency to be a few per cent drier.

The deposits of Kromdraai, the youngest Australopithecine deposit, indicates distinctly wetter conditions up to 35 to 40 inches per annum. The range of climate under which the Transvaal Australopithecine flourished was, therefore, from 20 to 40 inches, which in that region would probably indicate vegetation types which could all be included as Savannah. This rainfall range covers in Southern Rhodesia mopane savannah at the lower end and *Brachystegia* savannah at the upper end. Both are open woodland forms of about equal density, though the undergrowth in mopane tends to be much more open.

E. Remainder of Southern Africa

Angola and Portuguese East Africa are rather blank spaces archaeologically. Clark's distribution maps (1959) bring this out quite clearly, and it is not possible at this stage to amplify them. The remainder of southern Africa, therefore, will only be taken to include the Union of South Africa and South-West Africa; in both cases much archaeological work has been done.

The Chelles-Acheul is widely scattered topographically from coastal areas to interior plateau. In view of the deductions given earlier in this paper that Sangoan and later people were able to live under a variety of climates in the Bechuanaland-Southern and Northern Rhodesia region, there is no need to try to interpret the Vaal succession during this period. The critical period during which the Australopithecine existed has been dealt with elsewhere. The distribution maps given by Clark (1959) show clearly that Older Stone Age people were widely dispersed topographically. However the paleoclimatology for individual sites is interpreted, such a wide regional distribution must mean a wide climatic tolerance. Under present day conditions these sites range from over 60 inches per annum near Dondo in Angola to as little as 5 to 10 inches. While it is quite feasible to imagine Older Stone Age people existing quite easily under a climatic regime of 60 inches (as at Kalambo) and, applying the 50% increase probable at a maximum wet phase, even under 90, it seems extremely unlikely that they could exist in a climate of 5 to 10 inches per annum. This must indicate blown sand conditions with an almost total absence of surface water. Again applying the 50% increase at a wet maximum the climate would be changed to somewhere around 15. Under these conditions open grassland and sparse savannah would probably have stabilized the sand, and some water would be available, particularly in the wet season. Under these conditions it is conceivable that Older Stone Age people could have existed without impossibly long seasonal migrations.

On the basis of this speculative argument, the range of tolerance of Older Stone Age people is from about 15 inches per annum minimum to at least 80 to 90. This must have included botanical environments ranging from semidesert grassland to quite dense forest. If in later times Older Stone Age man was so environmentally tolerant, there seems no reason to doubt that Fauresmith-Sangoans and Middle Stone Age people were equally cosmopolitan.

The distribution maps for southern Africa given by Clark (1959) for all these cultural periods support such a suggestion. The distribution of Pebble Culture sites, few as they are, suggest far more environmental dependence as discussed elsewhere in this paper.

VI. CONCLUSIONS

From the review of Bechuanaland and Southern and Northern Rhodesia, it appears that at no time since the Acheulean at least, has the area been uninhabitable by man. Fresh evidence of the tectonic stability of the region has been given, and it has also been shown how little erosion has taken place during the Pleistocene. This is readily understandable in the plateau areas, but even in major river valleys, such as the Middle Zambezi, the relation of the oldest Pleistocene terraces to present river level below and later-Tertiary peneplain above, indicates a very minor degree of entrenchment. Throughout the Pleistocene, therefore, the topographic environment has been practically constant. The scenery of today was the scenery of our earliest Pleistocene ancestors, and erosion during this time has merely put the finishing touches to our present landscape.

Floral evolution has probably been so slow that even our modern vegetation would have been familiar to Older Stone Age man. The changes of environment in this region, therefore, have merely been the slow changes of climate, measuring thousands of years from the wet maxima to dry minima, during which vegetation had time to adjust itself.

An extensive treatment has been given to this region because it contains present rainfall zones ranging from about 15 inches per annum to about 60, about a mean value for much of the territory of about 25 inches.

The attempts so far made to evaluate southern African Pleistocene climatic changes quantitatively indicate less variation from the present climate than perhaps the words Pluvial and Interpluvial suggest. They also indicate that the present rainfall is about midway between the two extremes. It may be significant that eustatic sea levels suggest that we are midway between a glacial and an interglacial.

For the Australopithecine stage Brain's work suggests that rainfall varied between 70% and 140% of present rainfall. The evidence from Lochard in Older Stone Age times shows a possible range from 70%–160%. At the Victoria Falls in Late Acheulean to earliest Middle Stone Age the variation seems to have been from 60%–160%. The evidence from Khami Waterworks indicates a range from about 160% in the Gamblian maximum to about 80% in the Magosian.

These estimates are all based on geological evidence of various kinds, and there is a significant similarity in the range of rainfall variations suggested.

On palynological evidence, van Zinderen Bakker (1957–1961) has reached rather different conclusions. The only site within the southern Africa region from which quantitative conclusions have been drawn is Florisbad. Van Zinderen Bakker (*op. cit.*) concluded that the pollen evidence from "the oldest peaty layers, near the human skull" indicated 7 to 14 inches per annum at a date given by C-14 as 43,960 years ago. The present climate at Florisbad has about 28 inches. This is a decrease to 25% to 50% of present rainfall. At the other extreme he suggests an increase to 200% to 300% (in the Gamblian?) if the pollen diagram really represents alpine grassland.

The Bechuanaland-Southern and Northern Rhodesia region at the present day comprises rainfall zones of a wider range than individual sites disclose during Pleistocene pluvial-interpluvial cycles. The distribution of Stone Age sites in this region indicates that man could adapt himself in Acheulean and later times to conditions at least as diverse as those of the present day. Assuming that at wet maxima rainfall increased 50% and decreased 50% at dry maxima, then these people could adapt to conditions ranging from 90 inches per annum down to about 15. Below this figure surface water becomes an acute problem, and it is also near the limit below which sand begins to blow. For instance at Hopefield sand is blowing now with a rainfall of about 13 inches. But it seems fairly certain that at the time of Saldanha man there was a greater rainfall than this.

Not only has Acheulean and later man been cosmopolitan in regard to environment, but southern Africa is so vast that even at the worst periods of climate, there have been areas where survival was possible. Migrations due to climate must have taken place out of arid areas during extreme dry phases, but at no time can it have been impossible for hominids to find areas where their range of tolerance was exceeded.

Chellean time is very poorly known, but what has been discovered suggests, and it can do no more than suggest, that at that time hominids were much more tied to their climatic environment than later peoples. In Australopithecine times the evidence is, perhaps rather surprisingly in view of their age, better known. Pebble tool evidence has recently been under a cloud of mistrust, but must be used for want of any better evidence at present. On this basis their makers, Australopithecine or otherwise, do seem to have been restricted to relatively flat savannah country under conditions remarkably like the present day, with a range of annual rainfall of about 20 to 40 inches.

Much must await discovery and may modify tentative guesses, but for the moment as a working hypothesis in searching for more Australopithecine sites, this may be a useful guide. It suggests that the area most in need of exploration is the Lomagundi dolomite area of Southern Rhodesia. This is moderately dissected country, perhaps rather too dissected, and has a present rainfall of about 35 inches per annum. It is at present "unknown," but Dr. Brain and I are planning to start exploration for caves and breccias as soon as possible.

A second possibility is the Lusaka region of Northern Rhodesia, where breccias have already been discovered. However, all these have so far been found to be rather too recent.

If there is any validity in the attempt made in this paper to "bracket" the climatic tolerance of various stages of hominid evolution in southern Africa, then it seems unlikely that climatic changes, in themselves, have had much influence on cultural evolution. If, for instance, late Acheulean man was equally at home under climates ranging from 15 to 90 inches per annum, a relative increase in the lower limit or decrease in the upper one would not indicate a change so drastic that a completely new way of life would be forced upon him. Such changes would not act as a stirring mechanism inducing rapid population migrations. Pleistocene climatic changes were in any case so slow that they can hardly have affected the tempo of the inevitable "convection currents" which must occur all the time among unsettled and unspecialized hunter nomads. The difference between wet and dry seasons at any phase of Pleistocene climate produced a much greater environmental contrast than any pluvial-interpluvial cycle. People adapted to so many seasonal changes in the course of a lifetime would easily conform to slow shifts of mean climates in cycles of thousands of years.

In preparing this paper the writer has tried to avoid using any evidence which is not strictly geological. There are other lines of attack, and it would be most interesting to compare the results arrived at independently by quite separate lines of reasoning.

VII. ACKNOWLEDGMENTS

I would like to thank Mr. Roger Summers, my former colleague at the National Museum in Bulawayo, most sincerely for much helpful discussion.

Dr. C. K. Brain, Prof. Desmond Clark, and Prof. H. B. S. Cooke have also helped with much useful discussion.

Finally, may I thank the Wenner-Gren Foundation for their invitation to take part in the splendid symposium at Burg Wartenstein, where the discussion of other aspects of the problem of primate ecology in Africa was so stimulating.

BIBLIOGRAPHY

Bakker, E. M. van Z.

1957. "A Pollen Analytical Investigation of the Florisbad Deposits (South Africa)," *Proc. 3d Pan-Afr. Cong. Prehist.* (Livingstone 1955): 56–67.

1961. "Botanical Evidence for Quaternary Climates in Africa," *Symposium* (Port Elizabeth), South African Association for the Advancement of Science (in press).

Bond, G.

1946. "The Pleistocene Succession Near Bulawayo," *Occ. Pap. Nat. Mus. S. Rhod.*, 2:104–115.

1955. "The Excursion Handbook for S. Rhodesia," *Proc. 3d Pan-Afr. Cong. Prehist.* (Livingstone 1955): 1–30.

1957. "The Geology of the Khami Stone Age Sites: Southern Rhodesia, *Occ. Pap. Nat. Mus. S. Rhod.*, 3:44–55.

"Geology of the Kalambo Falls Prehistoric Site," (in press).

"Geology of the Middle Zambezi Basin," (in press).

Bond, G., and R. Summers

1934. "A Late Still Bay Hunting Camp Site on the Nata River, Bechuanaland Protectorate," *S. Afr. Arch. Bull.*, 9:89–95.

Brain, C. K.

1958. "The Transvaal Ape-Man-Bearing Cave Deposits," *Transvaal Mus. Mem.*, No. 11.

Clark, J. D.

1950. *The Stone Age Cultures of Northern Rhodesia.* Capetown: The South African Archaeological Society.

1954. "An Early Upper Pleistocene Site at the Kalambo Falls on the Northern Rhodesia-Tanganyika Border," *S. Afr. Arch. Bull.*, 9:51–6.

1959. *The Prehistory of Southern Africa.* London: Penguin Books.

1960. "Human Ecology during Pleistocene and Later Times in Africa South of the Sahara," *Cur. Anthrop.*, 1:307–24.

Cooke, C. K.

1957. "The Waterworks site at Khami, S. Rhodesia, Stone Age and Protohistoric," *Occ. Pap. Nat. Mus. S. Rhod.*, vol. 3, No. 21A.

Cooke, H. B. S.

1946. "The Development of the Vaal River and its Deposits," *Trans. Geol. Soc. S. Afr.*, 41:243–59.

1958. "Observations Relating to Quaternary Environments in East and Southern Africa," *Trans. Geol. Soc. S. Afr.*, Annexure to vol. 61. "The Pleistocene Environment in S. Africa," in *Ecology in South Africa* (W. Junk. publishers; in press).

Flint, R. F.

1959. "On the Basis of Pleistocene Correlation in East Africa," *Geol. Mag.* (London), 96:265–84.

1959. "Pleistocene Climates in East and South Africa," *Bull. Geol. Soc. Amer.*, 70:343–74.

Jones, N.

1945. "The Climate and Cultural Succession at Sawmills, S. R.," *Occ. Pap. Nat. Mus. S. Rhod.*, 11:39–46.

Jones, N., and G. Bond

1948. "The Lydiate Pebble Industry," *Occ. Pap. Nat. Mus. S. Rhod.*, 2:251–8.

Mason, R. J., A. B. A. Brink, and K. Knight

1959. "Pleistocene Climatic Significance of Calcretes and Ferricretes," *Nature*, 184:568.

Peabody, F. E.

1954. "Travertines and Cave Deposits of the Kaap Escarpment of South Africa, and the Type Locality of *Australopithecus africanus* Dart," *Bull. Geol. Soc. Amer.*, 65:671–706.

Söhnge, P. G., D. J. L. Visser, and C. van Riet Lowe

1937. "The Geology and Archaeology of the Vaal River Basin," *Union S. Afr. Geol. Surv. Mem.* No. 35.

Summers, R. F. H.

1960. "Environment and Culture in Southern Rhodesia: A Study in the Personality of a Landlocked Country," *Proc. Amer. Phil. Soc.*, 104:266–92.

Wayland, E. J.

1954. "Outlines of Prehistory and Stone Age Climatology in the Bechuanaland Protectorate," *Acad. Roy. Sci. Colon.* (Belgium), 25:1–47.

BABOON ECOLOGY AND HUMAN EVOLUTION

IRVEN DeVORE AND S. L. WASHBURN

THE ECOLOGY OF BABOONS is of particular interest to the student of human evolution. Aside from man, these monkeys are the most successful ground-living primates, and their way of life gives some insight into the problems which confronted early man. We have been concerned with an attempt to reconstruct the evolution of human behavior by comparing the social behavior and ecology of baboons with that of living hunter-gatherer groups, and applying these comparisons to the archaeological evidence (Washburn and DeVore 1961a). The following description of baboon behavior and ecology is based on field data collected during 200 hours of observation by Washburn in the game reserves of Southern Rhodesia in 1955, and on more than 1200 hours of observations by both of us in Kenya game reserves during 1959. The original study was financed by the Wenner-Gren Foundation for Anthropological Research, and the second trip was part of a study of the origin of human behavior supported by the Ford Foundation. Analysis of the field data is being completed under a National Science Foundation grant for the study of primate behavior. We wish to thank the foundations, and the numerous people who helped us in Africa—especially J. Desmond Clark, Stephen Ellis, L. S. B. Leakey, B. L. Mitchell, and B. Verdcourt.

CLASSIFICATION

This paper will primarily consider troop size, range, population density, and diet, but before discussing these topics we wish to give our views on the classification of these primates. Baboons are large, primarily ground-living monkeys of the family Cercopithecidae. As has been true of many of the primates, this group has been so divided that generic names have been applied to taxonomic groups which amount to no more than species. The most widely distributed baboon group occurs in the savanna and forest from the Tibesti Plateau in the north to Cape Town in the south, and across Central Africa from Dakar to the east coast. This group, the genus *Papio,* is usually divided into several species, including "chacma," "yellow," and "olive." There is no evidence that these forms are more than racially distinct, however, and "chacma" and "yellow" baboons occur in the same troops in the Rhodesias, although not in extreme form. It is not known whether the East and West African forms are distinct species, but it is likely that intermediate forms exist there, as they do between East and South Africa. When separated by long distances, individuals from these races appear to be quite distinct, as is the case of the "chacma" from the Cape, the "yellow" in Nyasaland, or the highland and coastal races in Kenya. But if intermediate forms exist and reproductive isolation cannot be demonstrated, these varieties are best considered races.

In West Africa there are two species of short-tailed forest baboons, the drill and the mandrill. These forest types differ no more from the widely distributed form of savanna *Papio* than the pig-tailed macaque (*Macaca nemestrina*) differs from the crab-eating macaque (*M. irus*). In North Africa and Arabia the desert baboon, *Papio hamadryas*, lives in country which is too dry and open for the other species. In summary, the genus *Papio* is divided into a number of races, and into at least four species, including the savanna species with several races, two forest species, and a desert species.

The gelada baboon, *Theropithecus gelada*, is very distinct from *Papio*. Its facial skeleton is constructed differently; it jumps and uses its tail differently; and it should probably not be regarded as a baboon at all. Today *Theropithecus* is confined to the mountains of Ethiopia, in the same region in which hamadryas occupy the lowland desert and "olive" baboons the savanna between the two extremes (Starck and Frick 1958). But *Simopithecus* is probably *Theropithecus*, indicating that the *Theropithecus* group formerly extended into East Africa. The practice of putting almost every fossil primate in a new genus nullifies the utility of the genus as a taxonomic concept. *Cercopithecoides*, for example, is a *Colobus* monkey. The increased understanding which the presence of such forms might contribute to the reconstruction of the ecology of Olduvai or Sterkfontein is lost by the multiplication of names which separate the fossils from similar living forms.

The African baboons are very similar to the Asiatic ground monkeys, the macaques. Both groups have forty-two chromosomes (Chu and Bender 1961), and their distribution does not overlap. The newborn are usually black, changing to brown. Skulls, teeth, and general physical structures are much the same. In social life and basic habits the two groups are very similar. In contrast to all other monkeys (both New and Old World), the macaques and baboons do most of their feeding on the ground. They can cross rivers and may live in dry areas, moving far from trees. Compared to other monkeys they are more aggressive and dominance-oriented, and their average troop size is considerably larger than any other species yet studied. These characteristics have enabled the baboon-macaques to occupy a much larger area than that of any other group of monkeys. It is an area very comparable to that utilized by *Homo* before the time of the last glaciation. Ground living, ability to cross water, an eclectic, varied diet, the protective troop, and aggressive males permitted the baboon-macaques to occupy this vast area with a minimum of speciation.[1] The contrast in the number of species between ground-living and tree-living monkeys emphasizes this point. There are more species in the genus *Cercopithecus* in the African forests than among all the baboon-macaques from Cape Town to Gibraltar to Japan. There are more species of langurs in Southeast Asia alone than species of *Cercopithecus*. Further, the most ground-living of the langurs (*Presbytis entellus*) has the widest distribution, and the same is true for the most ground-living vervet (*C. aethiops*). The taxonomic contrast between tree and ground monkeys is clearly seen in Ceylon where the island is occupied by one macaque, one dry country langur, and four forest forms (Phillips 1935). Apparently in Ceylon the rivers have been a major factor in isolating the langurs, but they do not form barriers for the macaques. The general relation between ecology and taxonomy in the monkeys appears clear: the more ground-living, the less speciation. There are many more adaptive niches in the forests than in the drier regions.

The men of the Middle Pleistocene, genus *Homo*, occupied the same range as the baboon-macaques but without speciation. Their way of life (based on tools, intelligence, walking, and hunting) was sufficiently more adaptable and effective so that a single species could occupy an area which ground monkeys could occupy only by evolving into at least a dozen species. This comparison gives some measure of the effectiveness of the human way of life, even at the level of Pekin and Ternifine man. Obviously, there is nothing to be gained by being dogmatic about the number of species of Middle Pleistocene men. Perhaps when many more specimens have been found it will be convenient to recognize two or three species, but the general form of this argument will still hold. There is no suggestion that any of the known fossil men (genus *Homo*) differ in size or form as much as a chacma baboon and a drill, or a crab-eating macaque and a pig-tail macaque. Even in its most primitive form the human way of life radically alters the relation of the organisms to the environment. As early as Middle Pleistocene times man could migrate over three continents without major morphological adaptation.

Australopithecus may have occupied an adaptive position midway in effectiveness between the ground monkeys and early *Homo*. Small-brained, bipedal toolmakers probably occupied larger areas than baboons, and without speciation. It is most unlikely that the East African and South African forms of *Australopithecus* are more than racially distinct. Robinson's suggestion that the jaws from Java called "Meganthropus" are closely allied to the australopithecoid from Swartkrans (Robinson, this volume) supports the notion that *Australopithecus* was already able to disperse widely with minimum biological change. The presence of small and large Australopithecoids in South Africa at the

1. *Cynopithecus* is a large black macaque, and there is no reason for putting this monkey in a separate genus.

same time suggests that their adaptation was much less effective than that of *Homo*. It may be possible to reconstruct more of this stage in human evolution with a more thorough study of the ecology of baboons, and by contrasting their mode of adaptation to that of man. With this hope in mind we will now consider the ecology of baboons in East Africa.

THE TROOP

Troop Size

Careful counts of more than 2,000 baboons showed a range in troop size from 9 to 185. Estimates by the Tanganyika Game Department, for troops in the Wankie Game Reserve (Southern Rhodesia), and carefully repeated counts in the Royal Nairobi National Park all give an average troop size of 36–42. The *largest* troop in Nairobi Park numbered 87. In the Amboseli Reserve, the *average* troop size was 80 and the largest troops numbered 171 and 185. The fact that troops are twice as large at Amboseli indicates the need to study several localities before generalizing. Table 1 summarizes the size of troops in the Nairobi Park and in the Amboseli Reserve.

TABLE 1
Troop Size

Nairobi Park	*Amboseli Reserve*
12	13
17	42
24	47
28	51 *
28	66 *
40	57
61	64
77	70
87	74
	78
Total 374	88
	94
	103
	171
	185
	Total 1203

* 51 and 66 sometimes combined to form a group of 117.

The smallest troops we observed numbered only 9 (Tsavo Reserve), 11, 12 (Wankie Reserve), 12 (Nairobi Park), and 13 (Amboseli Reserve). Three of these troops contained two adult males, and one only a single adult male; our data do not show the number of adult males in the fifth troop. These small troops are independent, functioning societies and not temporarily detached parts of larger troops. Baboon troops are closed social systems with a high degree of inbreeding. Often, adjacent troops can be distinguished from each other by the characteristic color patterns, length of hair, and form of face or tail. During both field trips, in over 1400 hours of observation, we saw only two individuals change from one troop to another. In these very small troops, inbreeding may be very important, and a whole generation may be the offspring of a single male.

The largest troops (103, 171, 185) were seen only at Amboseli, but for our study we selected open areas where the baboons would be visible as much of the time as possible. In areas with more rain the abundant vegetation supports more baboons, and large troops may be more common in these areas. For example, in the reserve of the north end of Lake Manyara (Tanganyika), and in the forested areas adjacent to the Athi River near Kibwezi (Kenya), we saw approximately one large troop per mile. This suggests that there were both larger troops and a much higher population density in these areas, but under conditions where continued observation was impossible.

The large troops may temporarily subdivide, and the troops of 88, 94, and 103 at Amboseli, and the troop of 77 at Nairobi, frequently split. When the troops of 77, 88, and 94 split, all the small infants and their mothers were in one

FIGURE 1
An adult male grooms an adult female while her two-year-old infant tries to nurse (Nairobi Park; DeVore).

section of the troop with the largest adult males. On one occasion troop 171 (Amboseli) was also seen dividing in this way. When all the individuals in a troop are together, there is a clear distinction between the large, dominant adult males, mothers, and infants occupying the center of the troop, and the other, peripheral, troop members around them (as described below under "Troop Structure"). The temporary divisions seen in these large troops are divided along these lines. Such a subdivision lasts for only part of the day and the troop reunites before nightfall. Another type of splitting, in which the troop divides into two sections with a normal distribution of males, females, and juveniles in each section, also occurs in some large troops. Troop 103 sometimes split into two troops of 66 and 37, each troop having a center, a periphery, and all the characteristics of a normal, independent troop. It seems likely that this kind of splitting represents the first stage in the formation of a new troop. Observations on troops 51 and 66 support this. These two troops stayed very close together; if one of the troops arrived at a water hole, the other was likely to appear, and, after using adjacent sleeping trees at night, they often followed the same route away from the trees the next morning. It is tempting to regard this situation as representing a large troop divided one stage further than was the case of troop 103. The reason for regarding 51 and 66 as two troops is that individuals within them did not shift; repeated counts showed that the membership of these two troops was constant, and sometimes they were entirely separate from each other, once for a period of days. It appears that large troops may become unstable, and that divisions occur in troops larger than 70 individuals which are

FIGURE 2
A play group of young juveniles around an adult male (Rhodesia; Washburn).

not seen in the small troops. If this division persists, and if the division contains a normal age-sex distribution, a new troop may result.

The division of troops, their large size, and the fact that they met at water holes made counting of troops at Amboseli very difficult. Troop 103, for example, was originally counted as three troops (one of 103, one of 66, and one of 37). Similarly, troops 51 and 66 seemed to form a troop of 117 on some days. A single count made at an Amboseli water hole might include only part of a troop, or a cluster of 400 baboons representing three adjacent troops.

The density of baboons in an area is related to the food supply, but the size of the troop itself bears no such simple relationship. The ranges of the smallest (13) and the largest (185) troops at Amboseli overlapped, and the size of Nairobi troops did not correlate with the different vegetation zones in the park. Social behavior, rather than ecology, seems to determine troop size.

FIGURE 3

(a) Animals grooming in the center of a troop are (left to right) a juvenile grooming an adult female, the mother of a small infant, a female grooming an adult male, another juvenile and a black infant (in grass). (b) The same group, but the adult male has turned to watch the black infant (Nairobi Park; DeVore).

Because adult males defend the troop from other animals, troop size is important to the individual's survival. Troop 185 contained over thirty large, adult males, compared to only two in troop 13 and one in the troop of 12 (Nairobi). Like troops of the much smaller vervet monkey (*Cercopithecus aethiops*), a small baboon troop yields to a large troop when they meet at water holes. When food supplies are limited, this gives a large troop an advantage over smaller ones.

Troop Structure

A detailed description of the social relationships within baboon troops is given elsewhere (DeVore 1962). Here we have emphasized those aspects of troop life which are adaptations to life on the ground. Baboons are intensely social, and membership in a troop is a prerequisite for survival. Most of a baboon's life is spent within a few feet of other baboons. Baboon troops are closed social systems, individuals very rarely change to a new troop, and the troop regards any strange baboon with suspicion and hostility.

Within the troop, subgroups are based on age, sex, personal preferences, and dominance. When a troop is resting or feeding quietly, most of the adult members gather into small clusters, grooming each other or just sitting (Fig. 1). Juveniles gather into groups of the same age and spend the day in these "play groups," eating, resting, and playing together. The most dominant adult males occupy the center of the troop, with the mothers and their young infants gathered around them, and the groups of young juveniles playing close by (Fig. 2). These dominant males, and the small black infants near them, seem to be greatly attractive to the other troop members. During quiet periods the other troop members approach the adult males and the mothers, grooming them or sitting beside them (Fig. 3). It is unnecessary for male baboons to herd the troop together; their presence alone insures that the other troop members will not be far away (Fig. 4).

Around this nucleus of adult males, mothers, and young juveniles are the more peripheral members of the troop—the less dominant adult males, older juveniles, and pregnant or estrus females. Estrus females and their consorts usually stay at the periphery of the troop. Although the juvenile play groups will not wander far from the troop's center, peripheral adults may leave the troop for short periods. While the center of the troop moves slowly along, the adult and older juvenile (subadult) males and adult females sometimes move rapidly ahead to a new feeding spot. This may separate them from the rest of the troop by a quarter of a mile or more, and they may not rejoin the troop for thirty minutes or an hour. Although peripheral adult males may make such a side trip alone, or in small groups, other troop members will not leave the troop unless accompanied by the males. Healthy "solitary males" observed during the early part of our study later proved to be troop members who had left the troop for a short while.

FIGURE 4
Females and juveniles feed near two adult males at Amboseli water hole (Washburn).

A baboon troop that is in or under trees seems to have no particular organization, but when the troop moves out onto the open plains a clear order of progression appears. Out in front of the troop move the boldest troop members—the less dominant adult males and the older juvenile males (Fig. 5). Following them are other members of the troop's periphery, pregnant and estrus adult females and juveniles. Next, in the center, comes the nucleus of dominant adult males, females with infants, and young juveniles. The rear of the troop is a mirror image of its front, with adults and older juveniles following the nucleus and more adult males at the end. This order of progression is invariably followed when the troop is moving rapidly from one feeding area to another during the day, and to its sleeping trees at dusk. A troop which is coming toward trees from the open plains approaches with particular caution. The tall trees in which baboons sleep are found only where the water table is near the surface, usually along a river or beside a pond. Vegetation is usually dense at the base of these trees, and it is in this undergrowth that predators often spend the day. The arrangement of the troop members when they are moving insures maximum protection for the infants and juveniles in the center of the troop. An approaching predator would first encounter the adult males on the troop's periphery, and then the adult males in the center, before it could reach defenseless troop members in the center.

Because they are in front of the troop by twenty to forty yards, the peripheral adult males are usually the first troop members to encounter a predator and give alarm calls. If a predator is sighted, all the adult males actively defend the troop. On one occasion we saw two dogs run up behind a troop, barking. The females and juveniles hurried ahead, but the males continued walking slowly. After a moment an irregular group of some twenty adult males was between the dogs and the rest of the troop. When a male turned on the dogs, they ran off. On another day we saw three cheetahs approach a troop of baboons. A single adult male stepped toward the cheetahs, gave a loud, defiant bark, and displayed his canine teeth; the cheetahs trotted away (Fig. 6). If baboons come upon predators while en route to their sleeping trees, the troop stops and waits while the males in the center move ahead and find an alternate route (the young juveniles and

FIGURE 5
Two adult males walk well ahead of the nucleus of a troop during progression (Nairobi Park; DeVore).

FIGURE 6
A young adult male displaying his canine teeth in a threat (Nairobi Park; DeVore).

mothers with infants stay behind with the peripheral adult males). Eventually the dominant males return, the original order of progession is re-established, and the troop proceeds along the new route. These behavior patterns assure that the females and young are protected in the troop's center.

The ultimate safety of a baboon troop is in the trees. When the troop is away from trees, the adult males are very important in troop defense. We saw baboons near such predators as cheetahs, dogs, hyenas, and jackals, and usually the baboons seemed unconcerned—the other animals kept well away. Lions, however, will put a baboon troop to flight. From the safety of trees baboons bark and threaten lions but make no resistance to them on the ground. The behavior of baboons when near trees contrasts strikingly with their behavior on the open plains. If the troop is under trees, it will feed on the ground within thirty yards of predators, including lions.

Ecology and Sex Differences

The role of the adult male baboons as defenders of the troop has been described. This behavior is vital to the survival of the troop, and especially to the survival of the most helpless animals—females with new babies, small juveniles, and temporarily sick or injured individuals. Selection has favored the evolution of males which weigh more than twice as much as females, and the advantage to the troop of these large animals is clear, but it is not obvious why it is advantageous for the females to be small. The answer to the degree of sex differences appears to be that this is the optimum distribution of the biomass of the species. If the average adult male weighs approximately 75 pounds and the average adult female 30 pounds, each adult male requires more than twice the food of a female.[2] If the food supply is a major factor in limiting the number of baboons, and if survival is more likely if there are many individuals, and if the roles of male and female are different—then selection will favor a sex difference in average body size which allows the largest number of animals compatible with the different social roles in the troop.

If selection favors males averaging 75 pounds, then it will favor females which are as much smaller as is compatible with their social roles. Since the females must travel the same distances, carry young, engage in sexual and competitive activities, there are limits to the degree of sexual differentiation, but the adaptive value of the difference is clear. For example, a troop of 36 baboons composed of 6 adult males and 12 adult females and their young (18 juveniles and infants) has a

2. The Game Department of Northern Rhodesia arranged for the shooting of some baboons which were given to Washburn for study. Weights of four adult males were 70, 72, 74, and 84 pounds. A subadult male weighted 62 pounds. Five adult females weighted 30, 30, 32, 32, and 36 pounds. A sexually mature female whose dentition had not fully erupted weighted 26 pounds. Five juveniles of assorted sizes altogether weighed only 65 pounds. The temporal muscle of one side averaged approximately 190 grams in the adult males and the corresponding muscle only 40 grams in the adult females. This combined with the difference in the canine teeth, means that the contrast in fighting ability between males and females is far greater than might be expected from the gross weight differences alone. In the subadult male, whose canine was fully erupted, the temporal muscle still weighed only 95 grams, or approximately half adult weight. Such an animal (although sexually, dentally, and osteologically mature) must wait for the completion of the growth of his muscles before participating fully in the life of the adult males.

biomass of some 1,000 pounds.[3] If the females also weighed 75 pounds each, 6 adult males and 6 adult females would alone total 900 pounds and have only one-half the reproductive potential of 6 adult males and 12 adult females. Because this would halve the number of young, it would greatly reduce the troop's chances of survival. Our data are not sufficiently detailed to analyze the actual distribution of biomass in the troops we observed, but our observations are compatible with the limited data on weights and the numbers of adult animals we saw. Viewing sexual differentiation in size as a function of the optimum distribution of biomass of the troop offers a way of understanding sexual dimorphism fundamentally different from the view which considers only sexual selection, dominance, and intratroop factors. Obviously, all factors should be considered. Adaptation is a complex process and results in compromises between the different selective pressures, but a distribution of biomass which doubles the reproductive potential of a species is so important that other factors may be minimized.

The importance of sex difference in body size is reinforced by social behavior and the structure of the troop. As described earlier, some subadult and adult males are peripheral in the structure of the troop. They tend to be first, or last, when the troop moves. They are the most exposed to predators and are, biologically, the most expendable members of the troop. Interadult male antagonism results in a social order which both protects females and young and reduces feeding competition with females and young. Without altruism, the dominance behavior of a small number of males keeps a feeding space available to subordinate animals.

Juvenile play prepares the adults for their differential roles. Older juvenile females do not engage in the serious mock fighting which characterizes the play of older juvenile males. In this "play" the males learn to fight, and by the time the canine teeth have erupted and the temporal muscles grown to adult size they have had years of fighting practice. Play, social arrangement, and structural sexual dimorphism all supplement each other, producing a pattern in which the females and young are relatively more protected than the large males. Sexual differentiation must be seen as a part of this whole complex social pattern which leads to the survival of troops of baboons.

3. In an average troop of 36 baboons there would be 6 males whose canine teeth were fully erupted, and 12 sexually mature females. Several of these females, however, would still be classed as "juveniles" if judged by their teeth and skeletons. Social maturity in the male is delayed beyond sexual maturity, and is attained only with full size, weight, and dentition (about the eighth year). Females become socially adult at sexual maturity, about the fourth year (Gillman and Gilbert 1946). The prolonged period of physical and social maturation in the male baboon explains the difference between the *adult*-male–*adult*-female ratio of one-to-two in a population in which the actual sex ratio is about one-to-one (see Carpenter 1958).

RANGE

On an average day a baboon troop leaves its sleeping trees at full daylight and moves rapidly to a spot where the animals feed intensively for two or three hours. In Nairobi Park this morning feeding period is often spent in a fig tree (if these are in fruit), along a watercourse, or out on the open plains. During the dry season in the Amboseli Reserve, feeding areas were usually at the edges of water holes. During the middle of the day baboons rest in the shade of bushes or trees, not far from the feeding place of the morning. The late afternoon is another period of relatively intensive feeding. It is often some distance away from the feeding area of the morning, and a different kind of food is usually eaten. If the morning was spent in a fig tree, the afternoon is usually spent eating grass on the plains; if the morning was spent on the plains, the afternoon meal often consists of the pods, buds, and blossoms of acacia trees. During such a day the troop completes an average circuit of about three miles in Nairobi Park, but this distance varies from a few yards on some days to six or seven miles on others. These figures refer to the distance between points on a map. As a troop meanders across a plain, however, the individuals actually walk twice as far as these figures indicate.

During the year a baboon troop moves over an area which probably averages

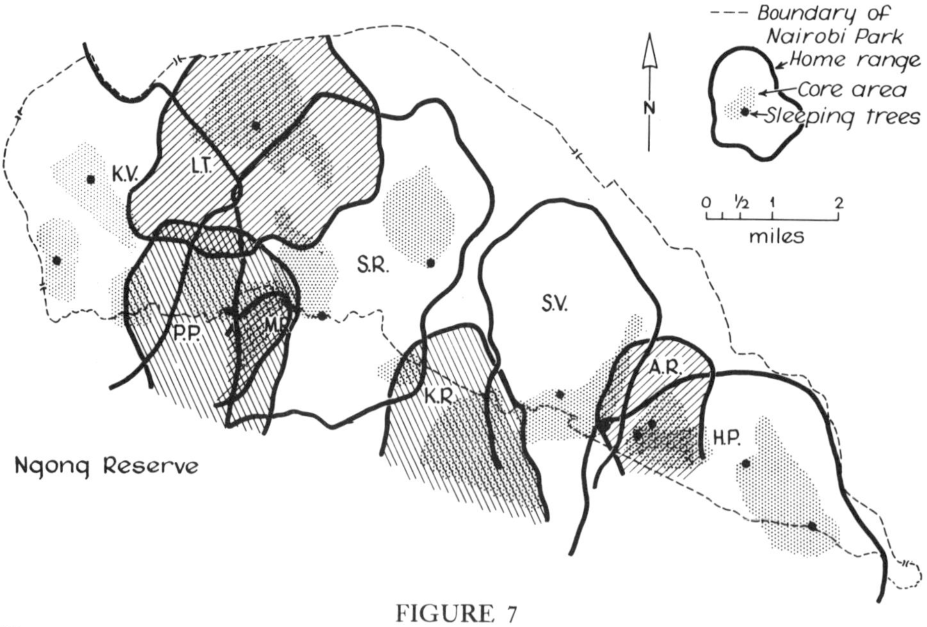

FIGURE 7
Home ranges and core areas of baboon troops in Nairobi Park, April to December, 1959. K. V.-Kisembe Valley, etc. (see Table 2).

about fifteen square miles in open savanna country, but which may be much smaller (for a small troop or a troop living in forest country). Even where the total animal range is as large as fifteen square miles, only parts of it are used frequently. These areas of frequent use may be called "core areas" (Kaufmann, in press). Figure 7 shows the total home range and the core areas in that range observed for Nairobi Park troops over a period of nine months. Although this study was largely confined to the forty square miles within the boundaries of the park, all but one of the nine troops in the park included the Ngong Reserve (to the south) in its range. The southern extent of these ranges into the Ngong Reserve is not known.

Table 2 shows the size and approximate annual range of the troops in Nairobi

TABLE 2
ANNUAL RANGE OF NAIROBI PARK TROOPS

Name of Troop	*Size*	*Range (in square miles)*
Mbagathi River	12	?
Lone Tree	17	9.2
Python Pool	24	7.0 (+)
Athi River	28	3.0 (+)
Songora Ridge	28	15.5
Sosian Valley	40	9.6 (+)
Kapio River	61	?
Kisembe Valley	77	11.7 (+)
Hippo Pool	87	13.8 (+)

Park. Only the annual ranges for the Lone Tree and Songora Ridge troops are complete, but range sizes shown for the other troops probably include about 80 per cent of their annual range. In general, a large troop contains more adult males and covers a wider range. Although both the Athi River and the Songora Ridge troops numbered twenty-eight, the range of the latter is about five times as large as the former. The Athi River troop had only one large adult male; the Songora Ridge troop had six. This gives an approximate measure of the additional range required by a troop with more large males. It also indicates that troops with more males can control a larger range.

Daily routines tend to keep baboon troops apart. Although Figure 1 shows that annual ranges overlap extensively, there is very little overlap of the core areas of adjacent troops. We saw no evidence that troops defend a part of their range as "territory" (Burt 1943), but in Nairobi Park one troop is seldom seen in the core area of another. The core area or areas of a troop contain sleeping trees, water, resting places, and food sources. A troop uses one core area and one grove of sleeping trees for many weeks at a time but may then shift suddenly to a new area. In 1959 throughout the dry season there were only two sources of water in Nairobi Park which also contained tall trees: the Athi River, which forms the southern boundary of the Park, and a water hole in the core area of the Lone Tree troop. All the troops except Lone Tree had at least one core area along the Athi River, which is the boundary between the park and the Ngong Reserve. At both Nairobi Park and Amboseli baboon troops usually slept in the tall fever trees (*Acacia xanthophloea*) which grow only where the water table is high (Fig. 8). Since the plants and fruit trees which baboons use for food also tend to be concentrated near water holes or along rivers, the core areas of a troop include food, water, refuge sites, and sleeping trees. Although a troop usually returns to the same sleeping place after its daily circuit, it also shifts from one core area to another over a period of weeks or months. The existence of alternative core areas serves to reduce contact between adjoining troops, and behavior patterns reinforce this distance. These spacing mechanisms, rather than defense of territorial boundaries, disperse baboon troops in an area.

FIGURE 8

Three adult males and three females (including two "consort pairs") enter the troop's sleeping tree at dusk; another adult male is napping above. This tree is typical of the fever trees along the Athi River which constituted the sleeping trees for most Nairobi Park troops (DeVore).

The population density of baboons in Nairobi Park is about ten per square mile. As a result of social factors and low population density, baboon troops are seldom within sight of each other. When a troop is living on one edge of its range, for example, its neighbor on that side tends to move to a portion of its range well away from other troops. However, where core areas of adjacent troops overlap (along the Athi River), troops may sleep only fifty yards apart without any display of aggression. Only one incident of intertroop aggression was seen when human intervention was not involved. Deliberately bringing two troops together by artificial feeding, however, can cause intertroop threats. The troop with the greatest number of adult males always won in these encounters, although a troop in its core area seemed to be more aggressive than one at the edge of its range.

By contrast to the infrequent encounters between baboon troops in Nairobi Park, troops at Amboseli were seen in close proximity every day. At the end of the dry season at Amboseli (September and October), baboon troops were tightly clustered around water holes. Figure 9 shows the location of these troops during this two-month period. The double water hole (center of the map) was used by troops 51, 66, 171, 88, 70, and 57. At this place a 100-yard crescent of vegetation contained two pools 50 yards apart, and it was here that more than 400 baboons from these troops might be seen together. We never saw any fighting between troops, and it soon became clear that there was a pattern to the various troops' use of the surrounding area. Troops 51, 66, and 171 used only the northern pool, and went north and west from the pools during the day (shaded area). The southern pool was used by troops 88 and 70, and probably by 57, and these troops ranged south and east from the pools. The area between the two pools was used by all, except that we have no record of troop 70 there. The shaded portion of Map 2 divides the region around the water holes according to its use by the different troops. The single exception is one observation of troop 171 at the water hole where troop 185 is shown.

Ordinarily this pattern of range utilization segregates the troops into clusters which recognize and tolerate each other at short distances without any sign of nervousness or tension. If a small troop is at a water hole and a large troop which also uses that water hole arrives, the smaller troop feeds slowly away. When troop 171 came once to the water hole usually frequented by 185, however, both troops paid close attention to the other. Adult males clustered where the troops were closest. The gestures, noises, and indications of nervousness were very different from the apparent lack of attention which is characteristic of troops normally frequenting the same water hole. By comparison with the behavior of troops in Nairobi Park, troop 171 was probably at the edge of its range, and troop 185 was occupying its core area.

As in Nairobi Park, the baboons at Amboseli slept in the tall fever trees around the water holes and marshes. The importance of trees as refuge from predators is illustrated by the ranges of the Amboseli troops. The marshy area north of

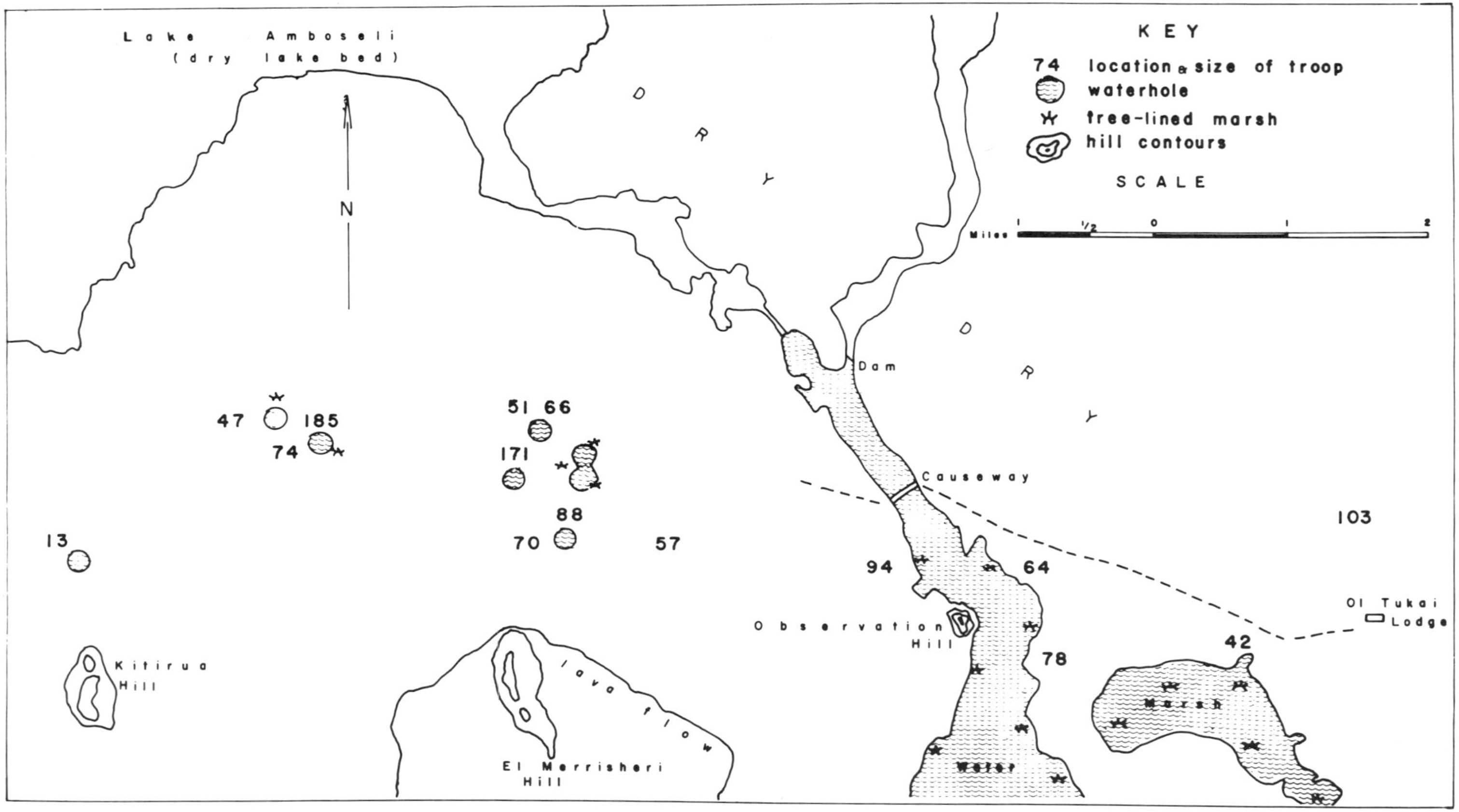

FIGURE 9

Size and location of 15 baboon troops at Amboseli National Reserve during September and October, 1959.

the causeway (Fig. 9) contained water and plenty of food, but there were no trees there. Despite the heavy competition for the limited food resources during this season, no baboon troop included this treeless area in its range. Lions were often seen near the marsh, but south of the causeway, where lions were seen even more frequently, there were trees, and three large troops (94, 64, and 78) lived in this area. Normally the adult males protect the troop, but against the largest carnivores the only safety is flight into the trees. In areas where there are lions, trees limit the distribution of baboons as much as does the availability of food or water.

In summary, baboon range is based on the existence of refuge sites as well as sources of food and water. These ecological factors control population density, but the interrelations of troops are based on behavior. Troop size, number of adult males in the troop, and frequency of contact between troops determine the outcome of intertroop relations. Territorial defense is not seen, but core areas of different troops tend to space troops apart within ranges which may overlap extensively.

In the evolution of human behavior, hunting is the best clue to the size of the range and the area which is defended from strangers (Washburn and DeVore 1961a). The pattern of core areas around water, within a larger range, described here for baboons, is analogous to the pattern of land use by primitive hunter-gatherers in savanna country today. The major difference between baboon range and that of human hunters is the vastly larger area which humans, like the other large carnivores, must control. The aggressive protection of the hunting territory by humans also contrasts with the behavior which spaces baboon troops apart. African bushmen and Australian aborigines range over a hunting territory of from 100 to 1200 square miles. A range of this size is far more comparable to the ranges of wolves, wild dogs, and large felines than to the small ranges of the nonhuman primates. Within these large ranges, camp sites near water sources correspond to the core areas of baboon troops. Access to the resources within the core areas of these hunters is rigidly controlled by social custom, religious sanction, and the force of arms. Interband relations between human hunters distinguish between "friendly neighbors" and strangers, a distinction which has an ancient, prelinguistic basis in primate behavior. The most striking difference between the social organization of baboons and human hunter-gatherers is the closed social system of the former and the rules of local exogamy which are usually found in the latter. Although formal rules of exogamy depend upon the presence of language, the exogamous pattern itself may have arisen during the shift to a hunting economy by men of the early and middle Pleistocene (Washburn and DeVore 1961a).

DIET

A more detailed description of baboon diet is in preparation; the following discussion outlines the range of baboon foods and the relation between food supply and the troop. Baboons, like the macaques of Asia, eat a wide variety of foods. Although the bulk of their diet is vegetable food, they will also eat insects, eggs, and an occasional small mammal. Most of the Nairobi Park, where baboon foods were collected,[4] is grassland with some scattered trees. *The Nairobi Royal National Park Guide Book* contains an excellent brief description of the flora and fauna of the park. Although the western edge of the park is dry semievergreen forest verging on woodland (the range of the Kisembe Valley troop, Fig. 7), observations were largely confined to the open grassland country, and the description of baboon diet which follows includes few food items from the forest habitat. A study of the foods eaten by baboons in forest areas would be necessary before the full range of baboon diet in this area could be known.

Vegetable Foods

The diet of baboons living in the savanna of Nairobi Park can be divided into: the vegetable foods which provide forage for them throughout the year, seasonal fruits, insects, and the live animals which they occasionally catch and eat. Grass is the baboon's single most important food. In ten months of observations, not a single day passed in which baboons were not observed eating grass, and for many weeks during the dry season, grasses composed an estimated 90 per cent of their diet. The portion of the grass eaten varies with the season. When the tassels contain seeds, these are "harvested" by pulling the tassel through the closed palm or clenched teeth. Most often, however, baboons pull up the grass shoots in order to eat the thick, lower stem at the base of the culm. Before eating the shoot, the dirt in the root system is carefully brushed away, and the roots themselves bitten off and discarded (DeVore and Washburn 1960). By the middle of the dry season, when grass shoots are rare, baboons concentrate on digging up rhizomes—the thick, rootlike runners of the grasses which lie from two to four inches beneath the surface (Fig. 10). Even after many weeks or months without rain, these rhizomes are still juicy, providing baboons with considerable water. The ability of baboons to shift to subsurface rhizomes and roots when surface vegetation is dry and sparse is one of their most important adaptations to the grasslands. It enables them to feed in an area which has been denuded of surface vegetation by the many ungulates with whom they share this habitat, and to find sufficient forage during long dry seasons. Digging these rhizomes out of the hard, dry soil with the fingers is a laborious task, and in

4. Identifications of plant specimens were made by B. Verdcourt, Botanist In-Charge, the East African Herbarium, whose assistance is gratefully acknowledged.

FIGURE 10

(a) An adult male digs for subsurface roots, tubers and rhizomes during the dry season (Nairobi Park; DeVore). (b) Close-up of a female's hands, showing rhizomes she has uncovered (Nairobi Park; DeVore).

the dry season baboons spend longer hours getting their food than they do during the rest of the year. The use of a simple digging stick or sharp stone would enormously increase their efficiency in extracting this food from the ground, but no baboon was ever seen trying to use a tool in this or any other way.

There are numerous plants on the Nairobi plains which have large, tuberous roots or bulbs, and the baboons are very adept at finding the tiny stem or leaf which indicates that such a root lies below. It may take as long as twenty minutes for a baboon to uncover a large root, and require a hole as large as 24 inches long, 8 inches wide, and 15 inches deep. Where the water table is high, along the rivers in Nairobi Park and around the water holes at Amboseli Reserve, the lush grasses attract many animal species, including baboons (Fig. 11). Not only is the grass more plentiful here during the dry season, but also the earth is softer and more easily dug and many water plants are found which grow nowhere else in the area. Baboons spend the majority of their time feeding in the grass near the water, but they will also wade into the shallow water to eat such plants as rushes and the buds of water lilies.

The baboon's usual diet is further extended by the various bushes, flowering plants, and shrubs of the savanna. In Nairobi Park they were seen eating the berries, buds, blossoms, and seed pods of such plants. Another very important source of food throughout the year is provided by the acacia trees. Probably the buds, blossoms, and beanlike seed pods of all acacias are eaten, but those of the fever trees (*A. xanthophloea*) are particularly important. Not only is this species used almost exclusively as sleeping trees, but when they are in the height of their bloom the baboons also usually feed in them for one or more hours before starting their morning round, returning in the afternoon for another heavy feeding period at dusk. Out on the plains the ant galls on the short whistling-thorn trees (*A. drepanolobium*) are constantly plucked for the ants inside, and extrusions of its sap are eaten as well (Fig. 12). Some edible portion—bud, flower, seed pod, sap—of one of the types of acacia tree will be available within a troop's range at almost any time of year, and acacias are second only to grasses in the quantity of food they provide for Nairobi Park baboons. In addition to the plants and trees which provide forage for baboons all year, certain seasonal foods may constitute the bulk of their diet for short periods. The most important source of these seasonal foods in Nairobi Park are fig trees. When large fig trees are in fruit, the baboons may also use them as sleeping trees.

The most important food sources in the park are the grasses, acacia, and fig trees, but despite the frequency with which they feed in these trees, baboons were never seen eating tree leaves. On the southeastern slope of Mt. Kilimanjaro, baboons were observed feeding on the forest floor, while vervets (*Cercopithecus aethiops* and *C. mitis*) fed in the lower branches of adjacent trees. Leaf-eating *Colobus* monkeys occupied the canopy of the same forest. Their ability to find

FIGURE 11
(a) Zebra, warthogs and baboons feed on the vegetation at an Amboseli water hole (Washburn). (b) Zebra, impala, warthogs, baboons and an Egyptian goose (left foreground) at an Amboseli water hole (Washburn).

FIGURE 12
An adult female gathers gall from an acacia tree (*Acacia drepanolobium*). The ants are eaten and the husk of the gall spat out. Note crouched stance (Nairobi Park; DeVore).

food both on the open plain and in the trees is a distinct advantage for the baboons. Although they compete with a wide variety of ungulates for their food on the plains at Nairobi Park, their only close competitors in the trees are the vervets. Vervets and baboons are commonly seen feeding in adjoining trees in the park and occasionally they occupy the same tree—the baboons on the lower branches and the vervets in the canopy.

In addition to the staple diet, other vegetable foods were frequently eaten when they were available. These included "kei-apples," croton nuts, sisal plants, mushrooms, and the produce of native gardens (potatoes, yams, bananas, beans, maize, peanuts, sugar cane, etc.). Since almost all cultivated plants in this area have been imported from the New World, it is clear that baboons are very eclectic in their food habits.

Insects

Baboons eat many types of insects when they can find them, but the climate of Nairobi Park with its dry season, its hot days and cool nights does not support a very heavy insect population. The most common insect eaten in the park is the ant living in the galls of the *Acacia drepanolobium* trees. The amount of ants eaten in this way, however, is very small compared to the grasses and plants eaten during the same feeding period. If the troop is walking slowly through an area strewn with large stones, some of these may be turned over and the ground beneath them examined carefully. Under such stones an occasional beetle, slug, or cricket will be found and is quickly eaten. Rarely, an ant nest is uncovered, and the baboon bends over and licks up the contents of the nest from the earth, licking additional ants from its hands and arms afterward. But the baboons' attitude toward insects is one of mild interest, and no troop was ever seen moving from its pathway to systematically turn over the stones in an area.

Besides the ants in acacia galls, a baboon most frequently eats the grasshoppers which it finds on the branches of the bushes or blades of grass where it is feeding. Young baboons are seldom able to capture grasshoppers, but an adult will move the hand cautiously and deliberately to within one or two feet of the insect, then grasp it very quickly in a movement which is usually successful. Not all insects encountered are eaten. When a rock is overturned, some beetles and centipedes are ignored while others are carefully selected. Too few instances were observed to be able to say whether such selection was by individual preference, or whether these insects were avoided by all baboons in the park.

Although insect food is minor in the over-all baboon diet, a very heavy infestation of "army worm" caterpillars in the park showed that for short periods insects can become the baboons' most important food. Beginning in early April, during the rainy season, army worms appeared in the park in large numbers. For about ten days the baboons ate little else. Feeding on the worms in a small area were: three baboon troops, totaling 188 animals; several troops of vervet monkeys, perhaps 75 in all; and a group of about 300 Marabou stork (*Leptoptilos crumeniferus*). The different baboon troops fed very near each other, and the other animals, without incident. All were gorging themselves on the caterpillars; several baboons were timed picking up 100 army worms per minute, and continuing at this rate for from 10 to 15 minutes without a break. The eating of insects, in addition to the extensive inventory of vegetable foods, further increases the dietary adaptability of the baboon.

Live Animals

On six, perhaps seven, occasions during the twelve months of study in Kenya and the Rhodesias, we saw baboons eating freshly killed animals. Twice they caught and ate half-grown African hares (*Lepus capensis crawshayi*). On the first occasion the male in possession of the hare was being harried not only by two more dominant males in this troop, but by a pair of tawny eagles (*Aquila repax raptor*) as well. The male in possession eluded his harassers and managed to consume most of the hare, the eagles retrieving scraps of viscera and skin. In his haste the baboon dropped the rib cage and a foreleg of the hare, with most of the flesh still attached, *but these pieces were ignored* by the other two baboons chasing him, despite their desire to obtain his catch.

Two or three times baboons were seen eating fledgling birds of some ground-nesting species, probably the crowned plover (*Stephanibyx coronatus*). On several occasions they chased fledglings some yards through the grass without catching them. We never saw baboons finding and eating eggs, but when offered a dozen guinea fowl eggs, they ate these without hesitation. Entire eggs were stuffed into the cheek pouches and the shell broken by the hand pressing the cheek against the teeth and jaws. More significant than the few instances of baboons' eating fledglings are the numerous times when baboons were seen feeding across a plain covered by bird nests without discovering the contents of a single nest. The same animals which are able to detect an underground root from only a tiny dried shoot on the surface will walk beside a bird nest six inches in diameter without noticing it. Furthermore, four species of weaver bird inhabit the park, and their nests are frequently clustered in the acacia branches where the baboons are eating, but no baboon was ever seen investigating such a nest, much less eating its contents (Fig. 13). The baboon's attitude toward food is clearly vegetarian. It is common to see a baboon troop completely mingled with a flock of guinea fowl without incident. The only eggs or fledglings which they seem to recognize as food are those which are literally stepped on as the troop searches for vegetable foods on the plains.

On December 14, near the close of the study, two very young Thomson gazelle (*Gazella t. thomsonii*) were caught and eaten by the adult males of a troop. The actual capture of the second gazelle was seen. An adult male baboon grabbed it, brought it above his head, and slammed it to the ground. He immediately tore into the stomach of the gazelle and began eating. Beginning with the most dominant males, five of the six adult males in the troop participated in eating this gazelle, and two hours later only skin, teeth and large bones remained. The viscera were eaten first, followed by the flesh, and finally the thin brain case was bitten open and the contents carefully scooped out with the fingers—bits of skull being pulled through the teeth and licked clean. The incisors, not the canines, were used in biting and tearing at the flesh (Fig. 14).

FIGURE 13
Although they frequently feed in acacia trees where weaver bird nests are abundant, baboons were never seen investigating the nests (Nairobi Park; DeVore).

These two Thomson's gazelle were apparently only a few days old, and were hiding in the grass some 150 yards from the herd of 38 with which they were no doubt associated. After the baboon troop moved on, two females from the herd of gazelle (of 35 females, 2 young, and one adult male) came over and paced nervously around the remains of the carcasses. It seems reasonable to assume that the discovery of these two young gazelle took place under circumstances very similar to those involved in the eating of the young hares, that is, that they were discovered accidentally in the grass. In fact, after the first gazelle had been found, and four of the males were pressing its possessor closely, the males passed within five yards of an African hare sitting in plain view. They clearly saw the hare but did not even walk over toward it.

FIGURE 14
(a) An adult male eats a newborn Thomson's gazelle. Note the use of incisors in biting (Nairobi Park; DeVore). (b) A male baboon bites into the brain case of another newborn Thomson's gazelle caught during the same afternoon as that shown in Figure 14a. Only the adult males had a chance to eat this meat (Nairobi Park; DeVore).

All these cases of flesh eating have one thing in common—they involve the eating of immature animals whose defense is to hide "frozen" in the grass, and in each case their discovery by the baboons seemed fortuitous. Nothing resembling a systematic search of an area or the stalking of prey was ever observed, nor was fresh meat eaten except when it was found alive or taken up immediately by a waiting baboon. Since baboons avoid lion kills when they are away from trees and other carrion is not eaten, the lack of interest shown by the male in the portion of hare which had been dropped (described above) may be due to their avoidance of carrion. It is also possible that baboons do not recognize as edible any meat which is not alive and easily caught. In either case it seems clear that their attitude toward other animals is not that of a predator, nor do the scores of other species with which they live peacefully so regard them.

The final instance of meat eating was observed in Amboseli Reserve. While watching baboons in an open area, we heard loud screeches and chattering in a tree where baboons and vervets had been feeding peacefully for the previous hour. When we approached the tree we saw an adult male baboon walking through the branches with a juvenile vervet dangling from his mouth, and the vervet troop had left the tree. The baboon consumed most of the vervet, carrying·the carcass in his mouth as he walked toward the troop's sleeping tree at dusk. This observation is in striking contrast to the many occasions when the two types of monkey were seen feeding peacefully together. During a brief aggressive interaction between the two species in Nairobi Park, DeVore saw an angry adult male baboon put a troop of vervets to rapid flight, and this case of meat eating may have been the incidental result of such a situation in the tree at Amboseli. Although Washburn saw baboons chase vervets quite frequently near Victoria Falls, he only once saw a baboon catch one. This was held in the mouth by the female who caught it. She was apparently bewildered by the situation and soon released it unharmed. In much the same way one of the fledglings DeVore saw eaten was actually caught by a juvenile baboon, which seemed puzzled by the object and quickly relinquished it to an adult male (who promptly ate it).

In summary, baboons may be described as very inefficient predators. Meat eating, to judge by the bewildered state of the female baboon who caught a vervet and of the young juvenile who caught a bird, would appear to be learned by each generation, and meat never becomes an important source of food for the whole troop. Only one baboon other than adult males (an adult female) participated in the eating of meat in any of the instances observed during the study. Accounts of meat-eating in captive baboons are contradictory. Kenya baboons kept near Nairobi Park ate meat readily, but Bolwig (1959) found that his captives refused it. In South Africa, where most reports of carnivorous baboons have originated, baboons are only now being systematically studied (Hall 1960, 1961), and we feel that the importance of meat in the baboon diet has been considerably overstressed. The usual reason given for the habit of meat-eating in South African baboons is that the hardship of drought creates the conditions under which it flourishes (e.g., Dart 1953), but when the two Thomson's gazelle were eaten in December the park was well into the rainy season, and the vegetable foods baboons ordinarily eat were more abundant than at any other time of year.

It would seem more reasonable to us, on the present evidence, to assume that meat has been a consistent but very minor part of the baboon diet throughout their evolutionary history. In localities where sources of animal protein can be obtained without danger, baboons apparently include these in their regular diet. At Murchison Falls, baboons are often seen digging out and eating crocodile eggs. Hall's description of the foods eaten by baboons along the coast of South Africa is very similar to the inventory of vegetable and insect foods discussed here, except that the South African baboons also eat marine foods such as mussels, crabs, and sand hoppers found along the beach. But baboons are ill fitted anatomically to be carnivores, and too great a dependence on meat eating could have been detrimental to their wide exploitation of the vegetable foods they depend upon today. By their utilization of a wide variety of plant and tree products, baboons have been able to spread over the African continent, and, together with the macaques, to cover most of the tropical Old World.

In the evolution of the human species, meat-eating played a very different role. We have suggested that the earliest hominids may have been living on a diet very like that of the baboons, that is, vegetable foods supplemented by an occasional small animal (Washburn and DeVore 1961a). The freedom to carry a simple digging implement in the hands would greatly enhance this adaptation. During the dry season in Africa, human hunter-gatherers are also very dependent on the subsurface roots and tubers sought by baboons. A digging stick greatly improves the humans' chance for survival during this period of food shortage, and it may be that the presence of baboon skeletons at Olorgesaille indicates the result of competition between baboons and humans over a limited food supply. It would be an easy step from killing baboons to protect a source of vegetable foods, to killing them for meat.

Scavenging

Scavenging has been regarded as an important phase in the evolution of man's carnivorous habits. It seems reasonable that a primate liking eggs, nestling birds, insects, and an occasional small mammal might add to this diet and develop more carnivorous tastes and habits by gleaning meat from kills. This theory seemed reasonable, and we made a particular effort to examine kills and to observe the relations of the baboons to them. Although we saw over a dozen recent kills (including gnu, giraffe, zebra, waterbuck, impala, Grant's gazelle, warthog, Masai cattle, and goat) and have thorough records on some, we were primarily looking at baboons. The subject of scavenging is so important, especially in the interpretation of the deposits in which *Australopithecus* is found, that a much more comprehensive study is needed. However, here are our tentative conclusions.

The scavenging theory is not supported by the evidence, and primates with habits similar to those of baboons could get meat by hunting far more easily than by scavenging. There are several reasons for this. The first is that most kills are made at night and are rapidly and thoroughly eaten. When the hyenas leave at dawn, the vultures locate the remains and clean the last meat from the bones. Some kills are made by day. We saw the remains of a gnu which a pride of ten lions finished in an hour. A pride of four lions (two not fully grown) killed a gnu one afternoon and ate almost all of it in one night. The vultures finished the rest, and the bones were undisturbed for three days. Many bones disappeared on the fourth night. Similarly, we saw two lions eat a warthog, three lions eat a Grant's gazelle, and five cheetahs kill and eat an impala. Only the meat of very large animals is left for long, and Africa is well supplied with highly efficient scavengers which leave little meat to tempt a primate.

Actually there are far fewer kills than might be expected from discussions of scavenging. In the part of the Amboseli Reserve which we studied intensively there were on the order of 100 baboons to one lion. The lions move over large areas, and the chances of a troop coming on a "kill" are very few. We saw a troop around a kill left from the previous night only once in Amboseli. It had been largely eaten, and the baboons appeared to take no interest in it. During nine months of observation in Nairobi Park, baboons were seen to pass near four kills and paid no attention to the few scraps of meat left on them. A Grant's gazelle carcass, presumably a leopard kill, hung in a fig tree where baboons ate and slept, but the baboons apparently ignored it. In addition, they did not attempt to eat fresh carrion when this was found. A further complicating factor is that when there is much meat left, the lions usually stay nearby, and the neighborhood of the kill is very dangerous.

In summary, the chances of a kill within the range of a baboon troop are very small; little meat is likely to be left; and the vicinity of the kill is dangerous. Most of the killing and eating is at night, and primates have neither the sense of smell of the hyenas nor the eyes of the vultures to locate the kill. As noted earlier, the baboons seem uninterested in dead animals. A slight increase in predatory activity against young animals would yield a far greater reward than scavenging, would be much less dangerous, and would represent a smaller change in habit. The use of a stick or stone for digging would increase the baboons' food supply more than any other simple invention. Perhaps in *Australopithecus* we see a form which had such a tool to exploit vegetable foods and which also used this tool as a weapon. If tools were being used at all, their use in the deliberate killing of small animals would be only a small change from the behavior observed in baboons. Once man had become a skilled tool-user in these ways, he could extend tool use to the hunting of large animals, to defense, and to driving carnivores from their kills. Scavenging may have become a source of meat when man had become sufficiently skilled to take the meat away from carnivores, but the hunting of small animals and defenseless young is much more likely to lie at the root of the human hunting habit.

DISCUSSION

In this paper we have tried to stress those aspects of baboon ecology which are of the greatest help in understanding human evolution. Obviously, man is not descended from a baboon, and the behavior of our ancestors may have been very different from that of living baboons. But we think that in a general way the problems faced by the baboon troop may be very similar to those which confronted our ancestors. At the least, comparison of human behavior with that of baboons emphasizes the differences. At the most, such a comparison may give new insights. Many topics have been summarized above, and in this discussion we will call attention only to a few major points.

The size of baboon troops may exceed that of hunter-gatherers and, their population density far exceeds that of primitive man. The human group differs in being exogamous, so that many local groups form the breeding population. We believe that this radically different breeding structure has exerted a profound effect on the later phases of human evolution and has long been a factor in preventing speciation in man.

The social structure of the baboon troop is important to the survival of the species. Survival depends on the adult males being constantly close to the other troop members. Roles in the troop are divided between the sexes, but these are in the context of a compact troop. With man, the hunters leave the local group, sometimes for days, and then return to their home base. Such a pattern is radically different from anything known in monkeys or apes. Hunting with tools basically changed the social structure of the band, the interrelations of bands, the size and utilization of range, and the relation of man to other animals.

Diet has already been discussed and we will not repeat here, except to point out that our opinion of the importance of scavenging has changed through observation of the actual situation at the kills. It is not enough to speculate that scavenging might have been important. One must estimate how much meat is actually available to a vegetarian, and how dangerous it is to get meat by scavenging.

Finally, we would stress that survival is a complex process, and that all the factors which lead to reproductive success must ultimately be considered. Varied diet, social structure, and anatomy, all are important, but their meaning only becomes clear as they are seen making possible the behavior of a population. Sex differences, peripheral animals, and range—each of these has meaning only in terms of the survival of groups. With the coming of man, every major category is fundamentally altered and evolution begins to be dominated by new selection pressures. Some measure of how different the new directions are may be gained from the study of the ecology of baboons.

BIBLIOGRAPHY

Bolwig, N.

1959. "A Study of the Behaviour of the Chacma Baboon, *Papio* ursinus," *Behaviour*, 14 (1–2):136–63.

Burt, W. H.

1943. "Territoriality and Home Range Concepts as Applied to Mammals," *J. Mammal.*, 24:346–52.

CARPENTER, C. R.
1958. "Soziologie und Verhalten freilebender nichtmenschlicher Primaten," *Handbuch der Zoologie*, 8(10)1–32.
CHU, E. H. Y., and M. A. BENDER
1961. "Chromosome Cytology and Evolution in Primates," *Science*, 133:1399–1405.
DART, R. A.
1953. "The Predatory Transition from Ape to Man," *International Anthropological and Linguistic Review*, 1:4.
DEVORE, I.
1962. "The Social Behavior and Organization of Baboon Troops." Unpublished Ph.D. dissertation, University of Chicago.
DEVORE, I., and S. L. WASHBURN
1960. "Baboon Behavior." 16 mm. sound color film. Berkeley: University Extension, University of California.
GILLMAN, J., and CHRISTINE GILBERT
1946. "The Reproductive Cycle of the Chacma Baboon, *Papio ursinus* with Special Reference to Problems of Menstrual Irregularities as Assessed by the Behaviour of the Sex Skin," *So. Afr. J. Med. Sci.*, 11 (biological supplement):1–54.
HALL, K. R. L.
1960. "Social Vigilance Behaviour of the Chacma Baboon, *Papio ursinus*," *Behaviour*, 16 (3–4):261–84.
1961. "Feeding Habits of the Chacma Baboon," *Advancement Sci.*, 17 (70):559–67.
KAUFMANN, JOHN H.
"The ecology and social behavior of the coati, *Nasau narica*, on Barro Colorado Island, Panama." *Univ. of Calif. Pub. in Zool.*, in press.
(NO AUTH.)
1957. *Nairobi Royal National Park Guide Book*. Nairobi: The Trustees of the Royal National Parks of Kenya.
PHILLIPS, W. W. A.
1935. *Manual of the Mammals of Ceylon*. London: Dulan and Co.
STARCK, D., and H. FRICK
1958. "Beobachtungen an äthiopischen Primaten," *Zoologische Jahrbücher*, 86:41–70.
WASHBURN, S. L., and I. DEVORE
1961a. "Social Behavior of Baboons and Early Man." In *The Social Life of Early Man*, S. L. Washburn (ed.), (Viking Fund Publications in Anthropology), 31, pp. 91–105.
1961b. "The Social Life of Baboons," *Scient. American* (June 1961), 204 (6):62–71.

OBSERVATIONS ON THE ECOLOGY AND SOCIAL BEHAVIOR OF THE MOUNTAIN GORILLA

GEORGE B. SCHALLER AND JOHN T. EMLEN, JR.

THE PURPOSE OF THIS PAPER is to present a brief summary of personal observations on the ecology and social life of the mountain gorilla (*Gorilla gorilla beringei*), and to a lesser extent on the chimpanzee (*Pan troglodytes schweinfurthii*) and orangutan (*Pongo pygmaeus*), as a basis for some comments on hominid origins and social behavior.

The gorilla data are based on a six-month ecological and distributional survey in the eastern Congo and Uganda from February to July, 1959 (Emlen and Schaller 1960), followed by fifteen months of intensive field work by Schaller in selected localities. The detailed observations on social behavior were made on a population of ten groups in an undisturbed section of the Virunga Volcanoes of Albert National Park, Congo, unless otherwise stated. Gorillas were there encountered 306 times and observed directly for 457½ hours. Most animals were recognized individually and several groups were habituated to the presence of the observer.

Two weeks were spent exclusively on chimpanzees, primarily in the Maramagambo and Budongo Forests of Uganda. Contact with orangutans was limited to 4 encounters with 8 animals in Sarawak over a two-month period (Schaller 1961). Full descriptions of gorilla and other ape behavior have been published elsewhere; this paper presents only those aspects which have the greatest interest from a comparative primate standpoint.

The study was financed by the National Science Foundation and the New York Zoological Society, with the latter institution acting as sponsor. Local sponsors of the expedition were the Institute of the Parks of the Congo, Makerere College in Uganda, and the Sarawak Museum. To these institutions, and the numerous persons who helped us, we are extremely grateful.

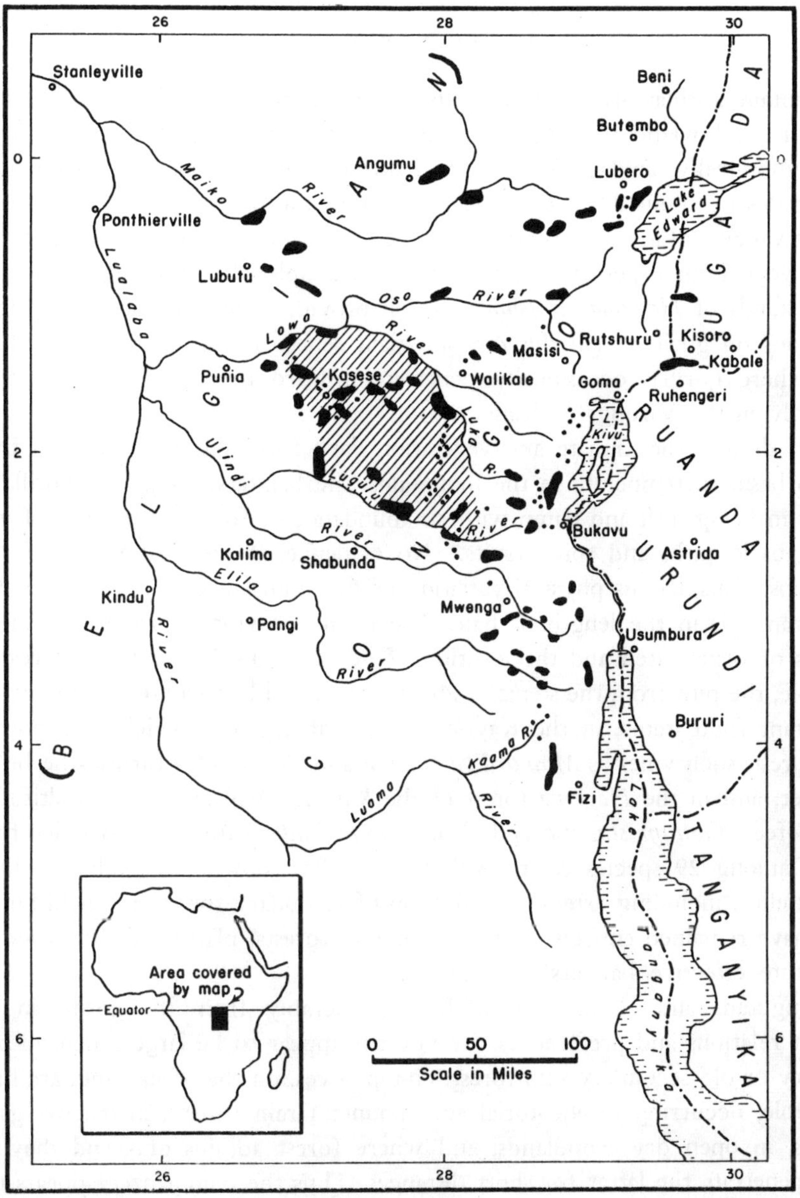

FIGURE 1
Geographic distribution of the mountain gorilla—1959. The black areas indicate the location and approximate shape of "gorilla areas" ranging in size from about 10 to 200 square miles each. The small dots represent records from outside these areas. The hatching marks a central region of fairly continuous but sparse population. All gorilla habitat west of about 28° 15′ E. lies in lowland rain forest; most of the habitat to the east of the longitude lies in mountain rain forest or in bamboo. (Reprinted from Emlen and Schaller 1960, with permission of the New York Zoological Society.)

ECOLOGY

Mountain gorillas are forest animals. Over the major portion of their range they inhabit lowland rain forest and at somewhat higher altitudes they occupy physiognomically similar mountain rain forest (Fig. 1). Bamboo, which occurs at altitudes of about 8,000 to 10,000 feet along the top of the western edge of the rift valley and in the Virunga Volcanoes, is a habitat of secondary importance. Other vegetation types are very local. For example, the rather open mountain forest stands of *Hagenia abyssinica*, which provided the best conditions for observations on gorillas, and the zone of giant senecios and lobelias above 11,500 feet, where gorillas occasionally penetrate upward to 13,500 feet, are found primarily in the Virunga Volcanoes.

Even though the climate and vegetation vary from distinctly tropical in the Congo basin to temperate in the mountains, the habitats utilized by gorillas are similar in being lush and damp with an abundance of food in the form of vines, leaves, bark, pith, and some fruits. (No evidence of meat-eating was found.) The most conspicuous physical variation of the animal over this extensive altitudinal range is in the length of hair. The main behavioral variations are in the species of plant eaten and the location of nests. In the lowland rain forest, for example, the pith from the stems of *Aframomum* and banana are two of the most important food items; in the Kayonza forest of Uganda, which is a mountain rain forest, such vines as *Momordica foetida* and *Basella alba* furnish the bulk of the diet; and in the *Hagenia* forest of the Virunga Volcanoes, at an altitude of 10,000 feet, *Galium simense* and *Peucedanum linderi* are the two chief forage plants among 29 species eaten. Although gorillas show considerable ecological adaptability, including extensive utilization of secondary forests and field borders, they have remained entirely within the humid forests; plains and dry woodlands constitute effective barriers to dispersal.

Orangutans and chimpanzees differ considerably from the gorilla in their habitat relations and preferences. Orangutans appear to be largely dependent on primary or old secondary rain forest; chimpanzees, on the other hand, are highly adaptable, occurring in equatorial and mountain rain forests, in narrow gallery forests, in open dry woodlands, and where forest adjoins grassland they may even penetrate the latter for short distances. Thus the chimpanzee appears to be the most adaptable of the great apes with regard to habitat.

LOCOMOTION

Gorillas are primarily quadrupedal and terrestrial. Extensive bipedal locomotion is rare in nature: we saw it only three times over distances ranging from 15 to 60 feet. Although the animals climb readily into trees to feed, nest, and for other activities, all of them, but especially the adults, do so with relative caution. They rarely run or jump around. When danger threatens they descend and flee on the ground, which to a lesser degree is also true of the chimpanzee but not the orangutan.

Chimpanzees are likewise quadrupedal, but only semiterrestrial. In agility and climbing ability they are only exceeded by the gibbon and siamang among the apes. They run and jump through the trees with great rapidity. One excited animal was seen to jump 30 feet from a nest to the ground, and leaps from branch to branch sometimes exceeded 8 feet. They may spend considerable time on the ground, especially in dry woodland habitats.

Orangutans rarely descend trees. They are fairly slow and careful quadrupedal climbers whose top speed usually does not exceed two to three miles per hour.

While it may be appropriate to class the apes as brachiators on anatomical grounds (Washburn and Avis 1958), it must be emphasized that in the wild only the gibbons and siamangs brachiate extensively; all of the great apes are essentially quadrupedal. We never saw gorillas brachiate in the wild, although they were occasionally seen to hang by their arms and reach for another branch while their legs hung free; chimpanzees have been reported to brachiate for short distances along a horizontal branch; and orangutans were observed to brachiate briefly at times but their main mode of progression was by quadrupedal climbing. Such commonly repeated statements in the literature as: "Only the anthropoid apes among the Primates are full brachiators, that is, their movements in the trees are exclusively arm-swinging movements and never quadrupedal" (Le Gros Clark 1960), should thus be modified.

GROUP SIZE, COMPOSITION, AND CHANGES IN COMPOSITION

The ten gorilla groups under close observation in the Virunga Volcanoes varied from 5 to 27 animals each, with some subsequent changes that raised the number of one group temporarily to 30. Average group size was 16.9 animals. It was noted, however, that groups in other areas tended to be on the average somewhat smaller, often consisting of only 3, 4, 5, and 6 animals each, a fact which seems to be at least in part correlated with habitat and predation by man.

Each group contained at least one silverbacked or adult male (about 10 or more years old), but some groups held 2, 3, or even 4 for varying periods of time. Adult and subadult females varied from only one in the smallest groups to as many as 12 in the largest ones. In addition most larger groups contained one or more blackbacked or subadult (?) males (about 6 to 10 years old) and a variable number of juveniles (about 3 to 6 years) and infants (0 to 3 years). Lone blackbacked and silverbacked males were commonly found, sometimes over 20 miles from the nearest group.

Gorilla groups are quite cohesive in that the members rarely drift far from each other. The diameter of feeding or resting groups is usually 200 feet or less, and only infrequently are the animals spread over 300 to 400 feet of forest.

Changes in size and composition occur quite commonly in some groups but only rarely in others. Two illustrations of the type of changes occurring in gorilla groups are given below.

1. Group IV was first observed on March 12, 1959, and detailed records of the composition of this group were collected periodically from August, 1959 to August, 1960.

Composition (August, 1959): 4 silverbacked males, 1 blackbacked male, 10 females, 3 juveniles, 6 infants (1 infant was born on March 12, 1959) = 24.

August 28 to August 30: A peripheral silverbacked male apparently left, and then rejoined the group.

September 9: A new silverback joined the group. The peripheral male of August left again but rejoined by September 22.

September 18–20: An infant was born.

Between October 2, 1959 and January 11, 1960: The peripheral male and the No. 2 male in the hierarchy left.

A new silverbacked male and two females, both with infants, were added. They probably represent a small group which joined.

An infant was born in late December.

April 24, 1960: A silverbacked male, who has been with the group at least since August, 1959, left.

April 25, 1960: An infant was born but died two days later.

Between May 1 and 15: A new silverbacked male joined the group and remained at least to May 24.

Between May 24 and August 12: The male who joined in early May left.

Thus in the course of 12 months at least 7 different silverbacked males associated with the group, but of the 4 present in August, 1959 only the dominant male remained one year later. Three infants were born and two females with infants were added. At the last encounter the composition was: 3 silverbacked males, 1 blackbacked male, 12 females, 3 juveniles, 10 infants = 29.

2. Group VII, which was studied in greater detail than any other group from October, 1959 to September, 1960, showed relative stability.

Composition (October, 1959): 1 silverbacked male, 2 blackbacked males, 6 females, 4 juveniles, 5 infants = 18.

Early February, 1960: An infant was born.

Between February 14 and March 16, 1960: An unknown female with infant joined the group, raising the group total to 21. (This was the only instance in which a single female was noted to join another group). No lone males associated with this group.

These examples indicate that established groups are fairly stable social units but that they may be augmented from time to time by the addition of individuals or other groups. Lone males may freely come and go in some groups, but apparently not in others.

No long-term observations on chimpanzees or orangutans were made, but the information obtained points to some differences. Whereas gorillas tend to remain in rather cohesive groups, chimpanzees are less closely knit. In one group of seemingly only two animals, the male and female were 300 feet apart. In another case a female with infant was apparently alone in a gallery forest. On the other hand, one temporary aggregation comprising 2 and perhaps 3 groups was seen, and probably at least 50 animals were spread over about a half mile of forest.

The orangutan forms small and seemingly unstable social groups. A perusal of the scanty literature and personal observation revealed that groups of more than 4 are rare, and that lone animals, both male and female, and groups of 2 and 3 are the most common, and even these appear unstable. For example, a female with small infant, which Schaller observed in Sarawak, fed alone in a tongue of forest. An examination of her nest site of the previous night revealed that a subadult had accompanied her, but it could not be found. Natives maintained they had also seen a male with the group two days previously, a statement which was corroborated by an examination of other nest sites. The most prevalent group combinations appear to be single adult females with one or 2 subadults; pairs with one male and female; and groups of 2 or 3 subadults.

HOME RANGE AND TERRITORY

Observations on peaceful interactions between distinct groups and the great areas of overlap of group range indicates that gorillas have no territory in the sense of an area exclusively held or defended against others of the same species. However, gorilla groups do restrict their activities to definite home ranges. The same range may be frequented by several groups in part; six different groups were seen in one small section of forest during the period of the study. Daily tracking of groups, sometimes for as long as 25 consecutive days, revealed average daily movements of a little less than half a mile, with variation from about 300 feet to over 3 miles. The route taken and distance traveled varied from day to day and did not follow a set pattern (Fig. 2). Home ranges of groups were on the order of about 10 to 15 square miles each.

SUBGROUPS AND JOINING OF GROUPS

Subgrouping or the temporary splitting of one group into two distinct units is a rare phenomenon in gorillas. Only one instance was noted, and in this the group split in two and united again later in the same day. Circumstantial evidence from trails and incomplete encounters pointed to several other instances of temporary splitting, but in any case it is an infrequent event.

Carpenter (1958) proposes that the agonistic territorial behavior which he observed between neighboring groups of several primate species is typical of primate societies, and on this basis suggests that merging of distinct groups does not occur. Overlapping of home ranges brings gorilla groups into close contact on occasion. In 12 definite meetings noted during the study, the groups usually passed each other slowly at distances of 300 feet or less, or sat near each other for varying lengths of time without actual interchange of individuals. However, partial or complete mixing of the two groups occurred at least 4 times. Once 2 groups remained near each other for 3 days before they were seen to mingle briefly (Fig. 2.) Most joining of groups was of brief duration, one day or less, but we have evidence that in one case two groups united for several months. Seemingly reliable data from an observer in Uganda indicates that the survivors

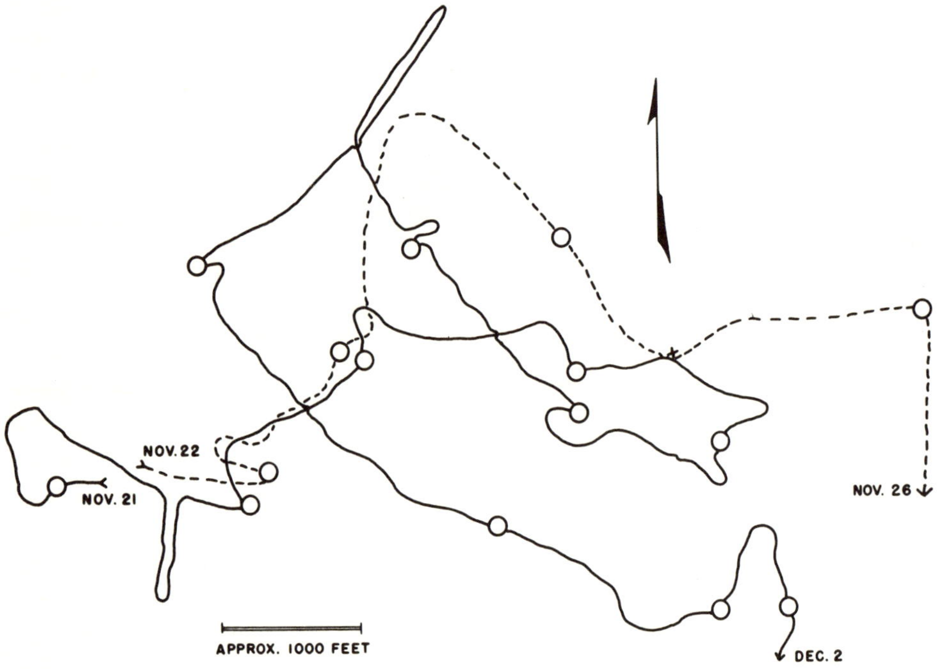

FIGURE 2

The daily movements of two gorilla groups on a steep slope of Mt. Mikeno in Albert National Park, Congo. Group VIII (solid line) was tracked from November 21 to December 2, 1959; group VI (dotted line) from November 22 to November 26, 1959. Circles indicate the night nest sites. The two groups nested near each other for two consecutive nights and they remained most of November 23 within sight of each other. They were observed to mingle briefly on November 25 (cross). The route of group VIII illustrates the typical daily meanderings of gorilla groups while frequenting a certain sector of the forest.

of a small group joined another one after the dominant male died. Agonistic displays in the form of threatening stares or bluff charges were seen between the dominant males of the respective groups on two occasions, but actual fighting was never observed.

INTRAGROUP RELATIONS

Although every gorilla has the possibility of interacting with every other member of its group, numerous factors tend to reduce the number of direct contacts between the animals. Forage is plentiful, competition for mates is apparently lacking, and numerous dominance interactions are simply avoided by circumventing situations which might produce them. In addition, adult gorillas possess temperaments which can best be described as self-contained or aloof, and physical contact with others is usually not sought. Thus, for example, mutual grooming was noted only .28 times per hour of observation, and dominance .23 times.

Dominance

Certain individuals commonly claim prerogatives in right of way along a trail and to a choice of sitting place and their right is recognized by others. Dominance was seen only once in competition for food, and it apparently did not feature in sex, although observations on the latter were limited to two copulations. In most dominance situations the subordinate animal merely received a look or a light touch with the back of the hand from the dominant one. Such signals are not disputed and no dominance fighting was ever observed.

To a large extent dominance is correlated with body size. All silverbacked males are dominant over all other animals; all females and blackbacked males are dominant over all juveniles and infants not in contact with their mothers; and among and between infants and juveniles size likewise seems to be the main correlate. If more than one silverbacked male is present in the group, there is a hierarchy among them. Females, however, seem to lack a definite hierarchy.

Leadership

The dominant male is also the leader, and, therefore, the focal point of the whole group. His leadership is almost absolute, for all other animals in the group, except sometimes the peripheral males, watch and respond to his every action—when he rests or nests, they do; when he travels, they follow. Differences in temperament and behavior of the dominant male thus affect the action of the whole group.

FIGURE 3
The dominant silverbacked male of group IV with six visible female and juvenile members of the group clustered near him. The dense tangle of herbs and vines in characteristic of the ground cover in the *Hagenia* forest. (Photograph copyrighted by George B. Schaller)

Sexual Behavior

Only two copulations and one invitation to copulate were observed. In one instance the female invited the male by presenting her rump, and once a female mounted a male briefly before copulation. All the males involved were subordinate individuals, and the dominant males never interfered aggressively. No instances of playmounting or any other type of sexual behavior were noted. The paucity of observations suggests that the sexual drive in gorillas is low. Chimpanzees, on the other hand, are more easily aroused to sexual activity if data from captive and semi-wild individuals are valid for comparison.

Mutual Grooming

Mutual grooming is rare between adult gorillas. No male was ever seen to groom another adult and only once was a male groomed by a female. Grooming between females is also infrequent, and they invite grooming by others only of areas of the body which they cannot themselves reach with ease. The invitation to grooming is indicated simply by presenting a certain part of the body to another animal. Thus mutual grooming among adults seems to have little or no social significance, merely a utilitarian one.

Most grooming activity involves a female and a younger animal. Females groom juveniles and infants readily; and occasionally a juvenile grooms a female, where the activity apparently functions in establishing social contact.

Mother-Infant Relationship

The development of one infant in the wild was traced from the day of birth to the age of 1½ years and that of several others for nearly one year. The ages of older infants and juveniles were estimated and are based primarily on comparative size. Infants are entirely helpless at birth and they must be supported by the arms of the female until about 3 months old, an age when they usually begin to ride on their mother's back for short periods. The first tentative crawling movements away from the mother occur between the age of 3 to 4 months and by 5 and 6 months other infants are actively sought in play. By one year of age, the infant wanders freely among the resting members of a group and it forages for its nourishment, with milk being now of secondary importance. By 1½ years it readily climbs around in trees high above the ground. By 2 years it may spend considerable time with other group members and even travel under its own power if the general movement is slow. Although the infant is more or less intimately attached to its mother to the age of about 3 years, the bond is not necessarily broken completely even then. Some juveniles frequently sit by a female, are groomed by her and sometimes sleep with her in the same nest until they are about 4 years old. This relationship persists in spite of the fact that lactation ceases by 1-1/2 or perhaps as late as 2 years and the female may have a new infant. In other words, a loose mother-child relationship continues well after the female has ceased to provide food and protection. Some male gorillas become independent of the group at about 6 years of age when they leave to lead a lone life in the forest, a habit never noted in females.

COMMUNICATION

Gorillas emit about 20 distinct sounds several of which apparently carry direct meaning or serve to attract attention to the performer and permit the communication of further information by means of gestures. Thus, a short, sharp *uh-uh* given by a male when females quarrel may cause them to subside. If, however, the male emits the same sound when the group is quietly resting or feeding all members first look at him and then face the direction which occupies his attention. Vocalizations in their daily routine are relatively infrequent and gestures usually serve to co-ordinate their actions and movements. The mere act of a

FIGURE 4
Four gorillas of group VI are in the branches of a *Hypericum* tree. A female with large infant sits in a crotch, a blackbacked male rests on the sloping trunk, and a female stands and beats her chest. (Photograph copyrighted by George B. Schaller)

male walking in a distinctive way in a certain direction suffices to indicate that he is leaving. There was no evidence of displacement or productivity, both considered by Hockett (1960) as distinctive of human communication systems; in fact, in their means of communication gorillas show no greater elaboration than many mammals. Early man could easily have co-ordinated his whole social existence by such simple means of communication as those employed by the gorilla. Even communal endeavors like game drives require only simple gestures as illustrated by the hunting techniques of wolves, hyenas and other carnivores.

NESTS

All great apes build nests, the orangutan exclusively in trees, the chimpanzee usually in trees, and the gorilla often on the ground. A detailed analysis of the nests indicates that the basic method of construction is the same in all three genera. Nests were not built in any special pattern, except that the branches were broken toward the animal to create a platform. They never included a constructed canopy, but gorillas were observed occasionally to nest beneath the leaning bole of a tree thus avoiding the rain. Gorillas never used the same nest two nights in succession. The breaking in of a few branches was an easy matter and usually required only one-half to 5 minutes.

Although gorillas nest both on the ground and in the trees, the percentage of ground to tree nests varies from area to area and is, at least in part, correlated with the availability of suitable trees in which to construct secure platforms.

FIGURE 5
Ground nest of a female mountain gorilla in the Virunga Volcanoes. The nest material consists primarily of the herb *Senecio trichopterygius*. Several sections of dung are visible on the nest rim and numberous flattened sections line the nest cup.

Location	*Total number of nests checked*	*Percent of nests on the ground*
Lowland rain forest (near Utu)	110	21.8
Mountain rain forest (Kayonza Forest)	179	53.6
Hagenia forest (Virunga Volcanoes)	2,488	97.0

As with other monkeys and apes, gorillas defecate wherever they are throughout the day and night, and this includes the cup of their nest on which they lie. The amount of dung deposited by a group per day is such that repeated use of a bedding area would soon make it unlivable. The first dwellers of caves or other creatures having a home base, whether they be ape or man-ape, must have learned to refrain from defecating until they had moved away from or to a certain part of the main living quarters, in the manner of various present-day rodents and carnivores.

CURIOSITY AND TOOL-USING

Gorillas are extremely curious to ascertain the nature of strange animate objects, and they approach a lone, quiet person sometimes very closely. But they seem to lack a comparable curiosity to investigate strange inanimate objects in their environment. A tin can, intestines of an antelope, a rucksack, paper—all were objects which lay directly in the path of gorilla groups on occasion without being handled. The animals showed little interest in fondling, pulling, or in tearing objects, a trait so prominent in chimpanzees in zoos.

Although Beatty (1951) and Merfield (1956) observed the use of tools by free-living chimpanzees, we saw no such behavior in gorillas. Young gorillas play with leaves and sticks and sometimes carry them around, and nesting material and food is occasionally transported by the animals for 15 to 20 feet, but adults showed no inclination to handle anything for the sake of manipulation alone. It should not be supposed, however, that quadrupedal gorillas have difficulty in handling objects, or in carrying anything such as a tool for long distances. Infants are supported by the female with one arm for at least 3 months and the female does quite well on her 3 limbs. The position of the fingers in quadrupedal locomotion is such that a stick or stone could be transported by the animals as easily as they sometimes carry a stalk of celery. Gorillas spend many hours daily on their haunches feeding or just sitting with their hands entirely free to manipulate objects. They can throw well with speed and accuracy in captivity. Gorillas, it would seem, are in many ways admirably pre-adapted to a life in which tool-using and hunting for meat could play a prominent role.

It may be conjectured that the gorilla's failure to develop tool-using is related to the ease with which it can satisfy its needs in the lush forest habitat. The forest is for these powerful vegetarians an evolutionary dead-end road in that there is no selective advantage for improvement of manipulative skills or mental activity along the lines which characterized human evolution. There is no reason to carry a tool if vegetable food is abundant everywhere, and no preparation of the food is required beyond stripping or shredding with the teeth and fingers. There is, it would seem, little selective pressure to try anything new or to improve on the old. Need for special manipulation involving tools might more likely arise in a harsh and marginal environment where selective premium is placed on mental activity and new modes of fulfilling bodily requirements. Man must have evolved in such a habitat, and of the apes today perhaps only the chimpanzee possesses the combination of need, potential, and habitat where the genesis of tool-using and hunting for meat might occur.

DISCUSSION

The descriptive and comparative data presented in this paper raise three points which bear on current views of social evolution in primates and man. All of these points emphasize the folly of premature generalization and the importance of a broad outlook in selecting the guidelines for theorizing.

1. The great apes present several contrasts in their social behavior and organization with what has been observed in other primates and what has come to be regarded by several scientists as the basic primate pattern from which man's social traits evolved. Territorialism is, for example, present in the gibbon (Carpenter 1940) but certainly not in the gorilla and probably not in the chimpanzee. The temporary joining of distinct groups occurs in gorillas and both Nissen (1931) and this study suggests that it happens also in chimpanzees. The relations between gorilla groups are quite peaceful as contrasted with those described for several other primate societies. Grooming between adult gorillas is primarily utilitarian, rather than intensely social or even of secondary sexual importance as in some of the monkey species which have been studied. Sex behavior, which is sometimes thought to be the main cohesive force in certain primate groups, has apparently no such encompassing function in gorillas. Thus, territorial aggression, group segregation, and sexual competition—three cornerstones of this hypothetical primate pattern—do not wholly apply to the gorilla. In fact, the question should be raised as to whether any present-day prosimian, catarrhine, or pongid pattern can, at this stage, be regarded as the one from which the human pattern developed. Caution should, therefore, be exercised in using data on the behavior of one or a few species as the basis for constructing theories of social evolution in primates, including man, as was attempted by Sahlins (1960).

2. Marked differences in social behavior exist among the apes. Social groupings vary from the almost groupless orangutans, where aggregations of 1, 2, or 3

animals are most common, to the "family" of 2 to 6 gibbons (Carpenter 1940), to the cohesive gorilla groups of 2 to 30 animals, and finally to the more loosely integrated chimpanzee groups which may aggregate up to 50 animals. Ecologically they vary from dependency on tall rain forest trees to adaptability to open, dry woodland. Temperamentally they vary from the exuberant chimpanzee to the reserved gorilla and the seemingly phlegmatic orangutan. The gorilla appears to have a low sexual drive, the chimpanzee a higher one. Information now available indicates that generalizations for the apes can be made only in very broad terms. A proper evaluation of similarities and differences must await further studies on all species.

3. The great apes, though structurally and mentally the closest living counterparts of man, are less like man in their adaptive responses and social behavior than many less closely related animals. Localization of behavior around a specific home or den is, for instance, prominent in many carnivores and rodents, as well as in birds and fishes. Social units based on a male-female pairing bond of considerable stability, and co-operative division of labor in the care of offspring are found in some carnivores and birds, but none of these attributes is markedly developed in apes.

All three of these points emphasize that social behavior and social systems in animals are more maleable in evolution than morphology, and that phylogenetic relationship as determined by structure is a relatively poor basis for inferring similarities of behavior. Distinctiveness of response to the environment has, in fact, been shown to possess marked survival value in the interaction of closely related and potentially competitive species. Accordingly, the search for evidence on the course of behavioral and social evolution followed by our prehuman ancestors might profitably be focused on ecological equivalence rather than phylogenetic relationship. The findings of archaeologists and paleontologists on the physical, biotic, and social environment of prehistoric hominids should be examined against a background of knowledge of behavioral adaptations in a wide variety of living animals. Such information, coupled with derived data on the sensory, motor, and integrative equipment of fossil forms, should provide a reasonably sound basis for constructing hypotheses on the evolution of human social behavior and social systems.

One theory of human social evolution suggests that the stable male-female relationship in human society is attributable to monogamy, the disappearance of estrus, and the persistence of sexual drive (Freedman and Roe 1958) as opposed to the violently hierarchial dominance behavior and intense competition for mates which is supposedly characteristic of nonhuman primate groupings (Sahlins 1960). An equally legitimate and logical theory can be advanced on what we know of gorilla behavior. Here the competition for mates is apparently held in abeyance not by physical dominance and suppression but by a simple reduction in the sex drive; the disruption of breeding groups is prevented not by fighting but by a general lowering of agonistic responses to subordinate and intruding lone males.

It is entirely possible that *Australopithecus* and other hominids lived in groups similar to those of gorillas, held together by social bonds in which sex and aggression played secondary roles.

It would be a mistake, however, to assume that all species of prehistoric man were alike in temperament, behavior, and social life. If the closely related chimpanzee and gorilla show differences, why not *Australopithecus* and *Paranthropus*? The basic social structure of the early hominids may have varied tremendously as a result not only of their mode of life, about which we know a little, but also of temperament, sexual drive, or aggressiveness, about which we know nothing.

Hypotheses on the paths of social evolution which our prehuman ancestors followed are beyond the reach of experimental proof and will always remain speculations. If, however, they are, to be satisfying to our sense of logic and effective guides for the interpretation of new discoveries, it is highly important that they be kept flexible and constantly in tune with pertinent information from all sources.

BIBLIOGRAPHY

Beatty, E. H.

1951. "A Note on the Behavior of the Chimpanzee," *J. Mammal.*, 32 (1):118.

Carpenter, C. R.

1940. "A Field Study in Siam of the Behavior and Social Relations of the Gibbon (*Hylobates lar*)," *Comp. Psych. Monogr.* 16 (5):1–212.

1958. "Soziologie und Verhalten freilebender nichtmenschlicher Primaten," *Handbuch der Zoologie*, 8(10):1–32.

Emlen, J. T., Jr., and G. B. Schaller

1960. "Distribution and Status of the Mountain Gorilla (*Gorilla gorilla beringei*)—1959," *Zoologica*, 45 (1):41–52.

Freedman, L. A., and Anne Roe

1958. "Evolution and human behavior. In *Behavior and Evolution* (Anne Roe and G. G. Simpson, eds.), New Haven: Yale University Press, pp. 455–79.

Hockett, C. D.

1960. "The Origin of Speech," *Scient. American*, 203 (3):88–96.

Le Gros Clark, W. E.

1960. *The Antecedents of Man.* Chicago: Quadrangle Books.

Merfield, F. G. with H. Miller

1956. *Gorilla Hunter.* New York: Longmans Green.

Nissen, H. W.

1931. "A Field Study of the Chimpanzee: Observations of Chimpanzee Behavior and Environment in Western French Guinea," *Comp. Psych. Monogr.* 8 (1):1–122.

SAHLINS, M. D.
1960. "The Origin of Society," *Scient. American*, 203 (3):76–87.
SCHALLER, G. B.
1961. "The Orang-utan in Sarawak," *Zoologica*, 46 (2):73–82.
1963. The mountain gorilla: ecology and behavior. Chicago: University of Chicago Press.
WASHBURN, S. L., and V. AVIS
1958. "Evolution of Human Behavior." In *Behavior and Evolution* (Anne Roe and G. G. Simpson, eds.), New Haven: Yale University Press, pp. 421–36.

ADAPTIVE RADIATION IN THE AUSTRALOPITHECINES AND THE ORIGIN OF MAN

J. T. ROBINSON

THE CHIEF PURPOSE of this paper is to speculate about the selective forces which brought the australopithecine group into existence, caused adaptive radiation within the group, and resulted in the origin of man.

AUSTRALOPITHECINE TAXONOMY

Although our prime concern in this discussion is with ecology, behavior, and functional anatomy, it seems advisable to discuss taxonomy first. Adaptive radiation cannot be discussed fruitfully if it is not clear what categories are involved in the group concerned; indeed whether adaptive radiation has actually occurred must first be decided by taxonomic analysis. From the literature it would appear that most workers regard the Australopithecine group as being virtually uniform taxonomically; some apparently go so far as to regard the known forms as members of the same phyletic sequence and therefore as exhibiting no adaptive radiation whatever. The literature also contains many statements that such and such are Australopithecine characteristics as though all known Australopithecines have the same characteristics.

Long and intimate acquaintance with almost all the known material has convinced me that the specimens fit into two quite distinct groups which are well differentiated in morphology and apparently also in ecology and behavior. Not only this, but it seems to me that the split between the two Australopithecine subgroups is far more fundamental than is that between one of these (*Australopithecus*) and the hominines. The other (*Paranthropus*) is quite aberrant as a hominid, whether it is compared with the contemporary *Australopithecus* or with the more advanced hominines. In order, therefore, to leave no doubt about the background, and the reasons for it, against which the later discussion occurs, taxonomy will be discussed here briefly, though it has been discussed elsewhere (Robinson 1954a, 1954b, 1956, 1961a, 1961b).

South African Australopithecines

South African Australopithecines are divided into two taxa, *Australopithecus africanus* Dart and *Paranthropus robustus* Broom (Robinson 1954a).

The practicing taxonomist is normally primarily concerned with identification and uses for this purpose good diagnostic "characters" which allow him to distinguish between closely related forms. The characters chosen must be selected and used with due regard to their ranges of variation. In this practical taxonomic sense there are a number of very good characters distinguishing *Australopithecus* and *Paranthropus*.

The first lower deciduous molar, for example, not only allows instantaneous recognition (even when considerably worn) of which group is being dealt with, but also serves to distinguish *Paranthropus* from all other hominids in which the tooth is known. On the other hand, the *Australopithecus* form of dm_1 is also found in all living and fossil hominines in which its nature is known, including

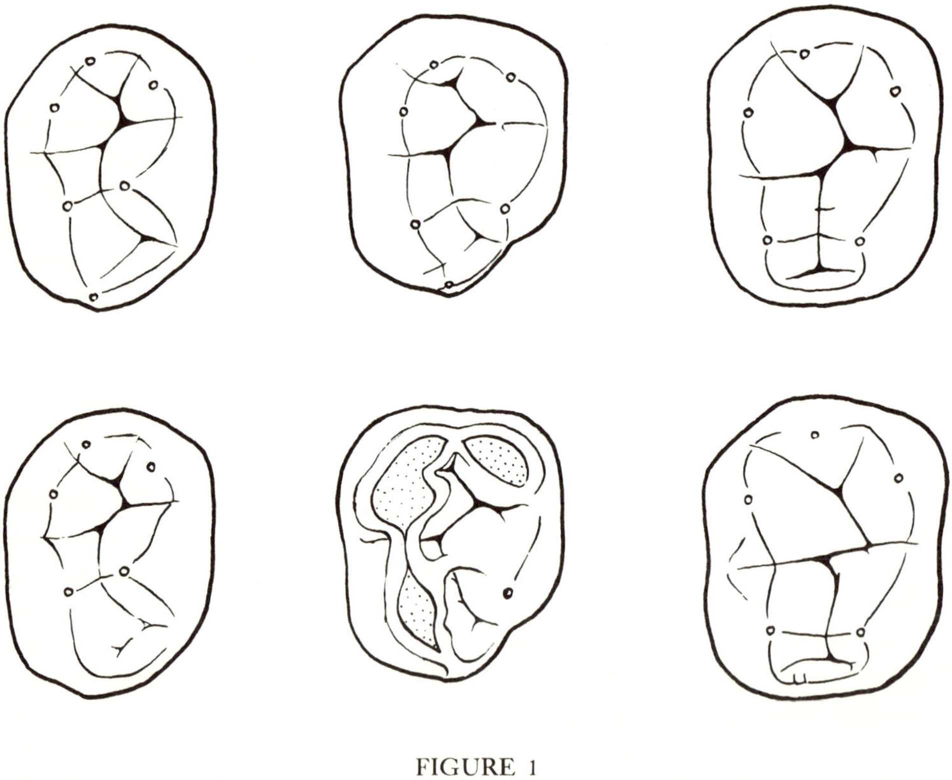

FIGURE 1

Upper row: examples of the deciduous first lower molar in *Paranthropus*, *Australopithecus*, and *Homo* (modern Bushman). Lower row: diagrammatic representations of the cusp and fissure patterns of the teeth in upper row. Twice natural size.

Pekin and Neanderthal man. The morphology of dm_1 thus serves to emphasize not only the distinction between *Paranthropus* and *Australopithecus*, but also the similarity between the latter and hominines. The permanent lower canine is another good diagnostic feature. The two Australopithecines can be separated without the slightest difficulty by means of this tooth; in *Australopithecus* the crown is large and highly asymmetric, while in *Paranthropus* the crown is small, more symmetric and with little relief on the lingual surface but the root is substantial. The large difference in proportion between the anterior and posterior teeth in the two forms is also a good diagnostic feature. In *Australopithecus* the canines and incisors are fairly large for a hominid and the postcanine teeth are of proportionate size. On the other hand in *Paranthropus* the postcanine teeth are larger than those of *Australopithecus*, as could be expected in a larger and more robust animal, but the incisors and canines are distinctly smaller than those of the latter. The condition seen in *Australopithecus* fits very well with that found in hominines; that in *Paranthropus* is quite aberrant and unlike that seen in any other known hominid. There are many other diagnostic features: the nasal cavity floor and its relation to the subnasal maxillary surface, the nature and shape of the palate, the shape and structure of the face and of the braincase, and others which cannot be discussed here because of lack of space.

In contrast to this practical, workaday taxonomic approach, there is a more inclusive and satisfying view which sees the animal as a member of a population in its natural environment rather than as a series of taxonomic characters. In such a view the isolated characters of the other approach are seen as parts of an integrated pattern. Viewed from this standpoint the differences between the two Australopithecine types are brought out even more clearly.

In *Paranthropus* it seems clear that the architecture of the skull and head in general is closely related to specializations of the dentition. The small anterior teeth, which in the maxilla are set in relatively lightly constructed bone and in the mandible in a more or less vertical symphysial region with no trace of chin, result in a relatively orthognathous face. The massive postcanine teeth, with strongly developed root systems, are set in massive bone. The area of support and the channels of dissipation of the forces generated by chewing are well developed. Examples of these are the thickened columns up either side of the nasal aperture, the enormously thickened palate anteriorly (over a centimeter thick in one adolescent where it can be measured opposite M^1), the pterygopalatine complex and the zygomatic process of the maxilla. The strongly developed musculature required to operate this massive postcanine dental battery has also affected the architecture of the skull in an obvious manner. Since in all known adults of both sexes in which the appropriate part of the skull is preserved a sagittal crest is present, the temporal muscles were clearly large in relation to brain-case size. The origin of the masseter, especially the superficial portion, is very clearly marked and extensive, as is the insertion on the broad and high ramus. The masseter must thus have been large and powerful. This was evidently true also of the pterygoid muscles in view of the relatively great development of the lateral pterygoid plate.

Maxillary prognathism is reduced by the relatively poor development of the anterior teeth. The support needed for the relatively massive postcanine dentition has resulted in a strongly stressed, hence completely flat, nasal area. The massive

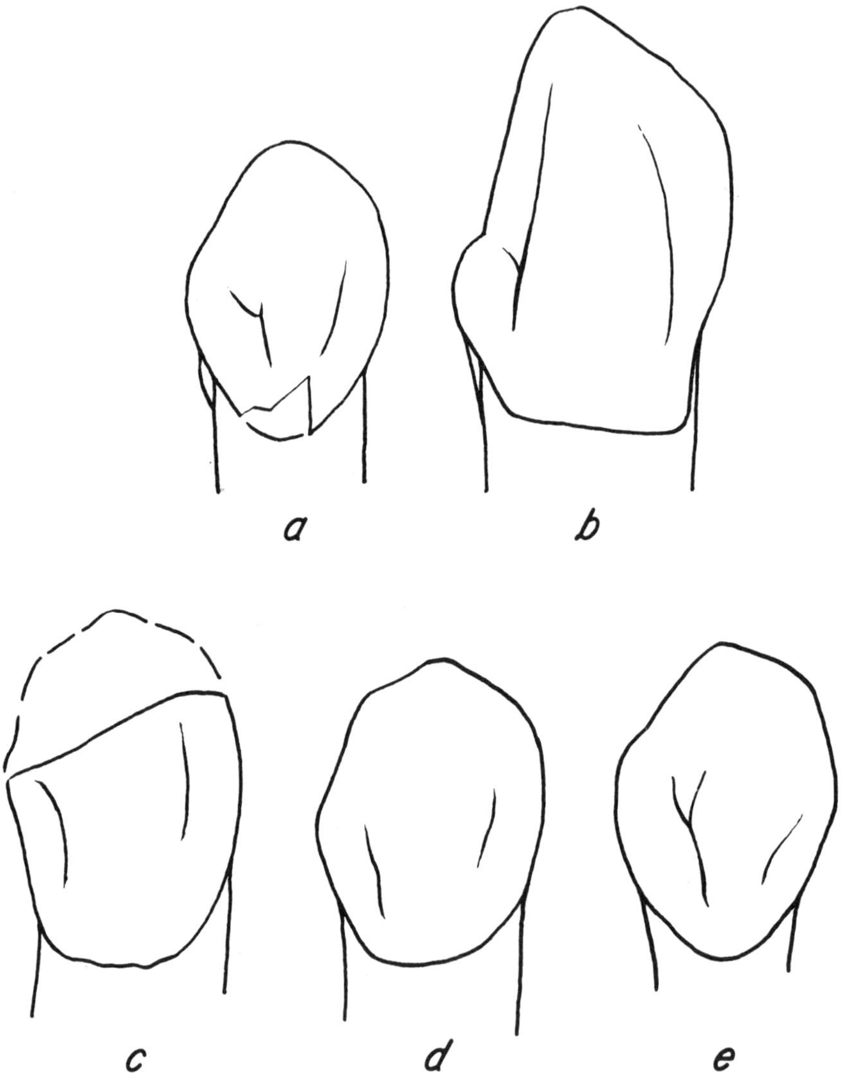

FIGURE 2

Mandibular canines of (a) *Paranthropus*, (b) *Australopithecus*, (c) and (d) *Homo* (Pekin man), and (e) *Homo* (modern Bantu). It will be recalled that (a) is from a very robust form while (b) is from a small and lightly built form; (c) and (d) after Weidenreich. Twice natural size.

chewing muscles are associated, among other things, with a strongly developed zygomatic region. These factors result in the typically wide and massive, but either flat or actually dished, face. The total lack of a true forehead and the comparatively great postorbital constriction make the brow ridges seem massive and projecting, though in actual fact they are not especially massive. The well-developed postorbital constriction, which in part at least is associated with the great development of the temporal muscle; the saggital crest, which is directly due to the relatively great size of the temporal muscles in relation to brain-case size; and the absence of a true forehead, result in a brain-case shape which is unique among hominids. The robustness of the jugal arch and the large attach-

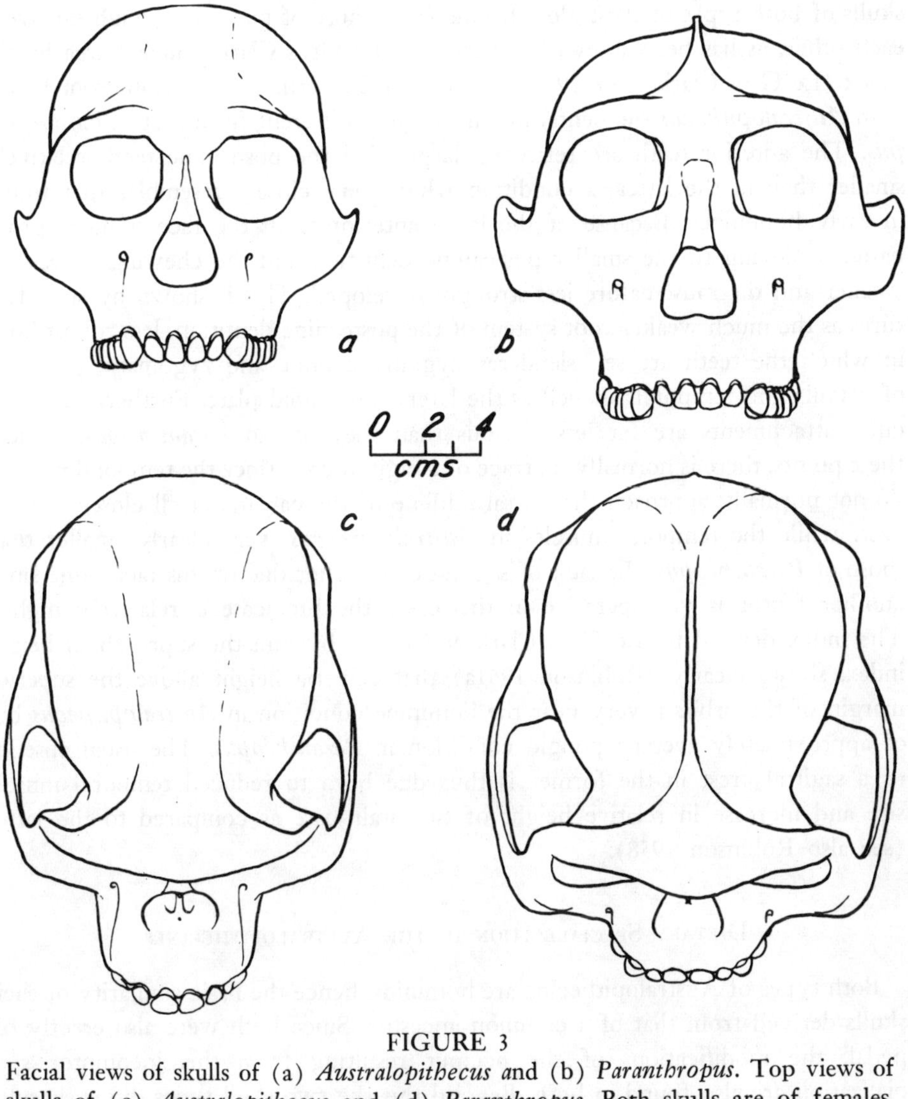

FIGURE 3

Facial views of skulls of (a) *Australopithecus* and (b) *Paranthropus*. Top views of skulls of (c) *Australopithecus* and (d) *Paranthropus*. Both skulls are of females.

ment area required by robust nuchal muscles cause the mastoid region to project laterally appreciably more than does the braincase above this region.

In *Paranthropus* the effect of the unusual dental specializations on the architecture of the skull—of the whole head, in fact—has been far-reaching. The result is a skull which bears considerable superficial resemblance to that of some pongids. That it is not a pongid skull is clearly evident, however, as a result of the effect of another important factor affecting skull architecture: erect posture. This has resulted in a very significant lowering of the relative height of the nuchal area of the occiput, which is quite differently oriented in the erectly bipedal hominids than it is in the quadrupedal pongids and all other terrestrial vertebrates. The altered orientation of the nuchal plane clearly distinguishes the skulls of both types of Australopithecine from those of pongids, though not from each other, as has been shown by the use of Le Gros Clark's nuchal-area height index (Le Gros Clark 1950, 1955; Ashton and Zuckerman 1951; Robinson, 1958).

In *Australopithecus* the dental picture is quite different from that in *Paranthropus.* The anterior teeth are relatively larger and the postcanine teeth relatively smaller than in the latter; a condition which very closely resembles that found in early hominines. Because of the large anterior teeth the face is more prognathous. Owing to the smaller postcanine dentition and the chewing forces are weaker and the musculature less strongly developed. This is shown by such features as the much weaker root system of the postcanine dentition, less robust bone in which the teeth are set, slenderer zygomatic bone and zygomatic processes of maxilla and temporal, as well as the lateral pterygoid plate. Furthermore muscular attachments are far less obvious than they are in *Paranthropus.* Besides these points, there is normally no trace of a sagittal crest since the temporal muscles do not normally approach the dorsal midline of the calvaria at all closely. However, while the temporal muscles in *Australopithecus* were clearly smaller than those of *Paranthropus,* the lack of sagittal crest is not due to this fact alone since another factor is yet operative in this case: the braincase is relatively higher. The index devised by Le Gros Clark and called by him the supraorbital height index, shows clearly (Robinson 1961a) that calvaria height above the superior margin of the orbits is very near the hominine condition in *Australopithecus* but of approximately average pongid condition in *Paranthropus.* The usual absence of a sagittal crest in the former is thus due both to reduced temporal muscle size and increase in relative height of the brain-case as compared to the latter (see also Robinson 1958).

Dietary Specialization in the Australopithecines

Both types of Australopithecine are hominids, hence the basic similarity of their skulls derived from that of a common ancestor. Since both were also erectly bipedal, the modifications of the occiput resulting from this locomotor specialization are also found in both. Beyond this the two skull types differ sharply.

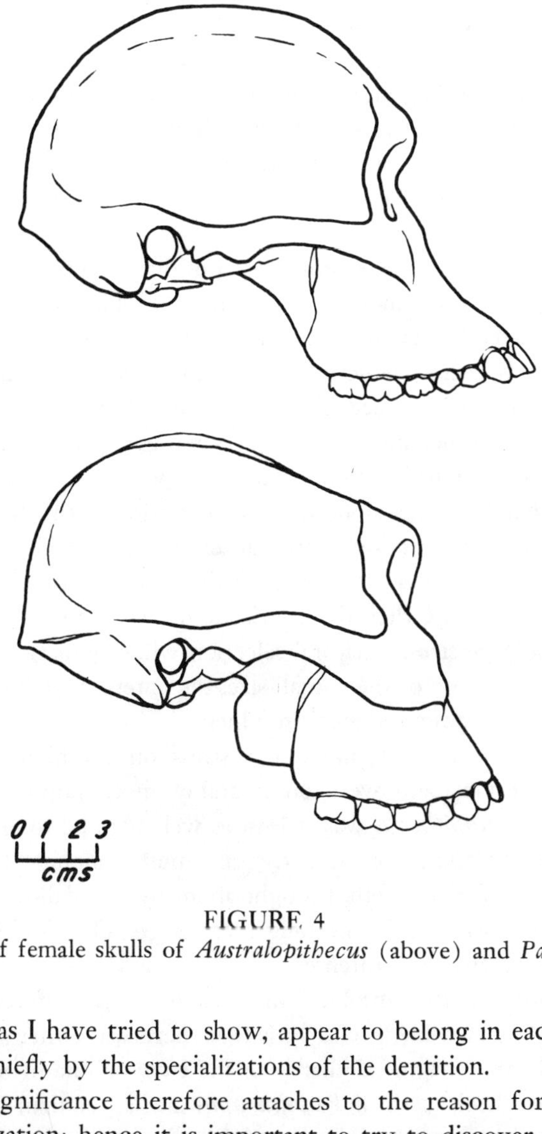

FIGURE 4
Side views of female skulls of *Australopithecus* (above) and *Paranthropus.*

The differences, as I have tried to show, appear to belong in each case to a pattern controlled chiefly by the specializations of the dentition.

Considerable significance therefore attaches to the reason for the differences in dental specialization; hence it is important to try to discover that reason. For reason there must be; it is not acceptable to say that all hominids have one sort of dentition except *Paranthropus,* which has quite a different type, but that there is no adaptive significance in this fact—it just happened that way without cause. The difference between the two dental types is very clearcut: on the one hand *Australopithecus* and all known hominines have a balanced pattern of tooth size and on the other *Paranthropus* alone has an unbalanced pattern, as it were, in which the size relation of anterior and postcanine teeth is quite different. It would therefore seem likely that there is also a fairly clearcut reason for the difference.

In *Paranthropus* the postcanine dentition is clearly very important. The tooth crowns are large, the enamel is thick, the occlusal surfaces large and of low relief and the root systems very well developed. The relatively great and flat occlusal surfaces and the massiveness of the postcanine teeth clearly point to a prime dietary function of crushing and grinding. The massiveness of the entire masticatory apparatus and the relatively rapid rate of wear of the teeth indicates a diet of tough material and one that probably needed much chewing and had a relatively low nutritive value. The anterior teeth either were of much less importance in relation to the diet than the postcanine ones, or the diet *required* small anterior teeth and that this feature thus represents positive adaptation to dietary needs. The latter seems doubtful since it is not obvious what manner of diet it could be that at once placed a premium on small anterior and massive posterior teeth. It seems more probable that reduction of the incisors and canines resulted from reduced need for them in respect not only of diet but also of other things such as behavior affecting other members of the same group or those of other groups in its environment. A fairly obvious and old explanation seems to be that in an erect biped the hands, either alone or in conjunction with objects used as tools, take over much of the function of the anterior teeth in obtaining food and in offense and defense. If this is the reason for the small anterior teeth of *Paranthropus*, as seems to me likely, then the cheek teeth were either being maintained at, or increased to, a large size by natural selection while reduction was occurring in the anterior teeth. In spite of their small size, the anterior teeth do not wear down rapidy. This again suggests reduced usefulness.

In *Australopithecus* there is much less stress on crushing and grinding and obviously the anterior teeth were considerably more important than those of *Paranthropus. Australopithecus* was at least as well adapted to erectly bipedal posture as was *Paranthropus;* consequently one must conclude that the tendency to reduction of the anterior teeth, brought about by bipedalism, was counteracted by some selective advantage for keeping them large. This could either have been diet or needs of defense and offense.

The implication of the absolute and relative size difference between the anterior teeth of the two forms clearly is that there was either a difference of behavior of considerable magnitude between the two, or one of diet, perhaps both. The fact that the cheek teeth of *Paranthropus* are larger than those of *Australopithecus* need not be due to anything more than the bodily size difference between them since the former is appreciably more robust than the latter.

The very great similarity in dentition and general skull structure between *Australopithecus* and the hominines suggests that they were basically similar in diet and behavior. That is to say that they were omnivores, eating both vegetable food and meat. This provides a logical reason for the relatively large anterior teeth since the need for these in meat-eating would be substantial until tool-making had reached a fairly advanced level. Since the teeth are unlikely to have been important for fighting in an erect biped with a relatively flat face, it seems likely that the canines of *Australopithecus* were large, like those of early hominines, because their diet included meat. This is perhaps supported also by the fact that this form is known mainly from periods of appreciable aridity when vegetable food will not have been plentiful, other than for grazers, hence the probability that animal food formed part of their diet. Finally, as appears later in this paper (also Robinson 1961b), the strong development of tool-using and of tool-making was very probably associated with the tendency for early members of the phyletic line leading to man to take to a certain amount of meat-eating. *Australopithecus* could very easily be representative of this stage. The hypothesis that *Australopithecus* was an omnivore that ate at least a moderate amount of meat does not therefore seem unreasonable.

The emphasis on crushing and grinding and lack of emphasis on the canine in the *Paranthropus* dentition seems inconsistent with a diet such as that suggested for *Australopithecus*. But these features are entirely consistent with a vegetarian diet, as is the massiveness of the whole masticatory mechanism and the much greater body size as a whole. This conclusion is supported also by the presence of grit in the diet, as suggested by damage to the teeth in the form of flakes of enamel detached from the edge of the occlusal margin. The flaking was clearly caused by the application of considerable pressure over a very small area—such as is produced, for example, by the stiletto heel of a lady's shoe as compared to the heel of a man's shoe—as would be the case in biting on a particle of grit but which would not be produced by biting on bone. Grit in the diet would seem to suggest roots and bulbs as part of the diet. The available evidence also points to the fact that *Paranthropus* is known only from wet climatic periods, both in Africa and the Far East. This again is consistent with a vegetarian diet.

The objection has been raised that it is not possible to tell, for example, what the diets are of various sorts of monkey by looking at the teeth, and therefore that reasoning such as the above is unsound. It would seem to me clear that distinguishing accurately between different sorts of vegetarian diet from dentitions which all are basically adapted to vegetarianism is a difficult task, especially as this example concerns the notoriously stable cercopithecoid dentition. But in the case of the Australopithecines the distinction is not between two forms of vegetarian diet but between a vegetarian and an omnivorous diet. That basic dietary adaptation *can* be recognized in many cases from teeth is a commonplace of mammalogy and vertebrate paleontology. In this instance one is faced with the fact that one form of Australopithecine fits so well in all relevant aspects into the picture presented by the hominines that it is not necessary to seek for it other explanations of basic adaptations of diet and behavior than apply to the hominines. But the other Australopithecine does *not* fit into this pattern, in respect of either the dentition or the skull morphology. So there must be some definite explanation of this difference. The dietary explanation seems entirely logical and, so far as I am aware, no evidence exists which is clearly inconsistent with such an explanation. On the contrary, independant confirmatory evidence seems to exist in the fact that *Paranthropus* appears to have shared the same territory with an early hominine. In the Swartkrans site "Telanthropus" and *Paranthropus* remains were directly associated. In Java the Sangiran site has yielded both "Pithecanthropus" and "Meganthropus" remains. According to von Koenigswald "Pithecanthropus" IV and the type mandible of "Meganthropus" came from the Black Clay (Putjangan beds) not far from each other, but "Pithecanthropus" II and III came from the later Kabuh conglomerate. According to Marks (1953) the "Meganthropus" mandible which he found in 1952 was not *in situ* but was lying on a slope of hard conglomerate and had enough matrix adhering to it to indicate satisfactorily that it had weathered out of the conglomerate. This conglomerate is one which von Koenigswald regards as a boundary bed (basal Kabuh) between the Kabuh and Putjangan beds. Therefore it would appear that "Pithecanthropus" and "Meganthropus" occurred synchronously at Sangiran, not merely at one time horizon, but over a significant period of time. As indicated elsewhere (Robinson 1953, 1955, 1961b) "Meganthropus" is fairly clearly a *Paranthropus*. The implication is clear that *Paranthropus* differed sufficiently in ecological adaptation from early hominines for them to be able to coexist in at least two places at opposite ends of the Old World. Continued coexistence is unlikely if the ecological requirements of the two were virtually identical but if their ecology was as different as is here suggested such coexistence is entirely logical and in no way remarkable.

The resistance to accounting for the differences between the Australopithecines by the dietary hypothesis seems to spring from the *assumption* that they represent a single phyletic line, hence that all known Australopithecines must, willy-nilly, be very closely related. The reason for such an assumption is not clear. Paleontology has provided much evidence to show that adaptive radiation usually follows the achievement of a new grade of organization. The emergence of the Australopithecine group represents the achievement of an important new grade of organization: that of the first known erect biped in evolutionary history. The new group must have arisen from a higher primate. Vegetarianism is the basic dietary adaptation of non-hominid higher primates, but meat-eating is an important part of hominine diet; hence it seems clear that dietary changes must have occurred. Some adaptive radiation in early Australopithecine history, particularly if it involved difference of diet, should consequently occasion no surprise. In later phases of hominid evolution adaptive radiation of any magnitude is very unlikely. This is because of the new adaptive mechanism, additional to the genetic one, which became available with the effective development of culture (*vide* Dobzhansky 1961). This more rapid, artificial adaptation results in a slowing down of natural adaptation, hence of speciation, owing to the fact that man can so easily adapt himself to different environments long before the genetic mechanism has the opportunity to do so. This does not mean that natural selection does not operate on man, but artificial adaptation modifies its action to a considerable extent. The Australopithecines in the initial phases of their evolution will have had no more than a rudimentary level of culture, hence the slowing down of the rate of speciation will not have applied. In the later phases of their evolution cultural development may have been sufficient to make its effect felt to some extent.

Some workers who disagree with the notion of significant adaptive radiation within the Australopithecines wish to regard *Australopithecus* and *Paranthropus* as successive members of the same phyletic sequence. If by this is meant that the Australopithecines were ancestors of the hominines, but through a sequence *Australopithecus*—*Paranthropus*—early hominine, then one is faced with the fact that the first member of this sequence is morphologically, and evidently ecologically too, much more like the last member of the sequence than it is like the middle one. Retrogressive steps *do* occur in evolution, but to have so many reversals in what would presumably have been a fairly rapid evolutionary sequence does not correspond with what is known of phyletic lines in other vertebrates where the evidence is more complete. It is worth considering a few of the reversals involved. *Australopithecus*, as already mentioned earlier in this paper, has the vertex of the skull significantly higher above the upper orbital margin than has *Paranthropus*. The supraorbital height index of the latter falls right in the normal range for pongids whereas that of the former agrees closely with early hominines. This is not a simple feature as it not only involves considerable alterations in skull architecture, but is doubtless also a reflection of expansion in the brain. The latter, it is now well recognized, is evolutionarily a very conservative organ. A very conservative organ thus developed to essentially the hominine condition, retrogressed to a pongid condition and then rapidly advanced to essentially the same hominine condition that it previously had, all in the space of a few hundred thousand years at most. Similarly dm_1 in *Australopithecus* has a structure also found in all hominines but not in *Paranthropus*. So a typically hominine condition has to give way to one which is unique among hominids and then return to identically the condition it previously had. At the same time the canines reduced from an early hominine size to a modern hominine size but then had to revert again to their previous size a short time later. The reduction in canine size would have occurred in a stage when meat-eating was being developed but cooking was not known and tools were not well developed. Furthermore, if my interpretation of the tool situation in the Sterkfontein Valley is correct, then a hominine was there present for a short while with *Australopithecus* before *Paranthropus* appeared.

In my opinion the known evidence indicates that the possibility of hominines having arisen from *Australopithecus* via *Paranthropus* is remote in the extreme and need not be considered seriously. If the latter, as known, is regarded as a direct lineal descendant of the known specimens of *Australopithecus*, then it means that Australopithecines were not ancestors of hominines, or if so, the *Paranthropus* group represents a side branch. It now appears that these two genera were contemporaries in Africa south of the Sahara since both are known from East Africa and apparently a little later in time both occur in South Africa with *Australopithecus* occurring in the latter area at about the time that *Paranthropus* was living in the former. If the latter is a lineal descendant of the known specimens of the former, then the transition must have occurred more than once in different places. However, the fact that all the known forms, wherever they occur in time or space, fall easily into one or the other of two very different groups, suggests that it is quite improbable that *Paranthropus* represents simply a later evolutionary phase of the known *Australopithecus* specimens.

Theoretical considerations thus indicate that adaptive radiation in the early stages of hominid evolution is by no means improbable. The available facts indicate considerable morphological difference between the two well-defined groups, more than can be found among living pongids for example, and imply that considerable differences in ecological adaptation were also involved. The ecological differences appear rather clearly to have involved differences of diet. At least differences in diet may logically be inferred from the morphological differences present; such a conclusion adequately explains the different morphology of the two groups and no alternative explanation has yet been advanced which does so. Furthermore, the dietary hypothesis is very fruitful in throwing light on the entire question of Australopithecine evolution and the origin from them of the hominines (Robinson 1961b, and later in this paper).

Non-South African Australopithecines

Australopithecines are at present known from two areas outside of South Africa: Java and East Africa. Coppens (1961) has reported a new Australopithecine find from another region, the Lake Chad basin. This specimen comes from Koro-Toro and apparently was coeval with an early Villafranchian fauna. The specimen has not yet been described in detail; if the dating and identification are correct, then this is the oldest of the known Australopithecines.

The Javanese form was first designated *Meganthropus palaeojavanicus* (Weidenreich 1945), but detailed analysis of the available information resulted in its being placed in the genus *Paranthropus* (Robinson 1953, 1955). The reason for this is that, with only the most trivial exceptions, the features of the known specimens fall within the observed range of the known *Paranthropus* material. Among these features are the massive mandible and the combination of small canines and enormously robust postcanine teeth. Although most of the crowns are missing from the 1952 mandible (Marks 1953), the roots are present in the sockets, including those of the left incisors and canines; along with those of the cheek teeth they reflect the characteristically *Paranthropus* condition. The conclusion that "Meganthropus" is a *Paranthropus* has been contested by von Koenigswald who has, however, produced no cogent evidence to refute it. Some of his main objections have been considered recently (Robinson 1961b). Not only is there so far no valid evidence differentiating the two groups of specimens, but "Meganthropus" exhibits some features which are diagnostic of *Paranthropus*. It is therefore reasonable to regard the former as a member of the latter genus.

Leakey (1959) reported the discovery of a good skull of a late adolescent Australopithecine from Olduvai. This form was regarded as new and named *Zinjanthropus boisei*. It has been shown, however, that the skull and dental characters, and their pattern of specialization, are typically those of *Paranthropus* (Robinson 1960). As in the case of "Meganthropus," the morphological differences held to validate generic separation from *Paranthropus* either disappear or become very slight if the observed range of variation of these features is taken into account.

In 1939 Kohl-Larsen discovered in the Laetolil beds, near Lake Eyassi in East Africa and in the same general region as Olduvai, a fragment of maxilla containing P^3 and P^4 and also an isolated upper molar. There were named *Präanthropus* (a *nomen nudum* since no species name was given) by Hennig (1948) and *Meganthropus africanus* by Weinert (1950, 1951), a view supported by Remane (1951). This matter has been considered at some length (Robinson 1953, 1955) and the conclusion drawn that (1) since one form is known only by mandibular and the other only by maxillary material, no evidence exists for placing them in the same genus; (2) since the East African specimen exhibits characters which all fall within the observed range of variation of the corresponding features of *Australopithecus*, the logical course is to refer the material to the latter genus. This is also the opinion of von Koenigswald (1957).

Recently Leakey (1961a, b) has announced the discovery of further material at Olduvai, including a juvenile mandible and two parietals from a Bed I horizon a little lower than that from which "Zinjanthropus" came. The mandible appears to have the characteristics of *Australopithecus* but the size of the parietals suggests that perhaps they belonged to a larger-brained creature than either of the Australopithecine types. Additional material is therefore needed before reliable conclusions can be drawn about whether the material from this horizon belongs to one or more forms and what its or their identity is.

We may conclude that

1. *Paranthropus* is a well defined genus consisting of a somewhat aberrant type of hominid whose morphological, ecological, and behavioral adaptations are quite distinct from those of all other known hominids. It is known from East and South Africa as well as Java; in the latter two places it is known to have coexisted at the same site with an early hominine.
2. *Australopithecus* differs clearly in morphological, ecological, and behavioral adaptations from *Paranthropus* but exhibits very considerable similarity in these respects to hominines. It occurs in South and East Africa but is not at present known from the Far East.

CULTURAL STATUS OF THE AUSTRALOPITHECINES

The level of culture achieved by the Australopithecines also concerns the subject of this paper. The relationship between Australopithecines and the stone industries found with them in the Sterkfontein Valley and at Olduvai has been discussed elsewhere (Robinson & Mason 1957, 1962; Robinson 1959, 1961a) and will not be dealt with here. The conclusion was reached that, despite commonly held opinion to the contrary, there is as yet no proof that either form of Australopithecine possessed a settled stone culture.

The evidence in fact favors the conclusion that the Australopithecines were no more than tool-users, employing whatever came to hand in the form of sticks, stones, bones, etc. This aspect of *Australopithecus* behavior has been dealt with at considerable length by Dart (e.g., 1957a, 1957b, 1958, 1960). In my opinion the evidence provided is enough to establish that this form was a tool-user, though it would appear that its osteodontokeratic prowess has been overrated considerably. This tool-using ability has been disputed by some authors. For example Mason (1961) holds that since a bone culture (due presumably to *Homo sapiens*) has been found in a Middle Stone Age (end-Pleistocene) deposit at Kalkbank and since early hominines were already in existence in Australopithecine times, therefore the Makapan Limeworks bone culture should be attributed to a homine that preyed on *Australopithecus* there. Washburn (1957) has also argued against the latter form having had a bone culture. His argument turns on whether the associated bones represent bone accumulation by the latter or by carnivorous animals such as hyaenids. Washburn and Howell (1960) accept the bone associated with the Olduvai *Paranthropus* as food remains of this vegetarian form and therefore as proof of predatory activity by it. However, in the same paragraph they state, "It is very unlikely that the earlier and small-bodied Australopithecines (i.e., *Australopithecus*) did much killing," without explaining why associated faunal remains are to be accepted as food remains of an Australopithecine in the one case but not in the other.

On theoretical grounds the probability that Australopithecines used tools would seem to be high. As is well known, tool-using of an indisputable character occurs among non-primate mammals, birds and even invertebrates. Australopithecines would appear in general to be no less well endowed than these other tool-users and in addition were possessed of the very great advantage of being erect bipeds with emancipated fore limbs. Furthermore, Australopithecines are very closely related to hominines, the supreme users and makers of tools. The probability that they used tools would thus seem very high. That bone was used would seem equally clear. Many hominines are known to have used bone for tools (indeed some still do) and incontestable evidence of the use of bone in Australopithecine times exists in the form of the bone tool from Sterkfontein (Robinson 1959) and the less perfect example from Olduvai (Leakey 1960), both of which were evidently used for working leather. Whether or not these were products of Australopithecine activity is not clear; but they do represent examples from two widely separated areas of the use of bone in Australopithecine times. That Australopithecines used bone is therefore a distinct possibility. It is interesting to note that some authors who do not accept tool-using for *Australopithecines*—in bone at any rate—nevertheless firmly believe that they were stone tool makers.

Acceptance of *Australopithecus* as a tool-user, in bone or anything else, therefore seems entirely reasonable. On general grounds it would seem probable that *Paranthropus* also used tools, though such activity may have been much more poorly developed in this vegetarian and may well not have included the use of bone.

THE ORIGIN OF THE AUSTRALOPITHECINAE

The Subfamily Homininae includes forms characterized morphologically by erect posture and large brain and behaviorally by relatively complex cultural activity. The latter feature appears to be dependant on the large brain since it seems that intelligence of the hominine caliber is not associated with brains smaller than an ill-defined lower limit in volume of the general order of 800 cubic centimeters.

The Subfamily Australopithecinae includes forms which have the erect posture, but not the large brain, of the hominines. Erect posture is more than adequately proven by the morphology of one virtually complete pelvis with most of the spinal column and a proximal portion of femur; three other adult innominate bones and two juvenile specimens; two proximal ends of femora and two distal ends, as well as a number of skulls showing the structure and orientation of the occiput. The pelvic morphology closely resembles that of hominines and differs sharply from that of pongids. A short, broad innominate is present which has the posterior part of the ilium expanded, hence there is a deep and well-developed greater sciatic notch; the iliac crest is in the form of a sinusoidal curve when seen from the top; the sacrum is broad; distinct lumbar lordosis is present and a femur with a strong lateral lean of the shaft from the vertical when the distal articular surfaces are placed on a flat horizontal surface with the shaft as nearly vertical as possible. The nuchal plane of the occiput has the near-horizontal disposition found in erect bipeds. Functionally the locomotor mechanism appears to be that of an erect biped. For example the arrangement of the origin and insertion of *gluteus maximus* are such that this muscle must have acted as an extensor of the thigh. *Gluteus medius* was evidently an abductor. A well-developed anterior inferior iliac spine suggests a powerful *rectus femoris*, and therefore probably *quadriceps* as a whole. This is a very important muscle in erect bipedal locomotion and in standing without additional support. A well-defined attachment area just below that for the direct head of *rectus femoris*, and a pronounced femoral tubercle, indicate a powerful ilio-femoral ligament strengthened and functioning in the manner of that in hominines. There is also evidence for the "locking" of the knee joint with the leg straight. The best available evidence for erect posture is for *Australopithecus;* that for *Paranthropus* indicates a basically similar condition, though perhaps not quite as well adapted to erect bipedalism as in *Australopithecus.*

The Australopithecines must have originated from some more primitive primate group, but it is not our aim here to inquire closely into what that group might be. The ancestral form may have been a member of the same early hominoid stock to which *Proconsul* belongs, as is commonly believed, or it may have been part of an independent line already quite distinct at the time the early Miocene East African pongids lived. *Amphipithecus* and *Oreopithecus* suggest that the hominids may have resulted from a line which was slow-rate for most of its history and which has been independent since the prosimian stage. The evidence is as yet too scanty for definite conclusions to be drawn.

Australopithecines appear to differ from pongids primarily in having erect bipedal posture, a primitive culture, and in the nature of the dentition. The differences between pongid and Australopithecine dentitions are most striking in the anterior teeth, especially the canines, incisors and P_3 as well as dm_1. The reduction in canine size, as was suggested already by Charles Darwin, probably resulted from the use of tools. Effective tool-using could only have become possible after erect posture had been acquired. The altered character of the incisors and canines in the early hominids may therefore have been a consequence chiefly of changed posture and locomotion. The differences between the pongid and hominid types of P_3 cannot primarily have been due to these changes, however, as the evidence clearly shows.

Since the improved cultural level and reduced canines of Australopithecines followed upon the change to erect bipedalism, the latter would appear to have been the key feature in the origin of the Australopithecines. This locomotor change represents a major adaptive shift which opened up entirely new evolutionary opportunities in this primate line as compared to all known previous ones.

The Origin of Erect Posture

The manner of origin of erect posture is, however, not clear. A critical part of the change centers around the shift in function of *gluteus maximus* from being primarily an abductor of the thigh to an extensor. The rest of the pelvic and thigh musculature of pongids and similar quadrupeds is very similar in function; but *gluteus maximus* functions very differently in the two groups and this difference is of profound importance in locomotion. The power provided by this muscle, particularly in the second half of a stride, is largely responsible for the efficacy of upright locomotion and is mainly responsible for the difference between the efficient erect locomotion of man and the far less efficient erect walking of a pongid, although the whole story is more complicated. It is readily apparent that a short, broad innominate, in which the breadth increase is mainly in the posterior part of the ilium, is a major cause of the change in function of *gluteus maximus*, since in such a case the origin of the muscle is placed well behind the acetabulum. This, coupled with the fact that the thigh is normally in at least a fairly extended position in erect bipeds, places the main line of action of the muscle behind the hip joint, hence contraction causes extension of the thigh, not abduction.

Higher primates commonly rear up on their hind legs under various circumstances normal to their way of life. The gibbon often does this in trees as part of locomotion, but apart from this the upright posture is probably mainly used for improving visibility, getting food or during play. The former is probably by far the most significant function, especially in the case of a mainly or entirely terrestrial animal living in the forest verge or in tree savanna where vision may be obstructed by shrubs and grass. It seems reasonable to suppose that members of a population in which the point had been reached where, in the erect position, *gluteus maximus* functions chiefly as an extensor would find it easier to use this posture or mode of locomotion.

Under these circumstances such animals would have the advantage of being able to keep alert about what was happening in their neighborhood rather more easily than in the case of pure quadrupeds. But this would certainly not be the whole of the advantage gained; the advantage conferred by having the hands free for manipulations such as tool-using is very obvious. It is now well recognized that even a small advantage is sufficient to allow selection to operate effectively. In this case the advantages would not be small. To start with, the advantage of freed hands would perhaps be comparatively unimportant, but the more erect posture was used, the more important it would become.

If erect posture and locomotion came to be used frequently in such a population, the nature of selection on the locomotor apparatus would alter considerably. Relatively minor changes only would at that stage be required to adapt fully to erect posture as the normal habitat. Rapid adaptation to erect bipedalism could thus be expected. It does not, therefore, seem especially difficult to see how natural selection would bring about a rapid readaptation, in respect of posture and locomotion, in a group in which the innominate had become sufficiently broad and short for the change of function of *gluteus maximus* to occur.

The difficult part to explain, as far as I am concerned, is the process which led to the changes in the innominate. Starting from the general pelvic type found in the prosimians and the arboreal monkeys, it is very difficult to see what manner of locomotory specializations could have brought about the required pelvic changes. Forms specializing in the direction of brachiating seem to acquire a pelvis which is long and narrow. This is the case in the living pongids, even though some are no longer primarily brachiators, and also in *Ateles*, a New World monkey which brachiates to an appreciable extent. The gorilla is now essentially a ground-dweller and when in the trees is mainly a quadrupedal climber (see Schaller & Emlen, this volume) and it also has a broadened ilium. The descent to the ground of an erstwhile brachiator might thus be regarded as suitable means for bringing about the required changes in the pelvis. This is, unfortunately, not what the case of the gorilla demonstrates. Not only is the pelvis quite clearly still of the pongid sort, but even the broadened ilium is not, as simple statistics alone might suggest, more nearly like that of hominids. All the increase in breadth is in the anterior part of the ilium, not the posterior, and appears to be related to the stout trunk of this animal. The greater iliac breadth therefore does not affect the function of *gluteus maximus*. Brachiators, whether modified for ground-dwelling or not, do not appear to offer any suggestion of tendencies in the required direction. Postulating that an arboreal form without brachiating specializations descended to the ground does not appear to help either. This experiment has also been performed, as it were, and the chacma baboon is an example of the result. Here again there is no evidence of a tendency for the innominate bone to shorten significantly or for the ilium to broaden in the direction of the sacroiliac articulation.

The known non-hominid primate locomotory specializations therefore do not appear to afford any help in explaining how an arboreal primate pelvis could have become modified to the point where changed muscular function could provide a basis for altered selection pressures causing adaptation to erect posture.

It might be felt that looking for the origin of the change in locomotor specialization is wrong; that behavioral changes might have led to erect posture. For example it might be supposed that the development of a proclivity for tool-using altered selection · so that the pelvic changes were brought about. This does not seem likely. In the first place, primates which have enough manual dexterity and grasping ability to use tools and which in captivity have been shown to be capable of a certain amount of tool-using, do not appear to make use of these abilities in their natural state. In some cases it would seem that even the most primitive use of tools would be very beneficial, but is not resorted to. An example is the baboon; a simple digging stick or other object would help considerably in recovering grass rhizomes, which are eaten to a considerable extent (see DeVore & Washburn, this volume). Such tool-using is used to very good effect by one of Darwin's finches of the Galapagos Islands and by the sea otter of the California coast with far less suitable equipment anatomically than the baboon has. In the second place, it is difficult to see how use of tools could bring about the required anatomical changes. If tool-using of the primitive sort under consideration was adopted by a quadrupedal apelike animal and in the use of tools it stood erect quite often, would this cause a selective situation favoring a short and broad pelvis? After all monkeys and apes commonly do stand erect, or even move in this position, and have presumably done so for millions of years without the pelvis changing in the required direction. Many other quite different animals also do this, bears and mongooses, for example. Only when the innominate had reached a state in which when the animal stood erect *gluteus maximus* functioned as an extensor of the thigh, would a situation exist where use of this posture would bring positive selection to bear in the direction of better adaptation to the bipedal locomotor habit. It is therefore very easy to see tool-using following adoption of erect bipedalism, but the latter being a consequence of tool-using seems very improbable.

The pelvic modifications were probably associated with changes which were not primarily concerned with locomotion, but which rendered the pelvis preadaptive for erect posture. That is to say that the changes responsible for making the change in function of *gluteus maximus* possible were being controlled by selection which was not concerned with erect posture. But when they had proceeded to a point where the change in muscular function was a possibility, a new situation existed in which selection for bipedality became a reality. A distinct adaptive shift would then occur and only after it had occurred was it possible to refer to the previous adaptive situation as being preadaptive for erect posture. However, I am unable to offer any explanation of what was causing the changes before the adaptive shift occurred.

Whatever the reason for the pelvic changes, it is a fact that they did occur, and once they had, a new adaptive trend came into being. According to the view presented here the process occurred in two phases: the first, during which it is difficult to see how selection for erect posture as such could have been operating, can in retrospect be regarded as the preadaptive phase; this was followed by the adaptive phase during which selection pressures were concerned directly with erect posture. This is, of course, typical of instances where a sharp adaptive shift occurs. In this case the threshold involved the changed function of *gluteus maximus*. Before this the pelvic changes represented a prospective adaptation; after the threshold had been crossed, adaptation to the new adaptive zone was rapid under the direct control of selection.

In respect of the preadaptive phase, it is of great significance that *Oreopithecus* had a somewhat shortened innominate with a relatively broad ilium (Schultz 1960). Not only is the ilium broad, but the increased breadth is posteriad, in the region of the sacroiliac articulation. Precisely the required changes required to bring about the changed function of *gluteus maximus* were therefore in progress in *Oreopithecus* and the early Pliocene horizon at which it occurred (Hürzeler 1958) would place it at about the right period in time. From the point of view of the pelvis, therefore, *Oreopithecus* is remarkably suitable as an Australopithecine ancestor in the stage before the adaptive shift to erect posture occurred. Some other skeletal features are also quite consistent with this view: the short face, canines of moderate size (substantial in males), compact tooth row, occlusal pattern of upper molars and fairly vertical chin region. If P_3 is indeed bicuspid in all cases, with the lingual cusp well developed, then this is additional and good evidence for this view since hominids are the only higher primate group so far known in which well developed bicuspedness in this tooth is normal. The cusp pattern of the lower molars is somewhat unusual: this may indicate that *Oreopithecus* was off the main line leading to the Australopithecines, or perhaps, only that this line was slower rate than others in which the dryopithecus pattern developed and therefore that the latter pattern was not yet fully in evidence. More information is needed about the nature of the upper limbs, which are known in the most complete skeleton. It is not inconceivable, therefore, that this form is, as Hürzeler (*ibid.*) has suggested, an early member of the ancestral line of the hominids.

The Nature of the First Australopithecines

The consequence of the adaptive shift involving the locomotor apparatus was to give rise to an erectly bipedal primate which was the first hominid and also the first Australopithecine. This form was probably a vegetarian. The reason for this conclusion is that vegetarianism, in its broadest sense, is characteristic of non-hominid higher primates and the hominid ancestors are therefore likely to have shared this characteristic. There is no reason to suppose that diet could have been an important factor in the locomotory changes. Furthermore, if the hominid ancestors had been meat-eaters from long before the Australopithecine stage, and therefore for a very considerable period of time, then it is likely that the dentition would have reflected this fact in adaptations to such a diet. This is not the case; hominid dentitions have no definite adaptations to a carnivorous diet. It thus seems fair to conclude that the first Australopithecine was an essentially vegetarian biped.

Since the conclusion has already been reached that *Paranthropus* is not as advanced in the hominine direction as is *Australopithecus* and had an essentially vegetarian diet, the possibility exists that the former has diverged less from the ancestral Australopithecine than has *Australopithecus*. The skull of *Paranthropus* is primitive for a hominid in some respects. There is no true forehead, the brow ridges are rendered prominent by a well-developed postorbital constriction and the vertex rises very little above the level of the upper orbital margins. This latter point is very well demonstrated by the supraorbital height index of Le Gros Clark (1950). The value of this index for *Australopithecus* (Sts. 5) is 61 (68 according to Le Gros Clark, 74 according to Ashton and Zuckerman, 1951). This approaches the figure for the index in *modern* hominines which, according to Ashton and Zuckerman, averages about 70 and ranges from about 63 to about 77. The value for a few specimens of Pekin man, determined from illustrations, appears to range from about 63 to 67 for the small sample. On the other hand the three great apes have mean values for this index which range from 49 for the orang to 54 for the gorilla, according to Ashton and Zuckerman. The figure for *Paranthropus* from Swartkrans is 50 and that for

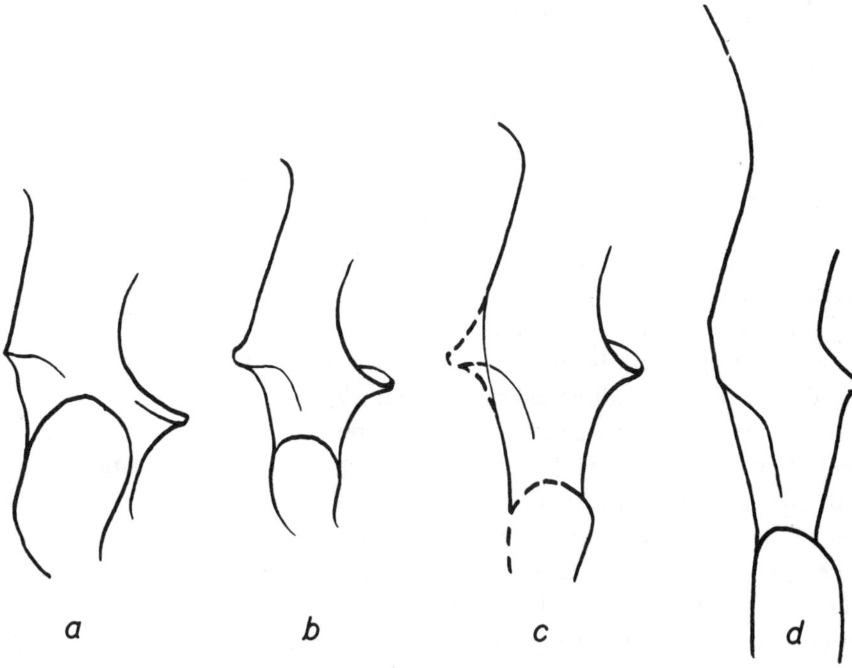

FIGURE 5
The ischial regions of the innominate bones of (a) *Homo* (Bantu female), (b) *Australopithecus* female, (c) *Paranthropus* (probably female), and (d) female Orang. The distance from the acetabular margin to the nearer edge of the ischial tuberosity in *Paranthropus* is half as much again than it is in *Australopithecus*, whereas the difference in size between the innominate bones of these two is much less than this.

Paranthropus from Olduvai, determined from photographs, appears to be just over 50. This feature reflects a significant feature of cranial morphology, which in turn almost certainly reflects some aspects of brain morphology. It is significant, therefore, that in these respects *Paranthropus* agrees with the pongids while *Australopithecus* exhibits a condition closely resembling that of the hominines. The vegetarianism of the former may also be interpreted most easily as a relatively primitive feature. If in fact the Australopithecine ancestors were vegetarians, then the simplest explanation of this condition in *Paranthropus* is that it was retained from ancestors. It is possible that it was a secondary condition; in the absence of evidence suggesting that this form was originally a vegetarian, then became an omnivore and then once more became a vegetarian, much the most economical hypothesis is to assume that it always had been vegetarian. There are still other significant features in which *Paranthropus* is more primitive than *Australopithecus*. In the former the ischium is well developed and the "bare area" between the acetabular margin and the area of muscular attachment on the tuberosity relatively much longer than in *Australopithecus*.

Also the total lack of separation between the anterior portion of the nasal cavity floor and the subnasal surface of the maxilla presents an almost ultra-pongid appearance in *Paranthropus* while *Australopithecus* has a considerably more hominine condition (see Robinson 1953, Figs. 4, 5, 6). Another feature seems at first glance also to fit into this pattern; the robust root systems of the premolars in which the fusion process has not gone as far as it has in *Australopithecus*. However it seems to me that perhaps this is partly due simply to absolute size and thus maybe to a heterogonic effect. The large size is, in any event, due to the dietary demands, and thus the less advanced root fusion really is part of the difference in diet and is not a separate difference.

There do not appear to be any features in which *Paranthropus* is clearly more advanced in the hominine direction than is *Australopithecus*. The small, hominine-like canines may appear to be an obvious exception to this statement. There is no doubt that the canines of *Paranthropus* match those of modern man better, in both size and structure, than do those of *Australopithecus*. However the canines of the latter form are very much more like those of *early* hominines and therefore are in fact better antecedents for the hominine canines than are those of *Paranthropus*. But the latter form cannot be held to be more primitive than *Australopithecus* in this respect. Here the combination of vegetarian diet and erect posture has allowed the canines to reduce and has thus produced what can only be regarded as a specialized condition. A somewhat similar condition holds for dm_1; the tendency in the deciduous molars and the permanent premolars of hominids as compared to the other higher primates, is toward greater molarisation. In *Paranthropus* dm_1 is clearly more molarised than is that of *Australopithecus*; hence it may be argued that in this respect the former is more advanced than the latter. This is undoubtedly a valid conclusion—in regard to degree of molarisation. But no hominines have the highly molarised *Paranthropus* type of tooth—they all have precisely the type of tooth that is found in *Australopithecus*. That is to say that where this tooth is known among hominines it is of the *Australopithecus* type. Again *Paranthropus* cannot be said to be more primitive in this respect. As in the case of the canines it is merely aberrant as far as the hominine line is concerned.

The position thus is that in some important respects *Paranthropus* demonstrably is more primitive than *Australopithecus*, judged from the viewpoint of hominines as the end-forms with which we are concerned. In some respects the former is merely aberrant, but in none does it appear to be more advanced than *Australopithecus*. There seems to be no alternative therefore to the conclusion that *Paranthropus* is the more primitive of the two Australopithecines. In view of this conclusion and the evidence which suggests that the forms coexisted in Africa but that *Paranthropus* probably survived longer in the Old World than did *Australopithecus*, it would seem clear that they belonged to separate, divergent phyletic lines. It would also seem clear that these lines had a common ancestor.

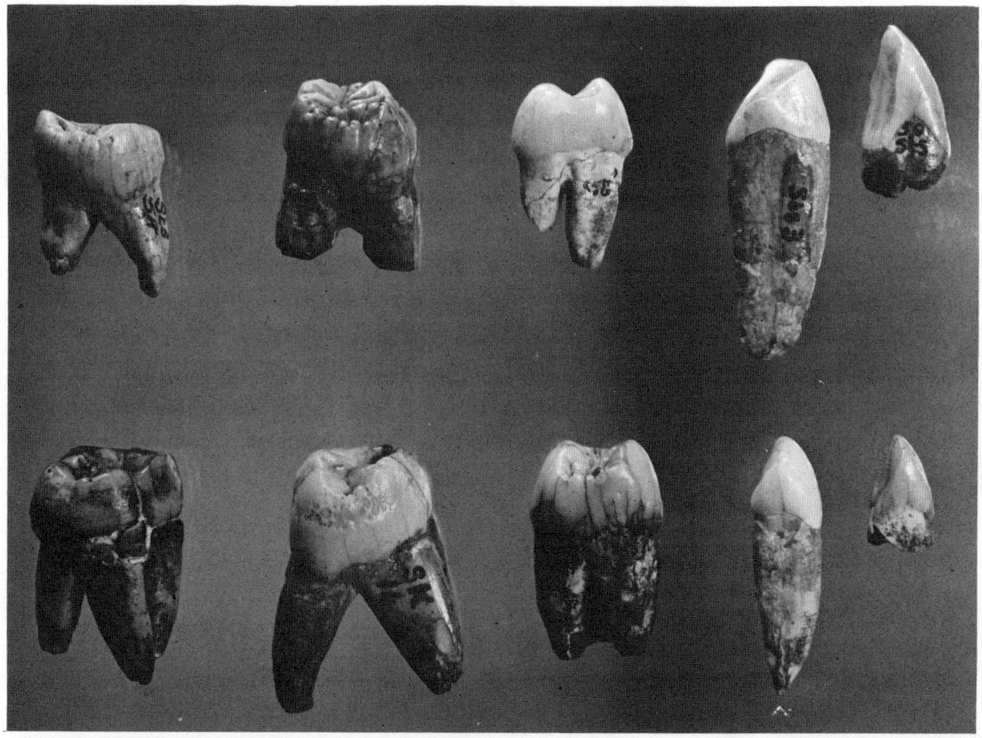

FIGURE 6
Teeth in the upper row belong to *Australopithecus;* those in the lower, to *Paranthropus.* The cheek teeth in both cases are maxillary, while the canines are mandibular. Note the greater development of the root systems in *Paranthropus;* these are average-sized teeth, while the second from the left in the upper row is a very large tooth for Australopithecus. The canines are in marked contrast to the cheek teeth—the canines of the small and lightly built *Australopithecus* are robust, while those of the large *Paranthropus* are small. Note hypoplastic enamel of the second tooth from left, lower row. This was presumably the result of a disease in childhood.

If the above conclusions are valid, then *Paranthropus* must be a less modified descendant of the ancestral Australopithecine than is *Australopithecus.* Not that the ancestral form will have had the exaggerated characters seen in the known specimens of *Paranthropus.* The canines, for example, will have been larger and therefore more in proportion to the cheek teeth. Body size is likely to have been smaller and therefore probably the skull will have been somewhat more gracile. This early *Paranthropus* will therefore have differed less from the known *Australopithecus* material than does the known, later, *Paranthropus* material. But it will nevertheless have been more nearly *Paranthropus* than *Australopithecus* because of diet, absence of forehead, pongid-like ischium, primitive nasal area, and probably many other things of which we are as yet unaware.

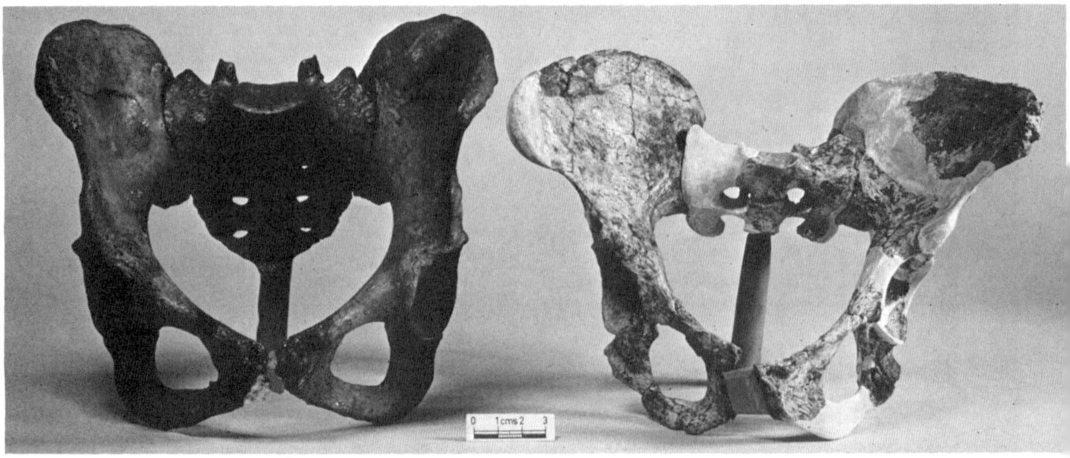

FIGURE 7
The pelvis of a female Bantu (left) and a female *Australopithecus* (right). In the latter, virtually nothing is missing except the lower portion of the sacrum. Owing to distortion during fossilization, the pelvis is no longer quite symmetric.

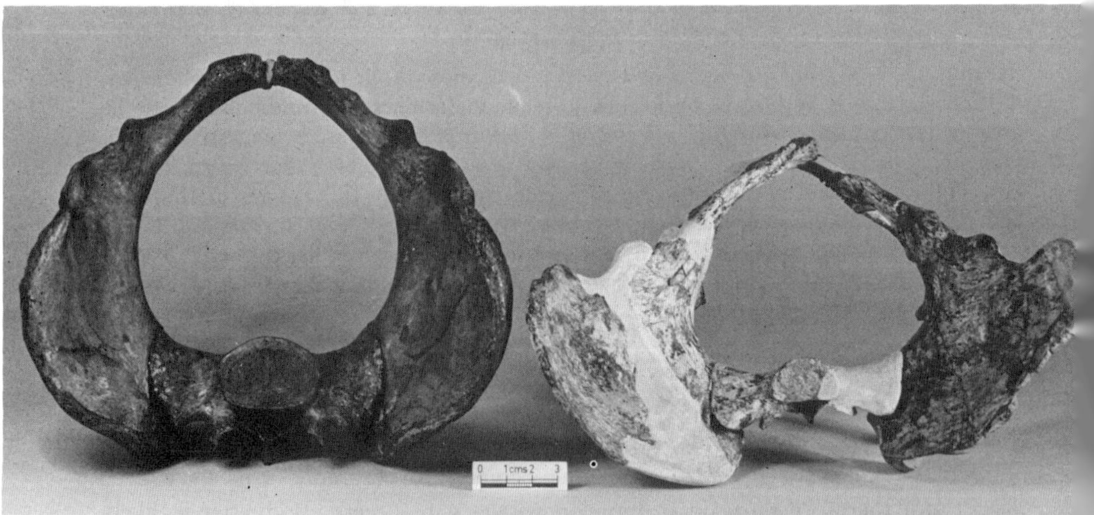

FIGURE 8
The pelvis of a female Bantu (left) and a female *Australopithecus* (right). The pelvic opening in the latter specimen is slightly smaller than it originally was and not quite symmetrical, owing to post-mortem distortion. This individual, of which a good deal of the skeleton is known, was about 4 feet in height and very lightly built. The pelvis is therefore relatively large.

ECOLOGY AND ADAPTIVE RADIATION

If *Paranthropus* represents basically the original Australopithecine stock and *Australopithecus* represents an adaptively different line evolving in a different direction, how did the latter line arise?

It seems unlikely that the earliest Australopithecines can have been as recent in age as the Pleistocene since the two phyletic lines were already well differentiated early in that period. On the other hand it seems logical to suppose that tool-using, tool-making, and increased brain size are virtually inevitable consequences of erect posture and that they will have followed the origin of the latter fairly rapidly in terms of the geological time scale. Consequently it is more likely that Australopithecines originated in the latter half of the Tertiary than in the earlier half; probably in the Pliocene, just possibly in the Miocene.

There is reason to believe that most of the Miocene was a period of expanding forests in Africa, but that the late Miocene and Pliocene was a time of desiccation and shrinking forests. The Kalahari sands of central and southern Africa throw some light on this matter. The original Kalahari sands overlie unconformably the Kalahari Limestone plain, which resulted from the African erosion cycle of early to mid-Tertiary times. However they predate the cutting of the Kalahari rivers into the Limestone in the lower Pleistocene. It would therefore seem that between the wetter period of the earlier Miocene and that of the early Pleistocene, considerable desiccation occurred during which the extensive deposits of Kalahari sand formed. These extend from fairly far south in South Africa right up into the Congo basin. The studies of botanists and of entomologists studying humicolous faunas support these conclusions in demonstrating marked forest expansion in the Miocene and equally marked recession in the Pliocene, leaving residual forests in a ring round the central Congo basin and in East Africa, and with a certain amount of expansion again in the Pleistocene. (See, for example, Mabbutt 1955, 1957; Cahen & Lepersonne 1952; also private communication from Leleup on humicolous faunas.)

One may conclude from this that suitable habitats for the vegetarian, original Australopithecines (*Paranthropus*) line will have become increasingly scarce through the late Tertiary. This will have been as true for other forms requiring forest or broken forest habitat and reasonably moist conditions, hence it could be expected that competition for such environments may have been more severe than usual. On the other hand grass savanna and other more arid environments will have expanded at this time, thus providing increased opportunity for animals adapted to, or capable of adapting to, such conditions.

The climatic changes in the desiccation process will not have been sudden. Australopithecines living in areas which subsequently became semiarid will have found that the dry season gradually became longer and drier. The critical time of the year, the latter part of the dry season, will gradually have become more difficult to cope with. It is reasonable to suppose that in these times of hardship insects, reptiles, small mammals, the eggs and nestlings of birds, etc., will have been eaten to supplement their diet. It is known that purely vegetarian primates will readily eat meat in captivity and that baboons, for example, will occasionally do so in the wild. Taking to a certain amount of meat-eating under environmental pressure could therefore occur fairly easily. As desiccation proceeded, such a deme will have found that it had to rely on the seasonal supplement to its normal vegetarian diet more frequently and to a greater degree. Under these circumstances it could be expected that population density will have dropped, probably to vanishing point in the most heavily affected areas. But it is probable that in at least some areas the creatures will have adapted satisfactorily to the altering circumstances and adopted a certain amount of carnivorousness as a normal part of their way of life. That is to say, the originally vegetarian diet will have become altered by the addition of a certain amount of meat-eating to an omnivorous diet.

It is quite clear, however, that such modifications to the environment will have altered the nature of selection acting on the group. Even an elementary level of tool-using will have had obvious advantages in the changing food situation. For the vegetarian part of their food, implements for digging will have made possible greater exploitation of the larger number of bulbs found in drier areas. Implements for bashing, hitting, or throwing, as well as digging, will have made capture and consumption of small animals much easier. Improved tool-using will thus have been favored by selection and any improvements in this respect will have improved adaptation, especially in respect of the carnivorous aspect of their diet. It is also obvious that improved intelligence will have been of great benefit in improving tool-using ability and dealing generally with the stresses of a somewhat hostile environment. Improved intelligence will consequently also have been favored by selection. Since there appears to be some relationship between intelligence and brain volume with regard to that portion of the range between the brain size of the larger pongids and that of the early hominines, it is probable that this part of the process of selection for improved intelligence will have been accompanied by an increase in brain volume. This would probably have shown up first as an increase in the size of the cerebral hemispheres; this would in turn have affected the brain-case by expanding the frontal region laterally and especially vertically. This is precisely what is seen in the brain-case of *Australopithecus* as compared to *Paranthropus.* Improved intelligence at this stage willl have improved tool-using ability so that progress in the direction of improved intelligence and improved use of tools will have reinforced each other and led to increasingly improved adaptation to the changing environment. This will have been especially true as far as meat-eating is concerned since improved tool-using will have made it increasingly easy to deal with the mechanics of capturing prey, penetrating the skin and removing the meat from larger animals, etc. Improved intelligence will have led also to improved hunting methods.

The changed environmental circumstances resulting from the known desiccation of a substantial part of Africa during the later Tertiary could therefore very easily have led to a second adaptive shift and the establishment of a second phyletic line in the Australopithecines. In this the introduction of a carnivorous element in the diet and an enhanced level of cultural activity were important features. *Australopithecus* is evidently precisely such a line. It is of interest that this form is present in the Sterkfontein Valley in the more arid periods, while *Paranthropus* is present only in the wetter periods (for climatic data see Brain 1958). The canines of *Australopithecus* are appreciably less reduced than those of *Paranthropus*, which suggests that the former genus arose from the latter well before the reduction of the anterior teeth in the latter had reached the stage seen in the known forms. Adaptation to an omnivorous diet which included an appreciable amount of meat-eating will have kept the canines as large as they originally were or perhaps even increased their size slightly.

THE ORIGIN OF HOMININES

Once the line adapting to drier conditions and altered ecology had become established, thus producing *Australopithecus* as we know it, its evolution would not have stopped there. The selection pressures operating, and entirely different from those controlling the direction of the *Paranthropus* line, would not cease to operate and it was virtually inevitable that adaptation would be carried well past the *Australopithecus* stage. The cultural situation would by then have become the vital factor. The need for tool-using in successfully adapting to the different way of life would, as indicated, place a high premium on intelligence. As this improved, presumably by an increase in size of the cerebral cortex so as to provide increased correlation and association areas, cultural facility also improved. When the modification of the brain had reached a point where hominine levels of intellectual ability began to appear, apparently when brain volume reached the general order of about 800 to 1,000 cu. cm., facility with tools reached the point where a characteristic hominine phenomenon appeared: the deliberate manufacture of tools for particular purposes. This provided still further scope for development and the increase in brain size occurred rapidly to about the modern volume. At this point it seems that correlation between brain size and intelligence is not especially close since all manner of other factors are involved. Cultural ability did not improve at the same rapid rate that applied to brain volume, presumably because there were many problems of communication and organization. But momentum gradually built up and the rate of change seems still to be increasing. "Telanthropus," from Swartkrans, was apparently an early member of this hominine stage. This form has now been included in the genus *Homo* (Robinson 1961a) in which, it seems to me, all hominines should go. Definitions of the genera *Australopithecus*, *Paranthropus*, and *Homo* are given in Robinson 1961b. From the Sterkfontein Valley have come, therefore, members of both the major lines of Australopithecine evolution as well as members of both stages of the *Australopithecus-Homo* stream. This is also true of the Olduvai region of Tanganyika.

CONCLUSION

The evolution of hominids thus seems to have involved two critical points or thresholds where adaptive shifts occurred. The first was an essentially anatomical change, the shift from quadrupedal to erectly bipedal locomotor habit. This freed the hands and opened up the possibility of becoming an efficient tool-user. However, with a vegetarian diet and suitable environmental conditions to provide the necessities for the way of life of the first Australopithecines, there would probably be comparatively little stimulus for any major change in the adaptive situation then existing.

The second critical point was the inclusion of meat as a normal part of the diet. This was a direct response to altered and altering environmental conditions. Increasing aridity over much of Africa, from which the vast majority of Australopithecine specimens are known, brought about the change in diet. The second adaptive shift was therefore primarily ecological. But the change in diet placed a premium on tool-using, that is on cultural activity, and on improved intelligence. The change in climatic conditions thus provided precisely the stimulus needed for rapid development of the potentialities present in the prospective adaptation represented by erect bipedal posture and freed hands. An inevitable consequence was the appearance of hominines with their relatively very high intellectual and cultural ability and the new means of adaptation, artificial adaptation, with the consequent effect on the pattern of speciation.

Once *Paranthropus* had given rise to the *Australopithecus* line it remained, as it were, in an evolutionary backwater. In the long run it was unable to hold its own and became extinct. *Australopithecus* evolved fairly rapidly, compared to *Paranthropus*, and soon the first hominines came into existence. This did not happen by the transformation of all demes of *Australopithecus*, since this form was still in existence after hominines had already appeared. But the competition between the two was probably such that as the hominines spread *Australopithecus* lost ground and became extinct. The whole process is graphically summarized in Figure 9.

The second adaptive shift, involving the incorporation of meat-eating in the diet, seems to me to have been an evolutionary change of enormous importance which opened up a vast new evolutionary field. This change in my opinion ranks in evolutionary importance with the origin of mammals—perhaps more

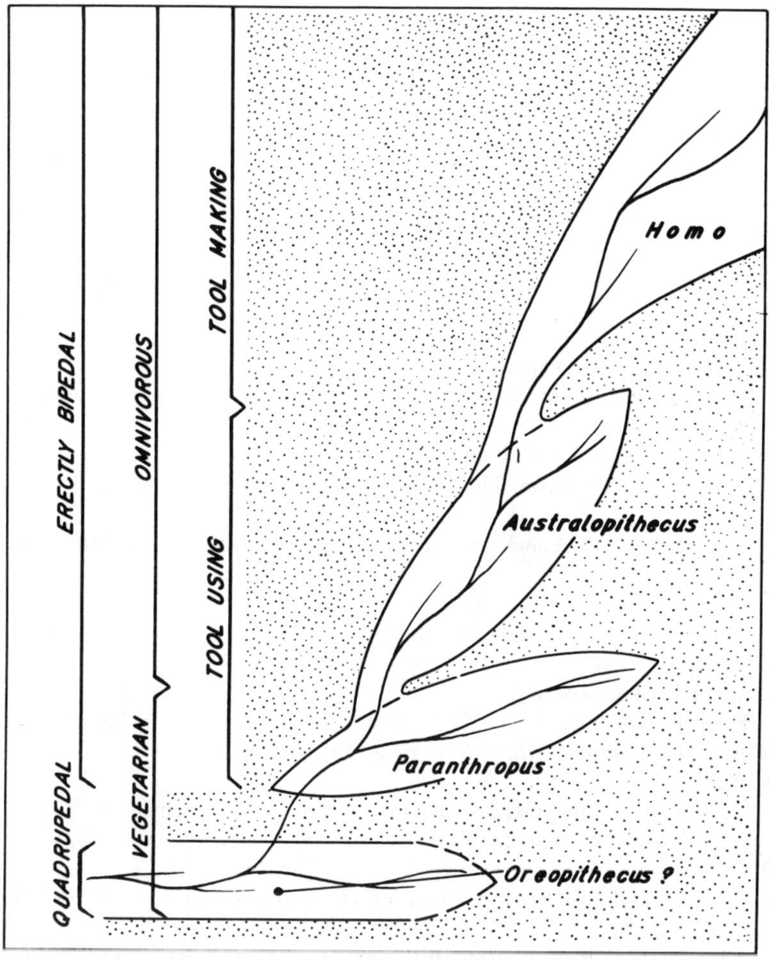

FIGURE 9
Diagrammatic representation of the more important adaptive zones occupied by the hominid evolutionary stream. The threshold between the quadrupedal and bipedal stages is a major one between essentially discontinuous zones. The second and third thresholds—change to omnivorous diet and tool manufacture—are of great importance but did not involve clearly discontinuous zones. It should be emphasized that this is not a family tree, but an adaptive grid.

appropriately with the origin of tetrapods. With the relatively great expansion of intelligence and culture it introduced a new dimension and a new evolutionary mechanism into the evolutionary picture, which at best are only very palely foreshadowed in other animals. Furthermore it brought into existence the first member of the animal kingdom able to colonize other worlds—or at the very least, able physically to leave this one permanently.

BIBLIOGRAPHY

Ashton, E. H., and S. Zuckerman

1951. "Some Cranial Indices of *Plesianthropus* and Other Primates," *Amer. J. Phys. Anthrop.*, 9:283–96.

Bartholomew, G. A., and J. B. Birdsell

1953. "Ecology and the Proto-Hominids," *Amer. Anthrop.*, 55:481–98.

Brain, C. K.

1958. "The Transvaal Ape-Man-Bearing Cave Deposits," *Transvaal Mus. Mem.*, No. 11.

Broom, R.

1950. "The Genera and Species of the South African Fossil Ape-Men," *Amer. J. Phys. Anthrop.*, 8:1–13.

Cahen, L., and J. Lepersonne

1952. "Equivalence entre le Système du Kalahari du Congo Belge et les Kalahari Beds d'Afrique Australe," *Mém. Soc. Belge Géol.*, Sec. 8, 4:1–64.

Coppens, Y.

1961. "Découverte d'un Australopithecine dans le Villafranchien du Tchad," *C. R. Acad. Sci.* (Paris), 252:3851–2.

Dart, R. A.

1957a. "The Osteodontokeratic Culture of *Australopithecus prometheus.*" *Transvaal Mus. Mem.*, No. 10.

1957b. "The Makapansgat Australopithecine Osteodontokeratic Culture," *Proc. 3rd. Pan-Afr. Cong. Prehist.* (Livingstone 1955): 161–71.

1960. "The Bone-tool Manufacturing Ability of *Australopithecus prometheus*,"

1958. "Bone Tools and Porcupine Gnawing," *Amer. Anthrop.*, 60:715–24.

Amer. Anthrop., 62:134–43.

Dobzhansky, T.

1961. "Man and Natural Selection," *Amer. Sci.*, 49:285–299.

Hennig, E.

1948. "Quartärfaunen und Urgeschichte Ostafrikas," *Naturw. Rdsch. Jahrg.* 1 (Heft 5):212–17.

Howell, F. C.

1959. "The Villafranchian and Human Origins," *Science*, 130:831–44.

Hürzeler, J.

1958. "*Oreopithecus bamboli Gervais.* A preliminary report," *Verh. naturf. Ges. Basel*, 69:1–48.

Koenigswald, G. H. V. von

1957. "*Meganthropus* and the Australopithecinae," *Proc. 3d. Pan-Afr. Cong. Prehist.* (Livingstone 1955): 158–60.

LEAKEY, L. S. B.
1959. "A New Fossil Skull from Olduvai," *Nature* (London), 184:491–93.
1961a. "New Finds at Olduvai Gorge," *Nature* (London), 189:649–50.
1961b. "The Juvenile Mandible from Olduvai," *Nature* (London), 191:417–18.

LE GROS CLARK, W. E.
1950. "New Palaeontological Evidence Bearing on the Evolution of the Hominoidea," *Quart. Jour. Geol. Soc. Lond.*, 105:225–64.
1955. *The Fossil Evidence for Human Evolution.* Chicago: University of Chicago Press.

MABBUTT, J. A.
1955. "Erosion Surfaces in Namaqualand and the Ages of Surface Deposits in the South-western Kalahari," *Trans. Geol. Soc. S. Afr.*, 58:13–30.
1957. "Physiographic Evidence for the Age of the Kalahari Sands of the South-Western Kalahari," *Proc. 3d. Pan.-Afr. Cong. Prehist.* (Livingstone 1955); 123–6.

MARKS, P.
1953. "Preliminary Note on the Discovery of a New Jaw of *Meganthropus* von Koenigswald in the Lower Middle Pleistocene of Sangiran, Central Java," *Indonesian J. Nat. Sci.*, 109:26–33.

MASON, R. J.
1961. "The Earliest Tool-makers in South Africa," *S. Afr. J. Sci.*, 57:13–16.

MAYR, E.
1950. "Taxonomic Categories in Fossil Hominids," *Cold Spring Harbor Symp. on Quant. Biol.*, 15:109–18.

REMANE, A.
1951. "Die Zähne des Meganthropus africanus," *Z. Morph. Anthr.*, 42:311–29.

ROBINSON, J. T.
1953. *Meganthropus*, Australopithecines and Hominids," *Amer. J. Phys. Anthrop.*, 11:1–38.
1954a. "The Genera and Species of the Australopithecinae," *Amer. J. Phys. Anthrop.*, 12:181–200.
1954b. "Prehominid dentition and hominid evolution." *Evolution*, 8:324–34.
1955. "Further Remarks on the Relationship between *Meganthropus* and Australopithecines," *Amer. J. Phys. Anthrop.*, 13:429–45.
1956. "The Dentition of the Australopithecinae," *Transvaal Mus. Mem.*, No. 9.
1958. "Cranial Cresting Patterns and Their Significance in the Hominoidea," *Amer. J. Phys. Anthrop.*, 16:397–428.
1959. "A Bone Implement from Sterkfontein," *Nature*, 184:583–5.
1960. "The Affinities of the New Olduvai Australopithecine," *Nature*, 186:456–7.
1961a. "The Australopithecines and Their Bearing on the Origin of Man and of Stone Tool-making," *S. Afr. J. Sci.*, 57:3–13.
1961b. "The Origin and Adaptive Radiation of the Australopithecines." In *Evolution und Hominization* (G. Kurth, ed.). Stuttgart: G. Fischer, pp. 120–40.
1962. "Australopithecines and Artefacts at Sterkfontein," *S. Afr. Archaeol. Bull.*, 17:87–126.

ROBINSON, J. T., and R. MASON
1957. "Occurrence of Stone Artifacts with *Australopithecus* at Sterkfontein," *Nature*, 180:521–4.

SCHULTZ, A. H.
1960. "Einige Beobachtungen und Mass am Skelett von *Oreopithecus*," *Z. Morph. Anthr.*, 50:136–49.

WASHBURN, S. L.
1957. "Australopithecines: The Hunters or the Hunted?" *Amer. Anthrop.*, 59:612–14.

WASHBURN, S. L., and F. C. HOWELL
1960. "Human Evolution and Culture." In *Evolution after Darwin*, Vol. 2, Chicago: University of Chicago Press, pp. 33–56.

WASHBURN, S. L., and B. PATTERSON
1951. "Evolutionary Importance of the South African Man-apes," *Nature*, 167:650–1.

WEIDENREICH, F.
1945. "Giant Early Man from Java and South China," *Anthrop. Pap. Amer. Mus. Nat. Hist.*, 40:1.

WEINERT, H.
1950. "Uber die Neuen vor- und Fruhmenschenfunde aus Africa, Java, China und Frankreich," Z. *Morph. Anthr.*, 42:113–48.
1951. "Uber die Vielgestaltigkeit der Summoprimaten vor der Menschwerdung," Z. *Morph. Anthr.*, 43:73–103.

HUMAN EVOLUTION IN MOROCCO IN THE FRAMEWORK OF THE PALEOCLIMATIC VARIATIONS OF THE ATLANTIC PLEISTOCENE

P. BIBERSON

INTRODUCTION

IN A SYMPOSIUM on primitive man and Pleistocene stratigraphy in the Mediterranean basin, held at Burg Wartenstein in 1960, I offered a picture of the evolution of the Moroccan Paleolithic within the framework of the Atlantic Pleistocene. In it, I summarized the chronology established by geographers, geologists, and prehistorians on the morphological, stratigraphic, and paleontological data, in order to incorporate with them certain archaeological discoveries (Biberson 1960a).

In that schematic synthesis, I gave a quick resume of observations allowing the reconstruction of the paleogeographic environment in which the evolution of the Paleolithic took place. The present symposium gives me the chance for greater precision, permitting to a certain extent a reconstruction of the successive ecological environments in which human evolution occurred in Atlantic Morocco.

The best information on the climatic variations during the succession of pluvial or interpluvial periods of the North African Quaternary should come from analyses of palynology. Lacking a collection of present-day Moroccan pollens, the study of the samples of diverse formations which I collected before 1956 has not yet been possible. In spite of this serious gap, by the concurrent use of lithic, granulometric, pedological, and paleontological data supplied by vertebrate and invertebrate fossils, a certain number of presuppositions can nevertheless be advanced which are only gratuitous hypotheses, even though they may be subject to individual interpretations by specialists.

If the human skeletal remains of Atlantic Morocco are still rare, stone industries are very well represented and permit us to trace the evolution of prehistoric cultures from the Lower Pleistocene to the Holocene. So it is possible to proceed, on archaeological and geological grounds, to comparisons of the greatest interest with other, more favored African countries and thus make up—varyingly according to the case—the gaps found in Morocco.

TABLE 1

Evolution of Lower Paleolithic Industries in Atlantic Morocco

MARINE CYCLES		CLIMATIC CYCLES			'CIVIL-ISATIONS'	STONE INDUSTRIES	Industrial Stages	TECHNIQUES	TYPE SITES
EUROPE	MOROCCO	EUROPE	EAST AFRICA	MOROCCO					
		Final Würm		Soltanian					
Neotyrrhenian	Ouljian						Aterian	Pedanculated tools	Superficial red sediments (Level A)
		Würm I	Gamblian	Presoltanian			Stage VIII	Levalloiso-Mousterian workmanship	Cap Chatelier (Level D_2)
Tyrrhenian	Harounian					Evolved Acheulian			
		Riss	Kanjeran	Tensiftian			Stage VII	Oval bifaces Evolved circular core	Sidi Abderrahman-Extension (Level E)
					Biface				
							Stage VI	Amygdaloid biface	Littorina Cave (Level D_0)
Milazzian	Anfatian				Civilization	Middle Acheulian	Stage V	Older subcircular core	Bear Cave (Level G_0)
							Stage IV	Primitive prepared core	Sidi Abderrahman-Extension (Level F)
		Mindel	Kamasian	Amirian			Stage III		S.T.I.C. pit (Level D)
						Older Acheulian	Stage II	Trihedrals	Sidi Abderrahman - Old Workings (Level M, Series 3)
							Stage I	Formless unprepared core	
									Sidi Abderrahman - Old Workings (Level M, Series 2)
Sicilian	Maarifian					Evolved Pebble Culture	Stage IV	Bi- and multidirectional working of pebbles	Sidi Abderrahman-Extension (Level G)
		Günz		Saletian	Trimmed pebble civilization		Stage III		Souk-el-Arba du Rharb
Calabrian	Messaoudian		Kageran			Older Pebble Culture	Stage II	Unidirectional working of pebbles	Deprez Pit (Casablanca)
		Donau		Moulouyian			Stage I		Tardiguet-er-Rahla
		Biber (?)	?	Lower Villafrauchian					

To clarify the sequence of ancient cultures within the framework of the Atlantic Pleistocene, Table 1 gives a schematic picture of the sequence of archeological stages within their paleogeographical framework.

It goes without saying that the correlations presented, based on the studies of various Moroccan geologists and my own observations, between the Moroccan chronology on the one hand, and that of the Alps, the Mediterranean, or eastern Africa on the other, are *very largely* hypothetical and by necessity subjective, given the present state of our knowledge. They are given only as approximations to establish, according to importance, the ideas of those who are not too familiar with purely local Moroccan terminology. This terminology is used here specifically to avoid serious errors that might otherwise arise from implying *a priori* correlations, always extremely dangerous when they are taken literally as is practically inevitable.

Only four human fossils from the Lower Paleolithic fit this stratigraphic and archaeological frame.[1] In the order of their age, on our scale of relative chronology, these are: the *Atlanthropus* of Sidi Abderrahman, Rabat man, Témara man, and Tangier man. At first glance this might seem pretty mediocre and yet few African or even European countries can boast of having revealed comparable evidence for periods as ancient. By a happy chance, these four finds are related to four successive stratigraphic levels and therefore to four stages of our evolutionary sequence of industries. On the other hand, certain morphological characters show that at least the most ancient of these fossil men are related to other early hominids of North Africa who are also associated with prehistoric industries. This allows, by archeological comparisons not altogether fictitional, serious presumptions on the nature of the makers in the oldest industrial stages in Morocco, in spite of the lack of paleontological data.

The most recent studies of prehistory emphasize the very African nature of the Moroccan industries, and, thereby, the great cultural unity of Africa during the oldest part of the Lower Paleolithic. It can be said that Pebble Culture ("Civilisation du Galet aménagé"), where its evolution could be studied in detail, developed everywhere by an extremely slow process and in an identical manner. The subtle transition between this Pebble Culture and the Hand Axe Culture ("Civilisation du Biface") appears, from what evidence can be found, to have been made in a similar manner. This unity is still evident in the Lower Acheulean, even where the variety of raw material introduces technological and typological nuances. The Middle Acheulean shows these nuances, but

1. While this paper was in press, we learned of the discovery of a Neanderthal skull in a pocket of red silt in a barytine mine in Jebel Irhoud, 60 km. southeast of Safi (Ennouchi 1962: "Un crane d'homme ancien au Jebel Irhoud (Maroc)," *C. R. Acad. Sci*, 254:4330–2). This makes five human fossils in Morocco. The petrographic character of the sediment and the associated fauna permit, after first report, the dating of the site to the Soltanian *sensu stricto*. Hence the skull would be contemporary with the Aterian. More investigations are needed, but the presence of a true Neanderthal in North Africa, clearly established for the first time, requires a modification of previous conclusions.

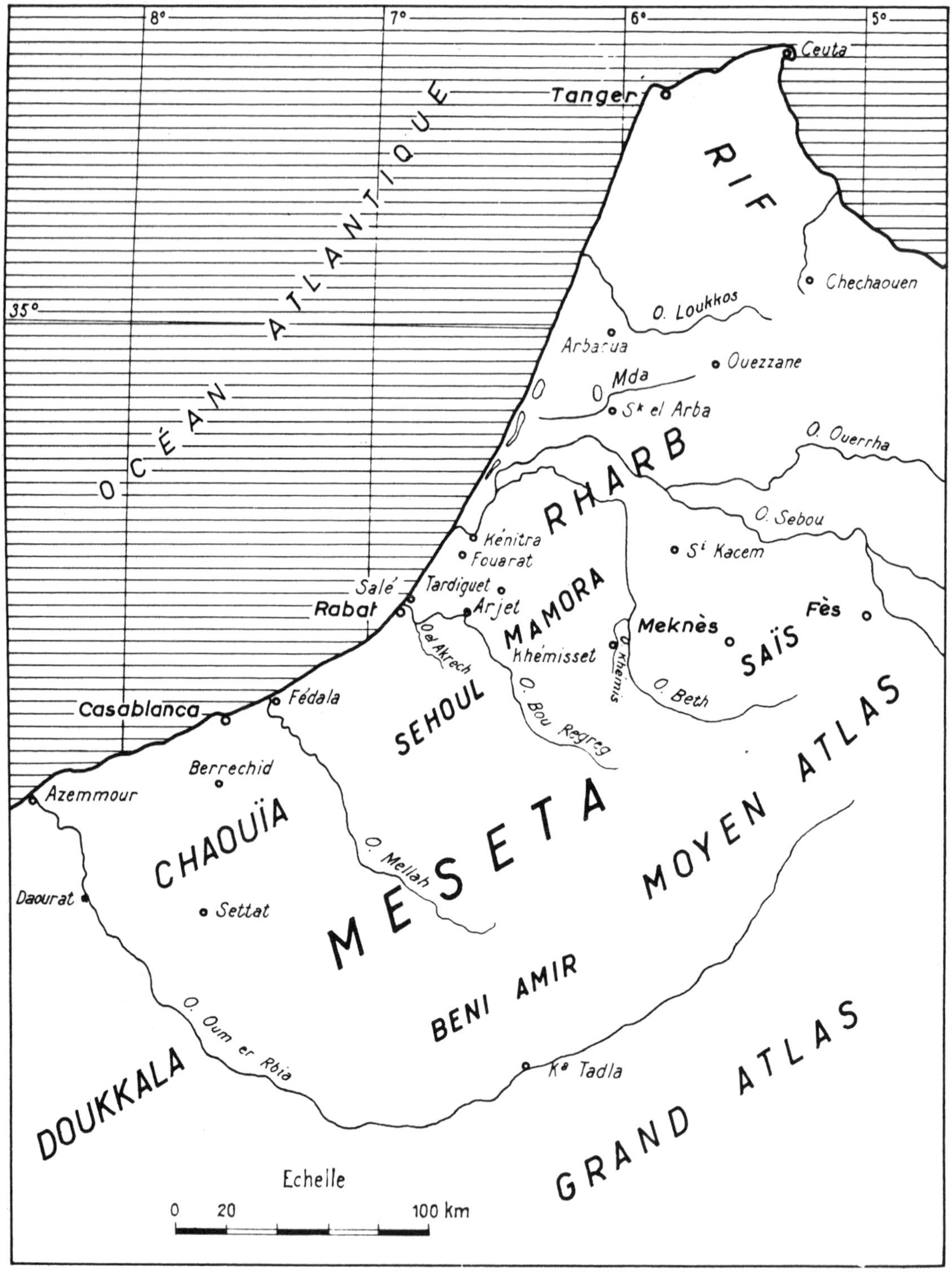

FIGURE 1
Principal sites of northern Atlantic Morocco.

apparent similarities are still very striking for industries widely separated in space (perhaps also in time). It is really only in the Evolved Acheulean that the divergences are affirmed and specialized industries are born, which creates true archeological provinces in diverse African regions.

From this, it may appear rash to extrapolate anthropological characteristics for these tool-makers. Nevertheless it can be done at least for the oldest periods if we keep to broad generalities.

We will now attempt to trace chronologically the various steps of human evolution in Morocco within the framework of the paleoclimatic variations of the Atlantic Pleistocene, keeping in mind that the attempt rests on still fragile bases requiring confirmation on many points by new observations.

I. LOWER PLEISTOCENE

A. The Lower Villafranchian

I shall not consider the Pliocene-Pleistocene boundary, which is still controversial (Biberson 1961a). We will begin with the Lower Villafranchian, although certain Moroccan geologists still place it in the continental Upper Pliocene (Gigout 1957).

In Atlantic Morocco, three ancient river estuaries have yielded mammal fauna of this period. These are: Fourarat (Choubert, Ennouchi, and Marçais 1948); the Oued El Akrech (Ennouchi 1951); and Daourat (Ennouchi, Gigout, and Marçais 1951).

The first is in the Rharb, a few kilometers south of Kénitra, composed of deposits of an ancient delta of Oued Sebou. The fauna is essentially *Anancus osiris* Aramb. and *Elephas africanavus* Aramb. Given the well-known subsidence of the Rharb plain, it is not surprising to find it at a very low altitude, almost zero.

The second, from the ballast-pit of Oued el Akrech near Argoub el Hafid, at an altitude of about 120 to 130 meters, is about 10 kilometers from the present coast, in the estuary of an ancient Bou Regreg. It has yielded *Elephas africanavus* Aramb., *Stylohipparion libycum* Pom., *Rhinoceros sp.*, and *Bos sp.*

The third is an abandoned quarry of Daourat, in the Oum-er-Rbia valley, 50 kilometers from the mouth, at an altitude of 180 meters. It has only supplied two molars of *Anancus osiris* Aramb., one of an adult, the other of a young individual. Although not very conglomeratic, it is still an estuary formation.

The lower fluvio-marine strata of the three sites have yielded marine molluscs belonging to the Astian facies of the Lower Pliocene regression (*Moghrébien* of Choubert and Ambroggi 1953). The association, at least in the first two, of mastodons and primitive elephants of the *E. planifrons* group and of *Stylohipparion*, convincingly demonstrates that, from the paleontological point of view, we must date them to the beginning of the Pleistocene. C. Arambourg particularly (Arambourg and Choubert 1957) compares them with the fauna of Ain Boucherit (Algeria) and of Lake Ichkeul (Tunisia).

The formations of the Lower Villafranchian are also represented by an important dune system which constitutes the morphologically dominant character of Atlantic Sahel, and which accompanies the regressive deposits of the Lower Pliocene; in the Moroccan interior diverse lacustrine formations are attributed to it, for example, in Saïs, between Fès and Meknès (Taltasse 1953).

If there was a climatic alteration in this earliest Pleistocene episode, nothing in the petrographic nature of the formations or in the composition of the fauna shows it. Morocco seems in that epoch, as in the Pliocene, to have had a humid and hot climate of subtropical type.

No hominid fossils have yet been found at this level, nor any certain prehistoric industry.[2]

B. The Moulouyian

The first periglacial terrace of the Moulouya Valley, incised in the morphological surface of the Lower Villafranchian, in eastern Morocco, has been accepted as typical of a new climatic episode which the geologists and geographers of Morocco named the Moulouyian. "Formations with periglacial character are very pervasive and extend widely along the piedmont, but deposits or glacial forms corresponding to this cycle have not yet been identified" (Choubert, Gigout, Joly, Marçais, Margat, and Raynal 1956).

In the Atlantic zone, specialists now attribute to it very characteristic formations, for a long time determined as Villafranchian and known previously under the name of "red formation of the Mamora" (Choubert 1950).

These are rubefied pebble and sandy sediment deposits of clear tint (corn-poppy red), more conglomeratic in the north of the Rharb plain where they may reach a depth of 200 m. in the brachysyncline of Oued Mda (Monition 1956), while toward the south in the Rabat region, there are finer sediments that hardly exceed 20 to 30 m. thickness (Choubert and Roche 1956). They cut into the sands of the Lower Villafranchian and most often cover the interdune depressions.

Unfortunately, these are generally acid formations without preserved fossils, so their fauna is, for the moment, unknown. This serious gap makes it difficult to correlate them with other fossil layers of the Algerian or Tunisian Villa-

2. However, it should be noted that in the pebbles of the basal levels resting directly on the Pliocene sands, in the layers of Tardiguet-er-Rahla (which will be discussed later), we collected (1957) some ten rounded pebbles longitudinally or transversally split, with fresh fractures, which resemble closely the poorest pieces of the earliest Kafuan of Uganda (C. Van Riet Lowe 1951). They did not appear to be decisive testimony of systematic human endeavor (Biberson, Choubert, Faure-Muret, and Lecointre 1960). Also, in the absence of fauna, it is not certain that these beds actually belong to the Lower Villafranchian; I would sooner attribute them to the Moulouyian (Biberson 1961a).

franchian in which this climatic cycle does not seem to have been separate from the Lower Villafranchian (Biberson 1961a).

In contrast, although still rare, well-characterized prehistoric industries have been recovered *in situ,* although in small quantities, in three stratigraphically well-dated sites of Atlantic Morocco. These are: the Arbaoua site, north of the Rharb plain; the Tardiguet-er-Rahla site in the Oued Tiflet Valley in the Mamora forest; and the ballast-pit of Douar Doum near Rabat (Biberson, Choubert, Faure-Muret, and Lecointre 1956, 1960).

The first of these is of no more than historical interest. Archaeologically, the second is valuable because the purity of its industry is assured by the absence of any possibility of redistribution. Stratigraphically, the last site is particularly interesting because it provides a revealing section (Biberson 1961b).

It happens that the summit of the Moulouyian beds was frequently enough affected by the phenomena of solifluction, of which we have several typical examples (Biberson 1961a); but it is this upper horizon that yields the first indisputable industries of Atlantic Morocco, which I made Stage I of the "older Pebble Culture."

The pebbles and even certain tools of this level are sometimes greatly altered by physico-chemical actions and even seem, according to certain specialists, to have been frost-cracked. If it appears difficult to admit periglacial phenomena at the latitude of Morocco, at so low an altitude and in such close proximity to the ocean, it must be recognized that these verified phenomena, along with the size of the transported materials, indicate, in relation to the Lower Villafranchian, a climatic alteration, probably toward greater coolness and certainly toward increased pluviality. The rubefaction of the deposits alone would certainly indicate an important climatic change, but we do not know what that meant for the environment in which the first known pebble-tool makers lived, or if the appearance of these primordial industries in Morocco had any relation to the deterioration of the climate and the arrival of this wet and cold period.

We are, then, reduced to conjectures about the physical characteristics of the makers of the "older Pebble Culture" of Morocco. Like C. Arambourg (following the recent colloquium in Paris on "L'Evolution des Vertebrés"), I would not be surprised if it is a question of prehominids related to the group of evolved Australopithecids. The recent discovery in Chad of an individual of this type, associated with fauna characteristic of the Lower Villafranchian in North Africa, reinforces this impression. But it will remain a mere hypothesis until new discoveries in Morocco proper provide additional data.

C. The Messaoudian

The Moulouyian terminates at the coast with the arrival of the first Pleistocene transgression of the Atlantic Ocean. I have called this the Messaoudian in

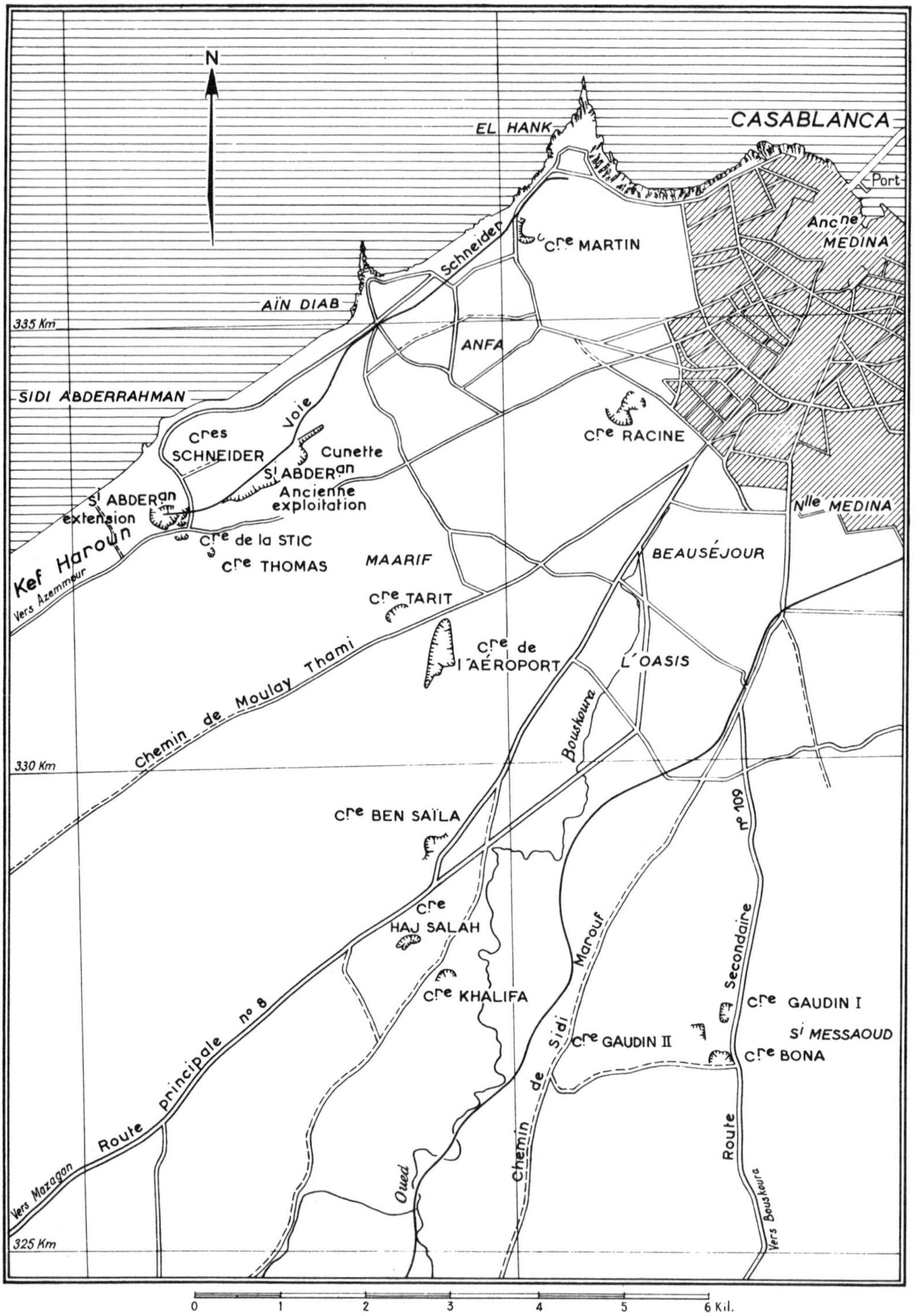

FIGURE 2
Map of locations of pits around Casablanca exposing the Pleistocene succession.

memory of the work of Neuville and Ruhlmann who first, in 1941, defined it after the study of quarries of Casablanca near the sanctuary of Sidi Messaoud (Biberson 1958).

The beach deposits of this marine cycle have been recognized at several points on the Casablancan outskirts, on an average of some 10 kilometers from the present shore, between 70 and 110 meters altitude. It seems that the maximum beach line reached on the reputedly stable part of the Moroccan Meseta (at Casablanca's latitude, for example) is found near the 100-meter contour.

The marine molluscan fauna is well known owing to the studies of G. Lecointre, who made it the type for his "ancient Quaternary I" (1953). It is characterized by the persistence of Pliocene species, notably *Ostrea (Gryphaea) virleti*, with all the transitional forms to *Ostrea (Gryphaea) cucullata*, var. *crassa* Lecointre, and the abundance of "Chilo-Peruvian" gastropods appearing in the final Pliocene of Morocco (Choubert's *Moghrébien*): *Acanthina crassilabrum* Lmk. or *Purpura plessisi* Lecointre, and *Trochatella trochiformis* Gmelin.

"Senegalian" species are abundant; I shall not cite the identifications G. Lecointre made on my finds (Biberson 1961a), but only point out the presence of *Mesalia brevialis* Lmk. and also of *Nerita* cf. *senegalensis* Gmelin, which is the species formerly described by Adanson under the name *Dunar*, a fossil until now unknown in Morocco.

This fauna has no "cold" character, contrary to that of the Calabrian or of the Sicilian of the Mediterranean. The abundance of *Ostrea cucullata* Born., whose present habitat is found mainly in the Indian Ocean, the Red Sea, and the Atlantic of the western African shores; as well as that of *Trochatella trochiformis* still living in the Cape Verde Islands; and of *Acanthina crassilabrum*, which seems not to survive today except on the shores of the Pacific in Chile and in Peru, seems certainly to indicate a marine littoral biotope of subtropical type.

Unfortunately, these ancient Moroccan beaches have yielded very few indications of the mammalian fauna of the period. I gathered fragments of tusks of a large proboscidean, unfortunately shattered, which are indeterminable.

In contrast, a few sites in certain quarries of Casablanca furnished pebble-tool industries that are very crude but similar to the archaic Pebble Culture of the Moulouyian. I made them Stage II of the Pebble Culture (Biberson 1961b). Nothing is yet known about the hominids, who seem, if they are judged by their tools, not to have been very different from those of Moulouyian.

This first positive movement of the ocean in the Pleistocene is very general on the shore of the Moroccan Meseta. It cannot be the result of a simple epirogenic deformation of the continent, but rather of a general climatic phenomenon causing a eustatic transgression. It corresponds, in effect, to an interpluvial period on the continent of which practically nothing is known unless it is that the aggradation period of the Moulouyian is replaced by a period of erosion. Nevertheless, the existence of interpluvial conditions at this period is consistent with the "warm" character of the littoral marine fauna.

D. The Saletian

Contrary to what is observed at the time of maximum transgression, the continental deposits banked against the beaches of the post-Messaoudian regression indicate a new pluvial period, the traces of which are clearly visible in many sites of the Casablancan outskirts and in numerous North Moroccan sectors.

The new climatic cycle which these phenomena of accumulation indicate has been called Saletian by geologists and geographers of Morocco, after the "site on the Salé plateau" on the right bank of the mouth of Oued Bou Regreg (Choubert and Roche 1956).

Here, the exploitation of ballast-pits and the profiles made by drainage trenches show an important gravel spread with very pronounced brown patina, which cuts into the Moulouyian of the "red Mamora formation" type. The coarse pebbles in the base levels often have traces of frost action at very low altitudes, indicating a cooler climate than in the Moulouyian. "In the high piedmonts and mountains these accumulations seem to correspond, in certain specified cases, to moraines. The coastal Saletian is characterized by a broad development of ferruginous concretions" (Choubert, Joly, Gigout, Marçais, Margat, and Raynal 1956).

One site in Casablanca is particularly interesting—the Schneider quarry of Maarif airport, which presents two well-separated archeological beds. The first is represented by a marine level at the base, precisely dated by its molluscan fauna (of *Acanthina crassilabrum* and *Trochatella trochiformis*) of the Messaoudian; it provided a collection of tools belonging to Stage II of the Pebble Culture. These beach deposits are cut into by continental formations of pebble beds and of pisolitic ferruginous sediments of the Saletian, which constitutes the second bed. This latter contains a new industry, quite different from the preceding, according to which I made Stage III the beginning of the Evolved Pebble Culture of Atlantic Morocco (Biberson 1961b).

The gravels and sediments of the Saletian are abundantly represented in the entire North Moroccan zone bordering the Atlantic. In the northern part of Rharb the beds are slightly tilted, indicating the effects of the last manifestations of the Riffian orogeny (Biberson, Choubert, Faure-Muret, and Lecointre 1960).

Their prehistoric sites are much more numerous than those of the Moulouyian, and the concentration of pebble-tools in them is considerably greater. The ballast-pits of the site named after Salé have yielded good collections, but the site of Souk-el-Arba of Rharb is the richest and appears the most homogenous. The resemblance to the Oldowan of Tanganyika is the most striking, and to the Algerian site of Aïn Hanech no less remarkable (Biberson 1961b).

The nature of the formations of the littoral Saletian is not favorable to the conservation of fossil remains. The study of continental molluscs collected here caused P. Jodot to say that the narrow calcareous bed from which they come must have formed during a humid period, in a climate comparable to that now of the Gulf of Cadiz, thus cooler than the present climate of Casablanca.

The only paleontological piece representing mammalian fauna which may be attributed to this horizon is a mandible of a primitive elephant from the *E. meridionalis* group, found long ago by J. Bourcart at the edge of the Salé plateau. C. Arambourg relates it to *Elephas recki* Dietr. of the Kageran-Villafranchian of Omo (Arambourg 1959). In spite of this paucity, Arambourg does not hesitate to equate the Moroccan Saletian with the Upper Villafranchian of Algeria, which appears plausible. From this perspective, the Moroccan fauna of the period must have been similar to that of Aïn Hanech and of Bed I of Olduvai.

Bearing the archaeological arguments in mind, we may believe that the toolmaker of Stage III of the Evolved Pebble Culture of Morocco might have been a close relative of *Zinjanthropus* discovered at Olduvai by L. S. B. Leakey (Leakey 1959).

Here again, in order to be precise, we must await the discovery of fossil evidence from Morocco, but the resemblance of the industries makes this hypothesis the most likely.

E. The Maarifian

A series of quarries in the Maarif quarter of Casablanca furnished precise data to characterize a new transgression of the Atlantic Ocean which attacked the consolidated formations of the Saletian. I called this series Maarifian (Biberson 1958).

Neuville and Ruhlmann isolated it in 1941 five kilometers to the northeast, but the interpretations they gave to the famous Sidi Abderrahman quarry profile, where they thought to have demonstrated the existence of two Pleistocene transgressions, were well criticized by J. Bourcart (1943), and their hypothesis was seriously shaken.

The opening of new quarries in the hinterlands of Sidi Abderrahman permitted the collection of abundant data which clearly proves the individuality of this Atlantic transgression. It is characterized by an abrasion platform and by ancient cliffs which are found 4 kilometers from the present coast, and whose base is between 55 and 60 m. In the high levels, significant erosion of the Messaoudian or the Saletian by this new advance of the ocean has been ascertained (Biberson 1961a).

The beach deposits of these sites yield a molluscan fauna which, by its composition, is not very different from that of the Messaoudian. The two "Chilo-Peruvian" gastropods *Acanthina crassilabrum* and *Trochatella trochiformis* are notable finds, but they are smaller than the specimens of the Messaoudian. These molluscs no longer seem to have been in their optimum biotope (Biberson and Lecointre 1956).

In contrast, the regressive beaches which can be followed up to one kilometer from the present coast in the Sidi Abderrahman sector, around 15 to 20 m. altitude, contain a curious mixture of relics of this "warm" fauna with a "nordic" fauna of *Purpura lapillus L.* and *Littorina littorea L.*, and certain sites even attest the complete elimination, at a given phase of the regression, of the "Chilo-Peruvian" fauna.

The retreat of the ocean was accompanied, on the continent, by the arrival of a new pluvial period (discussed in the next section) which explains the cooling of the coastal waters, the progressive elimination of the ancient "warm" fauna, and the invasion of the Moroccan coasts by a marine fauna from the northern Atlantic.

Very little is known about the ecology of the interior Moroccan regions during the interpluvial period corresponding to the maximum Maarifian transgression. Only the prehistoric industries indicate that this is the end of the Evolved Pebble Culture (Stage IV) showing the transition to hand-axe culture. Mixed with the pebble-tools, which are still in large majority (90 to 95% of all tools gathered), are mixed a few "poor hand axes," testifying that we are on the verge of another world (Biberson 1961b).

Unfortunately, nothing is known of the mammalian fauna of this period, nor of its tool-makers. The Lower Pleistocene ends, and the new data from the following continental cycle contrast sharply with the various periods of this Villafranchian *sensu lato*, summarized above.

Archaeological similarities with Olduvai industries of the first two levels of Bed II are striking. From the Moroccan littoral sites it is impossible to obtain a stratigraphy as detailed as that of the lacustrine levels of Olduvai; this is why it is probable that in and on the conglomerate at the base of the Sidi Abderrahman quarry, dating from the post-Maarifian regression, a complex of industries was collected that partially corresponds to the gap between Beds I and II of Olduvai and to Stages 1 and 2 of the "Chellean" of L. S. B. Leakey.

The very long duration of this Lower Pleistocene (Villafranchian *sensu lato*) which includes three humid periods and is estimated by some geologists at two million years (Choubert 1950-1959) must be taken into account.

II. MIDDLE PLEISTOCENE

A. The Amirian

The Middle Pleistocene begins with the first manifestations of a long pluvial period responsible, in the interior of Morocco, for the deposit of rose-colored "old silts." These are particularly well developed along the middle course of Oued Oum-er-Rbia, in the Béni Amir plain, hence the name given to this climatic cycle by geologists and geographers of Morocco. The periglacial formations must have been limited to the elevated parts of the mountains. "No traces of Amirian glaciation have yet been identified" (Choubert, Joly, Gigout, Marçais, Margat, and Raynal 1956).

This seems, then, to be a period less cold than the Saletian, but without a doubt at certain times very wet, given the abundance of rose-colored silts.

On the coast, several types of formations succeed each other in the course of this cycle. The series begins with limestone deposits banked against the regressive post-Maarifian beaches; they represent the lower Amirian. As the regression increased, the sand of the newly uncovered beaches was blown by the wind and an important dune complex was created, now greatly indurated. The calcareous sandstone is extensively quarried, especially near Casablanca in the Sidi Abderrahman sector. This windy period corresponds to the middle Amirian. A new period, favorable to the deposit of calcareous formations, developed during the Upper Amirian, while the dunes built up earlier were consolidated. There is now ample paleontological data to clarify the ecology of certain of these different periods.

The shells of continental invertebrates studied by P. Jodot tend to favor a temperate, humid climate for Casablanca in the lower Amirian, close to the present climate of Lisbon. The mammalian fauna supports this evidence. The hippopotamus and white rhinoceros are very abundant, indicating important water sources. *Elephas iolensis* Pomel, related to such Eurasiatic forms as *E. antiquus*, also appears. Morocco apparently experienced an extension of forest and prairie, at least in the littoral zone, and this biotope proved favorable to the life of various animal species having certain holarctic characters.

The "Great Dune" of Casablanca of the middle Amirian is relatively lacking in fossils. The terrestrial molluscs seem to have been disturbed in their development. According to P. Jodot, they may indicate a climate little different from the present littoral dune micro-climate of Casablanca, which he identifies as "warm-temperate," rather dry, and certainly windy. No mammals datable precisely to this period are known apart from the common hippopotamus, and tusk casts of an indeterminate elephant found in the consolidated dune.

More is known about the upper Amirian. The terrestrial invertebrates of certain calcareous deposits of the period indicate a change of climate to one similar to that now found in La Rochelle. This is the "coldest" indication supplied by the continental mollusks. The mammal fauna leads to analogous conclusions. Besides *Elephas iolensis*, again well represented, there appears for the first time in Morocco (so far as we now know) a bear related to *Ursus larteti* Bourg., but distinguished from it by several characteristics, notably its very large size. The presence of this forest fauna clearly indicates, it seems, the existence of a plant cover quite different from that supplied by the present flora.

The hand-axe culture developed during this very long pluvial period in five successive stages, stratigraphically and archaeologically distinct. The first three came in the lower Amirian, before the building up of the "Great Dune." They constitute the large industrial group which I named, in memory of Neuville and Ruhlmann, the Rahmanian or the "older Acheulean" of Atlantic Morocco (Biberson 1961b).

The typological and technological details of these subdivisions are not important here. On the other hand, it is important to note that this Casablanca Rahmanian has strong resemblances to the Algerian industry of Ternifine; they are alike particularly in the two initial stages (I and II). This should not be astonishing, since man frequented the shores of the lake of Ternifine surely much longer than the beaches of Sidi Abderrahman, which knew the vicissitudes of the sea and the deposit first of limestone and then dune formations during the lower Amirian. In my opinion, the industries of Ternifine must have accumulated at the bottom of the lake for a long period at the beginning of the Amirian; no doubt this prohibits their identification with one of the precise stages of the Moroccan Rahmanian, easier to separate stratigraphically. I suppose that exact comparisons with the diverse industrial stages of Bed II of Olduvai would lead to the same conclusions.

If isolated stages are not clearly comparable, the resemblances are nevertheless very striking, and it is not bold to think that the makers of this older Acheulean, whether they come from Olduvai, Ternifine, or Casablanca, must have been similar types of man. The human skull from Bed II of Olduvai can easily be associated, it seems, with the mandibles of the *Atlanthropus* of Ternifine, and we are thereby led to think that the tool-maker of the Moroccan Rahmanian must have belonged to the *Atlanthropus mauritanicus* Arambourg group.

One can also imagine that the new ecology created by the arrival of the Amirian pluvial, favorable to modification of the mammal fauna, also favored replacement of the more or less Australopithecoid hominids, creators of the Pebble Culture, by the *Atlanthropus* hominids of clearly "Pithecanthropian" form, makers of the hand-axe culture.

In contrast, we still lack sufficient evidence to compare hominids at the beginning of the Middle Acheulean, Stage IV of which manifests itself in the middle Amirian, and Stage V in the upper Amirian. We shall see below that Stage VI, which supplied human remains, leads us to think that it was still a question of *Atlanthropus*.

B. The Anfatian

In 1917, G. Lecointre discovered at the foot of the Anfa hill, in a southwest suburb of the city of Casablanca, a whole series of fossil layers characterized by the presence of two unknown marine mollusks of the ancient Quaternary; *Purpura haemastoma* L. and *Patella safiana* Lmk. (Lecointre 1926). Later he made this fauna the type of his "recent Quaternary" (1952-1953). In memory of these first determinations I named "Anfatian" the Atlantic transgression responsible for these beach deposits, well known today all along the Moroccan coast.

Neuville and Ruhlmann pointed out obvious traces at Sidi Abderrahman in their memoir of 1941. The renewal of work in this important quarry, and several years of systematic search in this sector, permitted me to take up again the detailed study of these marine formations, of which I recently gave a detailed analysis (Biberson 1955, 1956, 1961a).

The Anfatian, this third transgression of the Atlantic Pleistocene, is characterized by a coast line made up mainly of indented cliffs and sometimes, at the foot of the bays, by beaches of pebbles, gravel, or sand. The marine deposits accumulated in certain places to a relatively important thickness of between 20 and 34 meters altitude. Three horizons have been distinguished:

First, a lower one reaching an altitude of about 25 m., of scarce Chilo-Peruvian fauna (G. Lecointre calls the examples "dwarfed" *Purpura plessisi*), in which are mixed several still more numerous samples of "nordic" fauna. This obviously indicates ecological conditions of the regressive Maarifian.

Second, a middle horizon hardly surpassing 27 m., in which the Chilo-Peruvian fauna has totally disappeared, and in which, on the contrary, "nordic" fauna abound. The specimens of *Littorina littorea*, in particular, are in extraordinary abundance, although that species is unknown today south of the Gulf of Cadiz and does not reach the Mediterranean. The study of the shells done by G. Lecointre has brought that conchologist to conclude that these gastropods are found in their optimum biotope. The shells are well developed, more voluminous than those of the same species now living in the Channel. Their similarities with those from present sites extending from Portugal to the Baltic Sea allowed G. Lecointre to write that they are closest to examples of the Norwegian coasts. He concludes: "They indicate a very marked cooling of the sea, corresponding to a difference of 21 degrees latitude for the present seas" (Biberson and Lecointre 1956).

Third, an upper horizon up to 34 meters, from which the "nordic" fauna has completely disappeared, with a recurrence of a few surviving ancient Chilo-Peruvian fauna, and mostly a "quasi-cryptogene" explosion (according to G. Lecointre's expression, 1952) of two molluscs of the "Senegalian" fauna: *Purpura haemastoma* L. and *Patella safiana* Lmk.

One should stress the interest aroused by this replacement of fauna, not at a maximum regression which, normally marks the end of a marine cycle, but at a very advanced stage of a transgression close to the maximum. This biological phenomenon (another example was seen in the post-Maarifian regression) shows with what great prudence, for our thin Quaternary layers, the faunal evidence should be used.

There are very few mammal remains on the strictly Anfatian beaches. Those of the lower horizon belong more likely to the upper Amirian filling of the caves, redistributed by the transgression. Those of the middle horizon are

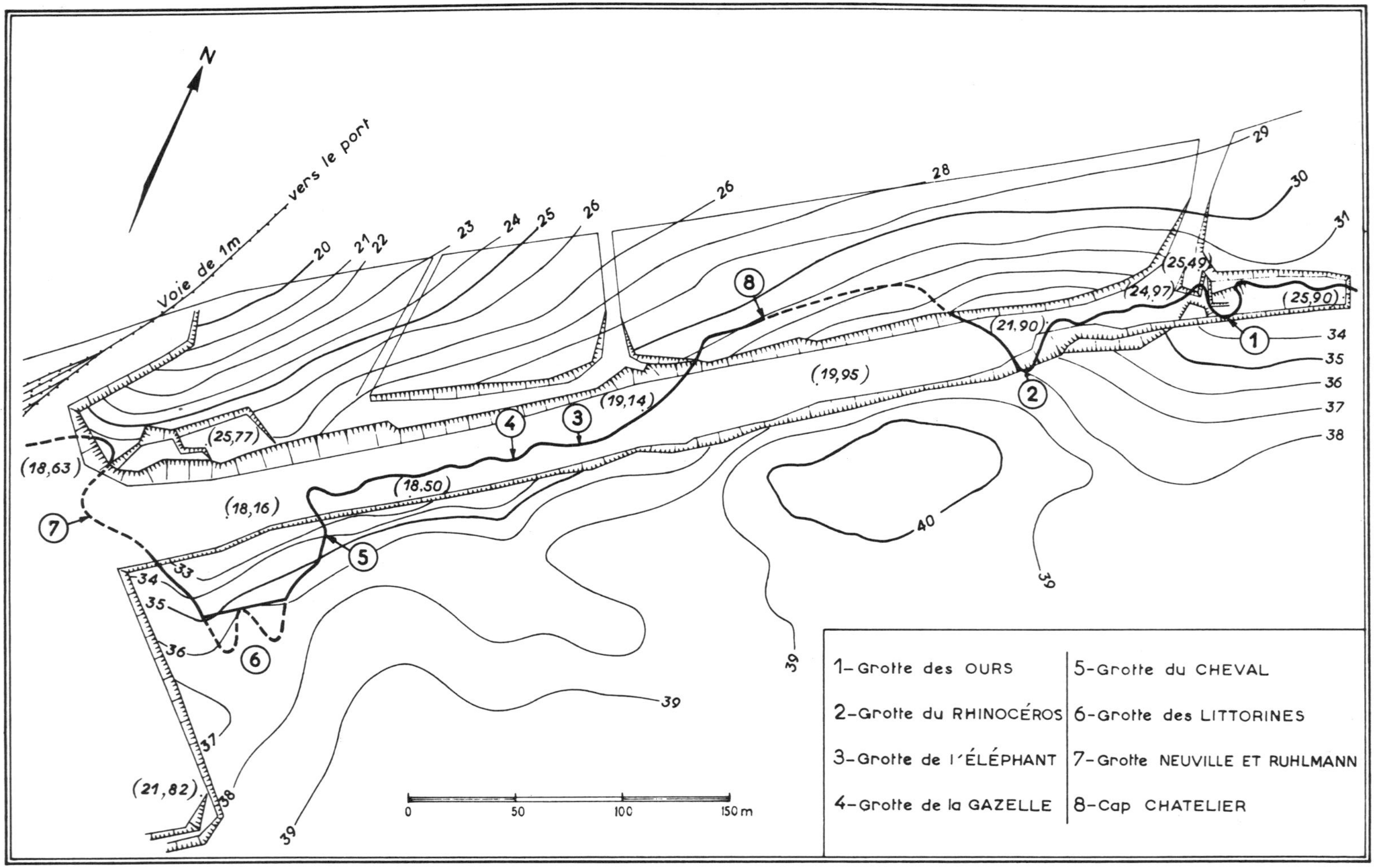

FIGURE 3
Reconstruction of the Anfatian cliff in the Sidi Abderrahman Cunette.

Elephas iolensis and *Ursus arctos bibersoni*, associated with hippopotamus and white rhinoceros. It seems, then, that Amirian mammal fauna survived. In the upper horizon, although one still finds hippopotamus and rhinoceros, the "woodland" elements disappear and, by contrast, antelope, gazelle, and zebrine equids are abundant, which seems to indicate the conditions to have been if not steppe, at least savanna. It is the fauna of the "Sudanese" type, well known as classic from the Middle Pleistocene of the Maghreb (Arambourg 1938).

The Anfatian beaches yield an abundant prehistoric assemblage more or less altered by the rolling sea, in which are found a mixture of all the stages, from the older Acheulean to the Middle Acheulean, and even pieces of the Pebble Culture (Biberson 1961b).

In particularly rich sites, it has been possible to establish, according to the degree of physical alteration, very good series, for example, in the Martin quarry, studied by M. Antoine (1930), and reviewed by Neuville and Ruhlmann (1941). After systematic searches in the "Grotte des Ours" and "Cap Chatelier" of Sidi Abderrahman, I was able to reach new conclusions (Biberson 1961b).

The industries contemporary with the Anfatian belong not to the "Abbevillian" as Neuville and Ruhlmann thought, nor to the "older Acheulean" as M. Antoine argued (1952), but clearly to the second part of the Middle Acheulean of which the last stage, Stage VI, even covers the Anfatian beaches, as discussed in the next section.

It is also in this Stage VI, at the beginning of the post-Anfatian regression corresponding to the arrival of a new pluvial period, that the first known human fossil in Morocco belongs. It has been identified as a descendant of *Atlanthropus* of Ternifine.

C. The Tensiftian

The geologists and geographers who established a classification of the continental Quaternary of Morocco seem to have had difficulties in choosing a type locality for the formations of the next to last pluvial (Riss) which succeeds the Amirian. They finally proposed the periglacial terraces that border Oued Tensift in the Marrakech region. Thus the name Tensiftian, given to this new climatic cycle.

In the interior of Morocco this pluvial does not seem to correspond to an important period of alluviation—the sediments are few and poorly represented—but rather to an epoch of intense chemical weathering, which would have been favorable to the development of "crusts" and of "powder-like calcareous tufas."

These same specialists estimate that the climate must have been more rigorous than in the Amirian: "The red formations are poorly developed or absent. In contrast, the contemporary deposits have a high proportion of pebble beds due to periglacial processes at altitude limits lower than those of the Soltanian (the subsequent climatic cycle). Just the same, the forms of erosion and of glacial accumulation (moraines, etc.) of this period extended further down" (Choubert, Joly, Gigout, Marçais, Margat, and Raynal 1956).

On the Atlantic littoral there is the same paucity of diluvial deposits. In the region of Sidi Abderrahman, in particular, this can be attributed first of all to the superficial redistribution of the Anfatian beaches, of thin formations of fine sediments beginning to wear away the foot of the ancient cliff, of dunes less important than those of the Amirian, and finally of "powder-like limestone" which formed in the interdunar hollows of the period (Biberson 1961a).

Sediments of the early Tensiftian, chiefly sandy at the base, more clayey at the summit, abound in fossils and have yielded prehistoric industries. At this level fauna of the Sudanese type has been identified, with white rhinoceros, antelope, gazelle, equids and ostrich, but neither hippopotamus nor elephant. In contrast, the oldest human remains of Morocco belong to this level.

A karstic cavity of the "Littorina Cave" of Sidi Abderrahman, which served as a hyena den, yielded two mandible fragments studied by C. Arambourg (Arambourg and Biberson 1955, 1956)—an early man with numerous archaic characteristics which belonged to the *Atlanthropus* of Ternifine. The associated industry is Middle Acheulean, more advanced than Stage V of the upper Amirian, which I made the type of Stage VI of the hand-axe culture of Atlantic Morocco (Biberson 1961b). Crude pieces on pebbles are still very numerous and the cylinder-hammer technique of wood or bone is far from being general. Nevertheless, certain flat bifaces of regular form and flakes worked into bifacial side scrapers indicate some improvement.

Following G. Choubert's interpretation of the stratigraphy of the Kébibat quarry (1959), the "Great Dune" of Rabat belongs to this cycle. The marine level of the base appearing at an altitude of only 7.50 meters, and the superposed dune able to have formed only in the event of an even more accentuated recession of the ocean, it is a question of a very advanced phase in the post-Anfatian regression. The maximum transgression is found around 30 m. and, therefore, of a period which can be considered as corresponding to the middle Tensiftian.

An important mammal fauna was collected in the "Rabat sandstone." Unfortunately the exact place of the paleontological pieces in the stratigraphy of the Pleistocene is not clearly established in all the fossiliferous quarries. Besides antelope, gazelle, and equids found in Casablanca, rhinoceros remains and, above all, *Elephas atlanticus* Pom., were found which are known to be related to the African elephants and which abound in the Amirian site of Ternifine. It is interesting to note the paleobiological considerations which C. Arambourg set forth concerning the ecological setting in which both *E. iolensis* and *E. atlanticus* must have prospered. The former, a prairie animal, if not of the forest, is opposed to the second, an animal of the savanna, if not of the steppe (Arambourg 1960).

Using these data, one can infer from the exclusive presence of *E. atlanticus* in this level that the climatic conditions in the middle Tensiftian, on the Atlantic littoral, were not as favorable to the spread of a forest flora as those of the

Amirian. This would agree with the estimations of the geologists, taken from petrographic and lithologic observations, on the aspect of the Tensiftian pluvial as colder but less humid than the Amirian pluvial.

Here lived the "Rabat Man" whose remains lay in the upper part of the "Great Dune" of Kébibat. These human fossils were studied by H. V. Vallois (1945, 1960), who saw in them the archaic characteristics of *Sinanthropus, Atlanthropus,* and Sidi Abderrahman Man. He concluded from his analysis: "The preceding description shows that the mandible and the teeth of the Rabat Man present a remarkable collection of primitive characteristics, many of which recall those of the Neanderthals, while others are close to the prehominians. Morphologically more archaic and chronologically more ancient than these Neanderthals, he must without a doubt be considered as a sort of African Pre-Neanderthal" (translated from Vallois 1960).

"Rabat Man" was not associated with a prehistoric industry, but if the dating of the site proposed by G. Choubert is accepted, he must be the maker of the evolved Acheulean. This new archeological stage first appeared in Stage VII of the hand-axe culture which was recovered in the "powder-like limestone" of the interdune hollows of the middle Tensiftian.

Several sites of this type in the Casablanca region have yielded assemblages showing considerable progress from Stage VI, which closed the Middle Acheulean. We find ourselves at the apogee of the Moroccan Acheulean, with good ovate or lanceolate hand axes, often very flat, well formed, and edges carefully shaped with wood. It is indicative that this remarkable assemblage is the work of a man who appears to us physically so far removed from *Homo sapiens.*

D. The Harounian

The Tensiftian climatic cycle ends, on the Moroccan littoral, with the arrival of a new Atlantic transgression. It is still difficult to appreciate the extent of the Tensiftian regression, so much so that the new advance has often been considered as "minor" and sometimes interpreted as merely a halt in the post-Anfatian regression.

Nevertheless, J. Marçais and G. Choubert isolated it long ago (1947) in the Marie Feuillet quarry in Rabat. M. Gigout also found it clearly in the section of Cap Cantin, north of Safi (1951). Recent observations in the Casablanca sector of Sidi Abderrahman seem to leave no doubt of its general occurrence.

An ancient cliff, the foot of which is found between 18 and 20 m., is clearly distinct from those of the Anfatian and the Ouljian, in the sector of the Kef Haroun rifle range—hence, the name I gave to this transgression (Biberson 1960, 1961a, 1961c). G. Choubert calls it Rabatian (1959), and M. Gigout "Episode of Kébibat" (1960). In my judgment, it corresponds to the Mediterranean *Eutyrrhenian* of Bonifay and Mars (1959), that is, to the "true Tyrrhenian" of Issel.

Unfortunately, the molluscan fauna of the Harounian beaches does not permit paleontological distinction of this marine cycle from the Anfatian which precedes it and from the Ouljian which succeeds it. In effect, the only characteristic fossils are those of the "Senegalian" fauna, *Purpura haemastoma* and *Patella safiana*, to the exclusion of the ancient Chilo-Peruvian and "nordic" fauna. The absence of *Strombus bubonius*, which seems to require a special environment (Lecointre 1926), has always greatly hindered the interpretations of paleontologists.

Nothing precise is known about the mammal fauna of the period; it is nevertheless probable that certain fossiliferous quarries around Rabat belong to this cycle. It seems that it is always the "Sudanese" fauna that perpetuates itself, and it is this relative uniformity of the fauna of "Rabat sandstone" that doubtless explains why the paleontologists are not prone to accept the data of relative chronology, which it would have been extremely interesting for us to have defined.

No proper prehistoric industry is known in the Harounian beaches. Only by deduction can one say that, during this period, the evolved Acheulean continued its development, but archeological data and precise stratigraphy are lacking, which considerably hinders comparisons with other African industries supposed to be contemporary. One must await a new pluvial period to find, in the continental levels subordinate to the Harounian formations, evidence of a new stage of hand-axe culture.

We are not much better informed, *a fortiori*, on the kinds of early men of this period. According to the fossils yielded by the continental levels of the Harounian transgression, we may be justified in thinking they were a type of *Atlanthropus*, as will be seen at the beginning of the following section which discusses the continental formations immediately above the beaches of the post-Harounian regression.

The Middle Pleistocene, including only two pluvial cycles, appears to be of much shorter duration than the Lower Pleistocene. At the most, it is estimated that the post-Sicilian regression (post-Maarifian in Morocco), that introduced it dates back 700,000 or 800,000 years.

III. UPPER PLEISTOCENE

A. The Presoltanian

Consequent to acknowledging the independence of the Harounian marine cycle, I was induced to separate the post-Harounian continental formations previously thought to belong to the upper Tensiftian, and attach them instead to the Soltanian (Choubert, Joly, Gigout, Marçais, Margat, and Raynal 1956). I called this first phase of the last Moroccan Pleistocene pluvial the Presoltanian. It is distinct from the second phase of this same pluvial, which constitutes the Soltanian *sensu stricto*, by the Ouljian transgression (Biberson 1961a).

The Presoltanian formations are well developed in the extension of Sidi Abderrahman, where Neuville and Ruhlmann described them (1941) as "reddish brecciated limestones."

This calcareous sediment contains the debris of the Anfatian ancient cliff begun by the sediment washed down in the lower Tensiftian; they are abundant on the Harounian beach deposits, which are coated with them. More clayey at the base, they become full of limestone at the top and change to a diffuse crust which can end in a calcareous flagstone of zoned structure. At "Cap Chatelier," still in the Sidi Abderrahman "cunette," they may reach a thickness exceeding 6 meters.

They have yielded a rich invertebrate and vertebrate fauna. The continental molluscs studied by P. Jodot indicated, according to that author, a temperate climate of the Lusitanian type, colder than the present littoral climate of Casablanca and closer to that of Lisbon.

All the mammal fossils indicate a fauna of the "Sudanese" type, which does not seem to have varied much since the Tensiftian. The only new species is *Sus scrofa algeriensis* Pom. which, so far as is known, appears in Morocco for the first time. An important microfauna of rodents, insectivores, and bats was collected, but it has not yet been studied and cannot be used for paleoclimatic conclusions.

The sum of the petrographic and paleontological data seems to denote a period of high rainfall but not of great coolness. Nevertheless, it should be kept in mind that the ocean has always had important thermal effects on the coast, and since the Presoltanian formations have not yet been isolated in interior Morocco, it is difficult at present to be categorical.

Prehistoric industries are very abundant. Neuville and Ruhlmann called them "Moroccan Micoquian" because of the abundance of small hand axes. Recent studies show that it is actually final Acheulean, very evolved, with numerous "Levalloiso-Mousterian" elements, which induced me to speak of "Levalloiso-Mousterian of Acheulean tradition" rather than *Micoquian*, a modifier which is hardly suitable. Numerous similarities with the Fauresmith of South Africa have also been revealed (Biberson 1961b).

During two seasons of systematic excavations in these levels, which covered several thousand cubic meters of material, the quarry of Sidi Abderrahman, although very rich in industry and fauna, has unfortunately yielded no human remains. In contrast, a mandible, discovered by Abbé J. Roche in the filling of the "Grotte des Contrebandiers" at Témara, between Rabat and Casablanca, seems to be datable to this period.

The cave, cut into the Harounian cliff, was refilled by the Ouljian transgressions. The filling is incontestably post-Ouljian, but distributed here and there are blocks of a derived breccia, very strongly consolidated. One of these blocks yielded the mandible of the "Témara Man." The pre-Ouljian breccia seems to indicate that the fossil belonged to the inter-Harounian-Ouljian stage, that is to the Presoltanian (Biberson 1961c).

This human fossil, studied by H. V. Vallois, has several remarkable characteristics. Some are archaic, close to those of *Atlanthropus,* Sidi Abderrahman Man, and Rabat Man; others are more evolved. But Témara Man cannot be allied with the Neanderthals, makers of the Mousterian in France, which seem nevertheless to have been their contemporaries in Europe (Vallois and Roche 1958).

B. The Ouljian

In 1949, M. Gigout defined a minor marine transgression, relatively recent, which he called the Ouljian, from the arabic word *oulja* denoting the littoral trough, at times a lagoon, which is most often found behind the recent dune of the present coast.

The cliff marking the maximum shoreline attained by this transgression is common in the morphology of the coastal zone (Guilcher and Joly 1954). Its foot is situated between 5 and 8 meters altitude. It can be clearly followed some tens of kilometers west of Casablanca. It is well outlined at the height of the quarries of Sidi Abderrahman, where it is distinctly separated from the Harounian cliff. It is pierced by numerous caves which, after the regression, frequently served as a habitat for man (Biberson 1961a).

As I said above, the molluscan fauna of the Ouljian is not in any way different from that of the Anfatian and of the Harounian. The most characteristic fossils are *Purpura haemastoma* and *Patella safiana,* which are still found on the present Moroccan coast. Nothing from the paleontological point of view distinguishes the Ouljian from prior transgressions.

The mammal fauna, still scarcely known, does not seem to have changed since the Presoltanian; but we must await the subsequent continental climatic episode to get positive data.

The prehistoric industries are not much better defined. The Ouljian beach levels yield numerous flakes of quartzite and sometimes of flint, very atypical, of more or less "Levalloiso-Mousterian" aspect which can be related just as well to the final Acheulean (Stage VIII of the hand-axe culture of Atlantic Morocco, proper to the Presoltanian) as to the older Aterian. Only one very characteristic pedunculate point was found by E. Bolléli in a consolidated dune linked to the post-Ouljian regression of the Contrebandiers beach, facing the cave where Abbé Roche found the mandible of "Témara Man."

It is known that in Algeria, in the Arzew region, an Aterian industry was found enclosed in the low *Strombus* beach (Camps 1955). It seems that this beach corresponds perfectly to the definition of the Neotyrrhenian by Bonifay and Mars (1959) and, in consequence, to the Moroccan Ouljian.

In Casablanca there are good grounds for thinking that the "quartzite Mousterian station of the plateau of the Martin quarry at El-Hank," described by M. Antoine in 1932, which yielded the oldest Aterian of Morocco, could be dated to the period on the continent corresponding broadly to the Ouljian transgression (Biberson 1961a).

It is only in the continental formations that accompany the post-Ouljian regression that one can find a parallel with the maximum of the great "Grimaldian" regression of J. Bourcart, data on which are becoming numerous and significant; but that involves a new climatic period called the Soltanian in Morocco.

C. The Soltanian

When the geologists and geographers of Morocco wanted to define the last Moroccan pluvial (Würm), corresponding to the well-known Grimaldian regression in the Mediterranean, they chose as a type locality the cave of Dar-es-Soltan, near Rabat (studied by A. Ruhlmann in 1951), because, they said, "the bulk of the filling of this cave is formed by thick sediments containing a complete succession of Aterian industries" (Choubert, Joly, Gigout, Marçais, Margat, and Raynal 1956).

This cave is breached in the Ouljian cliff, and its filling is unquestionably entirely post-Ouljian; it contains no series stratigraphically as complete as that of the Grimaldi cave, the base of which has yielded a "Mousterian with warm fauna" which belonged, it seems, to the beginning of the Würm and would parallel our Presoltanian. It follows that the Soltanian *sensu stricto* can only correspond with the second phase of the European Würm, after the Neotyrrhenian-Ouljian (Biberson 1961c).

On the other hand, in 1952 G. Lecointre stated, after having described the transgression of 20 meters (our Harounian): "That invasion is followed by an important regression called Grimaldian during which 'upper red silts' with Micoquian industry were formed. During this regression, which brought the sea much below present sea level, there was a transgression in Atlantic Morocco, the Ouljian of Gigout, carrying the shoreline to 7 or 8 meters, whose fauna seems to be identical to that of the beginning of the Tyrrhenian (i.e., Anfatian). The formation of the crust continues during this period, often coating the red sediments. The regressive movement then continued. No special name has been given to it, so I shall call it 'Grimaldian II.' This period is characterized by the deposition of formations covering the crust and whose surface bears only a few millimeters of this formation: dunes, terraces, etc. During this period, generally considered as equivalent to the great glaciation of the Würm, a cold pluvial climate came to North Africa. The fauna of the forests of Europe—deer, wild-boar, bear, etc.—invaded Berbérie. Man sought refuge in caves" (Lecointre 1952).

According to the geologists who defined the Soltanian, it is a question of "a relatively temperate pluvial, characterized by a widespread development of silts, comparable to the recent loess of temperate countries. These sediments run laterally, in the mountainous zones, to periglacial slope deposits above an altitude limit varying according to the regions. Soltanian glacial formations characterize certain elevated massifs" (Choubert, Joly, Gigout, Marçais, Margat, and Raynal 1956).

For a long time, the "Moroccan Würm" was considered essentially the period of formation of the "upper red silts." In reality, J. Bourcart showed that it was composed of more varied deposits (1943). Also, R. Raynal has since noted facies differences between the base and the summit of the red silts, and these granulometric data induced him to separate several climatic phases within this pluvial (Raynal 1956).

On the coast, the Soltanian presents a series of stratigraphic levels which are, from the base to the summit:

1. a rubified clay "soil" which is plastered against cemented beaches of the post-Ouljian regression;

2. dunes, most of the time well consolidated, parallel to the coast, at altitudes rather below the present level of the ocean; they sometimes form islets, almost islands or stony bars half closing the bays, which gives the coast an unexpected morphology of the "dalmatian" type (Guilcher and Joly 1954);

3. "allochthonous red silts," produced by the "dissolving" of the largely calcareous dune sands under the effect of physico-chemical actions, which were variously transported to fill the depressions; they often comprise lines of pebbles at the base and, towards the upper two-thirds, beds of gravel, which undoubtedly implies modifications in the sedimentary conditions denoting climatic variations;

4. finally a calcareous "crust" of less importance, which often seals all the deposits, especially near the coast (Biberson 1961a).

Thus, the analysis of the formations and depositional processes of the sediments indicates several climatic phases during this cycle, which may be parallel to the different phases of the Alpine Würm. But these correlations are still very hypothetical, and only the marked interruption of the Ouljian transgression can be, with any great truth, linked to that of the Neotyrrhenian of the Mediterranean, which is generally placed in the inter-Würm I—Würm II (Bonifay and Mars 1959).

The mammal fauna of the Soltanian *sensu stricto* has been known for a long time in North Africa in general and in Morocco in particular. C. Arambourg has described numerous specimens. The most striking is the arrival of species new to Africa and which are of Eurasiatic origin: *Rhinoceros merki*, *Ursus arctos larteti*, *Ursus arctos faidherbi*, *Sus algeriensis*, *Cervus* aff. *elaphus*, *Megaroides algericus*, etc. In 1938 he concluded his study on the fossil mammals of Morocco with these sentences: "The presence of bears, wild boar, and deer, is an index of the development on Morocco's soil of a wooded vegetation of which these animals are essentially the inhabitants. Further, the Eurasiatic affinities of these animals indicate an immigrant population, but one whose arrival was possible only when the local ecological conditions had become favorable.

"The extension of the forest which seems to mark the beginning of the Middle Paleolithic is due to the arrival of a humid and pluvial climate following the dry period. The local stratigraphy preserved traces of this humid period in the development of the eluvial red sediments of the residual decalcified clays, and of the products of slope wash that filled the caves where the characteristic Middle Paleolithic fauna and a Moustero-Aterian industry are generally found in mutual association" (Arambourg 1938).

In effect, the filling of the caves has given us both paleontological and prehistorical knowledge.

Three habitats of this period in Atlantic Morocco are particularly important: El Khenzira near Mazagan, Dar-es-Soltan near Rabat, and Mougharet-el-Aliya near Tangier. It is convenient to add, although outside the geographical limits of this paper, the "Grotte des Pigeons" of Taforalt, in the Oujda region.

Thanks to the successive archeological beds of the first two caves it has been possible to establish the evolutionary phases of the Aterian. The third has yielded remains of "Tangier man." The fourth, and the most recently excavated, has provided an exceptionally rich series of Ibero-Maurusian peoples of the "Mechta-el-Arbi race," and also afforded C-14 age determinations of greatest interest, of the chronological arrangement of the industries.

L. Balout demonstrated perfectly, in 1955, that the Aterian, which is *typologically* an evolved Mousterian with pedunculate points, cannot be prior to the older Mousterian of the caves of the Perigord and must be, at least partially, parallel to the Upper Paleolithic or Leptolithic of western Europe.

The human remains of Tangier, the only ones which seemingly can be related to this period, present, by comparison with Acheulean Man of the Maghreb, evolved characters, but also some others which would link them on certain points with "Rabat man" and "Témara man" (Vallois 1960). Unfortunately, as L. Balout wrote in 1955: "The human remains exhumed in Mougharet el Aliya are extremely deceptive—paleontologically because it is a question of a child, actually only a maxillary fragment; stratigraphically, because, except for an adult molar, all the human debris has been recovered either in an altered environment or even in a sieve; archeologically, because their relations with a "Levalloiso-Mousterian" (the interpretation proposed in 1953 by Howe and Stearns) rather than with the Aterian, are not demonstrated. In fact, "Tangier man," an illusory name, offers only a partial solution to the problem of man of the final Moghrebian Paleolithic, that is to say to "Aterian man" (Balout 1955).

Nevertheless, there seems to be no doubt, from the paleoanthropologic viewpoint, that the Tangier fossils cannot be attributed to *Homo sapiens,* nor even to *Homo neanderthalensis.* They seem to belong to the *Atlanthropus* stock seen at Ternifine with the older Acheulean, and to persist during the evolution of the

Middle Acheulean and the evolved Acheulean with the men of Sidi Abderrahman, Rabat, and Témara; but other discoveries are necessary in order to characterize the Aterian man properly.[3]

In contrast, we are today well informed on the "Man of Mechta-el-Arbi," first representative of *Homo sapiens,* who introduced Ibero-Maurusian industry to the Maghreb. This point, first established in Algeria (Arambourg, Boule, Vallois, and Verneau 1934), has just been solidly established in Morocco by the study of 170 individuals from the ossuary of Taforalt (Ferembach 1959).

What is of singular importance is that the dates of the Ibero-Maurusian occupation of the "Grotte des Pigeons" have been defined by C-14 analysis in two different laboratories as 12,000 to 10,000 years ago (Roche 1959), an age close to that of Magdelenian IV of the Saint Marcel cave (Indre).

Also, the Mellahian transgression (or Flandrian of Morocco) which ends the Soltanian and marks the extreme limit of the Upper Pleistocene, is similarly dated at only 6,000 years ago (Gigout 1959). The result is that the Ibero-Maurusian is still very clearly anterior to the Climatic Optimum of Europe, to which the last humid period of the Saharan Neolithic seems to correspond (Delibrias, Hugot, and Quézel 1957), and it must be dated from the final Soltanian, broadly equivalent with the end of the Würm of western Europe.

With the Mellahian we are at the beginning of the Holocene when current climatic conditions are established and only human behavior transforms the scenery and fauna to give us contemporary data of geography and history.

Thus we see the relative brevity of the Upper Pleistocene, which includes several climatic oscillations but, in reality, constitutes only one pluvial climatic cycle, as compared to the Middle Pleistocene and *a fortiori* to the Lower Pleistocene. Its duration certainly does not exceed 100,000 years.

CONCLUSIONS

From this rapid review of Moroccan Paleolithic man in the varied climatic environments of the Pleistocene, we may make some deductions, although the data they are based on require fuller multidisciplinary studies.

If no discovery in Morocco has yet provided fossil evidence of the makers of the Pebble Culture, comparisons of the paleontological and archeological collections suggest that they were closely related to *Zinjanthropus* of Olduvai, belonging to the Australopithecine group, a hypothesis which has just been reinforced by the recent discovery by Y. Coppens in the Chad. These very ancient forms must have evolved during a very long period of time—from the Lower Villafranchian to the end of the Saletian, that is, during the vast stretch of the Lower Pleistocene. The well-characterized industries belong only to the Moulouyian-Middle Villafranchian.

3. Cf. footnote 1 (p. 419). Discovery of Jebel Irhoud man may answer the question.

At the sudden change in climate that introduced the Middle Pleistocene, which we take as beginning with the Amirian-Mindel, one finds, if not a "faunal break" as was known in western Europe under the extreme influence of the glaciations, at least the disappearance of ancient animal species and the proliferation of new ones destined to continue their development over a long period. At the same time the hand-axe culture replaces the Pebble Culture and there is the appearance of *Atlanthropus* which belongs to the Pithecanthropian group.

How much is the arrival of the singular and well-defined climatic conditions of the pluvial Amirian responsible for these changes? It is difficult to say, but it is not impossible that faced with a new ecological environment, the most adaptable plant, animal, and human forms evolved and survived, while the others, too specialized, were condemned to disappear.

The recent discoveries of L. S. B. Leakey at Olduvai seem to show an analogous process with the beginning of the Kamasian of eastern Africa and attest in a very conclusive manner the perfectly coherent unity of the African continent in the Middle Pleistocene.

With the Upper Pleistocene, when Europe saw the appearance of Neanderthalers with the Mousterian industries, it seems that Morocco also saw the evolution of the old *Atlanthropus* human stock, responsible for the evolved Acheulean and maybe even for the Aterian.

Only at the extreme end of the Upper Pleistocene is *Homo sapiens,* bringing with him the Ibero-Maurusian industry, manifest in the Maghreb in the form of the Cro-Magnon-like race of Mechta-el-Arbi. C-14 datings of prehistoric habitats prove a time lag of tens of centuries between the arrival of the newcomers in Europe and in North Africa. Only the recency of the phenomena makes us see this difference, which is insignificant on the scale of the Pleistocene. If new methods of dating someday provide definitions of the earlier periods, similar facts on many diverse points of the ancient world may be verified. But the importance of Africa as center for the dispersion of species and even of the most ancient cultures will not be more obvious.

Another conclusion from the analysis of the Moroccan data is the general aspect of the simultaneous biological and cultural human evolution. It started slowly, because the Pebble Culture, attributed to beings close to the Australopithecines, developed uniformly during the long duration of the Lower Pleistocene. It was accelerated with the hand-axe culture at various stages, the work of still archaic Pithecanthropians; then it evolved all through the Middle Pleistocene, to emerge explosively in the Upper Pleistocene with the varied industries for which *Homo sapiens* is responsible.

BIBLIOGRAPHY

ANTOINE, M.

1930. "Notes de Préhistoire marocaine. III: Station chelléenne de la carrière Martin, près El-Hank," *Bull. Soc. Prehist. Maroc,* 4:57–117.

1932. "Notes de Préhistoire marocaine. V: Station moustérienne à quartzites du plateau de la carrière Martin a El-Hank," *Bull. Soc. Préhist. Maroc,* 6:23–46.

1952. "Les grandes lignes de la préhistoire marocaine," Special Publication of the *2d Pan-Afr. Cong. Prehist.* (Algiers 1952).

ARAMBOURG, C.

1938. "Mammifères fossiles du Maroc," *Mém. Soc. Sci. Nat. Maroc,* 46:1–72.

1955a. "Une découverte recente en paléontologie humaine, l'*Atlanthropus* de Ternifine (Algérie)," *Quaternaria,* 2:5–13.

1955b. "Récentes découvertes de Paléontologie humaine realisées en Afrique du Nord française (L'*Atlanthropus* de Ternifine-L'hominien de Casablanca)," *Actes. 3e Pan-Afr. Cong. Préhist.* (Livingstone 1955), pp. 186–94.

1959a. "Etat actuel des recherches sur le Quaternaire en Afrique du Nord," *Actes. 4e Pan-Afr. Cong. Préhist.* (Leopoldville 1959):255–77.

1959b. "Observations sur le Quaternaire littoral de la Méditerranée et du proche Atlantique," *C. R. Soc. Géol. France* (1959), No. 8: 209–10.

1960. "Au sujet de *Elephas iolensis* POMEL," *Bull. Archéol. Maroc,* 3:93–105.

ARAMBOURG, C., and L. BALOUT

1952. "Du nouveau a l'Aïn Hanech," *Bull Soc. Hist. Nat. Afr. Nord,* 43:152–9.

ARAMBOURG, C., and P. BIBERSON

1955. "Découverte de vestiges humains acheuléens dans la carrière de Sidi Abderrahman, près Casablanca," *C. R. Acad. Sci.* (Paris), 240:1661–3.

1956. "The fossil human remains from the Paleolithic site of Sidi Abderrahman (Morocco)," *Amer. J. Phys. Anthrop.,* 14:467–89.

ARAMBOURG, C., M. BOULE, H. VALLOIS, and R. VERNEAU

1934. "Les grottes paléolithiques des Béni Segoual (Algérie)," *Arch. Inst. Paleont. Humaine,* Mém. 13.

ARAMBOURG, C., and G. CHOUBERT

1957. "Les faunes de Mammifères de l'étage moghrebien du Maroc occidental," *Proc. 5th Cong. INQUA* (Madrid-Barcelona), in press.

BALOUT, L.

1955. *Préhistoire de l'Afrique du Nord.* Paris: Arts et Métiers Graphiques.

BIBERSON, P.

1955. "Nouvelles observations sur le Quaternaire côtier de la région de Casablanca," *Quaternaria,* 2:109–47.

1956. "Le gisement de l'Atlanthrope de Sidi Abderrahman," *Bull. Archéol. Maroc,* 1:38–92.

1957. "Nouveaux éléments sur la 'Pebble-Culture' du Maroc atlantique," *Proc. 5th Cong. INQUA* (Madrid-Barcelona), in press.

1958. "Essai de classification du Quaternaire marin du Maroc atlantique," *C. R. Soc. Géol. France*, No. 4: 67–9.

1960a. "L'évolution du Paléolithique marocain dans le cadre du Pléistocène atlantique," *Quaternaria*, 6:177–205.

1960b. "La place des Hommes fossiles du Maroc dans la chronologie du Pléistocène atlantique," *Proc. 6th Int. Cong. Anthrop. and Ethnol.*, Sec. 4, "Paleanthropologie et Origine de l'Homme" (in press).

1961a. "Le cadre paléogéographique de la préhistoire du Maroc atlantique," *Publ. Serv. Antiquités Maroc*, Mém. 16.

1961b. "Le Paléolithique inférieur du Maroc atlantique," *ibid.*, Mém. 17.

BIBERSON, P., G. CHOUBERT, A. FAURE-MURET, and G. LECOINTRE

1956. "Découvert d'instruments de la 'Pebble-Culture' dans les cailloutis villafranchiens d'Arbaoua," *C. R. Acad. Sci.* (Paris), 245:938–9.

1960. "Contribution a l'étude de la 'Pebble-Culture' du Maroc atlantique," *Bull. Archéol. Maroc.*, 3:7–53.

BIBERSON, P., and G. LECOINTRE

1956. "Progrès dans la connaissance du Quaternaire de Casablanca (Maroc)," *Bull. Soc. Géol. France* (Sér. 6), 6:855–66.

BONIFAY, E., and P. MARS

1959. "Le Tyrrhénien dans le cadre de la chronologie quaternaire méditerranéenne," *Bull. Soc. Géol. France* (Sér. 7), 1:62–78.

BOURCART, J.

1943. "La géologie du Quaternaire au Maroc," *Rev. Scient.*, 81:311–36.

CAMPS, G.

1955. "Le gisement atérien du Camp Franchet d'Espérey (Arzew)," *Libyca*, 3:17–56.

CHOUBERT, G.

1950. "Réflexions au sujet du Pliocène continental," *Notes et Mém. Serv. Géol. du Maroc*, No. 76, pp. 13–91.

1953. "Les rapports entre les formations marines et continentales quaternaires," *Actes 4e Int. Cong. INQUA* (Rome-Pisa):576–90.

1957. "Essai de corrélation des formations continentales et marines du Pléistocène au Maroc," *Proc. 5th Int. Cong. INQUA* (Madrid-Barcelona), in press.

1959. "Compléments à la note intitulée 'Essai de corrélation des formations continentales et marines du Pléistocène au Maroc," *Com. Géomorph. Périglaciaire Maroc.*

CHOUBERT, G., and R. AMBROGGI

1953. "Note préliminaire sur la présence de deux cycles sédimentaires dans le Pliocène marin au Maroc," *Notes et Mém. Serv. Géol. du Maroc*, No. 117.

CHOUBERT, G., F. JOLY, M. GIGOUT, J. MARCAIS, J. MARGAT, and R. RAYNAL

1956. "Essai de classification du Quaternaire continental du Maroc," *C. R. Acad. Sci.* (Paris), 243:504–6.

CHOUBERT, G., and J. MARCAIS

1947. "Le Quaternaire des environs de Rabat et l'âge de l'Homme de Rabat," *C. R. Acad. Sci.* (Paris), 224:1645–7.

CHOUBERT, G., and ABBE J. ROCHE

1956. "Notes sur les industries anciennes du plateau de Salé," *Bull. Archéol. Maroc.*, 1:9–37.

DELIBRIAS, G., H. J. HUGOT, and P. QUÉZEL

1957. "Trois datations de sédiments sahariens récents par le radio-carbone," *Libyca*, 5:267–70.

Ennouchi, E.

1951. "Découverte d'un Hipparion dans les environs de Rabat," *Notes et Mém. Serv. Géol. du Maroc*, 17:88.

1957. "Les Ursidés marocains," *Bull. Soc. Sci. Nat. et Phys. Maroc*, 37:201–24.

Ennouchi, E., M. Gigout, and J. Marcais

1951. "Un nouveau Mastodonte marocain a Daourat (Oum-er-Rbia) dans les couches fluvio-marines du Pliocène," *Notes et Mém. Serv. Géol. du Maroc*, No. 85: 147–54.

Ferembach, D.

1959. "Les restes humains épipaleolithiques de la grotte de Taforalt (Maroc oriental)," *C. R. Acad. Sci.* (Paris), 248:3465–7.

Gigout, M.

1949. "Définition d'un étage ouljien," *C. R. Acad. Sci.* (Paris), 229:551–2.

1951. "Etudes géologiques sur la Meseta marocaine (arrière-pays de Casablanca, Mazagan et Safi)," *Notes et Mém. Serv. Géol. du Maroc*, No. 86.

1956. "Recherches sur le Pliocène et le Quaternaire atlantiques marocains," *Trav. Inst. Sci. Chérif.* (Série Géol. et Géog. Phys.), No. 5.

1957. "Recherches sur le Quaternaire marocain," *Trav. Inst. Sci. Chérif.* (Série Géol. et Géo. Phys.), No. 7.

1959. "Age par radio-carbone de deux formations des environs de Rabat (Maroc)," *C. R. Acad. Sci.* (Paris), 249:2802–3.

Gigout, M.

1960. "Nouvelles recherches sur le Quaternaire marocain et comparaisons avec l'Europe," *Trav. Lab. Géol. Fac. Sci. Lyon* (n.s.), No. 6.

Guilcher, A., and F. Joly

1954. "Recherches sur la morphologie de la côte atlantique du Maroc," *Trav. Inst. Sci. Chérif.* (Série Géol. et Géog. Phys.), No. 2.

Howell, F. C.

1954. "Hominids, pebble-tools and the African Villafranchian," *Amer. Anthrop.*, 56:378–86.

1955. "The age of the Australopithecines of Southern Africa," *Amer. J. Phys. Anthrop.* (n.s.), 13(4):635–62.

1959. "The Villafranchian and Human Origins," *Science*, 130:831–44.

1960. "European and Northwestern African Middle Pleistocene Hominids," *Cur. Anthrop.*, 1 (3):195–232.

Leakey, L. S. B.

1951. *Olduvai Gorge.* Cambridge: Cambridge University Press.

1959. "The first Men. Recent discovery in East Africa," *Antiquity*, Vol. 33, No. 132.

Lecointre, G.

1926. "Recherches géologiques dans la Meseta marocaine," *Mém. Soc. Sci. Nat. Maroc*, No. XIV.

1952. "Recherches sur le Néogène et le Quaternaire marins de la côte atlantique du Maroc," *Notes et Mém. Sérv. Géol. du Maroc*, No. 99. 2 vols.

1953. "Le Quaternaire de Rabat-Casablanca et ses relations avec la préhistoire," *Libyca*, 1:13–15.

Monition, L.

1956. "La nappe artésienne de la plaine du Bou Agba (bassin de l'Oued Mda)," *Bull. Soc. Sci. Nat. et Phys. Maroc*, No. 7: 23–7.

Neuville, R., and A. Ruhlmann

1941. "La place du Paléolithique ancien dans le Quaternaire marocain," *Publ. Inst. Hautes Etudes Maroc.* (Collection Hespéris), No. 8.

RAYNAL, R.
1956. "Les phénomènes périglaciaires au Maroc et leur place dans l'évolution morphologique," *Biuletyn Peryglacjalny* (Lodz), No. 4: 143–62.
ROCHE, ABBE J.
1958. "Chronologie absolue de l'Epipaléolithique marocain," *C. R. Acad. Sci.* (Paris), 246:3486–7.
1959. "Nouvelle datation de l'Epipaléolithique marocain par la méthode du carbone 14," *C. R. Acad. Sci.* (Paris), 249:729–30.
RUHLMANN, A.
1951. "La grotte préhistorique de Dar-es-Soltan," *Publ. Inst. Hautes Etudes Maroc.* (Collection Hespéris), No. 11.
TALTASSE, P.
1953. "Recherches géologiques et hydrogéologiques dans le bassin lacustre de Fès-Meknes," *Notes et Mém. Sérv. Géol. du Maroc,* No. 115.
VALLOIS, H. V
1945. "L'Homme fossile de Rabat," *C. R. Acad. Sci.* (Paris), 221:669–71.
1960. "L'Homme de Rabat," *Bull. Archéol. maroc.*, 3:87–91.
VALLOIS, H. V., and ABBÉ J. ROCHE
1958. "La mandibule acheuléenne de Témara," *C. R. Acad. Sci.* (Paris), 246:3113–6.

VERY EARLY EAST AFRICAN HOMINIDAE AND THEIR ECOLOGICAL SETTING

L. S. B. LEAKEY

SUMMARY OF GEOLOGY

THE GEOLOGICAL SEQUENCE of Olduvai Gorge needs to be summarized briefly in order to see recent finds in their proper perspective.

The lowest deposit, which is clearly visible in part of the Gorge, is a massive lava up to 60 feet thick, but at one place, at the third transverse fault, underlying deposits, which seem to be lake beds, can be seen. They have not yet been explored.

The upper surface of the lava is very uneven although in some places fairly horizontal. In certain places it rises into small hills. In consequence Bed I, the lowest of the sedimentary part of the series, is of variable thickness. The hollows in the old surface were filled first (and relatively quickly perhaps) and then the rest of the area was covered more slowly. Bed I, as observed, varies in thickness from 100± feet to as little as 18 feet near site HWK, where the lava rises to its highest observed point.

The average thickness in the middle part of the Gorge near the junction of the main and side gorges is only 40± feet.

Bed I is thick and contains much very coarse material of volcanic origin to the east and becomes more and more fine grained as we proceed westward.

A number of oscillations in deposition led to temporary land surfaces which were inhabited by mammals, including Hominidae, at irregular intervals, but the main deposits of Bed I are lacustrine with fish, crocodile, and hippopotamus remains.

The major climatic, faunal and geological break occurs near the base of Bed II and not, as previously thought, between Bed I and Bed II. The fauna of the lowest part of Bed II and overlying the "marker bed" (B of Richard Hey) at the top of Bed I is of Upper Villafranchian age, and compares very closely with that of Omo in Abyssinia.

After this, there is evidence of a change of climate with aeolian sands being deposited, followed by severe channelling and deposition of earths and gravels. It is at this point that many of the animals which are characteristic of Bed I, such as *Dinotherium* and others, disappear.

Most of the land surfaces in Bed I and the lower part of Bed II represent short periods of recession rather than seasonal fluctuations; but, near the top of Bed I, there is a series of deposits, some seven feet in thickness, which may, perhaps, be regarded as representing genuine seasonal fluctuations such as those to be witnessed on shallow East African lakes today.

Bed II represents a renewed wet climate which lasted a very long time, and has a total thickness, almost everywhere, of from 60 to 80 feet. To the west it is composed largely of clay and silt and to the east has much coarse volcanic detritus. Bed III rests unconformably upon Bed II and unlike Bed II is not of lacustrine but of terrestrial origin. It is, in the main, bright red and in places contains torrential river gravels but is made up mostly of red soils. There are fairly thick gray terrestial beds and river sands in the lower half of Bed III. The thickness of Bed III varies with the distance from volcanic mountains in the east and south where there must, therefore, have been high land. This does not exactly coincide with the present-day volcanic ranges. In places Bed III fills valleys which are cut as deep as 60 feet into Bed II, in others 20 feet; but in most places the contact is more or less horizontal. Bed III eventually wedges out completely and disappears so that Bed IV rests unconformably upon Bed II to the west.

In due course Bed IV, which is again of lacustrine origin with fish, crocodile, and hippopotamus, was laid down over Bed III.

After Bed IV times there was major faulting so that the beds were dropped down in a series of steps to the east. The total throw is not measurable as the floor of the bottom step in the Balbal is filled with Gamblian sediments and we cannot therefore find the upper part of the main Olduvai series there. We know, however, by careful level-section measuring that the total throw is over 700 feet and perhaps nearer 800 feet.

FOSSIL FAUNA

The geological age of Bed I is shown by its fossil fauna of which we now have a large number of specimens in good condition and which have been found *in situ* at different levels. This fauna is Villafranchian but does *not* represent the lower Villafranchian which in East Africa is represented by deposits at Kanam East and Kanam West.

In this connection it is important to note that Hopwood's suggestion (in Leakey 1951) of a Middle Pleistocene age for Bed I at Olduvai was based upon very inadequate and sometimes wrongly identified data collected up to and including the year 1935. Since then, very much better material, in much greater quantity and more complete, has come to light. As a result the specimens which were used to provide data for Hopwood's list have had to be re-examined and of the species and genera, which were reported to be in Bed I, on the basis of the earlier material, only six are now quite certain. These are *Deinotherium bazasi, Stylohipparion sp., Equus olduvaiensis, Metaschizatherium hennigi,* and *Parmularius altidens.*

The supposed presence of such creatures as *Hippopotamus gorgops*, *Taurotragus oryx*, *Strepciseros strepciseros*, and many others, was based (so far as Bed I was concerned) on inadequate fragments and apparently inferred because better specimens of these occurred in the later and overlying beds.

Now that we have much better material we can say that what seemed to be a *Strepciseros strepciseros* is, in fact, a very large extinct Sitatunga-like animal, and similarly with other identifications.

When the new material has been fully studied, it is possible (but not probable) that some of the earlier listed animals will be put back into the Bed I level on the basis of surer foundations, but, at present, even the supposed presence of *Elephas recki* in Bed I is not proven, though it is superabundant in the upper part of Bed II and in Bed IV.

One well preserved specimen which was identified as *Elephas recki* and which is listed as having come from Bed I in 1931, came, in fact, from the lower part of Bed II, since the site at which it was reputed to have been found, Bed I, is not exposed.

While so much of the supposed faunal dating of Bed I has thus to be abandoned, it is not possible to give the details of the new faunal evidence which is now being studied (summer 1961), and which will be published as soon as possible.

It can, however, be said that the newly found fauna includes a saber-toothed cat, some very archaic pigs, a giant porcupine allied to that from Sterkfontein, a giant sitatunga-like antelope, and an okapi-like giraffid.

It is essential, moreover, to note that *it is not possible to talk, simply, of "the fauna of Bed I."* Bed I covers a very long period of time with a gradual change from very wet climate at the beginning to savanna conditions and then sub-desert ones.

Until the final faunal analysis has been made no more can be said, but the fauna of the upper few feet of Bed I is emphatically not the same as that from the middle and lower part of Bed I. It also differs, markedly, from the fauna found at the base of Bed II and higher in Bed II.

CLIMATIC INDICATIONS OF BED I

It is not possible, at this stage, to go into all the evidence in detail but it is quite certain that the rainfall must have been much heavier than it is today (combined with much greater cloud cover and consequently lowering of temperature and evaporation) before a lake of the type of which we have evidence at Olduvai could have existed over a very long period of time.

While the present Ngorongoro range of volcanic highlands did not yet exist, there must have been a range of higher land in this general region, and the Olduvai area lies on the west and lee side of the range, not on the rain shadow side.

On the Ngorongoro range today the rainfall on the two sides differs markedly.

That on the very wet east side is said to be nearly 60 inches a year. The whole of the run-off from this side goes into Lake Manyara by way of some thirty streams. Nevertheless, this mass of water fails to maintain Manyara as anything but an intermittent swamp and lake with high soda concentration.

In order to maintain a lake and a vegetational situation in which sitatunga- and okapi-like animals as well as very large ungulates could flourish, there must have been a very much higher over-all precipitation. This means, in turn, that there must have been a much heavier waterflow to the Manyara area in addition to the greater flow towards the drier Olduvai side. We can prove this to have been the case on the Manyara side by the very extensive deep water lake beds which occur many miles to the east of Lake Manyara today, while the Olduvai lake beds support the same idea to the west.

THE "ABSOLUTE AGE" OF BED I

Leakey *et al.* (1961) have published a preliminary report dealing with the age of Bed I on the basis of the potassium-argon dating method. The mean of the dates of the strata which straddle the land surface which contains *Zinjanthropus boisei* and the pre-*Zinjanthropus* juvenile, is 1,750,000 years. The dates for deposits at the top of Bed I average about 1,200,000.

This somewhat remarkable remote age does not mean that we have to place Bed I back into the Pliocene. It simply means that we have, in the past, greatly underestimated the length of time required for the Pleistocene. I envisage a date of well over 2,500,000 (perhaps nearer 3,000,000) for the lowest Villafranchian of Kanam West.

THE LIVING FLOORS IN BED I

We now have three stratigraphically different living floors in Bed I, all at the site known as FLK.

These are (1) site FLK I which gave us the *Zinjanthropus* skull; (2) site FLK NN I which gave us the mandible, parts of the skull, and certain portions of the post cranial skeleton of a juvenile, which is stratigraphically older than *Zinjanthropus*, although geologically more or less contemporary; (3) the site at FLK N I which has yielded quantities of stone tools but, so far, no hominid fossils.

Let us briefly consider these from the earliest to the latest. FLK NN I is apparently a living floor, but the material so far recovered from the area excavated is not very abundant. This may be due to the fact that the industry was much cruder and simpler, or it may be that we are only on the outskirts of a floor which, in places, may have had higher concentration of tools as well as better ones. Next year's work will answer this question. At present there is no way of telling for certain. On this floor there are fossil remains of many tortoises, many birds, a number of cat fish and also tilapia, together with some large mammals and many smaller ones. The stone-tool assemblage includes genuine Oldowan type tools, a number of natural stones with sharp edges and many bashers and unworked stones. There is also one beautifully made bone tool in the form of a "*lissoir*" which most strongly suggests the working of leather thongs. The hominid fossile remains includue parts of two parietals, a small piece of an occipital, part of the lower jaw, a clavicle, part of a scapula, parts of radius and ulna, parts of a hand, parts of a foot, and an upper molar, all of a juvenile of about 11 or 12 years old, together with some fragments of an adult (cf. Leakey 1961a, b).

The site FLK I is the living floor upon which the skull of *Zinjanthropus boisei* was found (Leakey 1959, 1960). I stress "was found" because it is no longer as likely, as it once seemed, that *Zinjanthropus* made the industry which is concentrated upon this living floor. The distribution over the living floor has been carefully recorded. There is a circular concentration of very broken-up bones resulting from the splitting of long bones to extract the marrow fat. Within the inner ring there is also a much greater concentration of stone tools and waste flakes than elsewhere on the floor. In the periphery are larger and less broken-up bones, broken ribs, broken jaw fragments, broken animal skull fragments, cannon bones of antelope, etc. These represent bones that had meat attached but did not have marrow in them, and which were therefore thrown away without being split open. In this peripheral area there is also a higher concentration of natural stones, which seem to have been brought to the camp site for one puropse or another. The details of the distribution pattern on the floor will be published in due course (cf. Clark 1961).

The stone tools include disc choppers, Oldowan choppers (so-called pebble tools, a name which I dislike as they were often made from lumps of rock which are not pebbles), hammer stones, bashers, retrimmed flakes, utilized flakes, cores, and a variety of unworked stone. There are no hand axes and no polyhydral stones. Much of the material used for tool-making was of relatively local origin—quartz, quartzite, and local lava—but some seem to have been brought in from very great distances, perhaps up to 45 or 50 miles away. These latter are only few in number.

The third site is FLK N I which lies immediately under the "marker bed" at the top of Bed I. Unlike the other two sites, which are simply thin land surfaces of about an inch thick and which do not represent an oscillating shore line, the deposit at FLK N I does seem to represent a lake which was drying up *with frequent seasonal oscillations*. The total depth of some six feet is full of bones and stone tools accumulated over a long period. Some were weathered before being fossilizd, others were not.

The fauna here contains antelope and gazelle remains in greater numbers, together with some hippo and horse and hipparion fossils, and very large numbers of small mammals, mainly rodents, but with some insectivores as well as small reptiles and birds.

The whole ecological picture is totally different from that presented by the fossils in the lower levels of Bed I and suggests conditions not unlike those of some of the drying-up lakes of the present day. At the top of Bed I the fauna becomes desertic and there is also the formation of "desert roses" below the then existing surface.

BED I CULTURAL LEVELS AND HOMINID REMAINS

The culture of Bed I is, as we have said before, confined to the Oldowan culture, but it now becomes evident that there may be a gradual evolutionary change in the Oldowan as we proceed from the bottom to top of Bed I. Studies are being made at present of the assemblages of FLK NN I, FLK I, and FLK N I, and when these are complete figures will be published.

While I will deal with the morphology of the fossil hominid remains in another session, I must briefly refer to them here in their cultural context.

Zinjanthropus boisei was found, as has been clearly stated, on a well-defined living floor about 20 feet below the upper or "marker bed" of Bed I. While I originally believed that the presence of a nearly complete skull on such a living floor probably indicated that the culture on that floor was made by the Australopithecine named *Zinjanthropus*, the position, as we shall see presently, is not nearly so clear now.

Certainly statements such as have been made by some of my colleagues in America, Great Britain, and South Africa that the Olduvai discovery of *Zinjanthropus* associated "with stone tools" "*proves*" that the South African Australopithecines were also stone-tool makers, cannot be regarded as more than the expression of a pious hope, not an established fact.

The pattern of distribution of bones and stone tools and waste flakes on the living floor of site FLK I makes it clear that the *Zinjanthropus* skull, like other larger specimens, was on the outskirts of the site and makes it possible that it was, like these other specimens, the remains of a meal. We do not yet know who may, or may not, have made the tools and eaten the meat upon this floor.

The discovery of parts of a "juvenile" or child at FLK NN I, also on a living floor in association with stone tools of the Oldowan culture, may be of very great significance in interpreting the evidence of who made the stone tools of the Oldowan culture.

The remains of this "child" include parts of the two parietals, a small fragment of occipital, parts of the lower jaw, one upper molar, the clavicle, part of the scapula, parts of the radius and ulna, and parts of a hand and foot. The jaw and its parietal indicates an individual who is clearly *not* an Australopithecine, in the accepted sense of that subfamily, but a Hominine. The parietals especially show a size comparable to a Pithecanthropine such as *Pithecanthropus II*, while the teeth have both a morphological pattern and indices which are clearly outside the known range of variation of corresponding indices in Australopithecines.

Similarly, the size of some of the parts of the postcranial skeleton speaks eloquently of something different. The morphology will be discussed later at another session.

Geologically *Zinjanthropus* and the pre-*Zinjanthropus* child are almost contemporary and must have overlapped. It seems, therefore, more likely that the type of hominid represented by the big-brained child may have been the maker of the Oldowan culture.

I should also report that we now have traces of teeth and bone of another hominid at site MK I. These fragments resemble the pre-*Zinjanthropus* juvenile rather than *Zinjanthropus*.

Turning very briefly indeed to the base of Bed II, it is clear that there is a major hiatus (in time) between the top of Bed I and the start of Bed II, and the fauna, once we get to Bed II, is fundamentally different. A very high proportion of new genera appears for the first time at this level, so far as we know at present.

Near the base of Bed II we have a living floor, at site BK II, with more than 9,000 artifacts showing the start of what we take to be the first true stage of the Chellean part of the Chelles-Acheul culture (Leakey 1957, 1958). From this site we have two human milk teeth which I do not consider to be Australopithecine, in spite of the claims of Robinson (1959) to the contrary (cf. also Dahlberg 1960, and Koenigswald 1960).

At the slightly higher level of site SHK II we have Chellean stage 2 with more than 6,000 tools resting on a limited living floor. No human remains have so far been found at this level, except a very weathered femur shaft.

Finally, I must briefly indicate the discovery of the skull of a hominine at the level of Chellean stage 3. This skull, as we shall see later, has superficial similarities with the Pithecanthropines but differs in points of detail.

It is not possible, at the present time, to give as much detail about the fossil hominid remains from Olduvai as I should like to do, since the detailed study is not yet complete. The following brief summary may serve a useful purpose.

A. Zinjanthropus Boisei

The many hours which I have devoted to examining this skull, since I wrote my preliminary report in *Nature* (Leakey 1959), have not materially altered the opinion which I gave at that time.

The *Zinjanthropus* skull, which is that of a young male, seems to represent a type of Australopithecine which is different from all the known South African representatives of this subfamily, whether they be *Australopithecus* or *Paranthropus*.

In certain characters (some of them very noticeable ones like the sagittal crest) *Zinjanthropus* recalls *Paranthropus*. In other characters the affinities of the Olduvai specimen lie with the genus *Australopithecus*. There are also a number of characters in which *Zinjanthropus* differs most markedly indeed from both the South African genera and approaches more closely to *Homo*. I refer, in particular, to the formation of the *foramen magnum* and the occipital condyles, the size and shape of the mastoid process, and the malar-maxillary region of the face.

I fully realize that the question of whether these differences justify generic, or only specific, differentiation, is one which depends upon point of view, but I believe that *Zinjanthropus* differs as much (or more) from the South African genera as gorilla does from chimpanzee. That is why I have accorded it distinct generic rank.

I should perhaps add that I believe that the supposed resemblance between *Zinjanthropus* and *Paranthropus* has been enhanced by Robinson's most recent reconstruction of his *Paranthropus*, a reconstruction which I cannot help but feel has been somewhat influenced by our photographs of *Zinjanthropus*.

I very much doubt if any of the *Paranthropus* skulls so far found are, per se, capable of giving rise to the reconstruction we have been shown of *Paranthropus*.

I would like here to issue a warning against the assumption that the tibia and fibula found near the *Zinjanthropus* skull are *necessarily* representative of the *Zinjanthropus* type of hominid. We have clear evidence of the presence of another type of hominid (see below) at the same time, and these limb bones could equally well represent this other type. We do not know.

As regards the cranial capacity of *Zinjanthropus*, all I can say is that it is somewhere in the region of 600 cc., perhaps a little over.

The age of *Zinjanthropus boisei* is something of the order of 1.75 million years and much older than other South African Australopithecines. I think it is likely that they represent later and over specialized groups that broke away from the earlier *Zinjanthropus* stock.

I do not however believe that any Australopithecines in East or South Africa gave rise to *Homo* and I believe that the presence in the Olduvai beds of a different type of Hominid at this remote date (see below) makes it increasingly necessary to discard the hypothesis that human evolution followed a simple series of stages, *Australopithecus-Pithecanthropus-Homo*, such as Le Gros Clark and Arambourg would have us believe.

B. The Pre-Zinjanthropus Juvenile

The fossil hominid remains from the site FLK NN I come from a geological level lower and older than that which yielded *Zinjanthropus*. These remains comprise parts of two parietals, parts of an occipital, the greater part of a lower jaw, parts of two clavicles, parts of a hand and of a foot, bits of a scapula, etc. There is also an upper molar.

Most of these remains belong to a single individual, a juvenile of an age corresponding to 11 or 12 years old today. Some of the bones, however, belong to an adult.

In my preliminary note on these fossils I stated that I did not consider that they represented an Australopithecine, but were much more a Hominine, and my subsequent studies have made me still more certain that this view is correct.

Although this juvenile was only about 11 to 12 years old at death, and probably therefore could expect at least a five per cent increase in brain growth, the parietals are far larger than those of the *Zinjanthropus* adult, or of any published South African Australopithecine, and, indeed, are almost identical in size to the parietals of *Pithecanthropus II.*

The teeth in the mandible, as well as the one upper molar, are metrically and morphologically quite unlike those of any published Australopithecine. The better preserved of the two clavicles is very like that of *Homo* and of a size comparable to many small living *Homo sapiens* today.

I am not competent to speak on the hand or foot bones.

The available evidence most clearly suggests that the juvenile does not belong to the subfamily Australopithecinae and that it reveals the contemporary presence of two different hominid types at Olduvai in upper Villafranchian times. One represented by *Zinjanthropus* is an Australopithecine, or a hominid which is heading for extinction, the other, represented by the juvenile, (and some other remains), seems to be much more closely related to the stock which eventually gave rise to *Homo*. That, at least, is my considered opinion.

It may be noted here, that an upper molar and bits of a skull of the same type as those of the pre-*Zinjanthropus* child have now been found at the *Zinjanthropus* site and level, proving that the two distinct types were contemporary and co-existent just as Pygmies, Bantu negroids, Europeans, gorillas and chimpanzees, are all contemporary and co-existent in the Eastern Congo today.

C. The Hominid Skull from Bed II at the Level of Chellean Stage 3 of Culture

A hominid skull cap of outstanding interest was found at site LLK Bed II, at the level of Chellean cultural stage 3, in December 1960 (Leakey 1961a). This skull is in process of being further cleaned of its matrix and I cannot say very much more than I did in my original preliminary report. I need only say that as the work of cleaning continues, nothing has been revealed to alter my original opinion that the resemblances between the skull and the Pithecanthropine type are more superficial than they are real. The supraorbital torus is more massive than any of which we have knowledge in other hominid skulls; much more so than in *Pithecanthropus* of either the Java or China forms, and different, morphologically, from Rhodesian man and the Eyasi skull.

It is unfortunate that this skull has (so far) no lower jaw, while the Ternifine (*Atlanthropus*) fossil hominid remains lack the parietals, the temporals and the occipitals. Comparison is therefore not possible.

There are those who will see in this new Bed II skull evidence to support the idea of a "Pithecanthropine stage" of hominid evolution, but I would urge caution.

It seems to me more likely that the Far Eastern *Pithecanthropus* and our East African skull have a common ancestor much further back, and that it was these African hominids (with some Pithecanthropine resemblances in a few characters) that gave rise eventually to *Homo*.

BIBLIOGRAPHY

CLARK, J. D.

1961. "Sites Yielding Hominid Remains in Bed I, Olduvai Gorge," *Nature,* 189:903–4.

DAHLBERG, A. A.

1960. "The Olduvai Giant Hominid Tooth." *Nature,* 188:962.

KOENIGSWALD, G. H. R. VON

1960. "Remarks on a Fossil Human Molar from Olduvai, East Africa," *Proc. Koninkl. Nederl. Atad. v. Wetenschappen* (Amsterdam), Ser. B, 236:20–5.

LEAKEY, L. S. B.

1951. *Olduvai Gorge.* Cambridge: Cambridge University Press.

1957. "Preliminary Report on a Chellean I Living Site at BK II, Olduvai Gorge, Tanganyika Territory," *Proc. 3d Pan-Afr. Cong. Prehist.* (Livingstone 1955): 217–8.

1958. "Recent Discoveries at Olduvai Gorge, Tanganyika," *Nature* (London), 181:1099–1103.

1959. "A New Fossil Skull from Olduvai," *Nature* (London), 184:491–3.

1960. "Recent Discoveries at Olduvai Gorge," *Nature* (London), 188:1050–2.

1961a. "New Finds at Olduvai Gorge," *Nature* (London), 189:649–50.

1961b. "The Juvenile Mandible from Olduvai," *Nature* (London), 191:417–8.

LEAKEY, L. S. B., J. F. EVERNDEN, and G. H. CURTIS

1961. "Age of Bed I, Olduvai Gorge, Tanganyika," *Nature* (London), 191:478–9.

ROBINSON, J. T. (and L. S. B. LEAKEY)

1959. "An Alternative Interpretation of the Supposed Giant Deciduous Hominid Tooth from Olduvai," *Nature* (London), 185:407–8.

ACHEULIAN HUNTER-GATHERERS OF SUB-SAHARAN AFRICA

F. CLARK HOWELL AND J. DESMOND CLARK

Our concern here is with the adaptations of human populations of sub-Saharan Africa during the later ranges of the Middle Pleistocene and into the earlier ranges of the Upper Pleistocene. No method of absolute age determination is as yet sufficiently refined to provide estimations in years for this span of time. However, there is a variety of less direct evidence to suggest that an appropriate order of magnitude of over 100,000 years, and probably 150,000 years, is a reasonable approximation. In Europe this time span would comprise a substantial part of the Great Interglacial (Hoxnian=Holsteinian) stage, the subsequent Riss or Saale glacial stage, the ensuing Last Interglacial (Eemian) stage, and the beginning of the early Würm glacial stage.

In sub-Saharan Africa there are obvious difficulties in correlation of local Pleistocene stages with that sequence established even in another nearby region. This is a consequence of poor preservation of mammalian fossils in many situations as well as the still inadequately understood climatic conditions of the Pleistocene in tropical and subtropical latitudes. The difficulties of intercontinental correlation are far greater and these we recognize. In terms of the general East African succession, established on lithological units and their contained mammal faunas, our concern extends from a time within the Kamasian-Kanjeran stage, through the Kanjeran and Kanjeran-Gamblian stages into the early Gamblian stage. We believe these designated intervals of Pleistocene time, best recognized in Eastern and Central, and parts of South Africa, to correspond *broadly* with the aforementioned European stages of later Middle and earlier Upper Pleistocene time. Direct correlations with Europe are tenuous, especially those based upon presumed but often still unproven climatic changes adjudged to correspond with, and to be conditioned by, glacial-interglacial conditions in northern latitudes. Such correlations are best established on mammalian faunal grounds and on pollen evidence; the available faunal evidence, and what little is yet available from pollen, does not conflict with the correlation suggested above. The correlation problem can best be settled by absolute age determinations, such as will eventuate from the refinement of the potassium-argon (K/A), and perhaps other methods, including radiocarbon for the terminal stages of the Acheulian.

Our several colleagues (Bishop, Bond, Grove, de Heinzelin, Monod) have dealt directly or indirectly with the field evidence which forms the basis for regional stratigraphic successions. This evidence also forms the basis for inferences into Pleistocene ecological conditions in sub-Saharan Africa. Hence it is pertinent to our own paper; but we will attempt to draw conclusions from the data without repeating the primary evidence from which conclusions have been drawn.

We have, rather arbitrarily, defined sub-Saharan Africa as that portion of the continent lying south of the parallel of 15°N. There are definite reasons for this choice. This parallel passes from just north of Dakar, south of Timbuktu, and across the great bend of the Niger river, through the northern reaches of the Chad basin, and eastward to south of the vicinity of Khartoum. It corresponds approximately with the northern limits of wooded steppe (*Acacia* and *Commiphora* spp.) or *Cenchrus*-type savanna in Rattray's (1960) recent grassland map. Actually west of 15° E. this vegetation type extends somewhat north of this parallel due to the increased rainfall regimen; and eastward of 15° E. the subdesertic steppe extends in uneven fashion southward of the 15th parallel (Keay 1959). Still it is a fairly useful boundary. Also there are corresponding broad equivalences with the summer (July) isohyet and, more closely, with the winter (January) 70° F. isotherm. Hence, in a broad sense, this delimits the southern (semidesertic) fringes of the Saharan region from the southerly-lying wooded grasslands so characteristic of northern equatorial, eastern, and southern Africa.

It should be emphasized that this boundary is inadequate for the later Middle Pleistocene range of time. There is abundant evidence, both geological and archaeological, to prove that climatic and hydrographic conditions in the Sahara were substantially unlike those prevailing at present. These had corresponding influences on the distribution of biotas, including early man, which are still sometimes not fully appreciated.

From the archaeological standpoint our concern is with that range of the Old Stone Age characterized by the Acheulian industry.[1] The discussion here

1. The terminology followed here, with a distinction between Pre-Chelles-Acheul and Chelles-Acheul (Chellian and Acheulian), is that laid down in 1947 by the Sub-Committee on Prehistoric Archaeology and passed as a general resolution (16:1) of the 1st Pan-African Congress on Prehistory (Nairobi). At that time the term Chelles-Acheul, prefixed with appropriate regional terms, was to be employed for the Earlier Stone Age, including later or local derivates like the Fauresmith and Sangoan. The term Pre-Chelles-Acheul was to be used as a general term for earlier lithic assemblages, with Kafuan and Oldowan employed to describe the earlier and later stages, respectively (16:6, 7). (Since that time the existence of the Kafuan, as originally described in the type locality and in the Kagera valley as well, has been seriously questioned by several workers, and the term is not now currently employed.) At the 3rd Pan-American Congress on Prehistory (Livingstone) it was resolved

is limited to only a portion of its extremely wide distribution, from Britain to the Cape and eastward to the Ganges flood plain. The evidence from sub-Saharan Africa provides maximum insight into rates of cultural change, composition of artifact assemblages, techniques of manufacture and implement typology, nature of occupation sites, and patterns of livelihood to a greater extent than any other part of the Eastern Hemisphere. Our colleague L. S. B. Leakey will provide a discussion of the still earlier stages within the Old Stone Age.

SPATIAL DISTRIBUTION

In a continent as vast as Africa, and in which archaeological exploration is still grossly inadequate, if not hardly begun in places, delimiting geographic distributions of Pleistocene human populations is nearly impossible. It is further complicated by the previously mentioned difficulties met with in effecting correlations between areas (other than on archaeological grounds, a risky venture involving circular reasoning). Isolated discoveries of certain types of implements, too frequently from surface exposures, and hence lacking in either geological or archaeological context, are scarcely useful. Various implements, as well as the techniques by which these were made, are time-transgressive to a considerable extent. Assemblages are necessary in order to make proper diagnoses and these are altogether too infrequent. We would suggest, indeed insist, that when collections are made by any interested party that these be as large as feasible under the prevailing field conditions, and that all representative components of the industry be collected to some extent (including small tools, cores, and waste products). If this procedure is followed such collections will be far more useful to the prehistorian.

For these reasons our remarks on this aspect of the problem are necessarily of limited value. Our colleagues at this conference, with firsthand experience in many areas personally unknown to either of us, can doubtless add much to this aspect of our paper.

There is an obvious need for adequate distribution maps of African archaeological sites. One of us has called attention to this elsewhere (Clark 1957b) and steps are being taken to compile such maps. In 1957 the South African Museums Association sponsored the compilation of an "Atlas of Prehistory" for southern Africa. In 1959 the Pan-African Congress and the Scientific Council for Africa South of the Sahara added their support to the project the scope of which has

---

(Resolution 6) that the terms Earlier Stone Age and First Intermediate, the latter to include Fauresmith and Sangoan Acheul-Levallois, etc., be followed.

More recently Mason (1961a, b) has suggested that the term Acheulian be employed to include the Chelles-Acheul (including the final Fauresmith phase), a procedure which merits consideration since similar terminology has been applied in northwestern Africa. However, this is a matter to be proposed for mutual discussion and formal action at the forthcoming Pan-African Congress on Prehistory and Quaternary Geology, rather than for individual decision.

now been extended to cover the whole continent. In a few years our task of the moment will be greatly simplified.

Some such maps, which surely contain errors and deficiencies, have already been prepared for parts of sub-Saharan Africa.[2] The most adequate (and recent) are those for the (Belgian) Congo (20 sites) (Mortelmans, no date), Northern Rhodesia (14+4) (Clark 1957 a,b, p. 423, Fig. 7b), and Southern Rhodesia (16+9) (Summers 1957, p. 399, Map 2); in a subsequent map (Summers 1960, p. 273, Fig. 6; also Summers and Cooke 1959) 57 sites in Southern Rhodesia are referred to the Earlier Stone Age, nearly all of which are Acheulian and nearly a third with *in situ* (geological context) assemblages. For the Guinea area of West Africa, Davies (1957, p. 593, Fig. 1) prepared a map of probable Acheulian sites in Nigeria (5+ Jos plateau localities), Togo (6), and Ghana (8). This distribution was subsequently revised on the basis of further field and museum experience for the area between 6° W.–10°E. and 4°–14° N. (Davies 1959, p. 205, Fig. 2). It includes Nigeria (5+2), Dahomey (0), Togo (5+3), Ghana (15+9), and Upper Volta (2), and the Ivory Coast (2+1). In nearly all cases, however, the finds are isolated occurrences and there is a real question as to whether these may not in fact represent the subsequent Sangoan industry which is both common and widespread over most of this area.

The distribution of sites in the Horn (the Somalilands and Ethiopia), which represent at the earliest a terminal manifestation of the Acheulian, has also been presented in a map by one of us (Clark 1956, p. 161; also in Clark 1952, opp. p. 158; also in Cole 1954, p. 201, Map 8). The Sudan sites (12), between the Second Cataract and the White and Blue Nile confluence, have been mapped (up to 1948) by Arkell (1949, p. 2; also in Cole 1954, p. 149, Map 5). Other maps, less complete but in some cases under revision, have been prepared for Uganda (O'Brien 1939, opp. p. 316; van Riet Lowe 1952, opp. p. 1; summarized in Cole 1954, p. 120, Map 3), and for Kenya and northern Tanganyika (Leakey 1931; summarized in Cole 1954, p. 137, Map 4, and p. 185, Map 7). There are no adequate published maps for many of the separate southern Africa territories; but, South-West Africa has been mapped (Fock 1958) and the monumental task of recording the sites in the Union of South Africa has been underway since 1958. A general map of the distribution of the earlier Old Stone Age in southern Africa (lying south of 8° S.) has been prepared by one of us (Clark 1959, p. 104, Map 9). Very schematic maps for the whole of Africa, and indicating general distributions but omitting specific localities, are presented in the book by Alimen (1955, p. 487, Fig. 151, and p. 488, Fig. 152; 1957, p. 417, Fig. 151, and p. 419, Fig. 152).

The map presented here (Fig. 1) attempts to show the distribution of the main Acheulian sites as these are known to us from published sources. Since, however, a considerable amount of material remains unpublished in the collections of museums, institutions, and private individuals, it is to be expected

2. The figures in parentheses refer to the number of Acheulian sites distinguished on each map.

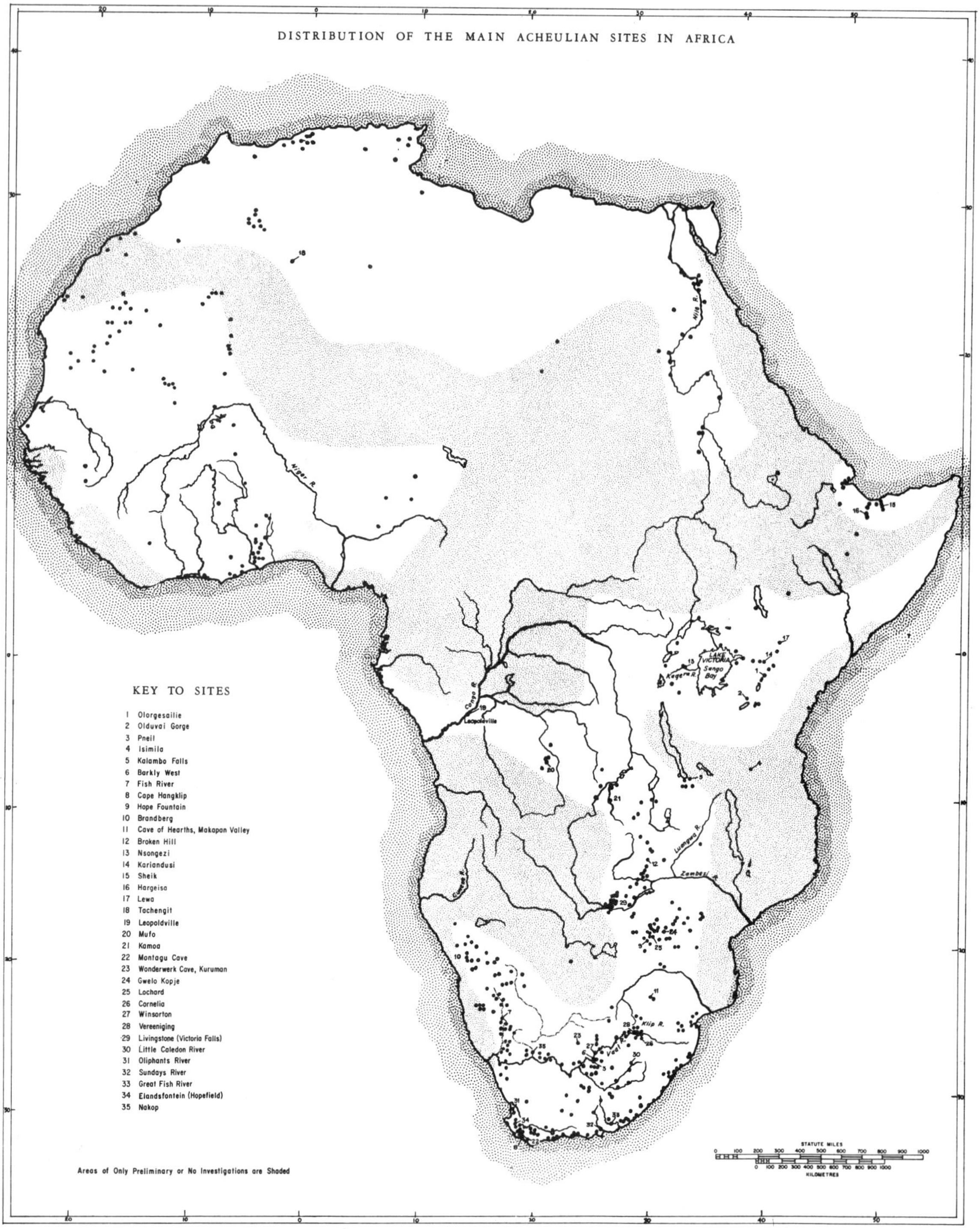

FIGURE 1
Distribution of the main Acheulian sites in Africa.

that were these also to be included, the number of sites would be greatly increased. Nevertheless, we believe that the general distribution is not materially affected by the incompleteness of the presentation. We had hoped that it would have been possible to differentiate the sites spatially, according to their situation (whether riverine, coastal, cave, lakeside, or open air occupation sites, etc.) and according to the time range involved. While, however, the character of some sites is well determined, the necessary information is lacking or uncertain for many others and it was, therefore, decided not so to distinguish them. The only cave occupations of Acheulian age that are known are those at Sidi Abderrahman (33° 25′ N., 7° 40′ W.), at the Cave of Hearths, Makapan (23° 45′ S., 29° 18′ E.), at the Montagu Cave (34° 0′ S., 19° 40′ E.), at the Wonderwerk Cave (27° 15′ S., 23° 25′ E.), and at an unpublished site in the Transvaal. The few coastal sites known show no adaptive specialization to a marine environment. The map reflects in particular the distribution of later and final Acheulian sites since these are much more numerous than those of the earlier stages which can be accurately differentiated only on clear stratigraphic grounds. Thus it is clearly impossible to differentiate the separate stages within the Acheulian industry except in very rare instances where basins of riverine or lacustrine sedimentation preserve long successions of human occupation. For these several reasons an exhaustive treatment would be both premature and and presumptive on the basis of the available data.

The very limited number of localities which preserve Early and Middle Acheulian industry in suitable contexts greatly limits any conclusions about spatial distribution. A general idea of distribution may be gained from occurrences in relative dateable contexts in the southern Cape, along the southeastern (Indian Ocean) coast of the continent, in the Vaal drainage system, in the Zambezi Valley and certain of its southern and northern tributaries, in some highland southern affluents of the Kasai, a southern tributary of the Congo, in the Olduvai basin and the adjacent Manyara trough, in the Olorgesailie basin north of the Magadi trough, in old northeastward (Kavirondo) and southwestward (Kagera) embayments of the Victoria basin, and in some tributaries of the Albert-Edward troughs of the Western Rift Valley. This distribution is in part fortuitous since it is (1) closely dependent upon persistent basins and major river valleys where Pleistocene deposition obtained and is exposed by subsequent erosion, and (2) related to the degree of commercial exploitation and/or archaeological investigation in the particular region the extent of which we have tried to show on the map.

The Later (and terminal) Acheulian is much more widely distributed south of the parallel of 15° N. This is surely in part a function of its more recent age; but also because it is related in many instances to a different cycle of river development and hence is more frequently preserved and exposed. In southern Africa, and also in eastern Africa perhaps to a lesser extent, the final manifestation of the Acheulian industry is found in the lowermost colluvial and other rubble deposits, at the base of Upper Pleistocene sediments of various sorts, exposed by recent sheet erosion and gully-cutting. It is most likely that this is a function of climatic change at the end of the Middle Pleistocene. It would correspond with the so-called Kanjeran-Gamblian "interpluvial" for which, at least in southern Africa, there is substantial evidence to indicate reduced rainfall conditions. One of us (Clark) has suggested that such conditions were generally unfavorable, led to cultural change and ecological readjustments, and led to population movements through newly opened corridors (Clark 1960b). The preceding damper "pluvial" (Kanjeran stage) was more favorable to cultural stability and relatively more settled populations. There are then some indications in the archaeological record to suggest perhaps both more widely distributed and also more numerous human populations in sub-Saharan Africa at the end of the Acheulian.

TEMPORAL DISTRIBUTION

In order to avoid lengthy discussion of local stratigraphy, which would be out of place here, we have chosen to illustrate a series of quite representative sections through river valleys and other basins of sedimentation. These sections demonstrate the stratigraphic occurrence of the Acheulian industry in eastern and southern Africa. The river valley occurrences include examples from: the

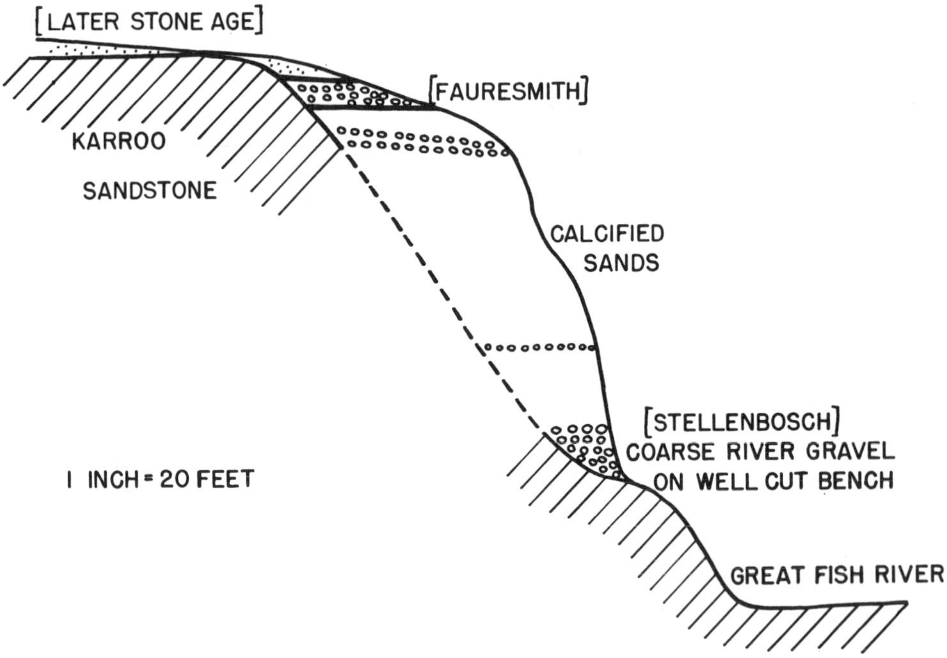

FIGURE 2
Sketch section of terrace deposit on left bank of the Great Fish River at Fort Brown, Cape Province. (After H. B. S. Cooke 1941)

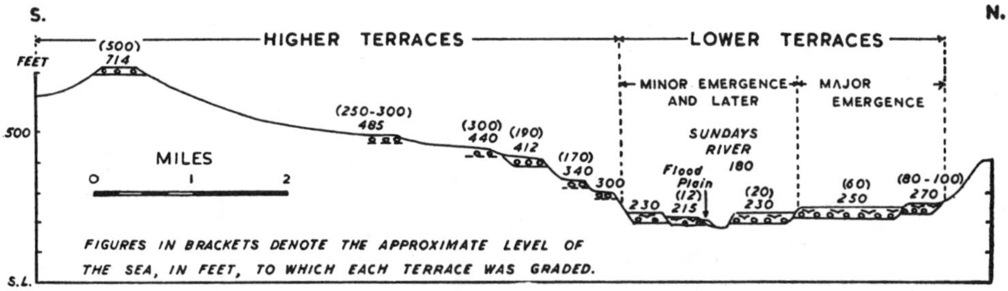

FIGURE 3
Generalized section across the Sundays River Valley at and in the vicinity of Kleingrasrug and Geelhouteboom, Cape Province. (After A. Ruddock 1957)

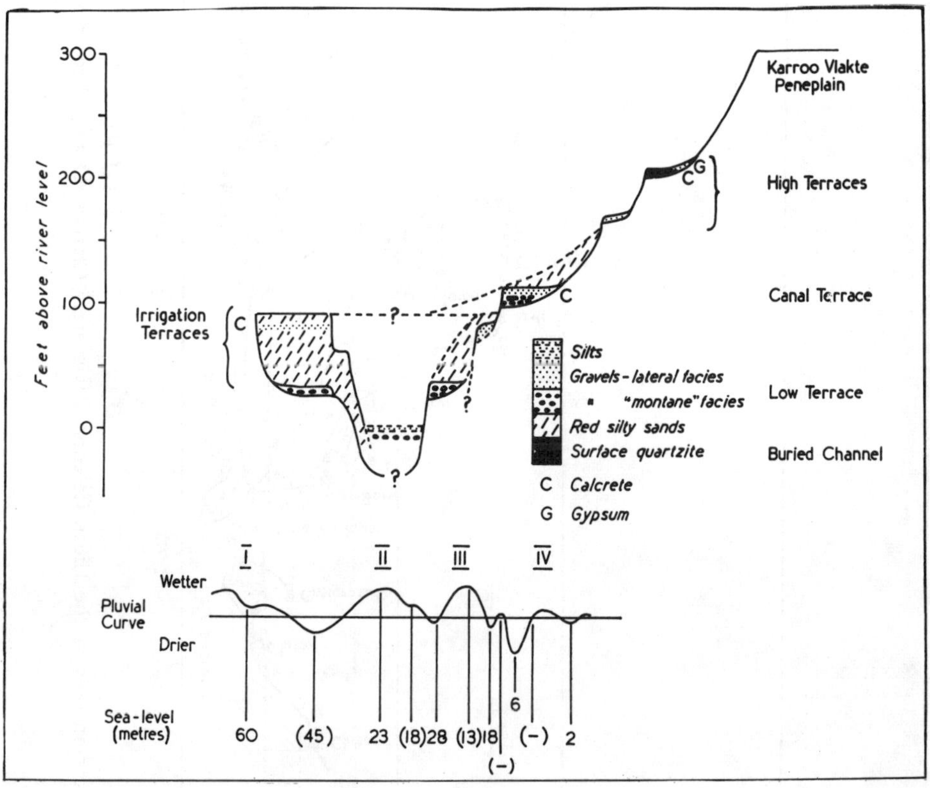

FIGURE 4
Composite section across the lower Olifants River Valley, Cape Province, with (below) a climatic interpretation of the terrace sequence. (After J. A. Mabbutt 1957)

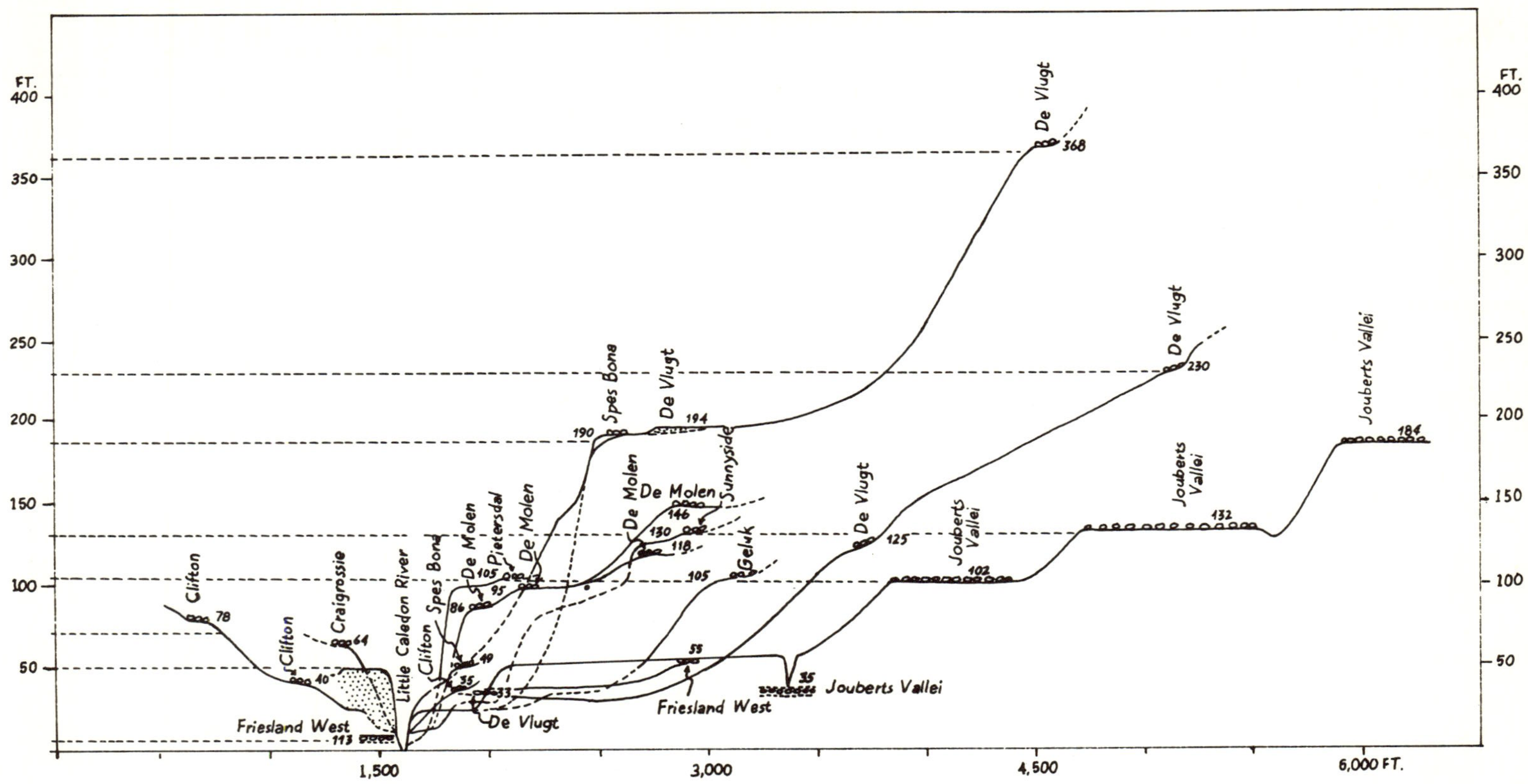

FIGURE 5
Combined sections across the Little Caledon River near Clarens, Orange Free State. (After D. J. L. Visser and C. van Riet Lowe 1955)

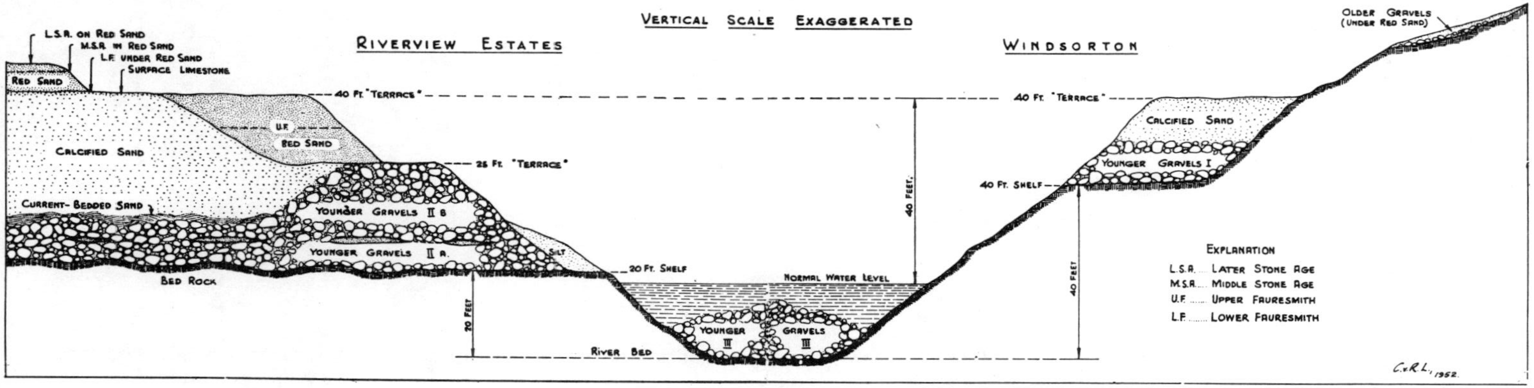

FIGURE 6

Composite sections of the Vaal River Valley, at Riverview Estates, and Windsorton, Cape Province. (After C. van Riet Lowe 1952)

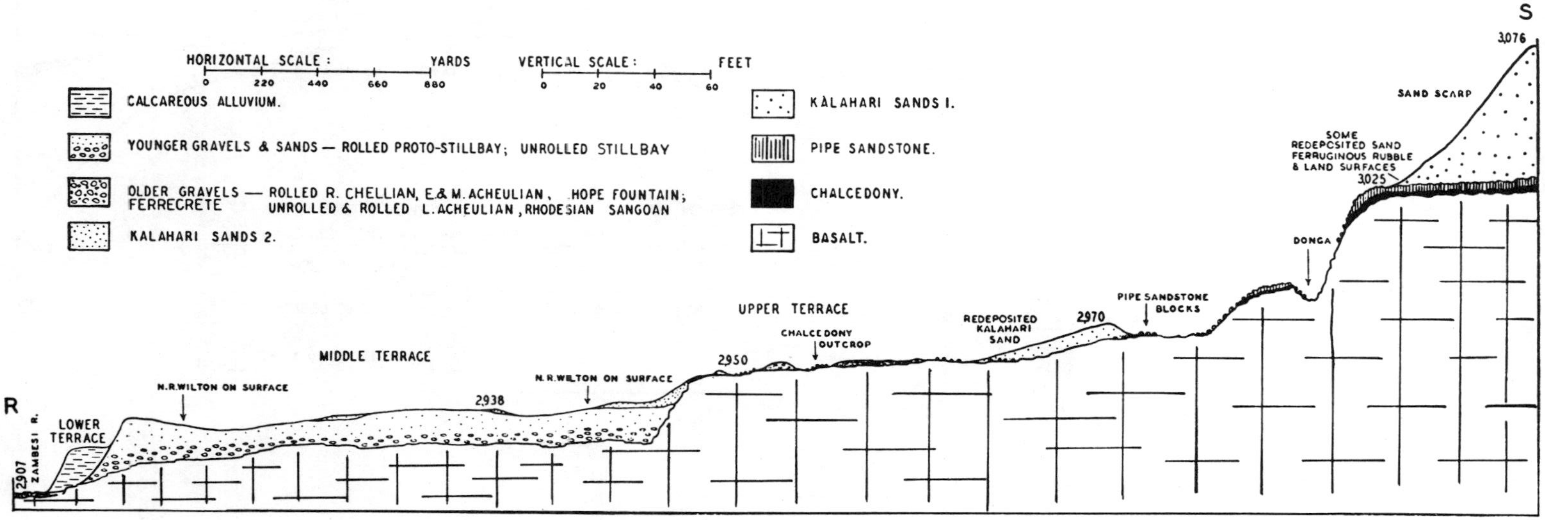

FIGURE 7

Section (north-south) across north side of upper Zambezi River Valley above Livingstone. (After J. D. Clark 1950)

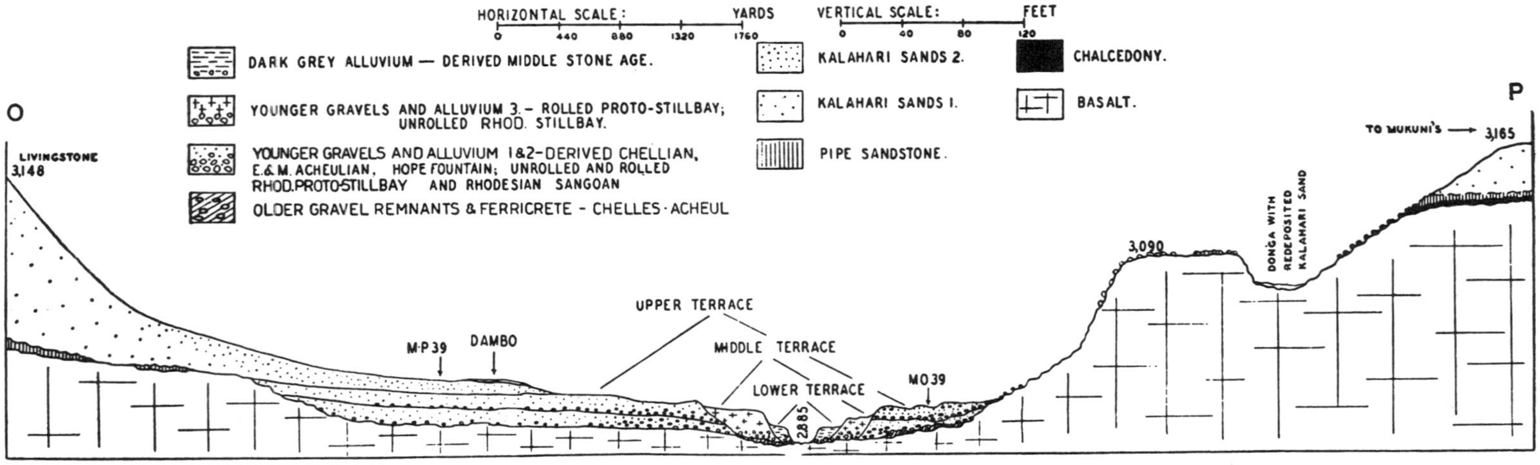

FIGURE 8
Section (nw-se) across Maramba River Valley, an upper Zambezi River tributary, near Livingstone. (After J. D. Clark 1950)

southern Cape including the Great Fish River (Fig. 2), Sundays River (Fig. 3), Olifants River (Fig. 4), the Central Plateau including the Little Caledon (Fig. 5) and Vaal Rivers (Fig. 6), the Middle Zambezi River of the Rhodesian plateau (Figs. 7–8), and the Luembe River of the southwestern Congo watershed (Fig. 9). One locality, Hopefield, is a pan site in southwestern Cape Province. Four localities, either in or adjacent to the Rift Valley System, represent open-air occupation places generally about former shallow lakes: Olduvai Gorge (eastern Serengeti Plain), Olorgesailie and Kariandusi (Eastern Rift Valley), and Kalambo Falls (southeast of the Tanganyika trough). Another is Isimila, a hill-locked basin in the southern Tanganyika highlands. The last, the Nsongezi locality, was adjacent to a former very extensive embayment of Lake Victoria (Fig. 12).

RIVER VALLEY OCCURRENCES

In sub-Saharan Africa, as in Europe and many other parts of the Eastern Hemisphere, the first studies on the Earlier Stone Age were concentrated in the major river valleys. These provide a Pleistocene succession marked by phases of aggradation and down-cutting. In a few cases in southern Africa efforts have been made to link the succession in the lower reaches of the rivers with coastal morphology and changes in Pleistocene sea levels. The evidence seems to favor the contemporaneity of the Middle to Late (but not latest or final) Acheulian with the regression from the higher or Major Emergence beaches along the southern coast; sea level fell from some 60(+) feet to approximately sea level and rose again to a level nearly two-thirds of the former, when the Minor Emergence beaches were formed. Investigations of this nature have generally lagged behind, however, and most effort has been concentrated on river systems in the interior plateau of the continent. Where mammalian fossils are included in aggradation deposits, correlations may be made with other areas on paleontological grounds; generally fossils are so uncommon, or wholly absent, as to preclude paleontological correlation. Hence, without ties to marine base levels and lacking fossils, it is only natural that most workers in sub-Saharan Africa have relied heavily on climatic interpretations of riverine depositional and erosional events.

In all the river systems of southern Africa, the Acheulian industry is found in the low-lying terraces above the present flood plain. Higher level terraces contain (usually rolled) early stages of the Chelles-Acheul industrial complex of early Middle Pleistocene age; the lowest terraces (where present) contain the Middle Stone Age industrial complex of mid- to late Upper Pleistocene age. The deposits making up these lower-lying terraces are sometimes complex, and cyclic fluctuations of rainfall have customarily been postulated to account for their presence. In some cases at least, e.g., Vaal River, it would appear that "they were deposited in a single and practically continuous cycle of downcutting, the present distribution being the result of checks and shifting of the channel" (Cooke 1946).

Especially significant in South African valleys is the common occurrence of substantial (up to 40–50 feet) accumulations of fine sediments, essentially silts and sands, over these gravel deposits. Fluvial activity was greatly diminished; and further evidences of climatic change is the extensive calcification of these fine sediments. The final manifestation of the Acheulian, generally termed Fauresmith, may frequently occur in and on the surface of these calcified sands and silts. In the Rhodesias, Angola, and southern Congo, a comparable phenomenon is the massive accumulation, up to 200 feet in places, of redistributed Kalahari Sands which mantle the ferruginized top of the Middle Pleistocene gravels. The Acheulian proper always occurs under these sands and the lower levels contain the Sangoan-Lower Lupemban culture, the contemporary and equivalent of the Fauresmith in the higher rainfall and thicker vegetation regions of tropical Africa.

The Vaal River, especially its lower reaches, has afforded a well-described geological and industrial succession (Sohnge, Visser, and van Riet Lowe 1937; van Riet Lowe 1945, 1952b, c; Cooke 1946). Table 1 records what is generally regarded as this classic sequence in South Africa. In point of fact the "sequence" is still far from being known or even substantiated as originally worked out. The recently renewed program of investigation of the Vaal River valley, with controlled excavations aimed at obtaining substantial and unselected collections of artifacts, will doubtless do much to clarify the situation.

The problem of distinguishing the Chelles-Acheul Stages 3–5 especially merits further study. This is largely due to the way in which collections have been made at most of the classic sites and also because of the geological complexities of the Younger Gravels 2–3 aggradation. Mason (1959, 1961b) has recently attributed all these collections to a "Later Chelles-Acheul" industry.

The question is whether these collections represent true stages or even artificially delimited parts of a cultural continuum; or, in the case of Stages 4 and 5, whether there is even any real age difference. There is some evidence, both from stratigraphic geological data and from quantitative studies on technology and typology of these collections by G. H. Cole (1961)[3] to support at least in part van Riet Lowe's differentiation of certain stages. The Homestead collection (Stage 3) does differ notably from those collections from other sites, in particular those generally attributed to Stages 4 and 5 (e.g., from Pniel and Larsen, both of which are fairly similar). At least some of the Fauresmith collections occur in post-Younger Gravels contexts, and hence may be dif-

3. Cole was able to examine portions of the Vaal River collections in the South African Archaeological Survey (Johannesburg). This included material referred to Stage 3 (Homestead, Riverview Estates; Canteen Koppie in part), Stages 4–5 (Canteen Koppie in part; Pniel Estates; Larsen Site, Riverview Estates), Earlier Fauresmith (Riverview Estates, Site VI) and Later Fauresmith (Newman's Pont Site, Riverview Estates).

TABLE 1

Geological and Cultural Succession in the Vaal River Valley

GEOLOGY	Fossil Mammals	CHELLES-ACHEUL STAGE	INDUSTRIAL CHARACTERISTICS
Redistributed red Kalahari-type sands		Upper Fauresmith (8)	Small hand axes (ovates); few cleavers; heavy Levallois flake component (retouched and plain triangular points, end scrapers, elongate backed flakes). Levallois flaking technique (as before with further refinement).
Sands and silts (10′-20′)		Middle Fauresmith (7) (in deposits overlying youngest gravels)	Small hand axes (lanceolates, ovates, cordiforms); rare cleavers; polyhedral stones; elongate triangular points; various scrapers; burins. Levallois flaking technique; rectangular, circular high-backed or flattish, and triangular cores.
Youngest gravels (in tributaries only)		Lower Fauresmith (6) (on eroded surface of calcified sands; in Youngest gravels	Small hand axes; mod. sized cleavers (end-struck); elongate, conv. flakes; scrapers; flake-blades. Levallois flaking technique: rectangular, circular high-backed or flattish cores.
Calcified sands and silts (40′-50′)		↑	
Current bedded sands			
Younger gravels[3] (3)		Chelles-Acheul (5) (on Y.G. 2 & 3, beneath and in calcified sands) ↑	Variety of hand axes (almond, pick-like, first cordiforms, lanceolates, ovates, limandes); cleavers (as in 3 and 4); various scrapers (end, side, hollow); burins; polyhedral stones. Levallois flaking technique: circular, biconvex, pyramidal, triangular cores.
Younger[2] gravels (2B) Discontinuous current-bedded sands		Chelles-Acheul (4) (in Y.G. 2 at Canteen Koppie; elsewhere in Y.G. 3)	Variety of hand axes and cleavers (of ⏢ section); often from end-struck flakes. Victoria West 2 flaking technique: large polygonal or circular high-backed (horse-hoof) cores.
Younger gravels (2A) Erosion		Chelles-Acheul (3) (rolled and unrolled in Y.G. 2)	Variety of hand axes (almond, limande, ovate, ovate-acuminate, pick-like); end and side struck cleavers, esp. U-shape, usually with ▱ section. Victoria West 1 flaking technique: fowl beak and formless cores.
Younger[1] gravels (1) Erosion		Chelles-Acheul (2) (only in derived condition in Y.G. 2)	Hand axes (almond-shaped, regular edges); end-struck (and first) side-struck cleavers. Cylinder-hammer technique. Asymmetrical cores

1. At Windsorton rests on 40-ft. platform (20 ft. above Vaal River) of Ventersdorp diabase; upstream at Vereeniging (above Klip River confluence) rests on 15–20 ft. platform (40 ft. above Vaal River) of Ecca shales (Karroo).
2. The stratigraphic significance of the Y. G. 2A and 2B distinction (van Riet Lowe 1952) is equivocable. Opposite Windsorton (Riverview Estate) rests on 20-ft. platform (few feet above Vaal River) of Ventersdorp diabase; downstream (near Barkly West) rests on 25-ft. platform (40–50 ft. above Vaal River) of Ventersdorp diabase at Canteen Koppie (north bank).
3. Opposite Windsorton (Riverview Estates) and downstream at Barkly West (Pniel Estates) rests on present rock floor of river, at or below present water level. In some places (Sheppard Island, Diamant, and Riverview Estates) Y. G. 2 and 3 inseparable and intergrade laterally (facies only).

ferentiated on geological grounds; but some other collections, especially the so-called "Fauresmith I," have always been essentially indistinguishable from the advanced Acheulian (=Later Chelles-Acheul). The later Fauresmith is also different, especially in the flake component; but, the basis of comparison is not always equivalent between collections. Also, the earlier stages of the Acheulian are very inadequately known; this is especially so in the case of Stage 2, there being a real break between the earliest Stage (1), which occurs in the high level Older Gravels (Breuil *et al.* 1948), and this advanced manifestation of the Acheulian.

The upper Zambezi provides another important succession (Clark 1950). It is briefly summarized in Table 2. The Early and Middle stages of the Acheulian are poorly known, being essentially found only in derived condition in the two aggradations of the Older Gravels. However, the Late Acheulian is known from undisturbed situations; and relatively large collections, including some from factory sites on the gravels, permit some estimation of the full tool kit. In this respect some comparisons may be made with the abundant assemblages from the undisturbed open-air occupation sites discussed below.

HOPEFIELD

Hopefield, also referred to as the Elandsfontein site from the farm on which it occurs, is situated on the edge of the "Sandveld Plateau" some 10 miles inland from Saldanha Bay and 300 feet above sea level. The site consists of a series of shallow fossil pans between parallel lines of old sand dunes capped by surface limestones. These fossil dunes were the product of dune invasion lasting from Middle into Upper Pleistocene times. The pans appear to have attracted game in considerable quantities as well as human settlement. The geology of the site has been studied by Mabbutt (1956) and its contents by Drennan (1954), Singer (1957), Boné (1961) and others, and furthe investigations are at present being carried out by R. R. Inskeep.

The well preserved fossil fauna is all dated by H. B. S. Cooke to the final phase of the Vaal-Cornelia stage (c.f. Cooke, in this volume). It thus all belongs to a single period and comes from a single horizon—a layer of nodular calcrete which apparently formed in the drying floors of the pans where the fossils accumulated. From this horizon also were recovered remains of Saldanha Man, a calvarium and mandibular fragment which belong to the physical type of *Homo sapiens rhodesiensis.*

The artifacts belong to three different ages referrable to the Fauresmith, Stillbay and Wilton-Smithfield cultures and are found eroded on the floor of the basin between the modern dunes. Although this has not yet been conclusively demonstrated there is little doubt that the greater part of the cultural material belongs to the Fauresmith stage and is contemporary with the fauna and Saldanha Man. The tools are mostly made from chalcedonic quartzite,

TABLE 2
GEOLOGICAL AND CULTURAL SUCCESSION IN THE LOWER REACHES OF THE UPPER ZAMBEZI RIVER VALLEY

GEOLOGY	CHELLES-ACHEUL STAGE	INDUSTRIAL CHARACTERISTICS
Calcification		
	Late Rhodesian Acheulian (rolled and unrolled in O.G. 2) both home and factory sites known.	Hand-axes (pear-shaped, pointed, ovate, limande, linguate), with plano-convex (on flakes), biconvex (on flakes and cores) and rarely lenticular cross-section, largely on flakes (side and especially end struck) but well rounded or unworked (pointed Micoque type); cleavers on flakes (side or end struck), U-shaped with ⏢ or biconvex section or subrectangular with pointed end and ⏢ section; small flake tools (scrapers); polyhedrals and spheroids. Tachengit flaking technique, no true cores known; fine wood technique for secondary flaking; some resolved flaking apparent.
Older Gravels (2) (10-20′ terrace)	Middle Rhodesian Acheulian (rolled in O.G. 2)	Hand-axes (pear-shaped, ovate, limande) usually biconvex cross-section, straight edges, usually rounded butts, on both cores and flakes; cleavers on flakes (side or end-struck) with ▱ or ⏢ section; retouched flakes. Tachengit flaking technique; wood technique for secondary flaking.
Erosion		
Ferruginization		
Older Gravels (1) (45′ terrace)	Early Rhodesian Acheulian (rolled in O.G. 1)	Heavy hand-axes (on cores) relatively symetrical with regular edges; cleaver on side-struck flake; wide angled, unfaceted flakes. Stone and wood technique for secondary flaking.

chert, and silcrete, but the source of this material is not yet known. A preliminary description of the tools has been given by Singer and Crawford (1958) and by Mabbutt (1956) and both of us also had the privilege of examining the collections in 1954 (F.C.H.) and 1959 (J.D.C.)

The assemblages would seem to indicate that the sites were both living and workshop camps since many flakes, polyhedral and round stone balls, and cores are present. The most significant tool is the hand axe, most of which are of small proportions, only 17 out of a total of 152 having lengths of over 6 in. Cleavers are rare (15 specimens only) and are not particularly well made; notable is the oblique guillotine-edged form. Also with this industry must be associated discs representing cores for the detachment of small flakes to be used as tools, and a number of small side scrapers and simple unifaced points closely resembling those found with the Acheulian at Kalambo, Broken Hill, and Isimila. The technique of the secondary retouch is predominantly "step flaking" and is thus typical of a Fauresmith age.

Nothing is as yet known of the settlement pattern but the fact that small concentrations of artifacts and broken bones of different animals occur together may perhaps indicate the presence of the food debris of the hunters, though there is a possibility that they might have come together from natural causes. If, as is believed, these pans were in existence at a time of drying climate then they may well reflect occupation at only one season in the year. From species counts and analysis of the animals present it should prove possible to determine whether the pans constituted a wet season dispersal area or a dry season concentration site.

SOUTH-WEST AFRICA

As in all but the peripheral parts of the Congo Basin and the Horn there is no certain evidence of human occupation of South-West Africa before Late Acheulian times. The industries are widely dispersed and occur in river gravels and sands, aggraded by torrential rains, consolidated by lime, and overlain by calcretes. Brain and Mason (1955, pp. 22–5) describe assemblages from Nakop on the Brak river where finished tools are assciated with much debitage including large proto-Levallois II-type cores. The tools are made in quartzite, are wind-abraded, and have a glossy patina.

Other important Acheulian collections from South-West Africa come from the terrace gravels of the Fish River in the south, from the Uis and Ugab Rivers in the Brandberg (Korn and Martin 1957, pp. 14–22), and from the middle reaches of the Cunene River on the Angola border. It is apparent that, even though the climate may never have been more than semiarid, a more abundant rainfall during later Middle Pleistocene times enabled Acheulian man to penetrate into the Kalahari and the fringes of the Namib Deserts.

SOUTHERN CONGO BASIN

By comparison with other parts of the continent, in this region the Acheulian is only poorly represented and this holds good for all the higher rainfall regions of Equatorial and West Africa. The industries represent a late phase and are contained in gravels situated only a few feet above the present streams. The most productive site is at Kamoa (Breuil 1944, pp. 143–444, 170–1; Mortelmans

1957, p. 17 and Plate 1) on the Zambezi-Congo watershed west of Kolwezi. Here hand axes of ovate and long ovate form are associated with U-shaped cleavers, trihedral picks, parallel-sided core-axe forms, flake tools, large and small factory waste, polyhedral stones and a single parallel-sided core-axe of Lupemban form. The site still remains to be described in detail but it would seem to represent a terminal Acheulian with forms already anticipating the Sangoan-Lower Lupemban culture.

In northeast Angola isolated hand axes of Acheulian type have been found lying on the land talus gravel at heights of 70 feet above the stream beds and buried by red redistributed sand of Kalahari type to a depth of up to 100 feet (Leakey 1949, pp. 44–7; Janmart 1953, pp. 37–8). In the valleys, assemblages of Acheulian hand axes have been found derived in the gravels of the buried channels. Acheulian tools have also been found by one of us (J.D.C.) in a fresh condition on a land surface overlying the latest accumulation of the Middle Pleistocene gravels at ±15 feet above the present rivers and covered by Laterite 2. This small but important assemblage comprises ovate and limande hand axes, a pick of Sangoan type, choppers, core scrapers, small flake tools, large and small utilized flakes, factory waste including pebble and discoidal cores, and polyhedral stones. Here again the industry would seem to be a terminal one already showing tendencies towards the Sangoan-Lower Lupemban, which directly succeeds it.

No Acheulian industries have been found in the Lower Congo or in the continuous rain-forest zones, but one Acheulian hand axe occurs, derived, at Leopoldville (Kalina Point) and here and elsewhere large and crude flake elements may represent a specialized variant of contemporary human activity.

LAKE BASIN OCCURRENCES

Two basins of internal drainage, Olduvai and Legemunge (Olorgesailie) afford the now classic Middle Pleistocene successions, with Chelles-Acheul assemblages in undisturbed contexts, in eastern Africa. The succession exposed at Olduvai Gorge must represent the key locality since its lower sediments (discussed in Leakey's paper) extend through the Middle into the Lower Pleistocene.

OLDUVAI GORGE

In his description of the hand-axe industry from Olduvai Gorge (3°0′S.–35° 20′ E.) Leakey (1951) delimited eleven stages within the local Chelles-Acheul.[4] The earlier stages (1–3) are restricted to Bed II, and are of earlier

4. There may easily be some confusion in stage designations since prior to the publication of the Olduvai Gorge memoir, the African Chellean material was treated as 5 separate stages, and the African Acheulian as 5 separate stages; these 10 stages plus the single stage of the Oldowan represent the 11 stages normally recognized at Olduvai Gorge.

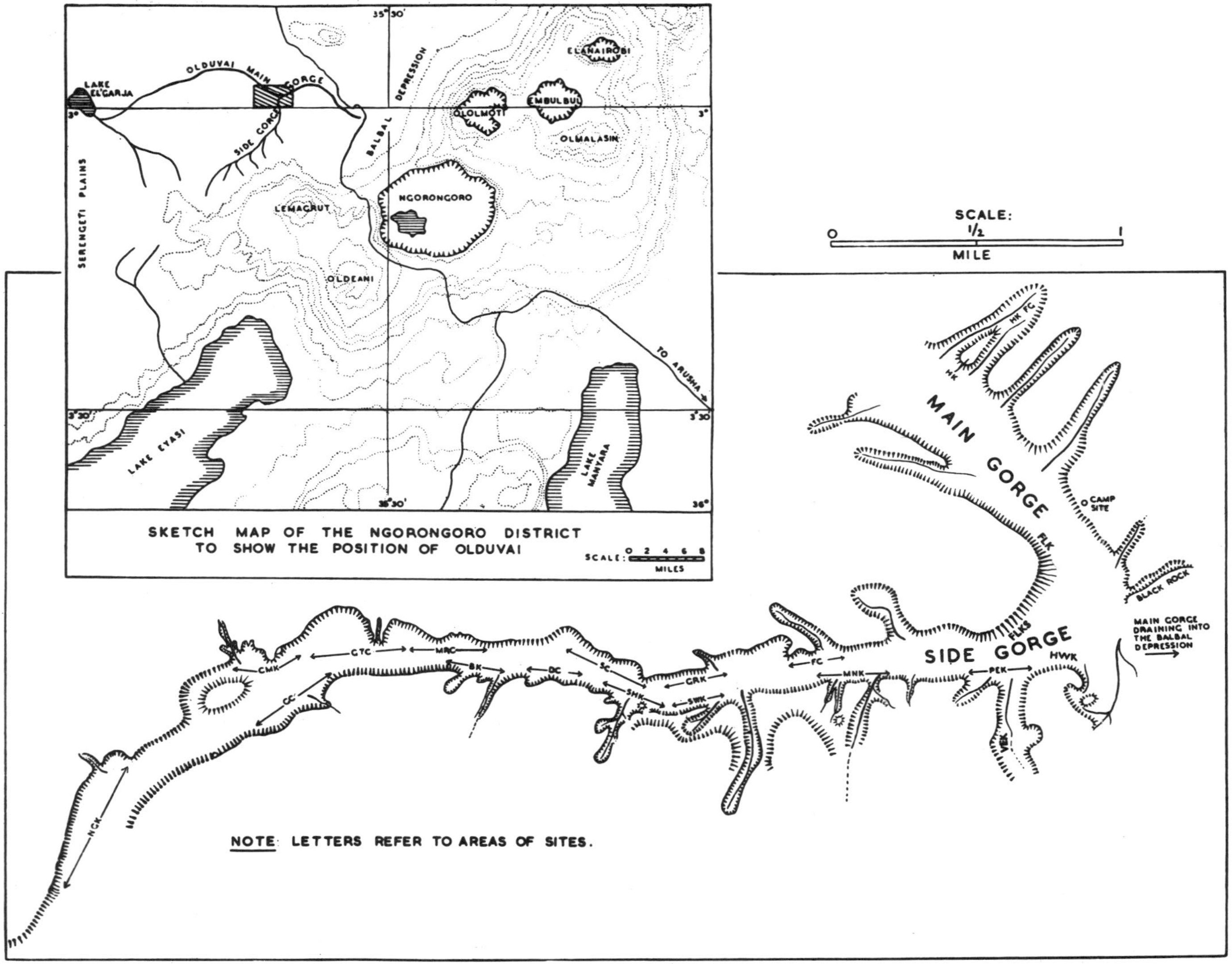
LAKE EL'GARJA
OLDUVAI MAIN GORGE
SIDE GORGE
SERENGETI PLAINS
BALBAL DEPRESSION
35°30'
3°
3°30'
36°
ELANAIROBI
EMBULBUL
OLOLMOTI
OLMALASIN
NGORONGORO
LEMAGRUT
OLDEANI
TO ARUSHA
LAKE EYASI
LAKE MANYARA
SKETCH MAP OF THE NGORONGORO DISTRICT
TO SHOW THE POSITION OF OLDUVAI
SCALE: 0 2 4 6 8
MILES
SCALE:
0 1/2 1
MILE
MAIN GORGE
SIDE GORGE
HK
HK FC
CAMP SITE
FLK
FLKS
BLACK ROCK
MAIN GORGE DRAINING INTO THE BALBAL DEPRESSION
HWK
PEK
VEK
MNK
FC
GRK
SWK
SHK
SC
DC
BK
MRC
GTC
CMK
CC
NGK
NOTE: LETTERS REFER TO AREAS OF SITES.

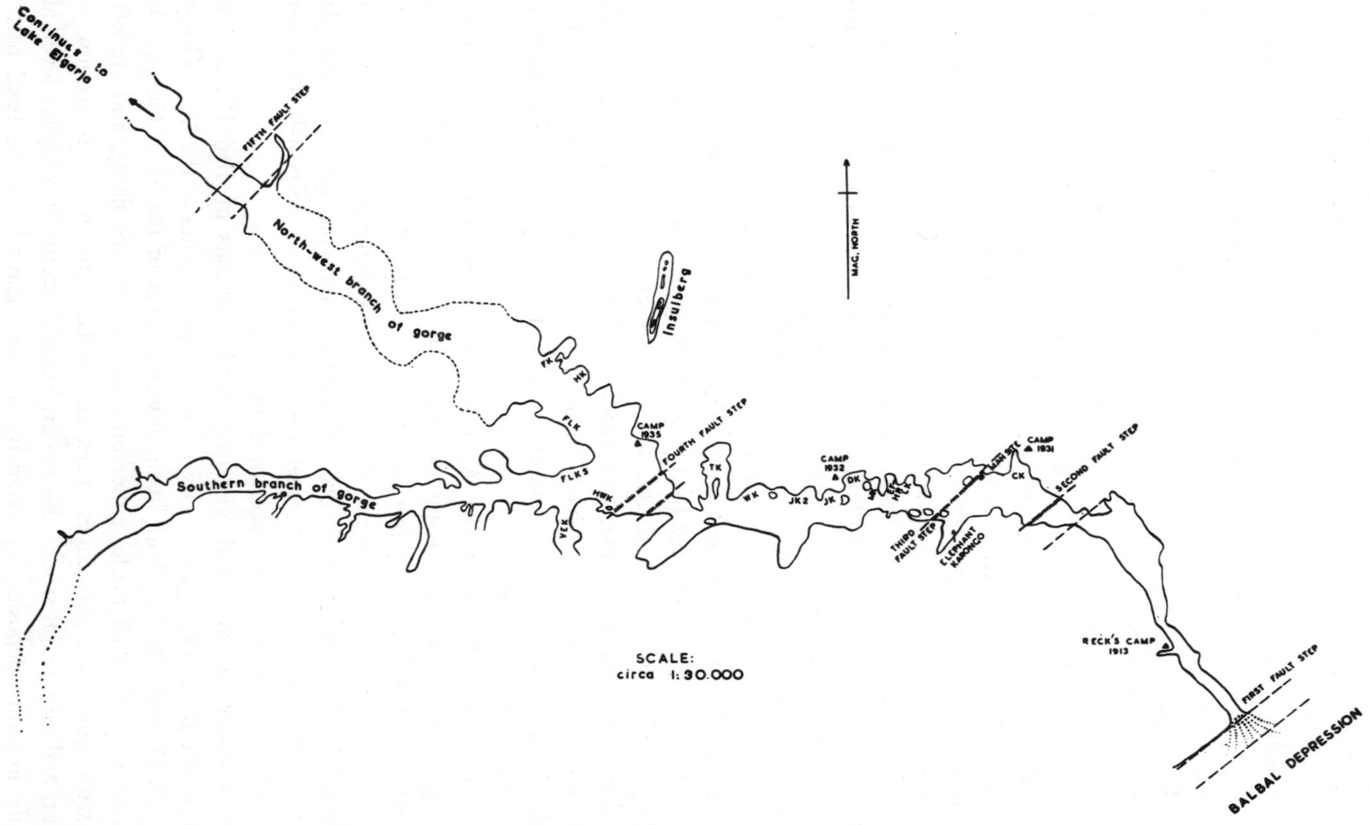

FIGURE 10
Location and plans of Olduvai Gorge. (After L. S. B. Leakey 1951)

Middle Pleistocene age (hence outside the scope of our discussion here). In the upper part of Bed II there seems to occur the first hand axes transitional to the true Acheulian, since these specimens exhibit "a cutting edge all round instead of only at the anterior end." The tools are also generally more almond-shaped and exhibit more extensive marginal secondary trimming. These (including 36 *in situ* specimens) have been referred to Stage 4, found at 10 feet below the Bed II–III junction at several sites (including FLK II, FLKS II, CK II, and HWK II) (Fig. 10). Stage 5 shows the first clear-cut evidence of use of the cylinder-hammer (wood) technique for secondary trimming. Implements (including 45 *in situ* specimens) referred to this stage occurred at the Bed II-III junction, also at several sites (including FLK II, VEK II, HWK II, TK II).

The big Red Bed (Bed III) which overlies Bed II at Olduvai contains rare fauna and Chelles-Acheul artifacts, but these are all in essentially derived condition (though some may be scarely rolled). Bed III is a terrestrial deposit, variable in thickness (usually 20 to 35 feet), which wedges out to the west (Upper Main Gorge) and south (upper Branch Gorge). It is comprised of fine, usually calcareous sediments, sandy clay, and silts, sometimes even ashy, with variable multiple horizons or lenses of channelled, more or less cemented, and often calcareous fine gravels and subangular pebbles (quartz and lavas). A number of isolated implements, collected from CK III, DK III, MK III, EF-HR III, MK III, and LK III, are referred by Leakey to Stage 6. The wood technique for secondary flaking is very well developed. Also cleavers, made on sideblow flakes, appear for the first time; and missile (polyhedral) stones are recorded. Since this material is largely derived, when the sediments were washed in from the uplands to the east, there are surely several still undistinguishable industrial phases actually represented.

The later stages (7–11) of the Acheulian all occur at various localities in several levels within the overlying Bed IV. This bed, which may be exposed to a thickness of well over 35 feet, is often cut into by the erosion, consequent upon faulting, which preceded the fluviatile and aeolian accumulation of sands and silts of Bed V in the Late Pleistocene. Bed IV is essentially deep water clays and marls, often calcareous and sometimes ashy, in places current-bedded sandstones, pebble beds, and even cemented gravel horizons.

Stage 7 occurs in the basal, generally sandy deposits of Bed IV, at VEK (lower, in the Branch Gorge), at HK (lower, upper reaches of the Main Gorge), and at CK IV and TK IV (both in the lower reaches of the Main Gorge). The hand axes are very well made, either on quartzite or lava flakes, and are often large with quite straight edges. Cleavers, usually biconvex in section, are frequent, and polyhedral stones also occur. Stage 8 occurs at a higher level, also generally in sandy deposits, separated by silts and clays from the basal part of Bed IV. The main sites are CK IV and MK IV, both in the lower reaches of the Main Gorge. Leakey has pointed out that the hand axes (which include ovates, long ovates, and some lanceolates) and cleavers, generally made on quartz and lava flakes, are notably smaller than in the preceding stage. Some "S-twist" ovates are present and the cleavers are often V-shaped rather than U-shaped.

Stage 9 occurs still higher, generally at the base of sandy deposits which form the so-called "minor red bed" within Bed IV. The most important sites, all in the lower reaches of the Main Gorge, include MK, LK, and JK. The JK site is a rich occupation area which includes considerable fauna. Its excavation has now been undertaken by Dr. Maxine R. Kleindienst in co-operation with the Leakeys' current program of excavations at Olduvai. There is a considerable variation in the hand axes, made on quartz or lava, within this stage; these include large finely made lanceolates and ovate-acuminates, ovates (some S-twist forms), pointed ovates, thick-backed knives, and a variety of small forms. Cleavers are common, generally U-shaped and with parallelogram section. Polyhedral stones also occur.

The uppermost sediments of Bed IV were generally eroded away during the pre-Bed V valley-cutting, or are inaccessible under a heavy mantle of vegetation. Stage 10 is represented by the HK site, a thin sandy band overlain by clays and silts, in the upper reaches of the Main Gorge. The "minor red bed" is not present here, but the evidence favors its stratigraphic position as overlying those previously mentioned. The HK site, partially excavated in the 'thirties, provided 552 hand axes and cleavers (total *in situ*, or already eroded out and lying below the site), as well as cores, waste flakes, hammerstones, and missile (polyhedral) stones. Quartzite was generally employed as raw material with the hand axes and cleavers being prepared on large flakes struck from big cores (several of which were recovered); only 17 large shaped tools were made of lava. Mammal bones, including much of a disarticulated hippopotamus, were associated with the assemblage, all of which is suggestive of a kill or scavenging site.

Some specimens (28 *in situ*), presumably from various occurrences in higher deposits of Bed IV, have been tentatively assigned by Leakey to a Stage 11. These include some beautifully made small hand axes and cleavers. However, further investigations are necessary to obtain larger *in situ* collections and to link these with the preceding Acheulian stages in the lower sediments of the bed.

OLORGESAILIE

The site of Olorgesailie (1° 30′ S., 30° 30′ E.) is situated in the floor (3,500 feet) of the southern Gregory Rift Valley. A lake existed over the present Legemunge Plain, north of the old volcano Olorgesailie (and probably also farther east and south as well) during the later Middle Pleistocene. It was enclosed in a trough (perhaps 50 square miles) formed by subparallel grid faults in the Rift floor during the earlier Middle Pleistocene. Erosion, following minor tilting and faulting, probably drained the trough at the beginning of the

Upper Pleistocene (Baker 1958). The present rainfall (18 to 20 in.) and small watershed are inadequate to explain the existence of the lake. A rainfall substantially greater than today, with repeated fluctuation in precipitation-evaporation ratios, is requisite to account for the lake and the multiple nonsequences within the succession of sediments.

The prehistoric site lies at the northern base of Mount Olorgesailie, probably a spur projecting along the southern shore of the old lake. It provides a truly unique succession of Acheulian occupation places in excellent stratigraphic contexts (Leakey 1946, 1952). The total thickness of sediments in the trough, diatomites with volcanic ashes, clays and marls, and only rare coarser sediments (sands and fine gravels), are exposed to at least 175 feet. The lake beds are cut by faults generally with throws of less than or no more than a few feet, except for one with at least a 40-foot throw. The Acheulian assemblages occur on old land surfaces (nonsequences) "sharply defined by lithological changes and evidence of erosion" (Shackleton, in Baker 1958). These are usually disconformities representing recessions of the lake margins; however, two such surfaces (3,8) are unconformities. A total of 14 such surfaces have been recognized at or near the main Olorgesailie site, now protected as a monument of the Royal Kenya Parks. Twelve (or 13, since one is split) of these have been excavated and afford mammal fossils and Acheulian assemblages. The fauna is broadly similar to that from upper Olduvai (Bed IV); and the assemblages correspond approximately with Chelles-Acheul Stage 9 (or later) from that site. The stratigraphic sequence at Olorgesailie is recorded in Table 3.

ISIMILA

A shorter sequence than that at Olorgesailie is exposed at the Isimila site (7° 53′ S., 35° 36′ E.) (5388′) in the southern highlands of Tanganyika (Howell 1960, 1961; Howell *et al.* 1961). Here a trough was left in an old, northeastwardly directed valley along the former Congo-Indian Ocean watershed. It was formed through end-Tertiary–Early Pleistocene faulting and tilting which left it with an underfit stream. Upon replacement of woodland vegetation by more open bush and scrub under drier climatic conditions at the end of the Middle Pleistocene, the stream was incapable of removing large quantities of mass wastage, landslides, and colluvial materials; consequently the drainage was blocked and a small basin came into existence.

It fluctuated in size between an open water pool or shallow pond in the old valley floor to a swamp or marsh. The Isimila Beds accumulated in the basin to a depth of 60 feet. The sediments were derived from the soil mantle on the adjacent hills and deposited in the basin in reverse order. These are medium- to fine-grained clastics, represented by level-bedded gray-green clays with bands of silts and consolidated sands. These are essentially lacking in organic matter or calcium carbonate related to typical lacustrine conditions. Clay-filled root

TABLE 3

GEOLOGICAL AND CULTURAL SUCCESSION AT OLORGESAILIE (BASED ON BAKER 1958, AND AN UNPUBLISHED CHART BY L. S. B. LEAKEY)

Bed No.	NATURE OF BED (SHACKLETON)	Land Surface (LSBL)	SEQUENCE IN MAIN AREA OF EXCAVATION (L.S.B.L.)
L14 L13 L12	Diatomite (14′) Diatomaceous clays (17′) Flakey diatomite with few grey ash stripes (22′) White diatomite, underlain by tuffaceous grey diatomite strong reddened at top (11.5′)	exp. ca. 1 mile w. of site ? 12	Hard compact sand and grit Acheul. ↑ Type B ? ↓
L11 L10	Impure diatomites with ash bands (19′-27′) Gravels, volcanic sands, pumice tuffs (3.5′-8′)	?11	Gravel with pumice (1.5′) and lava (1′) pebbles. Acheul. Type B
		10	Gravel (pebbley) and sand, depressions Acheul. Type A-B
L9 L8‴ L8″ L8′ L7	Grey bedded tuffs grading into diatomite (18′-21′) Dirty buff diatomaceous clays with manganese oxide film in cracks Crumbly clays, upper part locally reddened Reddened zone Massive marly clay with sporadic irregular concretions Small lenses of greenish sand and white diatomite	3½′-5′ 9	Fine bedded sands (3.5′) Gravel (¼′) Brown marls (3′) Red Bed (1′) Diatomites (3′) Marls (2′) Clays (2′) Sandy with calc. concretions Acheul. Type A
	Brownish buff concretionary bentonitic clay	5′-10′	Clays (1′) Sandy with calc. concretions (unconformable) Acheul. Type A
L6	Diatomaceous clay Green sand	8	Sand (¼′) Marls (1′) consolidated sandy gravel Acheul. Type A
	Yellowish-white bed Greenish sands (Green sand, Yellowish-white bed, Greenish sands: 0-1′)	7	White sand (½′) Sandy gravel Acheul. Type A-B
L5 L4	Dirty cream clays and marls (4′-8′) Grey pumice sands and gravels (3½′-7′)	6	Yellow marls (3′) and white marls (1′) Fossiliferous sand (3′) Grey clays (16′)
L3 L2‴ L2″ L2′	Bedded tuffaceous diatomite, greyish weathering (4½′-13′) Thin purple or greyish purple band (few inches) Crumbly impure diatomite Brownish laminated clay Pure white diatomite Brownish laminated clay Pale brownish buff clay (L2‴–L2′: 11′-14′)	? 4-5	Yellow clays (12′) Stratified clay and diatomite (1′) Hard white diatomites (½′) Stratified clays (1′) Grey marl (1½′) ashy undetermined Acheul. Yellow marls (½′) sandy Diatomites (¾′) undetermined Acheul. Marls (1¾′) soil (unconformable) Acheul. Type B
L1	Laminated creamy diatomite with bluish ash stripe a foot from the top Purplish buff clays (in part) ?	3 8½′-23′	Stratified white diatomite (2⅓′) Grey ash (Marker bed 2½″) Stratified white diatomite (2′) Acheul. Grey white clays (3½′) sands Type and gravels B
		? 1-2	Marls (3¾′) White diatomites (1¾′) sands and gravels (some boulders)
	Total observed thickness 175′		Hard white Kunkar bed (1′) Yellow marls (2½′) White marls (2′+) Observed thickness (base not seen) +80′

channels, some parallel to the bedding, apparently obtain from times of swamp vegetation (*mbuga* conditions). The beds are commonly mottled or stained black (oxidized Fe and Mn) indicative of alternative reduction and oxidation. The over-all geological evidence favors a wet-dry-wet succession, the time of Isimila Bed deposition and of Acheulian occupation occurring under essentially dry climatic conditions. The mammalian fauna, such as it is, equates broadly with that of Olduvai Bed IV or Olorgesailie; the Acheulian industry probably overlaps, but is generally later than that at the latter site. The Isimila Beds reveal traces of Acheulian occupation throughout, but the lower sediments are poorly exposed by erosion so that only the upper three layers of coarser sediments provide really substantial assemblages from *in situ* occupation places.

The stratigraphic sequence at Isimila is recorded in Table 4.

KALAMBO FALLS

An almost continuous cultural sequence from the end of the Acheulian up to the present day exists at the Kalambo Falls on the edge of the plateau at the southeast end of Lake Tanganyika (8° 30′ S., 21° 15′ E.). The Kalambo River, draining westward to the lake has cut a very deep but short gorge into the Rift escarpment and spills over the edge of the plateau in a spectacular fall 726 feet high. Immediately upstream from the falls the river winds through a small basin about two miles square bounded by quartzite hills some 600 feet high belonging to the Plateau Series. The quartzites and cherts of this series are well exposed in the Kalambo gorge and outcrop in the western end of the basin. They provided all but one of the several kinds of excellent raw material used for tools by the prehistoric inhabitants of the basin.

There is some evidence of high level boulder beds, but the bottom of the basin shows several cycles of cutting and filling with Pleistocene and later sediments. A number of clearly defined horizons representing land surfaces of varying temporality are sealed within these deposits and have furnished evidence of human occupation. These have been the subject of investigation during 1953, 1955, 1956, and 1959 (Clark 1954, 1960a, b, 1962; Bond, unpublished report). Some 3½ to 4 months were spent in extensive excavations during each of the last two field seasons and much cultural and organic vegetable material was collected. Pollen and macrolithic vegetable remains preserved in some of the deposits have shown that on two occasions the climate was probably both colder and wetter than that existing in the locality today.

The water level in the Kalambo basin was subject to several marked fluctuations from early Upper Pleistocene times onwards. This was probaby due to the blocking of the narrow spillway gorge by detrital material washed in from the valley sides and then the subsequent unblocking under semiarid conditions with torrential rainfall. When the water level was high swampy and open water conditions persisted in the valley, but when it was low the lake, or swamp, was

TABLE 4
Geological and Cultural Succession at Isimila

NO. OF BED		CHARACTERISTICS OF SEDIMENTS	INDUSTRIAL CHARACTERISTICS
		Korongo (gully) cutting	
		Development of Fe/Mn rich red soil mantle	? Magosian or L.S.A.
		Mottled clayey silt, silt and sands (colluvium)	
		Local erosion	
	Cyclic sedimentation	Mottled *mbuga* clays and silty clays, grading into clay, grading upward into transitional horizon (100- ? 300 cms)	post-Acheulian (Sangoan)
1 — 1a′	Cyclic sedimentation	consolidated sands (0-100 cms)	Late Acheulian open-air occupation sites
1 — 1a	Cyclic sedimentation	silty-clay (50-100 cms) consolidated sands (20-100 cms)	Type A = J 6-7 (u); H 9-J 8; K 14
1 — 1b	Cyclic sedimentation	silty-clay (20-50 cms) consolidated sands (10-60 cms)	Type A = K 6 A - B = J 6-7 (l)
2	Cyclic sedimentation	clay or silt (40-100 cms) consolidated sands (30-120 cms)	Type A = J 12
3	Cyclic sedimentation	clay or silty-clay (0-80 cms)	
3	Much lateral intergrading — no distinct cyclic sedimentation	consolidated sands (locally subdivisible into sand bands separated by finer sediments 200-400 + cms)	Type A = K 19 Type B = K 18 (2); Type C = H 15 (l)
4	Much lateral intergrading — no distinct cyclic sedimentation	clayey-silt or silt (10-80 cms) consolidated sands (locally subdivisible 80-160 + cms with mammal fossils)	
5	Much lateral intergrading — no distinct cyclic sedimentation	silty to sandy clay (20-150 cms) consolidated sands (160-270 cms)	
	Much lateral intergrading — no distinct cyclic sedimentation	silty-clay to silty sand grading down into older soil and claycrete (weathered bedrock (+ 400 cms)	
		Old Valley floor on mid-Tertiary surface (granite or grano-diorite)	

drained and the Kalambo River cut deeply into the sediments of the previous and earlier cycles. The short gorge is some 500 feet long, cut in hard quartzite that dips gently upstream and the gorge follows a strong joint direction. The shoulders of the gorge are approximately level with the top of the oldest of the later Pleistocene series known as "the Lake Beds." These cover an extensive area to the north and west and are most fine-grained at the northwestern end of the basin. Towards the flanks and eastward to where the river enters the basin they become coarser grained and pass into boulder beds. The base is nowhere exposed and could be about 30 feet or more below present river level. These beds accumulated in the apparently empty basin to an approximate height of 70 feet above the river and form an imposing scarp feature on the north side. In the central part of the basin they have been extensively removed by erosion and only the lower levels remain overlain by later land surfaces and aggradation. The succession at Kalambo Falls with carbon dating is summarized in Table 5.

It is in the lowest levels of the Lake Beds and from approximately 4 feet above and 2 feet below low-water level that the Acheulian industries are found. How far below water level the sediments are implement-bearing is not yet known. The fine, almost structureless, sands in which the latest Acheulian is found are sharply dissected by channelling and covered by current bedded sediments containing tools of the Sangoan culture. A disconformity exists, therefore, between the two, but the magnitude of this is not yet known. It would seem to be very slight and to represent only a short lived phase in the filling of the basin to the 70-foot level.

Acheulian occupation was found at five different levels on temporary living floors and portions of these were excavated on a horizontal grid pattern. These living floors rest on loose sand, either fine or coarse, and have no thickness, which points to only temporary occupation. They are adjacent to the water channels, and are separated from each other by a foot or so of sterile sand. They probably represent camps of only one season when the water level in the basin was naturally low. These sands and floors are interbedded with lenses of peaty clays containing much partially carbonized wood ranging from large tree trunks to small branches or twigs, leaves, seed pods, and fruits, reed stems and grasses, which have been preserved by the waterlogging of the lowest parts of the beds since the time of their original deposition.

The climate at the time when Acheulian man occupied the Kalambo basin appears, on the evidence of the vegetable remains (unpublished reports by E. M. van Zinderen Bakker, L. Chalk, and F. White), to have been cooler and wetter than it is today. Moist forest was present in places round the water (c.f. *Zyziphus*, *Ouratea*, *Cynometra* sp.). This was fringed by savanna trees (*Burkea*, *Cassia*, *Acacia*, *Brachystegia*, etc.). That the forest was wet and with closed canopy is shown by the high percentage of fern spores and the presence of *Cynometra*. The temperature appears to have been at first similar to that of the present day, but became cooler at the end of Acheulian times when

TABLE 5

Geological and Cultural Succession at Kalambo Falls, with Available Absolute Age Determinations from C-14

STRATIGRAPHY	NATURE OF SEDIMENTS	CULTURE	C-14 DATING
Sands & clays	Red talus & sand on upper slopes, grey swamp clays in valley. ±12′	Iron Age (Channelled Ware)	1,000 A.D.
	EROSION & DISCONFORMITY		
Red sand over land rubble on upper slopes. ±4′	Fine, unstratified sand with thin lenses of small land rubble.	Later Stone Age	
Sands & grits ±15′	Current bedded sands & grits with some fine clay veins, uncompacted.	Sterile	
Land surface 0-1″	Land rubble with tools.	Magosian	±7,550 B.C.
Sand 0-2′	Current bedded sands.	Sterile	
	EROSION & DISCONFORMITY		
Land surface ±1′	Much land rubble with tools.	Magosian	
Sands & clays c. 0-4′	Fine sands & peaty clays.	Sterile	
Land surface ±3″	Much land rubble with tools	Later Middle Stone Age	
Sands c. 0-3′	Current bedded fine & coarse sands.	Sterile	
Land surface ±2″	Much land rubble with tools.	Later Middle Stone Age (Upper Lupemban)	25,000 - 27,000 B.C.
	EROSION & DISCONFORMITY		
Sands, grits & fine gravels ±50′	Much sterile false bedding of sands & grits cut out by erosion with some temporary surfaces.	Sangoan on temporary surfaces passing up into early Middle Stone Age (Lower Lupemban).	}38,000 to 41,000 B.C.
	EROSION & DISCONFORMITY		
Sand 1-3′	Fine, pale, structureless sand.	Sterile	
Land surface 0-1″	Land rubble, tools, woodash & charcoal.	Acheulian*	
Sand 6″-1′	Fine & coarse pale sand, wood, ash & charcoal, etc. interbedded.	Sterile*	
Land surface 0-1″	Land rubble, tools, wood, etc.	Acheulian*	55,300 B.C.
Sand 1′-1′6″	Fine & coarse pale sand interbedded.	Sterile*	
Land surface 0-1″	Land gravel, clay, tools & wood.	Acheulian*	
Sand, clay & gravel 3-4′	Fine & coarse pale sands, clays & fine gravel, interbedded.	Occasional Acheulian tools*	

pollens of *Ilex*, *Curtisia*, Ericaceae, and *Podocarpus* became prominent, grass pollens increased, and the woodland cover was lighter. Such a vegetation is similar to that occurring today at altitudes of between 4,000 to 5,000 feet whereas the altitude of the Kalambo basin is between 3,500 and 3,200 feet. Edible fruits, which no doubt formed part of the diet of Acheulian man, included *Syzygium*, *Xylopia*, *Diospyros*, *Parinari*, and *Borassus*. All except the last are forest forms and, according to White (personal communication), suggest cooler conditions. If this is so, then it is unlikely that the *Borassus* was growing in the basin at the same time and may have been carried in by man, perhaps from the Tanganyika Rift.

An absolute radiocarbon date of 57, 300 ± 300 B.P. has been obtained by Groningen Laboratory from samples of *Ouretea* wood from Floor 6. This, and the botanical evidence, suggest that the Acheulian at the Kalambo Falls, which is very evolved technically and typologically, dates to the very end of the Kanjeran-Gamblian interpluvial and the beginning of the early Gamblian Pluvial, and would thus be the tropical equivalent of the Mousterian of the early Würm glacial stage of Europe. By 43,000 B.P. the Acheulian had been replaced by industries of the Sangoan-Lower Lupemban complex.

LOWER KAGERA VALLEY

In the earlier Pleistocene drainage from the eastern (volcanic) highlands still passed westward via the mature Kagera Valley (1° 0′ S.–30° 30′ E.) across the Uganda plateau to the Western Rift Valley. Rift faulting and associated tilting interrupted this drainage early in the Middle Pleistocene. Subsequently the down-warped Victoria basin extended its waters westward, as far as the Nsongezi area, 70 miles upstream of the present mouth of the 350-mile-long river. The Kikagati gorge to the west formed an effective barrier and/or an overflow (westerly) at certain times during the existence of the broad embayment into the Kagera Valley.

Wayland and his colleagues of the Uganda Geological Survey initiated investigations of this area in the 1920's, and between 1930 and 1939 more detailed work, including extensive pits, trenches, and the opening of paddocks, was carried out by Wayland (1934, 1935), Way (1937), and by O'Brien (and Solomon; 1939). Van Riet Lowe (1952) published an appraisal of much of the prehistoric material recovered through the activities of the Geological Survey; the results of the investigations by O'Brien (1939), during an eighteen-month period in Uganda in the mid-thirties, are considerably at variance with those reached by Wayland and van Riet Lowe. Wayland resumed his work, with much additional excavation, in the Nsongezi area in 1953–1954, but the results remain unpublished; one of us (Howell) saw something of this work, and the resultant implement collections, during a week's visit to the Kagera in October, 1954. Bishop (1958; Bishop and Poznansky 1960) made five safaris to the Kagera between 1956 and 1959, during which extensive levelling was carried out and the stratigraphy of many of the earlier pits and trenches was recorded. The results of these and other investigations into the Pleistocene of Uganda will be published in due course (Bishop, in press); we are most grateful to him for permission to utilize some of this data here.

The lacustrine-estuarine sediments of this old gulf are well preserved in the Nsongezi area (Fig. 12). The deposits underlie a broad flat which flanks the present meandering Kagera River as it now drains eastward to Lake Victoria. This Kafunzo flat (ex-100-ft. terrace) attains a height of 107 feet above present river level; it is somewhat less than 300 feet above the level of Lake Victoria. The deposits underlying the flat, termed the Nsongezi Series, reach at least 86 feet in thickness above the Karagwe-Ankolean phylittes of the old valley floor. Bishop has recognized three main lithological units in the Nsongezi Series:

III. Uppermost: Pale gray to buff clays and silts, with some gravelly horizons; below the flat these are weathered to mottled loams. Thickness: 21′–34′. Apparently indicate a change to shallower water, probably swampy conditions. Surface of Kafunzo flat, in some places, is rich in microlithic artifacts. Some horizons in the deposits yield patchy occurrences of Levallois flakes, etc.; this unit is widely exposed in the erosion gullies, which cut the edge of the Kafunzo Flat, in the tributary Orichinga Valley and yield a Lupemban II (Middle Sangoan) industry.

———slight unconformity———

II. Middle: Principally sandy beds, frequently clean and ferruginous; show false and slump bedding within normally parallel major beds which are interbedded with more silty and clayey horizons. (Within this unit, 60′–67′ above river level, is the MN artifact-rubble horizon.) Indicates a change from riverine to lacustrine, still-water deposition. Thickness: 35′–50′.

↑
probably grading into
↑

I. Lowermost: Basal boulder bed, resting on Karagwe-Ankolean phylittes. Thickness: 3′–10′.

Downstream of Nsongezi, especially in the Nyakanyasi area (Tanganyika), there are essentially only fine-grained sediments of deep-water origin (Spurr 1955). However, there is also silcrete formation and selenite crystals indicative of (periodical) shallow-water deposition. The role of eastward tilting or of climatic change in producing this and subsequent shallowing is poorly understood. At any rate there is substantial evidence to show that in the later Middle (and Upper?) Pleistocene, the embayment had a precarious existence and eventually disappeared except perhaps for isolated ponds and swampy pools.

Another formation in the Nsongezi area, the Kishen terrace (ex-30-ft. terrace) may be cut as a berm on re-exposed phylitte, the rock barriers hence acting

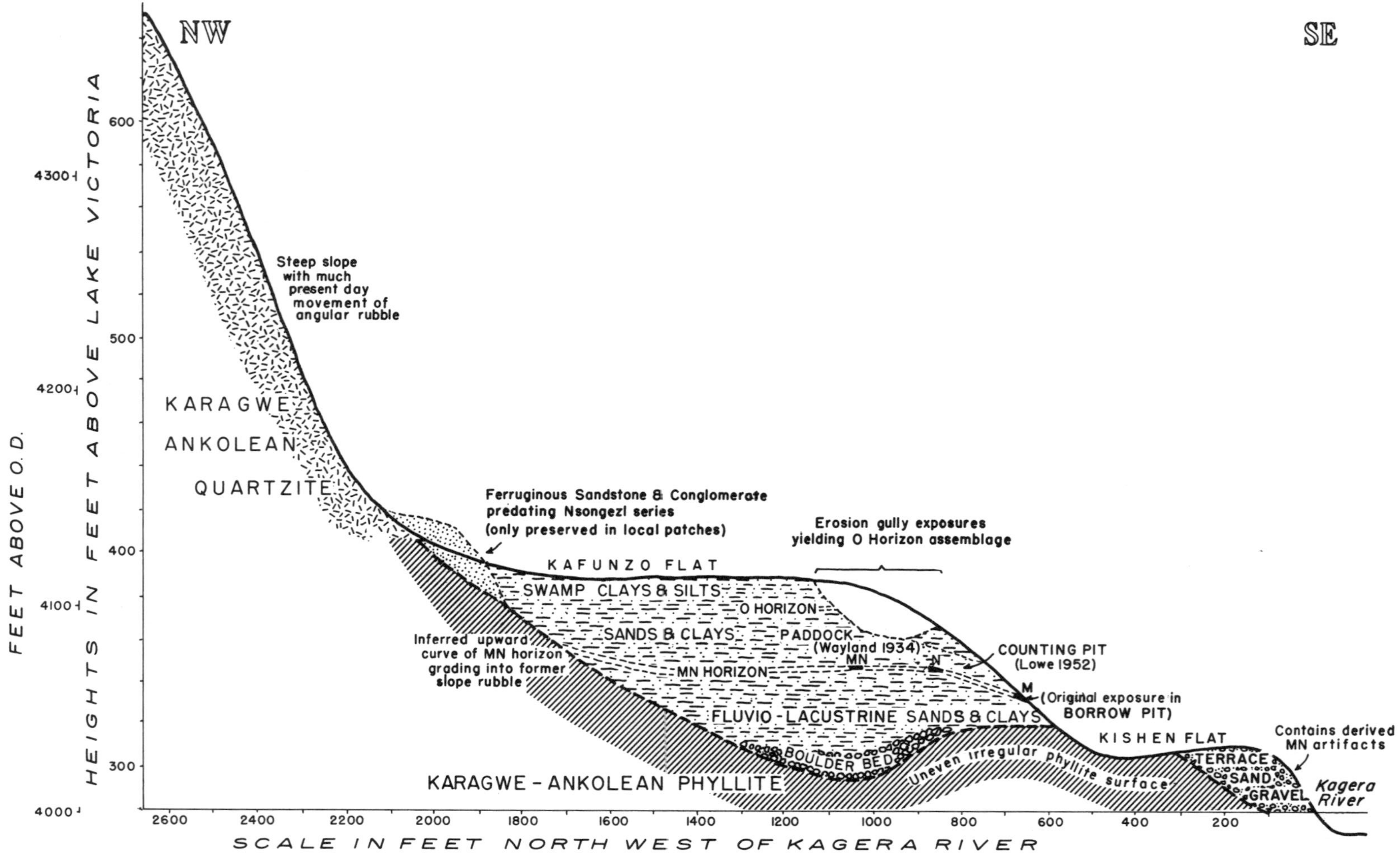

FIGURE 12
Generalized section across the northern margin of the Kagera River Valley at Nsongezi, Uganda. (Courtesy of W. W. Bishop)

as traps for gravel spreads, or represented by unconsolidated sands and gravels. It was thought by Wayland and van Riet Lowe to be older than the Nsongezi Series of the Kafunzo Flat. O'Brien and Solomon, on the other hand, thought these sands and gravels postdated the Nsongezi Series, although they differed in their interpretation as to how the deposits were formed. Bishop, in agreement with the latter authors as to the relative age of the sediments, regards them as a river deposit, not unlike the present load of the Kagera, largely derived from erosion of the earlier Nsongezi Series.[5] Pre-Nsongezi Series deposits are evidently present in the Orichinga Valley, for the Nsongezi Series there overlie northward-dipping banded and ferruginous sandstones and conglomerates which rest on phylitte bedrock.

The MN artifact-rubble zone in the middle unit of the Nsongezi Series is an extensive horizon traceable for over thirty miles. It is a relatively thin (1 to 2 feet) subaerial scree, locally ferruginized and composed of coarse angular blue gray quartzite rubble. Sometimes it is bifurcate (M *and* N) with less than a foot of fine sandy or gravelly sediment between two rubble zones.[6] The horizon is only slightly sloping, except where it has been affected ("cambered") by soil creep on steep present-day slopes, and generally has a very constant height above river level.

The MN horizon was intensively utilized and occupied by peoples of the later Acheulian. The industry represented there is in fact a Late Acheulian of proto-Sangoan facies.[7] There is a variety of minimally trimmed large tools on side-struck flakes, including numerous large to small hand axes (long and pointed ovates; some ovates; rare lanceolates, triangulars and cordiforms), many cleavers and cleaver flakes, generally U-shaped and some with pointed butts. There are also trihedral and elongate bifacial picks, discoids, steep and core scrapers, push planes, choppers, and polyhedral stones, many retouched flakes, and some flake tools (various scrapers, rare backed flakes). Some of these items

5. The age of the Kishen terrace is especially significant since van Riet Lowe (1952) thought it possible to trace the Chellian and Acheulian industrial succession from the artifacts contained in the deposits. These are practically all rolled and doubtless derive from the MN horizon. Hence, pre-MN implementiferous horizons are essentially unknown in the area.

6. This horizon was originally defined by Wayland (1934, 1935) as M, but was later found to be composite (hence MN) and van Riet Lowe (1952) defined its upper level on typological grounds. The levels which he thought to be distinct can now be traced, through intermediate trenches and pits, and shown to be the same horizon. It should be noted that Solomon (in O'Brien 1939) referred to a separate horizon, nearly twenty feet above M and sometimes implement-bearing (Levallois flakes, etc.), as N. It has not been recognized in the field by other workers.

7. Van Riet Lowe (1952), who believed up to 30 feet of sands and silts were deposited and then eroded between M and N horizon times, thought the M horizon contained Early Acheulian (comparable to Chelles-Acheul Stage 6 at Olduvai Gorge) and the N-horizon Late Acheulian (comparable to Chelles-Acheul Stage 7–9 at Olduvai Gorge). This conclusion is at variance with the geological evidence, as Bishop has demonstrated, and also with the archaeological evidence (Bishop and Posnansky 1960; also the observations of Howell at Nsongezi in 1954, including a small excavation into the M horizon in the Paddock 2 area).

are characteristic of the later Sangoan industry, and these are perhaps more common on the lower slopes of the MN horizon (according to Posnansky). Hammerstones are common as are waste materials, including much *débitage* and trimmed or smashed up rubble. The Tachengit flaking technique was employed to obtain large flakes from the quartzite rubble for the large shaped tools; these were generally made with stone-on-stone technique, but some cylinder-hammer (wood or bone) technique was employed for the better finished specimens. There are very numerous cores including block, formless, pyramidal discoidal, biconical, and struck varieties; and there is also evidence of some use of an early Levallois technique for production of flakes of predetermined shape.

The Nsongezi area and the adjacent section of the Kagera Valley is of extraordinary importance for prehistoric archaeological research. Very considerable geological background work has now been done, and many of the problems raised by the earlier fieldwork have been clarified or resolved by Bishop's more recent investigations. Not only is the MN a fantastically rich implement-bearing horizon, especially susceptible to quantitative studies, but there are also fossiliferous beds and peat deposits downstream of Nsongezi (near Nyabusora). Future detailed archaeological studies in the area, being undertaken by Glen H. Cole (in co-operation with the Uganda Museum) over an eighteen-month period in 1962-1963, should provide a much better understanding of the cultural changes manifested between the later Acheulian and the Sangoan industries.

CAVE SITUATIONS

Three cave sites are known in southern Africa, all situated south of the Limpopo River, which contain Acheulian occupation horizons.

The earliest known and first excavated (in 1919) of these is Montagu Cave, in the western Cape province (Goodwin 1929). It faces east, about 200 feet up in the side of a steep-walled valley cut in Table Mountain Sandstone. Its outer chamber was occupied successively by Acheulian peoples and, again in post-Pleistocene times, by peoples of the Later Stone Age. Its total depth of filling reaches some 12 to 17 feet. Three lower horizons (H, F, D), from one to three feet thick and separated by sterile horizons (G, E) of limey or phosphatic quartzitic sands, provided evidence of later Acheulian occupation, but without any trace of bone. The middle horizon (F) was probably a double occupation since implements occurred in two discontinuous concentrations. The basal level (H) contained only a few artifacts, but over 300 implements were obtained from each of the overlying Acheulian horizons. This site is frequently cited as an early example of the use of fire; however, Goodwin's description, based on the excavators' notebooks (and his own study of the assemblages) specifically mentions no charcoal in any level; in another place (*op. cit.*, p. 13) he claims "evidences of fire" without stating what these are (cf. Oakley 1955). Judging from some figures (no totals, but only counts of "better specimens") these are quite substantial assemblages. His figures include almond=long ovate (F=15, D =70); pear-shaped=pointed ovate and ovate acuminate, lanceolates (?) (F=62, D=15) and ovate-discoid (D=4) hand axes, cleavers (F=25, D=48) and numerous (biconical and discoid) cores (F=32, D=68). All are in Table Mountain Sandstone, the large shaped tools being made on large primary flakes from very substantial blocks of this raw material, some of which were also recovered in the excavations. Hammerstones (about a dozen each in F and D) from battered river cobbles were also present. Although Goodwin mentions no small tools or flakes having been recovered, these have been found during visits to the site by K. P. Oakley (in 1953) and M. R. Kleindienst (in 1960). Evidently then, as at so many other sites, these smaller implements were made on the occupation site. The artifact assemblages (in the South African Museum and the Musée de l'Homme) have recently been studied by Glynn Isaac and also by M. R. Kleindienst.

The second, and by far the richest, such site is the Cave of Hearths, situated in the Makapan Valley of the north-central Transvaal (van Riet Lowe 1938, 1943, 1948, 1954). Some 60 feet of hard brecciated earths fill the cave and provide a unique succession from late Acheulian, through four levels of the Pietersburg (local M. S. A. industry) and two levels of the L. S. A. (Smithfield) as well as the Recent African Iron Age (and even including traces of Pioneer settlers); ashy levels are found in the M. S. A. horizons and upward. The earliest, Acheulian occupation (some 33 feet) overlies a basal dripstone (stalagmite) surmounted by a "Basal Hearth," apparently a calcined guano deposit on the surface of which people of the Acheulian built fires (traces of their hearths are preserved) (cf. Oakley 1954, 1955). The site, its excavation, and the assemblages from it are still to be described in monographic form by its excavator, Dr. Revil Mason.

The last such site is the Wonderwerk Cave situated in the district of Kuruman, northern Cape Province (Malan and Wells 1943; also Malan and Cooke 1941). The upper levels contain Smithfield underlain by a probable Middle Stone Age industry (with Levallois technique). However, special interest atttaches to the site because the lower levels, about 18 inches of a red to yellowish-red earth above bedrock, yielded 11 small hand axes, assigned originally to the Fauresmith (final Acheulian). The specimens, in banded jasper, are isolated and only a small area was excavated to bedrock. Unfortunately, further excavations do not appear to have been undertaken, at least to our knowledge.

THE ACHEULIAN INDUSTRY

The Acheulian industry has been inadequately known through the collections made from riverine deposits. This is in spite of the fact that collections have sometimes been carefully made and even without the usual selection of only finely finished or unusual types of implements. However, the full (or at least relatively full) range of chipped-stone implements, cores, and waste materials can only be readily appreciated from the assemblages found in archaeological contexts, as left there by the peoples of the Acheulian, on the sealed-in, open-air occupation sites. As a consequence of the investigations already made at Olorgesailie, Isimila, and Kalambo Falls the detailed composition of Acheulian lithic assemblages is quite well known. Kalambo Falls, with its fortunate preservation of wood in waterlogged conditions, has also provided some useful evidence on artifacts in materials other than stone. As yet essentially nothing is known of worked bone in the African Acheulian.

The appendix to this paper contains a descriptive classification of the Late Acheulian (cf. Kleindienst 1961a). It was developed in the course of the Isimila work (by M. R. Kleindienst, G. H. Cole, and F. Clark Howell) and the attendant necessary comparative study of other Acheulian assemblages (by M. R. Kleindienst and G. H. Cole) including those from Kariandusi, Lewa (in part), Olorgesailie, Olduvai Bed IV (in part), Kalambo Falls (material collected through the 1956 season), and Broken Hill. However, Cole also studied several collections from sites in the Vaal River, and Kleindienst subsequently studied collections from Southern Rhodesia (including Lochard), and from South Africa (including Cornelia, on a tributary of the Vaal; and the Cape sites of Hangklip and Montagu Cave).

The main lithic artifact classes include shaped, modified, and utilized tools and waste products. The main categories in the shaped-tool class include cutting-edge tools, flake tools (scraping edges) (generally > 10 cm.) other uncategorized tools, heavy duty tools, and small tools (generally < 10 cm.). Some of these are illustrated in Figure 13. Within these categories various types of tools may be quite readily recognized and differentiated on the basis of kind and treatment of edges, secondary trimming, etc. The tool types also exhibit different shapes or forms which vary from site to site and between assemblages in the same site. The application of such a classificatory system, with proper definition, illustration, and quantification, permits another investigator to utilize fully the results of another's work. This is standard procedure in most archaeological studies, but it has been only within the last decade that it has been rigorously employed in Europe for Middle and Upper Paleolithic assemblages; and only in the last several years, in sub-Saharan Africa, for lower Paleolithic assemblages.

Figure 14 provides a graphic representation of the percentages of implement and artifact classes (in a composite assemblage) at six Late Acheulian sites. These include sites with geological context (Homestead, Vaal River), variously disturbed archaeological context (Kariandusi, Lochard), and primary or but slightly disturbed archaeological contexts (Olorgesailie, Isimila, Broken Hill). These are, of course, of different value since the single occupation at Broken Hill may represent but a few days, whereas Olorgesailie covers a range of

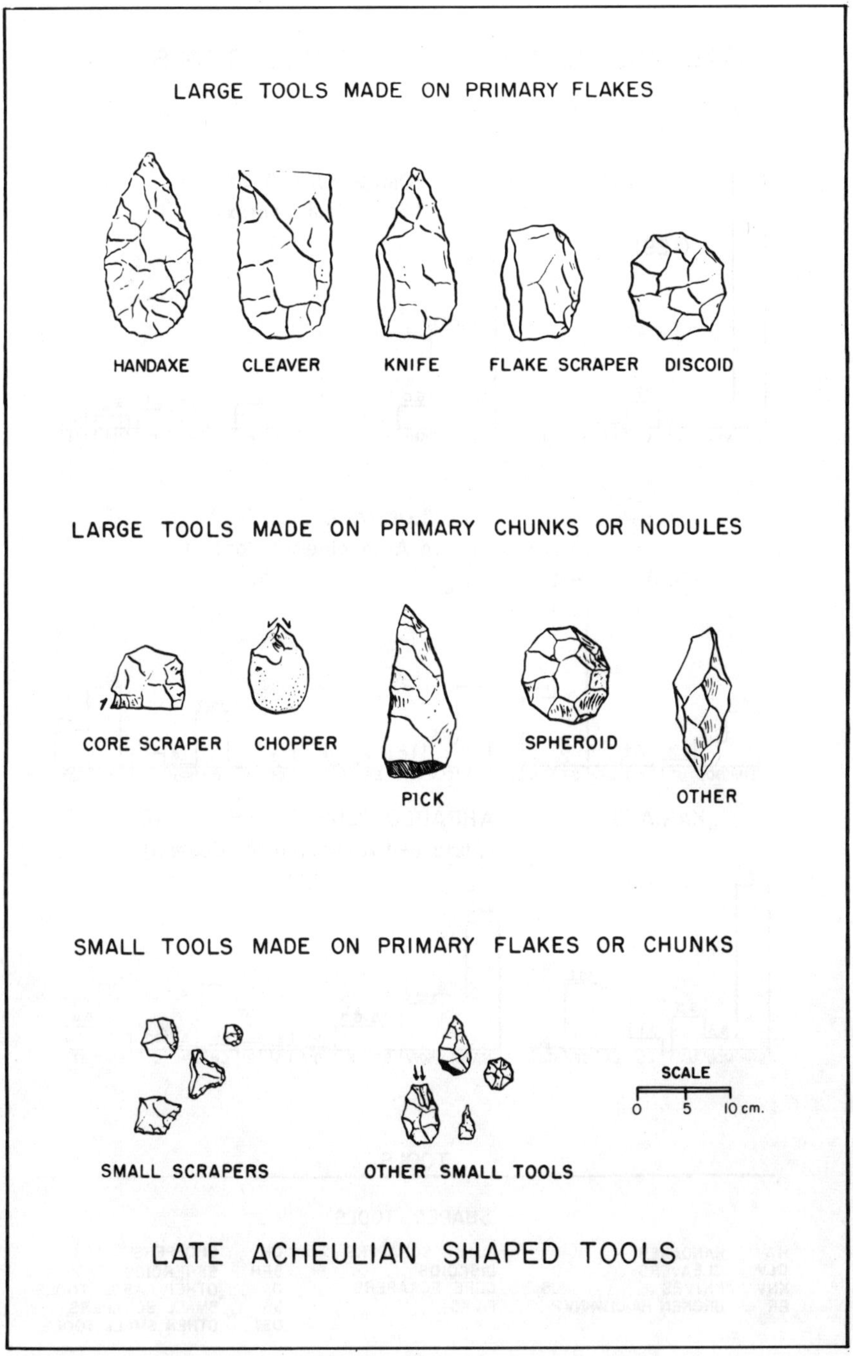

FIGURE 13
Some characteristic shaped tools of the African Late Acheulian. (After M. R. Kleindienst 1961b)

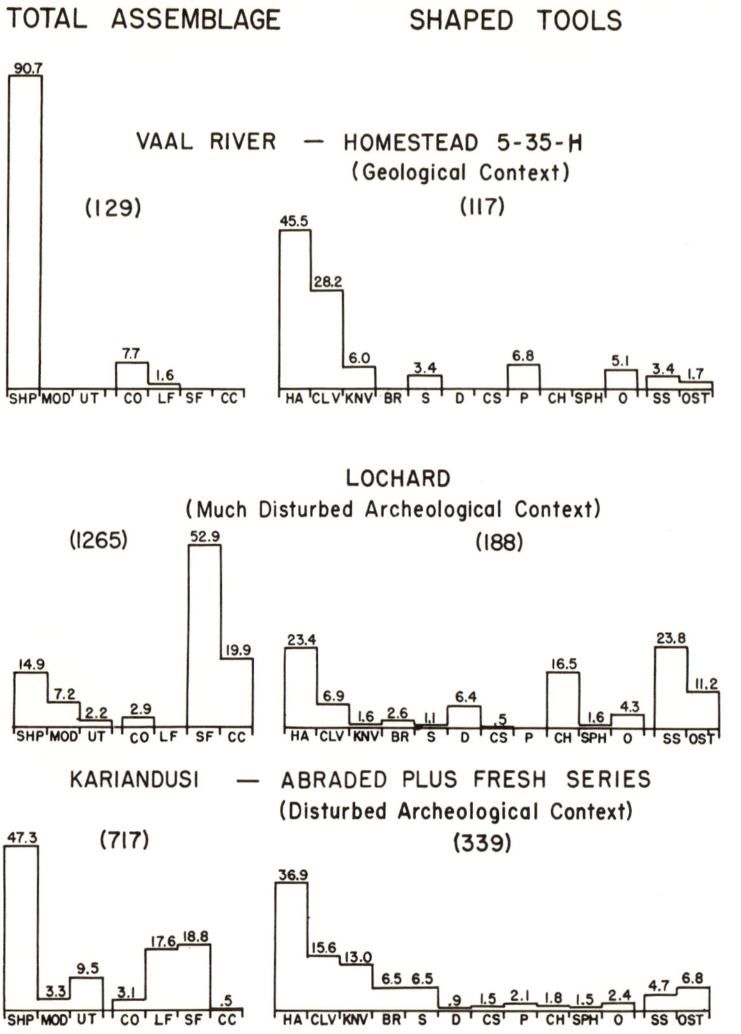

TOOLS

SHAPED TOOLS

HA	HANDAXES	S	FLAKE SCRAPERS	CH	CHOPPERS
CLV	CLEAVERS	D	DISCOIDS	SPH	SPHEROIDS
KNV	KNIVES	CS	CORE SCRAPERS	O	OTHER LARGE TOOLS
BR	BROKEN HA/CLV/KNV	P	PICKS	SS	SMALL SCRAPERS
				OST	OTHER SMALL TOOLS

FIGURE 14 Percentages of implement and other artifact classes in

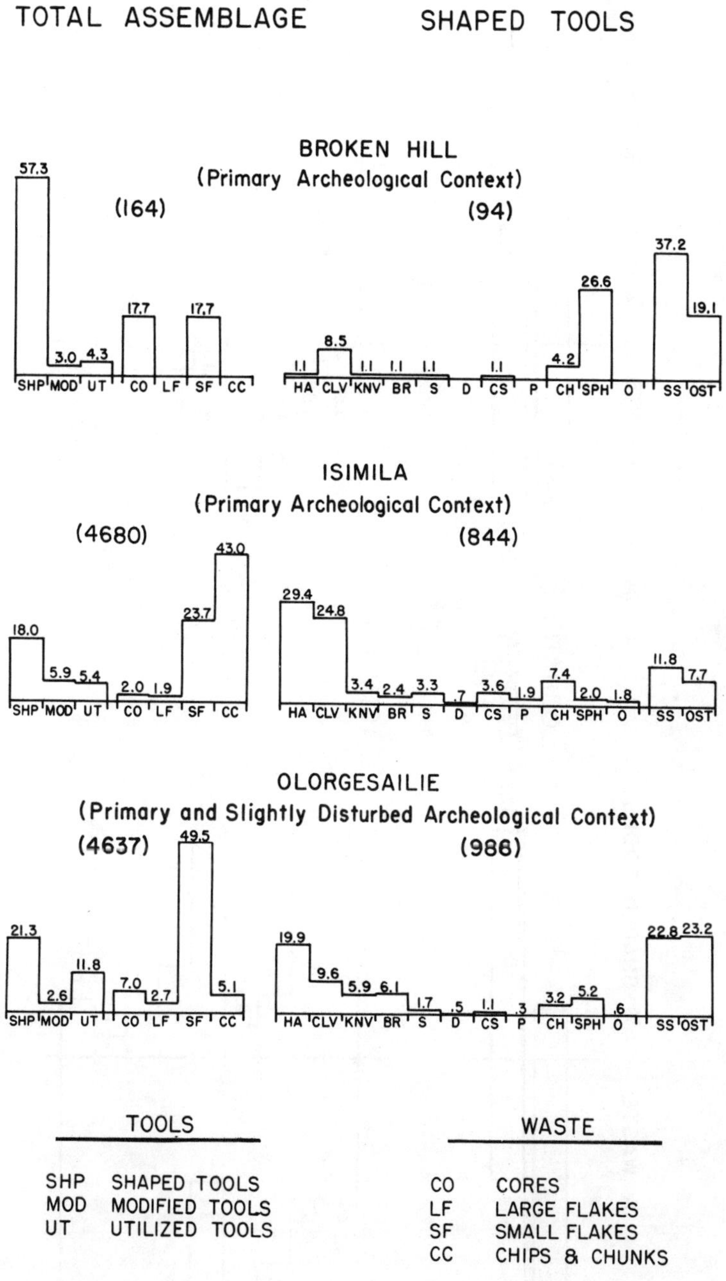

six (composite) late Acheulian sites. (After M. R. Kleindienst 1961b)

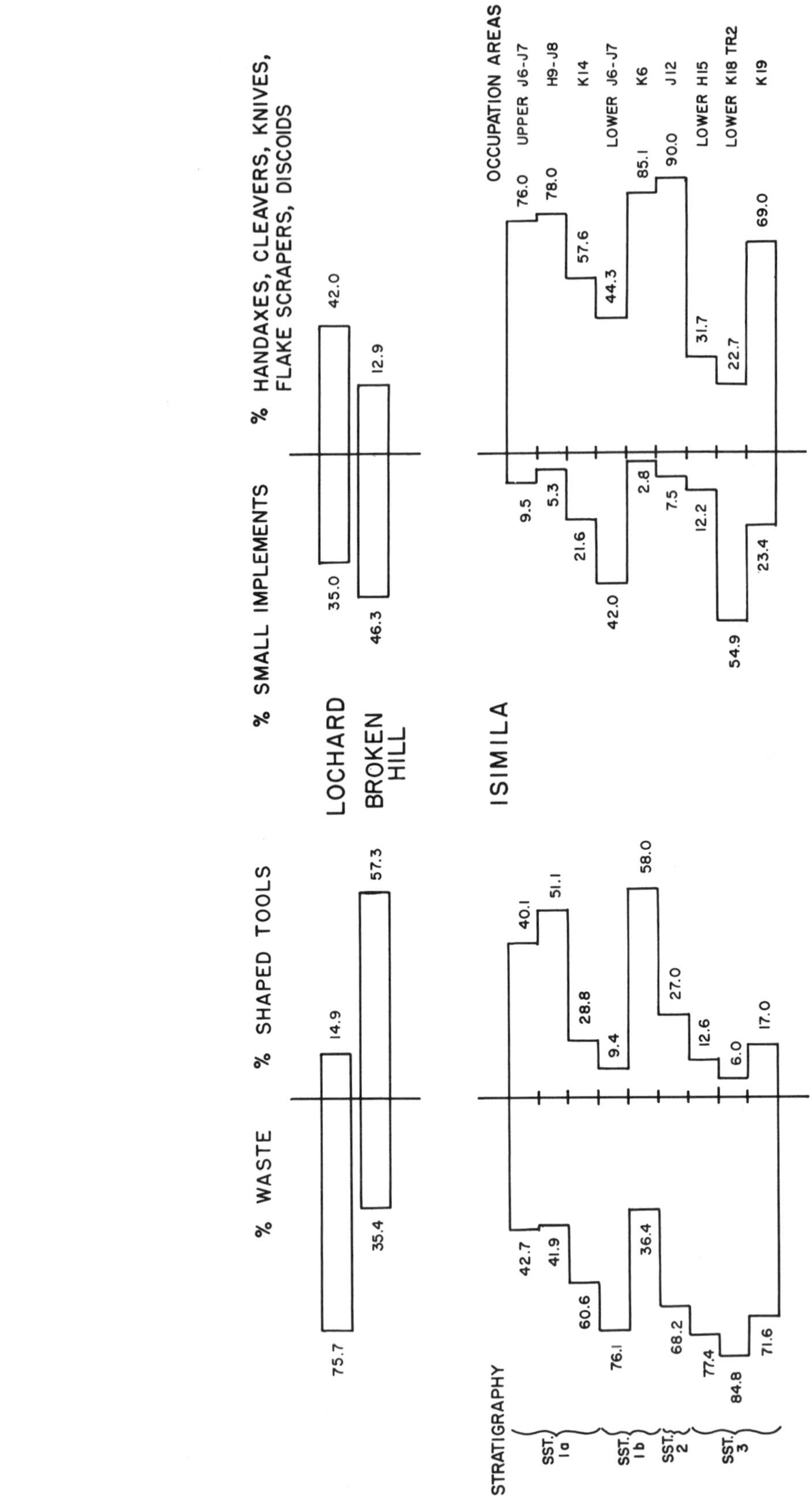

% WASTE
% SHAPED TOOLS
% SMALL IMPLEMENTS
% HANDAXES, CLEAVERS, KNIVES, FLAKE SCRAPERS, DISCOIDS
LOCHARD
BROKEN HILL
ISIMILA
STRATIGRAPHY
SST. Ia
SST. Ib
SST. 2
SST. 3
OCCUPATION AREAS
UPPER J6-J7
H9-J8
K14
LOWER J6-J7
K6
J12
LOWER H15
LOWER K18 TR2
K19
75.7
35.4
14.9
57.3
35.0
46.3
42.0
12.9
42.7
41.9
60.6
76.1
36.4
68.2
77.4
84.8
71.6
40.1
51.1
28.8
9.4
58.0
27.0
12.6
6.0
17.0
9.5
5.3
21.6
42.0
2.8
7.5
12.2
54.9
23.4
76.0
78.0
57.6
44.3
85.1
90.0
31.7
22.7
69.0

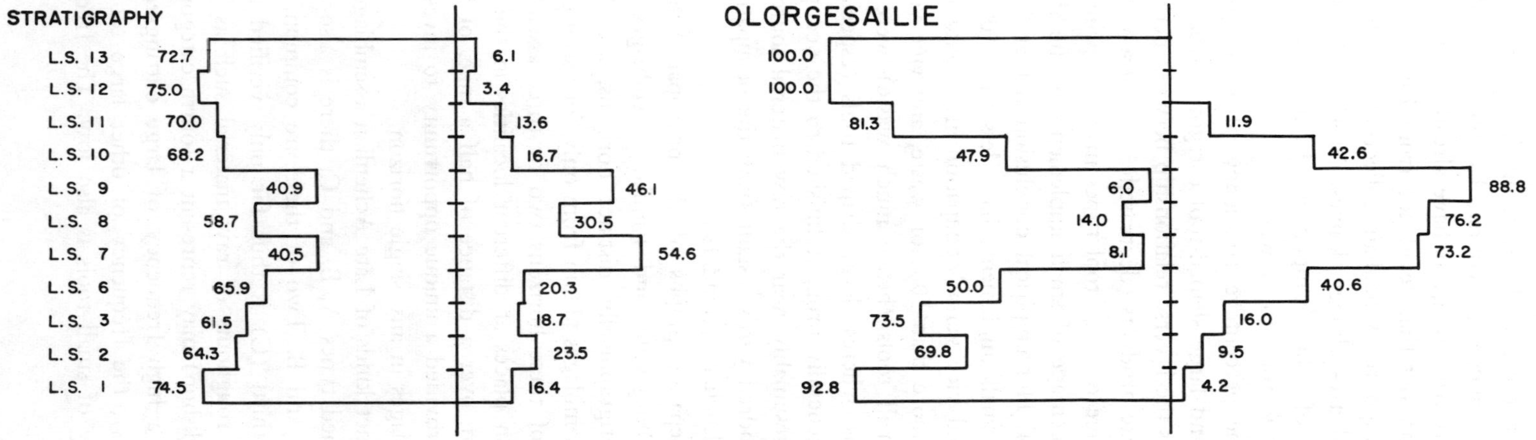

FIGURE 15
Percentages of waste products to shaped tools, and of small and large shaped tools in four Late Acheulian sites. (After M. R. Kleindienst 1961b)

thousands of years. Nonetheless, the comparison is deliberate and informative. The range of the components in the assemblage is much greater than usually stated (or believed). This is enhanced, of course, by selective collecting in geological contexts, usually where natural agents have already been a disturbing factor. Kleindienst (1961b) points out that "evidence from Isimila and Olorgesailie indicates that a large composite Acheulian collection (random sample) should include 40–60% of hand axes/cleavers/knives; 20–45% of small implements, and varying percentages of other components. In the total assemblage, it appears that the ratio is about ⅓ tools to ⅔ waste."

Figure 15 gives graphs of the percentage frequency of waste products to shaped tools, and of the small and large shaped-tool categories from four sites. The important conclusion here is the inverse relationship between percentages of shaped tools (increased) and waste products (decreased); it is closely "correlated with an increase in the percentage of large tool types made on primary flakes, and with a decrease in the percentage of small implements in the shaped tool category" (Kleindienst, *op. cit.*). Her empirical conclusion is that "occupation areas with more than 25% of small implements have less than 50% of large tools made on primary flakes and vice versa; occupation areas with more than 10% of small implements have more than 50% of waste, and vice versa." The evidence is now strong that small tools (hence much waste) were made on the spot at the occupation areas, whereas large shaped tools (especially when on large primary flakes) were generally brought finished to the occupation site having been made elsewhere, presumably near the raw material source. Such a source may be either a few hundred yards distant from the occupation site or as much as 40 miles away as at Lochard (Bond 1948).

Figures 16 and 17 present separate graphs of the principal Acheulian assemblages from the sites of Olorgesailie and Isimila. Assemblages from ten occupation areas, all from stratigraphically distinct horizons, are shown for Olorgesailie. The nine Isimila assemblages obtain from only four stratigraphically distinct horizons; but in three of these horizons two or three assemblages are known from separate occupation places at different localities in the particular horizon. The widespread erosion, over a distance of half a mile, of the upper three levels of the Isimila site provided a unique opportunity to investigate the lateral variation between assemblages in any single horizon.

Generally speaking three distinct kinds of Late Acheulian assemblages may be recognized. These may be termed types A, B, and C; there is also one more or less intermediate between A and B. Two of these are common, so far as current studies show, but the third (C) is thus far only verified at Isimila. The purely factory waste and roughouts of raw material such as at Gwelo Kopje and at Isimila (K 19 workshop) may represent a fourth category of site.

Type A is characterized by a high frequency of large cutting-edge tools, (hand axes, cleavers, and knives); the frequency of other large tools is consistently low; and the frequency of small tools is also low. The frequency of

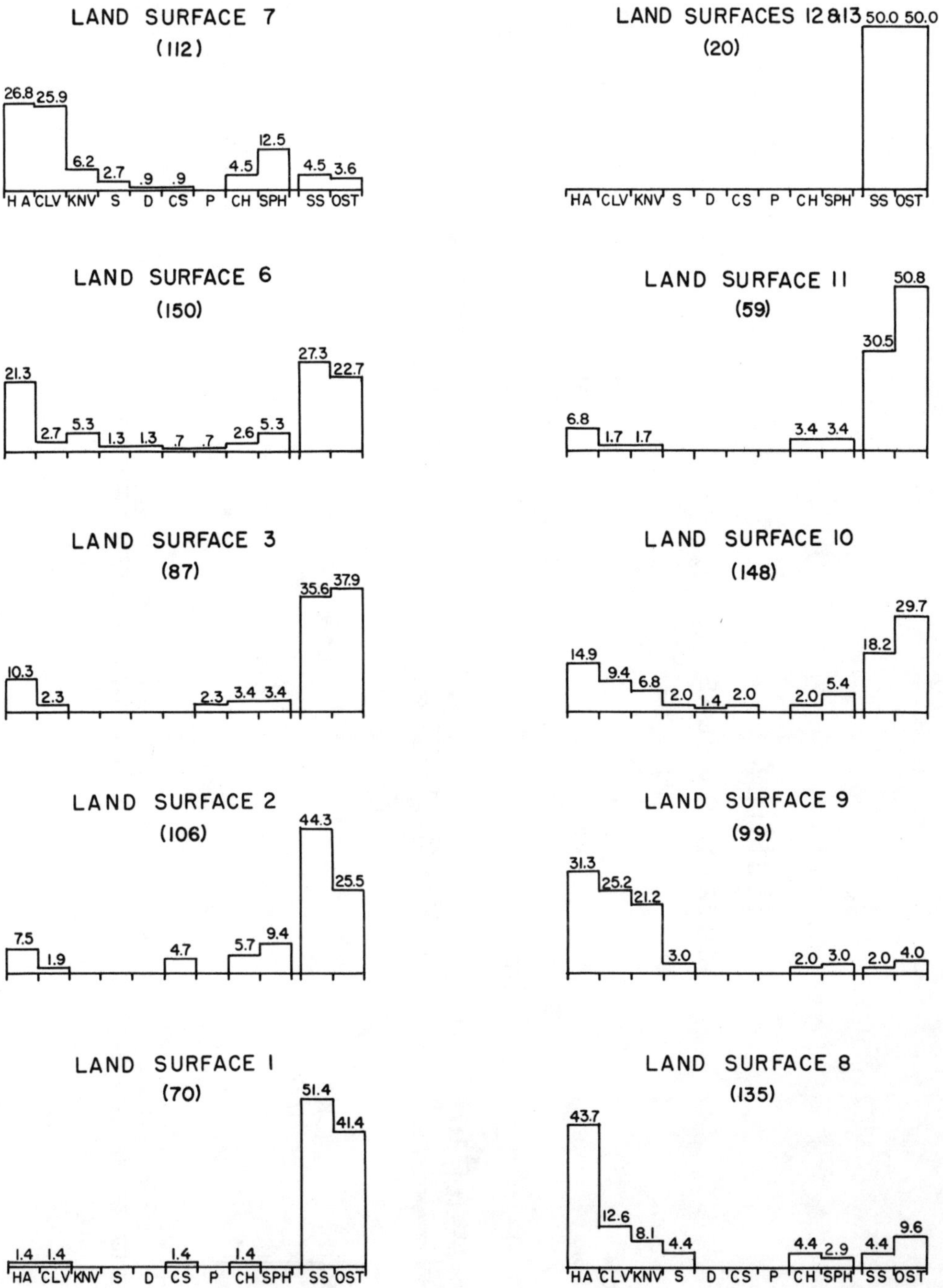

KEY

HA	Handaxes	P	Picks
CLV	Cleavers	CH	Choppers
KNV	Knives	SPH	Spheroids
S	Scrapers	SS	Small Scrapers
D	Discoids	OST	Other Small Tools
CS	Core Scrapers		

FIGURE 16
Composition of Acheulian assemblages from the Olorgesailie site.
(After M. R. Kleindienst 1961b)

ISIMILA

OCCUPATION AREAS

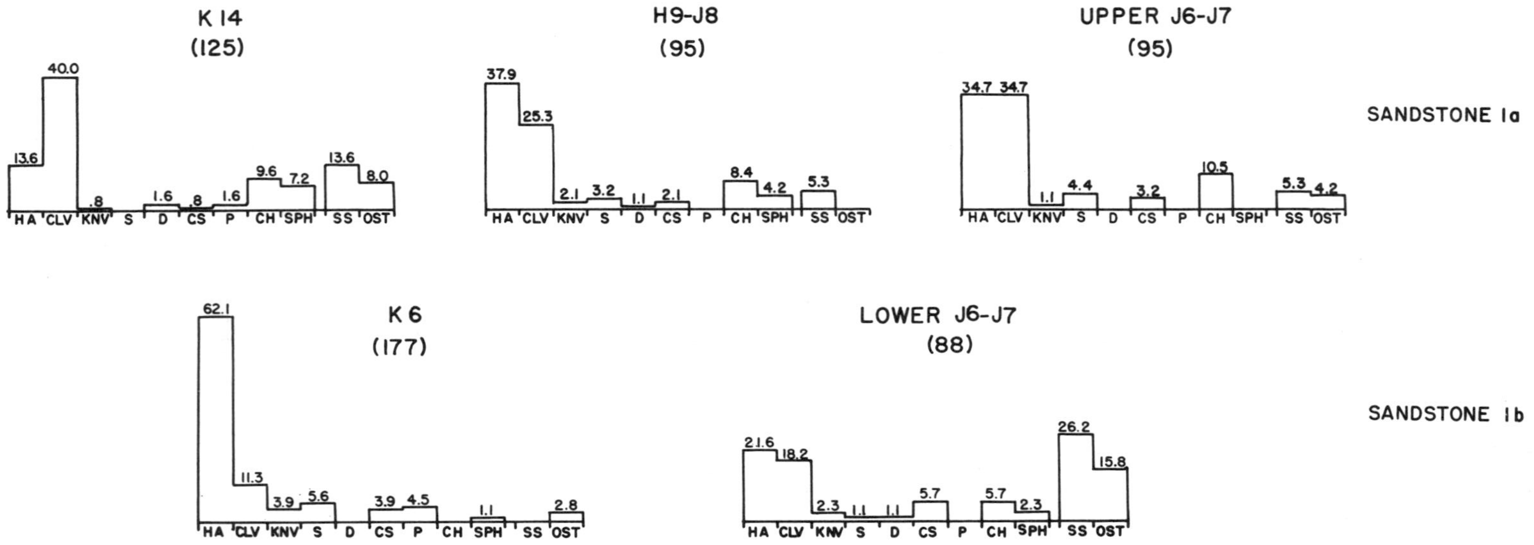

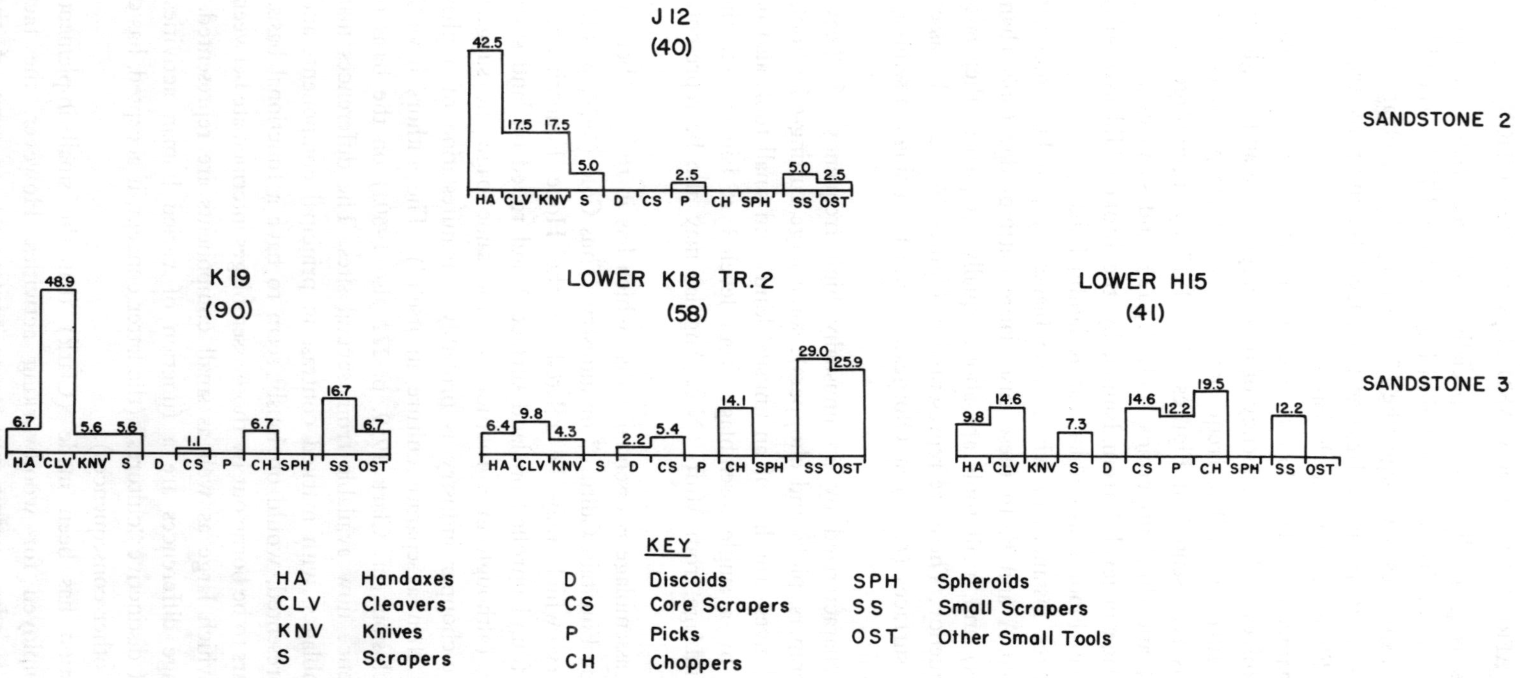

FIGURE 17
Composition of Acheulian assemblages from the Isimila site. (After M. R. Kleindienst 1961b)

waste products is generally low (especially small materials), although some large flakes are present. It is represented in Figures 16 and 17 by assemblages from three upper land surfaces at Olorgesailie (7, 8, 9), by six assemblages from Isimila (one from level 3, one from level 2, one from level 1b, and three from level 1a), and Acheulian Floors 4 and 5 at Kalambo.

Type B is characterized by a low frequency of large cutting-edge tools (hand axes, cleavers, knives); a low frequency of other large tools; and a high frequency of a variety of small tools. The frequency of small waste products is generally substantial. It is represented in Figures 16 and 17 by assemblages from three lower (1, 2, 3) and three uppermost (11, 12, 13) land surfaces at Olorgesailie, by one assemblage in level 3 from Isimila, by the Broken Hill assemblage, and by the Acheulian Floor 6 assemblage at Kalambo Falls.

There are three instances in which assemblages tend to be somewhat intermediate between A and B. In these cases large cutting-edge tools (hand axes, cleavers, knives) and small tools are about equally frequent; other large tools are in low frequency. These are represented in Figures 16 and 17 by assemblages from two land surfaces (6, 10) at Olorgesailie, and by a single assemblage from level 1b at Isimila.

Type C is characterized by an unusually high frequency of "heavy duty tools" (core-scrapers, picks and choppers); some large cutting-edge tools (hand axes, cleavers, knives—the last of an unusual kind) and small tools also occur. It is represented by a single assemblage from level 3 at Isimila, but the small Acheulian assemblage from Mufo (N.E. Angola) may also be representative of this category.

The type B assemblage is essentially that which has in the past been referred to as the "Hope Fountain Culture" or industry. Thus Cole (1954, p. 142) refers to "an industry which may be related to the Hope Fountain culture in Rhodesia . . . found unrolled on land surface 2 and rolled on land surface 6" at Olorgesailie (although in a footnote on the same page she suggests "this intriguing flake-chopper industry is probably a manifestation of a phase of a culture rather than a separate culture in itself"). The authors have reached the same conclusion (cf. Clark 1959, p. 222 also 1960a) on the basis of more adequate evidence now available from several sites. The differences manifested between assemblages, with marked contrast in principal components and major tool types represented, would obviously seem to have a functional basis. Moreover this appears to be borne out by those assemblages intermediate between Types A and B in which large as well as small components are represented. Hence these assemblage differences are a function of varied human activities rather than a mark of distinctive "cultures" (the latter concept, if accepted, has complex biological and other consequences).

The suggestion has been made (Clark) that the small implements were principally employed for woodworking activities. However, the factors of seasonality (e.g., dry- versus wet-season visitations) or even of the labor division by sex or age should not as yet be wholly excluded from consideration. The utility of other types of large implements is equally uncertain. The fact that the large shaped tools with cutting edges show little heavy usage, hence lacking deep nicks or battering, would indicate their employment on soft rather than hard resistant substances. The heavy duty tools (picks, choppers, etc.) on the other hand often show signs of heavy usage and hence the name applied to this category of implements. Only the further careful excavation of such human occupation sites in which more adequate context is manifested and in which wood and bone is preserved can throw further light on this baffling problem.

There are also differences between assemblages which can be explained most readily as stylistic in nature. This is particularly marked in the variable frequencies of certain forms (subtypes) of large shaped implements and is illustrated in Tables 6 and 7 which contrast the frequencies of forms of hand axes and cleavers in various assemblages at Olorgesailie and Isimila.

The tables require little special comment. At Olorgesailie there are more hand axes than cleavers on all the surfaces. They are largely of the ovate-long-ovate-lanceolate types. There is possibly an increase through time in the proportion of lanceolates and a decrease in the proportion of ovates and long ovates. Some forms occur only in very low frequency and/or only on certain surfaces; e.g., triangular forms on surfaces 6, 7, 8 and double-pointed forms only on surface 9. At Isimila, ovate and long ovate forms are about equally frequent, with lanceolates considerably less frequent. The assemblage of the K6 floor is unique in many respects including its variety of rare hand-axe shapes, higher frequency of lanceolates, and especially high-backed hand axes (also in its cleaver types). There are trends evident between the upper three sandstone horizons, with both ovates and long ovates decreasing in frequency, and lanceolates increasing slightly in frequency.

At Olorgesailie there are no clear temporal trends in cleaver types. In general divergent forms are practically absent, parallel cleavers are in low frequency, and convergent cleavers predominate. The proportion of convergent and parallel forms is similar on surfaces 7, 8, and 10; but on surface 9 nearly all cleavers are convergent. Ultra-convergent cleavers are found only on surfaces 7 and 8. Surface 8 is unique in the occurrence of three types of cleavers with pointed butts. At Isimila also there is no clear temporal trend. Generally speaking there are 3 to 4 times as many convergent as divergent cleavers, and 2 to 3 times as many parallel as convergent cleavers. Here is a regional contrast with Olorgesailie (where convergents predominate) versus Isimila (where parallel cleavers predominate). Farther south (Rhodesias, etc.) the frequency of divergent forms increases substantially.

TABLE 6

HANDAXE FORMS (%) AT OLORGESAILIE AND ISIMILA

Site-level	Lanceolate	Narrow lanceolate	Long ovate	Asymm. long ovate	Pointed long ovate	Ovate	Asymm. ovate	Pointed ovate	Ovate acuminate	Cordiform	Single shouldered	Triangular	Sub-triangular	Elongate triangular	Double pointed	Diamond	Truncated diamond	Limande	Butt untrimmed	Various	Total Number	
J 6-7 (u)	6.1		45.5	3.0		33.3		3.0	6.1						3.0						33	ISIMILA
H9	2.8		41.7	2.8	2.8	27.2	13.9	2.8	2.8			2.8									36	
K14	4.8	2.4	21.4	2.4		42.8	9.5	2.4	2.4		2.4				4.8			4.8			42	
K6	18.1	2.5	23.5	9.6	.7	19.6	1.8	2.5	7.5		2.1	.7	1.4	.3	1.4	.7	1.1	1.4	2.8	1.8	281	
J 6-7 (b)	10.5		36.8			31.6	10.5					5.3			5.3						19	
J12	5.8	5.8	17.6	5.8		52.9		11.7													17	
10	31.8		4.6			40.9	9.1					4.5			4.5		4.5				22	OLORGESAILIE
9	25.8		19.4	3.2		16.1			6.4						16.1	3.2	3.2	3.2		3.2	31	
8	20.3		18.6	8.5		25.4			1.7	1.7	1.7	15.2	1.7	1.7				1.7		1.7	59	
7	23.3		3.3	3.3		46.7	6.6					10.0				3.3				3.3	30	
6	25.0		15.6	3.1		21.9	6.2		3.1	3.1	3.1	9.4	3.1				3.1			3.1	32	
Total (%)	24.1		13.8	4.0		28.7	3.4		2.3	1.1	1.1	9.9	1.1	.6	3.4	1.1	1.7	1.1		2.2		

TABLE 7
Cleaver Forms (%) at Olorgesailie and Isimila

	Site-level	PARALLEL						ASYM. PARALLEL			CONVERGENT									DIVERGENT						
		Straight	Straight, pt. butt	Straight, sq. butt	Guillotine	Guillotine, pt. butt	Guillotine, sq. butt	Straight	Guillotine	Convergent-Guillotine	Straight	Straight, pt. butt	Straight, sq. butt	Guillotine	Guillotine, pt. butt	Guillotine, sq. butt	Ultraconvergent	Shouldered conv. str.	Shouldered Guillotine	Straight	Straight, sq. butt	Guillotine	Guillotine, pt. butt	Splayed	Side-cleaver	Total
ISIMILA	J 6-7 (u)	18.2			27.3						21.2			30.3								3.0				33
	H9	33.3		16.7	20.8		8.3				12.5			4.2									4.2			24
	K14	31.2	5.4	2.2	28.9	2.2	1.1	2.2	1.1		15.1			2.2		1.1			1.1	3.2	1.1	1.1		1.1	1.1	93
	K6	18.9	1.9	3.8	20.8	1.9	1.9				24.5	3.8		11.3	1.9					3.8		5.7				53
	J 6-7 (b)	12.5	6.3	6.3	43.7						6.3									6.3		18.7				16
	Sands 2	33.3			38.1	4.8					14.3			9.5												21
	Sands 3																									
OLORG.	10	28.6			7.1						42.9			21.4												14
	9	16.0	8.0		12.0						24.0	12.0		20.0	4.0							4.0				25
	8	5.9									47.1		5.9	29.4		5.9	47.1		5.9							17
	7	17.2			10.3						31.0			37.9			6.9	3.4								29

RAW MATERIALS AND TOOL MANUFACTURE

The nature and source of raw materials employed for stone artifacts in Earlier Old Stone Age times, and their relationship to tool manufacture, are matters which have either been generally avoided or, sometimes, badly misunderstood. Distinct types of rocks have quite different flaking qualities, some few being practically unsuitable for tool-making because of softness or lack of conchoidal fracture, whereas numerous others are more or less suitable depending, of course, upon availability and mode of occurrence. The presence and easy availability of especially suitable sources of stone for tool-making within those habitats favorable for occupation at that time must have been of tremendous importance in the distribution and ecological adaptation of Earlier Stone Age populations. However, it is necessary at any one time level to fully appreciate whether appropriate techniques of extraction, preparation, and actual tool manufacture were suitably developed to deal with a narrow or wide range of raw materials in regard to occurrence. Africa south of 15° N. affords an extraordinary variety of more or less suitable suite of rocks (flint being absent except in the westernmost Equatorial region). It is interesting and informative to see the ways in which these were used, often differentially, during the time range of the Acheulian. The excavation of undisturbed living sites in eastern Africa has provided clear-cut evidence for a fairly close correlation between the several classes of tools and the raw materials from which these were chosen to be made.

In the Cape region, quartzite, in the form of Table Mountain Sandstone (T.M.S.), was customarily employed. This is attested at a variety of coastal and inland riverine sites, and also in Montagu Cave. However, in the Vaal River drainage the area south of the river is underlain by rocks of the Karroo System. In the headwaters and bounding the river these are shales (lydianite) of the Beaufort and Ecca series, and tillites of the Dwyka series along the lower Vaal and the Hartz River down to the Orange confluence. North of the river are largely pre-Karoo rocks, including quartzites and Ventersdorp diabase. Although quartzite was still employed in the Vaal Acheulian, as it had been earlier, Ventersdorp diabase (andesite) and eventually indurated shale, in those regions where igneous dolerite intrusions occur, were predominantly exploited. The Younger Gravels of the Vaal, of later Middle Pleistocene age, are predominantly (90–95%) diabasic rocks, ranging in size from pebbles to large boulders; small (potato-size) quartzitic and other non-diabasic elements are much rarer. These contrast with the Older (so-called Red or "Potato" Gravels which are predominantly small non-diabasic elements. There were special flaking techniques, described by Goodwin (1933) and van Riet Lowe (1945), which permitted Acheulian peoples to remove very large primary flakes, destined for hand-axe–cleaver production, from such massive diabasic boulders after preliminary trimming into immense cores. In the final stages of the Acheulian indurated shale was being extensively exploited, presumably from natural outcrops, to the south of the river. In all events the large cutting-edge tools so characteristic of the Acheulian were being fashioned from large flakes, the initial form of which was more or less laid out by the initial trimming of the cores prior to primary flake detachment. Hence these are bifaces, in the sense of generally extensive bifacial workmanship, but are not core-bifaces (as is usually the case in northwestern Europe where flint was abundant).

At the Cave of Hearths quartzite was predominantly utilized, although felsite also rarely occurs in the Acheulian. Bond (1948) has admirably drawn attention to the use of a considerable variety of rock types, so long as there was conchoidal fracture and it was fine grained, among the Acheulian in Southern Rhodesia. These include rocks of the Basement Complex (aplite, quartz, granite, banded ironstone, hornblende schist, andesite, and assorted greenstones), the Karroo (fossil wood, silcretes, dolerite, and breccia), and Dyke rock (anorthosite, spotted norite), as well as chalcedony from the Kalahari System. He specifically mentions the probable transport of andesite at Lochard, from perhaps a distance of 15 miles, and perhaps Karroo silcrete from 40 miles away.

In the Zambezi Valley and its tributaries, chalcedony was commonly employed. Large rolled boulders were especially suitable and were trimmed with the so-called Tachengit technique to produce large primary flakes. However, quartz and quartzite, silicified sandstone, and basalt (especially for cleavers) were also utilized. Small tools are invariably made from chalcedony or silicified sandstone. There was either no or but slight transporting of raw materials over any distance by the peoples of the Acheulian (Clark 1950).

At Broken Hill a variety of raw material was employed with selection depending upon the particular tool category (Clark 1960). Quartzite and silicified limestone was fashioned into large cutting edge tools. Small tools were made predomniantly of quartz, but also of ironstone. Heavy duty tools like choppers and missile stones were of breccia and desilicified limestone. The same variety was evident at Kalambo Falls. Most choppers, core-scrapers, and picks, as well as some cores, are in quartzite; other cores are in silicified mudstone. Hand axes and cleavers are fashioned predominantly from hard quartzite or fine-grained felspathic quartzite. Some of the smaller hand axes are in chert and most of the flake tools are also made in this rock or more rarely from felspathic quartzite or quartz.

At Isimila a variety of raw materials were employed for tool-making. There was also selection among these for particular categories of tools. Mylonites, a cataclastic microcrystalline rock of the Ndembera Volcanics series, were principally employed for the large cutting-edge tools; however, quartz and quartzite examples are also recorded. Choppers, core-scrapers, stone balls and picks, the heavy duty tools, were normally fashioned on durable rocks such as granite, quartz, and, rarely, amphibolite. Small shaped tools were normally made on vein or (very rarely) crystal quartz.

The size of the large cutting edge tools indicates that very large blocks or nodules of mylonite were available to the peoples of the Acheulian. Several large cores, up to two feet in length, were also found eroded out on the surface. These were worked by the so-called Tachengit technique with a varying amount of preliminary, more or less radial trimming prior to the detachment of large primary flakes. However, some small proportion of such tools were also made on rock slabs rather than proper flakes. In some cases at least the mylonite occurred as relatively thin seams within a matrix of granite (see also below). Several varieties of mylonites are readily distinguished at the site, differing in flaking qualities, in color, and mineral composition. One variety was customarily used more or less to the exclusion of another in any particular assemblage, indicating either differential availability from time to time or, and this seems quite likely, preference and hence deliberate selection of a particular variety.

In the sites in and adjacent to the Eastern Rift Valley lava was widely used. Thus at Kariandusi, where both a rolled and unrolled Acheulian series are recognized, lava was generally employed in the former case with both lava and obsidian (in about equal proportions) in the latter. The small tools are in the same two materials, but several small tools are found in quartz and chalcedony. Lava was also normally used at Lewa. At Olorgesailie several sorts of trachytic lava was employed for all types of tools. No large cores have been found, however, perhaps suggesting direct flaking from lava outcrops. Other exotic raw materials include obsidian (some hand axes), quartz-quartzite, and chalcedony (chert); both chalcedony and quartz were used for some small tools and some missile stones are of quartzite. It is interesting that the nearest exposed obsidian source is southwest of Olorgesailie (at Ol Doinyo Nyegi), some 14 miles away.

In central equatorial Africa various quartzite rocks were widely used. The extraordinary exploitation of the angular rubble on the MN horizon scree is a striking example. In the Congo large blocks of *grès polymorphe* are an analogous phenomenon, and this material was widely available in those areas of the south and southeast where the Acheulian has been recorded to occur. Quartz and quartzite appear to have been preferred, however, at this time for the manufacture of the heavier tools and it is mostly the smaller implements that are made from *grès polymorphe*.

OCCUPATION SURFACES

In order to learn about the livelihood patterns of early hunter-gatherers, like the peoples of the Acheulian, it is essential to locate and investigate their occupation sites. The occurrence of stone artifacts in stratified geological contexts communicates essentially nothing about the behavior of extinct human populations. In Africa, and especially eastern Africa, occupation places have been preserved, and their careful excavation has provided some interesting information on life in the time range of the Acheulian industry.

In all the cases known thus far these occupation sites are situated adjacent to basins of deposition. The early human sites are relatively or wholly undisturbed having been sealed in by subsequent sedimentation due to periodic fluctuation of water levels, either as a consequence of local tectonics and/or climatic change. Hence the sites are situated on old land surfaces which generally represent minor non-sequences in the process of sedimentation. However, exposed erosional processes were at work on the surfaces to varying degress; consequently small elements could be and doubtless were disturbed. This would especially occur along the sloping margins of beaches where downslope creep and the activity, perhaps seasonal, of freshets were natural agents of disturbance.[8] From our own experience, and firsthand acquaintance with these sites, it is clear that each locality has its own special problems, a solution to which can only be reached through very careful excavation techniques and, of course, the co-operation in the field of experienced colleagues from other pertinent disciplines (geology, pedology, etc.).

Kariandusi (6,560 ft.) in the Nakuru-Elmenteita basin, was the first such open-air site ever excavated (Leakey 1931, 1936). It occurs in faulted (> 100 ft.) lake sediments of late Middle Pleistocene age.[9] These are over 150 feet thick and overlie unconformably quartz-trachytic lavas. The deposits are stratified tuffs (> 100 ft.), diatomites (with largely planktonic diatoms), and sandy beds below the occupation level; coarse to fine gritty bedded sandstones overlie the occupation level. The human occupation is in pebbly sandstone which grades up into a thick coarser blue-gray gravel. A rolled and abraded artifact series, largely in lava, and a fresh artifact series, both in obsidian and lava, are both represented in the Acheulian industry from the site. It seems likely, as Leakey has suggested, that this is a beach situation and the freshest tools represent the latest Acheulian occupation. At any rate this is a case of archaeological context which has been somewhat disturbed.

There is no direct evidence of the use of fire at Kariandusi.

At Olorgesailie there are multiple non-sequences representing old land surfaces. Up to seventeen have been reported and a dozen have been excavated into Acheulian occupation areas. Three of these surfaces (1, 10, and 12, excavated by Posnansky, 1959) reveal fully undisturbed or primary archaeological context; essentially all artifacts are in fresh, though rarely in mint, condition. Surface 10, underlain by fine-bedded sands and overlain by gravelly and gritty sands, is undulating and characterized by small hollows or depressions. The artifact as-

8. This is in fact evident today in the "Catwalk" area of the Olorgesailie site. G. H. Cole and M. R. Kleindienst made a quantitative study of the implements here which are derived from several Acheulian horizons. Small tools and flakes and cores were scarcely present in the area, other than at the break in slope beneath the low cliff from which the artifacts were being eroded; and especially some 30 feet away in and adjacent to a small watercourse these are abundant, although large implements are absent.

9. The very poorly preserved associated fauna includes *Elephas recki*, *Stylohipparion*, *Sivatherium*, *Pelorovis*, *Simopithecus*, and *Tapinochoerus*. It is broadly comparable with that from Upper Olduvai (IV) and from Olorgesailie.

semblages occur in concentrations, restricted both horizontally and vertically, in the depressions in association with much rounded small natural lava boulders. Leakey has suggested the stony rubble was deliberately accumulated and that these might represent windbreak floors. The question of rubble accumulation is noted again below.

On most of the other surfaces there is some evidence of generally minimal interference responsible for disturbed archaeological context or even partial geological context. This is especially the case on surface 9 where artifacts have weathered (though unworn) edges and the large shaped tools exhibit alignment of their long axes; both this fact and the low frequency of small tools would suggest water action, perhaps resulting in some sorting out of the small artifacts. Also on surface 6 the large implements are in quite fresh (though not mint) condition; but, the small tool component contains specimens ranging from fresh to heavily abraded and weathered; here also perhaps even partial geological context is suggested. On surface 8, possibly a flaking site, the artifacts are fresh, and only a few flakes are worn. And on surface 7 there is only little evidence of wear. On surfaces 2 and 3 the artifacts exhibit fresh to rather dulled edges which would suggest some slight disturbance. (The former reveals an irregular semicircle of natural stones (lavas) which one of us (Clark) has suggested might represent a sort of simple fish weir.)

There is no direct evidence of the use of fire at Olorgesailie.

At Isimila several sorts of artifact distribution occur on the occupation surfaces. Those occupation areas which exhibit both restricted lateral and vertical distribution, as at Olorgesailie surface 10, have been designated occupation floors. The assemblages occur at the base of the beds of coarser sands and silts; these may be on, penetrating slightly into, or a few inches above the usually well-differentiated contact with the underlying clay. The horizontal distribution is densely concentrated centrally and decreases laterally toward the margins of the occupation floor. These occur in all the main coarse sediment horizons as follows: Sands 1 (K6, K14, lower J6-7); Sands 2 (H20, Tr. 8); Sands 3 (H15, two floors; K18); and Sands 4 (G18).

There are other cases in which the artifacts, all quite fresh, are diffusely distributed vertically between one and three feet (50–100 cm.). However, the artifacts do appear to be horizontally concentrated. Two interpretations are suggested for this occurrence: either there was more prolonged occupation (compared to that at the occupation floors) during the accumulation of the coarse sediments, or there was a natural disturbance, or disturbance by man or animals, of the still unconsolidated sands. These occur in each of the three main upper horizons as follows: Sand 1 (upper J6–7, H9–J8); Sands 2 (J12); and Sands 3 (K19). Although a large number of artifacts are present at Isimila in Sands 4 these appear to be in geological rather than archaeological context (except for the partially excavated G18 occupation floor); this also holds for the basal sediments at the site.

The Isimila occupation areas contain varying quantities of non-artifactual stone or rubble which is more or less densely concentrated with the artifacts. In some cases there is very little rubble even within the same horizon. This is the case in some Sands 1 sites (K14, H9, and upper J6-7) and may merely reflect the periphery of an occupation area. However, in other cases, including the very same horizon, there occur dense concentrations of rubble. This is the case in the lower J6-7 and in the K6 occupational floors. In the former it comprised predominantly small (<10 cm.) chunks of quartz and quartzite with rare, somewhat larger blocks of granite material. In the K6 floor there were densely packed chunks of granite as well as schistose mylonite, generally of rather large size > 10 cm.). Two occupation floors in Sands 3 contained rubble, varying in size and concentration. In the K18 floor it was only moderately dense (compared with the K6 occurrence) and largely represented small chunks of granitic-gneissic (53.5%) and quartz-quartzitic (37.8%) rocks. In H15 the rubble was not especially close packed, but consisted of large (5-15 cm.) chunks of granitic rock.

This variation in the composition of rubble has been puzzling, but depends at least in part on types of raw materials used in tool manufacture. This is especially the case for the quartz and quartzite rubble. Where this material is abundant the small tool component, as well as related waste products, is substantial. The conclusion can only be that these small tools were fashioned on the spot, and the raw material brought there from elsewhere. This may also account for a certain amount of the granitic rubble where heavy duty implements, and also small cores, in this material are present.

However, there is clear evidence that large shaped tools were usually made elsewhere than on these occupation areas. Essentially none of the waste products which would result from the trimming of the large blocks of mylonite into cores for producing primary flakes are represented in the assemblages. This interpretation is confirmed by the discovery of an actual workshop (factory) site which was only recently exposed by erosion. Its full inventory reveals an abundance of such products along with flake roughouts, some simply finished or minimally trimmed large tools, but scarcely any well-shaped, large cutting-edge implements. Hence, occupation floors like K6, where well-finished, large cutting-edge tools are extraordinarily abundant in a massive and dense rubble concentration, pose a special problem. The conclusion can only be that the peoples of the Acheulian deliberately accumulated and concentrated quantities of stony rubble in certain occupation places. (This is also recorded on Surface 10 at Olorgesailie as mentioned before.) The purpose behind such activity is still quite unknown. There is no evidence of patterning in the occurrence of the rubble and no suggestion of its relationship to any sort of structure.

At Isimila there is no direct evidence of the use of fire.

Occupation material on the uppermost of the upper two Acheulian floors at Kalambo Falls is rich and generally well concentrated. That on the lower two floors is usually sparse and found in small, isolated accumulations. The implements and waste on the latest floors are all quite fresh, but some of that in the

KALAMBO FALLS : SITE A1

FLOOR 4 – ACHEULIAN

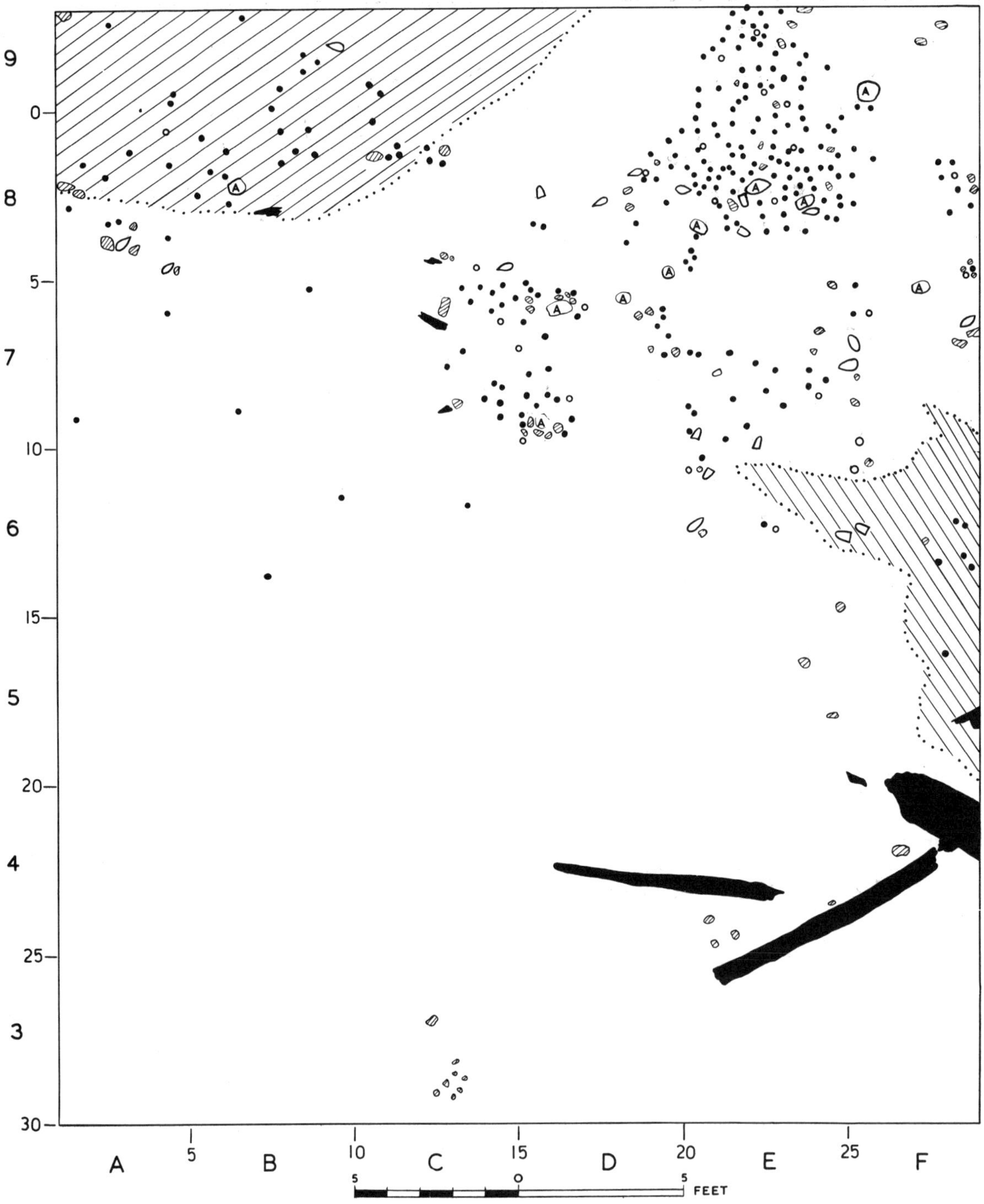

FIGURE 18 Horizontal grid

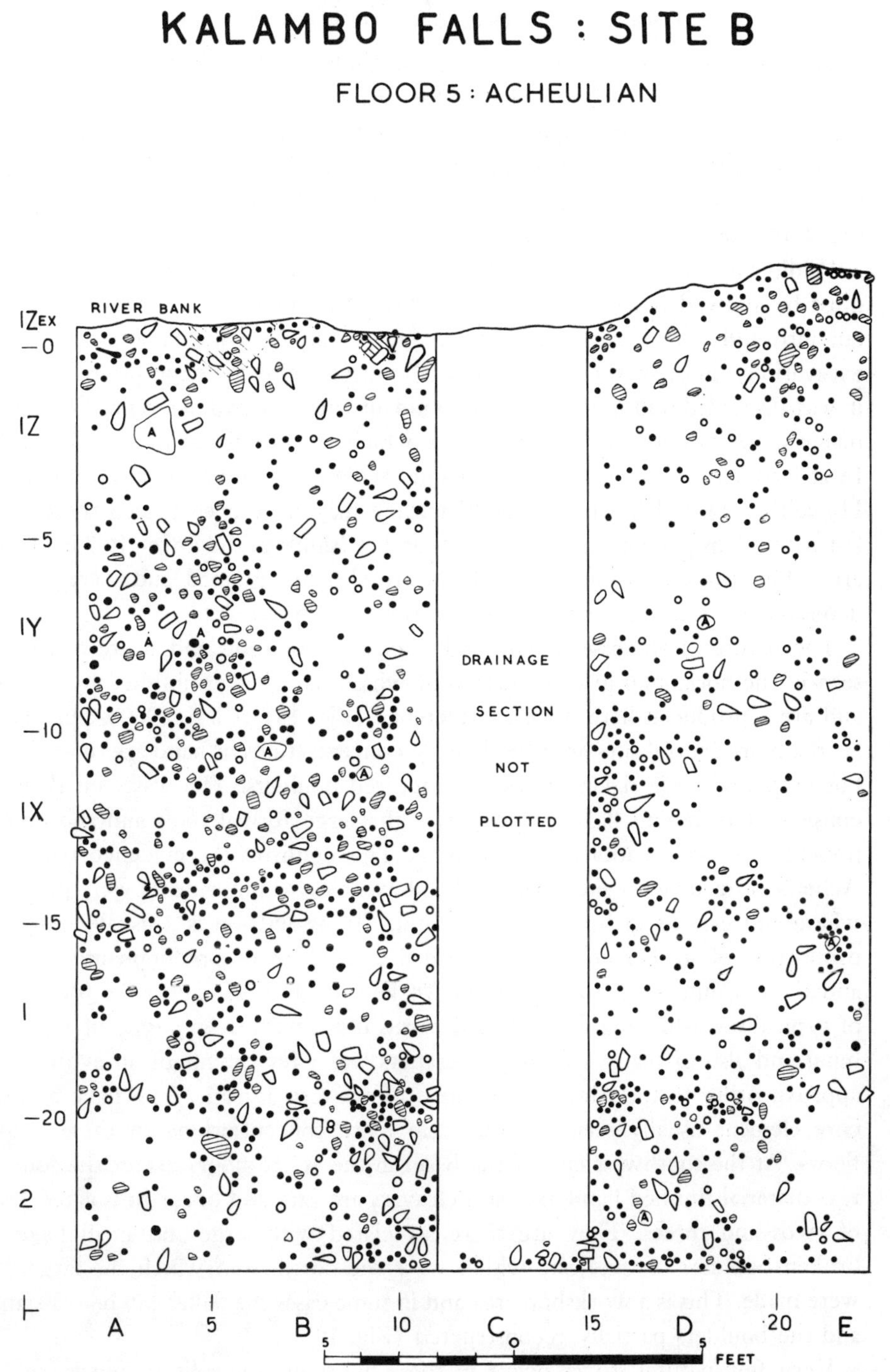

plots of two Acheulian occupation surfaces (Sites A1 and B) at Kalambo Falls.

lowest levels investigated has had the edges dulled by water action. The floors investigated in the most central parts of the basin are practically horizontal but those nearer the southwestern edge have a low angle of slope, as might be expected. It would seem that the Acheulian people camped by the fringing forest surrounding the open sand bars and lagoons on the edge of the water, as well as higher up the slopes nearer the source of the raw material.

While some slight displacement of material on the floors due to natural causes must be allowed for (as, for example, down the sides of the water channel that cut into and which marks the south edge of the Acheulian floors in the B2.59 excavation), most of the material must be in virtually the same position as when it was discarded and covered by the sand of the next season's rain. The cultural material lies horizontally on the floors which are composed of fine, loose sand. In at least two instances, however, hand axes were found vertically on edge (cf. Howell 1961) and it would seem that they may have been placed there by the last users. This phenomenon was also found at Isimila in the lower J6-7 occupation area. The sands overlying these floors are also fine-grained and were probably deposited by freshets fanning out into the open water.

The nature of the assemblages and association of tools and waste can be well seen in the horizontal grid plots, two of which have been published by one of us and are reproduced here for easy reference (Fig. 18) (Clark 1960b, pp. 316-17). At the more central sites are found heavy concentrations of hand axes and cleavers, but very few large flakes, cores, or fractured boulders. The waste on these sites consists of trimming flakes struck from the large formal tools and the numbers present suggest intentional resharpening. This confirms the evidence from other Acheulian occupation sites that the hand-axe–cleaver implements were made elsewhere and carried in for use at the camps. In association is a usually fairly low proportion of smaller cores, flakes, and flake tools made predominantly of chert, and these would seem to have been worked on the site. The flake tools consist of typically small, irregularly shaped forms often with steep angles of edge trimming and also of many well made scrapers and a few points of an evolved form approximating to the Mousterian forms of western Europe. The flake and small core element usually seems to occur in loose concentrations on these Type A floors. At the southwest end of the basin in the A2.56 floors nearer the source of raw material finished hand axes and cleavers are rare and occur in isolated groups of twos and threes. They are there associated with large and small flakes and broken boulders of quartzite which were the cores from which the larger tools were made. This is a workshop area and in some cases the flakes can be reassembled and the boulders partially reconstructed (Fig. 19).

Very few natural stones occur in these floors in contradistinction to those of the Middle Stone Age and it is likely that most of those that do occur were carried in by man though they show no clear signs of utilization. Some of the larger of these stones were used as anvils (Clark 1960b, p. 316) and the smaller ones could have served as smashing and pounding equipment. A few very well rolled

FIGURE 19
Views of Acheulian floor 5 (A56, B56) at Kalambo Falls to illustrate the different occurrences of finished tools, waste products, and workshop areas.

"pot hole" pebbles of quartzite occur though they were commoner on the Middle Stone Age floors where they were used in a special manner. G. Bond is unable to suggest a local source in the basin for these stones and it is possible that they might have been carried up from Lake Tanganyika some four hours walking distance to the west.

In one place on Acheulian Floor 5 (B2.G/H.3/5.59) was found a rough arc of stones, 23 in number, which it would seem may have been placed intentionally in position by man. The arc is close to the water channel cut into the floor and some slight natural displacement is likely to have taken place; but since no natural agency can be suggested capable of moving stones of the size of some of these under the circumstances in which they occur, human agency is the most likely explanation for their presence. The stones enclose a space some eight feet between the ends of the arc and 3 to 3 1/2 feet in depth in the center. They might mark the outline of some temporary shelter or windbreak. The ground within (con-

FIGURE 20
Club from Acheulian floor B2/59, Kalambo Falls.

cave side) these stones was remarkably free from tools and debitage which, however, occurs in the usual concentration on the outside (convex side) of the arc.

On the lower floors (6 and 7) the formalized hand axes and cleavers are more rare and the small-flake element increases. Moreover, there is little variety in the forms of the large tools present and the majority of these are only poorly made by comparison with those in the concentrations on the upper floors. Nevertheless, several very fine step-flaked scrapers and points occur together with the usual utilized and informal flake element.

As previously stated, wood and other vegetable remains are associated with these Acheulian floors. Some of the wood appears to have been shaped by fire and utilized as digging implements, knives, and clubs (Fig. 20). There is ample evidence for the presence of fire—logs with the ends charred, patches of consolidated ash, and charcoals all occur (Fig. 21). In at least one instance (B.1.5.56) ash and charcoals are associated with an oval patch of reddened clay suggesting

FIGURE 21
Close-up of Acheulian floor 5 at site B2/59, Kalambo Falls.

a hearth. Three oval-shaped patches of compressed and carbonized grass stems and woody plants, all filling shallow depressions and surrounded by sand, were found on two of the latest Acheulian floors. These are approximately 3 by 2 feet in area and 6 inches thick in the center. They may perhaps represent bedding in sleeping forms which had subsequently been burnt. At Kalambo Falls, therefore, good evidence exists that the Acheulian population were acquainted with and made use of fire.

Table 8 is included here to provide an appreciation of the density of artifact distribution on some occupation areas. Kleindienst (1961b) has recently summarized the evidence from Clark's (1959) work at Broken Hill, Posnansky's (1959) work in the higher level at Olorgesailie, and the work of Howell, Cole, and Kleindienst (1961) at Isimila. Generally speaking the occupation areas with Acheulian of Type A appear to have somewhat lower artifact densities. However, the most clear-cut difference is in the relationship between shaped tools and waste products, which is generally low in Type A Acheulian compared to the other facies.

The problem of *time* merits some brief discussion. In dealing with these open-air occupation sites the paleoanthropologist is fortunate to have slightly or wholly undisturbed archaeological context. But he is confronted with a real problem as to the length of time involved in any particular instance (cf. Kleindienst 1961b). How long was a particular land surface exposed during and after occupation and prior to being sealed in? What was the duration of visitation that produced the different occupation floors? Was this a matter of a few days, several weeks, or repeatedly over some months or a year? Were different occupation floors on the same horizon at Isimila occupied at the same moment in time by different bands, assembled for some specific purpose, or were these successively occupied at different times by the same or distinct bands of Acheulian folk? (Hence is this lateral variation in artifact assemblage composition a function of time [stratigraphic] or not? How do the vertically and horizontally restricted occupation floors compare temporally with the vertically diffuse occurrences? How are we to understand culture and culture change where some sites were repeatedly though intermittently occupied over some tens of thousands of years and others for but hundreds or tens of years and others only for days or weeks? The problem is really that of devising methods and approaches in order to deal with quite different *ordres de grandeur*. The matter of terminology for sites with archaeological context should also be re-examined. What do such terms as occupation site, living site, camp site, kill or butchering site, and workshop site really mean in terms of duration of visitation and the size of human group(s) involved? How may we deal with the problem of seasonal movements and range densities among such early hunter-gatherers? The discovery of early human occupation sites raises a whole series of problems of concern to the ecologically minded investigator.

Finally the question of subsistence should also be mentioned. Unfortunately

TABLE 8
DENSITY OF ARTIFACT DISTRIBUTION ON ACHEULIAN OCCUPATION AREAS
(after Kleindienst 1961b)

Occupation Area	*Type*	*Area (in sq. meters)*	*Density of shaped tools (per sq. meter)*	*Waste*	*Total pieces*
Olorgesailie 12 (2)	B	16.8	1.1	13.1	16.5
Olorgesailie (12) (1)	B	16.8	.2	4.5	5.2
Isimila K14	A	103	1.2	2.7	4.2
Isimila H9–J8	A	100	.9	.8	1.9
Isimila J6–7 (upper)	A	70	1.4	1.5	3.3
Isimila K6	A	61.8	2.9	2.0	4.9
Isimila J6–7 (lower)	A–B	70	1.3	10.9	12.3
Isimila J12	A	35	1.1	2.9	4.2
Isimila K19	A	26.3	3.4	15.8	20.5
Isimila K18	B	24	3.8	57.3	64.4
Isimila H15 (lower)	C	28.4	1.5	9.3	11.5
Broken Hill	B	33	2.8	2.0	5.0

we speak of Middle Pleistocene hunter-gatherers, with scarcely any knowledge of those species hunted and/or gathered. Here we are at the mercy of conditions of preservation on the occupation surfaces. Thus at Isimila, where acid conditions prevailed, bone was preserved in Sands 4, but where artifacts largely in geological rather than archaeological context seemed to be associated; only a single hippopotamus premolar and a skull of *Lepus* were found in any other occupation place (K14). We will never know whether bone originally existed in the upper levels of the site or whether its absence is merely a function of sediment acidity (and which also fails to preserve pollen).

At Olorgesailie there is associated fauna, generally broken up and fragmentary, on the occuption surfaces. It may provide some interesting evidence on the nature of the meat diet and preferences of peoples of the Acheulian. However, it will be essential to distinguish human prey from that of the various carnivores. A full inventory of the Olorgesailie fauna is not available to us, but the following remarks will suffice for the moment. Fauna is scarce on both land surfaces 6 and 8, but the former includes fish, the latter hippo, and both include some horse (*Equus oldowayensis*). On land surface 1 horse, giraffe, and antelopes were principally represented, and on land surface 2 mainly elephant and antelopes. Fauna is abundant on land surface 7 and includes especially the large baboon, *Simopithecus* (over 45 individuals represented) as well as some horse, hippo, and remains of fish. On land surface 10 wart hog (? species), hippo, and tortoise are represented.

The HK site at Olduvai has been referred to previously. Eventually the excavation of a large Acheulian occupation floor at Olduvai, where bone is extraordinarily well preserved, should provide some badly needed data on hunting and butchering practices and meat preferences. This work is currently in progress by Dr. M. R. Kleindienst. It will be particularly informative to have details on minimal numbers of individuals and the several age classes of the various species which may be represented.

Investigation of "wet basin" sites where vegetable remains are preserved is as yet the only way to fill the immense gap in our knowledge of Acheulian economies. As yet the Kalambo Falls and Lake Chila are the only sites where such remains are found in association and the results obtained are giving details of changing vegetation pattern, temperature, and rainfall conditions, never before available. As yet the use made by the Earlier Stone Age population of wood and its by-products and other vegetable materials is practically unknown. The scanty evidence that does so far exist suggests that such use is likely to have been considerable. Man's ability as a technician at this time, as evidenced by the utilization of many different kinds of rock for stone tools, cannot have failed to find expression in a wide range of other materials also and wet basin sites provide the only means of recovering those products of his handicraft made in perishable organic materials.

These sites also provide the only means of determining the extent of man's reliance on vegetable foods at this period. Circumstantial evidence already exists at Kalambo that certain fruits were eaten and it may be expected that tubers, roots, bark, and gum formed routine additions as well, no doubt, as honey. Until the part played by collecting as well as by hunting can be assessed, only a one-sided idea of Acheulian economy is possible; nor will we be able to advance much more in the determination of the uses to which the various tools were put by their makers. Although it has always been assumed that the Earlier Stone Age populations were primarily hunters it is more probable, from what we know and can infer of the environment they preferred, that some 75% of their food supply was derived from vegetable or other sources and only some 25% from meat.

The area covered by Acheulian sites usually would seem to have been fairly extensive. On the Kalambo Falls evidence, and also that from Isimila, an average estimate would be approximately an acre but, depending on the circumstances, both larger and small areas seem to occur. Where a locality, such as Power's Site at Pneil (Power 1955), was visited regularly over a long time the concentrations may cover an infinitely wider area, although some vertical division of industries might be possible here given careful excavation. Within most of the living floor sites examined it is possible to distinguish areas of heavy concentration and others where artifacts are sparse or quite absent. In addition to the problem of contemporaneity is that of determining what represents a unit of concentration. The wide horizontal areas uncovered at Isimila and the Kalambo Falls may, when the study of the assemblages is complete, give a partial answer but the whole answer can really only be obtained as a result of excavation of a complete occupation area. Such work is bound to be costly, but the investigation of one or two carefully selected key sites where faunal or vegetable remains, or preferably both, are present would add immeasureably to knowledge of this and other problems of Acheulian ecology.

The only open station sites that have provided evidence of fire are Nyabusoro in Uganda and the Kalambo Falls. At the former, burnt bone is said to have been found (Posnansky, personal communication) and at Kalambo the evidence is charcoal, ash, and wood. If, however, the latter site had not preserved the organic remains there would have been no evidence that fire had ever been associated. Worked stone showing fire-crackle is almost entirely nonexistent and the natural stones show no evidence of heat spalling. Quartz and quartzite pebbles lying on ground that is regularly burnt by bush fires and even specimens that have been intentionally heated in an ordinary temperature fire show little or no evidence of this other than some slight differential reddening of the cortex. It is, therefore, important to bear this in mind before rejecting the possibility that Acheulian man at Olorgesailie, Isimila, and other open sites used fire.

The problem of the concentration of natural rubble on some sites and its absence on others is also one requiring further investigation. On very gently sloping surfaces with almost imperceptible undulations and adjacent to a source of mechanically weathered material, land rubble tends to collect in the depressions. Such a phenomenon can be observed for example at Olorgesailie both in the prehistoric horizons and on the present-day surface as it can also at Kalambo Falls. Natural aggradation has first to be eliminated, therefore, before human agencies can be postulated. Where man is believed to have been the accumulating agent such rubble could have served several purposes. Some of it may have formed a means of defense in the sleeping places—ready to hand for throwing at wild animals coming too close to the camp. Alternatively if, as is believed, Earlier Stone Age man obtained some of his meat supply by driving and stoning, some of these concentrations may represent places where animals were cornered, killed, and butchered.

CONCLUSIONS

Most of what has been recorded in this paper refers to the Late or Upper Acheulian populations. Occupation sites of the Earlier and Middles stages are as yet very rare and none of these have been excavated by the horizontal grid so that nothing is known of the nature of the camping floors. From the apparently limited distribution of cultural material it is inferred that these Earlier Stone Age populations were comparatively restricted in distribution, but, by Late Acheulian times the wider distribution, the increased number of sites and the concentration of artifacts on them suggest both an acquisition of new territory in regions not previously occupied and an increase in overall population.

Acheulian industries would seem to be most concentrated in regions which today are grass and park savanna in East and South Africa and are invariably within easy distance of water or of places where it is evident that water would be available under a slightly increased rainfall. Lakes, pans, swamps, permanent and seasonal water courses, springs and deep water in limestone caves were all favored camping places. There is no evidence that the earlier Acheulian populations occupied country that is today arid or often waterless, on the one hand, or evergreen forest on the other. During the later stage, however, the populations spread into the semiarid and arid regions of the Horn, the Karoo, Kalahari, and fringes of the Namib as well as establishing themselves at favored localities in the Sahara. At the same time they made inroads on the peripheral parts of the Congo Basin into country that is today a Sudan savanna zone and penetrated along corridors into the forest itself opened by its recession from the main interfluves.

Occupation of arid regions can hardly have taken place unless accompanied by an amelioration of climate; such an increase in rainfall as has been previously postulated for the Kanjeran Pluvial. If Bond's (1955) and Brain's (1958) estimates of maximum and minimum rainfall intensity for Pleistocene "pluvials" and "non-pluvials" can be more generally applied in the continent it would seem that we should now think in terms of a 50% increase for the maximum of a pluvial and of a 25% decrease for the minimum of a dry period. Using such a basis and with the assistance of botanists and meteorologists it will be possible to reconstruct the general ecological conditions under which the Pleistocene population existed (cf. H.B.S. Cooke, this volume). Further work of the kind undertaken by Bond and Brain is needed, however, before it is possible to be certain that such quantitative estimates have general application.

If it was under "pluvial" conditions that Acheulian man penetrated the arid zones for the first time, the reverse must have been the case for his first occupation of forest country and we may, therefore, expect this to have taken place, not at the onset or height of a "pluvial," but during its decline and during the succeeding drier period. The typologically and technically final form of the Acheulian in the peripheral Congo savanna perhaps provides some confirmation of this.

In general, and on the basis of the above mentioned quantitative estimates it would seem that the Acheulian populations favored country with open rather than closed vegetation cover, supported by a semiarid climate. That is to say, much of the eastern and southern parts of the Central African plateau and the macchia vegetation of the southern Cape.

Edaphic gallery and fringing forests would have supplied many of the needs of the Stone Age populations and are likely to have been an important cause of their concentration on water courses. Their food sources would have been greatest and most easily come by at waterside camps. In addition, however, it must be mentioned that the stone for the making of tools which is very frequently found outcropping in such localities does not there become dehydrated and is thus easier to work properly than stone which has lost all its *eau de carrière*. This is an important factor recognized by all stone workers, both in ethnographical and other contexts, and it is easy to see how in prehistoric times such conditions could best be found in the immediate vicinity of water supplies.

These and many other problems of interpreting the ecosystems under which the Middle and early Upper Pleistocene human populations existed need close investigation and the aid of geologists, botanists, and zoologists, since reconstruction of the ecology is fundamental to an understanding of man's cultural achievements and social and economic attainments. This in turn rests upon team excavation work of selected occupation sites where faunal and/or vegetable remains are preserved in association with the cultural remains. The full occupational and technical significance behind any quaintitative analysis of the artifacts will only become apparent when this is examined in conjunction with the faunal and vegetable remains that formed an inseparable part of the camps of these hunting-collecting people.

The significance of the contributions to this symposium made by some of our colleagues in the behavioral sciences opens up new and extremely interesting possibilities for the interpretation of the faunal assemblages on the Acheulian living sites. Where such a high proportion of the biomass is provided by one or two large species, as it is in the Albertine Rift, the role of meat in the diet must have assumed a very different significance to that in moist forest or in semiarid veld, for example, where, in the one case, much of the biomass is in the trees and, in the other, consists of many smaller and swifter animals. It may be expected that the methods of securing the meat supply must have differed strikingly even within the small limits of variability permitted by the later Middle Pleistocene cultural level. Thus, our colleagues have provided us with most stimulating possibilities for further investigation, but we are only likely to be able to resolve some of the problems of interpretation with which we are faced by more excavation of selected living sites and a more critical and exact analysis of the total assemblages that they contain.

APPENDIX

Summary Classification of the Late Acheulian Assemblage (Eastern Africa)

I. Tools
  A. Shaped tools
    1. Implements characterized by *cutting edges* (some also with scraping edges), of any size, and having unifacial or bifacial trimming; usually exceed 10 cm. in length
      a. *Hand axes* (bifacial) and *pointed flakes* (unifacial). Forms = Lanceolate; narrow lanceolate; long ovate; asymmetrical ovate; pointed ovate; broad ovate; discoidal; cordiform; elongate lanceolate ("Larsen type"); ovate acuminate; single-shouldered ovate; single-shouldered narrow lanceolate; ovate with twisted point; triangular; elongate triangular; subtriangular; double-pointed diamond; truncated diamond; limande; elongate limande; asymmetrical limande; Micoque type; various; hand axes with cleaver-bit butts

b. *Cleavers* (bifacial) and *cleaver flakes* (unifacial). Forms = parallel sided (straight and guillotine bits); asymmetrical parallel sided (straight and guillotine bits); convergent sided (straight and guillotine bits); ultra convergent sided (straight and guillotine bits); asymmetrical convergent (straight or guillotine bits); shouldered convergent (straight and guillotine bits); divergent sided (straight and guillotine bits); splayed edges (straight and guillotine bits); burin-blow bit; side cleaver
c. *Knives*. Forms = pointed; end-and-side; disc; various
d. *Others* (elongate bifacial tools with cutting ends; round-bitted biface; round-bitted flake)
e. *Utilized flakes*

2. "Flake tools" characterized by *scraping edges*, unifacial or bifacial trimming; usually exceed 10 cm. in length
   a. *Side, or double side*
   b. *End, or double end*
   c. *Combined side and end*
   d. *Various*
   e. *Utilized flakes*
3. Other tools (uncategorized)
   a. *Chisel*. Forms = straight bitted; twisted bitted
   b. *Pushplane*. Forms = shallow-nosed; bitted
   c. *Various bifacial tools*. Forms = rectangular; elongate; discoid; end-notched; other
4. "Heavy duty tools" (various edge types and other forms; usually exceed 10 cm. in length)
   a. *Pointed: picks*. Forms = trihedral; bifacial; unifacial; flat-based, high back (unifacial or bifacial); flat-based beaked; point on cobble; block; spindle; core; various
   b. *Scraping edges*
      Core-scrapers. Forms = flat-based (steep, blunt); bevelled base (steep, blunt); keeled; double-edge
      Trimmed pebble or chunk
      (Utilized core)
   c. *Chopping edges*
      Choppers. Forms = pebble; chunk; flake; core; side; discoidal
      (Utilized core)
   d. *Other*
      Stone balls. Forms = missile; polyhedral; bolas
      (Hammerstones)
      (Anvils)
5. "Small implements," usually under 10 cm. in length
   a. *Scraping edges*. Forms = side, or double side; end, or double end; combined side and end; various; utilized flakes and chunks
   b. *Other* (Protoburin; burin; point, bifacial or unifacial; chisel-ended; elongate bifacial; bifacial with bevelled ends; discoid; borer; pointed; other)

B. Modified tools (trimmed and often utilized; large or small)

C. Utilized tools (identified only by signs of use; categorized by edge types)
   1. Hamerstones
   2. Anvils
   3. Utilized cores

4. Utilized flakes
5. Utilized chips
6. Others

D. Waste products (artifacts, but not tools; produced in manufacture of tools)
1. Cores. Forms = pebble; bashed chunk; formless; pyramidal; spindle; biconical; discoidal; struck (Levallois); double-struck; angle; large ("Tachengit")
2. Waste (flakes; chips and chunks)

BIBLIOGRAPHY

ALIMEN, [MARIE]-HENRIETTE
1955. *Préhistoire de l'Afrique.* Paris: N. Boubée et Cie.
1957. *The Prehistory of Africa* (rev. ed., translated by A. H. Brodrick). London: Hutchinson.

ARKELL, A. J.
1949. "The Old Stone Age in the Anglo-Egyptian Sudan," *Occ. Pap. Sudan Antiq. Serv.*, No. 1: 1–52.

BAKER, B. H.
1958. "Geology of the Magadi area," Degree Sheet 51, SW Quarter, *Geol. Surv. Kenya,* Report No. 42.

BISHOP, W. W.
1960. "A Review of the Pleistocene Stratigraphy of the Uganda Protectorate," Proc. Commission de cooperation technique en Afrique au Sud du Sahara (CCTA) Regional Committees for Geology, Leopoldville 1958 (1960), 91–105.
1962. "Pleistocene Stratigraphy in Uganda," *Geol. Surv. Uganda,* Report (in press).

BISHOP, W. W., and M. POSNANSKY
1960. "Pleistocene Environments and Early Man in Uganda," *Uganda J.*, 24:44–61.

BOND, G.
1948. "Rhodesian Stone Age Man and His Raw Materials," *S. Afr. Arch. Bull.*, 3:55–60.
1957. "The Geology of the Khami Stone Age Sites: Southern Rhodesia," *Occ. Pap. Nat. Mus. S. Rhod.*, 3:44–55.

BONÉ, E.
1958. "Préhistoire et Paléontologie du site néanderthalien de Saldanha-Hopefield," *Bull. Soc. Roy. Belge Anthrop. and de Préhist.*, 69:46–66.

BRAIN, C. K.
1958. "The Transvaal Ape-Man-Bearing Cave Deposits," *Transvaal Mus. Mem.*, No. 11.

BRAIN, C. K., and R. J. MASON
1955. "A Later African Chelles-Acheul Site near Nakop, South Kalahari," *S. Afr. Arch. Bull.*, 10:22–5.

BREUIL, H.
1944. "Le Paléolithique au Congo Belge d'après les recherches du Dr. Cabu," *Trans. Roy. Soc. S. Afr.*, 30:143–74.

Breuil, N., and J. Janmart

1950. "Les limons et graviers de l'Angola du nord-est et leur contenu archéologique," Companhia de Diamantes de Angola (Lisbon), *Publicaçoes Culturais*, No. 5:1–57.

Breuil, H., C. van Riet Lowe, and A. L. du Toit

1948. "Early Man in the Vaal River Basin," *Union S. Afr. Archeol. Surv.* (Archaeol. Series), 6:1–35.

Clark, J. D.

1950. *The Stone Age Cultures of Northern Rhodesia; with particular reference to the cultural and climatic succession in the Upper Zambezi Valley and its tributaries* (with a chapter by F. Dixey, appendices by H. B. S. Cooke, L. H. Wells, and G. Bond). Capetown: The South African Archeological Society.

1952. "Recent Prehistoric Research in the Somalilands," *Proc. 1st Pan-Afr. Cong. Prehist.* (Nairobi 1947): 146–64.

1954. *The Prehistoric Cultures of the Horn of Africa.* Cambridge: Cambridge University Press.

1954. "The Palaeolithic Hunters of Central Africa Some 100,000 Years Ago; Their Tools and Camp Sites Recently Discovered Near Lake Tanganyika," *Illust. London News,* May 29, 1954: 917–19.

1954. "An Early Upper Pleistocene Site at the Kalambo Falls on the Northern Rhodesia-Tanganyika Border," *S. Afr. Arch. Bull.,* 9:51–6.

1957a. "A Review of Prehistoric Research in Northern Rhodesia and Nyasaland." *Proc. 3d Pan-Afr. Cong. Prehist.* (Livingstone 1955): 412–32.

1957b. "The Importance of Distribution Maps in the Study of Prehistoric Cultures, and the Compilation of an Atlas of Prehistory for Southern Africa," *S. Afr. Mus. Assoc. Bull.,* 6:314–20.

1959a. "Further Excavations at Broken Hill, Northern Rhodesia," *J. Roy. Anthrop. Inst.,* 89:201–32.

1959b. *The Prehistory of Southern Africa.* London: Penguin Books.

1960. "Human Ecology during Pleistocene and Later Times in Africa South of the Sahara," *Cur. Anthrop.,* 1:307–24.

1962. The Kalambo Falls prehistoric site: an interim report. *Actes. 4e Pan-Afr. Cong. Préhist.* (Leopoldville 1959): 195–202.

Cole, G. H.

1961. "Culture Change in the Middle-Upper Pleistocene Transition in Africa South of the Sahara." Unpublished Ph.D. dissertation, Department of Anthropology, University of Chicago (microfilmed).

Cole, S. M.

1954. *The Prehistory of East Africa.* London: Penguin Books.

Cooke, H. B. S.

1941. "A Preliminary Survey of the Quaternary Period in Southern Africa," Union S. Afr. Bureau of Archaeology, *Archaeological Series,* 4:1–60.

1946. "The Development of the Vaal River and its Deposits," *Trans. Geol. Soc. S. Africa,* 49:243–60.

Davies, O.

1957. "The Old Stone Age Between the Volta and the Niger," *Bull. Inst. Franc. Afr. Noire* (series B), 19:592–616.

1959. "The Distribution of Old Stone Age Material in Guinea," *Bull. Inst. Franc. Afr. Noire* (series B), 21:102–8.

Drennan, M. R.

1954. "Saldanha Man and His Associations," *Amer. Anthrop.,* 56:879–84.

Fock, G. J.
1958. "Mapping of Prehistoric Sites," *S. Afr. Mus. Assoc. Bull.*, 6:13.
Goodwin, A. J. H.
1929. "The Montagu Cave: A Full Report of the Investigations of the Montagu Rock Shelter," *Ann. S. Afr. Mus.*, 24:1–16.
1933. "Some Developments in Technique During the Earlier Stone Age," *Trans. R. Soc. S. Afr.*, 21:109–24.
Howell, F. C.
1960. "Excavations at the Acheulian Occupation Site of Isimila, Iringa Highlands, Southern Tanganyika," *American Philosophical Society Yearbook,* 1959, pp. 481–5.
1961. "Isimila: A Paleolithic Site in Africa," *Scient. American,* 205 (October): 118–29.
Howell, F. C., G. H. Cole, and M. R. Kleindienst
1961. "Isimila: an Acheulian Occupation Site in the Iringa Highlands, Southern Highlands Province, Tanganyika" (geological section by E. G. Haldemann), *Actes. 4e Pan-Afr. Cong. Préhist.* (Leopoldville 1959): 43–80.
Janmart, J.
1953. "The Kalahari Sands of the Lunda (N.E. Angola), Their Earlier Redistributions and the Sangoan Culture," Companhia de Diamantes de Angola, Lisbon. *Publicaçoes Culturais,* No. 20.
Keay, R. W. J.
1959. "Vegetation Map of Africa South of the Tropic of Cancer," 24 pages and colored map 1:10,000,000. London: Oxford University Press.
Kleindienst, M. R.
1961a. "Components of the East African Acheulian Assemblage: An Analytic Approach," *Actes. 4e Pan-Afr. Cong. Préhist.* (Leopoldville 1959):81–111.
1961b. "Variability within the Late Acheulian Assemblage in Eastern Africa," *S. Afr. Arch. Bull.*, 16:35–52.
Korn, H., and H. Martin
1957. "The Pleistocene in South-West Africa," *Proc. 3d Pan-Afr. Cong. Prehist.* (Livingstone 1955):14–22.
Leakey, L. S. B.
1931. *The Stone Age Cultures of Kenya Colony.* Cambridge: Cambridge University Press.
1936. *Stone Age Africa: An Outline of Prehistory in Africa.* London: Oxford University Press.
1946. "A Prehistorian's Paradise in Africa: Early Stone Age Sites at Olorgesailie," *Illust. London News,* October 5, 1946: 382–3.
1949. "Tentative Study of the Pleistocene Climatic Changes and Stone Age Culture Sequence in N.E. Angola," Companhia de Diamantes de Angola, Lisbon. *Publicaçoes Culturais,* No. 4.
1951. *Olduvai Gorge.* Cambridge: Cambridge University Press.
1952. "The Olorgesailie Prehistoric Site," *Proc. 1st Pan-Afr. Cong. Prehist.* (Nairobi 1947): 209.
Lowe, C. van Riet
1938. "The Makapan Caves: An Archaeological Note," *S. Afr. J. Sci.*, 35:371–81.
1943. "Further Notes on the Makapan Caves," *S. Afr. J. Sci.*, 40:289–95.
1945. "The Evolution of the Levallois Technique in South Africa," *Man,* 45:49–59.
1948. "Cave Breccias in the Makapan Valley." In *Robert Broom Commemorative Volume,* Cape Town: Royal Society of South Africa, pp. 127–31.

1952a. "The Pleistocene Geology and Prehistory of Uganda. Part II: Prehistory," *Geol. Surv. Uganda*, Mem. No. 6.

1952b. "The Vaal River Chronology, An Up-to-date Summary," *S. Afr. Arch. Bull.*, 7 (28):1–15.

1952c. "The Development of the Hand-axe Culture in South Africa," *Proc. 1st Pan-Afr. Cong. Prehist.* (Nairobi 1947): 167–77.

1954. "The Cave of Hearths," *S. Afr. Arch. Bull.*, 9:25–29.

MABBUTT, J. A.

1956. "The Physiography and Surface Geology of the Hopefield Fossil Site," *Trans. R. Soc. S. Afr.*, 35:21–58.

1957. "Some Quaternary Events in the Winter Rainfall Area of the Cape Province," *Proc. 3d. Pan-Afr. Cong. Prehist.* (Livingstone 1955): 6–13.

MALAN, B. D., and H. B. S. COOKE

1941. "A preliminary account of the Wonderwerk Cave, Kuruman District," *S. Afr. J. Sci.*, 37:300–12.

MALAN, B. D., and L. H. WELLS

1943. "A Further Report on the Wonderwerk Cave, Kuruman District," *S. Afr. J. Sci.*, 41:258–70.

MASON, R. J.

1959. "Some South African Stone Age Cultures," *Nature*, 183:377–9.

1961a. "The Earliest Tool-makers in South Africa," *S. Afr. J. Sci.*, 57:13–16.

1961b. "The Acheulian Culture in South Africa," *S. Afr. Arch. Bull.*, 16:107–10.

MORTELMANS, G.

(n.d.). "Le Congo préhistorique." Colored map.

1957. "La préhistoire du Congo Belge," *Revue de l'Université de Bruxelles*, 2–3 (Jan.–April): 1–53.

OAKLEY, K. P.

1954. "Evidence of Fire in South African Cave Deposits," *Nature*, 174:261.

1955. "Fire as Palaeolithic Tool and Weapon," *Proc. Prehist. Soc.*, 21:36–48.

O'BRIEN, T. P.

1939. *The Prehistory of Uganda Protectorate* (with a chapter on the Pleistocene succession by J. D. Solomon, and an appendix on the mammalian fossils by A. T. Hopwood). Cambridge: Cambridge University Press.

PICKERING, R.

1960. "A preliminary note on the Quaternary Geology of Tanganyika," Proc. Commission de cooperation technique en Afrique au Sud du Sahara (CCTA) Regional Committees for Geology, Leopoldville 1958:77–89.

POSNANSKY, M.

1959. "A Hope Fountain site at Olorgesailie, Kenya Colony." *S. Afr. Arch. Bull.*, 14:83–89.

POWER, J. H.

1955. "Power's Site, Vaal River," *S. Afr. Arch. Bull.*, 10:96–101.

RATTRAY, J. M.

1960. "The Grass Cover of Africa." 168 pages and colored map, 1:10,000,000. *FAO Agric. Studies*, No. 49. Rome: United Nations Food and Agricultural Organization.

RUDDOCK, A.

1957. "A Note on the Relation Between Chelles-Acheul Implements and Quaternary River Terraces in the Valleys of the Coega and Sundays Rivers, Cape Province," *S. Afr. J. Sci.*, 53:373–7.

SINGER, R.
1957. "Investigations at the Hopefield Site," *Proc. 3d. Pan-Afr. Cong. Prehist.* (Livingstone 1955): 175–82.
SINGER, R., and J. R. CRAWFORD
1958. "The Significance of the Archaeological Discoveries at Hopefield, South Africa," *J. Roy. Anthrop. Inst.* 88:11–19.
SÖHNGE, P. G., D. J. L. VISSER, and C. VAN RIET LOWE
1937. "The Geology and Archaeology of the Vaal River Basin," *Union S. Afr. Geol. Surv.*, Mem. No. 35.
SPURR, A. M. M.
1955. "The Pleistocene Deposits of Part of the Kagera Valley, Bukoba District," *Geol. Surv. Tanganyika*, unpublished report AMMS/27.
SUMMERS, R. F. H.
1957. "Archaeology in Southern Rhodesia, 1900–1955," *Proc. 3d. Pan-Afr. Cong. Prehist.* (Livingstone 1955): 396–411.
SUMMERS, R.
1960. "Environment and Culture in Southern Rhodesia: A Study in the 'Personality' of a Land-locked Country," *Proc. Amer. Phil. Soc.*, 104:268–92.
SUMMERS, R., and C. K. COOKE
1959. "Archaeological Survey of Southern Rhodesia." Supplement to *Annual Report for 1958*, Commission for the Preservation of Natural and Historical Monuments and Relics.
VISSER, D. J. L., and C. VAN RIET LOWE
1955. "The Geology and Archaeology of the Little Caledon River Valley," *Union S. Afr. Geol. Surv.*, Mem. No. 47.
WAYLAND, E. J.
1934. "Rifts, Rivers, Rains and Early Man in Uganda," *J. Roy. Anthrop. Inst.*, 64:333–52.
1935. "The M-horizon. A Result of Climatic Oscillation in the Second Pluvial Period," *Bull. Geol. Surv. Uganda*, 2:69–76.

SOME ECOLOGICAL FACTORS EFFECTING HUMAN POPULATIONS OF SUB-SAHARAN AFRICA

J. HIERNAUX

I SHALL TAKE UP THE SUBJECT from the physical anthropology point of view, i.e., examine the influence of some ecological factors on the physical differences exhibited by the human populations of sub-Saharian Africa.

Human physical characteristics are conditioned by two groups of factors, heredity and environment, the relative importance of which differ according to the characteristic under consideration. Blood groups, for instance, are entirely conditioned by the individual's genotype; stature and weight are conditioned both by genotype and environment, the relative influence of genotype being more marked on stature than on weight.

Most biologists admit that environment does not influence genotype and cannot, therefore, even from generation to generation, directly affect the hereditary patrimony. It determines, rather, the degree to which the genetic potentialities of the individual inherited at the moment of conception are realized. From that moment, the environment acts in a differentiating manner: at first indirectly by affecting the mother, then, after birth, directly. Two genetically identical individuals—monozygotic twins—will differ as a result of mesological influences. Similarly, two genetically identical populations (i.e., presenting the same gene frequencies) will differ in the means or the frequencies of the characters on which the influences act.

The principal ecological factors in question are nutrition and pathology.

Nutritionists have published tables of the daily needs of the human organism in total calories, proteins, fats, and carbohydrates, in vitamins, and in soluble salts. These tables are certainly subject to considerable revision, since the determination of the optimum nutritional intake is extremely delicate. It cannot be universal, but has to be adjusted to each particular case, particularly in relation to the daily output of energy. Climate is undoubtedly one of the variables that intervenes in the optimal alimentary ration; others are morphology and physiology. But precise definition in these matters has not yet been achieved. It is possible that chronic malnutrition, by retarding growth, results in a reduction of size of the individual adult, and diminishes his energy needs, as Newman (1961) suggests. For example, the Scientific Council for Africa South of the Sahara

(1956) noted the rarity of signs of calcium deficiency among the Bantu, in spite of their meager intake of substances that control the calcium metabolism.

Whatever our knowledge in this field, it is certain that overfed populations are not found in sub-Saharan Africa, whereas undernourishment and specific deficiencies are frequent. Nutritional factors, therefore, play an important role there in the physical state of human groups and in the differences that separate them.

For example, two groups of Hutu from Rwanda, comparable from the genetic as well as the cultural point of view and sharing the same agricultural way of life, live the one above 2,000 m. altitude where the land is more fertile and better moistened, and the other below that altitude. They show highly significant mean differences of 7 kg. in weight, 31 mm. around the thigh, 3 mm. in the size of the head, and 18 mm. in the anteroposterior diameter of the thorax, all in favor of the population from the higher altitude (Hiernaux 1954).

Nutritional deficiency can occur from total calorie intake, or from one or more nutritional substances.

It appears that many African populations receive a number of calories which is below the optimum, at least at certain periods of the year.

For the majority of African agricultural groups, rations are greatly diminished between one harvest and the next. Some years these lean months are prolonged or intensified and famine sets in.

In Ruanda-Urundi, a land of high population density for Negro Africa, terrible famines that claim thousands or even tens of thousands of victims occurred periodically before the European administration introduced certain controls.

It is probable that with the two Hutu groups, a difference in the mean energy supplied is involved. This shows to what extent an African population can be modified by a quantitative variation in nutrition, other factors remaining constant.

If general undernourishment (perhaps periodic) plays an important role in Africa, deficiencies in specific nutritional substances have drawn the attention of specialists even more.

Protein deficiency is common among agricultural African populations, especially those whose diet consists mainly of vegetables that are poor in proteins. There is a strong contrast between the populations who eat mainly beans and peas, which are rich in proteins, and those for whom manioc, very poor in proteins, forms the basis of the diet. Among the latter, if no protein-rich foods are added to the manioc, protein deficiency becomes so serious that kwashiokor, a nutritional disease usually fatal without treatment, strikes a variable percentage, sometimes very high, of children. Those who do not die, or those who show only slight symptoms, are no less indelibly marked by it, and are far from showing, as adults, a physical development (and without doubt also psychical) corresponding to their genetic potential.

Since there are few marked individual differences in diet, whole rural populations carry the scars of a childhood protein deficiency.

Pastoral populations, or those that carry on agriculture and cattle-herding simultaneously, find a vital source of proteins in beef and milk. Of course, the meat must be eaten and the production of milk must be adequate. In Rwanda, where cattle is superabundant, cows have a primarily social value: a large part of the population never, or only a few times per year, eats meat. The milk yield is so poor that its consumption, although highly prized, is very low in a large majority of families.

The other domestic animals, goat, sheep, pigs, fowl, etc., rarely are a satisfactory source of protein, since they are either not raised in sufficient numbers, or their products are not part of the nutritional habits. Many populations milk neither goats nor sheep and consume no eggs; in Rwanda, for the Tutsi and the Hutu, eating mutton is considered a repugnant act, which only the Twa (pygmoid) do.

In numerous and vast regions game is actually too scarce to provide sufficient protein. The contribution of insects (termites, caterpillars, grasshoppers, etc.) is negligible.

Populations that have access to abundant fish have a great advantage from the point of view of animal protein consumption as compared to many other African groups, and their physical condition often reflects this.

No carbohydrate deficiency has been charged to Africa by nutritionists, but in certain populations, the low consumption of fat may result in specific troubles.

Vitamin deficiency abounds in many regions, particularly vitamin A, where the diet is poor in fat and vegetable carotenoids—palm oil saves vast areas of central Africa from this deficiency—and multiple vitamin B. It is probable that the intense rays of the sun compensate for the vitamin D deficiency of many diets, which would explain the rarity of rickets.

The dietary intake of soluble salts depends in part on the nature of the soil. Iodine deficiency, with its often dramatic physical and psychic consequences, appears where the soil is very poor in this element; calcium intake depends both on the composition of the water and the presence in the diet of foods rich in calcium.

Numerous pathological factors are also capable of influencing the average physique of the populations of vast African regions; mostly it is a matter of chronic parasitic diseases. Where not fatal, malaria, ancylostomiasis, bilharzia, onchocerciasis, to cite only a few, have the tendency to chronically debilitate their carriers and thus produce an effect similar to that of malnutrition. Very often parasitism and malnutrition are synergetic.

In certain biotopes favorable to *Glossina*, sleeping sickness is so severe and widespread that human occupation is all but impossible.

What kind and degree of differences can be created by the combined action of nutritional and pathological factors?

The preceding example shows impressive differences in one ethnic group. However, it is not a matter of extreme variations in environment. The Hutu at a moderate altitude are far from being the most disfavored of the African groups;

they consume significant quantities of vegetable proteins, and kwashiorkor is rare among them. The differentiating factor here seems to be, from the nutritional point of view, a difference of mean energy intake, plus a difference in the severity of malarial infestation, less strong in higher regions. More drastic differences in environment would certainly produce even more marked morphological differences within the same ethnic group.

Note that our two samples of Hutu do not differ significantly with regard to stature. It is probable, as Chamla, Marquer, and Vacher (1959) indicated when reviewing opinions on this subject, that qualitative nutritional differences can affect the average stature of groups having the same hereditary patrimony.

Several authors admit influence of the environment on the stature of African groups, basing their opinions on a statement by Torday and Joyce (1910) that the Kuba pygmoids in Kasai, who actually live in the savanna, would be bigger than those remaining in and suffering the detrimental effect of life in the forest. I had occasion to study this group of Twa and found them seriously undernourished and ravaged by kwashiorkor (Hiernaux 1954b). To date, they have not been compared with groups of like genetic composition, if there are any, and no environmental influence on their stature, in whatever sense it may be, can therefore be inferred from the existing data.

The important factors of direct influence of the environment on the average stature of African populations remain to be studied, and it is to be hoped that the various investigations will furnish us with significant data on the equally important problem of environmental influence on the rhythm of growth.

It will always be difficult to establish dependable hypotheses on the state of nutrition of the ancient African populations.

Given waters rich in fish, the fishing populations were assured of abundant animal proteins, which varied in the diet of herding populations living on milk and meat. With the same energy supply, the nutrition of agricultural populations would be very different if the basic food is cassava, millet, or bean.

Before the Neolithic, the economy of the large majority of African populations was based on hunting and collecting. This is still true for some groups, and they give us indications of the possible physical condition of those earlier populations. The example of the Bushman is misleading, because they have recently been pushed into arid regions where their very survival testifies to human adaptability and ingenuity. In contrast, many groups of pygmies still carry on, in the midst of the equatorial forest, an existence of hunters-collectors in which their symbiosis with natural resources does not seem to have been notably disturbed. This is the case with the Mbuti of the Ituri forest. Their state of nutrition, to judge from their muscular development, is excellent; they present no signs of avitominoses and do not seem to pay an excessive toll to chronic diseases. Nevertheless, the primary equatorial forest is not rich in game, and it seems to have repulsed African man during the major part of his evolution.

No doubt the hunting-collecting populations of the Stone Age enjoyed good nutrition and strength during the periods when climatic conditions did not make game scarce nor dried up wild vegetation. There was probably at this stage of human evolution a more rigorous correspondence than now between the density of human population and food resources.

In relation to the major theme of this symposium, let us now come to the problem of the influence of environment not only on individual realization of hereditary potentialities, but on the evolution of the hereditary patrimony of populations. This is expressed, when possible, in gene frequencies, which should not distract from the characters whose genetic determination is not yet entirely understood.

Mankind is certainly homozygous for a large number of genes, but it is variation that interests us here, i.e., those genes that present the possibility of several alleles. In certain cases, the various alleles show up in all human populations, in varying frequencies; in other cases, certain alleles show up only in certain populations (the sickle-cell anemia gene, for instance).

Four evolutionary mechanisms of the gene frequencies are generally recognized today: mutations, cross-breeding, random genetic drift, and natural selection. Let us see to what extent they are responsive to ecological conditions.

The importance of *mutations* to evolution is known; they create new genic material. Does environment effect their frequency? Although the rate of mutations has been greatly increased in the laboratory by physical (radiation) or chemical agents, the influence of mutative factors under natural conditions is still hardly understood. In certain regions the individual is exposed, by the nature of the soil, to more intense radiation (ten times more in one zone of Kerala in India than in the United States). It is possible that in African regions rich in uranium minerals (Katanga, for instance) similar increased radioactivity raises the rate of mutation. The importance of mutative agents in food, in the air breathed, and in tobacco is still ignored; it is suspected that fluctuations of energy, for instance variation in the temperature of the gonads, can be mutative.

The ecology of the rates of mutation in man is, then, only in its beginnings, although promising very important discoveries. Geneticists consider it possible that mutagens that act selectively on certain genes may be discovered (Stern 1960).

The influence of environment on *cross-breeding* is indirect: it aids or hinders coming in contact of different populations, which is the basic condition of cross-breeding. All mesological variations (climatic or other) that disrupt the balance of population and environment tend to provoke migrations, either because the new environment no longer permits of human occupation of the same density (at least where mastering natural forces is concerned), or because it no longer presents the conditions to which local populations are culturally adapted. The variations in extent of the primary equatorial forest and of the arid zones, and the degree of aridity of these latter, were certainly very important in causing the

African populations to shift about. More knowledge of man's past in the Sahara will shed light on this point.

Among the populations most closely tied to a biotope should be cited the herding populations, moving often and hardly sedentary. Over the years, they are likely to make long migrations occasioned by geographical variations—ecological conditions favorable to their flocks (abundance of food, availability of water, absence of *Glossina*). These migrations are generally followed by cross-breeding between neighboring populations, and this is habitually accompanied by acculturation. It may reach the point of the loss of individuality by one of the two groups (shepherd and non-shepherd). Such a reciprocal process with the tendency to ultimate fusion into one unique population, is happening today in Rwanda.

It is probable that desiccation of the pastoral Sahara (the bovid epoch of the rupestral paintings) brought about population movements that involved, in the long run, the greater part of Africa south of the Sahara.

Random genetic drift depends on the size of the population. This evolutionary factor of gene frequencies becomes more important in proportion as the size of the population decreases. It will therefore be influenced by the environment to the extent that the latter causes the formation of very small communities.

The pre-Neolithic hunting and gathering economy meant the scattering of small nomadic groups within a hunting territory. Ignoring the degree of endogamy of these groups, it is however probable that the conditions for random genetic drift often occurred. Certain environmental modifications must have reinforced the isolation of groups; others, on the contrary, must have facilitated genic exchanges or favored the reunion of many isolates into a collectivity less sensitive to random genetic drift.

If one considers the wide variations of gene frequencies observed in three generations in small isolated groups today (like the Dunkers of Pennsylvania studied by Glass, 1954), it must be admitted that random genetic drift played a very important role in the frequency of certain human characteristics, and that, in consequence, any contributing ecological factor influenced the anthropological map of Africa.

The thinning of an endogamous group is not alone in favoring random genetic drift; the same effect is achieved when a small group splits off from a larger stable population and establishes itself elsewhere. The daughter population can differ notably in certain ways from the parent population. This process must have occurred very frequently in Africa during the periods of expansion of certain groups. There is an example in the explosive growth, begun several millenia ago, of the Bantu-speaking populations (as defined by glottochronology) or that, archaeologically defined, of the dimple-based pottery population in the beginning of the Metal Age (two expansions that are perhaps linked, as I proposed hypothetically in 1959). The oral history of the Bantu tribes abounds in examples of emigration of very small groups splitting off from the mother tribe to found new ethnic entities of rapidly increasing size elsewhere. Their new habitat was often thinly populated, or not at all, and this discontinuity in certain African biotopes (the Congolese equatorial forest, for example) is in itself an important ecological fact.

All mesological factors favoring similar emigrations of small groups, then, may act on the evolution of their hereditary patrimony in permitting random genetic drift to come into play, and also, eventually, in bringing them into contact with other populations (for example, the small groups of pygmies which certain expansions of the agricultural Bantu assimilated, or have become symbiotic with, in the Congolese equatorial forest).

It remains to examine the influence of the environment on the fourth mechanism of the evolution of genic frequencies: *natural selection.* This occurs when the frequency of a gene or of a hereditary characteristic varies from one generation to the next because the carriers of this characteristic or gene contribute proportionately either more or less than others to the succeeding generation. Natural selection, then, depends on fertility and viability.

If the carriers of a gene are more fertile than those of other alleles, or if their offspring die less often before the end of their fertile period, the frequency of this gene increases from generation to generation, and the frequency of the allele or alleles diminishes.

Environment intervenes in natural selection when it is unfavorable to the fertility or the viability of the carriers of a gene, an assemblage of genes, or a hereditary characteristic.

Certain genes produce pathological characteristics that cause great fragility of the organism; hemophilia, for instance. All environments are unfavorable to the carriers of such genes, but not in the same way. A rude way of life, in a physical environment that subjects the organism to debilitating ordeals (nutritional deficiencies, famines, or malaria, for example) will cause more deadly malformations and hereditary diseases than an easy life in an environment where their lethality is lower thanks to a high level of medicine and hygiene.

Social reactions to the vitiated (often conditioned by the rigor of the environment) intervene in the severity of natural selection. While certain ethnic or social groups do the impossible to keep them alive, others do very little; some even kill them. If they reach adult age the society can still intervene by tolerating or hindering their reproduction.

Sociological and ecological factors hinder the action of a whole series of hereditary pathological characteristics, to determine the extent of selection.

The environment sometimes acts in opposite ways on the same gene, depending on whether it is a homozygote or a heterozygote. This certainly seems to be the case with the one responsible for sickle-cell anemia. This anomaly of the hemoglobin is determined by a gene S, allele of a gene A which determines normal hemoglobin. AA individuals are normal; SS individuals develop serious anemia in childhood, often fatal (it is so almost always in African countries where childhood is subject to a series of nutritional and pathological agressions; SS adults are very rare there). An intense selection, then, acts against the SS genotype. The heterozygotes AS are detected by a blood test; their genotypic constitution does not seem to handicap appreciably. On the contrary, it is admitted in a quasi general fashion today that they show a stronger resistance to malaria than AA individuals.

Bearing in mind the high mortality which malaria causes among children in many regions of Africa, it is conceivable that the result of the two selective actions in force, the one against the genotype SS by the almost total mortality in childhood and the other in favor of the genotype AS by its lesser susceptibility to malaria, is an increase in the frequency of the gene S, up to a certain rate of equilibrium, in malarial populations.

The sickle-cell anemia gene is present in almost all African Negro groups, in some populations of India and Arabia, in Mediterranean Europe, and in American Negroes. In Africa its frequency varies considerably from one population to another, and there is every reason to believe that the severity of malaria in past generations is one of the principal factors in this variability, along with cross-breeding and eventually of random genetic drift. All ecological variation (changing of local conditions or migration) that modifies the lethal nature of malaria (partially determined by the multiplication of the *Anopheles*) will, in a few generations, cause an evolution of the frequency of the sickle-cell gene in the populations which have it. The introduction of effective methods of prophylaxy and treatment or of eradication of the anopheles is a new example of human action on the intensity of selection and even on its direction: it can cause an inversion of selection, in making it act in favor of gene A in the population where previously it was favorable to gene S.

Likewise, another gene seems to confer protection against tertian malignant malaria, and its distribution in the ancient world coincides largely with the distribution of this disease. It is the sex-linked gene that determines a deficiency of glucose-6-phosphate dehydrogenase (Allison 1961). Although its presence gives relative protection against *Plasmodium falciparum,* it sensitizes the organism to the ingestion of certain antimalarials (like primaquine and pamaquine) and of the beans *Vicia faba,* which can cause serious hemolysis in the carriers of the gene. It is obvious that certain aspects of the environment have a pronounced selective action on the frequency of such genes, particularly, as with the sickle-cell anemia gene, those which influence malarial infestation.

Although hereditary characteristics, particularly those of simple genetic determination and of easily detected genotype like sicklemia, are studied by geneticists who wish to demonstrate obvious cases of natural selection, this evolutionary factor is also certainly active with regard to phenotypic characteristics of complex genetic determination, more frequently not clarified. Research on this action, often so important from the point of view of evolution, should not have to wait for a complete determination of genetic make-up.

The direct statement of natural selection in relation to human morphological characteristics has never, to my knowledge, been irrefutably made. In contrast, the hypothesis of natural selection is often formulated on the basis of a correlation between one or more variables of the environment and a morphological characteristic, or a complex of them. It supposes that this correlation is the reflection of an evolution of the frequencies of responsible genes toward an optimum adaptation to the environment. Such a hypothesis is reinforced when a physiological advantage of the characteristic in question is demonstrated in relation to the environment in which it is most frequently found. It is then easy to imagine that this physiological advantage is reduced mortality before the end of the reproductive period. The process of adaptive selection certainly seems to be the base of Bergmann and Allen's ecological "laws" which, to a certain extent, human populations follow. According to these laws, translated into terms directly usable by physiologists, the ratio of the body's mass to its surface tends to be lower in the biotopes where, at certain periods of the year at least, the mechanisms for eliminating body heat are put to the test. These mechanisms are eased when the ratio is low, for obvious physical reasons, and this can be realized by the cubic reduction of mass to the square reduction of surface; by a slender morphology, a low fat mass, or by a higher development of the limbs and extremities in relation to the trunk (the former, being columns of the lowest diameter, are much better radiators than the latter).

Newman (1953) found that, in the main, these ecological laws, established by zoologists, applied to the Amerindians who for about 15,000 years ranged over a vast continent with great climatic variation. (Perfect concordance cannot be expected, given the mobility of human populations). A distinct correlation appeared between mean stature and latitude; if Eskimos are small, their legs are very short in relation to the trunk.

On a world-wide level, Roberts (1953) found a correlation between weight and latitude. Malnutrition, so prevalent in the tropics, is certainly one factor, preventing any positive conclusion that the hereditary component of weight varies by natural selection in relation to the mean temperature of the coldest month, as is suggested by the data of Newman (1961) on the Amerindians.

The populations of sub-Saharian Africa all have, by various morphological means, a low mass-surface ratio, some, like the pygmies, by a great over-all reduction of size; others, like the Tutsi, the Hima, and the Masai, by a slender morphology with very long legs in relation to the trunk, a whole range of populations being found at various stages between these two extremes.

The hypothesis of evolution by adaptive selection in climates that severely test the mechanism of body-heat radiation at certain periods of the year at least, is insistent. It is supported by certain experimental verifications. Several works show the physiological superiority of the African over the European when it comes to intense physical effort in a hot climate (for example Wyndham, Bouwer, Devine, and Paterson 1952, and Thomson 1954). On the average, the African, even while perspiring less, maintains a smaller difference between his internal temperature (which itself remains low) and his skin temperature. The dissipation of heat is clearly more efficient in the African. The validity of these conclusions has, however, been doubted by several authors; a close critical analysis of the question was made by Barnicot (1960).

Without decisive experimental results, the hypothesis that the characteristics that lower the mass-surface ratio in Africans result from adaptive evolution can be retained. The morphological means used are diverse, as we have seen. There are few differences of body structure between human groups as striking as that separate the Tutsi and the pygmies. Why such an extreme difference? Several reasons are conceivable. First, adaptive selection can operate only on existing genetic materials. The end toward which it is oriented, in the case of the above hypothesis, not being narrowly allied to a given morphology, it is conceivable that it could, with the same efficacy, exaggerate either this or that characteristic according to the morphology of the original population. On the other hand, although almost all the sub-Saharian climates are hot, they can differ greatly in other ways—for example, humidity, ventilation, insolation. So it is probable that the selective pressures which acted on man in the equatorial forest were very different from those which acted in the Sahel. The hot, dry conditions of the African deserts seem to favor a slim morphology low in subcutaneous fat; Schickele (1947) showed the advantages of these characters for survival. In contrast, there is a certain association, in Africa as in Asia and America, between small size and ombrophilous equatorial forest. My observations on the Bantu tribes that emigrated to the forests of the eastern Congo show a tendency to dwarfing, their cross-breeding with pygmies not seeming to explain the phenomenon entirely (Hiernaux 1956). It is difficult not to bring in natural selection in the genesis of the pygmies, even if we have not been able to determine the ecological factor or factors responsible. In a like manner we sought in the hot steppe climate, dry and very insulated, the factors of the linear morphology of the Tutsi, Hima, or Masai shepherds. Again, adaptive selection is called upon to explain the concentration of dark skins in low latitudes and also, without being able to define the mechanism, the negative correlation between the nasal or facial index and the geographical latitude, found by Newman in the Amerindians.

Nevertheless, as with malformations, man can exert sexual selection of normal morphological characters. If certain characteristics are particularly prized, it is possible that their carriers find spouses or partners more frequently or more precociously, and produce a higher average number of children.

Such sexual selection could act in the direction of adaptive selection, or in the opposite direction, or on neutral characters in relation to adaptation. Here is a research subject of great importance; it has hardly been touched, and we know almost nothing of the role sexual selection can play in the genesis of the hereditary patrimonies of present Africa. It was evoked by Tobias (1957) with regard to the steatopygia of the Bushman; the latter considered steatopygia very seductive. Coon (1955) proposes to see in this character an adaptive trait, a hypothesis which Tobias (it seems) does not exclude. Maybe, he writes, steatopygia is an adaptive character associated with a hunting-collecting economy, with its hazards of periodic food restrictions, above all in fats and proteins, while Coon notes that steatopygia constitutes a fat reserve which does not impede thermo-regulation.

Brues (1959) suggested the existence of cultural selection according to which morphology adapts itself not only to the physical environment, but also to the dominant type of activity, determined by cultural habits. For pre-Neolithic populations, the kind of arms and hunting methods would be the factors to which such selection was sensitive. Tobias (*op. cit.*) attempts to apply these views to the Bushman and puts forth the hypothesis that their small size and light structure results from a selection with a view to adapting them to the long pursuits of animals superficially wounded by poisoned darts (implied by their hunting methods).

Ecological factors play a role, we have seen, in all the evolutionary mechanisms of hereditary patrimony.

If it has been necessary, for clarity of exposition, to study these mechanisms one by one, it must be remembered that they very often intervene simultaneously. The migration of a small group (from which random genetic drift) to a new biotope (from which reorientation of selection on the genes of the patrimony and on the mutants) being put in contact with new populations (from which crossbreeding) is an example of the habitual intricacy of evolutionary factors. In such conditions we must not hope to easily make clear the influence of environment on human evolution. The great complexity of environment and its interaction with human, moral, social, cultural, and sexual factors that equally play a selective role further complicate the question. All hypotheses based on an actual distribution must take into account the past of each population. As arduous as it is, the research of the relations between physical man and his environment are greatly worth the labor given to it; it concerns not only the past and present of man, but also his future.

BIBLIOGRAPHY

Allison, A. C.
1961. "Genetic Factors in Resistance to Malaria," *Ann. New York Acad. Sci.*, 91:711–19.

Barnicot, N. A.
1960. "Climatic Factors in the Evolution of Human Populations," *Cold Spring Harbor Symposia on Quant. Bio.*, 24:115–29.

Brues, A.
1959. "The Spearman and the Archer: An Essay on Selection in Body Build," *Amer. Anthrop.*, 61:457–9.

Chamla, M. Cl., P. Marquer, and J. Vacher
1959. "Les variations de la stature, en fonction des milieux socioprofessionnels," *L'Anthropologie*, 63:37–61, 269–94.

Coon, C. S.
1955. "Some Problems of Human Variability and Natural Selection in Climate and Culture," *Amer. Nat.*, 89:257–80.

Glass, B.
1954. "Genetic Changes in Human Populations, Especially Those Due to Gene Flow and Genetic Drift," *Advances in Genetics*, 6:95–139.

Hiernaux, J.
1954a. "Influence de la nutrition sur la morphologie des Bahutu du Ruanda," *Actes. 4e Cong. Internat. Sci. Anthrop. et Ethnol.* (Vienna 1952), vol. 1 (*Anthropologica*), pp. 157–62.
1954b. "Etat de nutrition des Kuba (Kasai)," *Zaire*, 7:719–27.
l'Afrique Centrale: Ruanda-Urundi et Kivu," *Ann. Mus. Roy. Congo Belge*,
1956. "Analyse de la variation des caractères physiques humains en une région de
Anthropologie 3.
1961. "Les débuts de l'Age des Métaux en Afrique Centrale," *Actes. 4e Pan-Afr. Cong. Préhist.* (Leopoldville 1959): 381–90.

Newman, M. T.
1953. "The Application of the Ecological Rules to the Racial Anthropology of the Aboriginal New World," *Amer. Anthrop.*, 55:311–27.
1961. "Biological Adaptation of Man to the Environment: Heat, Cold, Altitude, and Nutrition," *Ann. New York Acad. Sci.*, 91:617–33.

Roberts, D. F.
1953. "Body Weight, Race, and Climate," *Amer. J. Phys. Anthrop.*, 11:533–8.

Schickele, E.
1947. "Environment and Fatal Heat Stroke: An Analysis of 157 Cases Occurring in the Army in the U.S. During World War II," *Military Surgeon*, 100:235–56.

SCIENTIFIC COUNCIL FOR AFRICA SOUTH OF THE SAHARA
1956. *Nutritional Research in Africa South of the Sahara*. Publication 19, London: C. S. A.

STERN, C.
1960. *Principles of Human Genetics*. San Francisco and London: Freeman.

THOMSON, M. L.
1954. "A Comparison between the Number and Distribution of Functioning Eccrine Sweat Glands in Europeans and Africans," *J. Physiol.*, 123:225–33.

TOBIAS, P. V.
1957. "Bushmen of the Kalahari," *Man*, 57:33–40.
(n.d.) "Bushman Hunter-Gatherers: A Study in Human Ecology" (to appear as a chapter in *Ecology of South Africa*, H. B. S. Cooke, D. H. S. Davis, and B. de Meillon, eds.). Lochem, Netherlands: W. Junk.

TORDAY, E., and T. A. JOYCE
1910. "Notes ethnographiques sur les peuples communément appelés Bakuba, ainsi que sur les peuplades apparentées," *Ann. Mus. Congo Belge*, vol. 1.

WYNDHAM, C. H., W. v. D. M. BOUWER, M. G. DEVINE, and H. E. PATERSON
1952. "Physiological Responses of African Labours at Various Saturated Air Temperatures, Wind Velocities and Rates of Energy Expenditure," *Jour. App. Physiol.*, 5:290–8.

TRANSCRIPT OF DISCUSSIONS

SESSION I: JULY 10 (3:00–6:45 P.M.)

This session took the form of a general outline of Africa as a whole and by special regions, from the point of view of physical features and basic ecology.

COOKE began with a factual, geological outline of the whole continent stressing the importance of basins of sedimentation and, in Late Tertiary and Pleistocene times, of sediment traps formed by the Rift Valley system. Emphasis was laid on the fact that the effects of temperature and rainfall on vegetation can be considerably modified by basic geology, physical features, and the nature of the soil.

ARAMBOURG spoke on northern Africa. The area could be broadly considered under two headings:

1. The northwestern, Atlas Mountain, area, with an alpine climate.

Although mammalian fossils from the Upper Miocene here resemble the Pontian of Europe, they are definitely distinct species. Four to five thousand meters of marine deposits cover the Pontian-type fauna in this area and it is thus earlier than the Pontian in Europe, being of mid-Miocene age. It probably entered Africa from Asia.

Unfortunately the present political situation has made impossible the further investigation of Lower to Middle Miocene mammalian fossils in terrestrial beds in southern Tunisia.

2. The flatter area to the southeast of the mountains, including Hoggar, Tibesti, and the Mediterranean.

No mammalian fossils have been found in the continental deposits of Middle Pliocene age. The first fossil monkeys occur in the Nile valley area (Fayum).

At Moghara a mid-Miocene fauna occurs and about 400 km. south from the coast, marine and continental beds lying horizontal from the sea include all series from the Cretaceous upwards.

* Edited by Mrs. Betty Clark, with assistance of J. D. Clark and F. Clark Howell. One of the editors (F.C.H.) has added a few footnotes in proof, since further field and laboratory research has made it necessary to modify some of the statements made in the conference discussion two years previously.

New discoveries of Eocene fauna at Djebel Quoquin resemble that of the Fayum. At Zelten there occurs a rich Burdigalian fauna and a flora of all ages, including Eocene wood. Again, there are no connections with Europe, though possibly with southern Asia, and the fauna is specifically African.

Discussion

Moreau asked whether the Sahara had always been desert and was told that for most of the time it had. The Nubian sandstone, none of which is later than Eocene, interdigitates with Upper Jurassic and Eocene formations. In the western Sahara Paleozoic sandstones are found between Precambrian and Cambrian Basement rocks. There is evidence that the Nile shifted back and forth, and moist conditions may have occurred locally during the Pliocene. For example, the occurrence of gravels in the eastern Sahara may be evidence of drainage into the Nile valley.

Monod then spoke on Northwest Africa.

The country is generally of low relief with no mountain building and only a little gentle warping. Recent volcanics are very rare, although there are some of Pliocene age at Dakar and a recently discovered locality near Timbuctu. The Basement rocks have a sedimentary covering with, in the west, a large area of tillite. Most of the country consists of sandstones with karst topography in the western desert. There are a few limestones of Mesozoic and very early Eocene age. There is little hope of finding breccias.

No mammalian fossils have been found with the exception of the Villafranchian assemblages of Koro Toro (Chad basin).

The flora is not very diversified. Using the Yangambi nomenclature it was classified by T. Monod in 1957 (Les grandes divisions chorologiques de l'Afrique. Publication No. 24 CCTA/CSA, London, 1957.) Broadly there are three main divisions: tropical forests in the south, open woodland and savanna, and the desert (Yangambi-steppe). In some parts of the desert trees do exist, and in only certain areas can it be classified as climax. However, there is a great monotony in the vegetation; though there are many individuals, an area of western Sahara the size of France is characterized by only seven species.

de Heinzelin covered the Congo Basin area and the Western Rift.

Here the topography is largely of basin and swell formations, but elevation has been caused by epeirogenic movement at the coast.

During the Tertiary and Pleistocene the central area around Stanleyville was desertic as evidenced by sandy sediments over the Karoo rocks which contain Late Carboniferous and Jurassic fishes.

The ancient raised surface of the Buganda peneplain may not be related to the inselbergs that occur there.

Some of the Rift valleys are still in process of formation to a mature rift profile; the lithological formations must be considered as a unit and fitted into the composite picture.

This was followed by LEAKEY on East Africa.

In this area there occurs an unusual range of altitudes from sea level to 19,000 ft. so that some parts were always favorable for human occupation even at periods of great climatic deterioration.

Rainfall and vegetation are also very variable and the latter is greatly influenced by the porosity of the soil.

Considerable earth movement, both in the past and in post-Middle Pleistocene times, has taken place so that generalization over the whole area is impossible. Prolonged volcanism has also occurred from before the Miocene to not more than 800 years ago.

Since present-day climatic conditions vary in closely situated areas they may be supposed to have done so also in the past. Lake Victoria constitutes a special climatic area in East Africa since it receives convection rains.

The country is particularly favorable for the preservation of fossils, for example in the rifts. Conditions for their discovery are also good as faulting has caused the accessibility of older deposits. The Miocene beds around Lake Victoria and north into Uganda, consist of lake and terrestrial deposits, but no molluscs or fish have been found there. Near Somalia are Oligocene beds containing bone which should repay investigation. In the Rusinga area the beds containing remains of suids and elephant, and previously thought to be Miocene, are now considered to be Pliocene.

A revision of the published fauna lists from the Pleistocene beds is now necessary in the light of increased knowledge. The Omo fauna probably equates with basal Bed II at Olduvai.

BOND continued on south-central Africa.

Apart from the Luangwa and Nyasa rifts, the southern tip of the Tanganyika rift, and the depressed area of Zambezia and the Mozambique plain, this whole region is a plateau 3,000 to 4,000 feet high. There has been no earth movement since the end-Tertiary and the area has thus been stable.

Climatically the rainfall is high in the east (50″–60″), descending to the arid Kalahari in the southwest (15″). The vegetation is chiefly savanna and open woodland.

The vast area in the west that is blanketed by aeolian Kalahari sands has not yielded any fossil vertebrate remains. As a whole the area is not favorable to the preservation of fossil material. However, the Lake Nyasa region holds out hopes for the preservation of Pleistocene fossils and there may be some in the fanglomerates and other deposits in the Zambezi basin, but detailed fieldwork still remains to be done.

South of the true rift system the chances of finding Miocene deposits are slight.

Clark then spoke on Angola.

The country may be broadly divided into three regions:

1. a narrow, cool, desertic, coastal region;
2. an escarpment with high ranges and *Podocarpus* forest;
3. a gently warped Micocene plateau with inselbergs of Cretaceous and Jurassic age.

The region has been largely stable since the end of the Tertiary except for a slight movement at the end of the Pleistocene shown in the northward-draining rivers. Most of the area is covered by deep Kalahari sands so that the chance of finding fossils is slight, but good pollens are preserved.

The rainfall is high, but quickly drained soils affect the vegetation and restrict the spread of forest—an important fact for prehistoric man who made extensive use of the polymorphic sandstones of the Kalahari System for the manufacture of stone tools.

Massive breccias occur on the Chela massif. These appear to resemble the Upper Pleistocene breccias of Northern Rhodesia and have yielded fossils.

Discussion

Monod and Arambourg said that mammalian bones, including those of monkeys, had come from these breccias (some found by F. Mouta) and were at present being worked on in Paris.

Bond stated that Dr. C. K. Brain is finding fossil material in what may be related breccias in the Sinoia (Southern Rhodesia) area.

Clark said that some stone tools had come from the Chela breccias and that they appeared to be similar to the industry from Broken Hill.

Cooke concluded the session by describing South Africa.

There are three regions within this area that are over 3,000 ft., the highest being in Basutoland (11,000 ft). Uplift has been less on the west than on the east so that the escarpment is steeper in the southeast. The major drainage is to the Indian Ocean.

In Tertiary times the area as a whole was subject to erosion, but Tertiary sediments accumulated in the Kalahari basin formed by warping. The best potential fossil mammal deposits are buried beneath Kalahari sands. A small Miocene or Pliocene deposit at Walfish Bay, outside the Kalahari, was described by Stromer, but the material is now lost. (Robinson added that he considered this material to be post-Miocene on the basis of hyracoids similar to those from Sterkfontein. He said that another site in South-West Africa had also yielded bone similar to that from Sterkfontein, but that this was also lost.)

The major fossil-bearing localities in South Africa are:

The Vaal River terraces which have yielded mixed and derived material, largely of Middle Pleistocene age. It is also possible that deposits similar to those at Olduvai may be found to occur in the basin to the northwest of the southeastern massif. A minor pan in the closed basin of the Vaal near *Cornelia* has yielded a mammal fauna originally described by van Hoepen and now revised by Cooke.

The Florisbad hot springs have yielded, besides a human skull, cultural material dating from First Intermediate to Middle Stone Age times. The report on the pollens from this site produced by van Zinderen Bakker in 1955 is now being revised.

The Australopithecine caves. Similar limestone distributions occur in South-West Africa, the Chela massif, Sinoia, Broken Hill, and at the Roan Antelope in Northern Rhodesia. These should repay investigation, though in all the breccias so far examined no bones had been found. It is possible that water in reasonably close proximity is an essential as at Taung, Sterkfontein, and Kromdraai.

Hopefield, north of Cape Town, has yielded a fauna similar to Olduvai Bed IV.

It is also possible that fossil material may come to light in the Miocene shore deposits of the Mozambique plain.

SESSION II: JULY 11 (9:00 A.M.–12:30 P.M.)

TERTIARY FOSSIL MAMMALS

ARAMBOURG opened the session by speaking on the North African fauna. He said that the Eocene of the Fayum is distiinct from that of Europe. It includes primitive Proboscidia and large ungulates which are specifically African and different from those of Europe. The same horizons as in the Fayum occur in Libya with many species of fishes and two crocodiles (one long-and one short-nosed). On the coast occur sea snakes characteristic of the Fayum, and a marine crocodile.

Further north in the Fayum beds of Oligocene age occur with fossil forests and a fauna including *Paleomastodon* and monkeys. So far primates have not been found in the similar *Paleomastodon* horizons of Libya, but it is possible that they may occur.

At Zelten the fluvio-lacustrine beds containing the Burdigalian mammal horizons have yielded, besides fossil wood, Proboscidia, a small *Trilophodon* similar to *Mastodon pygmaeus* of Algeria, normal *Mastodon* but of a distinctly African and not European form, also ungulates and a rhinoceros of a form found in the Miocene in Europe and in Moghara.

New forms occurring in this fauna are:

1. a ruminant (*Prolibytherium magineri*) and forms of both *Sivatherium* and Giraffidae;
2. a carnivore of definitely African affinities (*Afrocyon*);
3. large suidae (*Libyochoerus*), again definitely African in form;
4. secondary elements including large, flightless birds (*Samornis rothschildi*) found in southern Tunisia may be related; two types of crocodile, the Nilotic and a long-nosed variety resembling the gavial, and a possible third form; *Euthecodon* and various other reptiles and fishes.

The Villafranchian follows without a break. The biotope of the region stretching from Tunisia to the Fayum does not seem to have changed appreciably from Eocene to Quaternary times.

Discussion

In reply to Moreau Arambourg said that the petrified forests occurred in the Burdigalian at Zelten. The sandy, Oligocene beds contained no mammals but much wood and the flora was determinable.

Leakey said that both short- and long-nosed crocodiles are present in the Oligocene of Kenya. The Cervidae recorded by Whitworth from Burdigalian were probably Giraffidae. Arambourg confirmed that there were no Cervidae in North Africa before the Würm and that these originated in Asia, but the problem required further study.

Bishop then spoke on the east-central African faunal sites.

He said that work was still in progress on the reconstruction of the pre-Pleistocene ecology of the area, that nine of the fifteen published areas had already been revisited and that two more, in Kenya and Uganda, would be revisited this year.

The sites fall into three groups.

1. lacustrine (not one large large lake but several small ones) with fine clays and silts and non-sequences in which the fauna occurs. The best are Rusinga and Karungu, but as yet no primates have been found.
2. Areas of occasional ponding, that is, water standing on the surface for short periods. Fine-grained, subaerial volcanics, representing periods when ash was falling, but no lava was flowing, present optimum condition for fossilization.
3. Coarse grits, gravels, and conglomerates with a little fine-grained volcanic ash which present poor conditions for fossilization, though occasional fossils are found.

A map showing the distribution of "Miocene" localities is given in Bishop's paper in this volume; the table of primates from these localities has been amended, following the data summarized in Whitworth (1958).

The most important sites for primate fossils are Rusinga (100 sites) and Songhor (a series of erosion gullies). At Maboko there are two faunas, but excavation is needed to produce adequate exposures.

Rangwa is the main volcanic center for Rusinga and being a carbonatite area it preserves fossils well. The lavas overlie lacustrine deposits and the fossil sites lie on the lower slopes. The ecology resembles that of Olduvai.

Tindoret is a much less dissected area than Rangwa. At Songhor nearby, there is a tremendous concentration of primate material.

Napak in Uganda (see Figure 2 of the background paper) exhibits highly calcareous grey ash and tuffs with redeposited calcium carbonate forming limestones which are overlain by further agglomerates and lavas. Much fossil wood is present, a few land mollusks, four carnivores, and 21 identified species of rodent which comprise a third of the total recognizable fragments. Five species of primate also occur.

The distribution of the fossils is not necessarily ecologically significant as it is essentially related to past volcanics. The ecology consisted of savanna with gallery forest in the valleys opening onto the volcanic slopes and shallow lakes with seasonal fluctuations. The importance of the sites is that several occur in similar geological settings and can be used for reconstruction purposes. There is also the possibility of obtaining potassium-argon dates.

Leakey then also spoke on East Africa.

He maintained that Dr. Bishop's argument was largely hypothesis and stressed that caution is needed in the interpretation of volcanic centers since the ages of the centers differ. Rangwa need not be connected with the Rusinga beds where the fauna is all Lower Miocene. The Tindoret eruptions are probably Pliocene or later and not Lower Miocene and so are not necessarily linked with Songhor. Since the potassium-argon dates are still being worked on it is, in any case, unwise to speculate as to age at this stage.

The ecology also probably varied, at least locally. For instance the Rusinga series (119 sites) covered a very long period of time. The primate and other bones there may have been the remains of crocodile larders as the deposits were actually formed in water and contain numerous crocodile remains.

Dr. Bishop's table, partly based on Whitworth, needs to be amended to include Leakey's own evidence and also that of Le Gros Clark.

At Maboko it is probable that the fauna of the channel is Villafranchian as it includes species too late in age to be Miocene; but, it is also possible that the river had cut through Pontian and Villafranchian beds, and in any case more work is required before definite conclusions can be reached.

The Songhor sites are terrestrial and contain a high percentage of *Galago* and a few *Proconsul* remains. The Kirimon group comprises fourteen sites, of which nine have yielded faunas from beneath the lava, and are late Miocene or early Pliocene.

Loperot, further north, is a difficult area to work as there is very little surface water. Losidok has yielded a juvenile *Proconsul africanus* and one *P. nyanzi* in addition to other forms which are not yet published and which do not appear on Dr. Bishop's list.

In Miocene sites fossil finding is difficult but further work, especially at sites like Songhor and Rusinga, would be well repaid.

The remains of true pongids have been found in Lower Miocene beds at Kiahera Hill and Rusinga. These forms are higher primates with a simian shelf, sectorial premolars, and long canines. They are associated with forest mammals and are contemporary with the *Proconsul* group.

A new site at Fort Ternan lies 8 to 10 miles east of Koru. Pleistocene deposits rest on the Koru Miocene series. Dr. Leakey stated that he had been watching this site since 1931. In 1959 a fossil bone was found in the hill-wash at the base of the limestone cliff and in 1961 from the same scree came: a small *Trilophodon,* more evolved than the Rusinga form; a pig, more evolved than the Rusinga pig but less so than those at Olduvai; skulls of ungulates beginning to show the growth of horns (there were no horns in the Miocene forms); carnivore and monkey remains; a form transitional between hippo and *Brachyotus.*

A proportion of these remains was articulated.

During June, fragments of a hominoid were found. These consisted of part of a maxilla showing a well-defined canine fossa, one lower premolar, possibly from the same hominoid, with an incipient second cusp, sectorial in form, and which is unlike *Proconsul.*

A preliminary potassium-argon date from biotites in the ash gives ±12 million years which puts the remains in the early Pliocene or late Miocene. The main Rusinga series is dated to 25 million years, with a small pocket dated to 12 million years (? Pliocene). Now that this rich Pliocene site exists with a hominoid, important discoveries may be imminent that will bridge the gap between the Burdigalian and the Villafranchian. More and similar sites may exist in the Kirimon area and perhaps also at Moboko.

Discussion

Leakey first asked Bishop why he considered Moroto to be Miocene. It might well be late Oligocene and older than the oldest sites at Rusinga.

Bishop replied that the site had only been included in a Tertiary map by a geologist who had found no fossils there. It needed re-examination and might prove important, as fragmentary mastodon remains had now been found there and two very eroded teeth, possibly of an anthracothere.

Cooke stated that Arambourg had found antelopes with incipient horns in the Upper Miocene at Wadi Hamamm. Moreover, during the Pliocene the evolution of gazelles and alcelaphines had been explosive in character.

In reply to Dr. Leakey's remarks regarding the Rangwa volcanics, Bishop said that on the evidence the Rusinga beds could have no other origin. The lavas dipped away from Rangwa, were thicker toward the volcano, and petrologically and stratigraphically they appear to be associated. Volcanic activity appeared to have been later in southern than in northern Tanganyika.

Howell quoted from a letter from Dr. T. Whitworth in which the ecology of the Kavirondo area in the early Miocene was described as "rather dry steppe to parkland with limited gallery forest and occasional shallow, often ephemeral lakes and scattered inselbergs, perhaps with thicker vegetation cover (a landscape fairly common in Africa today)."

de Heinzelin then spoke on the peneplains and the Albertine Rift.

The recent Lower Semliki Expedition had been interrupted by Congo independence and, although the specimens had been saved, the field notes were incomplete.

In Buganda there are two peneplains: the Buganda and the Kyoga (Fig. 1). The first is deeply weathered and caps some rises in the rift area; the second is lower and less deeply weathered.

The Semliki beds are now known to be very recent. In the west, across the Rift, there were thought to be three peneplains: I, the oldest (Cretaceous), II (Buganda), and III (Kyoga) which is pre-rift. But no evidence of I had been found and Lepersonne's findings are not accurate farther west.

Tectonic disturbance is very evident in the west and also in the Kaiso beds where there is an intricate stratigraphy. There is no definite break between this level and the typical facies with clay above (as in the classical facies on the shore of Lake Edward). A typical fauna with *Viviparus*, and *Melania* above, is found at the type locality. There is thus a definite sequence in the molluscan fauna.

The "Lower Miocene" white sandstones appear to rest on the downwarped Kyoga surface.

(At this point slides were shown with a commentary and it was explained

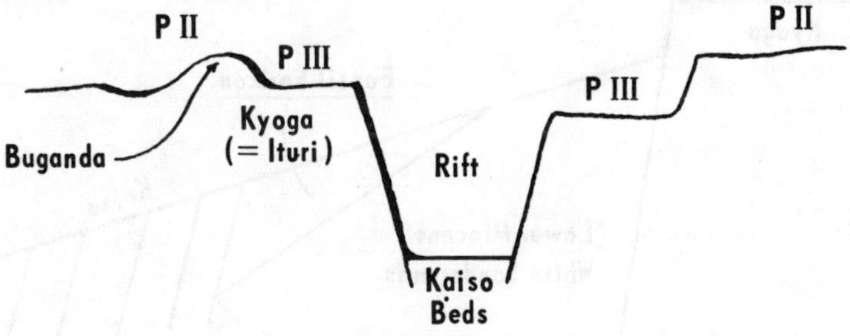

FIGURE 1
Erosion surfaces in eastern Equatorial Africa (schematic).

that fragmentary fossils occur in gullies and were eroded from exposures at the junction of the "Lower Miocene" Kaiso sediments at the important site of Ongoliba. Very little change is visible here in the beds and no major unconformity. Minor exposures of the white sandstones with fossils occur in the Mohari valley.)

de Heinzelin then tabled the section he had produced for the Pan-African Congress in 1959, at the request of Dr. Howell. He pointed out, however, that a change should now be made in this section since it is now certain that the oldest surface is post-Kyoga and not post-Buganda. This makes much more geological sense.

Discussion

Bishop agreed with de Heinzelin. He pointed out that if one started at Kyoga in Uganda there was no step up from the Kyoga surface which passes under Napak. The later Kyoga surface is thus clearly sub-Miocene. Previously, the subvolcanic surface had been considered to be the equivalent of the Buganda peneplain and this is manifestly not so. The Rift formation thus dates to post-Kyoga-Ituri times. In a borehole at Butiaba, 30 miles north of the type locality of Kaiso, it appeared that 4,000 feet of sediments occurred; and, in the center, 8,000 feet of sediments had formed on the Uganda side of the Rift since the down-faulting of the Ituri (Kyoga) surface. It is not possible to extrapolate from the coast (as had been done by L. C. King) because of the local modifications caused by rifting. He also considered it necessary to map completely the Kisegi series.

de Heinzelin said that the correlation of the various levels was complicated by tectonics and was a very complex system.

Leakey agreed and said that the picture of antiquity or modernity of molluscan fossils was not easy to see since older deposits contained modern-appearing forms;

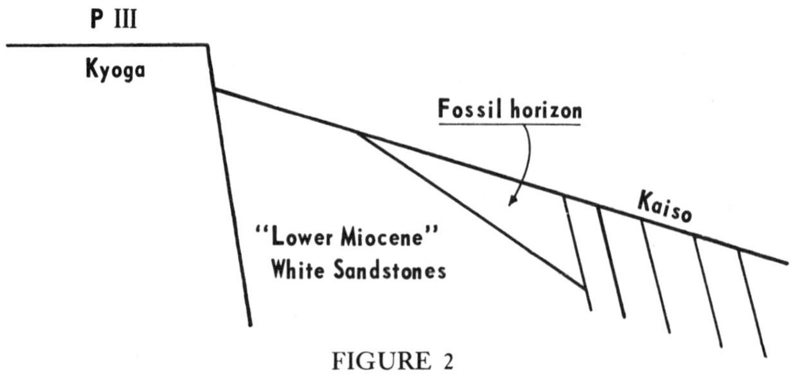

FIGURE 2
Schematic section through the Sinda and Mohari Beds.

he thought that Kaiso ecology must have affected the molluscan fauna. The east and west Rift faunas were different although the pictures were broadly similar.

DE HEINZELIN said that there was definitely a similarity between the Kaiso molluscan fauna and the Pontian of central Europe. The duration of Kaiso sedimentation was long and there must have been much faunal evolution, probably since pre-Quaternary times.

LEAKEY said that the classical Kaiso fauna came at the end of the sequence and being at the very end of the Villafranchian was later than the Bed I Olduvai fauna.

BISHOP said that the base of the Kaiso sequence showed conformity with the top of the Kisegi series; no basal Kaiso fauna had yet come from Uganda.

ARAMBOURG said that from the Upper Kaiso levels, not Wayland's original locality, had come very primitive *Stegodon*, *Anancus*, and *Elephas africanavus*. This suggested to him a basal Villafranchian age.

COOKE said that green shales, possibly of Miocene age, occurred in the bottom few hundred feet of the Butiaba borehole. DE HEINZELIN added that these were possibly of Karoo age. The Muzizi and Uruma sandstone represented a Karoo basin equating with the age of the schist in southern Tanganyika, i.e., Late Primary to Secondary.

COOKE cited a small collection of Miocene fossils from Karugamania which might have been derived from the "Lower Miocene" white sandstones. DE HEINZELIN said there did not appear to be any Miocene deposits from which this fauna could be derived although it had been defined as Miocene. It was almost impossible to separate the Lower Miocene from the Villafranchian in mixed assemblages such as this.

LEAKEY suggested separation on the basis of the degree of rolling in the same deposit (as at Maboko).

DE HEINZELIN said that the Buganda surface was probably much older than mid-Miocene and that the Kyoga, thought to be end-Tertiary, may also be older. So the whole question of the age of these beds must ultimately rest on the determination of the fauna.

SESSION III: JULY 11 (2:30–5:30 P.M.)

LEAKEY opened the session by speaking on Lower Pleistocene ecology, with particular reference to Olduvai Gorge.

The preliminary phases of work at Olduvai Gorge lasted from 1931 to 1949. From 1951 to 1958, work was concentrated on the base of Bed II. From 1958 to the present work has mainly been concerned with Bed I. Since the publication of the initial results, additional work has modified the original conclusions about the geology and the fauna but the cultural sequence, although considerably enlarged, is still substantially correct.

Bed I. This should definitely be regarded as distinct from and much older than Bed II.[1] In thickness it varies from ±100 ft. to ±12 ft. but this is to be considered by levels and not as a whole. The irregularity of the old lava surface had a major bearing on sites in the Gorge and on the formation of the peninsulas on which, for defense reasons, early man lived.

In the east, Bed I consists of coarse volcanic material; it runs under, and so is older than, the Ngorongoro range. To the west the sediments are finer. The contact between Beds I and II is more or less horizontal with very little evidence of erosion, although the surface of Bed I was subject to subaerial effects before the formation of Bed II.

The Fauna of Bed I. The fauna, as originally published by A. T. Hopwood, requires considerable revision due to more extensive discoveries. Only six of the original 21 genera defined by Hopwood on incomplete material can now be regarded as valid. Hopwood's suggestion that modern species are present is incorrect. Fifty-four mammalian species presently identified are all definitely archaic. It is not yet possible to define the Villafranchian boundary since complete details are not yet available.

The fauna of Bed I must be further divided rather than considered as a unity. The main part of the bed contains markedly wet-living fauna, succeeded by savanna forms with, at the top, desertic species.

The Fauna of Bed II. This varies, depending on the level. At the base it is very like that of Omo and includes some species carried over from Bed I. The main characteristic of the Bed II fauna is the very large size, i.e., giantism, of many of the species.

Potassium-argon dates. An average of five dates, made on volcanic rocks from samples taken by Evernden and Curtis of Berkeley, gives 1,750,000 years for the age of the sediments above and below those yielding the pre-*Zinjanthropus* child. Previous estimates of the length of the Pleistocene are thus too low and, including Kanam East and Kanam West (Lower Villafranchian), it may be as long as two million years.

Rainfall in the Pleistocene. In this area rainfall must have been higher than that of today to have maintained a lake the size of that which existed. The present-day rainfall maintains Lake Manyara only at a very low level. Pre-Acheul lake beds at Makunyuni Crossroads, probably equating with Bed I at Olduvai, seem to correspond to an increased Manyara rainfall.

Sites in Bed I. The date for the top of Bed I averages 1.23 million years. The

1. Geological studies, including many measured and levelled sections, by Richard Hey have recently shown that there is no major temporal break between Beds I and II. This is also borne out by the results of new (1962) potassium-argon dating by J. F. Evernden and G. Curtis (pers. comm.) [F.C.H.]

Zinjanthropus level is 20 feet below this, and this depth thus covers 500,000 years.

1. *Site FLK I.* This is limited in area and has a total extent of about 40 ft. and contains the *Zinjanthropus* level, a thin layer. *Zinjanthropus* is a ± 18-year-old individual whose teeth suggest a vegetarian diet. Culture was present in strong concentration in the center of the site and chipping areas can be distinguished. Much bone occurs, some broken to extract the marrow; and also larger, unbroken bones are scattered, more thinly, together with stone tools, around the periphery. The question arises as to whether *Zinjanthropus* was the maker of the culture.

2. *Site FLK NN I.* This is the level of the pre-*Zinjanthropus* child and is about two feet below the *Zinjanthropus* level at the same locality. There is a strong probability that the two fossils are broadly (geologically) contemporary. The culture includes naturally sharp stones brought in from a distance. Occupation occurs in a spread corresponding more to the edges than to the center of the *Zinjanthropus* occupation horizon. However, this may be a false impression created by the limited extent of the area excavated. A true bone-tool occurs showing signs of use, probably caused by leather working.

The hominid skeletal material includes parts of limb bones and part of an adult skull. The teeth of the pre-*Zinjanthropus* child show him to have been omnivorous in diet; his skull vault was larger than that of adult Australopithecines.

The problem to be considered is whether the pre-*Zinjanthropus* hominid made this or both the cultures; or whether the cultures were the work of *Zinjanthropus;* or whether either creature was a victim of the other, since both are fairly closely related in time.

3. *Site FLK N I.* This level is 14 feet above the *Zinjanthropus* level and consists of six feet of clays full of tools and broken bones. The deposits represent lacustrine sediments from a seasonally fluctuating lake into which occupation material had sunk.

4. *Site MK.* This is an older level which is at present being excavated and is yielding crocodile and other remains. The results will be published in due course.

Sites in Bed II. The time span between the top of Bed I and the bottom of Bed II is not more than 500,000 years.[2]

1. *Site BK II* is at the bottom of the bed and site SHK II is 10 feet higher up. Both contain Chellian tools, referred to Chellian stages 1 and 2, and including large polyhedrals, some the size of footballs. With the Chellian 1 were found two milk teeth of a possible hominine, but much later than Bed I, in fact well into the Middle Pleistocene.

2. *Site FLK II,* half way up Bed II, yielded a Chellian III living floor and some fossils, including a hominid skull. Clear suggestions of bolas occurred with some of the stones in groups of three.

2. See the previous footnote in regard to the lack of a substantial temporal hiatus between the top of Bed I and the base of Bed II. [F.C.H.].

Discussion

DeVore asked why the pre-*Zinjanthropus* child in a stratum two to four feet below *Zinjanthropus* should be considered to be contemporary. Howell thought it likely that, since 20 feet of Bed I is said to have covered a period of 500,000 years, the two to four feet that separated the two fossils would thus indicate a time difference of 50,000 to 100,000 years.

Leakey, in reply, said that Bed I showed no sign of a break similar to the hard marker beds at the top, and that while he was at present of the opinion that these two fossils were geologically contemporary the problem was still being studied.

de Heinzelin suggested a possible link between the desert period fauna at the top of Bed I and the mid-Kaiso fauna with spiny *Viviparus*.

Biberson pointed out that in the Villafranchian of Morocco three series exist: Lower, Middle (Moulouyien), and Final (Saletien) and that definite breaks occur in the deposits which may prove to be paralleled in other parts of Africa. In North Africa no industry is known to occur in the Lower but only in the Middle (and Later) Villafranchian.

Leakey pointed out in reply that the fauna of Olduvai Bed I is not the base of the Villafranchian, though Kanam East and Kanam West are also (Earlier) Villafranchian.

GIANTISM

Leakey, in reply to Moreau, stated that investigations were at present in progress into the possibility that the giant size of the fauna of Bed II is due to increased trace elements in the soil caused by increased rainfall. This giantism seemed to be greater in herbivores than in carnivores and was indeed a true giantism. Leakey emphasized that such an explanation was so far only very tentative. In support of the explanation the case was cited of three modern species of Kenya leopard with variation in size due to difference in diet. Mere size had not been used as the criterion for the indication of species in the Bed II fauna.

Cooke said that giantism occurred at several times in the Pleistocene and in fact only died out at its end.

Bourlière and Emlen pointed out the absence of true giantism in modern mammals. It was agreed that a variety of giant species at one time could imply a very favorable food supply.

WASHBURN, in citing instances of giant species in other areas contemporary with the Kenya fossils, suggested as a subject for discussion the relation of pluvial periods to increased food supply.

LEAKEY agreed that contemporary giant animals from elsewhere confirmed the evidence from Olduvai and said that in assessing this giantism the whole known mammal fauna of East Africa had been taken into account. He agreed with Cooke and Robinson that after bad climatic periods the fauna emerged in diminished variety and instanced the reduction of Asiatic fauna at the end of the Pleistocene and of *Afrochoerus* at the end of Bed IV. He felt, however, that Bond's suggestion that fauna actually diminished in size during bad climatic periods could not be proved in the present state of knowledge.

VULCANICITY

LEAKEY said, in reply to Moreau, that there was at present no evidence to show that the appreciable volcanic activity in Bed I and Bed II times was linked with Ngorongoro.

BISHOP suggested that the trace elements that might account for the giantism could, perhaps, have come from volcanic activity.

LEAKEY said that so far no specific source of the lava and ash of the Olduvai region had been determined. In reply to Cooke he said that age determination of the lava under Bed I was at present in progress and that when the results were available they might enable a dated type section of the Pleistocene sequence to be prepared. It was confirmed that there was no lava *over* the nearby Laetolil beds.

RAINFALL

It was agreed that increased rainfall would have been necessary to produce the wet conditions of the major part of Bed I times and again in Bed II times.

LEAKEY considered that these conditions in Beds I and II could only be explained by Pluvials in East Africa.

HOWELL was in agreement and that similar conditions applied particularly to Bed IV since this was in an area of limited drainage. He also agreed with Clark that an increase of about 50% over present-day rainfall would be adequate, and that this might also account for any possible decrease of temperature at that time.

SESSION IV: JULY 12 (9:30 A.M.–12:30 P.M.)

PLEISTOCENE HOMINIDS

COOKE opened the session on the Australopithecine sites.

General outline of the Australopithecine sites.

Sterkfontein lies some 30 miles west of Johannesburg and is situated on the southern rim of a basin in early Paleozoic dolomitic limestone.

Makapan is some 200 miles north on the same basin. The sites occur in old caves in the limestone formed by solution at the intersection of joint planes. In such caves the bone accumulations occur in the early stages of formation when the opening to the outside is small and material is washed in gradually. Secondary cave deposits also occur at Makapan. Subsequently, some of the deposits and even part of the roof material were eroded away. The date of the cave depends on the time when it was opened to the surface. The fillings date from the base of the Pleistocene and still continue.

Taung occurs in secondary limestone on the edge of the dipping surface formed by a series of carapaces, or delta deposits, of precipitated travertines, sometimes 200 feet high. The conditions necessary for the formation of such carapaces are fairly high precipitation and a reasonable reservoir of moisture. Chambers cut by erosion in the limestone are filled by soil and sand blown from the Kalahari region. The Taung child occurred in one such pocket of breccia, but its direct association with the fauna is not certain. The child occurred in the oldest pocket, sealed by a subsequent travertine, and Middle Stone Age material is present in a later breccia.

COOKE then showed slides on which he gave a commentary.

The caves with the most abundant fossil material are intimately related to the availability of water. Based on Brain's study of soil grains from a wide range of rainfall areas in Africa, the rainfall of these sites has been determined as:

Kromdraai—similar to the present day;
Swartkrans—drier, becoming wetter;
Sterkfontein Extension site—like the present, becoming wetter then drier;
Makapan—dry, becoming wetter;
Sterkfontein—declining rainfall.

There is abundant fauna at Sterkfontein and Makapan and before Leakey's revision of the Olduvai Gorge fauna the Makapan fauna could be compared to this. It seems to be Upper Villafranchian, though it is just possible that it equates with the lower part of Bed II.[3]

3. It should be noted that the "base" of Bed II, according to recent K-Ar determinations, is about a million years old; hence the possible equation of the Makapan Limeworks fauna with it does not make the former unduly "young." [F.C.H.].

ROBINSON then spoke on the Australopithecine fossils.

Sterkfontein. A hundred hominid specimens and a thousand associated faunal remains have come from the Type site. Stone implements have come only from the Extension site; also a bone tool, polished as if by use as a skin scraper, and with two broken facets.

Three breccias exist at these two sites. The top breccia is darker in color and is sterile. The section from the top downwards is end-Pleistocene material over Middle Stone Age, then a sterile layer overlying 2 to 3 feet of sterile breccia, all overlying the artifact-bearing breccia (Extension site) which overlies the basal breccia (Type site) containing the Australopithecine remains but no tools.

The best skull is that of Sts. 5; it is also the most prognathous. The occiput is in the hominid position indicating erect posture, there is human-type wear on the teeth, and the dental arch is hominine-like. *Australopithecus* would appear to have been about 4 feet tall and to have weighed 40 to 50 lbs. From his teeth his diet would seem to have been that of an omnivore. He would appear to have been an early hominid.

Swartkrans. This site is later than the Sterkfontein Extension site, and both *Telanthropus* and *Paranthropus* occur there. Morphologically these two forms differ markedly. The skull of *Paranthropus* has a sagittal crest confined to the top and central portion of the vault; it does not continue down the occiput. The dental alveoli form a nearly straight line across the front of the jaw instead of a curve. The mandible exhibits small canines and incisors but large molar teeth, on which the wear would seem to indicate a vegetarian diet. The creatures walked upright. The earliest instance of the alteration of the pelvic structure in the direction of erect posture would seem to be the case of *Oreopithecus.*

The suggested development of the South African fossils is from *Paranthropus* (vegetarian) to *Australopithecus* (omnivorous). The change to an omnivorous diet was probably caused by environmental pressure and contributary factors may have been improved intelligence and greater tool using. With the development of tool-making, *Homo erectus* and later still *H. sapiens* must have evolved.

LEAKEY then spoke on Olduvai Bed I.

After agreeing that *Paranthropus* and *Australopithecus* constitute two distinct genera he went on to discuss the pre-*Zinjanthropus* child. The first premolar puts it outside the possible range of variation of either *Australopithecus* or *Paranthropus.* The canine is large, like that of *Australopithecus,* and the teeth do not exhibit the extreme asymmetry of the *Paranthropus* teeth (cf. *Nature,* 191: 417-18, 1961).

Describing the morphology of the skull, Leakey said that although this was an eleven- or twelve-year-old child the interparietal width of the reconstructed cranium was as great as that of many adult East African Bantu skulls. The brain is much larger than that of the *Australopithecus* No. 5 (Sterkfontein) skull. All the characteristics point to this fossil having been a protohominine. For such a form to occur in Bed I in fairly early Villafranchian times (1,750,000 years ago) makes it more likely than *Australopithecus* to have been the ancestor of *Homo*.

On the other hand, *Zinjanthropus* is a true Australopithecine. He resembles both *Australopithecus* and *Paranthropus* though, while exhibiting features similar to both, he is not exactly like either. It seems most likely that *Zinjanthropus* represents the Australopithecine line of early Pleistocene hominids and the pre-*Zinjanthropus* child the *Homo* line. Such a hypothesis would make man's evolution similar to that of all mammals.

Arambourg then spoke on the Koro Toro fossil sites.

Fossil vertebrates had been known in the north Chad area since 1955. In 1960–1961, Y. Coppens had discovered a series of rich fossil bearing sites in the same area. There appear to be at least two Villafranchian levels, the older containing *Anancus osiris*, *Elephas africanavus*, and *Stylohipparion*.

At Fouarat (Morocco) the marine beds can be correlated with marine formations in the Mediterranean (Calabrian) and Europe. In Algeria at Aïn Boucherit there are again a Lower and an Upper Villafranchian, the latter containing a more evolved elephant. At Aïn Hanech the Upper Villafranchian is found with pebble tools and the fauna resembles that from Omo in East Africa.

At Wadi-Derdemy (Chad) the fauna in the later level contains *Elephas africanavus* and *E. recki*, and Stegodont (and at Koula an elephant more primitive than *E. africanavus*), two carnivores (a feline and a hyena), *Ceratotherium simum*, *Stylohipparion*, *Hippopotamus*, three species of pig (*Notochoerus* nov. sp., *Phacochoerus*, and a possible *Potamochoerus*), *Libytherium maurusium*, a giraffe, *Menelikia*, *Redunca*, *Hippotragus*, *Kobus*, an alcelaphine, a camel-like creature, three species of crocodile (*C. niloticus*, *C. cataphractus*, and *C. barbeaui*, nov. sp.) a gavial, *Trionyx*, and Siluridae and Serranidae.

In the Lower Villafranchian horizon of a nearby site, between Largeau and the frontier of Niger, was found a skull fragment— including the frontal region and the face—of a new Australopithecine form. In describing the skull Arambourg said that the degree of prognathism, the height of the forehead, and the slightness of the brow-ridges recalled *Australopithecus*, but that the development of the zygomatic arch and what is preserved of the dental series mark it as a female of the *Parathropus* group. However that may be, the skull appears to exhibit a greater cranial capacity than other Australopithecine forms and certain other features which may necessitate considering it, at any rate provisionally, as a distinct type of Australopithecine.

This find has opened up the possibility that the pebble-tool makers spread right from the north to the south of the African continent.

Monod spoke next on the stratigraphy of the Chad sites based on the work of Barbeau. He said that four major lacustrine extensions had been recorded separated by dry periods. These extensions were attested by the raised terraces of Chad at 4 m., 10 m., 20 m., and 30 m., the latter representing a considerable, northeastern extension of Chad and consisting of old white sterile sand, coarser green sands, and a little clay containing Villafranchian fauna. Coppens had also found two other faunas above the Villafranchian in the Chad area.

Discussion

Howell stressed the importance of the Koro Toro find because of its clearly defined association with the Villafranchian.

Butzer said the *Anancus osiris* from the Nile beds was probably associated with the 75 m. terrace and likely to be more recent than Sicilian, to which Arambourg replied that this occurrence of *Anacus osiris* was probably derived and must be regarded as Lower Villafranchian.

Cooke stressed the importance of the Suidae for dating purposes in this period particularly in association with elephants.

Washburn considered that the interpretation of human fossil material was not based sufficiently on the stratigraphical sequence of finds but too much on opinion. He considered that certain characteristics, for instance postorbital constriction, are merely related to face size. He said that since large-faced monkeys have low foreheads, and vice versa, the height of the forehead is not necessarily related to brain size. The ischium of the Australopithecines is very apelike so that it would not be possible for these creatures to be bipedal in a fully human sense. He suggested that human characteristics seemed to appear in the Middle Pleistocene. He could not agree with the opinions of Leakey and Robinson on the dietary habits of *Australopithecus* and *Paranthropus* since apes with a purely vegetarian diet nevertheless have large incisors so that small incisors might merely indicate hand feeding. Large muscles are required for pulling with the teeth, but relatively small muscles are needed to balance the skull so that there need be no correlation between muscle attachments and posture. Since both these Australopithecine forms were not far removed from the apes it seemed a pity to divide them into distinct genera.

Leakey said that the erect posture of *Zinjanthropus* was postulated on the occipital condyles. The wear on the molars indicated a rough diet, but the incisors were too small to have been of use for pulling though the muscles had nevertheless been large. Since *Zinjanthropus* was associated with a Villafranchian fauna and *Paranthropus* with a mid-Pleistocene fauna the former must be older.

Howell, however, was not convinced about the age of the Swartkrans fauna which probably, in fact, was Villafranchian.

Cooke said that the dating of these faunas would depend on the new faunal data from Bed I at Olduvai. The North African fauna, which can be correlated with the European evidence, is the only one in Africa so far where the Villafranchian boundary can be drawn, though it is still uncertain whether this is above or below the Cromerian. The Chad fauna is still not conclusively worked out and South Africa is too far away for the Villafranchian boundary there to be defined as yet. More faunal data from Bed I at Olduvai are needed though the potassium-argon dates are strongly in favor of Leakey's argument.

LEAKEY said that it was generally conceded that the *Paranthropus* level at Swartkrans was late in relation to Sterkfontein and Makapan. He also said that if there is any valid relation between tooth pattern and diet then *Zinjanthropus* was vegetarian and pre-*Zinjanthropus* omnivorous. This did not imply the descent of the one from the other, but merely the existence at some earlier time of a common ancestor.

ROBINSON thought that the varying sizes of the teeth of the Australopithecines contradicted Washburn's suggestion that large incisors can accompany vegetarian habits, as in apes. He suggested that the meat in the diet of *Australopithecus* probably counteracted any tendency to reduction of the size of the canines that might be expected in an otherwise more advanced form. Fossils should not be compared on isolated characteristics but as a whole and, even if size were discounted, *Australopithecus* and *Paranthropus* showed basic differences of adaptation.

WASHBURN thought that words must be used in a comparable sense when comparing tool use and adaptation and when comparing types. He suggested that "family" and "genus" were being given different connotations.

SESSION V: JULY 12 (2:30–6:15 P.M.)

MIDDLE PLEISTOCENE ECOLOGY

LEAKEY was invited to open the session by speaking in reply to the remarks made by Washburn during the discussion in Session IV. He said that he thought Washburn was arguing from one of two unacceptable premises, namely:

1. that man had originated from a chimpanzee-like pongid relatively recently in the Pliocene, or
2. that man was descended directly from an Australopithecine.

One must, however, work back from known fossils to the *Proconsul* group which, although they have been called pongids, are not, in fact, true pongids. It must be stressed that, whereas *Zinjanthropus* is quite clearly within the Australopithecine side branch, the pre-*Zinjanthropus* child with a very different morphology, definitely indicates a hominine. Therefore, there were two contemporary but different hominid branches already established in the Villafranchian. In the Lower Miocene of Rusinga a very definite pongid exists side by side with the *Proconsul* and *Sivapithecus* group. Part of a second such pongid fossil, complementary to the first, has since been obtained which again shows that in the Lower Miocene a true pongid and a non-pongid were present together. *Homo* origins must, therefore, be postulated as being in a group which is already showing tendencies in that direction.

HOWELL presented a table (Table 1) showing the probable relationships of the faunas in East Africa and those in South Africa. He stated that he did not agree

TABLE 1

TENTATIVE SCHEME OF RELATIONSHIPS OF EARLIER PLEISTOCENE SITES IN AFRICA

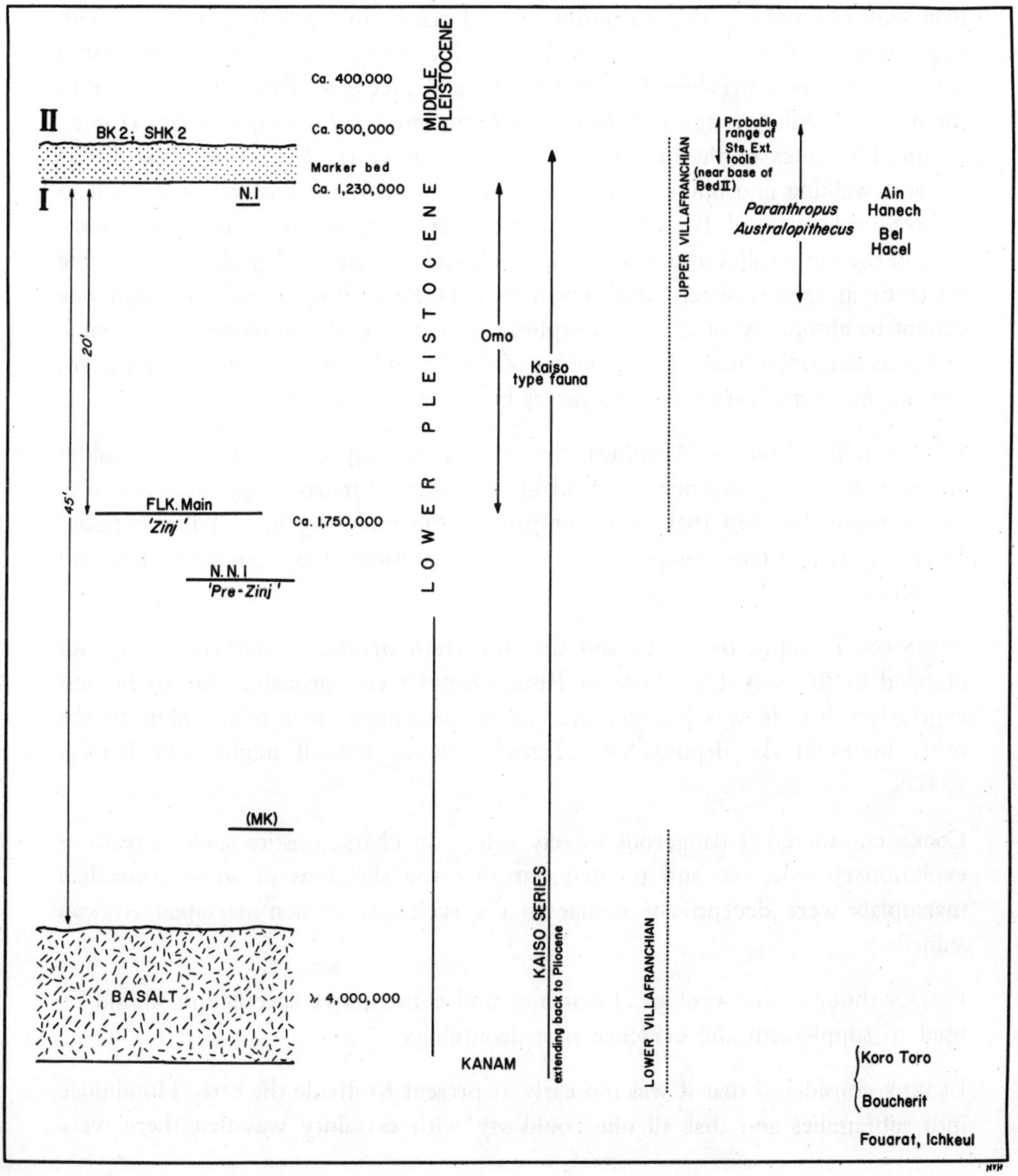

at all with Kurtén in regarding all Australopithecine fossils as of Middle Pleistocene age.

WASHBURN, in reply to Leakey, pointed out that different results could be obtained by starting from a different set of facts. By taking into account only evidence from bones one is ignoring the other physical features of man and the primates—for instance the blood groups of chimpanzees are closer to those of man than to those of other primates. At the end of the Pliocene great apes occurred with small teeth and the size of the canine varies greatly in for instance the gorilla.

ROBINSON pointed out that paleontology and taxonomy must be concerned with sequences. Le Gros Clark's supraorbital index shows that there is not just a simple correlation between forehead height and face size. *Paranthropus* fits into the pongid-monkey range and *Australopithecus* into the *Pithecanthropus* (*Homo*) group. The index of the height of the occiput above the Frankfurt plane differs in erect-walking and non-erect-walking forms and this analysis can be applied to *Australopithecus* and *Paranthropus*. Since they both occur at much the same time in the same valley there would appear to have been overlap. Sex is detectable on teeth in approximately the human way in *Paranthropus* and, although one cannot be absolutely sure, the dimorphism seems to be of approximately the same extent as in modern man. The structure of the first milk molar in *Australopithecus*, but not in *Paranthropus*, is the same as in modern man.

LEAKEY pointed out to Washburn that in paleontology one could only consider bones, as no other evidence was available. On skeletal morphology *Australopithecus* is nearer to man than is a chimpanzee. He said that he did not consider *Sivapithecus* as a true pongid and it might be considered as a source of hominid ancestry.

ROBINSON, in reply to Clark, said that the teeth of *Australopithecus* were not chipped in the way that those of *Paranthropus* were, probably due to his less vegetarian diet. It was not yet possible to determine their relationship to the other bones in the deposits and *Australopithecus* himself might have been a victim.

COOKE considered it dangerous to rely solely on characteristics such as teeth as evolutionary evidence and pointed out that the skeletons of some Australian marsupials were deceptively similar to the skeletons of non-marsupial African animals.

EMLEN thought that ecological evidence and constructive speculation should be used to supplement the evidence of paleontology.

LEAKEY considered that it was too early at present to divide the early Hominidae into subfamilies and that all one could say with certainty was that there were three, possibly four, distinct types, the *Paranthropus-Zinjanthropus* type, the *Australopithecus* group, and the pre-*Zinjanthropus* child. As in mammal evolution there must be many possible ancestral lines to *Homo* and not just one.

ARAMBOURG agreed that the same principles must be applied to human as to animal evolution. This had not been orthogenetic, but by progressive mutations. Since there are only a few individual fossils from scattered groups over a long period of time it is impossible to arrange these in a definite series and all that can be done is to indicate evolutionary changes.

HOWELL said that new information had complicated a picture hitherto simple because of lack of knowledge. The most important factor was the position of these fossils in their natural biotope.

CLARK then spoke on the Oldowan and Chellean floors and industries which he had been able to examine on Leakey's invitation. He said that the Oldowan floors in Bed I at Olduvai were the only ones of that age in a sealed and undisturbed state. Other sites where this culture occurs are either in river gravels and so disturbed, or incomplete as at Sterkfontein. This site is an instance of the importance of using the horizontal grid system for the uncovering of early Paleolithic sites.

On the *Zinjanthropus* living floor (FLK I), at FLK NI and on the dispersed floor at the top of Bed I, it was quite impossible that any of the material could have come there naturally so that all the stone and bone is in artificial association. In addition, the material is all fresh and the floors are flat so that no sorting has taken place due to runoff. The horizontal grid plot shows the existence of definite workshop and special occupation areas and a true cross section of the tool kit has probably been excavated. The floor must have been very quickly covered after the occupation and it is probable that only a transitory occupation is represented, though the number of natural stones brought in from a distance makes a longer occupation possible.

The faunal remains are not of complete beasts, but consist of parts of many animals, young and old and medium to small in size, and of birds. There had apparently been no attempt at any definite selection in the parts of the beasts that were present on the floors and this, therefore, suggests that scavenging may have been the main source of the meat supply. The bones are intentionally broken, but only those that would have contained marrow. Skulls are broken open to remove the brain, but the horn cores have rarely been separated from the top of the cranial vault.

A short film was shown illustrating experiments in the breaking of bones. Clark pointed out the characteristic fractures which were unlike the results of gnawing and cracking by carnivores. He said that the easiest way to break the bones was on an anvil or with a hammerstone and that once the ends were off the long bones, the shafts were easy to break, the resulting splinters resembling those of Dart's "Osteodontokeratic Culture." Fresh bone fragments are held together by sinew and so when found in the deposits such small fragments are often close together. A sharp instrument was needed to separate the limbs of a large animal before the bones were broken.

The culture from the *Zinjanthropus* floor impressed one as being near the beginning of tool-making. If the raw material used was flat in shape, flaking from one side only would produce the desired sharp edge; but bilateral flaking was needed if the material was nodular. Lava and quartz had been used, but no obvious selection had been practiced as apparently was the case with the pebbles on Dr. Biberson's Moroccan sites.

The industry consisted of:

1. Natural stones, none very large, which may have been missiles, or used for pounding food, bashing bones, or as a source of raw material. These stones did not often show signs of use.
2. Polyhedral and irregular stones showing signs of unintentional flaking and bruising.
3. Choppers and cores for the deliberate removal of flakes.
4. A great variety of flakes potentially useful as tools.

Napier has suggested that the hand of the pre-*Zinjanthropus* child and of *Paranthropus* was capable of a power grip but not of a precision grip. The number of small flakes would seem to suggest the possibility that the maker of the Oldowan Bed I culture could have had the precision grip also.

The industry from the top of Bed I looked more evolved than that of the *Zinjanthropus* level and was more like the industries of Bed II though not identical. This would seem to bear out Leakey's suggestion of a time lag between the two beds.

The industries from the lower levels in Bed II show considerable development with many flake tools, larger choppers and polyhedrals and hand axes, becoming more common in Stage 2 and 3. If Stage 3 was indeed the industry associated with the Chellian 3 skull, then it showed that man's tool-making ability had progressed considerably since Bed I and was clearly associated with, and no doubt the result of, increased brain size. The chopper-chopping tool element persisted for a very long time, covering the whole of the Chellian where, in the earlier stages, the chopper and the small flake tool are more typical than the hand-axe itself.

By Bed II times there was now also greater selection of raw material—mainly chert and quartz for small tools—and there was also definite evidence of hunting in the butchering of large animals as opposed to the postulated scavenging of the earlier levels.

Clark then went on to speak of the Sterkfontein Extension site which he said could not be described as a living site since no stone waste occurred with the tools. It was probably a place where man obtained water or drove game, and used and left his tools. These number 200 Robinson corrected this to 280) in mint condition and they are developed chopping tools. Mason contends that the industry resembles that of the 50-foot terrace of the Vaal, but more work is needed to substantiate this.

Of the Australopithecine bone assemblages Clark said that the same sorts of fracture are present as on the Bed I floors at Olduvai. There is apparently some intentional selection of the parts of the animals; few of the creatures represented are very large and all the assemblages are situated well within the caves suggesting the possibility of their having gravitated down a steeply sloping floor from higher levels nearer the entrance. Clark did not think that the "Osteodontokeratic Culture" was a culture in the sense intended by Dart, but there was no doubt that suitable broken pieces of bone may have been utilized and both Leakey and Robinson have recovered true bone tools of this kind.

The Sterkfontein Extension site was not considered to be as old as the top of Bed I but probably correlated with the base of Bed II.

Slides were then shown of Olduvai and Sterkfontein.

Discussion

Washburn, referring to the possible lack of precision in the grip of the pre-*Zinjanthropus* child, said that a student (G. Krantz) who had tried to make stone tools without using his thumbs had been unsuccessful.

Biberson said that it was likely that wherever early man found natural stones that could be used without further trimming they would be so used, since man is lazy and chooses the easiest way to solve his problems. He considered that the different ways of flaking the raw material, depending on the shape in which it occurred, implied a considerable process of thought on the part of a creature so primitive as the earliest hominids.

Leakey, in reply to Clark (above) said that there were many ways in which bone could be broken, each producing different effects. He said that the floor at the top of Bed I (FLK NI) was six feet thick. On the *Zinjanthropus* floor, almost complete large beasts did occur; it seemed unlikely that such large creatures had been scavenged almost whole. The natural stones were probably brought back as the occupants returned from forays and they seemed to imply if not lengthy at least continuous occupation. Few of the bones show signs of weathering though some on the NI floor above were heavily weathered. Leakey did not think that all the bones were broken for the marrow since some of those broken were not marrow bones.

The Oldowan-type chopper persisted up to the Neolithic, and it is characteristic of Bed I but after that it ceases to have diagnostic significance.

The Sterkfontein assemblage, which contains some flakes, would equate with assemblages from near the base of Bed II at Olduvai.

Robinson asked why man should come to Sterkfontein to obtain water and Cooke explained that in dolomitic country surface water was scarce and often only found in caves. He suggested that the carrying in to a site of foreign stones implied a bag and Leakey said that the skin of a freshly killed antelope makes an efficient bag.

Bishop suggested that during the time when *Zinjanthropus* lived, it was probable that the area as a whole was 250 feet lower than at present; lavas, quartz, and Basement rocks might then have been available nearer the site than is the case at present.

In reply to a question from Monod, Leakey said that the *Zinjanthropus* level was very flat so that as the waters of the lake rose the people simply moved further back from the water's edge.

Biberson said that the affinities of the industries in Morocco were partly African and partly European, but that local Arab names had been applied in the terminology. No tools occurred with the Lower Villafranchian faunas; there were a few split pebbles, but it was not certain that these were tools.

The Calabrian sites in the Mediterranean had a cold fauna and those in Morocco a warm fauna. A Saletien type site, similar to Aïn Hanech, yielded abundant tools, but no fauna. The Maarifien marine transgression at the beginning of the Middle Pleistocene has been tentatively correlated with the Sicilian. Later on the Maarifien caves were formed and these contain the remains of cave dwelling animals and stone implements. Later marine transgressions (Anfatien) further resorted these assemblages.

The Harounian transgression occurred at the end of the Middle Pleistocene. In the lowest levels 95% of pebble tools occur and only 5% of bifaces and in the later levels these percentages are altered to 60% and 40%, and also the first cleavers occur, as well as polyhedral stones, true bolas stones, and less crude hand-axes.

The new site at the "Cunette" of Sidi Abderrahman has yielded fragmentary jaw remains of an evolved *Atlanthropus* type. The Sidi Abderrahman site in the limestones has yielded a new Middle Acheulian industry.

(Thursday, July 13th, was a free day.)

SESSION VI: JULY 14 (9:30 A.M.–12:35 P.M.)

PRESENT-DAY ECOLOGY AND ITS RELEVANCE TO PLEISTOCENE STUDIES

The session was opened by Bourlière who spoke on the ecology of the Albert National Park. He emphasized that the data applicable to this area were not necessarily true for other areas. The Albert Park adjoins the Queen Elizabeth Park and the preserved area thus extends on both sides of Lake Edward from Lake Kivu to Ruwenzori. The environment is extremely varied and examples of most sorts of African habitats are found in the region. The lower altitude (1,000 m.) provides an open savanna environment. The following table (2) was presented, with additions by LEAKEY, to show the distribution of animal species in the different vegetation zones in Albert National Park and in Kenya. In this table a figure indicates the highest point at which an animal was observed, e.g., elephant at 3,700 m.

BOURLIÈRE made the following particular comments on portions of the table:

Hippopotamus. This is one of the most numerous mammals around Lake Edward, about 19,000 being present in Albert Park; 26 per kilometer of river in the gallery forest of the Rwindi River and 115 per kilometer of the Rutshuru River in the savanna and grass steppe. They can live in water at a very low temperature (nearly 0° C.) and their main basic need is grass on which they feed.

Okapi live only in lowland rain forest and do not cross the Semliki River.

Buffalo can be found throughout the region. Three forms occur: a plains form with black coat and long horns, a forest form with a yellowish coat and smallish horns, and an intermediate form, on the Virunga volcanoes, with a black coat and small horns.

Bushbuck is found almost anywhere, even in the lowland rain forest and in the mountain evergreen forest.

Duiker. Cephalophus nigrifons occurs from 2000–4270 m., the yellow backed species *(C. sylvicultor)* in the lowland rain forest, and *Silvicapra grimmia* in the savanna.

Waterbuck is generally met with in more or less wooded savannas, but is also abundant in the open and treeless plains of the upper Semliki.

Uganda Kob. Of all the ungulates this has the most definite ecological requirements—i.e., open savanna with shortish grass.

Topi. This species is less restricted to grass and steppe than is the kob.

Dendrohyrax. This animal extends from the alpine zone, where it is the staple food of the leopard, to the rain forest where it is abundant. It is known to climb bamboos and trees.

Lion occurs in small numbers in every environment.

Leopard also occurs everywhere, especially in the alpine zone.

Wild Dog, which was very abundant twenty years ago, has spontaneously disappeared since 1955.

In summing up, Bourlière concluded that these animals can be divided ecologically into a group having wide, fairly unspecialized requirements (such as elephant and buffalo) and a group with narrow requirements (such as the topi and, even more so, the kob).

TABLE 2

RANGES AND HABITATS OF SOME MAMMALS IN THE ALBERT NATIONAL PARK (BOURLIÈRE) AND IN KENYA (LEAKEY)

	Elephant	Warthog	Bush pig	Forest hog	Hippo	Okapi	Sitatunga	Bushbuck	Bongo	Buffalo
Alpine zone > 3600m.	↑3700m.			↑3750m.						↑4500m.
Ericetum (tree heaths)	○			○ K				↑3000m.		○ K
Hagenia belt	○			○						○ K
Bamboo belt	◐ K		K	○ K				○ K		○ K
Mountain rain forest	◐ K		K	○ K	K			○ K	K	○ K
Lowland rain forest	◐ K		○ K	○	↑2000m. ○ K 25/km.	○ clearings		○ K	○	○ K
Dry deciduous forest	◐ K			○ K	○ K			○ K		○ K
Thickets	◐ K	↑1400m.	○ K esp. in temporary natural clearings	○ K	○ K			◐ K		◐ K
Open savannas	○ K	○ K	K H where man has disturbed vegetation	○	● K 115/km.			○		◐ K
Grass steppe	○ K	○	H		● K 0°-35°-40° C. water temperature					○ K
Swamps	○ K		K		○ K H		○ K			○ K

	Duiker	Waterbuck	Uganda Kob	Topi	Dendrohyrax	Lion	Leopard	Crocuta crocuta	Wild dog
Alpine zone > 3600m.	↑4270m. nigriformis				◐	○ K	◐ K		
Ericetum (tree heaths)	○				○	○	○ K	○	
<u>Hagenia</u> belt	○				◐	○ K	○	○	
Bamboo belt	○				○	○ K	○ K	○	
Mountain rain forest	○	K			○	○ K	○ K	○	K
Lowland rain forest	◐ many species				○	○ K	○ K	○	
Dry deciduous forest	○	↑1700m. K			◐	○ K	○ K	○	K
Thickets	○	◐ K			K	○ K	○ K	◐	K
Open savannas	○ sylvicapro	○ K northern sector	○ especially short grass	○ K		○ K	○ K	◐	○ K
Grass steppe			●	○ K		○ K	○ K	○	○ K
Swamps						○ K			

KEY

○ – Animal found in environment

◐ – Animal more abundant

● – Favored environment

K – Leakey's East African distributions

H – Animal in contact with human settlement

↑ – Highest altitude of observed species

NYH

The table below (3) shows the total populations of the dominant ungulate species in the ± 1,200 sq. km. of open plains south of Lake Edward.

TABLE 3
TOTAL POPULATION OF THE DOMINANT SPECIES IN THE OPEN PLAINS SOUTH OF LAKE EDWARD (± 1,200 km^2)

	Elephant	*Buffalo*	*Topi*	*Uganda Kob*	*Hippo*
1931	± 150	± 2,000	± 10,000	± 15,000	?
			Decrease due (?) to restriction of fire		
1940	± 500		± 1,200	± 3,000	
1959	3.293	24,054	4,798	9,571	9,900
			recent increase due (?) to deforestation by elephants (?)		

BOURLIÈRE went on to say that the population density of a locality is determined by the amount of food available throughout the year, but more particularly at the end of the critical season, which in Albert Park is the dry season. The daily amount of fresh plant food needed by large African ungulates, when adults, is amazing. For instance, an elephant needs about 150 kg. of fresh food per day to keep its weight stable, a hippo ± 100 kg. and a buffalo ± 30 kg. So long as the food productivity of an area does not change, game does not usually migrate far. There is, therefore, only a little free exchange of individuals. In East Africa, where there is less surface water, animals may migrate extensively at the end of the dry season, but in Albert Park the most common cause of temporary groupings is around salt licks. It is probable that several trace elements may be scarce in the diet when grass dries up and all are important for health and ability to reproduce. This is probably why elephants lick lava so that they may absorb the chemicals washed out by rain.

Sometimes animals in the wild appear to suffer from mass death. In most cases this is due to epizootic causes such as rinderpest in buffalo or anthrax (?) in hippo. Volcanic eruptions are far less destructive. In 1957 and 1958, eruptions of the Gitebe, Mugogo, and Kitsimbanyi volcanoes caused the encirclement of small unburnt areas of forest where animals remained unharmed. It seems likely, therefore, that except for small species actually buried beneath the lava, volcanic eruptions do not cause large scale destruction of animal life. Many more animals are killed by toxic gases (36 to 44% CO_2) emanating intermittently from cracks in the ground. Birds, bats, insects, and small reptiles are instantaneously killed in these "masuku" as well as elephants, hippos, buffaloes, bushbucks, monkeys, lions, hyaenas, and baboons. Some of these places are true "elephant cemeteries."

BOURLIÈRE said, regarding the rate of disappearance of undisturbed skeletons in the wild, that data were available for one elephant killed in the Kagera National Park before it was proclaimed 25 years ago. Two years ago one long bone and fragments of the skull were still recognizable.

BOURLIÈRE said that the census of animals in Albert National Park in Table 2 of the background paper was a minimum count. It was based on the patrolling of all the open areas of the southern sector of the park on three consecutive days, once every two months, over a one kilometer grid. Every habitat appears to harbor its own characteristic species and to sustain different densities of population. Uganda kob, topi and buffalo rank first in steppes and low savannas when buffalo, waterbuck, and elephant are most abundant in tree savannas. The open plains of Albert National Park exhibit the highest biomass of herbivores per square kilometer anywhere in Africa (and probably of anywhere in the world), that is 24 tons per sq. km; Queen Elizabeth Park has 19.5 tons per sq. km. Farther east on the dry steppes of southern Kenya and Tanganyika the biomass is much lower: from 13 to 4.4 tons per sq. km. In Mauritania it is only 5–20 kg. per sq. km. and this is largely represented by the addax. Woodland biomass is more difficult to gauge, but a count done for Ghana over 250 sq. km. gave 5.6 kg. per sq. km. (3 species of duikers). The greatest mammal biomass in such an area is, of course, that of monkeys. In Czechoslovakia, in undisturbed, primaeval mountain forest, Turcek had calculated the biomass of animals and birds at half a ton per sq. km.

In East Africa the biomass of ungulates is unusually high, probably because the total animal population of an area is able to use all its vegetable products as many are specialized feeders. Physiology also plays some part in that wild ungulates breed younger and show better growth on a similar food supply than do domesticated forms. Of population turnover, that is the rate at which a population can maintain itself without changing its size, Bourlière said that large animals, such as elephant, effect turnover very slowly and smaller creatures more quickly. In Albert and Queen Elizabeth National Parks where 70% of the biomass is composed of three slow-turnover species, the other 30% is of small species with quick turnover. Farther east the smaller species with quicker turnover constitute a much greater percentage of the biomass. Obviously the best type of environment for primitive hunters would be one with a large ungulate biomass, but a high percentage of quick turnover animals.

DISCUSSION

EMLEN asked whether the figures given were for areas with no human disturbance.

BOURLIÈRE confirmed that this was generally so, but pointed out that animals, without human interference, can affect the vegetation as had been shown in Albert Park over the last 25 years. There, elephant and buffalo had increased to such an extent as to destroy parts of the trees in wooded savannas and the topi and the kob had thus been enabled to increase.

LEAKEY said that on the Serengeti Plains, in Ngorongoro Crater, and Nairobi National Park there had been progressive human increase over the last 30 years which had also meant a great increase in cattle, and these had never been included in any count of the wild ungulate biomass. Leakey said that in a drought zebra, wildebeest, and cattle are the first to be affected, and then hartebeest; but gazelle are hardly affected at all. Figures to demonstrate this phenomenon are in process of preparation. At the end of a very dry season or in a famine some deaths occur—usually of the newly born, yearling, or elderly animals.

BOURLIÈRE agreed that mass death of animals by famine can occur but more generally in northern areas (related to snow cover) rather than in tropical localities, although it can be a cause of death there also.

MOREAU suggested that the native fauna might be better able to utilize poorer grass for nutrition than cattle; and it is usually on the latter that the carrying capacity of land is based.

HIERNAUX suggested that the balance between the available food and the biomass in non-human populations was better kept than by man, particularly modern man, where the biomass is frequently unrelated to the food supply, resulting in famines.

COOKE quoted figures given by Knobel for the Kruger National Park and thought that these could be increased by 40–50%. Observations made on marked animals over two to three years had shown little movement of any kind, except among elephants which might move as much as 300 miles. The Gemsbok Park, also in South Africa, had six individual gemsbok (600 kg.) per sq. km., but this was not an official figure. These animals do not appear to need a water supply, but obtain their requirements from their food. This may be a significant point when considering Pleistocene animal populations at different periods. Cooke pointed out that the small gazelle population in Libya still remained selective in its feeding habits in a vegetation supported only by dew. This too may be an important point when considering the preservation of the vegetation of marginal areas.

LEAKEY said that in the Serengeti Plains wildebeest migrations had been observed since 1931 and bore no relation at all to food and water supplies, but seemed to be motivated by instinct. When no water was available in its customary calving area a group would return there to calve and the young would die. On the other hand zebra will move to find food. BOURLIÈRE pointed out that recent observations on wildebeeste made by Lee M. Talbot in the Serengeti-

Mara area do not support this "classic" point of view. To quote his recent paper: "The wildebeeste population movements appear to be determined by grass growth, which is a function of burning, grazing and rainfall, and availability of water. The location of calving appears also to be determined by these factors" (L. M. Talbot and M. H. Talbot, "Preliminary Observations on the Population Dynamics of the Wildebeeste in Narok District, Kenya," *East African Agricultural and Forestry Journal*, 1961, pp. 108–16).

LEAKEY went on to say that *Dendrohyrax* and *Procavia* (living in rocky areas) both climb trees and both occupy similar habitats though the one needs available water and the other does not. Hyrax bones in a prehistoric deposit are thus no good indication of the type environment. Even the black rhinoceros, which is said to eat only leaves, will also live entirely in grassland.

COOKE agreed with Leakey's remarks about the feeding habits of the black rhinoceros. He also cited instances of wildebeest surviving after drinking water so salty that bream and barbel were dying in it.

EMLEN said that there are physiological mechanisms in birds which enable them to deal with excessively salt water, but he did not know the wildebeest were able to do so.

SESSION VII: JULY 14 (2:30–6:15 P.M.)

The session began with a continuation of the discussion of Session VI after LEAKEY had added his East African data to Bourlière's table of the distribution of animal species in differing environments (see above).

LEAKEY, in amplification of the table, said that eland and buffalo were two of the most widely spread ungulates. In conditions of famine certain animals suffered by being reduced to feeding in unnatural positions on unaccustomed food, particularly kongoni, wildebeest, zebra, and giraffe (who also have difficulty in swallowing when eating saline earth because of the length of their neck). At times of climatic deterioration such animals suffer more. Some, which are normally nocturnal, will feed by day in times of drought as will the ant bear when termites are scarce. Wart hogs, that eat grass, sharing the habitat with grazing animals, will in times of drought eat thorn bush, first pushing off the thorns with their warts. Wart-hog molars can be misleading when found as fossils since pigs without warts have very similar molars.

In reply to a question from Cooke, LEAKEY said that bush pigs rooting in damp, muddy ground may be searching for roots. As regards wart hogs, Leakey said that there appeared to be a cycle starting with large numbers which became thinned by swine fever. The fever did not recur for several generations and the numbers increased again and then when the immunity had disappeared the swine fever recurred.

EMLEN referred to the table of animal distribution (see above) and suggested that it should be viewed horizontally as well as vertically because the absence of some species in a particular habitat may have a bearing on the presence there of others.

MOREAU next spoke on altitude as a factor in the distribution of birds and large vertebrates. He suggested, though he said that it was not conclusively proved, that with the exception of hyrax all animals with a wide altitudinal range appeared to be bulky. He instanced rhino that are common at 8,000 ft., especially on the wetter side on Kilimanjaro, and eland that have been observed on Kilimanjaro moorland. *Zosterops*, a large bird, exhibits increased wing length in relation to reduced minimum temperature and air pressure, and it seems likely that other animals may change also with altitude. At present it appears that smaller creatures show more habitat specialization than do large ones.

DISCUSSION

BOURLIÈRE said that, although more data are needed, it seems likely that some rodents have a wide altitudinal and habitat range, but it is not yet possible to say if they range from lower levels upwards or vice versa. It is also possible that shrews, that to some extent make their own microclimate, have a wider range than once was thought. EMLEN added that in the Rocky Mountains some shrews have a wide and some a restricted habitat. LEAKEY suggested that food habits—that is, the extent to which a creature was omnivorous—might have a bearing on range.

BOURLIÈRE said that in the Albert Park 135 kg. per sq. km. of rodents occurred in the same area as 24 tons of herbivores. The opposite was true in the southern European steppe country where 35,000 rodents were found in a square kilometer supporting 20.4 saiga antelope. He suggested that animal populations increased and decreased in waves. COOKE suggested that some such fact might help to explain the greater number of small animal forms in Later Stone Age as compared with Middle Stone Age sites.

MOREAU then continued with his remarks on the distribution of avifauna.

He said that birds seemed to be more specialized in range in that practically none are found both in forest and in non-forest habitats. There are also marked differences in birds found in tall, rank or short, sparse grass. Altitudinally 5,000–5,500 feet, where the mean annual temperature change is 18° C., seems to be the level-separating species. Distribution is also linked to temperature change, and in places near the coast which have a smaller daily temperature range, a 50% change in bird population occurs at 2,500–3,000 ft. The vegetation also changes at similar altitudes. At these levels birds adapted to lowland habitats fall out of the population and highland types take over, gradually falling out until only about six remain at the highest levels. Very few lowland birds are found in high country, although some birds which in Europe live in grassland occur in high land in East Africa.

Moreau said that on the whole there was little evidence of vertical movement of birds which seem less adapted to temperature change than do larger mammals. Birds have a wide range of food and in tropical Africa there is little evidence of seasonal migration though wide movements do occur in dry areas such as the Sahara. There have also been instances known of virus disease being carried by birds.

Monod then spoke on fauna distribution in desert regions. He stressed the great caution necessary before conclusions are drawn as to past ecology based on the comparatively few specimens usually found in a deposit. Monod divided his observations on desert fauna into five sections:

1. *Marginal distribution.* By this was meant the extreme range in the area inhabited by a species, which may be linked to a certain type of vegetation and climate. For instance the presence of elephants in the Afollé plateau in a Sahelian environment with acacias and euphorbia was very surprising, but was probably an extreme biogeographical limit. Wart hog had been recorded on the west coast below Cap Blanc in 1923 and hunting dogs in the Hoggar mountains in 1927. Tortoise is found at the limit of the Sahelian vegetation zone.

2. *Island distributions.* By this is meant the segregation of populations away from the main areas. This usually, but not always occurs in mountains. For instance, the Tibesti baboon occupies a mountain island situation and scimitar oryx has been found in the Spanish Sahara, which is not its main habitat. Thus, the bones of scimitar oryx in a deposit should not be taken as evidence of Sahelian conditions. The northern dormouse has been found on the coast in the same area, leopard in an extreme type of Sahelian environment with at most 50 mm. of rain a year, and there is an unconfirmed report of giraffe among dunes southeast of Adrar.

3. *Mountain distribution.* As soon as mountains are high enough to cause real biological zonation many cases of island-type isolation occur. The Sahara mountains do not demonstrate this phenomenon as well as others in Africa. Monod said that Professor Quezel had recently supplied him with new data from pollen analysis (there were now over 30 sites) bearing on past climate in the Hoggar. The last pluvial would seem to have had a cold Mediterranean steppe ecology changing to forests of conifers, oaks, and maples about 10,000–8,000 B.C.; thereafter becoming drier with pines and junipers to about 2,800 B.C., when replaced by the acacia Sahelian type of vegetation and subsequently to present day desertic condition. The possibility must be borne in mind that different types of environment may have existed in nearby regions. This may be brought about by wind direction and the 10° difference in orientation of

the pre-Neolithic dunes and the modern dunes above them would seem to provide evidence for such a change during the Pleistocene.

4. *Dependence on water.* It is well known that the presence of surface water does not determine the distribution of many desert species. Desert mammals will normally make use of any available surface water, some even digging for it in moist wadi beds, but many can live at great distances from water and some quite large mammals obtain their water exclusively from their diet, for instance *Addax* and fox which eat the watery rhizomes of *Philipaea lutea* often digging them out of the sand. In the western "Empty Quarter," which is roughly 1,000 by 500 km. in size, there is never any surface water, but several species of birds and animals live there—for example fennec, jackals that follow the herds of mendez antelope and live on the young, the newborn, and the placentas, while the *Addax* themselves subsist almost entirely on *Aristida pungens.*

5. *Aquatic life.* In the desert this constitutes a true island type of distribution. Water occurs usually in isolated pools, but even when there is a small system, as in the artificial canals of the Algerian Sahara, it is still isolated. Crocodiles, fishes, crustaceans, mollusks, and even a *Medusa* all occur. Mass mortality takes places if the water overflows and subsequently dies up, or if the salinity increases as happens in coastal lagoons. There are many purely temporary rain-water pools which contain specialized invertebrate fauna whose eggs are preserved in the mud when the adult dies as the water dries up. The mollusk *Melania* has a wide distribution and lives in the western Sahara and the Tibesti mountains. As a fossil it is found near Timbuctu, Accra, and Dakar, but not in the Niger. There is no apparent physiographic barrier to its spread there though the fact that it is viviparous may have in itself been a limiting factor controlling its distribution.

Discussion

DeVore cited an instance of fish said to have been carried to the Galapagos Islands in the feathers of ducks and so to have restocked the water after a volcanic eruption.

Bourlière, in reply, said that fish eggs might have been carried in the mud on birds' feet, but actual fish would not have been transported. In connection with such forms of transport he mentioned the case of pollen found in Tierra del Fuego from plants that did not occur there and which must have been wind-blown. Such possibilities must be borne in mind when assessing the value of pollen analysis for the determination of past ecology.

Leakey mentioned that *Medusa* had been known to appear suddenly in pools in East Africa that had previously been dry for long periods. Monod added that *Medusa* had a wide distribution. It occurred in permanent water in the Sahara, and was endemic in Niger, Tanganyika, and Uganda, and sporadic in many places.

BOURLIÈRE, referring again to plants, said that seeds could be dormant for a considerable time and that after 1,000 years in peat in Korea and Mongolia, 80% of poppy seed had germinated. Acidity was important in the germination of many seeds and this was often provided by their passage through the bodies of animals.

COOKE, in drawing attention to the isolated islands of vegetation in southern Africa, suggested that soil conditions must play a large part in determining what seeds became established. He thought too that the vegetation islands attract certain animals unusual in the surrounding area.

BUTZER stressed the importance played by wind in the distribution of pollen and the significance of this for pollen analysis. It was also essential to have a very wide knowledge of modern pollens before making inferences as to past ecology from fossil pollens.

PRIMATE ECOLOGY AND BEHAVIOR

At this point a film on baboons in Kenya made during the course of fieldwork by DeVore and Washburn was shown, and was followed by slides showing certain aspects of baboon behavior.

DISCUSSION

WASHBURN explained that the areas where baboons are easiest to see are not those where they prefer to live as they like to be near trees. Studies had been made at both Amboseli and the Nairobi National Park in Kenya.

DEVORE, in reply to Emlen, said that the baboons appeared to live in social symbiosis with impala and waterbuck.

LEAKEY said that the Olduvai baboons had a daily movement to water of seven or eight miles, whereas Washburn had mentioned three to four miles.

WASHBURN said that his observations had shown there was great variability and adaptability.

COOKE said he had seen baboons on the 1,000-foot Lebombo escarpment plucking young leaves for food and turning over stones, but that he had not seen them eat grass as Washburn and Devore said. LEAKEY thought that diet depended on season and availability.

WASHBURN agreed, but said that baboons do indeed eat grass, though rather unobtrusively and trained observation is sometimes necessary to detect it.

CLARK mentioned how the Valley Tonga tribe in Northern Rhodesia collect grass seeds by scooping them with the hand off the stem into a container and how he had seen baboons doing the same thing, but directly into the mouth.

WASHBURN closed the discussion for the day with the observation that arboreal species seemed to have the least worn dentition and the smallest jaw muscles, and also mentioned some dental characteristics of chimpanzee and *Australopithecus.*

SESSION VIII: JULY 15 (9:30 A.M.–12:50 P.M.)

The session was opened by DEVORE who spoke on baboon behavior. He began by setting out in tabular form the home ranges of the baboon troops he had observed in the Nairobi National Park.

Troop	*Range (sq. mi)*	*Size of Troop*	*Observations*
S.R.	15.5	26	6 adult males
H.P.	13.8 +	87	
K.V.	11.7 +	77	
L.T.	9.2	17	2 adult males (one very old)
S.V.	9.6 +	40	
A.R.	2.9 +	28	1 adult male—area largely
P.P.	7.0 +	24	overlaps others

DeVore said that there did not appear to be any defended territory and the home range varied very considerably being centralized round a "core area." One troop may have several core areas and a much larger range than the one they normally use. The home ranges of several troops can overlap, often considerably, though there is very little overlapping of core areas. Sleeping places are in trees, often acacia or fever trees, and near water.

A very common activity among baboons of all ages and both sexes is that of grooming, though it is most common among adult females. Presenting, normally characteristic of females in estrus, may also be practiced as an aspect of social behavior.

In the Nairobi National Park the country was park savanna-grassland. Where open patches with similar vegetation occurred on the edge of the forest these were also used by the baboons. The diet of the baboons included a very wide range of foods and in the dry season 80% of the day was spent in feeding. The diet can be described under three main headings:

1. *Vegetable foods available all the year round.* Grass constitutes 90% of this vegetable food. The culm or base of the stem is cleaned with the hand and eaten, or the seeds may be drawn straight into the mouth. The juicy rhizomes are also eaten—sometimes the animals will dig as deeply as 2–3 inches to find them—so that baboons are able to live in overgrazed areas. They will also dig sometimes quite large holes for roots, bulbs, and tubers. When a troop feeds along the river bank in general it is to procure grass rather than water. The seed

pods of various acacias are eaten (c.f., *A. clarigera* and *A. drepanolobium*) as are tree galls and the buds and leaves of fever trees, which trees are also favored sleeping places.

2. *Seasonal vegetable foods.* In the Nairobi Park such foods included figs, croton nuts, the imported *Kei* apple, sisal of which the shooting center stem was eaten, *Euphorbia candelabrum*, and the leaves and vine of *Cucumus postulatus* (but not the fruit, probably because it may be poisonous). Native gardens were also popular feeding grounds. The adult males ate first. They were able to distinguish unwholesome food. A drugged banana, offered to an adult male was immediately refused though it was accepted and eaten by an infant.

3. *Insects and small animals.* The insect population of the area was not heavy and so did not provide a great proportion of the diet. Sometimes the animals would turn over rocks looking for insects, they would eat the insects in tree galls and occasional grasshoppers and beetles. During a plague of army worms, however, the baboons fed almost exclusively on them. On six occasions they were seen to eat meat and on one occasion a bird. They did not look for nests, but would rifle them if they came across any. All the animals they were seen to eat were young, and were encountered while the baboons were looking through the grass for their normal food. Adult baboons recognized that meat was edible, but the young did not. On one occasion the baboons caught a Thompson's gazelle, banged it on the ground and consumed everything except the hair and the large bones. They do not, however, eat meat found dead, however fresh it may be, and although they are constantly adding new foods to their diet they do not always appear to recognize meat when they see it. Their basic adaptation would thus appear to be to vegetarianism. The inhabitants of the *Zinjanthropus* level of Bed I at Olduvai would appear to have taken a step further from vegetarianism than have the baboons.

WASHBURN then spoke on primate behavior and some of its pertinence to an understanding of early human behavior.

A table (Table 4) of the size of all baboon troops studied indicated the variations in troop size. It was pointed out that the size of the troop varied from 12 to 185 and seemed to be determined by social factors. The number of individuals per square mile depended, however, on the available food supply. When a troop reached about 80 individuals, subdivisions began to appear within it and this may be the manner in which new troops are formed.

Although troops may intermingle around a waterhole there seems to be very little interchange of individuals. A single troop shows considerable signs of inbreeding, the members exhibiting related characteristics as if they were siblings. When baboon-type mating patterns were replaced by human-type mating patterns, greater gene flow was released and one would thus expect man to show racial rather than specific differences. In baboon troops considerable sexual selection occurs in that the females are usually all impregnated by the larger males.

TABLE 4
BABOON POPULATIONS IN KENYA

Troop Counts in Nairobi Park	*Size:*	*Remarks:*
1. Mbagathi	12	
2. Lone Tree	17	2 killed in April 2 born in September
3. Python Pool	24	
4. Athi River	28	adult male changed to Sosian Valley troop
5. Songora Ridge	28	3 born in December
6. Sosian Valley	40	lost 3 adults and an infant; gained a male
7. Kapio River	61	gained about 4 infants
8. Kisembe Valley	77	in Oct.–Nov.
9. Hippo Pool	87	
Total	374	
Average troop size:	41.5	Population density: about 10 per sq: mile

Troop Counts in Amboseli Reserve

1. 7th Hole, No. 2	13	8. South	70 (?)
2. Corner	42	9. Grey No. 2	74
3. 7th Hole	47	10. Wastelands	78
4. Grey	51 ( )	11. Dark	88
5. Normal	66	12. Observation	94
6. Halfway	57	13. Emali	103
7. Causeway	64	14. Big	171
		15. Big No. 2	185
Average troop size:	80	Total	1,203

Consort pairs of baboons might live apart from the troop for short periods of hours and/or days when the female was in heat. In some troops two or more divisions seemed to exist and these might separate for varying periods. Such divisions sometimes consisted of the dominant males, the females, and the young in one group and the remainder in the other, or they might represent a more natural cross section of the whole group.

Washburn said that he had observed the baboons' reactions to eleven kills, but they exhibited little interest. There were not a great many kills in the area and those which occurred were rapidly consumed by the carnivores, vultures, and hyenas. So scavenging would seem a rather uncertain way to obtain a meat diet. An animal that had already developed a taste for meat-eating might scavenge, or young animals could be killed fairly easily to supply a need for meat.

Washburn considered that the ground-living primates, baboons, macques, etc., formed a remarkably uniform group, both in structure and behavior, and probably also psychology. It would seem that the more ground living the group the wider and more diversified the areas it can occupy, in much the same way as Middle Pleistocene man, and the more likely differences are to be racial rather than specific. Usually not more than one kind of the ground-living form occupies one area.

In tree-living forms, however, the opposite would seem to be the case and they tend to form species rather than races, as for example in the guenons (*Cercopithecus*).

Discussion

Emlen, DeVore, and Bourlière drew attention to the wider range of speciation among ungulates and their greater numbers in plains conditions as opposed to forests.

Leakey considered that many of the ungulate species in the plains might in fact be regarded as races. Cooke suggested that life in marginal areas where creatures were more limited in range and habitat made for speciation. Leakey, however, considered that the significance of the wider range was in its bearing on the food habits of the animal. Emlen suggested that the less speciation and greater raciation of primates might be a result of their wider food range.

In reply to Clark, Washburn said that the gelada was not really a baboon and constituted an exception to his suggested hypothesis. The present classification of these primates, which inhabit different ecological settings, was: *Papio doguera, P. anubis, P. papio, P. cynocephalus, P. chacma, P. hamadryas; (Drill); (Mandrill); Macaca; Theropithecus gelada.*

These forms had a wide distribution, but Washburn said that in his experience no two occupied the same area at the same time.

Leakey, on the subject of scavenging, said that early man may well have eaten young animals found in the bush. He considered scavenging to be quite feasible since lions often leave their kill for several hours and a lion is not dangerous when it has just fed. In the Olduvai area a small family could scavenge meat at least twice a week from kills and it is also possible to catch and kill adult small antelopes with one's bare hands, as he knew from experience. Therefore, a creature that desired meat, unlike the predominantly vegetarian baboons, could definitely obtain it. At Olduvai the proportion of adult animals in the floors increases from the pre-*Zinjanthropus* level to Bed II.

Monod referred to information given by P. Ishac that villagers in the East Chad Republic take parts of lions' kills and in the dry season provide the lions with water.

Clark quoted the practice, common in savanna country in Africa, of sending Bantu children out in the early morning to observe where the vultures are flying so that the group may scavenge meat from the kills. [*Note.* In connection with baboon behavior attention was drawn to a paper by K. R. L. Hall on "Cape Baboons" which instanced the eating of crabs and mussels. Baboons had also been seen to eat clay and DeVore confirmed that he had seen this at Lake Magadi.]

Schaller then illustrated and described some aspects of the ecology and behavior of the mountain gorilla. Schaller said that there were two gorilla populations, one in West Africa and the other, which he had studied, in the Central African mountain region. His study had been made in the valley overlooking the Virunga volcanoes, which is open country with a good gorilla population—three per square mile. This Central African gorilla group has a fairly wide range of habitat and can be observed from Lubero near the north of Lake Edward south to Fizi and in a wide area in between and to the west. Although they do not stay normally at the higher levels these animals have been observed from an altitude 1,500 ft. up to 13,500 ft. They do not live in the thick primary or very old secondary forest lacking an understorey, but in old garden patches or similar clearings where the secondary forest is beginning to regenerate. Gorillas will not cross rivers by swimming, but prefer to go dry shod so that a river can be a barrier to their movement unless they go up to the headwaters, where it is narrower, to cross.

For sleeping places they prefer the drier areas, such as slopes, but will go down to the valleys to feed. Each night a new nest is built; it is always left fouled. The nest may be made on the ground, as in *Hagenia* forest, or in a strong tree in the lowland rain forest. They do not like to stay in thick rain forest or bamboo forest.

As to diet, Schaller said that gorillas are very selective feeders, and most of their foods are bitter to the taste. They are enabled to live at such varying altitudes because a certain number of their favored food plants overlap from area to area, so that the change is made gradually, as it were. One group may frequent areas as much as 5,000 ft. apart. They only eat certain parts of a plant and have been observed to eat the inner white pith of banana and *Afromomum* and some *Senecio* and the leaf base of *Lobelia*. In all, parts of some 20–30 plants are eaten. The animals have also been seen to scrape their incisors on ground with a high sodium and potassium content, but they do not seek out licks. Young captive animals will eat anything offered to them, but older individuals will not. Water is often scarce and the gorillas drink very rarely, apparently obtaining water largely from their diet.

The daily cycle as observed by Schaller consisted of wakening about an hour after sunrise, feeding for two to three hours, resting in a tight group until 2:00 P.M., feeding again for three to four hours, and then sleep.

The female gorilla weighs about 150–200 lbs. and the biggest wild shot male weighed 480 lbs. Age has been computed on the records of zoological gardens. It seems that the female reaches sexual maturity at 6–7 years and the male at about 10 years though he is not fully adult until 2–3 years later. By 12 a male has a grey back and an old animal is all grey except for the arms. Hair may be shorter in tropical rain forest than in colder areas. The gorillas' facial characteristics are very variable.

Schaller said that the average home range is about 10 miles, but this may be altered every few months. There is no daily movement to and from an area as with baboons. The average group is 17 in a normal range of 5–27 individuals, but groups of one male, one female, and one young are fairly common. A new group is probably begun by such a pair leaving a larger grouping. Group behavior is determined by the personality of the dominant male, many changes occurring in the group of a friendly male, but few in that of an aggressive male. It is probable that all the older animals in an area recognize one another. They will come close to and observe humans but will not touch their possessions. There is no clearcut evidence of tool use among these gorillas, but in West Africa they have been observed to crack nuts with a stone after first laying these on a rock, and also to put a stick into a bees' nest to suck off the honey. In Tanganyika chimpanzees were seen (by J. Goodall) to put a stick into a termite hill and to eat the adhering insects.

At birth the young are quite helpless and even up to a month old are unable to hold on to the mother. Unlike baboons, gorillas groom very rarely; there is no adult intersexual grooming, though occasional grooming of a subadult by a female occurs.

Locomotion is entirely quadrupedal. They squat and sometimes stand up, but they rarely move bipedally. They are good at climbing trees though adult males do not do so very often.

Fighting is rare, though some scars, usually round the face, can be observed. A show of agression is usually sufficient. This may be because in most areas there is no competition for food and if a lone male joins a group he seems to have a right to the females. There are thus probably no reasons for fighting. Sexual behavior is infrequent and occurs only at the invitation of the female.

Chimpanzees share most of the forests with the gorillas, but are more abundant in areas where there are no gorillas. Chimpanzees usually nest in trees.

Discussion

Leakey suggested that the main differences between the behavior of baboons and gorillas could be attributed to the fact that the former are ground dwelling and the latter are forest-adapted.

Schaller agreed. He went on to say that sometimes the gorillas' daily movement is in excess of what is needed to obtain food. It seemed that the size of groups could be determined by the extent of the food supply or even by disease. No sick animals had been observed other than one case of skin rash and a common cold, but in West Africa these animals are subject to tick-borne fevers, yaws, and leprosy. Death may also occur as a result of accident. LEAKEY referred to a dead gorilla which he had examined in East Africa and which had had no parasites, internal or external.

EMLEN said that the gorilla troops moved about as a compact unit, perhaps to protect the young from preying leopards, but solitary males often wander as much as two miles from the group. SCHALLER said that the females and very young always stay with the dominant male. In reply to Hiernaux, who asked whether the behavior of chimpanzees could be considered more juvenile than that of gorillas, EMLEN said that a comparative study of the two creatures was at present in progress in California.

SCHALLER said that the life span of a gorilla was probably about 25 years. The female begins to reproduce at about 7 and has one infant about every four years, which means not more than four or five in a lifetime. An infant stays with the mother until it is about three years old. A 20% mortality is fairly likely.

The session closed with the showing of a short film of gorillas taken in the field by Schaller.

SESSION IX: JULY 15 (2:30–6:10 P.M.)

(Continued discussion on baboon and gorilla behavior.)

The first part of the discussion has been reported under the subjects touched upon. Those taking part were BIBERSON, BOURLIÈRE, CLARK, DEVORE, EMLEN, DE HEINZELIN, LEAKEY, MONOD, SCHALLER, and WASHBURN. The main points that emerged were:

1. *Diet.* Gorillas' interest in sodium or potassium rock, though variable, did not seem to be seasonal. They seemed to eat woody pith and not the insects beneath the bark. The groups observed by Schaller had not been seen to drink, but water was scarce in their area. Baboons show no interest in dead meat. In the animal kingdom as a whole cannibalism is rare.

2. *Co-operation for mutual assistance.* Among gorillas little was observed apart from the mother-infant relationship; and on one occasion an adult male carried a juvenile to safety. Baboon and macaque males protect infants and sometimes also larger juveniles. If a troop is hunted the older males stay in the rear to fend off enemies. Lagging members of a troop are encouraged and dead infants are sometimes carried with the troop. Food is never shared.

3. *Survival of injured individuals and the stability of the population.* Injured individuals are observed among primate groups, but not usually among other animals. Ninety-seven per cent of gibbons' skeletons examined by A. H. Schultz showed healed fractures. Infant mortality among baboons would have to be of the order of 66% to keep the population stable, but populations always

fluctuate, probably around a mean number. Among injured individuals the incidence of survival probably depends on the chances of death in the local environment; an individual that becomes separated from the group is generally lost. Among ungulates the injured seem to die, though elephant and wild dog protect injured animals.

4. *Disease.* Very little disease was observed and it is not known if these creatures are subject to epidemics. One gorilla was known to die of gastroenteritis and both baboons and rhesus monkeys are known to suffer from arteriosclerosis.

5. *Dominance.* Among baboons the *alpha* male is supported by a group of dominant males who assist him to keep his position. Only the older males in the dominant group are able to breed.

6. *Bipedalism.* Ape and human-type walking differ, the ape having the knee bent; the human pelvis and femur are a different shape from those of apes. It was thought possible that there might have existed a stage intermediate between ape-type bipedal locomotion and human walking which still permitted substantial freedom of the hands.

7. *Right-handedness.* Experiments on chimpanzees carried out 25 years ago showed more right-handed than left-handed individuals. On baboons and gorillas data were insufficient, but right- and left-handedness and ambidextrousness had all been observed. Much experiment was at present in progress on many animals and birds. From a study of stone tools early man is said to have been right-handed.

Bourlière then suggested the following table to correlate the various phenomena affecting early man and other animals:

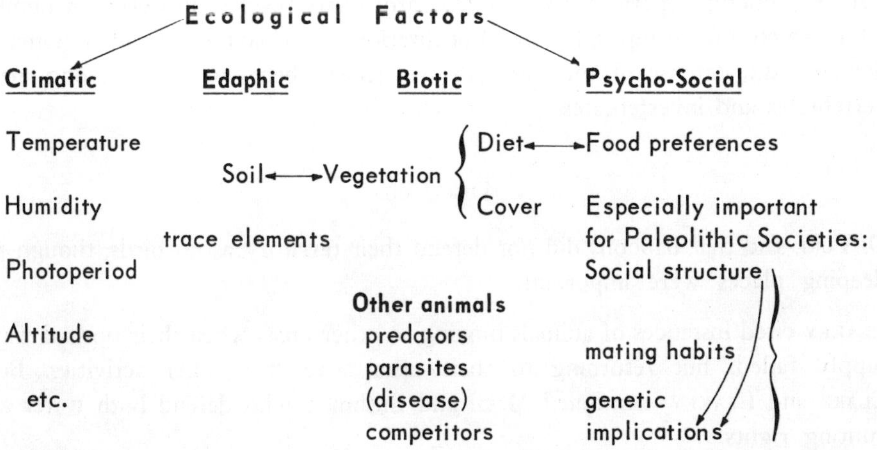

Emlen then suggested that the study of the psycho-social factors was what was really meant by ethology and he thought that this could be applied to a consideration of proto-hominid behavior. The information obtained from prehistoric sites relates to the structure and tools of the actual creatures, to the climate, etc. The picture of social behavior to be built up from this may be helped by studies of present-day primates, vertebrates, or even lowlier forms. To date parallels for human behavior had usually been sought among other primates, but since these are more closely related species they may conceivably have evolved greater divergence to escape too close ecological competition. A study of other animals, less closely related to man, might be illuminating. For instance early man was a hunter and it might be helpful for the understanding of his behavior to consider that of carnivores, such as lion. Bipedal locomotion was practiced to raise the eyes above the level of the surrounding vegetation and various animals jump into the air for the same reason (for instance the wild dog, jack rabbit, coyote, etc). One can suppose, therefore, that height above the ground was useful to early man and may have influenced development of the ability to walk upright. Some animals sit up on the haunches to free the hands, for instance squirrels and some monkeys. Many creatures use tools such as sea otters, various kinds of birds, and even spiders.

On the subject of language and communication Emlen said that this was not much better in gorillas than in birds, but in the past, situations may well have arisen when man's survival depended on improved communication, which must then have been highly selected for. It need not at first have been true language—for instance bees communicate by the dances performed over the hive—but as human groups grew larger and communication became more important it must have developed.

Social groupings similar to the human male-female relationship are found among various carnivores, a few rodents, and some birds. Troop size, for instance, among American quail or laboratory rats reaches an optimum number after which the group splits up. Localization of behavior around a home or den and the defense of this area are widespread behavior traits among many vertebrates and invertebrates.

Discussion

DeVore said that baboons did not defend their territory as do birds, though the sleeping places were important.

Leakey cited instances of animals moving to other areas when their normal water supply failed, but returning to their usual areas for other activities. Both Clark and Leakey mentioned Masai and Bushmen who defend both water and hunting rights.

In reply to Bourlière, Schaller said that gorillas did use facial expression as a means of communication, but usually in conjunction with gesture or sound; vocalization was particularly important when the troop was scattered.

Leakey considered that the range of noises possible to a baboon was, on his own experience, not so varied as that of a dog.

Bourlière drew attention to the importance of the fact that in baboon troops, for instance, the part played in reproduction by the males and the females differed and a difference also occurred in solitary species as opposed to social species. A baboon female matures at 4 and may reproduce soon afterwards. The male, on the other hand, matures at 5 or 6 but, owing to his position in the hierarchy cannot reproduce until about 10 and his expectation of life is 15.

[From 4:30–5:00 P.M. de Heinzelin showed a film illustrating Ruwenzori environments. This film had been taken during an expedition to study the existing glaciers.]

ECOLOGY AND HUMAN POPULATIONS

The session then continued with Hiernaux speaking on ecological factors and their effects on human population.

Hiernaux considered that it was necessary to add to Bourlière's table, given earlier in the session, the following psycho-social factors which greatly affect man, particularly modern man:

1. *Economic factors* influencing food production, population density, and movement from rural to urban areas, thus affecting gene flow. The breaking up of isolates and the formation of larger population groups greatly increase vigor (heterosis), at least in the first generation.
2. *Political factors.* A strong central power permits greater population density in a given environment, for instance in Ruanda Urundi.
3. *Technical improvements* that affect food production.
4. *Aesthetic ideas,* particularly those which influence the choice of a mate.
5. *Moral and religious ideas,* for instance polygamy and monogamy.
6. *Modern medical practice* may change or override the direction of natural selection, for instance by the suppression of malaria.

These factors all influence the biological development of man, though in modern man more widespread hybridization has probably reduced the factors of differentiation.

In large breeding populations without hybridization the gene frequencies remain constant, but in small populations they fluctuate and genes may eventually be lost. Fluctuation and genetic drift play a very important role in breeding populations of less than 100. During most of the period of his evolution man must have been subject to genetic drift since until Neolithic times he probably lived in small groups favorable to it and small isolates probably more easily led to raciation than larger modern populations. Modern man also has more cultural means of influencing natural selection.

Hiernaux went on to say that where a group is isolated and subject to selective pressure (for instance pygmies in equatorial forest in Africa and possible pygmyization in India, according to Gusinde) raciation seems to occur. The length of time needed to produce pygmies is not known. The shortest, those of the Ituri area, do not, as is usually supposed, live deep in the forest, but near the eastern and northern margins. In the Central Congo forest there are no true pygmies, only pygmoids of shortish stature who may have been brought there by the Bantu. All have traditions of having come there, but there is no fossil evidence to show the stature of the Neolithic inhabitants. It is possible that the pygmies came at the beginning of the present millennium and that they did not penetrate deep into the forest where the biomass is low, until forced by pressure from outside. Many forest cultures seem to depend on fish.

Hiernaux said that the rate of evolution of different characters varies and may be influenced by selective pressure—for instance the advantage of the morphological change to produce a larger brain in early hominids. As one feature changes others also change—for instance the cephalic index drops with increase of stature—so that correlation of such differences must be reckoned with when considering fossil hominids. In addition one favorable change may be accompanied by other, necessarily related but unfavorable changes. An important line of investigation, Hiernaux considered, would be into the change in variables throughout growth and the way in which natural selection can operate by arresting growth at some point. An example of this in humans is the lengthening of the growing period which retards the time of development of some genes and causes different morphological proportions as in the pedomorphological features of Bushmen.

Discussion

Bourlière emphasized both the importance of population size and the marriage system on human biological development. In the pygmy populations of the world the gene frequency varied, but each resembled that of the surrounding population, thus suggesting a recent move of the pygmy groups into a forest habitat. The smaller over-all size may be in response to the reduced need for evaporation from the body surface, but the sweat gland pattern of pygmies had not yet been investigated.

Washburn said that Hiernaux's remarks had raised all the major problems of physical anthropology today. With regard to the protection against malaria afforded by the sickle cell, Washburn said that an increase in gene frequency from 0.2 to 0.8 would protect 32% of individuals from disease, leave untouched 64% and 4% would still die of it. He thought that if *Homo sapiens* was protected in this inefficient way against malaria he could not have orginated in Africa.

Bourlière, Cooke, and Leakey immediately pointed out that there was no evidence to show that malaria existed in prehistoric times in Africa. Hiernaux said that the disease could spread very rapidly since, due to population movement, it had spread up the mountains in Ruanda Urundi during the last 20 years.

WASHBURN pointed out that, comparatively speaking, pygmies have very large incisor teeth. This means that if stature and face height are reduced and tooth size is not, than the nasal index appears unusually large. This is a point to remember when considering fossil hominid remains.

(The discussion was continued in Session X.)

SESSION X: JULY 16 (10:00 A.M.–12:55 P.M.)

(Continued discussion on pygmies.)

The discussion covered seven main topics:

1. *The rate of adaptation of pygmies to forest conditions.*

DE HEINZELIN said that extensive microlithic industries in quartz with glass beads and iron occurred outside the Ituri forests, but within the forests there was only much older material.

HIERNAUX said that there had been Neolithic culture in the north. After it had died out an interval had elapsed before agriculturalists had tried to live in the forest. The present true pygmies did not seem to have been there long, but nevertheless longer than the smallish Bantu Iron Age people who also live there. It was possible that the Batwa of Virunga were a mixture between the true pygmy and the Bantu Negro.

CLARK said that the archaeological evidence suggested that during the late Pleistocene considerable specialization had taken place in the southern Congo and that there seemed to have been two sorts of cultural adaptation.

COOKE asked if the pygmies could have developed their short stature elsewhere and merely used the forests as a retreat.

LEAKEY said that some of the skulls from Elmenteita A, B, and F people and from Njoro River (Kenya) appear to have been pygmy and others hybrid. Such a situation would arise today in an area where Bantu and pygmy peoples were mixing. Small size seemed to go back to the Middle Pleistocene and would be an advantage when fleeing through undergrowth, for instance in forest margins.

GROVE then asked if the small size of West African cattle was due to diet or natural selection. BOURLIÈRE pointed out that to deal with heat the surface of the body and not the weight should be increased, hence the hump and dewlap of the small cattle. Color also plays a part and it has been found that some types of European cattle thrive better in the tropics on account of their color as well as rangy build.

HIERNAUX said that where short and tall occurred in the same population it was necessary to study the effect of stature on selection of mates. Useful information on physiology and temperature regulation could also be obtained from a study of the amount of sweat produced by acclimatized individuals of different body area.

COOKE drew attention to such a study which had been made by Windham (Transvaal and Orange Free State Chamber of Mines) in connection with deep mining.

COOKE, DEVORE, HIERNAUX, LEAKEY, and WASHBURN agreed that social and cultural conditions influence the selection of a mate and thus the perpetuation of desired qualities in a population, but the data on these phenomena are poor.

WASHBURN also suggested that if physiological grounds point to the production of a certain type which does not in fact result then other factors must be investigated. For instance study has shown that Eskimos do not in fact have any notable physiological adaptations to cold; their adaptation is cultural. Tropical heat may be countered by wearing no clothing or, as in the case of the Arabs, much clothing.

2. *Baboon and Gorilla diets.*

LEAKEY pointed out the need to investigate baboon diet in different areas as in fact they might be eating very similar foods in different habitats. He agreed with DeVore that the most important factor was that baboons exploit most of the food possibilities of an area.

EMLEN said that monkeys and gorillas were more specialized eaters than baboons. Of 100 food plants collected that are eaten by gorillas, about 30 occur in each habitat, with some overlap between habitats.

BOURLIÈRE said that the important factor in diet was the caloric value of the plants eaten. For instance, Tonkin peasants who eat over 100 species of plant obtain 90% of their daily calories from rice. It is possible that the physiology of animals enables them to use parts of plants that the human body cannot utilize.

It was agreed that the study of the feeding of captive animals was important. But EMLEN showed how it might also be misleading since the most successful attempt at feeding a captive gorilla was on meat which in the wild the creature never eats.

WASHBURN suggested that this might imply that diet cannot be judged from tooth patterns.

3. *Teeth as indicators of diet.*

WASHBURN said that in primates there is no simple correlation between cusp pattern and diet. HOWELL instanced the variation in cusp patterns of various species and races of gibbons studied by J. Frisch. The significance would seem to be in the presence of large molars and small incisors and canines.

ARAMBOURG said that variations within a species are so common that no conclusion can be reached on the basis of a few isolated fossil pieces.

HOWELL, LEAKEY, and ROBINSON agreed on the importance of the study of the gibbons' dentition but considered that it should be combined with that of skull morphology.

COOKE said that it was necessary to consider the whole dentition and not merely individual teeth. He said also that the chewing pattern of mammals can be detected in the morphological characteristics of the jaws as well as in the teeth.

BOURLIÈRE said that experiments with rats had shown that the size and weight of the masseter and temporal muscles were connected with the male sex hormone and that this might be important in a study of fossil human skulls.

WASHBURN said that minor differences in diet in omnivorous animals could not be detected from the teeth. However, it was likely that plains-living forms needed stronger teeth to deal with harder foods. The addition of a relatively small proportion of meat to the diet of a hominid would probably not show in the teeth and he felt that the only answer to such a question would come from archaeological studies.

CLARK, LEAKEY, and ROBINSON agreed with Washburn's last remarks. ROBINSON said that the general tooth pattern, and not the cusps alone, of *Australopithecus* was similar to that of early and modern man, whereas that of *Paranthropus* was not. This difference could only be explained by diet.

HOWELL suggested that as *Australopithecus* seemed to be only about a third the weight of *Paranthropus*, if body size were enlarged his dentition might resemble in size that of *Paranthropus*. This still left unanswered the basis for the notable reduction in the anterior teeth and the changes in cusp pattern of the posterior teeth.

ROBINSON considered that the important point was the respective proportions of the posterior and the anterior teeth in the same dentition. *Paranthropus* had small front teeth and large back teeth with a heavier skull to carry larger muscles. So the back teeth are important in determining diet. In *Australopithecus* the wear is on a different part of the first molar and the tooth pattern is more like that of man (hominines). There was little time difference between the two forms, and he did not think that the smaller front teeth of *Paranthropus* could be a change caused by different methods of obtaining or dealing with the diet which he regarded as largely vegetarian. HOWELL pointed out that if the *Paranthropus* form was primarily responsible for the fauna at Olduvai he could not be entirely vegetarian.

LEAKEY considered that the teeth of *Paranthropus* and *Zinjanthropus* reflect the same food habits, but that the pre-*Zinjanthropus* child is more like *Australopithecus* in its teeth, though larger in stature.

4. *Co-existence of distinctive hominids.*

LEAKEY pointed out that the very near contemporaneity of *Paranthropus* and *Australopithecus* in the same area must have meant that they occupied differing niches in the same environment.

ROBINSON agreed. He cited also *Paranthropus* and "*Telanthropus*" (at Swartkrans) and *Meganthropus* and *Pithecanthropus* (at Sangiran), drawing the inference that one was probably vegetarian and the other omnivorous.

HOWELL did not consider that this explanation could be strictly applied at Olduvai, or to *Australopithecus* and *Paranthropus* since there was a stratigraphic distinction involved. He agreed with the Swartkrans example, but pointed out the field evidence was still inadequate to decide the case in Java.

5. *The makers of the industries at Olduvai and in the South African caves.*

LEAKEY said that the association of a skull and tools in any horizon cannot at present be taken to prove that the skull in question necessarily represented the maker of the tools. But if two contemporary types occur, of which one in another area did not have tools associated, it is reasonable to assume that the other, with the bigger brain, was the tool-maker. More data are needed, though at present it seems as if the pre-*Zinjanthropus* child was hominine and so might have made the tools.

HOWELL and ROBINSON agreed that from the Sterkfontein evidence (at least at the Main Site) there was no clear association of *Australopithecus* with the tools. *Australopithecus* is associated with tools at the later Sterkfontein Extension site, but there was disagreement as to which hominid this was, as several possibilities were presented.

CLARK pointed out the need for care in excavating cave sites where the occupation material is usually at the front and the bones at the back of the cave.

COOKE suggested that before man practiced intentional burial, human bones occurring in caves might be just part of the deposits. He said that 70% of the animal skulls occurring in South African caves were either very old or very young and this made them difficult to determine.

WASHBURN said he had always regarded *Australopithecus* as a tool *user* and thought that this must have been necessary for the protection of his group. The tools he used were probably of wood and certainly need not have been stone. At the same time a human upright carriage, as opposed to a gibbon-type walk, human hands, smaller canines, and a smaller face must have been acquired so that *Australopithecus* was probably at the end of a long line of tool users.

LEAKEY suggested the use of horns or antlers as the first tools, rather than wood. It was agreed that the core of factual evidence in the so-called Osteodontokeratic culture should not be entirely overlooked.

6. *Transition from quadrupedalism to bipedalism.*

ARAMBOURG said that separation of the lines of man and ape probably began in the Oligocene (c.f. *Oreopithecus*) and asked Washburn what precisely he understood by being "semi-bipedal" which Arambourg would think was a contradiction in terms.

WASHBURN replied that in structure this was not possible, but in habit one could be perhaps semi-bipedal. For instance an ape living on the edge of a forest could walk bipedally on the ground or could take refuge in trees and would thus be behaving in a semi-bipedal way. Tool-using would increase the need for upright carriage while still allowing the early hominid to take refuge in trees when necessary. Balance in the gibbon-type of walking is achieved by the position of the toes and is unlike locomotion in man. The session closed by Washburn's realistic and much appreciated demonstration of gibbon-type bipedal locomotion across the conference table.

(The afternoon of Sunday, July 16, was free.)

SESSION XI: JULY 17 (9:30 A.M.–12:40 P.M.

PAST CLIMATES

BUTZER introduced the subject of paleoclimatology, with special reference to the subtropics (Tables 5, 6, 7).

TABLE 5
GENERAL CIRCULATION PATTERNS IN LOWER AND HIGHER LATITUDES OF NORTH AFRICA AND EUROPE

Climate Type	*Subtropical Belt*	*Middle-Higher Latitudes*
Warm, dry	High pressures	High pressures, westerly and southwesterly flow
Warm, moist	Jet-stream branch	Southerly flow or high pressures
Cool, moist	Jet-stream branch	Northerly or westerly flow
Cool, dry	High pressures	High pressures of south-westerly flow

TABLE 6
SIMPLIFIED SCHEME OF PALEOCLIMATES IN THE MEDITERRANEAN AND SAHARA

Paleoclimate	*Stratigraphic Unit*
Temperate/warm, generally dry but with moist thermal maximum	Postglacial
Cool, generally dry	Main and Late Glacial
Cool, moist	Early Last Glacial (Würm)
Warm, moist	Late Interglacial
Warm, generally dry	Last Interglacial (Eem)

TABLE 7
LOCAL LATE PLEISTOCENE AND HOLOCENE STRATIGRAPHY

The Balearic Islands	*Northeast Africa*	*Climatic History*
(Minor aeolianites) Rendzinas	Dunes	Post-Subboreal
(2 m. sea level) Rendzinas	Dunes	SUBBOREAL
(4 m. sea level) Alluvial silts	(4 m. sea level) Local alluvial gravels and rendzinas	ATLANTIC
(Flandrian Transgression) Rendzinas	Dunes	BOREAL
" " "	Dunes	
" " "	Alluvial gravels	
(Major Aeolianite) interrupted by tufa crust and rubefaction zone	Dunes	MAIN LATE GLACIAL Predominantly dry
Colluvial beds		
Rubefaction	Burozem	EARLY GLACIAL
(Major aeolianite)	Alluvial gravels	Cool, pluvial maximum followed by variable desiccation
Minor colluvial bed		
(Major Aeolianite)		
Colluvial beds, alluvial gravels, tufas		
(2–3 m. sea-level) reduced thermophile fauna, tufa crusts	Local tufa deposits in Cyrenaica	LATE INTERGLACIAL
Local aeolianite, terra rossa	Terra rossa	Warm, pluvial conditions
(4–2 m. sea levels) thermophile fauna	(4–2 m. sea level) thermophile fauna?	
Colluvial beds		INTERGLACIAL
(8–6 m. sea levels) common (?) fauna, terra rossa?		Variably moister and drier than now
Colluvial beds		
(10–12 m. sea level) thermophile fauna, terra rossa	(10–13 m. sea level) thermophile fauna	

He began by saying that past climatic fluctuations cannot be explained in terms of simple correlation of glacials and pluvials. The system must be pieced together from the evidence at present established and based on a differentiation of warm and cool pluvials.

The Warm Pluvials. These correspond to interglacials in the north temperate zone, and were characterized by the development of red soils, *terra rossa,* in the Mediterranean area. Present-day Mediterranean soils are fossil or relict. The last intensive red soil formation is probably of Mindel age and is preserved, for example, on hill tops in northern Italy. Such soils are not developing today. Later soils show considerable contrast and have much less humic content. Intense weathering would have been necessary to produce the *terra rossa;* hence considerable rainfall is postulated with a climate at least as warm as the present. Since modern parallels for these soils are lacking it is difficult to determine exactly the type of vegetation which they supported. They have yielded no pollen evidence.

The western Sahara has similar, though not so extensive, *terra rossa* soils. Egypt has rubefied soils, though not true *terra rossa,* and which were developed prior to the last pluvial. Red soils similar to those in the Mediterranean occur in Senegal and on the Hoggar mountains; the latter are less well dated.

The Neolithic Pluvial, or subpluvial, phase. Butzer said that this was dated to 6,000–3,000 B. C. and was first recognized in rock drawings showing a savanna-type fauna in the central highlands from west to east of the Sahara. The lack of parallel drawings to the north (with one exception) and south suggest a people locally confined. The drawings are very naturalistic. A 50–100 km. shift in the desert-savanna border would not imply catastrophic climatic changes, but the ecological importance of minor climatic shifts would be very great. Egyptian dynastic reliefs of the Fourth or Fifth Dynasty in the middle of the third millennium B.C. depict trees, and tree remains occur in the Egyptian deserts associated with Badarian industries.

Pollen analysis, the work of Quézel, is giving a general picture of the genera present though not in order of importance. No spectra have yet been published and it has not been possible to date most of the 33 samples studied. The samples come from Neolithic "Black" soils which are not a good stratigraphic horizon, and interpretation is difficult, but the vegetation seems to have been essentially woodland of modern Mediterranean type. The species have been grouped according to altitude: Mediterranean woodland above 2000 m., an intermediate parkland between 1500–800 m., and desert vegetation below 700 m.

The Cool Pluvials. Butzer described the cool pluvials as periods of aggradation in the Mediterranean area (Catalonia, Balearic Islands, the Levant), Egypt, and Cyrenaica.

In the Mediterranean today there is no closed vegetation mat so that an increase in moisture would cause wide erosion.

Butzer said that during the cold pluvial light brown steppe soil, known as burozem, was developed in Egypt. In the western Sahara pollen evidence occurs with the Aterian-Mousterian in gravels of Late Pleistocene date. Some species are the same as in the Neolithic, including some acacias and a large series of temperate (moist) species including hazel, fir, and lime.

In the southern Sahara at the Atbara mouth gravel deposits occur similar to those of the last pluvial in Egypt; and a Mousterian industry occurs on the clay plain between Omdurman and Khartoum. In Egypt it is difficult to separate wadi terraces indicating local moist climate, and Nile terraces showing a greater flow of the river itself. Conditions in the Sebilian period in Egypt were as they are today. Dunes formed in the Balearic Islands and the Senegal valleys in the regressions of glacial times.

The pluvials seem to have been localized and relatively short lived. Aridity in the Sahara (and stagnation of ice farther north) was also caused by changes in the circulation. Some of the stratigraphic evidence and probable general correlation is summarized in Table 7.

DE HEINZELIN then spoke on his and other work in the Western Rift Valley. He said that the large body of sand, composed of aeolianites, and stretching through Angola to the Kalahari, is divisible into three units: (1) the oldest, pre-Quaternary sand related to the Kalahari; (2) the Selonga and related sands; and (3) the Yangambi plateau sands.

Aeolian sand implies lack of vegetation due to poor rainfall.

The modern fauna of the Western Rift is lacustrine like that of Lake Victoria or Upemba. The Kaiso beds give the first indications of climatic change, the gypsum deposits of which point to increased evaporation. De Heinzelin summarized the Kaiso series as:

Kaiso IV: with *Melania brevissima;* probably correlating with Omo and Olduvai I, even perhaps dating to the beginning of the Middle Pleistocene.

Kaiso III: fauna (invertebrate) the same as evolved II;

Kaiso II: with *Viviparus carinatus*, gypsum deposits, some artifacts at Kanyatsi (J. de Heinzelin, in press).

Very early Kaiso.

He said it was important to determine to what extent the climatic changes in South Africa, Olduvai, and Kaiso could be correlated. In the Upper Pleistocene there was a broad extension of terrace deposits with boulders, colluvium, and red and orange soils, belonging to the high terraces of the Semliki and implying higher rainfall. Reworked Acheulian material in these terraces indicated a post-Acheulian, probably Sangoan, age for this period of increased precipitation. There also seemed to be a connection between the red soils and the glaciers. The red soils seem to be related to the moraines of Ruwenzori glaciers and are displaced by faulting 2,500 m. up the western Ruwenzori. Thus,

4,000 m.: very old moraines of wide extent over old land surface;
4,500 m.: lowest extent of modern glacier;

4,800 m.: present snowline;
5,000 m.: present glaciers, definitely retreating (as in Europe).

Below the 4,500 m. level glaciers, the moraines are new and recent, for example Lac Gris, about A.D. 1850-1600, and corresponding to the "Little Ice Age" of Europe. Other moraines—Lac Vert, 4,150 m., Lac Noire, 3,750 m. and others at 3,000 m. are also recent—perhaps a thousand years old.

Ruwenzori does not appear to be very old and only reached its present height comparatively recently. When Ruwenzori was at a lower altitude and an ice cap developed on it there must have been a deterioration of climate of the order of 6°–9°C. Such a change is borne out by deep sea cores from the Atlantic.

A change of world mean temperature implies changes of circulation and vegetation, etc. It might even include a change in radiation which could perhaps have been a cause of the giantism in mammal fauna, as recorded at Olduvai.

Biberson here observed that changes in temperature recorded in deep sea cores were minimal and implied much greater variations in continental interiors.

Bond then spoke on the paleoclimates of the stable, southern African block. He said that southern Africa had enjoyed latitudinal stability since the Jurassic. The fact that all the Cretaceous and Tertiary marine beds lie at the present coast line indicates that the shape of this part of the continent was practically constant. Apart from a very gentle movement that altered the direction of some rivers there have been almost no later Tertiary tectonics and no volcanicity. The climate of much of the Pliocene seems to have been arid, but this aridity decreased in the Quaternary. Again there was great stability and little change of topography which again reduced the factors making for climatic change. However, it also removed one of the opportunities for Pleistocene sedimentary traps so that most of the deposits are thin, scattered, and largely unfossiliferous.

The kinds of evidence giving indications of past climates during the Quaternary are:

River action. "Cut and fill" gravels which are difficult to interpret, particularly in major rivers where action at a given place does not necessarily reflect the climate at that particular place.

Sands. Not all of these are of aeolian origin and are usually only present near a source, for instance the Kalahari. These may be associated with river or other deposits; where redeposition is concerned the surface texture rather than the shape of the grains permits determination of whether the agency was wind or water. The lower limits of rainfall for aeolian redeposition may be 15″; and the post-Sangoan and post-Magosian sands at the Victoria Falls indicate about that rainfall.

Dambo deposits. Comparison should be made between these and modern conditions in the area. On hard, crystalline rocks, even with a high rainfall, rivers may cut wide, shallow valleys and be unable to carry away the sediment run in from the sides. Vegetation also plays a part. A study at present being made of deposits with some Acheulian industry and modern dambo conditions seems to suggest considerable rainfall differences and clues as to past climate may be obtained from this work.

Chemical alteration of existing deposits. This includes, for example, silica, calcium, or iron depositions. They may have only local occurrence, but they can indicate rainfall, for example 18-20″ for post-Acheulian calcification at Lochard. If such alteration can be dated climatic conditions can be fixed at a definite time in the past.

Special environments. 1) *Cave breccias.* The quantitative work of C. K. Brain has indicated that rainfall of between 20″–40″ is necessary to produce these formations. 2) *Colluvial hill washes.* Analyses based on the average weight of felspar grains transported by sheetflood and of the rainfall above which granite weathering seems to be mechanical (i.e., 35″) can give indications as to rainfall.

BOND said that the total rainfall of the Quaternary seemed to vary from about 60% to 160% of present rainfall. This is probably a reasonably accurate estimate since it was arrived at by three different methods.

Still more study of the present day is essential to an understanding of the past.

Bond said he had not adduced any evidence to indicate temperature, but to some extent this accompanies rainfall change. He gave the following comparative figures:

	Lochard	*Khami*
Post Acheulian—minimum rainfall not below:	20″	18″
Today:	22–25″	22–25″
Late Acheulian—maximum rainfall not over:	40″	40″

COOKE then circulated a revised version of a table (Table 8) from his background paper (see below) modified by inferences concerning new evidence from Olduvai Bed I.

DISCUSSION

DE HEINZELIN and LEAKEY pointed out that "Kaiso" must be regarded as a formation and not a bed.

BISHOP said that until all the new evidence was studied he preferred not to subdivide the Kaiso. The lithological similarity between the type site at Lake Albert and Lake Edward makes tentative correlation possible, as does the molluscan fauna. New evidence from studies by Arambourg shows no change in the molluscan fauna between the type locality and the new sites to the south.

TABLE 8

PROVISIONAL CORRELATION OF MAJOR PLEISTOCENE DEPOSITS IN AFRICA

Possible Age	*South Africa*	*East Africa*	*North Africa*
Recent	Deposits with Later Stone Age cultures and no extinct faunal elements	Deposits with Neolithic and Mesolithic cultures and no extinct faunal elements	Deposits with Neolithic and Capsian associations
	Vlakkraal and other deposits	"Upper Gamblian" beds and Eyasi Beds	Deposits with Micoquian and Aterian cultures
Upper Pleistocene			
	Florisbad and Chelmer	"Lower Gamblian" Beds	
Middle to Upper Pleistocene	Hopefield	Olduvai Bed IV +	
	Cornelia	Olduvai Bed III to IV	Lac Karar etc.
		Semliki Beds	
Middle Pleistocene	Vaal "Younger Gravels"	Olduvai Beds II and III	Ternifine
Cromerian or uppermost Villafranchian	Kromdraai		
	Swartkrans	(unconformity)*	
Upper Villafranchian	Sterkfontein extension		
	Makapansgat, Taung and Sterkfontein	Olduvai Bed I	Ain Hanech
		Laetolil	Bel Hacel etc.
		Omo	
		Upper Kaiso	
		Lower Kaiso	Lac Ichkeul,
Lower Villafranchian		Kanam	Ain Boucherit, Koro Toro, Fouarat etc.

* See footnote 1, in which it is noted that recent geological work at Olduvai Gorge has shown that Beds I and II constitute a conformable series [F.C.H.].

COOKE suggested that at present Kaiso should be left broadly within the Villafranchian and that no attempt should as yet be made to relate it to Bed I.

He also felt that no very definite correlations should yet be made based on fossil fauna. He pointed out that as new material comes to light many older "species" are discarded. Faunas come from within deposits laid down under local conditions, not necessarily typical of the Pleistocene as a whole. The fauna itself is broadly typical African and since it has a wide range can only very occasionally be used to indicate ecological settings. Ungulates and carnivores range on the whole from desert fringe to forest fringe though there are a few, like the okapi, that suggest a definite environment, in this case forest. With regard to the South African Australopithecine deposits the most typical picture is one of a general savanna environment with occasionally drier steppe animals, but the limits are never such as would exclude creatures living in local thicket or bushy areas. This is entirely consistent with the geological inference that the climate of these deposits was like that of today, but perhaps slightly drier. The ecology of the *Paranthropus* deposit shows a higher proportion of thicket-dwelling creatures and the drier elements are absent. Mammal evidence, therefore, should only be used to lend support to geological indications. Bed I at Olduvai was wet in the lower parts and drier in the upper as reported earlier by Leakey, and this is supported by the faunal evidence. Bed II showed general Serengeti plains conditions within the range of general East African environment.

Cooke then went on to speak of palynology which he thought had so far produced very little reliable work and had given no absolute dates. The geological effects produced by varying climatic changes depend largely on the vegetation cover, so that it is difficult to use major stream deposits as climatic indications. A complete vegetational mat is rare in Africa and so 40–50″ of rainfall usually causes the stripping of the surface soil and its deposition in hollows. Wide seepage prevents too great a run off, however, so that silting results if the drainage is not too steep.

Decreasing rainfall causes devegetation, which in turn allows torrential rain to cause great erosion and torrent deposits. Still greater devegetation causes erosion too great for transport, and redeposition then occurs accompanied by calcification of colluvial material and torrent fills. The cycle is complete with bare open surfaces of material and then is recommenced.

Cooke showed slides of maps postulating the changes in vegetation likely to accompany given intensities of rainfall change (cf. the deduced climatic curve for the Australopithecine deposition, H. B. S. Cooke, this volume).

If the present rainfall of southern Africa were cut by 50% to 60% the Kalahari vegetation area would be almost desert and open to wind action. East Africa would suffer much less apart from a reduction of forest and temperate grassland. On the other hand if the present rainfall were increased by 140% to 150% the general changes would be less dramatic; there would be an extension of *Brachyste-*

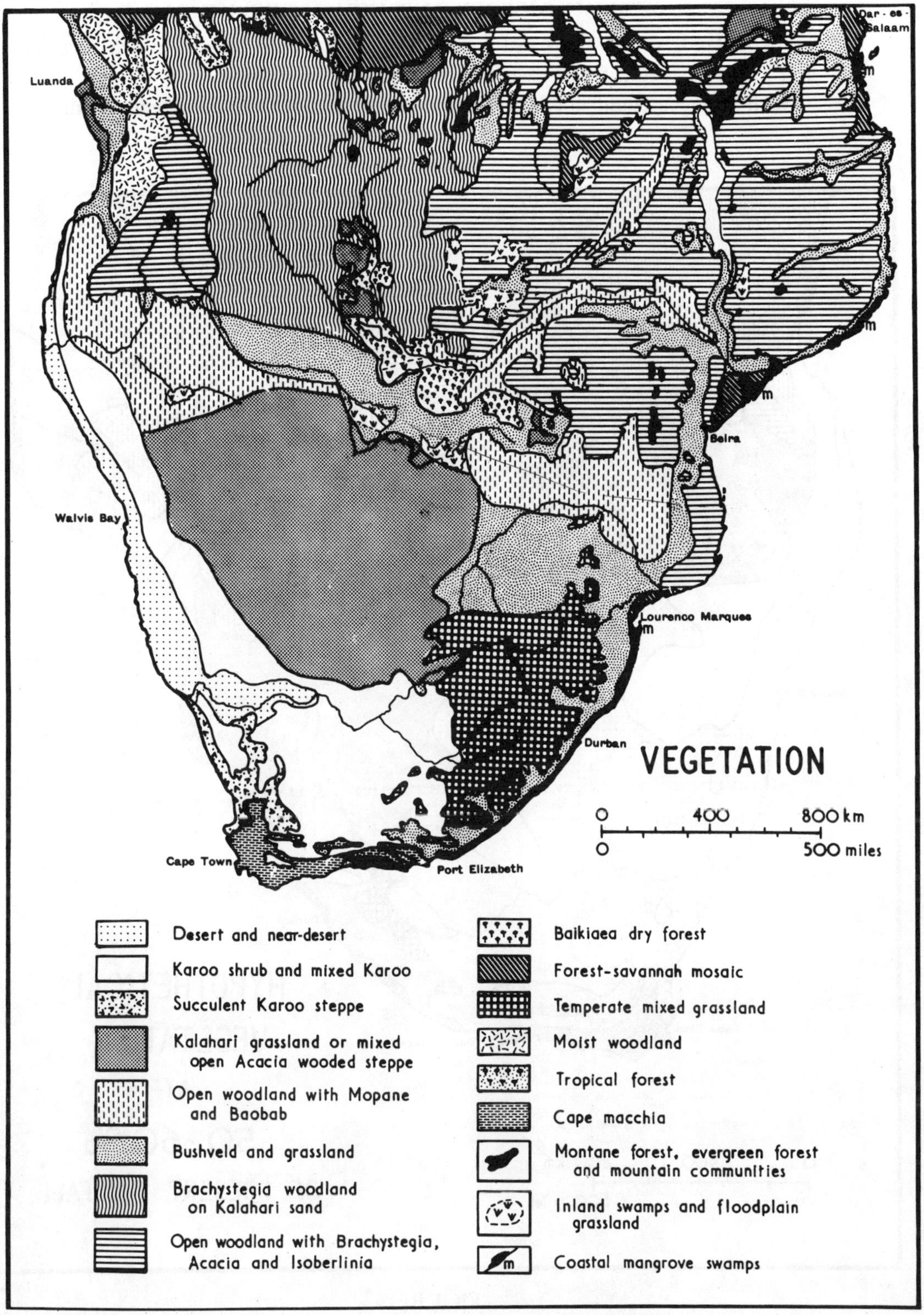
Luanda
Dar-es-Salaam
Walvis Bay
Beira
Lourenco Marques
Durban
Cape Town
Port Elizabeth
VEGETATION
0 400 800 km
0 500 miles
Desert and near-desert
Karoo shrub and mixed Karoo
Succulent Karoo steppe
Kalahari grassland or mixed open Acacia wooded steppe
Open woodland with Mopane and Baobab
Bushveld and grassland
Brachystegia woodland on Kalahari sand
Open woodland with Brachystegia, Acacia and Isoberlinia
Baikiaea dry forest
Forest-savannah mosaic
Temperate mixed grassland
Moist woodland
Tropical forest
Cape macchia
Montane forest, evergreen forest and mountain communities
Inland swamps and floodplain grassland
Coastal mangrove swamps

FIGURE 3

gia woodland, and swamps and forest vegetation would be very much favored, and hence so would forest-dwelling mammals and avifaunas.

Such changes could not be synchronous throughout the continent as rainfall,

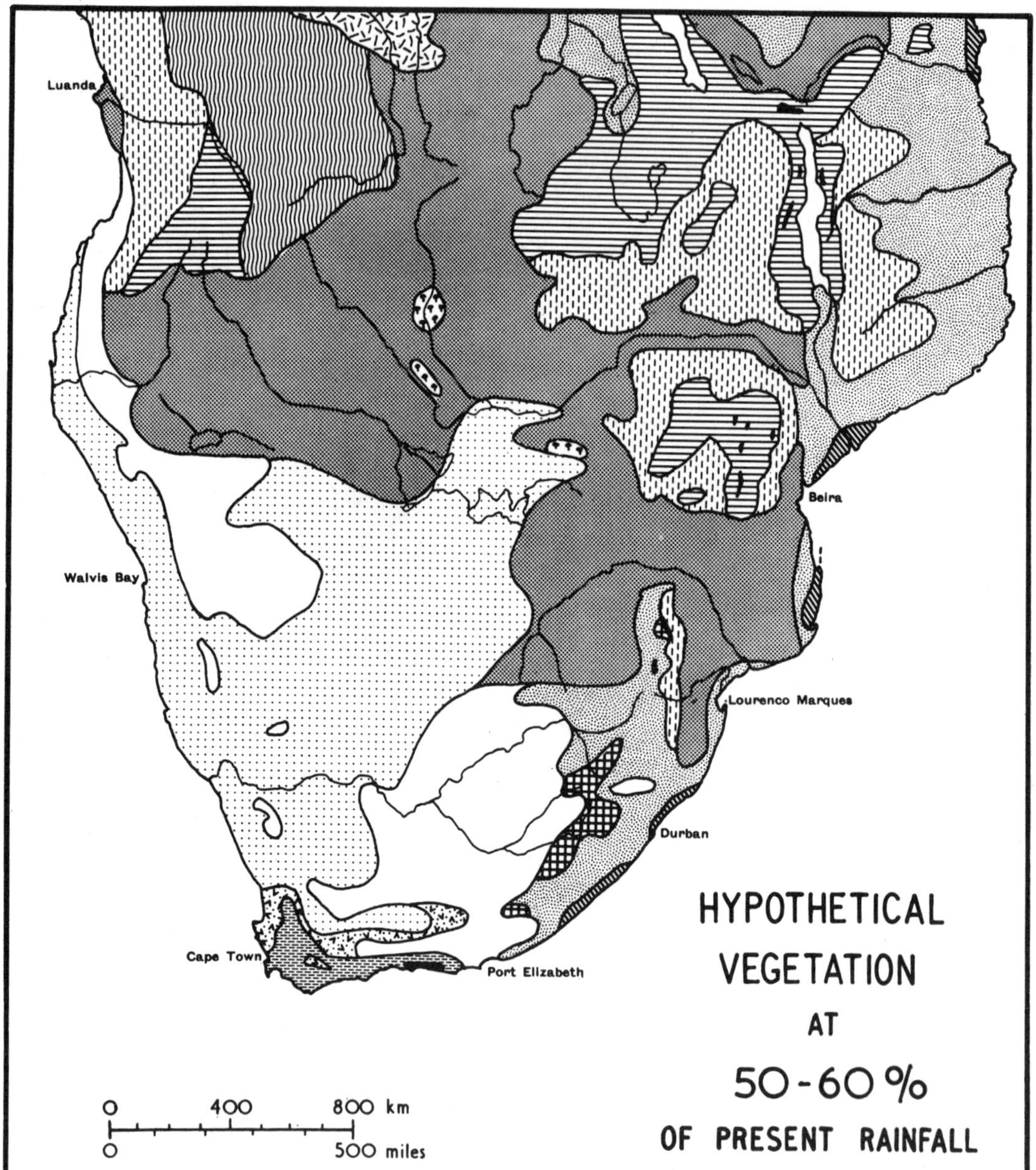

FIGURE 4

and thus vegetation, is affected by the movement north or south of the anticyclonic belts.

He then showed slides of diagrams and maps which he had prepared in an

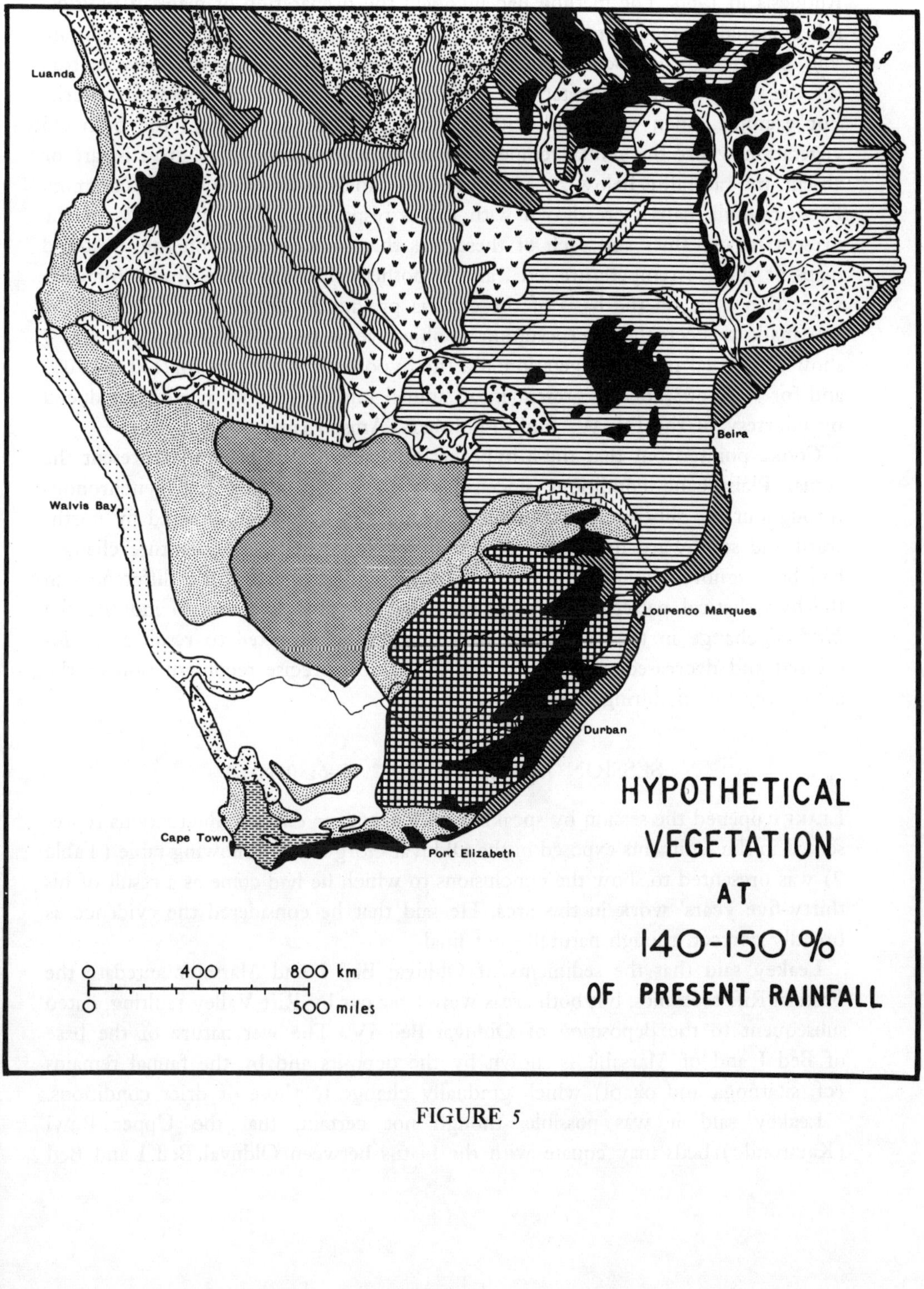

FIGURE 5

attempt to gain a visual impression of the possible changes in vegetation pattern which might have resulted from changes in rainfall of the order of magnitude discussed earlier by Bond and himself (see papers, this volume). He had tackled the whole of southern Africa (south of 6 1/2° S.) to supplement the similar essays made by Clark in 1957 for Northern Rhodesia and Summers for Southern Rhodesia in 1960. The method had involved the preparation of maps of physical relief, actual surface temperature, temperature range, frost distribution, rainfall, soils, etc., which had been used in order to derive more or less objective factors for the recognition of vegetational boundaries as delimited by botanical criteria. For example, open woodland with mopane and baobab replaces the bushveld vegetation where the annual temperature is normally above 70°F., frost is rare or absent, the rainfall is below 20–25 inches, and poorly drained clay soil is common. Fresh rainfall maps were drawn with values respectively 60% and 140% of the present mean values and each of these was then superimposed on the unaltered maps of surface temperature, etc., and hypothetical values were drawn for each vegetation type on the basis of the criteria previously determined. Every effort was made to keep the interpretation as objective as possible. Figures 3, 4, and 5 show the actual present vegetation and the hypothetical patterns for 140%–150% and for 50%–60% of the present mean annual rainfall (the maps are reproduced by courtesy of the Dr. W. Junk, Publishers, Amsterdam).

Cooke pointed out that these hypothetical maps are unlikely to represent the actual Pleistocene conditions since changes were probably not synchronous throughout the subcontinent as rainfall would also have been affected by northward and southward movement of the anticyclonic belts. Temperature changes had been ignored but these would probably tend to increase the differences in the hypothetical patterns. He emphasizd that his objective was to indicate the *kind* of change in the vegetation which might be expected to result from increased and decreased rainfall and not to claim a precise reconstruction of the actual vegetation during the Pleistocene.

SESSION XII: JULY 17 (2:30–6:30 P.M.)

LEAKEY opened the session by speaking on Pleistocene climatic fluctuations represented in the sediments exposed in the Olduvai Gorge. The following table (Table 9) was presented to show the conclusions to which he had come as a result of his thirty-five years' work in this area. He said that he considered the evidence as broadly correct, though naturally not final.

Leakey said that the sediments of Olduvai Bed I and Marsabit antedate the Eastern Rift tectonics, but both areas were later cut by Rift Valley faulting, dated subsequent to the deposition of Olduvai Bed IV. The wet nature of the base of Bed I and of Marsabit is shown by the deposits and by the faunal remains (cf. sitatunga and okapi) which gradually change to those of drier conditions.

Leakey said it was possible, though not certain, that the Upper Rawi (Kavirondo) beds may equate with the hiatus between Olduvai Bed I and Bed

TABLE 9
EAST AFRICAN STRATIGRAPHY

Stage	Division	Deposit	Climate
HOLOCENE			
MAKALIAN		Nakuran	Wet
		Makalian 2.	Dry
		Makalian 1.	Fairly wet
GAMBLIAN		Hiatus -	Dry
		Gamblian 3	Wet
		Gamblian 2	Wet
		Gamblian I	Wet
KAMASIAN-KANJERAN		Olduvai Bed IV	Gradually drying Wetter
		Olduvai Bed III (Red earth formation)	climate like today
		erosion	
		Olduvai Bed II	Wetter
		Hiatus - Rawi Beds ??	
VILLAFRANCHIAN	UPPER	Olduvai I & Marsabit	Dry Drying Wetter
	LOWER	Kanam?	No climatic evidence

II. There was no deposition then so that the lake must have disappeared and the changed fish fauna of the Rawi beds suggests similar conditions.

The lake deposits of Olduvai Bed II times point to a climate wetter than that of today, and the giant fauna suggests different vegetation conditions. After Bed II the climate was drier, but not desertic. Ravines and valleys were cut, often as much as 60 ft. deep, into Bed II before the formation of Bed III of which the climate was similar to that of today. Later, during the deposition of Bed IV, a lake was again formed, pointing to a renewed wetter climate.

There were thus three wet phases and two less wet phases. After deposition of Bed IV major faulting and disturbance occurred. It may have had some effect on the formation of lakes and rivers. In the same way the major faunal break could have been due to volcanicity and not necessarily to increased precipitation.

After the high level basin of the Gamblian I pluvial the levels dropped, but there is no disturbance visible in the beds. The levels subsequently rose again and, as this was a closed drainage basin, the cause must have been increased rainfall.

The Gamblian III beach is not as marked as those of the earlier phases and it was followed by wind-blown sand in an area where forest had previously persisted.

The Makalian Wet Phase in the Mesolithic was not so great as a pluvial and some drying occurred again before the Nakuran Wet Phase. Gamble's Cave has failed to provide charcoals from which a C-14 date could be obtained. When Apis Rock is further excavated charcoals may well be found there.

CLARK then presented some evidence of Upper Pleistocene climatic fluctuations as recorded at Kalambo Falls and in northern Angola.

Clark thought that the discontinuous distribution of fauna in eastern and southwestern Africa could possibly be explained by the existence of a corridor of drier country during the Last Interpluvial and Early Gamblian dry periods, which facilitated the movement of such animals as ostrich, oryx, wildebeest, Grant's gazelle, giraffe, and brown hyena. The sand redistributions in the Congo Basin would seem to support such a hypothesis.

Clark said that pollen from Kalambo Falls seem to show that in Acheulian times the climate was colder, with greater humidity and reduced evaporation. Such conditions are now dated to 40,000–12,000 B.P., i.e., to the greater part of the Gamblian Pluvial. In the later Middle Stone Age (=later Gamblian) *dambo* conditions prevailed with a cooler climate and a low percentage of deciduous woodland (only 4% of *Brachystegia* as against present-day predominance). A quantity of fern pollen occurs and also a few forest forms. The macroscopic evidence from the Kalambo Falls shows a percentage of moist evergreen forest near water and mist forest on the slopes in Late Acheulian times. There are also some savanna-woodland trees. There is evidence to indicate that the *Podocarpus* forests were some 300 miles nearer Kalambo at this time. This would suggest a lowering of temperature of about 5°C.

Evidence from Mufo in Angola (2,100 ft.) points again to a cooler climate during the Early Lupemban (formerly Middle Sangoan) culture, and low values for moist forest forms until the upper layers of the containing deposit. He considered that all of the redistributed Kalahari-type sands in this region did not date to the dry period between the "Kanjeran" and the Gamblian. There is cultural evidence showing the presence of man in the later and final Middle Stone Age.

The Last Interpluvial, although enough drier that the forests retreated into the valleys, was not much drier than today; but, the actual dry season may have been longer. Deeply weathered lateritized crusts, which had formed on the lower slopes, and cooler temperatures prevented the forest from re-establishing itself during the Gamblian Pluvial. Its place was taken by montane forest and grassland; and sand was blown and washed over the surface and into the valleys. The average rainfall today is 58″ but a decrease of 25% and increased dry wind velocity might have caused the sand to move in. This sand is unstratified and its age can only be deduced from the artifacts found in it. Sand can also move by slumping caused by extreme saturation at the base.

The *chanas* in Angola (3,000′) carry no trees and, according to Mendonça, never have done so, unlike the Nigerian plateau with 70″ of rainfall which GROVE said now carried only grass because of deforestation by human occupation.

DISCUSSION

COOKE and LEAKEY did not consider a dry corridor necessary for the spread of giraffe, etc., to the movement of which present-day vegetation would be no barrier. They thought that the discontinuous distribution of some kinds of animals could be the result of the isolation of pockets of an earlier, wider distribution. BISHOP asked the precise meaning of the term "laterite." CLARK replied that he had been using the term already applied to the Angola formations, which in fact had more the appearance of ferricrete.

COOKE defined laterite as aluminum oxides accompanied by great chemical alteration; and ferricrete as iron oxides merely cementing existing deposits. Great seasonal change and 40″ of rainfall were not necessary for the formation of ferricrete, impeded drainage being sufficient.

WASHBURN pointed out the importance for primate development and possibly for genetic isolation, of an edge of the forest environment which would have been increased by fingers of gallery forest.

MOREAU felt that on the basis of the present avifauna, connections must at one time have existed between East and South Africa and COOKE pointed out that even a drop of 3° in temperature would affect evaporation and hence forest spread. (He pointed out that temperature had not been indicated on the rainfall maps he had shown at the end of Session XI.) Cooke also said that the upper levels of Kalahari sand that supported vegetation were somewhat looser than the primary sand and permitted the upward movement of drainage water and thus supported tree growth, although conditions could exist where rain water percolated so far that it was lost to vegetation.

CLARK said that evidence showed that montane forest at the Kalambo Falls was 1,500′–2,000′ lower than at present on at least two occasions in the past.

DE HEINZELIN agreed with Clark's remarks on sand movement in valleys for the later periods of deposition, but he did not think that they could account for the thicker, more aeolian and chiefly pre-Quaternary sands of, for instance, the Kalahari, Selonga, and the Yangambi plateau. He also considered that climate could affect the type of vegetation growing on the same very thick sands as at Yangambi and in the Kasai.

COOKE stated that once a few trees had become established on the sand, woodland would spread, but that any stripping, naturally or humanly induced, would be rapid. The top of the sands quickly became bleached by the humic acids from vegetation and the original color was that found at lower levels.

LEAKEY asked de Heinzelin how he could be certain that the Yangambi sands were pre-Quaternary because early cultural material was not at all widely distributed in the sand and for instance in Angola had only been discovered as a result of large scale commercial operations.

DE HEINZELIN replied that culture does occur, but only in the latest redistributed sands.

COOKE said that since all these sands originated in sandstone formations of aeolian origin they already exhibit desertic characteristics which make distinguishing between them even more difficult.

In reply to a question from Moreau, CLARK said that the aeolian sand of Gamblian date filling the valleys in the Katanga, and referred to by Mortelmans, represented a very minor redistribution at the very end of the Pleistocene dry period. It occurred also in Angola and the Upper Zambezi but it could not be compared in extent to the main redepositions of the Last Interpluvial.

The session continued with a discussion of the Chelles-Acheul industry. LEAKEY spoke on (and illustrated by slides) its earlier stages, especially in Bed II at Olduvai. He said that the climate at the base appeared to have been stable with oscillations represented farther up in the bed. In the Chellean industry hand axes at first formed only insignificant proportions among the Oldowan-type choppers and other artifacts, but increased both in quantity and quality by Chellean 3 times. At the BKII (Chellean 2) site there was a much higher level of culture than occurs in Bed I with imported material for stone tools and lumps of red ochre. The fauna of this level implies lush vegetation: *Deinotherium* (also found in Bed I), *Pelorovis, Bularchus,* various pigs, etc. An increase in antelopes (alcelaphines), etc., implies an increase of savanna conditions higher up in this bed. Two very large milk teeth had been found in the BK II site, but whether of an Australopithecine or a hominine was not yet certain. With the Chellean 3 occurred a well-preserved hominid (hominine) skull. The living floors of the Chellean Stage are found on the disconformities within the deposits which make up Bed II.

Subsequently, slides were shown of various later Chelles-Acheul sites, including Olorgesailie, Isimila, Kalambo Falls, Broken Hill, and localities in Angola and South Africa, with observations by Leakey, Howell, and Clark on the industries and their occurrence in the containing deposits.

CLARK, referring to some recent suggestions for renaming the Chelles-Acheul "Acheulean" because of its very long range, advocated adhering to the terminology agreed upon at the 1947 Pan-African Congress on Prehistory unless this was changed by international agreement. He said that the Acheulean culture is normally considered to consist of hand axes, cleavers, smaller flake tools, and a variety of less formalized tools together with some worked wood and bone. The range of raw material is considerable, although fine-grained rocks were usually utilized for making smaller tools. Sometimes material was carried in from a distance, as at Olduvai and Bembezi, but Clark considered that it was the *form* in which the raw material occurred and not its nature that was important and that occasioned variations in technique.

LEAKEY thought that raw material was very unimportant once the basic technique had been mastered. He pointed out that in any case it was local material that was normally used and even small tools were sometimes made of poor material. Materials brought in from outside were, in any case, not always used for toolmaking. He did not think that the trade in obsidian in East Africa appeared until late. He agreed with Clark's opinion that the terminology agreed for the Acheulean in 1947 should only be changed by international agreement and not unilaterally.

BIBERSON dissented, however, and thought that a general term, such as "handaxe culture" would be preferable. He deplored the use of the same word for a culture and for a technique.

HOWELL also agreed that any change should only be made by agreement.

ARAMBOURG agreed with the view expressed by Biberson and suggested the term "La civilisation du biface." He believed, on paleontological grounds, that this could also be connected with a definite human type. BIBERSON pointed out that even if the word "civilization" had a different connotation in English, as Howell had brought out, one must use terms to suit the particular language being employed.

CLARK brought up the subject of distribution maps now being prepared for the "Atlas of Prehistory" project. He admitted that these would be incomplete since many areas still awaited exploration, but the maps would have value in that they would show the distribution as it is known at present and also which areas are unknown.

Clark said that many of the early Upper Pleistocene sites were without depth and the scatter was slight. In final Pleistocene times in Angola (Lupembo-Tshitolian culture) there was a great concentration of population on the interfluves overlooking the forests. In Angola also the raw material was largely *grès polymorphe.*

On the whole, seacoast and some river valley cultures specialized in choppers and chopping tools, the savanna cultures in flakes, and the grassland cultures in points. The Middle Stone Age showed advances over the Earlier Stone Age. Pigment indicated the expression of an aesthetic sense, ritual burial began to be practiced, there was a greater specialization in tools, and intentionally shaped bone tools occurred. There were also regional distributions to be recognized. The question of a common language was raised. There was now a greater concentration for food on a few larger animals, and improved hunting methods may have enabled man to cause the extinction of some overspecialized forms. Near permanent settlement now took place on forest fringes, around lakes, and along some savanna rivers, and the facies of the industries altered during this period.

COOKE considered that Clark had tried to combine evidence from too many varied sources for his exposition of Pleistocene culture. He questioned also whether it was permissible to outline the distribution of Acheulian man in relation to *modern* vegetation patterns since these may not have obtained then.

CLARK replied that the period cannot in any case be considered as a unit, but only as a series of fluctuations. He thought that on the whole the climate was, as he had indicated, at times wetter than today since man was enabled to live in such places as the Sahara and South-West Africa.

LEAKEY said that it was not possible to make generalizations for Africa as a whole and that all one could with certainty say of the needs of Acheulian man was that he required water, stone, game, and a climate that was not too cold. In the present state of knowledge Leakey did not think that any useful purpose could be served by distribution maps.

WASHBURN disagreed and considered that such a correlation was necessary to stimulate and guide further research.

COOKE thought that generalization should only follow after careful study of distribution and reconstruction of the ecology. He thought, however, that it would be a help to know where, in the areas thoroughly explored, the Acheulian did not occur.

CLARK said that in some specialized areas it can now be said with certainty that the Acheulian does not occur and in any case he considered that broad, tentative conclusions could now be drawn (see Howell and Clark, this volume).

BIBERSON said that although more work was still needed it can be said that the Acheulian occurs in the Sahara in three stages (early, middle, and late).

HOWELL said that the best basis for comparison of Chelles-Acheul industries for sub-Saharan Africa was the succession at Olduvai Gorge, but until detailed divisions had been worked out for Bed IV there it was not possible to make effective comparisons. It seemed to him that the earliest Acheulian in the Sahara was Middle, though Biberson did not agree.

SESSION XIII: JULY 18 (9:30 A.M.–12:30 P.M.)

HOWELL opened the session by suggesting the following scheme (Table 10) of correlations of the later Acheulian sites.

LEAKEY contended that work at Hopefield must be finished before its position in the table can be adjudged. He thought that in assessing Acheulian culture, too much emphasis was placed on typology and not enough latitude allowed for group preferences or style differences. HOWELL agreed that this was perhaps inevitable as a consequence of the existing methodology.

CLARK suggested that, since in Acheulian times culture contact could be ruled out, intended use could probably be a cause of difference in tool form.

LEAKEY pointed out that in order to obtain a complete picture of a culture a whole living area should be completely excavated. This had never been done and even though expensive was essential. He thought that a further reason for the difference of tool forms could be seasonal changes of habitat.

TABLE 10

TENTATIVE SCHEME OF TEMPORAL RELATIONSHIPS OF ACHEULIAN SITES IN SUB-SAHARAN AFRICA

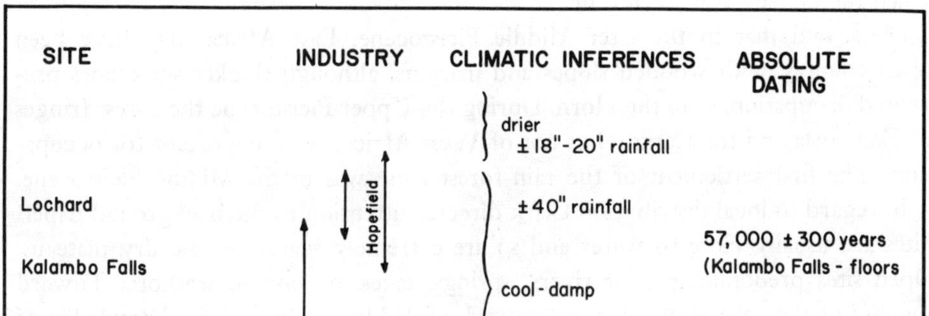

Clark then introduced the discussion on the later Middle Pleistocene cultures.

The tentative correlation table shown below was circulated (Table 11).

The tool forms of this stage included formalized large tools (hand axes, cleavers, polyhedrals) and informal tools, mainly small flakes of various kinds and wood for spears, digging sticks, etc. Again the range of raw material was wide though sometimes careful selection was practiced.

Clark divided the distribution broadly into regional and local. Sites, he said, were most common in open grass and park savanna (e.g., in the East African grassland areas) and then in the Central African savanna (though not in the West African) and in lowland coastal areas. Sites also occurred in semi- and true desert, though not until much later in South-West Africa and the Horn, in the latter case the thicket vegetation may have been a contributory cause restricting distribution. Sites of this period were least common in rain-forest fringes. In sites in the first two types of country (East and Central African grassland) the full range of industries was found, fairly early forms occurred in sites of the third type (semi- or true desert, for instance Nakop and Sahara sites), while in the last there occurred final forms already transitional to the Sangoan when the climate was drying up.

Clark said that in the later Middle Pleistocene, East Africa may have been open country with wooded slopes and margins, although thicket sometimes prevented occupation, as in the Horn. During the Upper Pleistocene the forest fringes of Equatoria and the Congo, but not of West Africa, were important for occupation. The first settlement of the rain-forest zone was in the Middle Pleistocene.

In regard to local distribution Clark directed attention to the background paper. Sites are usually close to water and so are extremely scarce on the dry plateaus. Open sites predominate near rivers, springs, lakes, or on the seashore. Toward the end of the succession caves were used, probably as a result of a knowledge of fire-making. During the interpluvials the populations seem to have moved to the mountains and forest fringes. It was necessary for these people to live close to water as carrying equipment was poor and proximity to water also meant proximity to food supply. Trees, used as a refuge at night, could also be found near water, and stones taken from streams made a better raw material. Some hunting methods, such as the driving of animals could also be better employed near water or swamps.

Discussion

Leakey said that stones from rivers were easier to work, not because they contained "*eau de carrière*," but, since they had not been exposed to temperature change, had not developed small cracks. Biberson and de Heinzelin agreed with this, but thought that the stone must not be too cold.

Clark then continued by saying that the information as to the nature of the Middle Pleistocene settlements and their contents was at present based on evi-

TABLE 11

Scheme of Late Pleistocene Successions and Industries in Africa and Europe

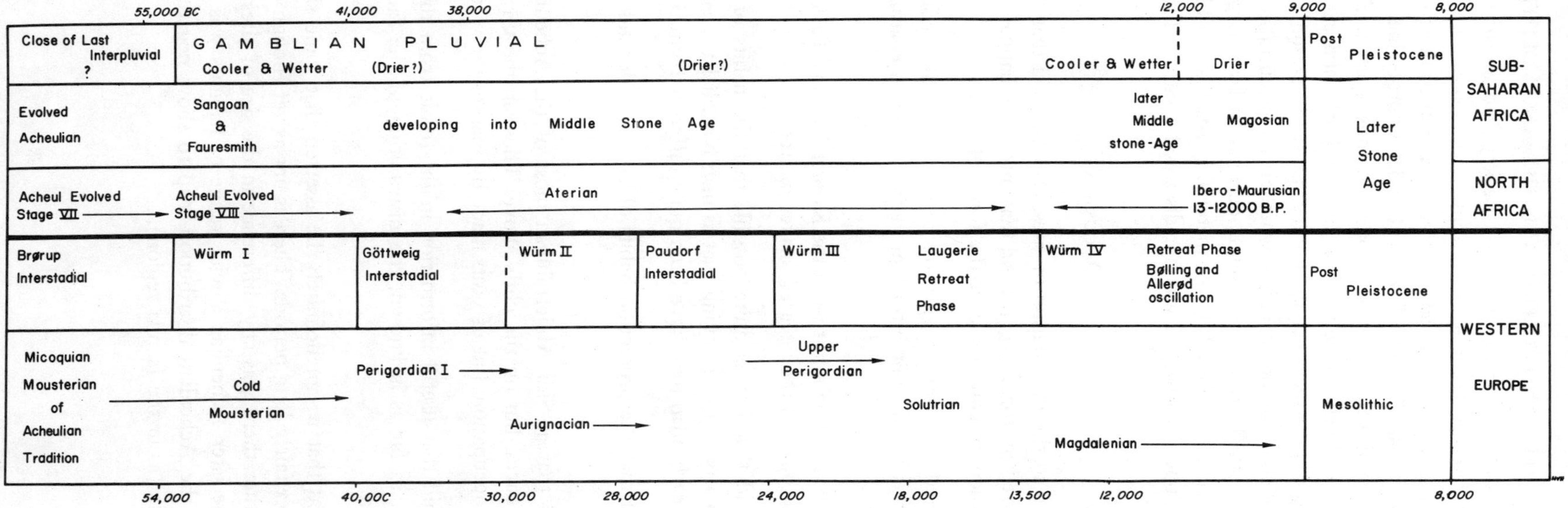

dence from only four or five sites. As Leakey had already pointed out, there were no completely excavated areas. Some sites covered large areas, widely extended horizontally and of no special thickness. Sometimes there were large concentrations of cultural material as at Isimila, Olorgesailie, and Kalambo, but within the general scatter smaller concentrations existed. Some sites, such as Broken Hill and Gwelo, were very small. The horizontal extent and the lack of thickness causes one to suspect temporary, probably seasonal occupation. There is also the difficulty of determining whether a concentration represents several occupations by small groups or a single occupation by a larger population. The amount of movement suffered by the material is often negligible.

The later Upper Pleistocene was not discussed (but cf. Table 11).

Discussion

With regard to the distribution of sites, DeVore said that the greatest amount of food occurred on forest fringes and that this was a habitat favored by baboons as it enabled them to enter or leave the forest at will.

Leakey referred to Clark's suggestion that these people slept in trees, but considered that this would not afford protection from predators, many of which climbed trees.

Cooke in this connection said that in the Kruger National Park Africans preferred to sleep in the open as they feared snakes in trees.

Washburn said that it was quite possible that the habits of animals may have changed and Clark agreed, saying that as man himself has changed, the effect on carnivores of early man may have been quite different from that of modern man.

Emlen added that the protection habits of creatures depended on the nature of the danger.

Biberson said that at Sidi Abderrahman most of the Acheulian tools were not found in the caves, but on the cliffs above. Those in the Grotte des Ours were not a sign of occupation, but of visits made by man to the water there.

Washburn said that useful information might result from the compilation of a map showing, as far as is known, the distribution of the fauna at the time of Acheulian man.

Howell stated that it can definitely be asserted that some sites were occupied repeatedly over fairly long periods. There is not as yet enough evidence, however, to show whether there was any difference in the size of sites at drier or wetter periods. In the later Acheulian a wide range of habitats was used and it would be useful on the Acheulian distribution map to show negative evidence where this is available (as in the Isimila region).

Bourlière pointed out that the increase of plant biomass which necessarily follows a change from dry to wetter conditions does not affect in the same way herbivorous, omnivorous, or carnivorous mammals. One important fraction of the metabolizable energy ingested as food appears immediately as heat and is lost for an organism which lives in a warm climate. This fraction, known as the specific dynamic action (SDA), or calorigenic effect, or heat increment of the ration, depends not only on the level of nutrition of the animal but also of the composition of its diet. Proteins exhibit the greatest calorigenic effect (as high as 40% of their caloric value) in mammals, when fats and carbohydrates exhibit lesser SDA (5%). Herbivorous and omnivorous mammals will therefore have an increased advantage over meat eaters. This phenomenon could well explain the remark made by Dr. Leakey at a previous session that, among mammals, giantism is always greater in herbivores than in carnivores.

Cooke said that before Acheulian times in South Africa the sweeping clear of the watercourses down to bed rock must have been very widespread since Acheulian hand axes are frequently found resting on the bedrock.

Bond said that in Southern Rhodesia the headwaters of the Bembezi was the only gully area with a concentration of Acheulian tools. He said that even in Kalahari sand areas valleys had been cut to bedrock so that raw material was available. In the middleveld, although there was no Acheulian, there was no lack of later cultures.

Biberson said that in southern Morocco the raw material was obscured in much the same way as by the Kalahari sand in Southern Rhodesia.

Clark pointed out that in areas where no stone is found one should not overlook the probability that perishable raw materials were still used.

Grove suggested that man may have had important effects on his environment, for instance by burning vegetation and thus causing soil erosion and the scouring of valleys.

Cooke, Moreau, and Leakey pointed out, however, that natural fires were in any case very frequent in Africa.

Monod said that in the Sahara the Acheulian was concentrated near ancient lake beds and that scattered sites also occur on plateaus capped by quartzitic sandstones, which were used as raw material, and where water was probably present at that time. The Spanish Sahara is not yet explored but must contain Acheulian material. A few scattered sites are known in the Upper Guinea forest.

Biberson said that where the *regs* in the northern Sahara permit tool-making, enormous spreads of final (and older) Acheulian occur. The *regs* are datable and Chellean spreads occur on the upper ones and Acheulian (final, wood technique) on the middle; but there is no Sangoan. He also said that Mlle. Alimen has found stratified sections with Acheulian in Algeria and it is also found in Morocco. So the Sahara obviously had human occupation much earlier than evolved Acheulian.

CLARK said that at the end of Acheulian times when the climate was drying the sites moved to higher areas.

HOWELL said that Isimila, where an occupation area with vertically dispersed cultural material was known, might be an exception to the statement that Acheulian sites had no depth. But the particular area could have been boggy and the apparent depth of material might have resulted from tools' having sunk into ground.

CLARK said, in reply to a question from Schaller, that food was not usually carried into the living place. LEAKEY disagreed. He thought that the large numbers of tools on these sites were partly to be explained by the need to have them ready for instant use if necessary. It was finally agreed that a smallish animal would probably be carried into the camp, but that if a large animal were killed the group would probably camp round it.

HOWELL thought that at this period there was no clear evidence of projectile weapons. So killing would have to be done at close quarters. In any event some of the tools would be required for dealing with vegetable foods.

BIBERSON cited Bordes as considering the Mousterian and Levalloisian to be seasonal expressions of the same culture. He also thought that the concentrations of tools would not be so large if raw material were scarce since under those conditions a tool would be resharpened and reused.

BOURLIÈRE wondered if there might not have been direct competition between man and the giant baboons in East Africa. LEAKEY suggested that it might have been similar competition between man and herbivores that first drove man to become omnivorous. He agreed that hand axes could be used for digging pits or roots or for cutting up meat, but pointed out that most of those found were new and sharp and so could not have been used for digging. HOWELL and CLARK agreed. LEAKEY suggested that the groups of fine obsidian hand axes at Kariandusi might even have been used for barter.

CLARK said that the large proportion of hand axes in the Acheulian industries in South Africa might be artificially high and due to selective collecting in the past. But COOKE did not think that this could always explain the small variety of tools found on open sites.

LEAKEY suggested that there might have been spears, probably for stabbing, not throwing, and that these might have been tipped with horn, which does not survive.

CLARK thought it possible to distinguish from the fauna on a site whether it was a "kill site," when the fauna would tend to represent one large beast, or whether it was a "home site," when several smaller creatures would occur. He pointed out that hand axes used in experiment to cut up animals show little wear. They might also have been used to cut footholds for climbing trees. LEAKEY disagreed.

DEVORE suggested that they might have been used for digging in soft substances, for instance for water lily roots, much used as food by modern hunting peoples. LEAKEY again thought that horn was more likely to have been used for this purpose and to have left no trace.

CLARK said that some Acheulian living floors showed special features such as the line of stones at Olorgesailie that might be a fish weir, and the rubble on some floors perhaps for throwing at game. He said that at Kalambo there was a possible sleeping place lined with vegetable material and what might be the base of a windbreak.

COOKE suggested that the layers of stones at Isimila and Olorgesailie resembled the bases of temporary huts made by travellers in Swaziland. HOWELL recommended the usefulness of a study of the remains on an archaeological site with a view to finding modern ethnographic parallels.

CLARK then went on to say that the occurrences of fauna at these Acheulian sites can vary considerably. At Olduvai, concentrations of the remains of large animals occur, and the later Acheulian has been found with a broken up hippo; in the base of Bed II occurs the undisturbed *Deinotherium;* at Isimila is a hippo with some portions missing and at Olorgesailie and Nyabusora are found fish remains. At the Kalambo Falls site, seeds and other parts of plants are found which point to occupation during stages of the dry season (all illustrated with slides).

At the end of the period caves began to be occupied after the use of fire became known. Evidence of fire came from several sites, for example Nyabusora, Wonderwerk, Kalambo Falls, the Cave of Hearths, and the Montagu Cave. At the Kalambo Falls site, if the charred wood had not been there no evidence of fire would have remained. It is possible that the minor red bed at Olorgesailie may have formed due to bush fires. Clark thought it seemed evident that fire was used later in Africa than in Europe, perhaps because the climate was warmer.

SESSION XIV: JULY 18 (2:30–4:05 P.M.)

CLARK was asked by the Chairman (Howell) to continue with the remarks he had begun in the morning on later Middle Pleistocene cultural adaptations. The other participants were invited to interrupt where points were made which they might like to question.

CLARK began by saying the Late Acheulian industrial units could probably be divided into four main types:

1. Assemblages containing a high proportion of large cutting-edge tools (such as hand axes and cleavers) and a small proportion of other tool types;

2. Assemblages consisting of few or even no large tools and considerable numbers of smaller tools. Such assemblages were at one time considered to belong to a separate culture: the Hope Fountain.

LEAKEY interrupted at this point and said that the work carried on by himself and his wife in East Africa left him convinced that the Hope Fountain was in fact a separate entity and that Clark should not speak as if all archaeologists were in agreement on this point. On land surface 3 at Olorgesailie there were four sites, three of which contained purely Hope Fountain assemblages and the fourth, which was largely Hope Fountain, had three hand axes associated. These assemblages could not be considered as Acheulian (s. str.).

HOWELL said that on Sands 3 at Isimila there occurred, all on the same horizon, an Acheulian site of type C (that is, large numbers of "heavy duty" tools and a very few small tools), two living floors (type B) not of this type, and a workshop floor.

CLARK said that at Broken Hill an Acheulian level occurred with, two feet below, a "Hope Fountain" horizon with no hand axes, but that the flake tools in both cases were identical.

LEAKEY, while agreeing that flake tools, scrapers, etc., occur in both types of assemblage, thought that the retouch in each case was entirely different and that Hope Fountain tools were unmistakable. He agreed with Howell that two different kinds of assemblages occurred on the same level and that this might, as Clark had suggested, be the result of two different occupations by the same group. Or it might indicate different clans with different traditions or perhaps guilds of stone workers and wood workers, but that one did not know at present.

HOWELL invited Leakey to contribute a footnote to his own and Clark's joint paper describing why he disagreed with their conclusions as to the validity of a Hope Fountain culture. Leakey agreed to do so.

DE HEINZELIN said that this was precisely the same kind of problem as that of the Mousterian in Europe.

BIBERSON said that the same problem occurred in Tunisia.

LEAKEY admitted that the facts were not in doubt but that it was the explanation with which he disagreed.

CLARK suggested that the cruder tools might be the results of efforts at toolmaking on the part of juveniles, perhaps turned out at certain seasons to fend for themselves (as are young Bantu boys). He also said that the large heavy tools might indicate a thicket environment.

LEAKEY agreed that the difference was probably due to environment and said that it need not be a time difference but could be spatial.

Bishop suggested that the fact that the Sango Hills were islands 10 to 15 miles off the shore of Lake Victoria in final Acheulian times must have had a bearing on the paleoecology since it appeared to imply ability on the part of Sangoan man to cross water. In reply to Leakey, who disagreed that the Sango Hills could have been islands, Bishop explained that he had now mapped the lake deposits and that the conclusion was unavoidable.

Howell suggested that the Sangoan people might have been stranded there when the hills became islands. He also agreed that more work should be done on the rich Kagera sites. He said that van Riet Lowe's typological work had been rather divorced from stratigraphy. At Nsongezi the MN horizon—a single horizon—was flat, like the present-day terrace surface, at 65–67 feet above the present river with a rich spread of artifacts up to two feet thick. There were two similar horizons about 31 miles downstream where rich beds occurred in bare sediments and 400 feet of uptilting would have to have taken place to give lacustrine beds above and below the implement floor. Thirty feet higher in the sequence is another quite rich horizon with late Lupemban. Howell commented on the general ecological and population significance of large concentrations and discontinuities of population at the end of the Pleistocene. Clark and Leakey agreed.

Washburn suggested, however, that if man, who was less dependent on water than the baboon, was able to support only a sparse population, he must have been a less successful creature than the baboon and he wondered if this could have been due to man's greater dependence on a meat than on a vegetable diet.

Howell added that available sources of raw material were also necessary for human groups.

Washburn asked if the heavier tools were for working wood at the end of Acheulian times or for obtaining vegetables.

Clark said that the heavier Sangoan tools were usually considered as woodworking tools.

Leakey thought, however, that they should rather be regarded as tools for use in a wooded environment since "woodworking" implied a much finer tool.

Clark pointed out, however, that the working edge on these Sangoan tools was of the same type as that on the later core-axes and on tools right up into the Neolithic, but agreed that by "woodworking" he included also the by-products of wood such as bark.

Clark then spoke of some of the special features on Late Acheulian living floors. He suggested that the line of stones on floor 2 at Olorgesailie, while it might possibly be natural, could be a fish weir.

For the concentrations of rubble at Isimila and Olorgesailie there might be several explanations (there was virtually no rubble at the Kalambo Falls):

(1) that these were natural stones which had gravitated down the slope; (2) that they were sleeping places, though this was thought unlikely; (3) that they were animal kill places, having been used as missile stones; (4) that the hollow filled with stones at Olorgesailie might have been a cooking place where fire was used even if not made. To the objection that none of the stones showed fire crackle or reddening Clark pointed out that at Kalambo, where evidence for fire was incontestable, no signs were visible on the stones.

LEAKEY objected that had there been fire, at least charcoal, which is almost indestructible, would have remained. He also thought that the 10-foot-diameter hollow at Olorgesailie had been deliberately filled with rubble (which could not have got there naturally) but he did not know for what purpose.

HOWELL said that at Isimila patches of rubble occurred interspersed with areas covered with fully finished stone tools. The extensive erosion (especially of the upper sand horizons) permits one to walk over the site and to observe the general extent of occupation and the horizontal discontinuities between them.

CLARK then described elipsoid hollows at the Kalambo Falls site filled with compressed grass, reed, and woody material—definitely not hearths—and suggested that these might have been sleeping places.

LEAKEY drew attention to the current African practice of putting down a layer of grass on which to skin a dead animal.

COOKE suggested grass for use in straining water from water holes. He also drew attention to the natural carbonization of wood in peat bogs, which might be mistaken for the effects of fire.

CLARK pointed out that charred wood, charcoal, and consolidated ash all occurred at Kalambo.

The plot of an arc of 25 stones, part of a circle roughly 9 feet in diameter, from the Kalambo site was then shown and it was agreed that this could be the base of a windbreak.

Clark also described roughly spherical, but flattened lumps of black clay containing carbonaceous material, surrounded by a layer of lighter clay without such material, that had been found at the Kalambo Falls site. No satisfactory expanation of their purpose was offered and geophagy was not considered very likely as analysis had not revealed any appreciable quantities of mineral salts.

Clark then went on to say that evidence of fire occurs only at the very end of Acheulian times in Africa and comes from four sites, of which two were open sites: Nyabusora, where charred fish and animal bones suggest cooking; and Kalambo, where charred wood suggests woodworking. He agreed with Leakey that lightning could not be entirely discounted as a source of the fire but said that there did seem to be unmistakable signs of a hearth. The other two sites were in caves, one the Cave of Hearths, where signs of fire have been found throughout the 30 feet of deposit, the bottom two feet of which consist of guano burnt as a result of the fires of Acheulian man.

Cooke said that there were signs of a hearth which had been used for a considerable period in this cave. He asked Clark how the culture of the Cave of Hearths compared with that of Kalambo.

Clark replied that he had not had an opportunity to examine it properly, but he thought it would be later.

The other site where fire is reported to have been found dating from Late Acheulian times is the Montagu Cave.

The consequences of fire-making must have been important.

Man would now have semipermanent homes, and society and community life could develop.

Hunting techniques could be improved, such as firing grass to drive game. Clark agreed with Leakey that man must have done this with wild fire before he knew how to make it himself but the practice could not then have been so widespread.

Man's tools would be improved and he would be able to use a greater variety of materials.

He could live in country which had hitherto been unacceptable.

Fire afforded protection, not merely light as Leakey contended, and the fire was usually made between the sleeper and the entrance to the cave or other shelter.

Fire was also of value for cooking food.

Bourlière pointed out that the value of fire in regard to cooking food was very great indeed since many foods, both meat and vegetable, could be eaten and digested cooked but not raw.

Clark summed up his remarks by saying that the general impression of Acheulian man was of temporary occupation until fire enabled him to take up more permanent quarters. By bringing and keeping people together, fire also probably helped in the development of true language. Clark agreed with Leakey that effective communication must have existed long before final Acheulian times, but that language as such probably developed then.

In closing, Clark said that it was clear that teamwork between archaeologists and scientists of other disciplines was necessary if valuable evidence was not to be lost through ignorance.

Leakey made the alternative suggestion of a wider training for archaeologists in other disciplines before they went into the field.

Biberson said that his evidence in Morocco confirmed much of what Clark had said. Most of his sites were factory sites but some were living sites; and until the final Acheulian, in spite of very careful search, no signs of fire had been found at all. With the final Acheulian (stage 8) at Sidi Abderrahman signs of

TABLE 12 Climatic Change and Radiocarbon Dates in Sub-

(After J. D. Clark 1962. Actes du IVe Congrès Panafricain de Prehistoire et de l' Etude du Quater-

WESTERN EUROPE

Radio Carbon Dates	Climatic Stage	Main Cultural Division	Cultures
A.D. 0- B.C.	POST GLACIAL	NEOLITHIC MESOLITHIC	Various
8000	LATE WURM	UPPER PALAEOLITHIC	Magdalenian
10,000-8850	ALLERØD OSCILLATION		
10,500	LATE WURM		
	BØLLING OSCILLATION		Solutrian
11,500	MAIN WURM STAGE		
ca. 15,000	MAIN WURM:		
21,000	LATE PHASE		
25,000			PROTO-MAGDALENIAN
26,950	PAUDORF OSCILLATION		Aurignacian; Perigordian III-V
	MAIN WURM: EARLY PHASE		
28,720	GOTTWEIG INTERSTADIAL		Persisting Mousterian; Perigordian I
39,950	EARLY WURM STAGE	MIDDLE PALAEOLITHIC	Mousterian

naire, Sect. III. Pré- et Protohistoire. Tervuren, Belgium: Musée Royal de l'Afrique Centrale)

SUB-SAHARAN AFRICA

Radio Carbon Dates	Climatic Stage	Main Cultural Divisions	Cultures for which radio-carbon dates exist
A.D. ± 1080		IRON AGE	Channelled and Stamped Wares
B.C. 140	NAKURAN WET PHASE	LATER	Late Wilton
1300			Various Neolithic cultures Wilton/Smithfield
	DRIER	STONE	
3400			Wilton
4500			Nachikufan I
	MAKALIAN	AGE	
6000	WET PHASE		Ishangian
7550			Late Magosian
9100	DRIER	SECOND	Lupembo-Tshitolian
9600		INTERMEDIATE	Latest Pietersburg
12,500		MIDDLE STONE AGE: UPPER	Final Lupemban
13,100	MAIN		Later Pietersburg
17,000			Mazelspoort II Variant
		MIDDLE	
25,000	GAMBLIAN		
			Rhodesian Lupemban
28,000	STAGE		Mazelspoort I Variant
		LOWER	
37,000			Hagenstad Variant
38,000			Sangoan (38,000–40,500)
40,500	DRIER	FIRST	
41,000			
	OSCILLATION	INTERMEDIATE	
57,300	EARLY GAMBLIAN STAGE	EARLIER STONE AGE	Evolved Acheulian

fire occurred in the form of cracked and reddened stones, but no charcoals had been found on any site in Northwest Africa.

(Wednesday, July 19, was a free day.)

SESSION XV: JULY 20 (9:30 A.M.–12:30 P.M.)

PAST CLIMATES (2)

Discussion was directed to the nature and mechanism of the changes in Pleistocene climate and the implication of these changes for the different areas.

DE HEINZELIN said that work of a purely theoretical nature had been done by Bernard on the application of Milankovitch's theory to Africa.

BIBERSON said that there seemed to be a striking connection between glaciations and pluvials, but that this might apply to warm or cool pluvials.

LEAKEY thought that humidity and soil were more important for plants in Africa than was temperature. BIBERSON added that since it was the effect of the ecology on man that was to be determined temperature was of secondary importance. BOURLIÈRE, however, pointed out that for naked men without fire temperature could well be of importance and a change of 6°–10°C. might be very significant.

MOREAU considered the presence of Ericaceae at Kalambo significant since now it is not found below 1,700 m.

HOWELL suggested starting with the known dates from Kalambo and Mufo, and the above table was used as a basis for the discussion.

DE HEINZELIN added remarks on glaciations and said that there was a world-wide parallelism, both now and historically, between the retreat of glaciers.

BUTZER thought that evidence for the Atlas should be obtainable on the same basis as that for Europe. BIBERSON agreed. He then went on to say that he did not think Bishop was correct in considering that without exact dates the evidence was too theoretical, and he thought it important for the correlation of North Africa both with Europe and with other parts of Africa.

HOWELL agreed with Bishop that exact dating was more important. LEAKEY said that at present absolute dating is uncertain and apt to vary with the method used.

BIBERSON and BUTZER referred to changes in sea level during small-scale glacial fluctuations and pointed out that oscillations were going on even at the present day.

HOWELL said that even general correlations between Europe and Africa were of importance.

MOREAU, speaking again of climate, said that for early man it was not the actual amount of rain that mattered, but when it fell. The movement of the equatorial zone 5° north or 7°–8° south of its present position would have a great effect on vegetation and so on man.

BIBERSON agreed that such a movement would have a global effect on glaciations and on pluvials. For North Africa, he said, it was important to be able to correlate glaciations in Europe and pluvials in Africa.

DE HEINZELIN said that deep-sea cores from beneath the equator indicate temperature changes of 6°C. corresponding with glaciations in northern Europe.

ARAMBOURG said that while there were four glaciations during the Quaternary in Europe, that only in the last one did the fauna move south to the Mediterranean, even though it was a warmth-loving fauna. He considered that this might be significant when considering the correlation of glaciations and temperature. He disagreed with Bourlière's suggestion that the animals might not in fact have been warmth-loving and that movement was because of inadequate food supply. The Quaternary fauna was all African in its affinities and most likely to have been warmth-loving.

MOREAU said that the flora and fauna of Mount Cameroon was of East African species and that to explain this one must postulate one or more periods of the Pleistocene when the climate of the areas was similar and when the montane forest was widely extended to ensure a continuous distribution. This could have been made possible by the northward movement of the equatorial belt and since the floras and avifaunas are so very similar with no subspecific distinctions, this must have occurred fairly late in the Pleistocene.

MONOD said that the Gulf of Guinea islands and massifs in West Africa also have some East African flora. In the same areas, as also in East Africa, there is also some Palaearctic flora. It is, of course, possible that this derived directly from the Palaearctic region and not via East Africa. There is evidence on St. Thomé Island for species' having been introduced by wind action and this could have happened elsewhere.

Monod suggested that Pluvials might have created extensive lakes in the west, due not to local precipitation but to rainfall at a distance (for instance Lake Chad), so that the lake sediments in the Spanish Sahara do not necessarily indicate a greater rainfall but may be due to drainage from a distance.

HOWELL said that the important part of rainfall was its reliability or lack of it (cf. East Africa) and this would have been accentuated by shifts of the equatorial zone.

CLARK said that the evidence for greater cold at Kalambo and in Angola was based not only on the presence of cold-loving pollen, but also on the absence of warmth-loving varieties.

GROVE spoke of the importance of dune evidence in determining a sequence which can be widely applied in the Chad area; he instanced Barbeau's recent work.

BIBERSON said that correlations had been made using molluscs between Morocco and Lisbon and on present habitats these indicated a change of temperature of the order of 21°C. Lecointre had made similar correlations between the English Channel and Norway.

BUTZER mentioned a species that exists now in Algeria and disappeared from Majorca during the Würm when the climate there must have become colder.

The discussion was reopened by BOURLIÈRE who suggested that during the Pleistocene the large ungulates with a very wide distribution might have been rather like the present-day white-tailed deer in the deciduous forests of North America. This species has a very wide distribution and great variation in brain size and body size, but all the varieties can reproduce.

ARAMBOURG said that such an analogy could not be applied to Pleistocene hippos which were definitely a morphological series.

HOWELL said that recent changes in altitude in East Africa, for instance Ruwenzori, are of importance and where exact figures on uplift do exist these are of great significance so far as the creation of barriers or the opening of corridors favorable to dispersal is concerned. Some figures of the magnitude of uplift are presented in Table 13. BISHOP and LEAKEY pointed out, however, the great local variability in Africa and its great size compared with Europe.

TABLE 13
RIFT VALLEY TECTONICS

Lake Edward
End Lower Pleistocene faulting more than 100′
Grid faults with up to 60′ throw
Rejuvenation of the Rift faults within the past 70,000 years
Lake Albert
End-Lower Pleistocene faults, in places more than 1,000′
Ruwenzori uplift continuing into the Upper Pleistocene
Kagera Valley
Tilting (and possible warping); applies to the whole of western Uganda
Similar movement went on during the whole of the Pleistocene to produce a total uplift, at the swamp divide 150 miles west of the Lake, of about 2,000′
It is also important to remember the effect of volcanic activity on environments.

GROVE suggested that the uplift might have been accompanied by local subsidence.

MOREAU said that sub-Saharan West Africa seemed to have been omitted from the discussion. He said it had suffered little tectonic movement. The dry dune system there probably oscillated some 200 miles south, spreading savanna conditions and leaving three refuge areas for primates (cf. the work of Booth) and the fauna was thus fragmented into small groups. It has been suggested that changes in sea currents may explain the Dahomey Gap in that they may have affected rainfall in West Africa. Davies has suggested that in West Africa, Stone Age tools may now be covered by equatorial forest.

Monod said that a semiarid zone must have extended farther south towards the Gulf of Guinea, but he could not see where the refuge areas might be from whence the forest was subsequently recolonized as the mountains (2,000 m.) were not really high enough to constitute refuges.

Moreau suggested that the West African forest must also have advanced northward.

Bourlière suggested that it would be of interest to consider the increased size of the mammalian fauna in relation to paleoecology and to see whether there was, in fact, a widespread increase in size.

Bishop then outlined the successions in the Western (southern part) and Eastern Rifts (Tables 14, 15).

TABLE 14
Geological Succession, Western Rift (Uganda)
Lake George-Edward and Kaiso Areas
(Bishop 1960, etc.)

	Gully erosion and gully fill minor step faulting	
LATER PLEISTOCENE	25′ to 30′ terrace gravels	Explosion craters—ashes and tuffs with paleosols
	erosion 90′ terrace gravels	
	.... form of present lakes established	
MID-PLEISTOCENE	Semliki series lake deposits—with paleosols at top	
	.... rift faulting	
EARLIER PLEISTOCENE	Kaiso series lake deposits	
	.... rift faulting	
PRE-PLEISTOCENE	Kisegi lake beds, etc.	
	.... rift faulting	
PRE-CAMBRIAN	Long period of erosion Basement Complex	

TABLE 15
GEOLOGICAL SUCCESSION, GREGORY RIFT (KENYA)
OLORGESAILIE-MAGADI AREA
(Baker 1958, p. 9.)

LATER PLEISTOCENE	Lake Magadi Evaporites erosion High Magadi Lake Beds
	minor faulting-erosion
MID-PLEISTOCENE	Olorgesailie Lake Beds; land surfaces with Acheulian assemblages)
	grid faulting-erosion
EARLIER PLEISTOCENE	Oloronga Lake Beds minor volcanic vents Plateau trachytes, etc.
	rift faulting
PRE-PLEISTOCENE	Earlier volcanics rift faulting-erosion
	long period of erosion
PRE-CAMBRIAN	Basement Complex

DE HEINZELIN mentioned the volcanoes near Ishango Bay and that it seems possible that volcanic explosions might have changed ecological conditions and fauna in that area. The Nilotic fauna was greatly reduced in species, and in size, and recolonization was slow. At the same time crocodiles disappeared from most of the area, no doubt due to the changed chemistry of the water.

LEAKEY agreed, and then referred back to Moreau's remarks on the connections between East and West African fauna. He supported the hypothesis of a northward vegetation movement, but felt that the problem needed further study. Some animal species, for instance reptiles, had spread in both directions.

BOURLIÈRE then summed up the discussion by saying that it had been agreed that there had been a real change of climate over the whole continent, possibly accompanied by a temperature change of the order of 6°C. It was also agreed that the uplift in the East had had no counterpart in the West where the vegetation zones were regular. In the East the broken-up vegetation zones had probably been of great importance for hominid development allowing as they did the possibility of isolation with its greater chances of evolution.

DE HEINZELIN said that the movement of the equator and the repercussions to the north and south should also be considered.

LEAKEY thought that the effects would be different in the different zones. He also agreed with Biberson that tilting and movement must have affected the fauna before the Upper Pleistocene, probably in the Villafranchian. Such movement would probably be gradual at first and only later more violent. In 1959 in Kenya and Tanganyika, displacement of the order of 10 feet due to earth movement was recorded.

SESSION XVI: JULY 20 (2:35–6:15 P.M.)

Some aspects of the mode of life of modern hunter-gatherers, and of agriculturalists on savanna belt margins who in bad seasons are forced to return to hunting-gathering, were then considered. Changes in mode of life brought about by new political situations urgently require that field studies be initiated before these are further altered or disappeared entirely.

BOURLIÈRE opened the discussion by asking:

1. How do modern hunter-gatherers feed themselves, both quantitatively and qualitatively, and how does this vary with seasonal change?
2. How much variation in reproduction and evolution is determined by famine periods?
3. What foods do people use in times of famine?

Some data was available on these matters, as CLARK and HIERNAUX pointed out, but many aspects of the ecology of such groups was still unknown and further studies were badly needed.

CLARK first discussed the Hukwe Bushmen of the Zambezi-Mashi area.

The group of about one hundred lived in open woodland and moved between five waterholes about 20 miles apart over an arc of some 100 square miles.

The main foods were: the oil-rich nuts of *Ricinodrendron rautenanii*, which are available all the year round, either on the tree or fallen to the ground; roots (not identified); caterpillars, particularly off the *Baikiea plurijuga* tree; lizards; snakes; mice; honey, which is particularly prized; and game animals (roan, zebra, kudu, waterbuck, duiker, and giraffe), the most preferred being the eland. They were not seen to eat salt, but may have obtained it from a trading store.

Clark said that a large bull eland he shot lasted a group of 30 Bushmen for two to three days. Everything was eaten, including the skin. The meat was either cooked and then eaten or sundried after being carried back to camp. He had unfortunately not noticed how the bones were dispersed. He had, however, seen a few skulls and horns behind the camp, though skulls were not kept as proof of prowess. The people did not seem undernourished. Clark said he had not obtained any information about food distribution to individuals, but he had noticed that very few old-looking individuals were in evidence. The Valley

Tonga (BaWe) people will, in time of famine, deprive the old of food and thus cause their death, but he did not know if the Bushmen did this.

For hunting the Hukwe Bushmen used nowadays chiefly spears since the Barotse Parent Chief had made the use of poisoned arrows illegal. They exchanged honey and skins (which they cleaned with mandibles, sand and brains) with the Bantu for iron which was beat out cold. Bone, preferably of giraffe, was made into needles. The technique of scraping utilizes a stone, rather than a metal rasp or scraper, and with the scraper at right angle to the material.

LEAKEY emphasized the great need for a study of the daily food habits of such a group and of the debris left in their camps. A regular periodic check should also be kept on the debris after the group has moved away. It is also necessary to analyze the food plants, both raw and cooked, to assess the nutritional value.

BOURLIÈRE, after being told by CLARK that he thought that the Bushman infants were suckled for two years, instanced Tonkin women who suckle infants for 5 to 6 years. He emphasized the necessity for children in a primitive society to be accustomed gradually to adult food. HIERNAUX agreed.

He went on to speak of the Ituri pygmies. They now live in symbiosis with Bantu agriculturalists so that there are no longer any true Ituri pygmies. They do still live largely by hunting and gathering, using nets to capture driven game. HIERNAUX said that no nutritional study has been made of the Twa, pygmoids of Ruanda Urundi. He said that pygmies appear to be better nourished than the surrounding Bantu, and on the whole people living on protein are better nourished than agriculturalists and hence are more fertile. For this reason the pygmy women were sought after as wives by the Bantu around them. The pygmies in the Lake Albert area are beginning to practice agriculture and to trade with the surrounding Bantu. Nutritionally the trade products are very important—for instance banana beer, which is more nutritious than bananas. They also obtain iron by trade. The Tutsi and the Hutu only eat their cattle if these die, and such protein foods as insects are unacceptable to them. Hiernaux thought that the Sandawe and Tindiga of Tanganyika, who were perhaps in symbiosis with the Bantu, probably lived a life more akin to that of prehistoric peoples than did the Bushmen whose culture must have deteriorated after they had been driven into an unnaturally inhospitable habitat. In part of Ituri, Hiernaux said, witch doctors used a plant to increase milk supply and that this appeared to be efficacious. The plant is not identified.

DISCUSSION

LEAKEY said that among the Kikuyu also, goats and cattle are wealth and not food, and although meat sacrifice is frequent and is always eaten, this is not "food."

HIERNAUX agreed that to understand a people's nutrition a thorough study must be made. He cited the Tutsi who state that they have only one meal a day and yet are eating continually.

LEAKEY said that in time of famine, meat-eaters suffer less than agriculturalists as game is then easier to kill.

CLARK said that on Iron Age sites in southern Africa very few domestic animal bones occurred, no doubt since cattle were considered as wealth and wild meat only as food. Such sites dated roughly from 1400–1800 A.D.

MONOD mentioned that when collecting people catch fish, they will eat not only the fish but everything they catch.

BOURLIÈRE agreed and described the contents of the collecting bags of the Ituri people which included worms, toads, and snakes. The snakes were cut up immediately and the head discarded so that archaeologists should not expect to find any of the important parts. Anything not usable was discarded on the spot. Bourlière then said that Cooke had told him of two organizations in South Africa doing work on the Bushmen. These were the Kalahari Research Group of Witwatersrand University which had been doing a very complete study for the last five years and would probably continue for the next five, and the Bechuanaland Administration which kept a full-time research officer living among the Bushmen to study their diet, habits, etc. It might be profitable to prepare lists of desired information which these people might collect during the course of their work. Bourlière thought it would also be useful if anthropologists provided data on the nutrition of the tribes they studied.

LEAKEY, in reply to a query on hunting methods, mentioned a published description of a game drive at Karamojo (*Uganda Journal of Wild Life and Sport*). Referring to weapons, he said that the discovery of harpoons in a deposit did not necessarily imply that fishing was practiced since harpoons could be used for spearing game, as by the Tindiga.

MONOD said that both harpoons and fish hooks occur in the Sahara and there is one tribe that ceremonially collects the animals stranded on islands when floods occur.

CLARK said that the Bon and Ribi peoples of Somalia use nets, dogs, and poisoned arrows for hunting. In reply to Bourlière who asked if there was any evidence for the date at which poison was first used, he said that he thought arrows with microliths would be ineffective without it. LEAKEY disagreed, provided the microliths were hafted in such a way that they stayed in the wound and caused the animal to bleed to death. He did, however, think that groups of microliths found together on a site might be those which had been cut out in the piece of poisoned flesh. In reply to Bourlière he said that no game pits of Pleistocene date had been found but BIBERSON mentioned some instances on the banks of the Rhine and CLARK said that modern Bushmen using only a digging stick dig pits to catch elephants.

BIBERSON asked how driving animals into mud would be effective in catching hippo who often live in liquid mud.

LEAKEY explained how in the Ruvuma River hippos are driven down the path to the swamp and impale themselves on stakes set up in the way and so bleed to death.

With regard to the introduction of the hunting dog there did not seem to be any evidence as to the date of its introduction. CLARK said that a possible dog occurred with the Iron Age at Mumbwa Cave. COOKE added that dog occurred with the Iron Age at the Cave of Hearths. BIBERSON said there was Neolithic evidence for the dog in the Sahara and MONOD added that its age in the Middle East was about 7,000 years.

DE HEINZELIN said that evidence from Ishango, dating to more than 8,000 years B.P., suggested that the earliest semisettled groups might have been fishers on the edge of lakes.

MONOD then spoke on some aspects of nutrition in the Saharan peoples.

He said he would consider chiefly the southern Sahara and the Sahelian regions where the resources of Recent and Neolithic times were better known than those of Acheulian times, which indeed might have been very different.

North of Chad the people were exclusively hunter-gatherers and there were more food plants than at first sight might appear. Racially these people are not a well-defined group and are probably not remnants of prehistoric groups.

Some of the plants in the steppe or semidesert are normally eaten only by special groups such as slaves or children. Very few plants are poisonous (only three are known that poison cattle) and in these regions of light, sandy soil it is relatively easy to dig things up. During famines such substances as powdered bone and datestone, wood, and skin are eaten.

A book published by Storey in 1958 (*Some Plants Used by the Bushmen in Obtaining Food and Water*) shows conditions similar to those of peoples living in the Saharan steppes.

Monod then described the food of these people. He said that several kinds of mushrooms and truffles are eaten. These do not necessarily appear every year, but only sporadically. Rhizomes are eaten, but only in times of famine are very many species eaten. Water lily bulbs and seeds are eaten and the galls on roots. An extensive number of leaves are eaten, particularly those of plants that grow after rain and produce a mucilaginous substance, for example *Portulaca* and Cruciferae. Seeds of *Cucumis* sp. are eaten ground into flour and mixed with date flour. They are slightly oily and when mixed with water produce a nutritious and filling substance. Wild grains are collected in baskets; P. Fuchs states that a woman in Sahel, southeast Tibesti, can collect ten camel loads of grain in a season. Grain is also collected from the nests of certain ants. There might be some connection between the use of wild grain and the saddle quern.

Rains usually occur annually so that wild fruits, albeit of a lowly kind, as for instance jackal dates, are a seasonal possibility and many species are eaten. As they are all well known and have native names, they may have been a source of food for a long time. Three or four species of *Acacia* with edible gums occur. These are usually used as medicine, but are also eaten by the lower strata of the population. Sweet foods are rare. Nowadays the date is the chief one. Sweet nodules occur on tamarisks, produced by the "scale insect," and these are also eaten.

Today *Sesamum indicum* is cultivated for oil. Fish are eaten and widely exported. Other creatures eaten include tortoises, possibly snakes and crocodiles, lizards, the tail of *Uronyastix,* and birds which are sometimes caught in snares or by rod and line (cormorants).

Among mammals, rodents, jerboas, hyrax, and ungulates, particularly *Addax* (gazelle), are eaten. Rock paintings also depict giraffe, elephant, and other large animals. *Addax* hunting is practiced by several people on foot with dogs and spears or on camels on expeditions lasting up to thirty days. Dried *Addax* meat is almost an industry the unit being an *Addax* skin stuffed with the dried meat. For eating, the meat is ground between two stones and mixed with rice gruel. The work of Brouin gives detailed descriptions of the hunting methods of various such peoples. Traps are widely used, the most common being a circular "wheel" trap with radiating spokes that surround the leg of a victim. Nets of gazelle tendon are also used.

The midden material on sites shows that various mollusks and water crustaceans were eaten in Neolithic times. Maggot worms are made into loaves and dried in the sun by the Dauda and a mixture of grasshopper flour and date flour is a favorite food of travellers.

Water always presents a problem. It is often obtained from the stomach contents of the *Addax* by filtering, but in earlier times a network of lakes in this area must have provided a more favorable environment.

The introduction of cattle did not displace hunting. Agriculture could only be established where it was possible to irrigate, but desert living people benefit from trade with agriculturalists.

He stressed the need for more work on the nutrition of these areas and lists of queries might be prepared for investigation by the Commission on Urgent Anthropological and Ethnological Research.

BOURLIÈRE pointed out that under conditions such as those described by MONOD groups must always be small so that present-day numbers do not represent a diminution of population. He said that basic food requirements for a person of middle size were: (1) 2,000 calories a day; (2) amino acids found only in certain animal proteins; and (3) mineral salts to provide trace elements necessary to the body and to ameliorate the taste of unpalatable food.

Two other problems were reconsidered. These were a restatement of the known facts concerning the variations of the Chad basin and the problem of the curious paleodune formations; and the distribution of certain African freshwater fish with implications for reconstruction of the paleoecology of the Pleistocene.

GROVE opened the discussion on the Chad basin. He said that the barrier from the Logoni River across northeastern Nigeria, similar to the barrier beaches of Mexico, southern France, and the southern Baltic, might be Villafranchian in age. If there were humid periods in the past they must have caused the filling of a lake to 330 m. above sea level, judging by the formations (Lake Chad is now 280 m. above sea level). But the age of such a lake needed to be determined. J. Barbeau (and others) had shown that there are four distinguishable levels; if the series were truly alternating with the four sets of dunes this could be a useful basis for work. Barbeau's four levels fit in with the distinguishable shorelines and the four levels worked out by Faure for an area farther west. However, Grove thought that the barrier was the result of a single build-up and not of several.

COOKE considered the barrier probably to be Neolithic. LEAKEY thought that it could not be Villafranchian, but might be Upper Pleistocene.

GROVE said that if Leakey were right then this presupposed a wetter climate and less evaporation at that time which was important.

MONOD suggested that a higher lake level might have been simulated by the sinking of the Chad basin.

MOREAU thought that Grove's suggestion of a wetter and so cooler Late Pleistocene climate could explain the connection between the East and West African flora and fauna.

GROVE said that to obtain useful evidence from the dune system a greater knowledge of present-day wind direction was needed. The fossil dunes did not seem to suggest any change in direction.

MONOD said that dune formation was still going on. Two parallel areas of longitudinal dunes (*cordon lourd*), now in a Saharan climate had once been in a Sahelian climate, and between them is a veritable "*boulevard de préhistoire*," but the formation of these dunes is still little understood. He agreed that if rainfall increased in the southern Chad basin it would probably improve the rainfall of the whole area to the west. He thought that the *cordon lourd* dunes were formed in a less arid climate than the present because of the sands of which they are composed. They must have been pre-Neolithic since that material is found on top of them, but it is not possible to say whether they are post-Acheulian because that material has only been found in the corridors between the dunes.

Butzer said that conditions in Libya were similar. Grove thought that the humidity at the time of the Acheulian must have affected the whole of the Sahara.

Cooke then considered the distributions of certain African fish, in particular two species of *Hydrocyon* and four of *Protopterus*. These showed isolated distributions not apparently caused by modern phenomena. He thought it very important to know more about the formation of the continent in Pliocene times for an understanding of the later periods. If the Pliocene drainage systems were lower and the East African central mass were 1,000 m. lower than now, pressure systems would break down and wide extensions of weak anticyclone belts would result in a very different climatic situation. Following such Pliocene conditions, volcanic development in the east would have had great effects on the climate and ecology of the region.

Butzer considered that the important factor was the reduced thermal gradient from the pole to the equator with absence of polar ice. This could have caused a breakdown of the pressure system without postulating a lower elevation for the continent.

Leakey thought Cooke's hypothesis untenable on the evidence of the fauna.

Cooke said that even were his hypothesis correct East Africa would still have been the highest, coolest, and thus most favorable region of the continent. Mount Kenya shows evidence of glaciation, but since other mountains do not they must have been lower.

SESSION XVII: JULY 21 (9:30 A.M.–12:30 P.M.)

Bourlière summarized the main themes of the past two weeks under three main headings (see Table 16.)

Climatic changes. It had been shown that these could be broadly correlated and that they were mainly concerned with alternation of wetter and drier periods. Changes in the average yearly temperature which had also been shown to have occurred in the African Pleistocene were of importance.

Physiographic changes. The Pleistocene uplift of the Central African highlands, tectonics and changes in river drainage systems are all important for early human evolution since they are factors making for isolation.

Differences between East and West Africa. Not only today, but in the Pleistocene also, these two areas presented striking differences in their microenvironments.

These past changes had important implications for human evolution. Such changes could act on early human groups, both at the individual and at the population level, but one should not overlook the kind of effect on the individual that also influences the group, for instance fecundity. Bourlière said that in con-

TABLE 16
ECOLOGICAL FACTORS AND MAN

I. *PLEISTOCENE ENVIRONMENTAL CHANGES.*	
(a) CLIMATIC.	Rainfall humid / dry } periods Average yearly temperature (6–10°C lower)
(b) PHYSIOGRAPHIC.	
Factors favoring isolation	Uplift 1000 m. Volcanism Changes in river drainages and lake sizes
(c) EAST-WEST CONTRASTS.	
II. *POSSIBLE INFLUENCES OF THESE ENVIRONMENTAL CHANGES ON MAN.*	
(a) INDIVIDUAL LEVEL.	Size and body build Fecundity, time of sexual maturity and differentiation in sexes.
(b) POPULATION LEVEL.	Population units { size of breeding populations; isolation Overall population density (population cycles?)
(c) CONTINENTAL LEVEL.	—Rainforest—no humans until recently. —Forest edges—rich, most important at the beginning. —Open habitats—lakes and shallow water, important at bottoms of population curve. —More niches in the East—where isolated groups could grow up and evolve.

sidering prehistoric hunter-gatherer groups with a probable low level of nutrition, the possibility must be faced of seasonally oestrous females as is the case with other mammals. This is said to have been recorded among Eskimos and possibly pygmies. Nothing is known of Bushmen in this respect. Among normally well-nourished European women the menstrual cycle became interrupted in concentration camp conditions. On the other hand difficult conditions do not have this effect on peoples in India so that it is possible that the human female can adapt to poor environmental conditions. It was at the population level that greater isolation of units also occurs, giving more chance of genetic drift. Cultural development can also be affected if the units become isolated in differing environments. Rapid population increase frequently results from the colonization of empty environments, as exemplified, for instance, by the introduction of European rabbits into Australia. When the carrying capacity of the environment is reached a sudden drop occurs, followed by a series of diminishing increases as the environment recovers until a stabilized state is reached. In natural conditions populations are thus never constant. In early man the population graph would have probably followed similar lines, but the introduction of new technological advances such as tool-making or fire-using would have widened the possibilities of early men and so allowed their populations to increase, as did the Neolithic revolution in the Near East and Western Europe. In wild mammals, population fluctuations are generally most spectacular in simplified biotic communities such as those of desert or tundra. The same phenomenon might well have occurred in populations of early men, when climatic conditions altered the environment making it unfavorable. It is at such periods that the effects of "the survival of the fittest" should be seen most clearly. In more favorable environments larger population groups act as a buffer to some extent.

The most favorable environment for early man was probably one with a large biomass of small and medium sized ungulates.

Discussion

Hiernaux began by saying that in Ruanda Urundi there was a well-known population and people recognized that famine would occur roughly every seven years. Hiernaux thought that the further man moved from his primitive beginnings the more environmental changes acted on him indirectly through his culture. The Neolithic and Industrial revolutions were probably only two of a series through which man has passed and in the early times a major climatic change could have had the same effect. Evolution from an *Australopithecus* to a *Homo sapiens* stage was dependent on adjustments in the brain. Periods of drastic psychological stress, such as could have been caused by great changes in an environment to which early man had become adapted, necessitated new adaptation, not so much to the climate directly but to its effects on man's culture. He said also that he would like to add to Bourlière's table above (Table 16) the factor of the time of sexual maturity and the difference in male and female.

Leakey considered that the greatest revolution in human history was that of tool-making itself which at once would have given man access to a vastly increased food supply. This would have had a great effect on his evolution into a true human (hominine) since by making tools man, so to speak, domesticated himself.

Washburn agreed that tool-making must have had such a great effect. Domestication implied rage control which permitted creatures to live together. With smaller adrenal glands go smaller brow ridges and faces, even in laboratory rats, and man's adrenal glands are smaller than those of a chimpanzee. *Homo sapiens* may have had smaller adrenals from the time he first appeared.

He thought it possible that true speech may have come to man with the bigger brains of the Acheulian since when brain cortex size is reduced the first ability lost is that of speech.

He also pointed out that man with his greater mobility draws his mates from greater distances than do other primates and thus extends the available gene pool. In the short run it is existing frequencies that can be built up to allow a more rapid working of natural selection.

EMLEN agreed with Hiernaux's point that as the Pleistocene went on cultural effects became more important than climatic change per se and his culture made man increasingly less dependent on environment. He added that ability depended both on the cortex area and the size of the brain.

DE HEINZELIN noted that the first evidence of man's counting ability was the marks on bones associated with the Mousterian and the Upper Paleolithic.

WASHBURN said that the development of the brain seemed to have followed and not preceded the use of tools. He said that although brain development and not sheer size determined ability, size was ultimately important since below a certain size the necessary development was not possible. Physical anthropologists are obliged to try to determine brain development from brain cases and not from the actual brains and the range of fossils available is in any case small and may not be truly representative.

LEAKEY referred to an examination carried out by the Royal College of Surgeons of the brains of outstanding men of intellect, sport, etc., the conclusions of which were that there did not appear to be any correlation between brain size and ability. Leakey thought that brain size relative to body size was important.

ARAMBOURG agreed and said that he considered human evolution to be basically cerebral development.

COOKE thought that the short time of human development should not be overlooked. He pointed out that man was the only creature that had gone through such tremendous cortical expansion in so short a period.

BOURLIÈRE agreed that accelerating progress was a *sine qua non* of human evolution. Both he and LEAKEY pointed out, however, that this progress had not been gradual but rather by saltations. BOURLIÈRE asked if more precise information could not now be produced based on the methods of detailed plotting of living floors.

LEAKEY said that this should be quite possible, but it would entail the excavation of a complete floor which necessitated both trained personnel and considerable funds. He instanced floor BK II at Olduvai which he said was a thin floor quickly covered. In the area excavated over 9,000 artifacts had been collected and it seemed possible that this represented occupation by a large group for a short time but before one could be definite on this better techniques were necessary and more information on such subjects as modern Bushman occupation was needed. It was not possible to estimate the number of individuals by trying to decide how many tools each would need, Leakey said, although he could say that one good Oldowan chopper would cut up one and possibly two Thomson's gazelles. Obsidian is a good material and does not blunt too easily, but even so estimates are not possible as to how many would be needed by one hunter for say, one month. It is also not necessary, as Clark had suggested, to use tools to cut meat off bones since with a knowledge of the beast's anatomy most of it can be removed by pulling.

HOWELL said that the main need now was to locate and exploit good sites, and especially those where preservation was optimal and hence the range of inference maximized. Such potential sites should be preserved from molesting until the money and personnel are available to investigate them properly.

COOKE felt that further work was needed on living sites. He suggested that a comparison of the number of stone tools with the number of fossil animals might provide information on the ecological setting. He also thought that to consider human evolution adequately a greater study must be made of Pliocene ecology. At no time did it seem that the whole continent had become impossible for man to live in. The beginning of the differential development of man and the higher primates must have gone back even before Villafranchian times. It seemed not unlikely that man first began to develop in forest fringe areas which in the Pliocene may well have been restricted so that it is quite possible that the beginnings of human evolution could be found there.

LEAKEY, on the basis of SHK II and BK II fossil faunas, thought that little could be determined in the way suggested by Cooke since the number and age of individual animals on sites varied considerably. He said that statistics were in preparation for the Olduvai occupation floors, but that the process was necessarily a very slow one. He agreed that one should not have a closed mind as to the possibility of man's evolution from Australopithecines, though he thought that the pre-*Zinjanthropus* adult when found might turn out to be more hominine than Australopithecine. He thought that in the Pliocene we may already expect man to show an established pattern of behavior in relation to his ecology.

CLARK and HOWELL both continued to stress the practical needs of such paleoanthropological studies in Africa. These included:

1. the planned excavation of more home sites, particularly in the Sahara, South-West Africa, and in the Kagera Valley of Uganda;
2. a more detailed analysis of the material from such sites. These would include:
   a. a study of the degree of variability of the industries and the relationship of this to raw material;
   b. assessment of the importance of natural stones;

c. assessment of the amounts of meat and vegetable food required by groups of varying size;
d. notation of the exact nature of the faunal remains—number of species, age of individuals, selectivity, etc.;
e. experiment and comparison in bone-breaking techniques and the setting up of bone-weathering controls;
f. observations of present-day hunter-gatherer occupation sites;
g. more pollen analysis and C-14 dating; and, finally,
h. the collection and preservation of absolutely everything on such a site.

CLARK said in order to obtain a better appreciation of past ecology it would be necessary to:

1. analyze fossil fauna from several sites in an area to obtain a better picture of the probable ecology;
2. assess the effects of fire on vegetation;
3. assess the probable biomass existing in an area at a given time to be able to give an approximate size for the human groups;
4. compare nutrition patterns of present-day hunter-gatherers with probable meat and vegetable supply at any time in the past which should help in the estimation of the size of human groups; and
5. study the effect of changing climatic conditions on ecological zones and thus on human groups. Thus,
   Pluvials encourage isolation so restricting gene flow and making for cultural specialization; and
   Interpluvials encourage movement, gene flow, contact, and cultural variability and experiment. For instance the Sangoan was not an influx of people, but a natural development from the Acheulian to fit a particular environment (with which HOWELL and LEAKEY agreed).

LEAKEY and MONOD considered that to do adequate pollen analysis in Africa one should first carry out an exhaustive study of existing species.

COOKE replied that van Zinderen Bakker had already published three volumes of an atlas of African pollens and that two more were in preparation as well as slides for comparison. He felt that fuller use should be made of such trained personnel. Similarly with paleontological material: more centralized facilities should be available and fuller use should be made of existing experts.

BOND and MONOD stated that Carbon 14 laboratories would shortly open in Salisbury and Dakar and would need samples for dating.

BIBERSON said that in the early Acheulian of Morocco most of the fossil animals were young. He said that a human group had persisted in North Africa for a very long time. He listed the human remains associated with Acheulian at Ternifine, Middle Acheulian at Sidi Abderrahman, and also those from Temára and Tangier. ARAMBOURG said that *Atlanthropus* was associated with bifaces. As he had a simple mentality he did not require complex tools. Similarly with *Pithe-*

canthropus, *Sinanthropus*, the Mauer jaw, and the Steinheim skull which belong to the same group. He could not agree with Biberson that although Swanscombe was associated with the Middle Acheulian it was not a Pithecanthropoid.

LEAKEY could not agree that the Acheulian was a simple culture. He was sure that if everything on a site were collected then even in Europe, where the fallacy still persists, the hand-axe culture would be found to include flake tools as well as bifaces as a basic component and in addition it already was known to include bone, ivory, ochre, etc.

CLARK and HOWELL endorsed Leakey's remarks and advocated the improvement of archaeologists' techniques so that nothing on a site would be missed.

HIERNAUX agreed, but asked how a man who was an amateur in most of the specialties involved would know what and how to collect. He said that a list was needed of technical instructions and of experts who would deal with the various materials collected.

BIBERSON said that Mlle. A. Laming's *La Découverte du Passé* gave details of some of these techniques.

BOURLIÈRE said that on the whole the environment determines the vegetation, animals select their environment and, very largely, man makes his.

BIBERSON, CLARK, HOWELL, and LEAKEY immediately replied that in the very early stages man did not make his environment, but adapted to it and in different places different environments produced similar adaptations. For instance, the Fauresmith was an adaptation to open country in South, Central, and Northern Africa.

FINAL SESSION: JULY 21 (2:30–4:30 P.M.)

A joint letter by Leakey and Robinson was presented to the participants. It stressed the urgent need to have casts, made by the new plastic method now available in the United States, of the best preserved hominid fossils in South Africa and Kenya. The symposium supported the views expressed and moved to ask for the necessary funds to initiate such a program.[4]

The participants expressed their thanks to the two chairmen, Howell and Bourlière, and to the director and staff of the Wenner-Gren Foundation for a most stimulating and enjoyable symposium.

4. This project has now been completed, and has also been expanded to include hominid fossils from Europe and Asia. The work has been carried out with support from the Wenner-Gren Foundation with the advice of the American Institute of Human Paleontology. The casts will be made available in due course [F.C.H.]

ADDENDUM

A Tentative Paleogeographic Map of Neogene Africa *

Two versions of this map were communicated to some of our colleagues in November 1961 and March 1962, respectively.

It appears, from correspondence and compliments received, that the "Pre-Rift" map is unanimously considered the most useful and provocative. It is probably also the most coherent, although it is only a vague indication of the age of the proposed reconstruction.

The indefiniteness of the age of this reconstruction leads to difficulties if North Africa is included, for here topographical modifications were more rapid and important due to alpine orogeny.

Should we adopt *Burdiglian, Vindobonian, Sarmatian, or Plaisancian* contours? Paleogeographic reconstructions of the Mediterranean area are partial and difficult to integrate.

Faced with these inconsistencies, we decided to make the 30th parallel the northern limit of the map. A geologist specializing in the geology of Maghreb and Libya could probably propose valuable complementary reconstructions.

COMMENTARY ON THE MAP

Note: this map is *inexact*. It is only a suggestion of a possible stage in the evolution of Africa. It was intended to suggest a stage in the morphology and hydrography of the African continent (excluding North Africa) at the end of the Tertiary, prior to the major faulting of the Rift Valleys.

It could be that features used as evidence were not, actually, contemporary. The document is, in a certain way, a "telescopic view" of paleogeographic evolution around the Miocene-Pliocene transition.

Nevertheless the map gives an idea of the modifications undergone by the great river basins, the displacement of fall lines and sea coasts.

In relation to the stratigraphic sequence of continental African formations, the stage reconstructed here occurs in approximately the following way:

- Previous to the tectonic fractures of the rifts; these do not yet disrupt the primitive drainage basins
- Almost contemporary with East African deposits which are called "Lower Miocene"

* Compilation and commentary by J. de Heinzelin, based on additional information received from W. W. Bishop, G. Bond, K. W. Butzer, H. B. S. Cooke, A. T. Grove, and T. Monod.

• Previous to the formations of the Kaiso sequence and to the formations containing "Villafranchian" type fauna.

• Almost contemporary with the end of the extension of peneplain III, Ituri peneplain, and Kyoga peneplain.

• After the true Kalahari Sands.

• About contemporary with the Salonga Sands (these two sand formations are here considered to be of subdesertic origin).

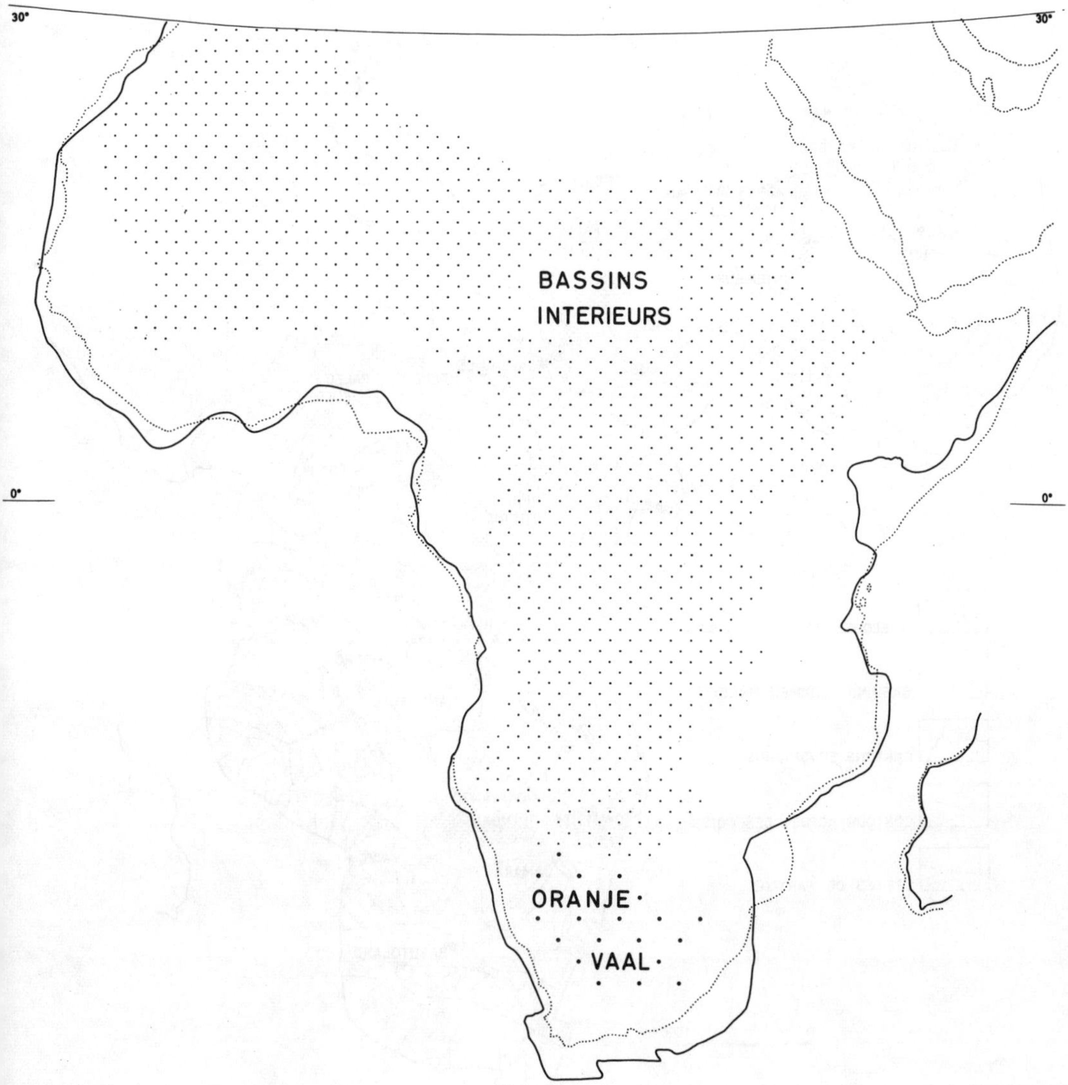

FIGURE 6

The arguments in support of this reconstruction are of several types:

1. Geomorphology: persistence of ancient peneplain surfaces the length of swell structures; persistence of depressions forming many interior basins; persistence of a flexure which is particularly noticeable along the western and southern coasts.

2. Stratigraphy: continental deposits filling the intracontinental basins.

FIGURE 7

3. Pedology: accessible data on the subject of present distribution of lateritic crusts, of eolian sands, and of alluvial formation.

4. Character of present hydrographic systems, of which certain characteristics are clearly abnormal and bear witness to ancient captures.

5. Biogeographical distribution of certain species and associations, such as fresh-water fish, small mammals, and forest associations.

MAJOR FEATURES OF PRE-RIFT AFRICAN PHYSIOGRAPHY

1. An entirely peneplain continent, with very few mountainous massifs, except the Atlas, which, perhaps, was not very different from its present height (ca. 4,000 m.). Notable are some of its residual massifs with their present heights (these being probably greater than their past height through the accentuation of epirogenic uplift).

Loma Mountains, Sierra Leone	2,000 m.
Mount Nimba, Guinea	1,750 m.
The Jos Plateau and hills	1,700 to 1,900 m.
Mount Ganga, Cameroon Plateau	1,650 m.
Ahaggar, in the Sahara, pre-Cambrian basement, minus thickness of younger volcanic formations	1,800 m.
N'Dele Plateau	1,530 m.
Benguela Ridge	1,300 to 2,300 m.
Mountains North of Karamoja Lomariti, Loliba, Langia, Lotuka	3,000 m.
Katanga	1,600 m.
Damara	2,400 m.
Basutoland, Monts'-aux-Sources	3,600 m.

2. Large epirogenic deformations of the swells and interior basins. Drainage predominantly toward the interior basins. These basins are, from north to south:

(Atlas and South Atlas basins, Murzuch basin, not illustrated)
Djouf-Aouker, Aouker-Hodh, interior Niger, Paleo-Chad, Bahr-el-Ghazal or Paleo-Sudd
Interior Congo
Cuanza-Cassanje, Lulua, Lufira, Moero, Bangweulu
Cunene-Ovambo, Etosha-Okavango-Ngami

It is not necessarily true that lakes filled these depressions. They could also have contained an interior hydrographic system, or sandy desert or semidesertic accumulations, or swamps. Thus we use dotted lines for their contours.

A part of the Vaal and Orange basins may have also functioned as interior basins.

3. The coasts have been more or less modified, essentially by the play of continental flexure, as far as the fracture of the continental plateau in certain cases.

There are some drowned coasts (Guinea, South Africa) and others uplifted. At many points in the western and southern coast the fall line is very close to the sea coast.

It appears that the relief, the hydrography, and the distribution of the erosion surfaces of Guinea and South Africa cannot be explained without postulating a much greater oceanic extension of the continent. The reconstructed sketch is imaginary.

The western coast of Africa shows a zone of almost continuous peripheral continental flexure. Perhaps some faults affected the continental slope (shelf).

The eastern coast, on the contrary, indicates a more faulted structure, in connection, probably, with the older opening of the strait of Mozambique.

4. The great submarine canyons presently known are all relatively young, due to the play of continental flexure and turbidity currents.

Canyons on the Guinea coast: drowned

Senegalese canyon of Cayar, Abidjan canyon

Congo canyon: drainage (with cataracts) of the interior basin of the Congo crossing a schistose-sandstone plateau, after being captured by a small coastal river

Cunene Canyon: diverted to the north in relation to the present mouth; drainage of the Cunene basin.

No known canyons exist along the eastern coast.

5. Many ridges are persistent traits of the topography: Guinean Ridge, Ubangi-Chari Ridge, lengthened by the Haut-Uele Ridge (the latter becoming the Congo-Nile divide).

6. The reconstruction of the coast and the hydrography of East Africa (Kenya, Abyssinia, and Somalia) is highly conjectural.

It is supposed that the Lake Victoria region drained toward the east. The topography has been completely modified by the rift tectonics and the eruption of enormous sheets of lava. The thick cover of volcanic deposits on the Abyssinian plateau hides all the earlier structures.

7. Arabia was an integral part of the African continent before the opening of the Red Sea. The beginning of a marine tectonic rift occurred during the Miocene in the Suez area. There is very reason to believe that the drainage of Abyssinia and the Sudan flowed toward the northeast and Arabia. The reconstruction of that drainage, notably that of Wadi Halfa, is still conjectural but it is certain that it has few connections with the present Nile Valley.

8. Most of the great modern rivers are composites. Many represent outlets toward the ocean of otherwise endorrheic basins: Niger, Congo, Cuanza, Cunene, Nile, Zambezi (perhaps the Orange and the Vaal).

BIOGEOGRAPHICAL INFERENCES

1. Three great more or less isolated permanent massifs could have served as a forest reservoir in equatorial Africa: the Guinean ridge, (with the highlands of Loma, Nimba, Jos, Ganga, etc.), Gabon ridge, and that of Ituri-Sudan. Their small mammalian faunas are of relatively different origin.
2. The Abyssinian region was separated from tropical Africa and communicated, rather, more with Arabia.
3. The interior basins as well could have been centers of speciation, especially for the fresh-water forms. A separate place must be given to the Bahr-el-Ghazal depression (comprising the modern Lake Rudolph). A slightly elevated sill separates it from the ocean. From its privileged position, we can conceive the following double role:
   a. A possible stage for the penetration of brackish aralopontian species;
   b. fresh-water speciation of marine species.
4. Interconnection between interior basins, captures, and disruption of hydrographic basins by subsequent rifts explain the distribution of fresh-water faunas. Thus, notably, the case of the Malagarasi where species from the Congo basin are found.
5. The differentiation of high mountain floras in Central Africa is subsequent to the state of the continent suggested here. The massifs which could have served as links for Macronesian and European forms are reduced in number and distance. The Karamoja mountains and the N'Dele plateau could have been particularly important in this regard.

BIBLIOGRAPHY

CAHEN, L.
1954. *Géologie du Congo belge*. Liège: Vaillant-Carmanne.
CAPOT-REY, R.
1953. "Recherches géographiques sur les confins algéro-libyens," Trav. Inst. Rech. Sahar. (Algiers), 10:33–73.
Carte géologique internationale de l'Afrique au 1:5,000,000.
CASTANY, G.
1951. "Etude géologique de l'Atlas tunisien oriental," Regence de Tunis, *Annales des Mines et de la Géologie*, No. 8.
FURON, R., *et al.*
1958. "Esquisse structurale provisoire de l'Afrique," (Congrès Géol. Int. Assoc. des Sérvices géologiques africains).
GROVE, A. T.
(n. d.) "The Benue Valley," Min. Nat. Res., Northern Nigeria.

Grove, A. T., and R. A. Pullan
1961. "Some Aspects of the Pleistocene Paleogeography of the Chad Basin," in this volume.
Mouta, F.
1954. "Esboço geologico de Angola. Noticia explicativa," Lisbon.
Sadek, H.
1959. "The Miocene in the Gulf of Suez Region (Egypt)," United Arab Republic—Egyptian region. Geol. Survey and Mineral Research Department.
Sys, C.
1958–59. Carte des sols du Congo belge et du Ruanda-Urundi, 1:5,000,000. Bruxelles: INEAC.
Termier, H., and G.
1959. *Atlas de Paléogeographie*. Paris: Masson et Cie.
Urvoy, Y.
1942. "Les bassins du Niger," Mem. Institut française de l'Afrique noire, No. 4.
Wills, L. J.
1951. *A Paleogeographical Atlas*. London-Glasgow: Blackie and Sons Ltd.

INDEX